PATTY'S INDUSTRIAL HYGIENE AND TOXICOLOGY

Fourth Edition

Volume I, Parts A and B
GENERAL PRINCIPLES

Volume II, Parts A, B, C, D, E, and F
TOXICOLOGY

Third Edition

Volume III, Parts A and B
THEORY AND RATIONALE
OF INDUSTRIAL HYGIENE
PRACTICE

PATTY'S INDUSTRIAL HYGIENE AND TOXICOLOGY

Third Edition
Volume III, Part A
Theory and Rationale
of Industrial Hygiene
Practice: The Work Environment

ROBERT L. HARRIS
LEWIS J. CRALLEY
LESTER V. CRALLEY
Editors

CONTRIBUTORS

E. W. Arp, Jr.
W. A. Burgess
K. A. Busch
L. E. Christenson
W. C. Cooper
L. V. Cralley
L. J. Cralley
P. M. Eller
P. E. Enterline

R. L. Fischoff
R. L. Harris
J. L. Holtshouser
J. Koscelnik
N. A. Leidel
J. R. Lynch
M. W. Lyzen
M. H. Munsch

R. L. Patnoe
D. J. Paustenbach
S. M. Rappaport
B. D. Reinert
B. Rogers
R. B. Weidner
M. R. Zavon

A Wiley-Interscience Publication

JOHN WILEY & SONS, INC.
New York / Chichester / Brisbane / Toronto / Singapore

This text is printed on acid-free paper.

Copyright © 1994 by John Wiley & Sons, Inc.

All rights reserved. Published simultaneously in Canada.

Reproduction or translation of any part of this work beyond
that permitted by Section 107 or 108 of the 1976 United
States Copyright Act without the permission of the copyright
owner is unlawful. Requests for permission or further
information should be addressed to the Permissions Department,
John Wiley & Sons, Inc., 605 Third Avenue, New York, NY
10158-0012.

Library of Congress Cataloging in Publication Data:
Theory and rationale of industrial hygiene practice / [Robert L.
Harris, Lewis J. Cralley, Lester V. Cralley, editors].—3rd ed.
     p.   cm.
  At head of title: Patty's Industrial hygiene and toxicology,
Volume III.
   Includes bibliographical references and indexes.
   Contents: 3A. The work environment.
   ISBN 0-471-53066-2 (V. 3, Pt. A: acid-free paper)
   1. Industrial hygiene. 2. Industrial toxicology. I. Harris,
Robert L., 1924– . II. Cralley, Lewis J., 1911–   . III. Cralley,
Lester V. IV. Patty's Industrial hygiene and toxicology.
RC967.T48   1993
613.6'2—dc20   93-23747

Printed in the United States of America

10 9 8 7 6 5 4 3 2 1

Contributors

Earl W. Arp, Jr., Ph.D., Ashland Oil, Inc., Ashland, Kentucky

William A. Burgess, CIH, Harvard School of Public Health, Boston, Massachusetts

Kenneth A. Busch, M.S., Cincinnati, Ohio

Lynne E. Christenson, Ph.D., Department of Anthropology, San Diego State University, San Diego, California

W. Clark Cooper, M.D., Berkeley, California

Lester V. Cralley, Ph.D., Fallbrook, California

Lewis J. Cralley, Ph.D., Cincinnati, Ohio

Peter M. Eller, Ph.D., CIH, Division of Physical Science and Engineering, NIOSH, Cincinnati, Ohio

Philip E. Enterline, Ph.D., Graduate School of Public Health, University of Pittsburgh, Pittsburgh, Pennsylvania

Robert L. Fischoff, M.S., CIH, CSP, Program Manager, Process Safety Technology, I.B.M. Corporate Headquarters, Bethesda, Maryland

Robert L. Harris, Ph.D., Department of Environmental Sciences and Engineering, University of North Carolina, Chapel Hill, North Carolina

Joseph L. Holtshouser, CIH, CSP, Manager, Industrial Health Services Management, Goodyear Tire and Rubber Company, Akron, Ohio

Jacqueline A. Koscelnik, J.D., Reed, Smith, Shaw and McClay, Pittsburgh, Pennsylvania

Nelson A. Leidel, Sc.D., Atlanta Georgia

Jeremiah R. Lynch, Senior Environmental Health Scientist, Exxon Chemical Americas, East Millstone, New Jersey

Maria W. Lyzen, R.N., M.S., UAW—CM Center for Health and Safety, Auburn Hills, Michigan

Martha Hartle Munsch, J.D., Reed, Smith, Shaw and McClay, Pittsburgh Pennsylvania

Richard L. Patnoe, Ph.D., Boulder, Colorado

Dennis J. Paustenbach, Ph.D., CIH, DABT, Vice President and Chief Technical Officer, McLaren/Hart, ChemRisk Division, Alameda, California

Stephen M. Rappaport, Ph.D., Department of Environmental Sciences and Engineering, University of North Carolina, Chapel Hill, North Carolina

Bruce D. Reinert, Los Alamos National Laboratory, Los Alamos, New Mexico

Bonnie Rogers, DrPH, COHN, FAAN, School of Public Health, University of North Carolina, Chapel Hill, North Carolina

Robert G. Tardiff, Ph.D., A.T.S., Vice President, Health Services, EA Engineering, Science and Technology, Silver Spring, Maryland

Robert B. Weidner, J.D., Cincinnati, Ohio

Mitchell R. Zavon, M.D., Lewiston, New York

Preface

Industrial hygiene is an applied science and a profession. Like other applied sciences and professions, such as medicine and engineering, it is founded upon basic sciences such as biology, chemistry, mathematics, and physics. Drawn from these other sciences is a foundation of theory that supports the practice of industrial hygiene. The rationale for industrial hygiene practice is based on the profession's purposes, that is, the recognition, evaluation, and control of work-related health hazards.

This volume retains the fifteen chapters on theory and rationale from the Second Edition, although all of these chapters have been either rewritten or updated and expanded. Four new chapters have been added. Our abilities to monitor exposures, both in terms of the numbers of measurements that are feasible and the variety of techniques from which we may choose, have increased the availability and complexity of exposure data. A new chapter on interpretation of levels of exposures has been added. The field of risk analysis has developed enormously in the past decade, and its principles are increasingly influencing industrial hygiene practice, particularly in standards setting and litigation. We have added a chapter on risk analysis in industrial hygiene. As industrial hygiene continues to develop as a profession, we must be aware that others recognize it as a profession as well, with all of the expectations that attend such recognition. Society's expectations may have malpractice and other types of litigation among their consequences. We have added a chapter on professional liability and litigation. Finally, other professions are not without their industrial hygiene hazards. To illustrate the need for, and application of, industrial hygiene in other professions that in the past have not been subject to in-depth industrial hygiene scrutiny, we have added a chapter on occupational hazards in archaeology. This, we hope, will encourage industrial hygienists to examine other professions we may have neglected, and within which the recognition, evaluation, and control of health hazards may help to bring needed protection to fellow professionals and their co-workers.

Industrial hygienists know that variability is the key to measurement and interpretation of workers' exposures. If exposures did not vary, exposure assessment could be limited to a single measurement, the results of which could be acted upon, then the matter filed away as something of no further concern. We know, however, that things change, that we must be alert to recognize new hazards, we must continue to evaluate new and changing stresses, and that we must monitor the performance of controls and from time to time upgrade them. The science of industrial hygiene is in continuous change as well. This volume represents the theory and rationale for industrial hygiene practice in the work environment as it is understood by its chapter authors in 1992–1993. Improvements and changes in theory and practice take place continuously and are generally reported in the professional literature. Because of the changes that will take place after its publication, it will be prudent to have information in this volume interpreted by professionals who stay abreast of new developments in industrial hygiene.

ROBERT L. HARRIS
LEWIS J. CRALLEY
LESTER V. CRALLEY

Raleigh, North Carolina
Cincinnati, Ohio
Fallbrook, California
December 1993

Contents

PATTY'S INDUSTRIAL HYGIENE AND TOXICOLOGY

Third Edition

Volume III Part A
THEORY AND RATIONALE
OF INDUSTRIAL HYGIENE
PRACTICE: THE WORK ENVIRONMENT

CHAPTER ONE

Rationale

Robert L. Harris, Ph.D.
Lewis J. Cralley, Ph.D.
Lester V. Cralley, Ph.D.

1 BACKGROUND

The emergence of industrial hygiene as a science has followed a predictable pattern. Whenever a gap of knowledge exists and an urgent need arises for such knowledge, dedicated people will gain the knowledge.

The harmful effects from exposures to toxic substances in mines and other workplaces, producing diseases and death among workers, have been known for more than two thousand years. Knowledge on the toxicity of materials, the hazards of physical and biologic agents, and of ergonomic stressors encountered in industry, and means for their evaluation and control were not available during the earlier period of industrial development. With few exceptions, the earliest attention given to worker health was in applying the knowledge at hand, which concerned primarily the recognition and treatment of illnesses associated with a job.

However, the devotion of prime attention to the preventive aspects of worker health maintenance through controlling job-associated health hazards became quite evident if the best interest of workers was to be served in preventing occupational diseases. Not until around the turn of the twentieth century, though, did major effort begin to be directed toward the recognition, measurement, evaluation, and control of workplace environmental health stresses in the prevention of occupational diseases.

The aim of this chapter is not to document or present chronologically the major

Patty's Industrial Hygiene and Toxicology, Third Edition, Volume 3, Part A, Edited by Robert L. Harris, Lewis J. Cralley, and Lester V. Cralley.
ISBN 0-471-53066-2 © 1994 John Wiley & Sons, Inc.

past contributors to worker health and their relevant works or the events and episodes that gave urgency to the development of industrial hygiene as a science and a profession. Rather, the purpose of the chapter is to place in perspective the many factors involved in relating environmental stresses to health and the rationale upon which the practice of industrial hygiene is based, including the recognition, measurement, evaluation, and control of workplace stresses, the biological responses to these stresses, the body defense mechanisms involved, and their interrelationships.

The individual chapters of this volume and its Part B companion volume cover these aspects in detail.

Similarly, it is not the intent of Parts A and B of this volume to present procedures, instrumental or otherwise, for measuring levels of exposure to chemical, physical, biological, or other stress agents. This aspect of industrial hygiene practice is covered in detail in *Patty's Industrial Hygiene and Toxicology, Volume I, General Principles*. Rather, attention in Parts A and B of this volume is devoted to other aspects of workplace exposures such as representative and adequate sampling and measurement, variations in exposure levels, exposure durations, interpreting results, the rationale for control, and the like.

2 INSEPARABILITY OF ENVIRONMENT AND HEALTH

Knowledge is constantly being developed on the ecological balance that exists between the earth's natural environmental forces and the existing biological species, and how the effects of changes in either may affect the other. In the earth's early biologic history this balance was maintained by the natural interrelationships of stresses and accommodations between the environment and the existing biological species at each particular site. This system related as well to the ecological balance within species, both plant and animal.

Studies of past catastrophic events such as the ice ages have shown the effects that changes in this balance can have on the existing species. The forces that brought on the demise of the dinosaurs that lived during the Mesozoic Era are uncertain. Most probably major geological events were involved. Studies have also shown that in the earth's past history a great many other animal species, as well, have originated and disappeared.

The human species, however, has been an exception to the ecological balance that existed between the natural environment and the evolving biological species in the earth's earlier history. The human ability to think, create, and change the natural environment has brought on changes above and beyond those of the existing natural forces and environment. These changes have had, and continue to have, an ever-increasing impact upon the previous overall ecologic balance.

The capacity of humans to alter the environment to serve their purposes is beyond the bounds of anticipation. In early human history these efforts predictably addressed themselves to better means of survival, that is to food, shelter, and protection. As these efforts succeeded, humankind was able to devote some of its energy to gaining knowledge concerning factors affecting human health and well-being. Thus evolved the medical sciences, including public health. In some instances these efforts resulted in intervention in the ecological balance in the control of disease. In other situations the environment may have been altered to make desirable resources available, for example, in the damming of streams for flood control and developing hydroelectric power. This type of alteration of the localized natural environment and the associated ecological systems may have an impact by developing additional stresses in readjusting the existing ecological balance.

Of more recent impact on health have been the stresses of living brought on by activities associated with personal gratifications such as life-styles as well as those associated with an ever more complex and advancing technology in almost all areas of human endeavor.

The quality of the indoor environment is receiving increasing attention in relation to good health. This applies to the study and control of factors giving rise to psychological stresses associated with living or working in enclosed spaces, as well as air pollution arising from life-styles, building designs, and materials and activities.

The advantages associated with changes in the environment for human benefit and improving the essential quality of living should be assessed for their cost-effectiveness as well as their potential to produce deterioration of the environment and concomitant new stresses. That humans, for optimum health, must exist in harmony with their surrounding 24-hr daily environment and its stresses is self-evident.

For better understanding of the significance of on-the-job environmental health hazards, an overview of the 24-hr daily stress patterns of workers is helpful. This permits a perspective in which the overall component stresses are related to the whole of workers' health.

Our habitat, the earth and its flora and fauna, is in reality a chemical one, that is, an entity that can be described in terms of tens of millions of related elements and compounds. It is the habitat in which the many species have evolved and in which a sort of symbiosis exists that supports the survival of individual species. The intricacy of this relationship is illustrated in the recognition that copper, chromium, fluorine, iodine, molybdenum, manganese, nickel, selenium, silicon, vanadium, and zinc, in trace amounts, are essential to human health and well-being. All, however, are toxic when absorbed in excess, and all are listed in standards relating to permissible exposure limits in working environments. Some compounds of several of these elements are classified as carcinogens. It is most revealing that some trace elements essential for survival are under some circumstances capable of causing our destruction. Thus, the matter of need or hazard is a question of "How much?"

The environment is both friendly and hostile. The friendly milieu provides the components necessary for survival: oxygen, food, and water. On the other hand, the hostile environment constitutes a combination of stresses in which survival is constantly challenged.

Although numerous factors are obviously involved in the optimal health of an individual, stresses arising out of the overall environment, that is, the workplace, life-style, and off-the-job activities, are substantial. The stresses encountered over full 24-hr daily periods have an overall impact on an individual's health. Any activity over the same period of time that can be stress relieving will have a beneficial effect in helping the body to adjust to the remaining insults of the day.

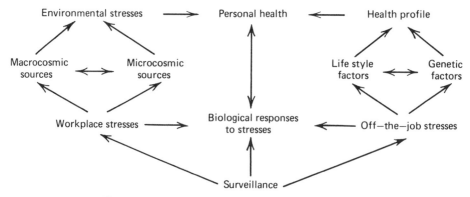

Figure 1.1 Inseparability of environment and health.

The inseparability of the environment and its relation to health is presented graphically in Figure 1.1.

An environmental health stress may be thought of as any agent in the environment capable of significantly diminishing a sense of well-being, causing severe discomfort, interfering with proper body organ functions, or causing illness or disease. These stresses may be chemical, physical, biologic, psychologic, or ergonomic in nature. They may arise from natural or created sources.

2.1 Macrocosmic Sources

Macrocosmic sources of environmental stress agents, such as the hemispheric or global quality or state of air, soil, or water, are those emanating from the sun or from extensive geographical perturbations and are capable of affecting large geographical areas. Examples of natural stress agents are ultraviolet, thermal, and other radiations from the sun, major volcanic eruptions that release huge quantities of gases and particulate material into the upper atmosphere, the changing of the upper air jetstream and other factors that influence climate, and the movement of the earth's tectonic plates resulting in earthquakes and tidal waves.

Examples of human-created macrocosmic stresses include interference, through release to the atmosphere of some synthetic organic compounds, with the upper atmosphere ozone layer that shields the earth from excessive ultraviolet radiation; the burning of fossil fuels that increases the carbon dioxide level in the atmosphere and alters the earth's heat balance and surface temperature; and the destruction of forests and pollution of oceans that inhibit the biosphere's oxygen-producing capability.

These macrocosmic stresses may act directly upon individuals through such conditions as excessive exposure to ultraviolet and thermal radiation or indirectly by influencing the earth's climate—sunshine, rain, and temperature—thus affecting vegetation and habitability.

2.2 Microcosmic Sources

Microcosmic sources of stress agents are those emanating from localized areas and generally affecting a single region. These are most commonly at the regional or community level and may also include the home and work environments.

Examples of natural sources of microcosmic stress agents are pollen, which gives rise to sensitization, allergy, and hay fever; water pollution from ground sources having high mineral or salt content; and the release of methane, radon, sulfur gases, and other air contaminants from underground and surface areas. It is noteworthy that human evolution has taken place in the presence of natural macrocosmic and microcosmic sources.

Human-created microcosmic sources of health stress agents at the community level include noise from everyday activities such as lawn mowing, motorcycle and truck traffic, and loud music; air pollution from motor vehicle exhaust, release of industrial emissions into the air, emissions from refuse and garbage landfills, toxic waste disposal sites, spraying of crops, and life-style; surface water pollution through the release of contaminants from home, community, agricultural, and industrial activities into streams; and seepage into ground water of contaminants from landfills, agricultural activities, and from industrial and other waste disposal sites.

Stress agents from these sources may cause direct responses, such as the effects of noise on hearing and toxic exposures on health, or indirect responses such as acid smog and rain affecting vegetation and soil quality.

Regarding microcosmic sources, it is noteworthy that segments of industry are taking seriously their obligations and opportunities for protection of the health and well-being of both their employees and the communities in which their plants operate. In late 1988 the Chemical Manufacturers Association (CMA) adopted an initiative identified as Responsible Care: A Public Commitment. Each CMA member company, as a condition of membership, will manage its business according to the following listed principles (CMA, 1988):

To recognize and respond to community concerns about chemicals and our operations.

To develop and produce chemicals that can be manufactured, transported, used, and disposed of safely.

To make health, safety, and environmental considerations a priority in our planning for all existing and new products and processes.

To report promptly to officials, employees, customers, and the public information on chemical related health or environmental hazards and to recommend protective measures.

To counsel customers on the safe use, transportation, and disposal of chemical products.

To operate our plants and facilities in a manner that protects the environment and the health and safety of our employees and the public.

To extend knowledge by conducting or supporting research on the health, safety, and environmental effects of our products, processes, and waste materials.

To work with others to resolve problems created by past handling and disposal of hazardous substances.

To participate with government and others in creating responsible laws, regulations, and standards to safeguard the community, workplace, and environment.

To promote the principles and practices of responsible care by sharing experiences and offering assistance to others who produce, handle, use, transport, or dispose of chemicals.

These principles pledge good practices in controlling microcosmic sources, and, through those principles applying to employees, endorse good industrial hygiene practices. The Responsible Care initiative bodes well for community and worker health, as well as for the long-term financial health of the industry.

2.3 Life-Style Stresses

The life-style of individuals, including habits, nutrition, off-the-job activities, recreation, exercise, and rest, may have beneficial effects as well as stresses that exert a profound influence on health. Extensive knowledge has been developed, and continues to expand, on the influence of habits such as smoking, alcohol consumption, and use of drugs on health. Stresses from such activities may have additive, accumulative, and synergistic actions or may exert superimposed responses on the effects of other exposures arising in places of work. These off-the-job agents can be causes of respiratory, cardiovascular, renal, or other diseases, and may create grave health problems in individuals over and above any effects of exposures encountered in workplaces.

Knowledge is available on the deleterious effects on the health of their offspring caused by smoking, alcohol consumption, and drug use by women during pregnancy; effects include malformation and improper functioning of body organs and systems, low birth weight, and so forth.

The benefits to health of good, adequately balanced nutrition, that is, vitamins, minerals, and other essential food intake, are gaining increased attention in relation to general fitness, weight control, prevention of disease, supporting the natural body defense mechanisms, recovery from exposure to environmental stresses, and aging. Conversely, malnutrition and obesity are associated with many diseases or dysfunctions, and may have synergistic effects with exposures to other stresses.

Recreational activities are important aspects of good health practices, releasing tension brought on through both off- and on-the-job stresses. Conversely, many recreational activities may be harmful, such as listening to excessively loud music, which may lead to hearing decrement, frequent engagements in events and schedules that interfere with the body's internal rhythmic functions, failing to observe needed precautions while using toxic agents in hobby activities, and pursuing activities to the point of exhaustion.

Both exercise and rest are important for maintaining good health. Exercise helps in maintaining proper muscular tone as well as weight control. Exercise, however, should be designed for specific purposes, maintained on a regular basis, and structured to accommodate the body physique and health profile of the performing individual. Otherwise more harm than benefit may result. Rest provides time for the body to recuperate from physical and psychological stresses.

The hours between work shifts and during weekends provide time for the body to excrete substantial portions of some chemical agents absorbed during a work shift or a workweek. The significance of these nonexposure recovery periods, of course, depends on the biologic half-time of the chemical agent of interest. Work patterns that disturb this recovery period, as in moonlighting, may have an especially deleterious effect if similar stresses are encountered on the second job. Likewise, smoking, alcohol consumption, and use of drugs may impair the body's proper recuperation from previous stresses.

2.4 Off-the-Job Stresses

Workers may encounter a host of stresses outside places of work. These may be chemical, physical, biologic, or psychologic in nature and usually are encountered during the 8- to 10-hr period between the end of the work period and the beginning of sleep time, and on weekends. Many off-the-job activities, if performed in excess and without regard to necessary precautions, are capable of producing stress and injury. Hobby and recreational activities may account for a substantial portion of this time. Hobby participants may encounter environmental exposures from activities such as soldering, welding, cleaning, gluing, woodworking, grinding, sanding, and painting, which are experienced by workers on the job but that are well controlled on the job. The home hobbyist, however, often does not have available the appropriate protective devices such as local exhaust ventilation, protective clothing, goggles, and respirators. The same hobbyist may neglect other good safety practices; he or she may ignore precautionary labels, take shortcuts to save time, fail to use the basic principles of keeping toxic materials from the skin, neglect thorough washing after skin contact, and may smoke or eat while working with such materials.

The off-the-job gamut of health stresses is wide and formidable. To deal with these stresses satisfactorily requires that a degree of accommodation be reached based on judgment, feasibility, personal options, objectives, and other factors. It is evident that although the components of the total sum of environmental health stresses must be considered on an individual basis, no single component can stand alone and apart from the others.

2.5 Workplace Stresses

Places of work may be the most important sources of health stresses if workplace operations have not been studied thoroughly and the associated health hazards have not been eliminated or controlled. This was evidenced during the time of early industrial development when little information was available on methods for identifying, measuring, evaluating, and controlling work-related stresses. During this period workplace exposures were often severe, leading to high prevalence of diseases and excess deaths.

Beginning about the turn of this century, and especially since the 1950s, management, labor union, government, academic, and other groups have taken substantial interest in worker health and in the control of exposures to stresses in places of work. This, along with the setting of standards of exposure limits, has created a broad support for expanding knowledge and practice to prevent illnesses and diseases associated with on-the-job exposures. This volume focuses on the theory and rationale for recognizing, evaluating, and controlling on-the-job health hazards.

2.6 Biological Response to Environmental Health Stresses

The human body consists of a number of discrete organs and systems derived from the embryonic state, encased in a dermal sheath, and developed to perform specific functions necessary for the overall functioning of the body as an integral unit. These organs are interdependent so that a malfunction in one may affect the functioning of many others. As an example, a malfunction in the alveoli, which hinders the passage of oxygen into the blood transport system, may have a direct effect on other organs through their diminished oxygen supply. Similarly, any hormonal imbalance or enzyme aberration may affect the functioning of many other body organs. Once absorption has occurred, toxic agents may selectively target one or more of the organs. Each organ has its own means for accommodating to, or resisting, stress and adjusting to injury, and its own propensity for repair.

Although the organ structures and functions of several experimental animal species have many similarities to those of the human body, with some more similar than others to those of humans, care has to be taken in extrapolating research data on any one experimental animal species to the human. Similarly, research data obtained in isolated systems such as cell cultures have to be cautiously interpreted when applied to even the same cells in the whole integrated human body organ system.

2.7 Body Protective Mechanisms Against Environmental Stresses

The human body, in coexisting with the hostile stresses of the external environment, has developed a formidable system of protection against many of these stresses. This is accomplished in a remarkable manner by the ectodermal and endodermal barriers resisting absorption of noxious agents through inhalation, skin contact, and ingestion, and supported by the backup mesodermal and biotransformation mechanisms once absorption has occurred. These external and internal protective mechanisms, however, are not absolute and can be overwhelmed by a stress agent to the extent that they are ineffective, with resultant disease and death. Also, these protective mechanisms may become impaired in various degrees from insults associated with life-styles and other daily activities. In studying the effects of specific stresses on health, it is important to be aware of the body's protective mechanisms. Suitable control of a specific stress should supplement the body's protective response to that stress.

2.8 Coaptation of Health and Environmental Stresses

To survive, the human body must live in balance with the surrounding environment and its concomitant stresses. Since these stresses, singly or combined, are not constant in value even for short periods of time or over limited geographical areas, the body must have built-in mechanisms for adjusting to differing levels of stresses through adaptation, acclimatization, and other accommodating mechanisms. There is a limit, however, to which the body can protect itself against these stresses without a breakdown occurring in its protective systems.

An extremely thought-provoking concept on associations between environmental stresses and health decrements has been presented by Theodore Hatch (Hatch, 1962, 1972). His concept examines associations between stresses and the human body's adjustments, compensations, and finally breakdown and failure, in response to them. The concept is particularly useful in understanding the effects of multiple risk factors when they include those of both occupational and nonoccupational origin.

If humans are to have freedom of geographic movement and of living in highly hostile environments that produce environmental insults greater than can be handled by the body's protective and adaptive mechanisms, some means of protection other than accommodation by the body must be provided. Extreme examples of the need for such protection are living and working in confined spaces where the immediate human environment is under absolute control against outside catastrophic stresses as in the cases of space travel and submarine activities.

More typical of this coaptive relationship is exposure to ultraviolet radiation from the sun. It is obvious that avoidance of all ultraviolet radiation from this source is impractical. In addition to the ozone layer of the upper atmosphere shielding the earth from major levels of ultraviolet radiation emitted by the sun, and the body's own protective mechanisms such as skin pigmentation, further accommodation is reached through the use of special clothing, skin barriers, eye protection, and a managed limitation to exposure.

At high altitudes where the partial pressure of oxygen is diminished from that to which the human body may be accustomed at lower elevations, the body acclimates in time by increasing the number of red blood cells and hemoglobin that carry oxygen to the tissues. When the availability of atmospheric oxygen decreases below the limit of acclimation, further accommodation may be provided externally through the use of supplemental oxygen supply.

The human body has a number of regulating mechanisms to keep its temperature within normal limits when it is exposed to excessively high or low environmental temperatures. This permits living in a limited but wide range of environmental temperatures. Further accommodation to extremes in environmental temperatures may be provided through special clothing, protective equipment, and living and working in climate-controlled structures.

Where excessive exposures to environmental stresses exist in a workplace, emphasis is placed on their elimination or on lowering them to acceptable levels. Stress levels should be lowered through recognized engineering or administrative control procedures to the point that the body defense mechanisms can adequately prevent injury to health. Some situations may not be amenable to engineering or administrative control and may properly require the use of personal protective equipment or other control strategies.

There is a limit to what can be done to alter existing environmental stresses from natural macrocosmic sources. Thus, these become ubiquitous background stresses upon which other exposures from microcosmic sources such as community and industrial pollution, off-the-job activities, life-styles, and on-the-job activities are added. Rationally, then, it is primarily the stresses created from predominantly microcosmic and a few macrocosmic sources that are amenable to control. These must be kept within acceptable limits to permit humans to avoid health damage from the stresses of an increasingly complex technologic age.

3 INDUSTRIAL TECHNOLOGICAL ADVANCES

In the early history of humankind, the many activities associated with living were at the tribal level where emphasis was placed upon survival, that is, on procuring adequate food, protection, and shelter. The tribes, many of whom were nomadic, were undoubtedly aware that they lived in accommodation with their environments. This would have been evidenced through the appropriate use of clothing, safe use of fire, and observing climatic patterns.

It was inevitable that the nomadic way of life would give way in most instances to a more settled life-style in which cooperative efforts for food, shelter, and protection were more dependable than those based on individual or small group effort and ingenuity. During this transition period accommodations to the natural elements were made easier through more permanent shelter and a more organized pattern of living.

The next advancement in accommodation came through the realization that increased production could be attained through specialization of work pursuits wherein a designated work group devoted its principal effort to the making of a single commodity such as clothing, pottery, or tools, or the growing of foods. Each group shared its commodity in exchange for the products of other groups. This type of trade evolved to cottage-type industries that related primarily to the community level of commerce. Even at that level of production many of the health stresses associated with different pursuits were intensified over those of nomadic living in which every person was a sort of jack-of-all-trades. This was especially true where the operations tended to be restricted to confined and crowded spaces.

As means for communications and transportation improved, trade increased between adjoining communities, and the search continued for ways of producing commodities in increased volume with less manpower. Similar types of production operations tended to expand and to be concentrated within the same housing structure. This led to increases in the health stresses of the whole workforce in instances where the stress agents were cumulative in intensity and response.

This trend toward industrialization continued and intensified with the advent of the steam engine. Developments such as the steam engine gave rise to the industrial revolution, which brought about larger factories and new production techniques, along with increased health risks. While in earlier times of cottage industries there were relatively few health risks in any one workplace, the new technology and industrialization led to more complex patterns of exposures.

Since the advent of the industrial revolution technological advances and their application to production have expanded at an ever-increasing pace. In the latter half of the twentieth century the application in industry of knowledge gained through space technology and other such research has rapidly expanded into the current high-technology electronic age.

The advent of this high technology and its application to production is having its effect upon both the nature of employment and the concomitant health stresses. While in the past workers needed only special instructions to perform most job operations—and this will continue for some time—the move into higher technology has created a demand for highly trained employees for many job positions. This trend can be expected to increase dramatically. Computers, word processors, video display terminals, lasers, microwave, and other electronic equipment are becoming commonplace in industry. Robots, which require sophisticated management and control, are taking over many repetitive operations such as in painting and metalworking.

The nature and extent of associated health stresses are becoming more complex with the advent of high-technology industry. At the same time, the health effects of these stresses are becoming more detectable with more sophisticated measurement and diagnostic tools.

The urgent need for knowledge concerning the effects of exposures to health stresses associated with an ever-expanding industrial technology, along with the methodology for their recognition, evaluation, and control, gave rise to the science of industrial hygiene. This science must keep attuned to the ever-changing applications of technology in industry.

4 EMERGENCE OF INDUSTRIAL HYGIENE AS A SCIENCE AND PROFESSION

Science may be defined as an organized body of knowledge and facts established through research, observations, and hypotheses. As such, a science may be basic, as exemplified by the fundamental physical sciences, or it may be applied in the sense that the principles of other sciences are brought to bear in developing facts and knowledge in a specific area of application.

During the early history of industrial development the lack of knowledge on the effects of health stresses associated with industrial operations and how these could be controlled, with the concomitant massive exposures to toxic materials in places of work, led to many serious episodes of illness, disease, and death among workers. An example is the high incidence of silicosis and silicotuberculosis that existed a century ago among workers in hard rock mines, the granite industry, and in tunneling operations, wherever the dust had a high free silica content.

During this early period the major effort on behalf of workers' health was to apply the knowledge at hand, which related primarily to the recognition and treatment of occupational illnesses.

It was not until around the turn of this century that specific attention began to be devoted to the preventive aspects of industrial illnesses. Scientists and practitioners, including engineers, chemists, and physicists, began to apply their knowledge and skills toward the development of methods and procedures for identifying, measuring, and controlling exposures to harmful airborne dusts and chemicals in workplaces. At that time there were no recognized procedures for carrying out these activities.

After trying various potential air-sampling procedures, the impingement method was judged the most adaptable one at that time for collecting many of the airborne contaminants such as particulates, mists, some fumes, and gases. The light-field microscopic dust counting technique was developed for enumerating levels of mineral dust in the air; and conventional wet chemical analytical methods available at the time were adapted for measuring quantities of chemical agents in these samples.

Even during this early period of development of airborne sample collection and analytical procedures, scientists realized that the exposure patterns that existed were more complex and complicated than the instruments for sample collection and analysis could define. These scientists also knew of many of the deficiencies associated with the measurement procedures being developed. They were aware that the data being collected represented only segments of the overall exposure patterns. They believed, however, that these segments could be used as indices that would represent overall exposure patterns so long as the production techniques and other operational factors remained the same. It must be remembered that at that time information was not available on respiratory deposition and dust size. It was imperative to them, and rightly so, that some method, with whatever deficiencies that may have been incumbent, be developed for indexing airborne levels of contamination in workplaces, both for estimating levels of exposure and for use as benchmarks in determining degrees of air quality improvement after controls had been established.

Since the earliest instruments for collecting airborne contaminant samples were nonportable, the collected samples represented general room levels of contamination, and the results depended on where in the workplace the samples were taken.

The procedures described above for airborne sampling and analysis in workrooms, as primitive as they may seem today, served well during that period of time. They accounted for the drastic reduction in massive exposures that existed in many work sites and were the methods and procedures upon which future refinements would be made.

These early scientists showed that it was feasible to lower the massive workplace airborne contamination levels that often existed at that time; and by relating the data to the health profiles of workers, they observed that the lowering of exposure levels also lowered the incidence of the associated diseases. Thus began the first field studies that were to have a profound influence on the development of the earliest permissible exposure limits and in developing the rationale upon which the practice of industrial hygiene is predicated. The insights developed in these field studies were to be further substantiated through laboratory and clinical research.

The development of the hand-operated midget impinger pump during the 1930s was a decided improvement over the standard impinger pump since it was portable and permitted movement about a workplace while airborne samples were being taken. Samples of particulate or gaseous agents could be taken with glass impingers or fritted glass bubblers near workers as they moved about their tasks. This new worker exposure data demonstrated that workers often had higher levels of exposure than those indicated by the general room airborne levels.

Other instruments, such as the electrostatic precipitator and evacuated containers, came into use during this period. In the late 1940s the paper and membrane filter methods for collecting some airborne particulate samples came into use. The filter sampling procedure was found to be superior in many respects to the impingement method. The method did not fracture particles or disperse agglomerates, which often accompanied impingement collection, and could be performed in ways that permitted direct microscopic observation of particles and gravimetric measurement of samples.

Insights also began to emerge on the importance of particle size in relation to deposition and retention of particulate material in the respiratory tract. Electron microscopes and membrane filters made it possible to study particles of submicron sizes.

A surge of improved and sophisticated techniques for quantifying workers' exposures to health stress agents took place in the 1960s. This applies both to sample collection and analytical techniques in which much lower levels of exposure to specific agents could be determined. There also began a dramatic increase in toxicologic and epidemiological studies by government, industry, universities, and foundations, directed to obtaining data upon which to base exposure standards as well as good industrial hygiene practices.

Another major advancement in developing better methods for studying occupational diseases occurred at midcentury. Toxicologic and other studies had revealed that lowering the level of exposure and extending the exposure duration changed the dose–response pattern of many toxic agents. As an example, a high level of exposure to airborne lead produces an acute response over a relatively short period of time. In contrast, lowering the exposure level of this agent and extending the exposure time shows a different dose–response pattern, a chronic form of lead poisoning. Thus, in studying the effects of exposure to health stress agents it is important to obtain relevant dose–response data over an extended period of exposure time.

One method of obtaining relevant health profile data on workers is through study of their medical records. Another method is through the study of causes of death among worker populations using death certificates located through Social Security, management, retirement system, and labor union records. Such studies have revealed that a lifetime of work exposure to an agent, or an extended observation period of 20 or more years from time of initial exposure, may be necessary to fully define the wide range of dose–response relationships. This may be especially true for carcinogenic and other long latency types of exposure response.

A more recent advancement relates to chronobiology, the study relating to the body's internal biological rhythms and their effects on organ functions, and so forth. The workweek schedule can have a direct effect on these rhythmic patterns and health. Also, there is some evidence that the rate of absorption of toxic materials and reactions to stress may relate in some way to an individual's chronobiology.

The establishment of professional associations to support the interests and growth of the profession has played an important role in developing industrial hygiene as a science. In the United States in the 1930s the American Public Health Association had a section on industrial hygiene that supported the early growth of the profession. The American Conference of Governmental Industrial Hygienists was organized in 1938. The American Industrial Hygiene Association was organized in 1939. The American Board of Industrial Hygiene was created and held its first meeting in 1960. This board certifies qualified industrial hygienists in the comprehensive practice of industrial hygiene and in the past has certified in six additional industrial hygiene specialties as well. Industrial hygienists certified by the board have the status of Diplomates and as such are eligible for membership in the American Academy of Industrial Hygiene.

The American Academy of Industrial Hygiene has developed and adopted a 15-point Code of Ethics for the professional practice of industrial hygiene. The code addresses industrial hygienists' responsibilities to the profession, to workers, to employers and clients, and to the public. It is noteworthy that the Code of Ethics specifies that the primary responsibility of an industrial hygienist is to protect the health of employees (i.e., workers) and that responsibility to any employer or client is subservient to that to workers.

The American Industrial Hygiene Association administers a laboratory accreditation program with the objective of assisting those laboratories engaged in analyses of industrial hygiene samples in achieving and maintaining performance levels within acceptable ranges.

The American Industrial Hygiene Foundation was established in 1979 under the auspices of the American Industrial Hygiene Association. The functions of the foundation are carried out by an independent Board of Trustees. The foundation provides fellowships to worthy industrial hygiene graduate students, encourages qualified science students to enter the industrial hygiene profession, and stimulates major universities to establish and maintain industrial hygiene graduate programs.

In the United States a number of occupational health regulations were established in the early 1900s with emphasis, in several, on listing limits of exposures to a relatively few agents. These regulations were effective at the local, state, and federal levels depending on governmental jurisdiction.

The Social Security Act of 1935 and the Walsh–Healy Act of 1936, had an immense impact in giving increased stability, incentive, and expanded concepts in the practice of industrial hygiene. These acts stimulated industry to incorporate industrial hygiene programs as an integral part of management. They also stimulated broad-base programs in industry, foundations, educational institutions, insurance carriers, labor unions, and government that address the causes, recognition, and control of occupational diseases. These acts established the philosophy that the worker had a right to earn a living without endangerment to health and were the forerunners for the passage of the Occupational Safety and Health Act of 1970.

Passage of the Occupational Safety and Health Act of 1970, which has the purpose of assuring "so far as possible every man and woman in the nation safe and healthful working conditions" had a very broad bearing on the further development and practice of the industrial hygiene profession. These enabling acts, and the regulations deriving from them, have been substantial factors in the broad recognition of industrial hygiene as a science and a profession. It has been necessary to expand the profession in all of its concepts and technical aspects to meet its expanded responsibilities.

Other industrialized countries have had similar experiences in the professional recognition and growth of the science relating to the recognition, measurement, evaluation, and control of work-related health stresses.

4.1 Definition of Industrial Hygiene

The American Industrial Hygiene Association defines industrial hygiene as "that science and art devoted to the recognition, evaluation, and control of those environmental factors or stresses, arising in or from the workplace, which may cause sickness, impaired health and well-being, or significant discomfort and inefficiency among workers or among the citizens of a community." Because the science and practice of industrial hygiene continues to evolve, the association is reviewing this definition for possible revision.

By any definition, however, industrial hygiene is an applied science encompassing the application of knowledge from a multidisciplinary profession including the sciences and professions of chemistry, engineering, biology, mathematics, medicine, physics, toxicology, and other specialties. Industrial hygiene meets the criteria for the definition as a science since it brings together in context and practice an organized body of knowledge necessary for the recognition, evaluation, and control of health stresses in the work environment.

In the early 1900s the major thrust in the control of workplace health stresses was directed toward those areas in industry having massive exposures to highly toxic materials. The professional talents of engineers, chemists, physicians, physicists, and statisticians were those largely used in these programs. As industrial technology advanced, the complexity of workers' exposures also increased, along with an increase in the professional knowledge and skills needed to study the new health effects and to develop the methods for recognition, evaluation, and control of the new environmental stresses. The need for new knowledge and skills continues now with the advent of high technology in the electronic and allied industries. Factors such as improper lighting and contrast, glare, posture, fatigue, need for intense concentration, tension, and many other stresses arise in the operation of computers, word processors, video display terminals, and laser, microwave, and ionizing radiation equipment, which are becoming commonplace in industry. Thus, concerns for the health of employees above and beyond that of toxicity response arise. The study and control of these nonchemical stress agents point to the need for the occupational health nurse, psychologist, human factors engineer, ergonomist, and others to join the professional team in studying the effects and control of the ever-widening list of health stress agents in places of work.

In the early practice of occupational health nursing, emphasis was placed on such activities as the emergency treatment of traumatic injuries stipulated in written orders of a physician and in maintaining records and information relating to physical examinations and the like. With the current advanced training of occupational health nurses, this limited role has been found to be wasteful of professional talent and resources. Occupational health nurses are often the first interface between workers and pending health problems and are in a position to gain information on situations and health stresses both on and off the job that may, unless addressed, lead to more serious responses. Occupational health nurses have increasingly become members of the multidisciplinary team needed in the recognition of job-associated health stresses. Similarly, psychologists, in the study of effects of strain, tension, and similar stresses, and human factors engineers and ergonomists in designing machines, tools, and equipment compatible with physical and morphologic limitations of workers, are examples of other professionals joining the multidisciplinary team studying the effects and control of the increasingly complex health stresses associated with advanced industrial technology.

The complexity of the multifaceted professional effort needed for carrying out the responsibilities of professional practice in the protection of worker health is further illustrated in the 28 technical committees of the American Industrial Hygiene Association and the 7 different specialty areas of certification that have been used in the past by the American Board of Industrial Hygiene.

4.2 Rationale of Industrial Hygiene Practice

The practice of industrial hygiene is based on the following observations, experiences, and rationales:

1. Environmental health stresses in the workplace can be quantitatively measured and expressed in terms that relate to the degree of stress.
2. Stresses in the workplace, in the main, show a dose–response relationship. The dose can be expressed as a value integrating the concentration or intensity, and the time duration, of the exposure to the stress agent. In general, as the dose increases the severity of the response also increases. As the dose decreases the biological response decreases and may at some time and dose value exhibit a different kind of response, chronic versus acute, depending on the time duration of the stress, even though the total stress expressed as a dose–response value may be the same.
3. The human body has an intricate mechanism of protection, both in preventing the invasion of hostile stresses into the body and in dealing with stress agents once invasion has occurred. For most stress agents there is some point above zero level of exposure that the body can tolerate over a working lifetime without injury to health.
4. Levels of exposure of workers to specific stress agents should always be kept within prescribed safe limits. Regardless of their type, all exposures in the workplace should be kept at lower than prescribed levels as are reasonably attainable through good industrial hygiene and work practices.
5. Some stress agents may cause serious biological responses among a few workers at such low levels that exposures should be controlled to levels as low as

reasonably achievable regardless of any higher regulatory limit. An example of such an agent is one having genotoxic properties.

6. The elimination of health hazards through process design and/or the use of nonhazardous substitute materials should be the first objective in maintaining a safe workplace. When this is not feasible, recognized engineering or administrative controls should be used to keep exposures within acceptable limits. In some cases, when feasible engineering and administrative controls are insufficient, supplemental programs such as the use of personal protective equipment or other control practices have application.

7. Surveillance of both the work environment and workers should be maintained to assure a healthful workplace.

4.3 Elements of an Industrial Hygiene Program

The purpose of an industrial hygiene program is to assure a healthful workplace for employees. It should include all the functions needed in the recognition, evaluation, and control of occupational health hazards associated with production, office, and other work. This requires a comprehensive program designed around the nature of the operations, documented to preserve a sound retrospective record, and executed in a professional manner.

The basic components of a comprehensive program include the following:

1. Coordinated technical activities capable of detecting occupational health stresses in any part or process of the establishment.

2. Capability to conduct or obtain measurement and evaluation activities for the assessment of occupational health stresses anywhere in the establishment.

3. Capability to determine the need for and to obtain and maintain effective engineering and administrative controls necessary for safe and healthful workplaces throughout the establishment.

4. Participation in the periodic review of worker exposure and health records to detect the emergence of insufficiently controlled health stresses in the workplace.

5. Participation in research, including toxicological and epidemiological studies designed to generate data useful in establishing safe levels of exposure.

6. Maintaining a data storage system that permits appropriate retrieval of information necessary for the study of long-term effects of occupational exposures.

7. Assuring the relevancy of the data being collected.

An integrated program is capable of responding to the need for the establishment of appropriate exposure controls, both for current needs and for those that may result from technological advances and associated process changes.

The almost universal availability of high-capacity and powerful desk-top computers that has taken place over the past decade has greatly facilitated the conduct of industrial hygiene programs. A great amount of industrial-hygiene-related software for record keeping, technical reference (e.g., regulations, safety data sheets, etc.), sampling data analysis, exhaust ventilation design, and other such industrial hygiene functions has become available from commercial sources or through professional journals, professional associations, and individual industrial hygienists. The American Industrial Hygiene Association has a Computer Applications Committee whose mission is to provide a forum for advancing the use of computer applications by occupational and environmental health professionals. Among other activities this committee reviews and reports on available software.

The industrial hygienist at the corporate or equivalent level should report to top management. His or her responsibility involves appropriate input whenever product, technological, operational, or process changes, or other corporate considerations, may have an influence on the nature and extent of associated health hazards. When new plants or processes are planned, the corporate industrial hygienist should assure that adequate controls are incorporated at the design stage.

5   HEALTH HAZARD RECOGNITION

An important aspect of a responsive industrial hygiene program is that it is capable of recognizing potential health hazards or, when new materials and operations are encountered, to exercise expert judgment in maintaining an adequate surveillance program until any associated health hazards have been defined. This should not be a problem in cases involving operations, procedures, or materials for which adequate knowledge is available. In such cases it is primarily a matter of application of available knowledge and techniques. In operations and procedures involving a new substance or material for which relevant information is limited or unavailable, it may be necessary to extrapolate information from other kindred sources, to use professional judgment in setting up a control program with a reasonable factor of safety, and to incorporate an ongoing surveillance program to further define health hazards that may emerge. In some instances it may be necessary to undertake toxicological research prior to the production stage to define parameters needed in setting up the control and surveillance program.

One of the basic concepts of industrial hygiene is that the environmental health stresses of the workplace can be quantitatively measured and recorded in terms that relate to the degree of stress.

The recognition of potential health hazards is dependent on such relevant basic information as:

1. Detailed knowledge of the industrial process and any resultant emissions that may be harmful
2. The toxicological, chemical, and physical properties of these emissions
3. An awareness of the sites in the process that may involve exposure of workers
4. Job work patterns with energy requirements (i.e., metabolic levels) of workers
5. Other coexisting stresses that may be important

This information may be expressed in a number of ways depending on its ultimate use. A very effective form is a material process flowchart that lists each step in the process along with the appropriate information just noted. This permits the pinpointing of areas of special concern. The effort in whatever form it may take, however, remains only a tool for the use of the industrial hygienist in the actual identification of the stresses in the workplace. In the quantification itself, many approaches may be taken depending on the information sought, its intended use, the required sensitivity of measurement, the level of effort and instruments available, and the practicality of the procedures.

Aside from the production workplace with the attendant toxicological, physical, and other related health stresses, a new area of concern is rapidly gaining special attention where employees may be subjected to a high degree of stress from tension, physical and mental strain, fatigue, excessive concentration, and distraction such as may exist for operators of computers, word processors, video display terminals, and other operator-intensive equipment. Off-the-job stresses, life-style factors, and the immediate room environment may become increasingly important for such workers. The recognition of associated health stresses and their evaluation require a special battery of psychological and physiological body reaction and response tests to define and measure factors of fatigue, tension, eyestrain, deficits in the ability to concentrate, and the like.

6 EXPOSURE MEASUREMENTS

Both direct and indirect methods may be used to measure exposures to stress agents. Table 1.1 illustrates direct and indirect measurements of chemical agents.

Table 1.1 Methods for Measuring Worker Exposure to Absorbed Chemical Stress Agents

Direct	Indirect
Body dosage	Environment
Tissues	Ambient air
Fluids	Interface of body and stress
Blood	Physiological response
Serum	Sensory
Excreta	Pulse rate and recovery pattern
Urine	Body temperature and recovery pattern
Feces	Voice masking, etc.
Sweat	
Saliva[a]	
Hair[a]	
Nails[a]	
Mother's milk[a]	
Alveolar air	

[a] Not usually considered to be excreta.

6.1 Direct Measurements

To measure directly the quantity of a chemical agent actually absorbed by the body—fluids, tissues, expired air, excreta, and so on—must be analyzed for the agent per se or for a biotransformation product. The quantification of a body burden of the agent requires information regarding the biological half-time of the agent or its metabolite as well. Such procedures may be quite involved, since the evaluation of the data at times depends on previous information gathered through epidemiological studies and animal research. Studies on animals, moreover, may have used indirect methods for measuring exposure to the stress agent, necessitating appropriate extrapolation in the use of such values. The current adopted list of Biological Exposure Indices published by the American Conference of Governmental Industrial Hygienists lists 61 determinants (the agent or a metabolite) for 33 compounds and gives notice of intent to adopt 5 more determinants for 5 compounds. Fourteen additional compounds are under study by the committee for establishment of biological exposure indices.

One decided advantage of biological monitoring coupled with information on the biological half-time of an agent or its metabolite is that exposure can be integrated on a time-weighted basis. Such integration is difficult to estimate through ambient air sampling when the exposure is highly intermittent or involves peak exposures of varying duration. Conversely, biological monitoring may fail to reflect adequately the influence of peak concentrations per se that may have special meaning for acute effects. Urine analysis may also provide valuable data on body burden in addition to current exposures when the samples are collected at specific time intervals after exposure. In general, quantitative body burden interpretation of analytic values for a biological specimen requires knowledge of biologic half-time for the agent of interest and an appropriate exposure-sampling time sequence and schedule.

Sampling of alveolar air may be an appropriate procedure for monitoring exposures to organic vapors and gases. An acceleration of research in this area can be anticipated because of the ease with which samples can be collected.

The use of biologic specimens, particularly for purposes of research, ordinarily requires the informed consent of each study subject who provides a sample. This is clearly necessary for an invasive procedure, such as blood sampling, and may apply even to the collection of excreta and exhaled air. Invasion of privacy may be at issue, for example, detection of alcohol consumption, in the analysis of exhaled breath or other excreta.

6.2 Indirect Measurement

The most widely used technique for the evaluation of occupational health hazards is indirect in that the measurement is made at the interface of the body and the stress agent, for example, the breathing zone or skin surface. In this approach the stress level actually measured may differ appreciably from the actual body dose. For example, all the particulates of an inhaled dust are not deposited in the lower respiratory tract. Some are exhaled and others are entrapped in the mucous lining of the upper respiratory tract and eventually are expectorated or swallowed. The same is true of gases and vapors of low water solubility. Thus, the target site for inhaled chemicals is scattered along the entire respiratory tract, depending on their chemical and physical properties. Another example is skin absorption of a toxic material. Many factors, such as the source and concentration of the contaminant, that is, airborne or direct contact, and its characteristics, body skin location, and skin physiology, relate to the amount of the contaminant that reacts with or is absorbed through the skin. For chemical agents the sampling and analytical procedures must relate appropriately to the chemical and physical properties, such as particle size, solubility, and limit of sensitivity of analytical procedures, for the agent being assessed. Other factors of importance are weighted average values, peak exposures, and the job energy requirements, which are directly related to respiratory volume and pulmonary deposition characteristics.

Exposures to physical agents such as noise and ionizing radiation are almost always measured by indirect methods such as dosimetry or assessment of work area intensity levels.

The indirect method of health hazard assessment is, nevertheless, a valid one when the techniques used are the same or equivalent to those relied on in the studies that established the standards.

7 ENVIRONMENTAL EXPOSURE QUANTIFICATION

Procedures for measuring airborne exposure levels of a stress agent depend to a great degree on the reasons for making the measurements. Some of these are: (1) obtaining worker exposure levels over a long period of time on which to base permissible exposure limits, (2) compliance with standards, and (3) performance of process equipment and controls. It is essential that the data be valid regardless of the purpose for which they were collected and that they be capable of duplication. This is a key factor in establishing exposure limits to be used in standards and in fact-finding related to compliance. Since judgment and action will in some way be passed on the data, validity is paramount if the data are to be used as a bona fide basis for action.

7.1 Long-Term Exposure Studies

In epidemiologic studies in which the relationship between a stress agent and the body response is sought, ideally the stress factor would be characterized in great detail. This could require massive volumes of data suitable for statistical analysis and a comprehensive data-collecting procedure so that a complete exposure picture may be accurately constructed. The sampling procedures and strategy should be fully documented, including number and length of time of samples, their locations, their types, that is, personal or area samples, and should be adequate to cover the full work shift activities of the workers. Any departure from normal activities should be noted. These are important since the data may be used at a later date for a purpose not anticipated at the time of sample collection.

This ideal situation is seldom the case in epidemiologic studies. In research on long-term health effects the typical epidemiologic study involves use of surrogates for exposures, or efforts to retrospectively reconstruct exposures. The collection of valid retrospective data may be extremely difficult. If available at all, actual sampling data may be scanty; the sample collection and analytical procedures used in the past may not have been well documented as to precise methodology, and may have been less sensitive and efficient, or may have measured different parameters, than do current procedures; sampling locations and types may not be well defined; and the job activities of the workers may have changed considerably even though the job designations may be the same. Other factors that need to be considered in securing retrospective data relate to contrasting past and current plant operations, including changes in technology and raw materials, effectiveness of exposure control procedures and their surveillance, and housekeeping and maintenance practices. In many instances an attempt to accommodate these differences has been made through broad assumptions and extrapolations with an unknown degree of validity and without expressing the limitations of such derived data.

The effect of national emergencies may significantly change the nature and extent of workers' exposures to stress agents. The experience during World War II is an example. The number of hours worked per week were increased substantially in many industries. Control equipment was allocated to specified industries and denied to others. Local exhaust ventilation systems at times became ineffective or completely inoperable due to lack of maintenance and replacement parts. Less attention was given to plant maintenance, housekeeping, and monitoring procedures. Substitute or lower quality raw materials had to be used in many instances.

Although the major impact of World War II upon levels of exposure to harmful agents occurred from around 1940 to the early 1950s, the effects of many of these exposures may not have shown up among members of that work force and its retirees until decades later.

Thus, expressing exposure levels in the past for more than a few years may be only extrapolated guesses unless factors such as the above can be clearly examined and the data validity established.

7.2 Compliance with Standards

In contrast to the collection of data for epidemiological studies, data collected for the purpose of compliance with standards, as they are now interpreted, may require relatively few samples if the values are clearly above or below the designated value for the agent of interest. If the values are borderline, evaluation may call for a more comprehensive sampling exercise and may be a matter for legal interpretation. The nature and type of samples taken should meet the criteria upon which the standards were based. Scientifically, though, the data should be adequate to establish a clear pattern with no one single value being given undue weight and should meet data analysis requirements.

7.3 Spot Sampling

The exposure of a worker may arise from a number of sources, including the ambient levels of the agent in the general room air, which in turn may be influenced by ambient levels of the agent in the community air, leaks from improperly maintained operating equipment such as from pump seals and flanges, the inadequate performance of engineering control equipment, and the care workers take in performing job operations. Spot sampling can easily detect the effects of any one of these factors on the overall worker exposure level and point the direction for further exposure control action.

8 DATA EVALUATION

The evaluation of the intensity of a physical agent or of airborne levels of a chemical agent to determine compliance with a standard or to determine specific sources of the stress agent are generally uncomplicated and straightforward. The evaluation of environmental exposure data that serve as a basis for determining whether a health hazard exists is more complicated and requires a denominator that characterizes a satisfactory workplace. Similarly, the use of environmental exposure data for establishing safe levels of exposure or a permissible exposure level, as in epidemiologic studies, requires their correlation with other parameters such as the health profile of the workforce.

As pointed out earlier, the early field studies of the 1920s and 1930s showed that when the very high exposures of workers were lowered, there was a corresponding lowering of the related disease incidence in workers. These and other studies gave support to the dose–response rationale upon which the practice of industrial hygiene is primarily based, that is, there is a dose–response relationship between the extent of exposure and severity of biological response to most stress agents and in which the response is negligible at some point above zero level.

There is great difficulty, however, in determining lower levels of exposure to a specific agent and its effect on the health of workers over a working lifetime. Often this is done through extrapolation of other data or by trying to estimate past exposures. In the lower range of the dose–response region, the incidence of disease from exposure to an agent may be so low that it approaches the level for that disease in the community outside the industry under study. This results in part from exposure of the general population to stress agents such as smoking, alcohol consumption, drug use, hobby activities, community and in-house pollution, and the like. These incidental stresses may be similar in magnitude to those on the job or may be additive to, accumulative, or synergistic with stress from on-the-job exposures. For various reasons, often including limited study population size, even well-controlled studies may not be sensitive enough to give data that can be reliably extrapolated to the lowest dose–response region for lifetime exposure.

The problems of estimating dose–response of human populations at low levels of exposure to hazards has given rise in recent years to a new scientific field of endeavor called risk analysis. There is not yet unanimity of opinion among scientist in the field on the most appropriate models and estimating procedures for all types of risk situations, including those involving lifetime exposures to low levels of health stressors.

It is known that the body has protective mechanisms to guard against the effects of low levels of exposure to many environmental agents. For a great number of agents encountered in the industrial environment, data on dose–response relation-ships support the industrial hygiene rationale that the level of exposure does not have to be zero over a lifetime of work to avoid injury to workers' health. Some agents, however, such as those having genotoxic properties (i.e., being able to di-rectly damage genetic material in cells), and perhaps some associated with hyper-sensitivity, may not have a threshold of biologic response, and the lowest achievable level of exposure for workers should be required.

Evaluation of data from exposure stresses relating to tension, fatigue, annoy-ances, irritation, decrements in ability to concentrate and discern, and the like are often subjective and may also involve the personal background, traits, habits, and so forth of those being stressed. Such stresses may require that attention be given to personal behavior for proper definition and control. Evaluation of such situations generally must be done by specialists other than an industrial hygienists.

9 ENVIRONMENTAL CONTROL

The cornerstones of an effective industrial hygiene program can be described as:

1. Proper identification of on-the-job health hazards
2. The measurement of levels of exposure to such hazards
3. Evaluation of all exposure data in context with work schedules and job de-mands
4. Environmental control

In essence, the success of the entire program depends on the success of the control effort. The technical aspects of the program must encompass sound practices and must be related both to workers and to the medical preventive program.

The heart of a control program must rest with process and/or engineering controls properly designed and properly operated to protect workers' health. The most effec-tive and economic control is that which has been incorporated at the stage of process design and production planning, and which has been made part of the process. With new processes this can be accomplished by bringing industrial hygiene input into the bench level, pilot plant, and final stages of process development. It is neither good industrial hygiene practice nor sound economics to neglect exposure control in process design with the intention of adding supplemental control hardware piecemeal as indicated by future production or to comply with regulations.

Although engineering and administrative means should be used wherever feasible to achieve control of exposures, the need may exist for the judicious use of personal protective equipment under unique circumstances, for example, during breakdowns, spills, accidental releases, and some repair jobs. Personal protective equipment should be used only sparingly and under appropriate circumstances, and never as a substitute for more reliable and effective engineering or administrative controls.

Increasingly, engineering controls are being supported with automatic alarm sys-tems to give an alert when controls are malfunctioning and excessive air contami-nation or physical agent intensity is occurring.

The control of stresses associated with high technology in the operation of equipment such as computers, word processors, video display terminals, microwave equipment, and the like requires a different engineering approach from that used in the control of toxic stresses. Providing optimal lighting, adjusting equipment to the operator's stature, and maintaining an overall general room compatibility with tasks are required when such stresses are encountered. Additional considerations including special rest periods and designated exercises may be appropriate.

10 EDUCATIONAL INVOLVEMENT

In the late 1930s very few universities in the United States offered programs leading to degrees in industrial hygiene at either the undergraduate or graduate level. In 1992 more than 30 colleges and universities offer programs leading to undergraduate or graduate degrees in industrial hygiene. This reflects the enormous growth in the profession that has taken place over the past 50 years. The passage of the Occupational Safety and Health Act of 1970 had a major impact on this growth.

The American Industrial Hygiene Association has increased in membership from 160 in 1940 to more than 9000 in 1992. In the 1982–1992 decade the number of Diplomates in the American Academy of Industrial Hygiene more than doubled from 1865 to 4253; in the spring of 1992 there were also 725 industrial hygienists in training.

Professional organizations such as the American Industrial Hygiene Association, the American Conference of Governmental Industrial Hygienists, and the American Academy of Industrial Hygiene offer excellent opportunities for the interchange of professional knowledge and the continuing education of industrial hygienists. These professional organizations invite participation through technical publications, lectures, committee activities, seminars, and refresher courses. As an example, the American Industrial Hygiene Conference and Exhibition of 1992 listed 457 technical papers covering a wide range of subjects. The same conference offered 103 professional development courses for the purpose of increasing knowledge and skills in the practice of industrial hygiene. This participation by experts in the many facets of the profession enhances the overall performance of practitioners and permits industrial hygienists to keep abreast of new technology in the recognition, measurement, and control of workplace stresses. These associations and the American Academy of Industrial Hygiene support the profession in its fullest concept. In return, practicing industrial hygienists are obliged to keep involved in the educational and knowledge sharing process by making professional information available to others who have interest in, and responsibility for, the health and well-being of workers.

Industrial hygienists should take active roles in educating management concerning environmental stresses in places of work and the means for their control. Alert management can bring pending processes and production changes to the attention of industrial hygienists for study and follow-up, thus avoiding inadvertent occurrences of health problems.

The educational involvement of workers is extremely important. Workers have the right to know the status of their working environments, the stresses that may be deleterious to their health if excessive exposures occur, and of the control systems that have been instituted for their protection. Knowledgeable workers are in a position to enhance their own protection through the proper use of control equipment, the proper response to administrative controls, and, where it is needed, to the proper use of personal protective equipment. When a control system malfunctions, a worker is often the first to observe it and can call it to the attention of management. In cases of spills and leaks, or equipment breakdown, a worker who is knowledgeable of the hazardous nature of the materials involved can better follow prescribed emergency procedures. Industrial hygienists are in an excellent position to participate in worker protection education programs.

11 SUMMARY

Gigantic strides have been made during the past four or five decades in characterizing and controlling environmental health hazards in places of work. In many industrial plants where comprehensive industrial hygiene programs are in effect and exposures to work hazards are well controlled, the off-the-job stresses such as smoking, alcohol consumption, drug use, and hobby activities may have greater effect on workers' health than do their on-the-job stresses.

Industrial technology changes rapidly. As technology changes and new technology is applied, new and different on-the-job health stresses emerge. Industrial hygienists must stay abreast of these changes and with the procedures for their recognition, evaluation, and control. It is vital that the techniques used in measuring occupational health stresses cover all the relevant components of each stress and that these are incorporated into any resulting control program. The practice of industrial hygiene rests on having valid data, on proper judgment in evaluating these data, and on effective follow through.

The science and profession of industrial hygiene has a vital role in industry. A well-implemented comprehensive industrial hygiene program leads to a healthful workplace.

The following chapters are devoted to the theoretical basis and rationale for the science and profession of industrial hygiene.

REFERENCES

CMA (1988). *CMA News*, Vol. 16, Number 8, November 1988, Chemical Manufacturers Association, Washington, D.C.

Hatch, T. (1962). "Changing Objectives in Occupational Health," *AIHAJ*, **23,** 1–7.

Hatch, T. F. (1972). "The Role of Permissible Limits for Hazardous Airborne Substances in the Working Environment in the Prevention of Occupational Disease," *Bull. World Health Organization*, Geneva, **47,** 151–159.

CHAPTER TWO

Measurement of Worker Exposure

Jeremiah R. Lynch

1 INTRODUCTION

This chapter explains why workplace measurements of air contaminants are made, discusses the options available in terms of number, time, and location, and relates these options to the criteria that govern their selection and the consequences of various choices. In addition, this chapter will discuss industrial hygiene exposure assessment methods in the broader context of exposure assessment as it is used outside the workplace.

A person at work may be exposed to many potentially harmful agents for as long as a working lifetime, upward of 40 years in some cases. These agents occur singly and in mixtures, and their concentration varies with time. Exposure may occur continuously or at regular intervals or in altogether irregular spurts. The worker may inhale the agent or be exposed by skin contact or ingestion. As a result of exposure to these agents, they come in contact with or enter the body of the worker, and depending on the magnitude of the dose, some harmful effect may occur. All measurements in industrial hygiene ultimately relate to the dose received by the worker and the harm it might do.

Early investigators of the exposure of workers to toxic chemicals encountered obviously unhealthy conditions as evidenced by the existence of frank disease. Quantitative measurements of the work environment to estimate the dose received by the afflicted were not needed to establish cause-and-effect relationships and the

Patty's Industrial Hygiene and Toxicology, Third Edition, Volume 3, Part A, Edited by Robert L. Harris, Lewis J. Cralley, and Lester V. Cralley.
ISBN 0-471-53066-2 © 1994 John Wiley & Sons, Inc.

need for exposure remediation measures. At the same time the ability of these early industrial hygienists to make measurements was severely limited because convenient sampling equipment did not exist and analytical methods were insensitive. Pumps were driven by hand, equipment was large and heavy, filters changed weight with humidity, gases were collected in fragile glass vessels, absorbing solutions spilled or were sucked into pumps, and laboratory instrument sophistication was bounded by the optical spectrometer. To collect and analyze only a few short-period samples required several days of work. The probability of failure due to one of many possible equipment defects or other mishaps was high. Consequently, few measurements were made, and much judgment was applied to maximize the representativeness of the measurements or even as a substitute for measurement.

Changes in working conditions, in technology, and in society have changed the old methods of measurement.

- With few exceptions, workplace exposure to toxic chemicals is much below what is commonly accepted as a safe level.
- As a consequence of the reduction of exposure, frank occupational disease is rarely seen. Much of the disease now present results from multiple factors of which occupation is only one.
- Workers have the right to know how much toxic chemical exposure they receive, and this often results in a need to document the absence of exposure.
- Technology provides enormously improved sampling equipment that is rugged and flexible. This equipment, used with analytical instruments of great specificity and sensitivity, has largely replaced the old ''wet'' chemical methods.

As a consequence of these changes in the workplace and advances in technology, it is now both necessary and possible to examine in far more detail the way in which workers are exposed to harmful chemicals. Personal sampling pumps permit collection of contaminants in the breathing zone of a mobile worker. Pump–collector combinations are available for long and short sampling periods. Passive dosimeters, which do not require pumps, are available for a wide range of gases and vapors. Systems that do not require the continual attention of the sample taker permit the simultaneous collection of multiple samples. Data loggers can continuously record instrument readings in a form easily transferable to a computer. Automated sampling and analytical systems can collect data continuously. Sorbent-gas chromatograph techniques permit the simultaneous sampling and analyses of mixtures and, when coupled with mass spectrometers, identify obscure unknowns. Sensitivities have improved to the degree that tens and hundreds of ubiquitous trace materials begin to be noticeable.

At the same time the demands placed on our information-gathering systems are greater. Now we must not only answer the question: ''Is exposure to this agent likely to harm anyone?'' but provide data for many other purposes. Worker exposure must be documented to comply with the law (Corn, 1976). Employees are demanding to be told what they are breathing, even in the absence of hazard (USPHS, 1977). Epidemiologists need data on substances not thought to be hazardous to relate to possible future outbreaks of disease (Rappaport and Smith, 1991). Design engineers need contaminant release data to relate to control options (Lipton and Lynch, 1987). Process operators want continuous assurance that contaminant levels are within normal bounds. Management information systems that issue status reports when queried require monitoring data inputs. Data needs are so pervasive that there is a tendency to monitor without a clear idea of what the data will be used for or whether it will meet the need. An overall purpose of this chapter is to suggest the objectives that need to be considered in an analysis of the value of exposure measurement.

2 OBJECTIVES OF EXPOSURE MEASUREMENT

The central question that must be asked before measuring exposure is: "What use will be made of the data?" That is, what questions will the data answer or what external information need will be satisfied? The collection of data should be looked on as part of a decision-making process. If no decision is to be made, or if nothing is to be done differently either in the short or long run as a result of the data collected, regardless of the result, then why collect the data? In some situations a correct control decision may be perfectly clear without any measurements, although measurements may serve to reinforce the decision or to convince others. In other cases it is difficult to see how any obtainable data will aid in making decisions or that the cost of obtaining the data needed exceeds the cost of making the wrong decision. To make decisions under uncertainty, as is usually required in industrial hygiene, the techniques of decision analysis (Raiffa, 1970) are useful. These techniques also permit the calculation of the value of information that can then be compared with the cost. While the cost of information for an identified decision is important, it is also useful to consider what other questions will need to be answered or what other information will be needed. The resources available for the measurement of exposure are usually limited so data must serve several purposes. Some of those purposes are discussed in the following sections.

2.1 Hazard Recognition

As a starting point for a complete assessment of the risk to health posed by an occupational environment, it is necessary to know the substances to which workers are exposed. Systematic recognition of all possible hazards requires inventories of materials brought into the workplace, descriptions of production processes, and identification of any new substances by-products or wastes. However, these sources of information may not be enough to identify all substances, particularly those present as trace contaminants or substances generated by production process, either inadvertently or as unknown by-products. To complete the identification of all substances present, before going to the next step of evaluating exposure and risk, it may be necessary to make some substance recognition measurements. Since these measurements, which are typically made by such techniques as gas chromatography— mass spectrometry (GC–MS), are not intended to evaluate exposure, they may be area rather than personal samples and may be large-volume samples for maximum sensitivity.

2.2 Exposure Evaluation

The most common reason for measuring worker exposure to a toxic chemical is to evaluate the health significance of that exposure. These evaluations are usually made by comparing the result with some reference level. Traditionally, the threshold limit values (TLVs) for airborne contaminants of the American Conference of Governmental Industrial Hygienists (ACGIH, 1991) have been used to represent safe levels or "conditions under which it is believed that nearly all workers may be repeatedly exposed day-after-day without adverse effect."

The introduction to the TLV list goes on to discuss the classes of workers who may not be protected by the limits and notes that: "These limits are not fine lines between safe and dangerous concentrations nor are they a relative index of toxicity. They should not be used by anyone untrained in the discipline of industrial hygiene." The reason for this last stipulation is that industrial hygiene training covers the caveats that apply to the TLVs so that appropriate judgment can be used in their application to take into account their uncertainty and to properly describe to workers and management what results are to be expected from using the TLVs.

In 1970 the U.S. Occupational Safety and Health Administration (OSHA) was given the responsibility for setting legally enforceable permissible exposure limits (PELs) for U.S. workplaces (OSHA, 1970, 1992). The ACGIH TLVs were the source of the original PELs and were one of the sources of a later update. Other countries have also used the TLV as a basis for their standards, but more recently they have established their own standard-setting mechanisms.

Unfortunately, established limits like the OSHA PELs, ACGIH TLVs, and AIHA Workplace Environmental Exposure Limits (WEELs) cover only a small fraction of the chemicals that occur in industrial workplaces albeit the most common ones. Where there is exposure to a substance for which there is no established or recommended limit, it may be necessary to develop a supplemental standard for use in a particular plant or company. Standards for substances whose toxicology is not well known are generally set to avoid acute effects in man or animals and, often by analogy to other better-documented substances, at a level low enough to make chronic effects unlikely. When very little is known about a substance, it may only be possible to estimate a lower level at which it is reasonably certain that no adverse effects occur and an upper level at which adverse effects are likely. The width of the gap between these two levels is a zone of uncertainty that needs to be considered in evaluating the results of exposure measurements.

2.3 Control Effectiveness

When changes in equipment or processes are made that affect the release of substances that are contributing to worker exposure, measurements of the magnitude of that change may be needed. These measurements provided empirical data on control effectiveness to confirm design expectations or to use as a basis for the design of other modifications. In the simplest case, before-and-after measurements are made when a new control, such as a local exhaust hood, is installed on a contaminant release point. From the results of these measurements, it is possible to predict the reduction in worker exposure, which can be confirmed by subsequent exposure measurement.

Unfortunately, the situation is rarely that simple. Most worker exposures are caused by multiple release points, creating a work environment of complex spatial and temporal concentration variations through which the worker moves in a not-altogether-predictable manner. Furthermore, interaction between several release points or other factors in the environment may confound the results. A needed improvement such as a new exhaust hood may seem to be without effect because the building is air starved, or a poor hood may seem to function well because of exceptional general ventilation. The time of contaminant release may depend on obscure and uncontrollable process operation factors. As with measurements made for other purposes, control evaluation studies must be carefully designed and are likely to consist of a series of factorial measurements analyzed by statistical methods.

2.4 Model Validation

Various physical models have been developed to predict contaminant concentrations in a workplace (Jayjock, 1988; Ryan, 1991; USEPA, 1991). These models may describe near-field dispersion from a source such as a valve leak or an open tank, or the mixing of a contaminant in an enclosed space. The models can be used to predict exposure to a chemical that has not yet been manufactured, to estimate past exposure, and to extend the range of exposure measurement to situations not measured. As physical models, they are based on first principles (heat and mass transfer) and an empirical observation (Gaussian dispersion). The models are likely to be accurate when the assumptions made in developing the models are true. However, these assumptions are rarely perfectly true, and so it becomes important to know how sensitive the model is to deviations from the assumptions. To learn this, it is necessary to validate the model by comparing the model with actual exposure measurements over a wide range of conditions. The number of physical models in use has increased more rapidly than validation studies, so there are a number of unvalidated models.

Statistical models based on exposure data coupled with factors, such as job tasks, believed to be associated with exposure are also used to estimate worker exposure (Nicas and Spear, 1992). Some statistical technique such as regression analysis is used to combine the data so that the factors contributing to exposure may be used as independent variables to predict unknown exposures. Since the model is developed from the data, there is no need for validation as long as the model is applied within the range of the measurements used to create it. Additional measurements may be used to extend the model, increase confidence in the output, or to replicate the model in apparently similar circumstances.

2.5 Methods Research

Industrial hygiene methods development research hypotheses often take the form: ''Will sampling and analytical method A give the same result as method B?'' If we are unable to reject this hypothesis in a carefully designed experiment, then we accept that methods A and B are equivalent within our limits of error and given the bounds of the experimental conditions (Thompson et al., 1977; Donaldson and Stringer, 1976; McCammon and Woodfin, 1977; Horowitz, 1976).

Sampling and analytical methods development research usually requires extensive laboratory work, but in most cases field testing is necessary because completely realistic environments with interferences usually cannot be generated in laboratory chambers, and the difficulties of making field measurements introduce errors that may affect one method more than another. For these reasons, most practicing industrial hygienists tend to distrust assertions of equivalence that are not backed up by field data. To be credible, experiments of this kind should clearly define the range of concentrations and conditions over which the equivalence has been tested. Personal versus area equivalence of coal mine dust measurements made in long-wall mines should not be assumed to hold in room-and-pillar mines. Manual versus automated asbestos counting relationships based on chrysotile do not apply when counting amosite fibers.

Enough data should be collected not only to determine whether the methods are correlated but also to determine the ability of a measurement by one method to define the confidence limits on a prediction of the result that would have been obtained by the other method (Fig. 2.1). Often a high correlation coefficient is obtained when many pairs of measurements have been made, indicating that the two methods are certainly related; yet the scatter is such that one method may only be used as a

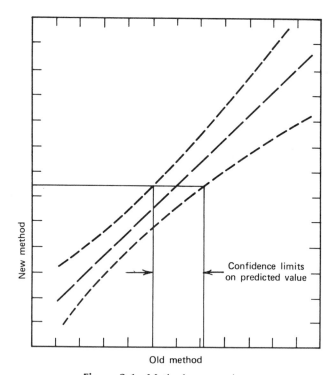

Figure 2.1 Methods comparison.

predictor of the other method to within an order of magnitude. The design of experiments for the purpose of measuring method equivalence is discussed elsewhere in this series. Considerations such as the environmental variability related to location, time, and numbers, which are described below, should be taken into account, to maximize the range of conditions over which the equivalence is evaluated without introducing so much error that the relationship has no predictive ability.

2.6 Source Evaluation

Measurements can be made to evaluate the magnitude or trends in the generation of air contaminants at the source. These measurements are made to detect leaks, inadvertent loss of control, or other events that may cause a change in the amount of contaminant released into the workplace. Patterns of contaminant generation may change because of loss of temperature control in a vessel or tank, dullness of a chisel in a mine or quarry, or bacterial contamination of a cutting oil. These events cannot always be detected by changes in process parameters, and it may be unacceptable to wait until they show up in exposure measurements because the margin for error in a control system designed for every strict standard, such as vinyl chloride, may be too small, or as in the case of hydrogen sulfide, because the consequences of overexposure, even for a short period, are too serious.

Automatic leak detection systems may be installed where it is important to instantly detect any leak, or leaks may be detected by periodic manual surveys that check spot concentrations near pump and valve seals, flanges, and so forth. Various environmental laws administered by the U.S. Environmental Protection Agency (USEPA) require the monitoring of equipment leaks for reporting purposes under the Superfund Amendments and Reauthorization Act (SARA) Toxic Release Inventory and for the control of fugitive emissions of volatile organic compounds (VOC) or volatile hazardous air pollutants (VHAP). These emissions are the same emissions that are the sources of most worker exposure in chemical plants and oil refineries. The control of these emissions will reduce worker exposure so leak detection and repair (LDAR) programs required for environmental reasons also serve occupational health objectives. Substantial work has been done to develop and evaluate methods for this purpose (CMA, 1989; USEPA, 1988, 1990).

2.7 Epidemiology

Occupational epidemiology is concerned with the relationships between occupational factors and disease trends. The kinds of conclusions the epidemiologist may draw from a study depend on the kind of data used in the study. If the epidemiologist only knows the industry or place of employment of the individuals in the study group, then the conclusions can only be in the form that a disease excess may be associated with work in that industry or establishment. This is a scientifically sound conclusion, but it does not lead the managers and industrial hygienists responsible for worker health to the chemical agent, if any, which may be causing the disease. To come closer to useful causal relationships, the epidemiologist needs to know to what substances each worker was exposed. With data on the degree of exposure it is possible for the epidemiologist to detect dose–response relationships that can aid in confirming a causal relationship between an agent and a disease. If the degree of exposure is accurately known over time, a dose–response relationship, which can be used to estimate safe levels of exposure, may be calculated.

Ideally, the epidemiologist would like to have measurements of the exposure of all workers to all substances over time from the beginning of employment. In the National Coal Board study of coal workers pneumoconiosis in the United Kingdom, exposure and health status were measured over a long enough period so that the results could be used as a basis for the present coal dust standard in the United States. Major prospective studies such as this, however, are rare. Historically, studies of causality and dose–response for chronic diseases of occupational origin used employment within an industry and/or job as a surrogate for exposure. This resulted in major misclassification of employees and/or of exposure categories with the result that causal relationships were obscured and significant associations not found (Stewart and Herrick, 1991). As a result, most modern occupational epidemiologists conducting retrospective studies enlist the aid of industrial hygienists to reconstruct exposure by making use of whatever exposure measurements are available. In addition, they will use data on plant, process, and maintenance events that influence exposure and employee recollection of exposure conditions. Typically, this assembly of information is converted into a job time–exposure matrix by one or a panel of several experienced industrial hygienists. Validation studies may be used to compare judgments with known exposures. These methods are being continuously improved so that the degree of reduction in misclassification now depends largely on the effort put into retrospective exposure assessment (Stewart et al., 1991; Rappaport, 1991).

Prospective exposure assessment for epidemiology is hindered by the fact that most industrial hygiene measurements are made for purposes other than epidemiology (i.e., hazard evaluation, problem solving) and do not represent the exposure of the whole population. Harris (1993) has addressed the question of what the plant industrial hygienist, who has day-to-day responsibility for worker protection, can do to provide for future epidemiology, without displacing more immediate and urgent work. These actions include taking more samples chosen to represent the exposure of all groups of worker, including those not at risk of overexposure by current standards, and to represent exposure to substances not presently known to present a special risk. Obviously, the development of such a program is complicated by our inability to predict what workers or substances will be of interest in future research. It may be that a few ''fingerprint'' samples, analyzed by such detailed methods as capillary GC with mass spectroscopy, will be the best choice to generally characterize the kinds of exposures that are occurring. In addition to more sampling, the collection and preservation of process, maintenance, and job history data will greatly benefit future exposure reconstruction. When the resources are available and the need for future epidemiology is evident, exposure estimation schemes may be used to fill in the gaps between what is measured and what is not. These schemes rely on experience based on observation to place workers in exposure categories over time.

In deciding what data to collect for use in future epidemiologic studies, it should be known if the data will meet the need. Close, early cooperation between industrial hygienists and epidemiologist is needed to avoid expensive data collection programs that fulfill no need.

2.8 Illness Investigations

When an employee has a frank occupational disease, such as lead poisoning, confirmed by both physical findings and analysis of biological materials, or is known to have been overexposed based on analysis of biological materials, the industrial hygienist should determine the cause (source) of the overexposure. When the conditions that led to the overexposure still exist, they may be evaluated by measurements made after the event. It is also possible to evaluate overexposure that resulted from past episodes that were not observed and evaluated when they occurred. A history of past exposure opportunities can be constructed and used to estimate those that led to the present case. In some instances, it may be necessary to reenact or simulate an event to measure what may have happened—being careful, of course, to ensure that all participants are protected. Exposure, obviously, need not always be by inhalation and may include off-the-job activities.

A much more difficult investigation is the search, in an occupational setting, for the cause of an outbreak of illness or complaints of illness that may or may not be of occupational origin, or even if related to occupation may result from factors other than exposure to toxic substances. Marbury and Woods (1991, p. 306) relate that "since the early 1970s, outbreaks of work-related health complaints have occurred in large numbers in a wide variety of nonindustrial workplaces such as hospitals, schools, and office buildings. In some cases, careful evaluation of the building or the affected population has revealed an agent responsible for the outbreak. In most cases, however, no specific etiologic agent can be identified as its cause."

The assessment of indoor air quality and its relation to work-related health complaints involves environmental and personal monitoring for such known agents as environmental tobacco smoke, carbon monoxide, nitrogen dioxide, formaldehyde, and volatile organic compounds (VOC). In addition, carbon dioxide can be used as a surrogate for the absence of fresh (outdoor) air. Indoor air can also be a vector for the transmission of infectious disease organisms such as *Legionella*. Beyond the measurement of air contaminants, the resolution of indoor air problems involves detailed knowledge of human responses, building system performance, and factors affecting the service of the building air handling plant. Indoor air quality assessment is discussed in detail by Samet and Spengler (1991) and Nagda et al. (1987).

2.9 Legal Requirements

Section 6b7 of the Occupational Safety and Health Act provides for "monitoring or measuring employee exposure at such locations and intervals, and in such a manner as may be necessary for the protection of the employees" (OSHA, 1970). Under this act, OSHA has responsibility for establishing exposure limits (PELs) in the working environment. Elsewhere in the act OSHA is required to set standards that require monitoring to be performed. Responsible and effective implementation of the congressional intent behind these provisions require that OSHA devise a scheme requiring monitoring where it will be of value in the protection of worker health and not elsewhere. The ideal regulation should not apply to the majority of establishments, which have no conceivable hazard resulting from the substance being regulated and should apply requirements of increasing strictness to other establishments as the significance of the hazard in the establishment increases, ultimately calling for measurements of sufficient frequency to ensure that the potential for harm is fully assessed in the few establishments where exposures are great enough to create significant risk. Further, this should be done by a regulatory scheme that is easy to understand and implement.

This sorting of workplaces by level of risk can be done by prescribing a series of thresholds or triggers that lead to increasingly stringent requirements for a decreasing number of employers. First, all employers who do not have the substance present in the workplace should not be required to monitor. Although the "presence" of a material seems a simple enough criterion that everyone would interpret in the same way, the extreme bounds of interpretation, which are of concern in legal arguments, include the presence of as little as a few molecules of a gas or a single asbestos fiber. As analytical techniques become more sensitive, almost everything is to be found almost everywhere, at least at the level of a few molecules. What is needed in a regulation is an exclusion, such as a percentage concentration in a liquid, below which the presence of a substance is of no health significance.

For employees who work in a place where a substance is present above the excluded level, the next step should be to determine whether there is any possibility that the substance is released into the workplace such that workers may be exposed. The setting of this threshold must reflect consideration of the conditions under which the substance is present and the consequences of release. Thus, nuclear reactor decay products are continuously monitored against the possibility of leaks even when they are hermetically sealed. Such high toxicity materials are not released into the workplace except under very rare emergency circumstances. On the other hand, cadmium released into the workplace as a result of silver soldering should be monitored, but it is not necessary to measure exposure to cadmium where cadmium-coated auto parts are stored. Since no simple "potential for release" trigger has yet been devised, there is some regulatory error (employers included who should not have been, and vice versa) at this decision point. A further step is needed, therefore, before a full monitoring program with its consequent expense is mandatory. One step is to use a small number of measurements of the exposure of the maximum risk employee under conditions when the exposure is likely to be the greatest. If the results of these measurements are sufficiently below the PEL so that it is possible to be confident that the PEL is not likely to be exceeded, then further monitoring to demonstrate compliance with the PEL is not necessary. However, any change in circumstance that may increase the risk is cause for a reevaluation. This initial measurement scheme, however, is not appropriate with a highly toxic substance such that continued vigilance must be maintained against the possibility of leaks or other inadvertent releases.

In cases where it has been established, by means of an initial measurement or data from other sources, that significant exposure is occurring, possibly over the PEL on occasion, a regular program of periodic monitoring should be required. The frequency of monitoring should relate to the level of exposure and should consider trends between measurements that might lead to conditions with unacceptable consequences.

Monitoring programs undertaken to meet legal regulatory requirements also meet the needs of worker protection if the PEL is appropriate and if unregulated substances and risk situations are evaluated by other means.

2.10 Routine Monitoring

Many employers attempt to meet most of the purposes of exposure measurement through a simple routine monitoring program. Schemes for the logical stepwise analysis of data to arrive at decisions regarding monitoring have been developed by the U.S. National Institute for Occupational Safety and Health (NIOSH) (Leidel et al., 1977), European Council of Chemical Manufacturers' Federations (CEFIC, 1982), the West German Federal Ministry of Labor (BMA, 1979), and the American Industrial Hygiene Association (Hawkins et al., 1991). The CEFIC occupational exposure analysis flowchart is shown in Figure 2.2. At the start a chemical inventory of products, by-products, intermediates, and impurities is assembled and annotated with data about hazards, limit values, regulations, and standards. The hazard (potential for exposure) is assessed based on the process, equipment, material volume, temperature, pressure, ventilation, work practices, and precautions. This information is analyzed to determine where and when substances may be released into the workplace and what exposures are possible as a result. The exposure status of workers in work areas identified by this analysis is then initially assessed using such a priori information as earlier measurements or computed concentrations based on comparable installations, work processes, materials, and working conditions. A compliance evaluation based on this computation is now made if possible; if not, exposure measurements are made of the maximum risk (most exposed) employee for the job function under study. If the results are out of compliance, exposure reduction measures are taken and the process repeated. If conditions are in compliance but greater than a decision level (DL), then an occupational exposure monitoring protocol is developed and implemented. If exposures are below the DL, the process stops. The DL, which is expressed as a fraction of the occupational exposure limit, is based on judgment. In general, it would not be greater than 0.5, would usually be 0.25, and may be as low as 0.1 of the occupational exposure limit (OEL) for special circumstances such as carcinogens. Unlike the OSHA action level, the DL is used only for monitoring decisions, not for decisions involving training, medical examinations, and so forth.

The above scheme is one approach to the general problem of designing employee exposure monitoring programs that fit the need. Several investigators (Harvey, 1981; Rappaport et al., 1981a,b; Tuggle, 1981, 1982; Rock, 1982) have commented on the limitations of various other schemes and have proposed alternatives. In general, exposure assessment strategies fit the pattern shown in Figure 2.3.

3 SOURCES OF WORKER EXPOSURE

The core concern of industrial hygiene is the prevention of disease arising out of the workplace. Toxic substances cause disease when some amount, or dose, enters the body or comes in contact with it. Workers are exposed to toxic substances by in-

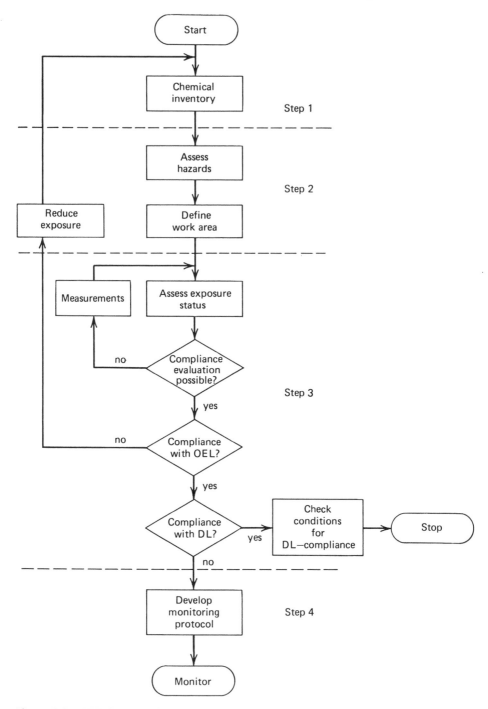

Figure 2.2 CEFIC occupational exposure analysis chart. The occupational exposure level (OEL) and decision level (DL) used depends on the substance being evaluated.

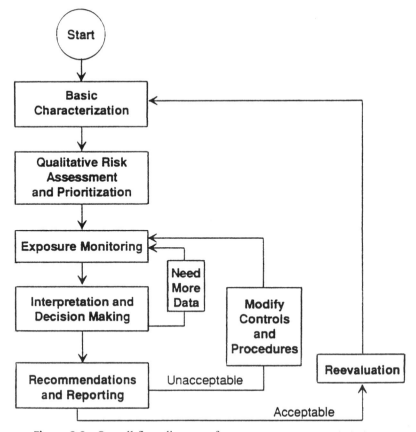

Figure 2.3 Overall flow diagram of an exposure assessment strategy.

halation, ingestion, skin contact, and even, under rare industrial circumstances, ingestion. To accurately assess the total exposure of a worker, it is necessary to understand the sources and characteristics of exposure.

3.1 Production Operations

Industry is generally thought of in terms of continuous repetitive operations that generate air contaminants to which workers are more or less continuously exposed. Paint is sprayed on parts passing continuously in front of a worker. Dust is generated by a foundry shakeout on a continuous casting line every few seconds. Fumes seep steadily from cracks in aluminum smelting pot enclosures. Welders join structural members on a production basis, with only short breaks between welds. Operators watch controls in the midst of chemical plant air contaminated by fugitive releases. In some cases, such as a grinder cleaning sand from a casting, the concentration of the contaminant is closely related to the work performed and is probably higher in the workers breathing zone than it would be several feet away. Other workers, such as dorfers and creelers in a cotton spinning mill, are exposed to dust released from hundreds of bobbins, and their own activities, short of leaving the workplace, have little effect on their exposure. All these continuous exposures are actual rather than potential and present the least difficulty in evaluation.

3.2 Episodes

Much of the exposure workers receive occurs as a result of events or episodes that occur intermittently. Glue is mixed on Wednesday. The shaker mechanism breaks and a mechanic must enter the baghouse. Coke strainers preceding a pump need to be dumped when the pressure drop becomes excessive. A drum falls off a pallet and ruptures on the floor. Samples of product are taken every two hours. A pressure relief valve opens. A packing gland bursts. A reactor vessel cover is driven from its hinges by overpressure in the vessel.

These exposure events can be periodic or they can occur at irregular intervals. They may be planned and predictable or altogether unanticipated. Some events are frequent and result in small exposures, whereas others may be catastrophic events causing massive exposures and even fatalities. As a class, episodic exposure events result in a significant fraction of the exposure burden of many workers and may not be ignored. While their evaluation is extremely complex, and often only broad estimates of the probability of an unlikely event and the consequent risk may be available, it is useful to consider episodes separately from continuous operations since they may be missed altogether by routine monitoring programs. Some possible ways to "catch" unpredictable events include continuous, instantaneous real-time monitors (usually area monitors), discussions with workers and supervisors on the details of processes and events, and a nonthreatening incident reporting system. The analysis of episode data will be discussed later.

3.3 Dermal Exposure

When relating exposure to biological effects, either in an epidemiologic study or because of an outbreak of illness that might be related to occupation, the total dose is the relevant quantity, not merely the dose received by inhalation. Amounts of a substance entering the body by any route may contribute to the total dose.

Many substances, especially fat-soluble hydrocarbons and other solvents, can enter the body and cause systemic damage directly through the skin when the skin has become wet with the substance by splashing, immersion of hands or limbs, or exposure to a mist or liquid aerosol. Some substances, such as amines and nitriles, pass through the skin so rapidly that the rate at which they enter the body is like that of substances inhaled or ingested. The prevention of skin contact to phenol is as important as preventing inhalation of airborne concentrations. A few drops of dimethyl formamide on the skin can contribute a body burden similar to inhaling air at the TLV for 8 hours.

The "skin" notation in the TLV list identifies substances for which skin absorption is potentially a significant contribution to the overall exposure. While the TLV list offers no precise definition of a "significant contribution," Fiserova-Bergerova et al. (1990, p. 630) have proposed the following two criteria for the skin notation.

- "Dermal absorption potential, which relates to dermal absorption raising the dose of nonvolatile chemicals or biological levels of volatile chemicals 30% above those observed during inhalation exposure to TLV-TWA only."
- "Dermal toxicity potential, which relates to dermal absorption that triples biological levels as compared with levels observed during inhalation exposure to TLV-TWA only. (This toxicity criterion may not be valid for TLVs based on irritation or discomfort.)"

Note that the skin notation relates to absorption or toxicity via a skin route and not to the potential to cause skin damage and dermatitis. These later effects are important, however, and should be indicated on such hazard data sources as material safety data sheets (MSDSs).

For some substances with low vapor pressure, like benzidine, skin absorption is the most important risk. Benzene, on the other hand, though absorbed through the skin, is absorbed at such a low rate that skin contact probably contributes little to the body burden. To judge the degree to which skin absorption is contributing to exposure, it is necessary to consider both the rate of absorption and the degree of contact. Clothing wet with a substance that remains on the worker for prolonged periods provides the maximum contact short of immersion. On the other hand, the poultice effect does not occur on wet unclothed skin; thus, evaporation can take place and the result is less severe. Contact with mist that does not fully wet the clothes or body but merely dampens them is not as severe as being splashed with the bulk liquid. Some substances, which by themselves do not significantly penetrate the skin, can have their absorption significantly enhanced by the presence of vehicles in solutions or mixtures. The TLV list notes that direct skin contact with the liquid substance is probably of greater importance than contact with the vapor. This follows from the fact that the gas in contact with the skin contains mostly (>99%) air for typical occupational vapor concentrations. Human and animal studies have estimated that dermal absorption of vapor accounts for between 0.1 and 10 percent of total uptake including inhalation for nonpolar vapors such as benzene, xylene, and others. These facts lead to the general conclusion that the absorption of vapor through the skin is not a problem. An exception appears to be 2-butoxyethanol where dermal uptake from ambient air was found to be appreciably higher than respiratory uptake, particularly in hot, humid conditions (Johanson and Boman, 1991). This is important because clothing that protects against liquid contact will not necessarily prevent vapor contact.

Protective clothing that is impervious to the substance will reduce absorption to nil on protected areas (Oxley, 1976). However, a leaky glove that has become filled with a solvent is providing contact with the hand equivalent to immersion. Barrier creams are often used to prevent dermatitis but may not always prevent skin absorption (Lauwerys et al., 1978). Although skin contact and absorption must be considered as contributing to the dose for many materials, few quantitative data on rate of absorption are available, and these are in a form that is difficult to apply in an industrial setting (Fukabori, 1976a,b). Furthermore, such factors as part of the skin exposed, sweating, and the presence of abrasions or cuts can cause order of magnitude differences (Cronen and Stoughton, 1962).

Dermal exposure assessment is a complex matter that has not received much attention in the practice of industrial hygiene. Estimation of dermal exposure involves exposure scenarios and pathways, contact duration and frequency, body surface area in contact, and substance adherence. The amount of contaminant that crosses the skin barrier and enters the body is influenced by the properties of the substance and the properties and condition of the skin at the exposed site. Models of dermal absorption have been developed (Berner and Cooper, 1987; Flynn, 1990). These models start with the partition coefficients of the substance, its mole weight and solubility, and the diffusivity of the compound in the lipid and protein phases of the skin. Complex theoretical models have been simplified and combined with empirical models to yield an estimate of the chemical specific dermal permeability constant. The dermally absorbed dose (mg/kg day) can then be calculated from this constant and the skin contact area, exposure time, frequency and duration, and body weight. Since these quantities are rarely known, they must be assumed. When conservative assumptions are used, the conservatisms may compound to yield a very conservative estimate of the dermal dose.

3.4 Ingestion

Although ingestion is an uncommon mode of exposure for most gases and vapors (Key, 1977), it cannot be ignored in the case of certain metals such as lead. Indeed, for workers exposed in pigment manufacture and use, ingested lead, either coughed up and swallowed or taken on food, though less well adsorbed than inhaled lead, constitutes a significant fraction of the total burden. Spot tests (Weeks, 1977) and tests with tracers have shown that materials present in a workplace tend to be widely dispersed over surfaces, thus support the essential rule prohibiting eating and smoking where toxic substances are present.

3.5 Injection

Toxic substances may pass through the skin by intentional or unintentional injection. Opportunities for all kinds of materials to enter the body by injection connected with drug abuse or therapeutic accident are obvious. Bulk liquids may also break through the skin and enter the bloodstream without the aid of a needle when driven into the body as high-velocity projectiles released from high-pressure sources. Airless paint spray and hydraulic systems (LeBlanc, 1977) often use pressures in this range, and such pressures often occur inside pipes and vessels in chemical plants. Inadvertent cracking of a flange under pressure can cause the traumatic introduction of a toxic substance into the body. Solid particles may also enter this way; and if soluble or radioactive, they may cause damage beyond the initial injury.

3.6 Nonoccupational Exposure

In addition to the exposure to toxic substances a worker receives at work, an increment of dose may also be delivered during nonworking hours. Air (Sterling and Koboyashi, 1977), water (Wigle, 1977), and food all contain small amounts of toxic substances that may also be found in the occupational environment. As a rule, the nonoccupational dose is at least an order of magnitude below the occupational dose, just as community air standards are much lower than TLVs, and many substances present in industry are very rare in the community. Yet, in certain cases, these pollutants have an impact. The consequence of arsenic exposure resulting from copper smelting in northern Chile is difficult to assess since arsenic poisoning is endemic there as a result of naturally occurring drinking water contamination. Chronic bronchitis resulting from air pollution in the industrial midlands of England is confounded with the lung diseases of coal miners. The cardiovascular consequences of urban carbon monoxide exposure may be not unlike those of marginal industrial exposures. The benzene exposure of smokers probably exceeds typical present-day occupational exposure.

Wallace et al. (1988) have shown that indoor exposure to chemicals are often higher than outdoor although the compounds involved may be different. This fact, together with the observation that people spend 80–90 percent of time indoors (Ott, 1989; Lioy, 1990) makes indoor exposure important. The sources of indoor exposure in homes are passive smoking, cooking, off-gasing of furniture and building materials, perchloroethylene from dry-cleaned fabrics, household chemicals and pesticides, release of VOCs in water from showers and washing machines, and infiltration of outdoor air pollutants. Office buildings have similar sources plus office machines (Hodgson et al., 1991).

Many workers have hobbies or other leisure activities that result in exposure to chemicals and physical agents (McCann, 1979; Helper et al., 1979; Hart, 1987). Some of the specific hazards are lead frits in ceramic glazing, solvents in coatings, benzidine compounds in fabric dyes, cadmium in silver solder use for making jewelry, plastic hardeners and curing agents, silica flour used in lost wax casting, welding fumes in metalworking and auto repair, and noise from loud music and target practice. Lead exposure on police and presumably private firing ranges is significant. Epoxies are used in home workshops, and garden chemicals contain a variety of economic poisons. Although data are scarce, it seems likely that leisure activity exposure to toxic substances rivaling high but permissible work exposure is rare. Yet, significant overexposure does occur as evidenced by the fact that cases of illness have been reported as a result of such exposures.

Intentional exposure to solvents for their narcotic effects is well documented (Poklis and Burkett, 1977; Oliver, 1977; Korobkin et al., 1975; Hayden et al., 1976; Warriner et al., 1977; Weisenberger, 1977). Glue sniffing, gasoline sniffing, ingestion of denatured alcohol, methanol, and even turpentine have been reported. This category of chemical abuse results in exposure and often damage far beyond that normally encountered in industry. The industrial hygienist must be alert to the possibility that a case of disease may be related to intentional addictive exposure.

4 CHARACTERISTICS OF EXPOSURE AGENTS

The physical and chemical properties of a substance are important because of the effect they have on exposure measurement quite apart from their effect on the magnitude of the exposure and its consequences. The most important properties of any substance that affect sampling are those that determine whether it can be collected on a sampling medium and treated or removed in a manner that permits analysis. Many vapors are rapidly adsorbed and desorbed from one or another of a variety of solid sorbents, but some high boiling materials are very difficult to desorb and some gases do not adsorb well. Similarly, most dust particles are collected efficiently by membrane filters with 0.8 μm and even larger pore sizes; however, fresh fumes may pass through and hot particles can burn holes. Beyond these apparent physical and chemical dependencies of exposure measurement methods, there are some less obvious complications that frequently occur.

4.1 Vapor Pressure

Since liquids with high vapor pressure tend to evaporate completely when they are aerosolized, it is rarely necessary to measure them as mists. Liquids with very low vapor pressure may be present only as mists and must be measured with methods appropriate for aerosols. The situation can be very complicated for a liquid of intermediate volatility, high enough to produce a saturated vapor 10 percent or more of the TLV, yet low enough that mists will not quickly evaporate. If samples are collected on filters, only the mist will be caught and part of this liquid may be lost by evaporation into the air flowing through the filter during sampling, particularly if samples are collected over long periods. This effect undoubtedly results in a significant loss of the lower molecular weight two- and three-ring aromatic compounds present in coke oven effluents, when sampled by the usual filter methods.

When mixed mist–vapor atmospheres are sampled with a solid sorbent, the vapor is likely to be caught and retained efficiently, but the charcoal granules and associated support plugs do not constitute an efficient particulate filter (Fairchild and Tillery, 1977). As a consequence, compound devices with a filter preceding a sorbent tube have been developed. Despite severe sampling rate limitations, these systems are satisfactory when the sum of aerosol and vapor concentrations is to be related to a standard or effect. If, however, the aerosol and vapor mist are to be considered separately, perhaps because of different deposition sites or rates of adsorption, the single sum of the two concentrations is not enough. Furthermore, it cannot be assumed that the material collected on the filter is all the aerosol, since part of the liquid evaporates and is recollected on the charcoal. Since the reverse cannot be true, in cases when the vapor is more hazardous, this assumption is "safe" (i.e., protective). Size-selective presamplers that collect part of the aerosol and none of the vapor may approximate the respiratory deposition–adsorption differential and so deliver a sample weighted toward the vapor to a biologically appropriate degree. However, even here the possibility of evaporation from the cyclone, elutriator, impactor, or other component must be acknowledged. Several investigators have studied this problem and suggested ways of dealing with it for airborne styrene (Malek et al., 1986; Geuskens et al., 1990) and paint spray (Chan et al., 1986).

4.2 Reactivity

Most toxic substances are stable in air at the concentrations of interest under the usual ranges of temperature and pressure encountered in inhabited places to the degree that they may be sampled, transported to the laboratory, and analyzed without significant change or loss. Some few substances are in a transient state of reaction during the critical period when the worker is exposed. In the spraying of polyurethane foam (Peterson et al., 1962), isocyanates present in aerosol droplets are still reacting with other components of the polymer while the aerosol is being inhaled, thus must be collected in a reagent that halts the reaction and yields a product that can be related to the amount of isocyanate that was there. Similarly solid sorbents such as charcoal can be coated with a reagent to react with the substance collected to yield a new compound, which will be retained and can be analyzed.

Unwanted reactions can also occur on the collecting media. For example, a substance that will hydrolyze may do so if brought in contact with water on the solid sorbent, particularly if a sorbent like silica gel, which takes up water well, is used. Substances that may coexist in air and not react or react only slowly because of dilution may react rapidly when concentrated on the surface of a collection medium.

4.3 Particle Size

The route of entry, site of deposition, and mode of action of an aerosol all depend on some measure of the size of particle, usually but not always the aerodynamic equivalent diameter. Exposure measurement methods must discriminate among different size particles to sort out those of greater or lesser biological effect. The most common instance of such size selection is the exclusion of particles not capable of penetrating into the terminal alveoli when measuring exposure to pneumoconiosis-producing dust. Such "respirable mass sampling" by cyclone, horizontal elutriator, or various impactors is discussed at length elsewhere in this series and is not covered in detail here. However, penetration into the smallest parts of the lung is not the only division point in size selective sampling. Cotton dust particles probably do not produce the chest tightness response by penetration into the deep lung. Rather, histamine may be released in large airways (bronchioles), which can be reached by larger particles; thus a different size selective criterion (50% at 15 μm) and device (vertical elutriator) are used (Lynch, 1970a). Even larger particles may enter the body by ingestion, if they are caught in the ciliated portion of the bronchus, transported to the epiglottis, and swallowed.

At some size particles are so large they cease to be "inhalable," thus should be excluded from exposure samples (Vincent and Mark, 1981). It is difficult to select a criterion on which to base size selectors for "inhalable dust" samplers. Particles having falling (setting) speeds greater than the upflow velocity into the nose could be said to be "noninhalable" except that some individuals breathe through their mouths. These particles turn out to be quite large, thus have falling speeds that cause them to be removed from all but the most turbulent or recently generated dust clouds. Because of their size, of course, they represent a mass far out of proportion to their number. By using the general tendency of open face filter samples to undersample large particles when pointed down (Davies, 1968; Agarual and Liu, 1980), it is possible to use this vertical elutriation effect to discriminate against noninhalable particles in most cases.

To respond to the dependence of biological effect on particle size, the ACGIH (1985, 1991) has established three classes of TLV for airborne particulate matter.

- Inspirable particulate mass. This is the quantity of particulate matter that should be measured for those materials that are hazardous when deposited anywhere in the respiratory tract. This class includes some fraction of particles as large as 100 μm.

- Thoracic particulate mass. These are the particles that are deposited anywhere within the lung airways and the gas exchange region. The sample for thoracic mass collects particles with a median aerodynamic diameter of 10 μm, which is similar to the cotton dust sampler.

- Respirable particulate mass. These particles are hazardous when deposited in the gas exchange region of the deep lung. They include the traditional pneumoconiosis-producing dusts like silica and are collected with a particle size selector that passes 50 percent at 3.5 μm.

4.4 Exposure Indices

Most toxic substances encountered in the workplace are clearly defined chemical compounds that can be measured with as much specificity as we please. Benzene need not be confused with other compounds, and other closely related compounds (e.g., ethyl benzene) have different toxicology. For some toxic substances we tend to think of an element (lead) in any one of a number of possible compounds. Thus specificity is defined in terms of the element rather than the compound. Often, however, the situation becomes more complex because the compound containing the element of interest has a significant effect on its uptake, metabolism, toxicity, or excretion (Roy, 1977). Although the influence of the chemical structure containing the element probably varies even among similar compounds, for simplicity the compounds are usually grouped as organic/inorganic (lead, mercury), soluble/insoluble (nickel, silver), or by valence (chromium).

The problem of differentiating the several classes of compounds of a toxic element in a mixed atmosphere adds complexity to sampling method selection, and it is sometimes necessary to make, and clearly state alongside the results, certain simplifying assumptions. It is commonly assumed when measuring lead exposure in gasoline blending, for example, that all the lead measured is organic. Similarly when measuring the more toxic soluble form of an element, the ''safe'' assumption may be made that all the element present was soluble.

The greatest complexity occurs when toxicity is based on the effects of a class of compounds or of a material of a certain physical description. Some polynuclear aromatic hydrocarbons (PNA) are carcinogens of varying potency, and they usually exist in mixtures with other PNAs and with compounds (activators, promotors, inhibitors) that modify their activity. Analysis of each individual compound is very difficult and when done does not yield a clear answer, since given the complexity of the mixture of biologically active agents and their interactions, a calculated equivalent dose would have little accuracy. In these instances it is common to measure some quantity related to the active agents and to base the TLV on that index. For PNAs a TLV has been based on the total weight of benzene- or hexane-soluble airborne material. While this limit may be appropriate for coal tar pitch volatiles for which it was developed, it may not work for other PNA containing materials. Crude oil and cracked petroleum stocks may contain PNA; but, whereas the coal dust particles mixed in with coal tar pitch volatiles are not soluble in benzene, almost all of the petroleum-derived materials admixed with PNAs are soluble in benzene. For example, a heavy aromatic naphtha (HAN) may or may not contain PNAs depending on the manufacturing process but is completely soluble in benzene. Thus a measurement of the benzene-soluble fraction of a HAN aerosol will reveal nothing about the PNA content. Alternate indices include the single carcinogen B(a)P, the sum of a subset of six carcinogenic PNAs (Table 2.1), or 14 or more individual PNAs (Schulte et al., 1974).

Asbestos is another toxic substance for which the parameter of greatest biological relevance is difficult to define (Zumwalde and Dement, 1977; Merchant, 1990; Esmen and Erdal, 1990). In the early studies when measurements were made with an impinger, few fibers were seen and consequently a count of all particles present was used as an index of overall dustness. More recently the TLV has been based on counts of fibers longer than 5 μm as seen with a light microscope. Since most fibers present are usually shorter than 5 μm and too thin to be visible under a light microscope, it has been suggested that counts of all fibers seen by an electron microscope would provide the most meaningful estimate of risk. Long fibers may be more dangerous; however, short fibers will dominate the count; thus some adjustment may be necessary.

Lippmann (1988) has observed that the several diseases caused by asbestos fiber inhalation (asbestosis, mesothelioma, lung cancer) are associated with differing parameters of fiber exposure. Based on this, the asbestos exposure indices shown in Table 2.2 are recommended.

Table 2.1 Carcinogenic PNA Subset

Benz[a]anthracene
Benzo[b]fluoranthene
Benzo[j]fluoranthene
Benzo[a]pyrene
Benzo[e]pyrene
Benzo[k]fluoranthene

Table 2.2 Summary of Recommendation on asbestos Exposure Indices

Disease	Relevant Exposure Index
Asbestosis	Surface area of fibers with: Length >2 μm; diameter <0.15 μm
Mesothelioma	Number of fibers with: Length >5 μm; diameter <0.1 μm
Lung Cancer	Number of fibers with: Length >10 μm; diameter >0.15 μm

Source: From Lippmann (1988).

Byssinosis appears to be caused not by cotton itself but by inhalation of cotton plant debris dust baled with the cotton. The total dust airborne in a cotton textile mill is mostly lint (Lynch, 1970a), and indices have aimed at excluding these "inert" cellulose fibers by collecting a "lintfree" fraction by use of a screen or vertical elutriator. More relevant indices could include plant debris only or the specific biologically active agents if known.

Indices are used where a group of compounds interact to produce a biological effect, or where the active agent is unknown or unmeasurable. In choosing an index we try to maximize biological relevance with a method of measurement that is practical to use. On the one hand very simple parameters like gross dust or total count are easy to use but include much irrelevant material. On the other hand counting fibers by scanning electron microscope or detailed analyses of individual PNAs is difficult, expensive, and not likely to be undertaken frequently. In making the choice, it is important to remember that the primary objective is the protection of the worker. Very exact and highly relevant methods, though scientifically satisfying, may be so tedious that very few samples are taken, and because of the larger variability of worker exposure, the true accuracy of the exposure estimate is lower than it would have been if many samples had been taken by a less specific method.

When the overall contaminant level has been reduced and with it the level of the biologically active agent, the level of all correlated indices will be lower. The danger of less relevant, more indirect indices is that serious systematic bias may occur, particularly when an index from the workplace where the health effect relationship was estimated is used in other quite different workplaces. The carcinogenic risk of roofers using asphalt is far lower than for coke oven workers at the same level of exposure as measured by benzene solubles. Byssinosis patterns may be different in mills that garner old rags or process linters than in raw cotton card rooms. In using a measurement that is not perfectly specific, it must be remembered that the result obtained has a less than direct connection with the biological process, and the stronger the effect of extraneous factors, the more care is needed in interpreting the result.

4.5 Mixtures

Industrial workplaces rarely contain only one airborne contaminant, although it is uncommon for there to be several toxic substances each at or near its TLV. The measurement problems caused by the presence of gases, vapors, or dust—some in even higher concentrations than those of primary interest—are discussed below. The question of what to measure and how to interpret the result when a worker is simultaneously exposed to several agents involves their biological mechanism and can be answered only by considering the mode of action of the substances in the body. Possible effects of mixtures can be divided into three broad categories (ACGIH, 1991).

- *Additive*. When two or more hazardous substances act on the same organ system, their combined effect, rather than that of any one by itself, should be the criterion for overexposure. For additive substances the ratios of the concentration C_n of each substance to its limit T_n are summed:

$$\frac{C_1}{T_1} + \frac{C_2}{T_2} + \cdots + \frac{C_n}{T_n}$$

When this summation exceeds one, then the TLV for the mixture as a whole is considered to be exceeded.

- *Independent.* In some cases there is good reason to believe that the major effects of the components of the mixture are not additive but act on altogether different organ systems; for example, an eye irritant and a pneumoconiosis-producing dust, or a neurotoxic substance with a liver poison. In these cases the TLV is exceeded when one of the components exceeds its TLV.
- *Synergistic.* It is possible that components of a mixture, or a workplace exposure plus a life-style risk (tobacco, alcohol), may act together synergistically or potentiation may occur. This action usually occurs in high rather than low doses. The interpretation of exposure data where synergism or potentiation may occur needs to be evaluated on a case-by-case basis using the scientific evidence that such action may happen and at what levels it happens.

Although possible interactions of substances are extremely complex, it has been custom to accept the simplifying assumption that "in the absence of information to the contrary, the effects of the different hazards should be considered as additive."

Hydrocarbons used as fuels or solvents are usually a mixture of a large number of individual aliphatic compounds and their isomers, often so numerous that analyzing for each individual compound and comparing the result with an individual limit is impractical. Indeed, since TLVs exist for only a few of the compounds present, limits must be stated for the mixture by considering the compounds as a class. Gasoline (Halder et al., 1986), for example, may contain aliphatic and aromatic hydrocarbons and such oxygenate additives as methyl tert butyl ether (MTBE). In view of the differences in toxicity of the several substances and variations in content, calculation of a "TLV" for gasoline is a complex matter (Runion, 1977; McDermott and Killiany, 1978; WGD, 1991). One difficulty in expressing a TLV for any vapor mixture is that it has been customary to state TLVs for gases or vapors in parts per million. Analytical results, typically from gas chromatographs, emerge initially as the weight in milligrams of each fraction present, from which the concentration, in milligrams per cubic meter, can be calculated. To convert this to parts per million, it is necessary to know the molecular weight, which will not be known for unidentified homologues or for the mixture as a whole. For this reason, it is preferable to state TLVs for vapor mixtures as a weight concentration (mg/m^3) rather than as a volume fraction (ppm).

Certain combinations of substances present a far more complex situation than can be described by either independent or simple interaction. Benzo[*a*]pyrene and particulates, with and without sulfur dioxide, carbon monoxide and hydrogen cyanide, ozone, and oxides of nitrogen, and some other mixtures result in complex interactions that may cause effects beyond those predicted by the merely additive case.

4.6 Period of Standard

The concentration of industrial air contaminants varies with time, and recordings such as that in Figure 2.4 are typical. In general, where the concentration is varying with time, the height of the maximum and depths of the minimum are greater as the period of the measurement is shortened. If it were possible to make truly "instantaneous" or zero-time measurements, the peaks and troughs would be very great indeed. Real measurements using continuous reading instruments do not show quite such wide extremes because the response time of the instrument causes some averaging, thus preventing true zero-time measurements. Even so, instruments with short response times, such as those with solid-state sensors, show wide variation from the average, and even those with relatively long response times such as a beta adsorption type of dust monitor, still reveal peaks and valleys that are more than double or less than half the average. As longer and longer period measurements are made, the extremes regress toward the average and obviously, if we define our average over an 8-hr period, a single integrated sample over that period would show no extremes, but only the average. However, even daily averages have highs and lows compared to monthly or yearly averages in all but perfectly nonvarying environments, which do not occur in the real world.

Given the variance of the universe of instantaneous concentrations and the probability distribution of the concentration over one averaging time, it should be possible to determine the probability distribution of the concentration over any other averaging time. The mathematics of this relationship has been addressed (Spear et al., 1986; Preat, 1987), and, as expected, the arithmetic mean of the exposure distribution is constant and independent of averaging time, and the variance decreases with increasing averaging time. For lognormal distributions, however, while the arithmetic mean is independent of averaging time, the geometric mean decreases with averaging time as the geometric standard deviation increases (Rappaport, 1991). These changes depend in part on the degree of autocorrelation of instantaneous concentrations.

The relationship of averaging time and environmental variance has the consequence that measurements made over different integrating times have a relationship to each other that is a function of the variance. For example, if a given value of the concentration had a 50 percent (or 90%) chance of occurring over one averaging time, there would be a unique higher and lower pair of values that had an equal chance of occurring over one averaging time, and there would be a unique higher and lower pair of values that had an equal chance of occurring over a shorter averaging time. Since not all values of environmental variance are equally likely but rather tend to be in the range of geometric standard deviations of 1.5 to 3 or 4, it is possible to say in some cases that certain values for different averaging times are inconsistent with each other. Thus, it is very unlikely that an industrial environment

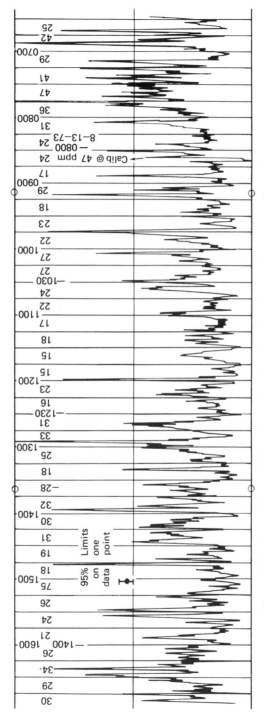

Figure 2.4 Actual industrial hygiene data showing intraday environmental fluctuations. Range of carbon monoxide data on chart is 0 to 100 ppm.

51

would be so constant that the 100-ppm, 8-hr limit for ethyl benzene could be exceeded without exceeding the 125-ppm, 15-min short-term exposure limit (STEL). The effect of setting short-term limits close to long-term limits is to force the effective long-term limit down. Thus to avoid exceeding a 125-ppm, 15-min STEL, it will probably be necessary to achieve an 8-hr average value of much less than 100 ppm of ethyl benzene perhaps even lower than 50 ppm. There may be valid toxicological reasons (Turner, 1976; Lauwerys, 1983) for setting a short-term limit based on, for example, acute irritation, and a time-weighted average (TWA) that is aimed at preventing some chronic effect; however, it should be recognized that the two are not independent, and when they are set outside the range of approximately equal likelihood, holding concentrations below the limit for one averaging time means holding them far below the limit for the other averaging time; thus one limit is in effect forcing the other.

In recent years, the TLV committee of ACGIH has been aware of this issue and has revised and removed a number of STELs that had no independent toxicological basis and were close to the 8-hr TWA. Based on typical distribution of exposures, the TLV committee now recommends for substances that do not have STELs. "Excursions in worker exposure levels may exceed three times the TLV–TWA for no more than a total of 30 minutes during a work day and under no circumstances should they exceed five times the TLV–TWA provided that the TLV–TWA is not exceeded." (ACGIH, 1992, p. 5)

In terms of sampling strategy, the significance of limits for different averaging times that are statistically inconsistent is that since one has a relatively greater likelihood of being exceeded than the other, regardless of the absolute likelihood of either, there is an opportunity to devise schemes that emphasize measurements to detect the likely event and use these measurements and knowledge of variance derived from them to draw inferences about the less likely event.

5 SAMPLING STRATEGY

The term *sampling strategy* as used here means the assembly of decisions about how to make a set of measurements to represent exposure for a particular purpose. The measurements should yield a data set that is logically and statistically adequate to satisfy the objective of the measurements as discussed in Section 2. An optimum strategy is selection of elements under the control of the exposure assessor that most efficiently achieves the objective given the physical circumstances and environmental variability (HSE, 1989).

5.1 Environmental Variability

An important factor in the design of any measurement scheme is the degree of variability in the system being observed. This variability has a primary effect on the number of samples to be taken and the accuracy of the results that can be expected. When the system being observed is the exposure of a worker to a toxic substance in a workplace, variability tends to be quite high. During the course of a day there are minute-to-minute variations and daily averages vary from day to day.

A typical recording of actual intraday environmental fluctuations appears in Figure 2.4. Highly variable environmental data of this kind, which are truncated at zero, generally have been found to be best described by the lognormal rather than the normal distribution (Fig. 2.5). This two-parameter distribution (Aitchinson and Brown, 1963) is described by the geometric mean (GM) and the geometric standard deviation (GSD), which is the antilog of the standard deviation of the log-transformed data (Lynch, 1976). A rough equivalence between GSDs and the more familiar coefficient of variation (Table 2.3) is valid up to a GSD of about 1.4. Figure 2.6 illustrates the consequence of various values in terms of spread of data. Models and data derived from community air pollution measurements (Bencala and Seinfeld, 1976; Larsen, 1971; Kalpasanor and Kurchatora, 1976; Stern, 1976) have been useful in studying the in-plant microenvironment. Some studies of occupational environmental variability (Leidel et al., 1975) have found lognormal distributions of data with GSDs in the range of 1.5–1.7, with few less than 1.1 and as many as 10 percent exceeding 2.3. Rappaport (1991) summarized various studies with GSDs ranging from 1.2 to 9.3 with most under 3.0.

One can speculate on the probable causes for this variability in worker exposure. Fugitive emissions, which are like frequent small accidents rather than main con-

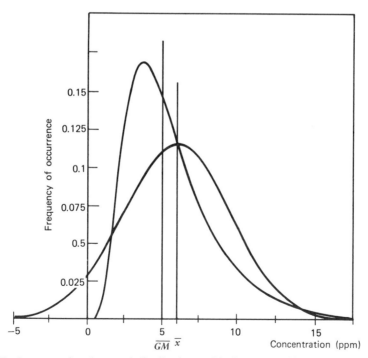

Figure 2.5 Lognormal and normal distributions with the same arithmetic (mean) and standard deviation: GM—geometric mean.

Table 2.3 Log and Arithmetic Standard
Equivalence

Geometric Standard Deviation (GSD)	Coefficient of Variation (CV)
1.05	0.049
1.10	0.096
1.20	0.18
1.30	0.37
1.40	0.35

sequences of the production process, occur almost randomly. Production rates change. Patterns of overlapping multiple operations shift irregularly. Distribution of contaminants by bulk flow, random turbulence, and diffusion is uneven in both time and space. Through all this our target system, the worker, moves in a manner that is not altogether predictable. These and other uncertainties are the probable causes of the variability typically observed, but the situation is far too complex to pinpoint each individual source of variation and its consequences.

Variance can be divided into within-worker variance and between-worker variance. While within-worker variance is a result of changes in the exposure of a worker from day to nominally similar day, between-worker variation is at least theoretically

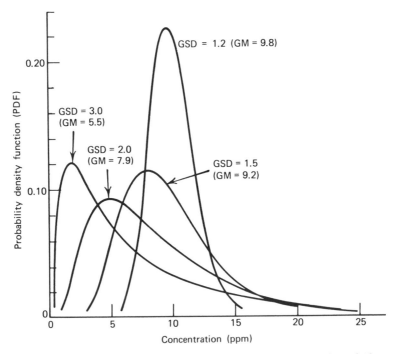

Figure 2.6 Lognormal distributions for arithmetic mean concentration of 10 ppm.

controllable by the way workers are grouped for sampling purposes. Since it is not usually practical to measure the exposure of all workers all the time, it is customary to group workers who are relatively homogenous with respect to exposure. In some cases, this can be done by observation, but when the between-worker variance is seen to be large with respect to the within-worker variance, it may be necessary to use sampling data to group workers.

Sampling schemes may deal with the variability from all sources as a single pool and derive whatever accuracy is required by increasing the number of samples. Alternatively one can postulate that a large part of the variability is due to some observable factor or factors and by means of a factorial design account for this portion of the variance, leaving only the residual to be dealt with as error. No hard and fast rules can be made regarding the choice except that it seems logical to expect the factors identified to account for a statistically significant fraction of the variance (F test) if it is to be worthwhile to sample and analyze the data in this manner. Shift, season of the year, wind velocity, and rate of production are some of the factors affecting worker exposure that may be worth singling out to ensure that part of the gross variance can be assigned to an accountable cause. Even if successful, it is likely that the residual or error variance will still be large, much larger than the variance due to measurement method inaccuracy, and will exert a major influence on our decisions.

5.2 Purpose of Measurement

The development of a sampling strategy requires a clear understanding of the purpose of the measurement. Rarely are data collected purely for their own sake. Even when data are collected because of a demand by others, such as the government or employees, the use of the data should be considered.

What questions will be answered by the data? What decisions could depend on those answers? For example, do we want to know whether these workers are overexposed? And if so, will a decision to take control action follow? Is the control likely to be a minor change in a work practice or an expensive engineering modification? Or is it intended to combine the data with those of health effects to answer the question: What level of exposure is safe? (Campbell, 1976). Will it then be decided to modify the TLV or establish a new PEL by regulation? Or are the numbers to be assembled to answer the question: What level of control is presently being achieved in industry? And may this answer lead to new decision regarding what control is feasible? Last, will workers use the result to find out whether their health is at risk and, as a consequence, decide to change jobs or seek changes in the conditions of work? Often there are many questions that need answers and thus data are collected for multiple purposes. As the discussion that follows indicates, the purpose of the data determines the design of the measurement scheme. All too often data intended for multiple purposes turn out not to be suitable for any purpose. Thus it is usually necessary to focus on the prime need to be sure the strategy will meet this requirement; then, if possible, minor adjustment or additions can be made to meet other needs, if this can be done without losing the main purpose. The optimum sampling strategy is that which combines the choice of method and sampling scheme with respect to sampling location, time, and frequency so that we are confident that the data are adequate for the decisions that follow.

5.3 Location

The most common purposes of measurement of exposure of workers is to estimate the dose so as to prevent or predict adverse health effects. These health effects result from a substance entering the body by some route, and it is possible to estimate the dose by measurement of the substance or a by-product on the way in or in the way out. The use of biomonitoring methods and biomarkers for the reconstruction of internal dose is discussed below. However, due to the limitation in these methods and the need to know the source/time pattern of the substance intake, industrial hygienists most often estimate inhaled dose by measurement of the concentration of a substance in inhaled air. Although air samples are sometimes collected from inside respirator facepieces, it is generally not possible to sample the air being inhaled directly. Therefore, the location of the sample collector inlet in relation to the subject's nose and mouth is important. We categorize sampling methods in terms of their closeness to the subject and the point of inhalation as personal samples, breathing zone or vicinity samples, and area or general air samples. A personal sample is one that is collected by a sampling device worn on the person of the worker, which travels with the worker. Breathing zone samples are those collected in the envelope or "breathing zone" around the worker's head, which is thought, based on observation and the nature of the operation, to have approximately the same concentration of the contaminant being measured as the air breathed by the worker. Area or general air samples are the most remote and are collected in fixed locations in the workplace.

Obviously, personal samples are the method of estimating dose of the inhalation route preferred since they most closely measure inhaled air. OSHA (1982) enforcement operations reflect a longstanding belief that personal sampling generally provides the most accurate measure of an employee's exposure. However, even personal samples may not sample exactly the air being breathed as indicated by the fact that even a few inches difference in the placement of the filter head of a personal sampler has been reported to make a significant difference in the concentration measured (Chatterjee et al., 1969; Butterworth and Donoghue, 1970) particularly when dust comes from point sources or is resuspended from clothing. For uniformly dispersed aerosols, however, there appears to be no bias between forehead or lapel versus nose locations (Cohen et al., 1982).

If personal sampling cannot be used, some other means of estimating exposure must be accepted. Breathing zone measurements, made by a sample collector who follows the worker, can come close to measuring exposure. However, this intrusive measurement method may influence worker behavior and the inconvenience of the measurement will limit the number of measurements and therefore reduce accuracy, as discussed below.

When fixed station samplers are used, knowledge of the quality of the relation between their measurements and the exposure of the workers is necessary if worker exposure is to be estimated. An experimental design that collects large numbers of pairs of measurements of quantities, that are in any way related, will yield a significant correlation coefficient. The important question in the use of general air measurements is: What confidence can be placed in the estimate of worker exposure? Not only is the regression line important, but also the width of the bounds on the confidence limits of a predicted exposure value from some set of fixed station measurement as shown in Figure 2.1.

In studying the relation between area and personal data with respect to asbestos the British Occupational Hygiene Society concluded (Roach et al., 1983, pp. 26–27):

> The relationship between static and personal sampling results varies according to the characteristics of the dust emission sources and the general and individual work practices adopted in a particular work area.
>
> (i) When identical sampling instruments are deployed simultaneously at personal and static sampling points and the distances between them are reasonably small, at least two-thirds of the personal sampling results obtained in a given working location are higher than those obtained from static sampling.
>
> (ii) The differences found between the two types of result tend to be particularly great where the static sampling points are relatively remote from dust emission points, as, for example, when "background" static testing is adopted.
>
> (iii) In certain cases, results from personal sampling may be lower than those from static sampling, owing to factors such as the positioning of the sampling point with respect to air extraction systems.
>
> (iv) The correlation coefficient between the personal and static measurements is statistically significant but, even so, no consistent relationship of great practical utility could be found in the limited data available.

Many attempts have been made to estimate worker exposure from fixed station air sampling schemes (Breslin et al., 1967; Sherwood, 1966; Baretta et al., 1969; Calabrease, 1977). The static sampling arrangements ranged from manually operated fixed stations in some studies to computer-based automated monitoring systems in others. In some studies the time a worker spent in an area was taken from observations of work patterns. Leidel et al. (1977) analyzed these studies and concluded that exposure estimates based on general air (area) monitoring should only be used where it can be demonstrated that general air methods can measure exposure with appropriate accuracy. Linch and Pfaff (1971) on the other hand conclude that "only by personal monitoring could a true exposure be determined."

Although worker exposure measurements are most often used in relation to health hazards, not all measurements made for the protection of health need be measurements of exposure. When it has been established that an industrial operation does not produce unsafe conditions when it is operating within specified control limits, fixed station measurements that can detect loss of control may be the most appropriate monitoring system for workers' protection. Local increases in contaminant concentration caused by leaks, loss of cooling in a degreaser, or fan failure in a local exhaust system can be detected before important worker exposure occurs. Continuous air monitoring (CAM) equipment, which detects leaks or monitors area concentrations, is often used in this way. All such systems should be validated for their intended purpose.

5.4 Period

Free of all other constraints, the most biologically relevant time period over which to measure or average worker exposure should be derived from the time constants of the uptake, action, and elimination of the toxic substance in the body (Roach, 1966, 1977; Droz and Yu, 1990). These periods range from minutes in the case of fast-acting poisons such as chlorine or hydrogen sulfide, to days or months for slow systemic poisons such as lead or quartz. In the adoption of guides and standards, this broad range has been narrowed, and the periods have not always been selected based on speed of effect of half-life. It has been observed (Campbell, 1976, p. A4), however, that "transient exposures of a shift or less are unlikely to affect the risk of chronic disease provided that the following conditions are met: (1) the damage induced by the agent is reversible; (2) elimination of the toxicant is first-order; (3) the mean exposure is less than one-quarter to one-eighth of that which corresponds to the threshold burden; and (4) the exposure distribution is stationary and adequately described by a log-normal distribution."

When these conditions are met, measurement of the long-term average (multishift) exposure is much more efficient than measuring single-shift "peaks." However, where standards have been developed or have been interpreted as single-shift limits, an increase in the averaging time is in effect an increase in the standard, and so it may be necessary to lower the numerical value of the standard to remain risk neutral. For most substances, a time-weighted average over the usual work shift of 8 hr has been accepted since it is long enough to average out extremes and short enough to be measured in one work day. Several systems have been proposed for adjusting limits to novel work shifts (Brief and Scala, 1975; Mason and Dershin, 1976; Anderson et al., 1987).

Once the time period over which exposure is to be averaged has been decided for either biological or other reasons (Calabrease, 1977; Hickey and Reist, 1977), there are available several alternate sampling schemes to yield an estimate of the exposure over the averaging time. A single sample could be taken for the full period over which exposure is to be averaged (Fig. 2.7). If such a long sample is not practical, several shorter samples can be strung together to make up a set of full period, consecutive samples. In both cases, since the full period is being measured, the only error in the estimate of the exposure for that period is the error of sampling and analytical method itself. However, when these full-period measurements are used to estimate exposure over other periods not measured, the interperiod variance will contribute to the total error.

It is often difficult to begin sample collection at the beginning of a work shift, or an interruption may be necessary during the period to change sample collection device. Several assumptions may be made with respect to the unsampled period. It may be assumed that exposure was zero during this period, in which case the estimate for the full period could be regarded as a minimum. Alternatively, it could be assumed that the exposure during the unmeasured period was the same as the average

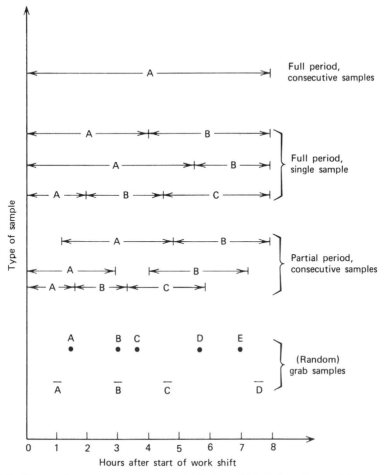

Figure 2.7 Types of exposure measurements that could be taken for an 8-hr average exposure standard.

over the measured period. This is the most likely assumption in the absence of information that the unsampled period was different. However, it is difficult to calculate confidence limits on the overall exposure estimate since the validity of the assumption is a factor, there is no internal estimate of environmental variance, and the statistical situation is complex.

When only very short period or grab samples can be collected, a set of such samples can be used to calculate an exposure estimate for the full period. Such samples are usually collected at random; thus, each interval in the period has the same chance of being included as any other, and the samples are independent. This sampling scheme of discrete measurements with a day is analogous to a set of full-period samples used to draw inferences about what is happening over a large number of days. In both cases the environmental variance, which is usually large, has a major influence on the accuracy of the results.

Short-period sampling schemes can be useful with dual standards. For example, if a toxic substance has both a short-term, say 15-min limit and an 8-hr limit, 15-min samples taken during the 8-hr period could be used to evaluate exposure against both standards. This involves some compromise, however, since samples taken to evaluate short-period exposure are likely to be taken when exposure is likely to be at a maximum rather than at random. Statistical techniques for evaluating exposure with respect to dual standards are also available (Brief and Jones, 1976; Rappaport et al., 1988). As discussed earlier, when a dual standard is inconsistent, so that one limit is more likely to be exceeded than the other, sampling schemes that evaluate exposure with respect to the limit more likely to be exceeded can be used to provide some confidence about the other limit.

The traditional method of estimating full-period exposure is by the calculation of the *time-weighted average*. In this method the workday is divided into phases based on observable changes in the process or worker location. It is assumed that concentration patterns are varying with these changes and are homogeneous with each phase. A measurement or measurements, usually shorter than the length of the phase, are made in each phase, and the exposure estimate E is calculated according to the formula:

$$E = \frac{C_1 T_1 + C_2 T_2 + \cdots + C_n T_n}{8}$$

where C_n = concentration measured in phase n
        T_n = duration (hr) of phase n ($\Sigma T = 8$)

Figure 2.8 represents this procedure graphically. Although the exposure estimate itself is simple to make, calculation of the confidence limits on this estimate can be

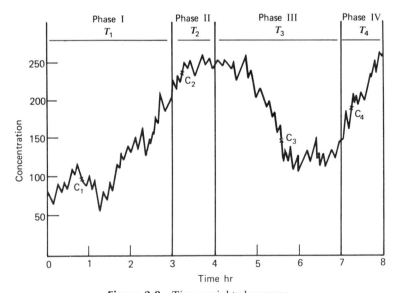

Figure 2.8   Time-weighted average.

very complex. Each phase must be treated separately. A set of samples must be collected to determine the mean and standard deviation for the phase. These data must then be combined in a manner that weights the variance to obtain an error estimate for the whole. This complex calculation does not include, however, any consideration of the imprecision in selection of phase boundaries. Given the number of samples required in each phase to provide adequate error estimates and the lack of confidence in the end results due to the several layers of assumptions, an equal number of grab samples collected at random over the whole period or stratified by task may be more efficient.

The recommendation that samples should be taken over a full shift to determine employee exposure to toxic substances has been questioned (Douglas, 1977). It is maintained that it is possible in some instances to characterize a worker's exposure with a few short period samples. There are arguments to support both sides; ultimately, however, the issue can be decided in each individual case based on the answer to two questions: Can the risk of error in the decision to be based on these measurements be calculated from the data? Is this risk acceptable, given the consequences of the decision?

When an averaging time longer than a full shift is needed, that long-term average (LTA) is usually calculated from some number of full shift samples. For example, the coal mine dust standard is based on the average of five full-shift measurements. Single-sample means of measuring multiple shifts, while excluding nonshift periods, are not widely used.

When workplace measurements are made for purposes other than the estimation of worker exposure, different considerations apply. While a single 8-hr sample may be an accurate measure of a worker's average exposure during that period, the exposure was probably not uniform, and the single sample gives no information on the time history of contaminant concentration. To find out when and where peaks occur, with the aim of knowing what to control, short period samples or even continuous recordings are useful. Similarly, when a control system is evaluated or sampling methods are compared, measurements need be only long enough to average out system fluctuations and provide an adequate sample for accurate analysis. As in the case of the decision on location, the purpose of the measurement is a primary consideration in the selection of a time period of a measurement.

5.5 Frequency

By increasing the number of measurements made over a period of time or in a sampling session, the magnitude of the confidence limits on the mean result or on the fraction of periods that may exceed the limit can be reduced. With smaller (tighter) confidence limits it becomes easier to arrive at a decision at a chosen level of confidence or to be more confident that a decision is correct. In Figure 2.9 decisions are possible in cases A and C, but not in case B. By collecting more samples, it might be possible to tighten the lower confidence limit (LCL) in case B1, for example, permitting the conclusion that these data do in fact represent an overexposure. The choice of the number of samples to be collected rests on three factors: the magnitude of the error variance associated with the measurement, the size of a dif-

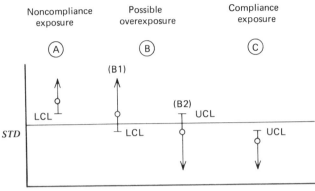

Figure 2.9 One-sided confidence limits.

ference in results that would be considered important, and the consequence of the decision based on the result.

The error variance associated with the measurement depends in most cases on the environmental variance. An exception is the rather limited instance of evaluating the exposure of a worker over a single day by means of a full-period measurement. In that case the error variance is determined by only the sampling and analytical error and confidence limits tend to be quite narrow. Usually, however, our concern is with the totality of a worker's exposure, and we wish to use the data collected to make inferences about other times not sampled. There is little choice; unless the universe of all exposure occasions is measured, we must "sample," that is, make statements about, the whole based on measurement of some parts.

As discussed earlier, the universe has a large variance, quite apart from the error of the sampling and analytical method. In terms of our decision-making ability, the error of the sampling and analytical method may have very little impact. In Figure 2.10 the inner pair of curves define the decision zone for an environmental coefficient of variation equal to .60 with no sampling or analytical error; the outer zone includes a typical detector tube error having a coefficient of variation of .25 (Lynch, 1970b). Even the relatively large error of one of the less accurate methods results in only a slight contribution of analytical variability to the total variability of measurements as shown in Figure 2.11 (Nicas et al., 1991).

The American Industrial Hygiene Association has addressed the issue of appropriate sample size (Hawkins et al., 1991) and recommends in the range of 6–10 random samples per homogeneous exposure group. Fewer than 6 leaves a lot of uncertainty and more than 10 results in only marginal improvement in accuracy. Also, it is usually possible to make a reasonable approximation of the exposure distribution with 10 samples although a rigorous goodness-of-fit test often requires 30 or more. Since the confidence interval is very sensitive to the sample estimate of the GSD, Buringh and Lanting (1991) have recommended using an assumed GSD (of approximately 2.7) with small sample sets.

Figure 2.12 also illustrates the effect of sample size or our ability to arrive at a conclusion. These curves give the number of grab samples required in order to be

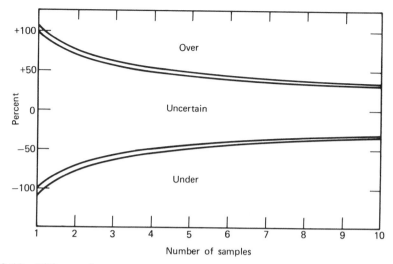

Figure 2.10 Difference between mean measured concentration and TLV required for a decision at the 95 percent level versus number of samples averaged.

confident that an overexposure did not occur for several typical levels of environmental variance. As can be seen, the difference between the mean and the standard necessary to achieve confidence in the conclusion decreases sharply as the number of samples used increases from 3 to 11. An important conclusion is that for a fixed sampling cost and level of effort, many samples by an easy but less accurate method may yield a more accurate overall result than a few samples by a difficult but more accurate method due to the effect of increasing sample numbers on the error of the mean in highly variable environments.

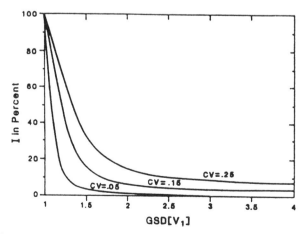

Figure 2.11 The percent contribution of analytical variability to total variability in measurements of 8-hr TWAs (Nicas et al., 1991).

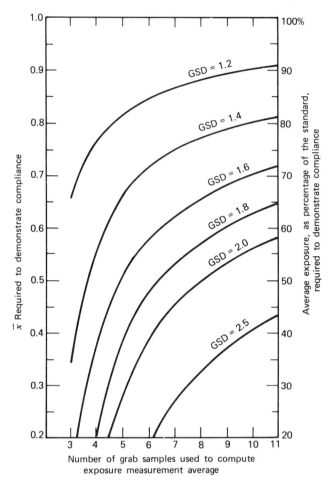

Figure 2.12 Effect of grab sample on compliance demonstration: GSD = variation of grab samples.

In selecting a sample size, note that it is possible to make a difference statistically significant by increasing the number of samples even though the difference may be of small importance. Thus given enough samples, it may be possible to show that a mean of 1.02 ppm is significantly different from a TLV of 1.0 ppm, even though the difference has no importance in terms of biological consequence. Such a statistical significance difference is not useful. Therefore, in planning our sampling strategies, we should first decide how small a difference we would consider important in terms of our use of the data and then select a frequency of sampling that could prove this difference significant, if it existed.

The consequences of the decision made on the basis of the data collected should be the deciding factor in selecting the level of confidence at which the results will be tested. Although the common 95 percent (1 in 20) confidence level is convenient because its bounds are two standard deviations from the mean, it is arbitrary, and other levels of confidence may be more appropriate in some situations. When measurements are made in a screening study to decide on the design of a larger study, it may be appropriate to be only 50 percent confident that an exposure is over some low trigger level. On the other hand, when a threat to life or a large amount of money may hang on the decision, confidence levels even beyond the three standard deviations common in quality control may be appropriate. To choose a confidence limit, first consider the consequence of being wrong and then decide on an acceptable level of risk.

Since sampling and analysis can be expensive, some thought should be given to ways of improving efficiency. Sequential sampling schemes in which the collection of a second or later group of samples is dependent on the results of some earlier set are a possibility. This common quality control approach results in infrequent sampling when far from decision points but increases as a critical region is neared. Another means of economizing is to use a nonspecific, direct-reading screening method, such as a total hydrocarbon meter, to obtain information on limiting maximum concentrations that will help to reduce the field of concern of exposure to a specific agent.

Few firm rules can be provided to aid in the selection of a sampling strategy because data can be put to such a wide variety of uses. However, the steps to be followed to arrive at a strategy can be listed:

1. Decide on the purpose of the measurements in terms of what decisions are to be made. When there are multiple purposes, select the one or several most important for design.
2. Consider the ways in which the nature of the environmental exposure and of the agent relates to measurement options.
3. Identify the methods available to measure the toxic substance as it occurs in the workplace.
4. Select an interrelated combination of sampling method, location, time, and frequency that will allow a confident decision in the event of an important difference with a minimum of effort.

5.6 Biological Monitoring

Exposure can be measured after it has occurred by reconstruction of internal dose (USEPA, 1992). If the internal dose can be estimated and the connection between the applied dose and the internal dose (pharmacokinetics) is known, it is possible to work back to the applied, external dose. The internal dose can be estimated by means of biomonitoring or biomarkers provided the relationship with the dose can be established and interfering substances or metabolites are absent.

Biological monitoring can sometimes be used to estimate dose by measuring the substance itself in blood or urine, a metabolite of the substance, a biological effect of the substance, or the amount of the substance or its metabolite bound to target molecules. Various methods and their uses are discussed by Baselt (1980), Lauwerys

(1983), Zielhuis (1978), and Kneip and Crable (1988). Biological monitoring methods are available for only some chemicals. While biological monitoring often can provide very direct information about past exposure, the relationship with exposure is weak or nonexistent for most chemicals.

6 MEASUREMENT METHOD SELECTION

The assortment of tools available to the industrial hygienist for measuring worker exposure to toxic substances is much larger now than a decade ago, although there are still many gaps. This section describes the choices available with respect to the location at which the measurement is made, time period or averaging time of the measurement, ability to select certain size aerosols and reject others, and degree to which human involvement can be lessened by automation and computer analysis. The attributes of the various sampling methods, which are important in deciding if they are appropriate for a particular sampling strategy, are summarized. The detailed techniques for sample collection and analysis are covered elsewhere in this text and in other references (NIOSH, 1991; OSHA, 1991).

6.1 Personal, Breathing Zone, and General Air Samples

Section 5 described the locations at which samples could be collected with relation to the air actually inhaled by the worker and defined the terms *personal*, *breathing zone*, and *general air*. Personal samplers are devices that collect a sample while being worn on the worker's clothing with the sample inlet positioned as close to the mouth as practicable, usually on the lapel. One kind of personal sampler consists of a pump on the worker's belt or pocket and a sampling head containing the sorbent tube, filter, or other collection medium, clipped to the lapel close to the nose and mouth. These pump-type active samplers have been largely replaced for vapor sampling by passive personal samplers that collect the air contaminants by diffusion onto a solid sorbent. When initially introduced, many questions were raised about the accuracy and reliability of passive dosimetery, and many studies have been performed. These studies have been reported by Berlin et al. (1987) and reviewed by Harper and Purnell (1987). In general, passive samplers, when properly designed and used, are as accurate as active samplers. Since passive sampling is less expensive and more convenient than powered sampling, more samples can be collected, so overall accuracy, including environmental variance, is better.

Personal samplers of both types are worn by the worker, so they must be lightweight, portable, and not affected by motion or position. These restrictions tend to limit the size of sample that can be collected. In the case of passive dosimeters, the geometry of the diffusion channel and the diffusion constant of the substance result in an equivalent sampling rate that is usually quite low, thus requiring long sample periods. Personal sampling with wet collectors, such as impingers or bubblers, is difficult due to the danger of spillage and glasswear breakage. Spill-proof impingers have, however, been used. Gravitational size selectors such as horizontal elutriators, which are affected by position, have not made successful personal samplers, although inertial devices such as cyclones or impactors are satisfactory.

When limitations of weight, complexity, wet collecting media, or position prevent successful personal sampling, it is still possible to make an approximate measurement of worker exposure by collecting the sample in the "breathing zone" of the worker. This vaguely defined zone is the envelope of air surrounding the worker's head that is thought to have approximately the same concentration of the contaminant being measured as the air breathed by the worker. Breathing zone samples may be collected by a fixed sampler with the inlet near the nose of a stationary worker or, in the case of a mobile worker, by carrying the sample collecting equipment and holding the sample inlet near the worker's head, while moving around the work site with the worker. The obviously awkward and time-consuming nature of this kind of sampling limits its usefulness.

When measurements of workers' exposures are not needed or indirect estimates are adequate, the concentration of a contaminant in the general air of a workplace may be measured at some fixed station. Many of the equipment limitations imposed by personal and breathing zone sampling systems do not apply to general air samplers. Portability is not critical, so electrical components may be either battery or line powered. When line power is available, powerful pumps may be used to provide enough vacuum for critical orifices to obtain precise flow control or to operate high-volume vacuum sources capable of collecting very large samples. Wet devices and both horizontal and vertical elutriators are practical. Very large samples, which may be needed to obtain sufficient sensitivity for trace analyses, may be collected on heavy or bulky collecting media. Multiple samplers of different types may be arrayed close to each other to provide sampler comparison data. Most new methods of measurement of worker exposure were first tested in fixed station arrays, where they were compared with older methods before being adapted to personal sampling.

6.2 Short- and Long-Period Samplers

Available sampling methods have limited flexibility in the period of time over which the sample can or must be collected. Some methods are inherently grab samplers, although the increased interest in long-period samplers has led to their adaptation. Most detector tubes (AIHA, 1976) were originally intended to produce a result after a few pump strokes of up to several hundred milliliters. The interval between strokes could be lengthened, but instead of increasing the sampling period, this produces an average of several short samples taken over a longer time. Automatic systems have been developed as fixed station samplers that can extend the low range of some tubes by repeated pump strokes spread out over a long period. In the continuous flow mode, short-period detector tubes have been recalibrated for use at very low flow rates over long periods, and special tubes have been developed (Jentysch and Fraser, 1981; Keane, 1981) as long-term samples for up to 8-hr flow rates of 5–50 cm^3/min. Other inherently short-period methods are those that use a liquid, particularly a volatile liquid, in a bubbler or impinger. As the air passes through the sampler, the collecting medium evaporates, eventually to dryness, with loss of sample, unless terminated in time. Usually, without the addition of liquid, sampling periods of in excess of 30 min are impractical with wet devices.

Vessels that collect a whole air sample are convenient grab samples and some can be adapted for longer period sampling. Canister-based methods (McClenny et al., 1991) used for measuring pollutants can maintain constant flow rates for time-integrated continuous sampling over 24 hr or more. Many low flow rate personal sampler pumps have an air outlet fitting that can be used to fill a bag. Allowing for possible contamination due to the air passing through the pump, long-period samples may be collected. When the bag is carried in a sling on the worker's back, personal samples may be collected.

Greater time flexibility is available with solid sorbents. Even though there is a fixed volume of air that can be sampled at a concentration before breakthrough occurs, the freedom to use a wide range of low flow rates permits personal and fixed sampling over periods of 8 hr or more. Other air contaminants, competing for active sorbent sites and particulates adding to the resistance flow, limit the maximum volume of air that can be sampled in much the same way as breakthrough.

Systems that tend to be suitable for long-period samples only are usually those where the sampler can barely collect the minimum amount of material required by the sensitivity of the analytical method. A personal sample for respirable quartz using a 10-mm cyclone size selective presampler, operated at 2 L/min or less, will sample less than a cubic meter of air in 8 hr and thus collect less than 100 μg of quartz at the TLV. Shorter samples will confront the serious sensitivity limitations of most methods for quartz (NIOSH, 1991). The same problem occurs with personal samples for beryllium and for detailed analyses for multiple compounds such as PNAs.

High-volume fixed station or even personal particulate samplers using large size selectors and filters, where necessary, allow shorter sampling periods that are still longer than "grab samples." Even when analysis is not necessary and only gross or respirable weight is being measured, analytical balance limitations prevent very short samples for particulates. There are, however, various "instantaneous" mass monitors based on beta absorption (Lilienfeld and Dulchunos, 1972) or the piezoelectric effect (Sem et al., 1977) that are capable of making a measurement in as little as one minute.

6.3   Size Selection

When it is necessary in sampling for particulates to include or exclude certain size particles, limitations are created by the nature of the size selecting devices available. The unsuitability of elutriators as personal samplers due to their orientation requirements has already been mentioned. All size selectors make their stated cut only at a predetermined flow rate, which must be held constant over the period of the sample against changing filter resistance and battery conditions (Bartley and Brewer, 1982). Whereas cyclones and impactors tend to compensate for flow rate changes and elutriators compound the error, all need pumps that not only sample a known volume over the sampling period but do so at a known constant flow rate. Such pumps are usually larger and heavier than low-flow pumps, which need only sample a reliably known volume, and approach the limit of practicality as personal samplers. In addition to the requirement for constant flow, pump pulsation must be damped out if it upsets the size selector.

Isokinetic conditions usually must be established when sampling for particulates in high-velocity streams (Fuchs, 1975). In stacks, particles with high kinetic energy due to their weight and velocity can be improperly included or excluded from the sample under nonisokinetic conditions. However, workplace sampling is generally done at low ambient air velocities of less than 300 ft/min. Further, air velocity and direction is usually continually changing. Under these conditions, isokinetic sampling of the kind used in stack sampling is not necessary or practical. However, consideration must be given to the effect of the velocity and direction of the air at the filter inlet. Open-face filters may oversample when face up or undersample face down. Some undersampling may be desirable to avoid collection of large, noninhalable particles. Davies (1968) has developed theoretical relationships for filter inlet performance. For asbestos fiber sampling in accordance with NIOSH methods 7400 or 7402 (NIOSH, 1991), a conductive plastic cowl is attached to the filter inlet to improve the distribution of fibers on the filter surface and to reduce charge effects.

6.4 Continuous Air Monitors

Continuous air monitors (CAMs) with fixed sensors in critical locations connected to alarms and remote indicators are commonly used for carbon monoxide, hydrogen sulfide, chlorine hydrocarbons, and other acutely toxic or explosive gases. These systems may use passive sensors, they may pump the air through a detection cell located at the point of collection, or they may pump contaminated air to a remote analyzer. Sequential valving arrangements allow one remote analyzer to be coupled to many sample lines. Provision for automatic calibration and zeroing may be included.

Systems that make measurements automatically at a number of locations and gather the data at a central readout point have been installed in a number of plants. Adaptation of these systems to the estimation of worker exposure has led to the development of complex computer-based monitoring systems (Fig. 2.13). Since the contaminant sensors do not detect the presence of a worker, some data on worker location must be added to estimate exposure. Time and motion studies that yield percentage of time in various measured locations could be used with the daily average fixed station measurement for that location to calculate a weighted average exposure. The drawbacks are that time/activity distributions exhibit considerable variation, even under routine conditions, and the most significant exposures often occur during nonroutine periods. Also an assumption is made that the concentration at a time is independent of the worker's activity.

Alternative schemes provide a device that reads a card carried by each worker to signal the computer of each entry and departure from a monitored area. Time in the area can be multiplied by the general or weighted average concentration from the sensors in that area as measured while the worker was present. This situation is analogous to estimating exposure from fixed station sampler measurements. An even more elaborate system could place small transmitters on each worker that are tracked

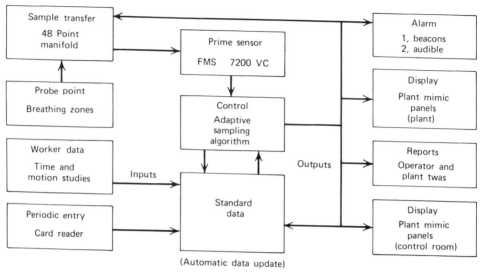

Figure 2.13 Vinyl chloride monomer (VCM) monitoring system. (Courtesy of Eocom Corporation)

automatically by a sensing network, and these detailed worker location data are combined with fixed station measurements for various locations to estimate exposure.

Continuous monitoring may also be done with remote sensing systems such as those that use Fourier transform infrared (FTIR). These systems project a beam of electromagnetic radiation in the infrared region over a path through the potentially contaminated air volume to a receiving telescope and spectral analyses either directly or via a mirror. These systems can sense and measure a large number of gases or path lengths of hundreds of meters and give real-time readout. While not directly applicable to worker exposure measurement, the use of remote sensing devices as fugitive emission detectors is a useful addition to closed process chemical plant control.

6.5 Mixtures

Toxic substances seldom occur by themselves. Mixtures not only cause difficulties in estimating the biological consequences of exposure, but also complicate exposure measurement. The other substances present in an air mixture need to be considered, even when only one component of the mixture is being measured and the other components are far below toxic effect levels, because of the effects of the other substances on the sampling systems. Charcoal is a very useful sampling medium because it will absorb and retain so many substances. However, because of this property, it is possible for all the active sorption sites to be occupied by other materials. The substance being measured may then break through long before the recommended sample volume, based on collection of the pure substance, has been passed through the tube.

An analogous case is the measurement of a low concentration of asbestos fibers in an environment containing a great deal of other, nonfibrous dust. To collect enough fibers to give the fiber density necessary for an adequate count without counting an unreasonable number of fields would result in the collection of so many grains that the fibers would be obscured. Thus the sample volume is limited by the total dust present and low fiber density is compensated for by counting a large number of fields. Overload from other airborne substances can also result in plugging of filter samplers, particularly when liquid accumulates on the surface of membrane filters and "blinds" the pores. The use of thick depth filters of glass or cellulose fibers provides greater capacity, although at the possible sacrifice of some efficiency.

Where mixture of substances present in the workplace are additive, an exposure index E is calculated as discussed earlier in this chapter. A difficulty with this approach is how E is distributed and where the variance in E comes from. Kumagai and Matsunaga (1992) has proposed a model that splits the variance into specific factors, which cause one component of a mixture to vary and common factors that effect all components. This model, which has been empirically validated, offers the generality that E is lognormally distributed and so all previously proposed methods for evaluating occupational exposure can be applied.

6.6 Method Selection

The purpose of this chapter is to explain the considerations that go into the choice of a sampling strategy (location, period, frequency) and the selection of a method that will permit the accomplishment of that strategy. Table 2.4 summarizes the degree to which the most common methods possess the attributes of importance in selecting various sampling strategies. The column headings are explained as follows:

6.6.1 Method

The methods listed are sampling methods, but the ratings of attributes that follow assume the usual range of analytical methods that can be applied to the size and type of sample collected.

6.6.2 Sampling Period

By "short" is meant essentially grab samples while "long" means 8 hr or longer in a single sample.

6.6.3 Ability to Concentrate Contaminant

Sampling methods that extract contaminant from the air and collect it in a reduced area or volume are potentially able to improve analytical sensitivity by several orders of magnitude. However, the concentrating mechanism (filtration, sorption) may introduce errors.

Table 2.4 Sampling and Analysis Method Attributes

Method	Sampling Period	Ability to Concentrate Contaminant	Ability to Measure Mixtures	Time to Result	Intrusiveness	Proximity to Nose and Mouth
Personal sampler/solid sorbent						
Sorption only	Medium to long	Yes	Yes—gases	After analysis	Medium	Very close
Sorption plus reaction	Medium to long	Yes	No	After analysis	Medium	Very close
Personal sampler/filter						
Gross gravimetric	Medium to long	Yes	Yes—particulate	After weighing or analysis	Medium	Very close
Respirable gravimetric	Long	Yes	Yes—particulate	After weighing or analysis	Medium	Very close
Count	Medium to long	Yes	Yes—particulate	After counting	Medium	Very close
Combination filter and sorbent	Medium to long limited	Yes	Yes	After analysis	Medium	Very close
Passive dosimeter	Long	Yes	Yes—gases	After analysis	Low	Very close
Breathing zone impinger/bubbler						
Analysis	Medium-limited	Yes	Yes	After analysis	High	Close
Count	Medium-limited	Yes	Yes—particulate	After counting	High	Close
Detector tubes						
Grab	Short	MA	No	Immediate	High	Close
Long period	Medium to long	NA	No	Immediate	Medium	Very close
Gas vessels						
Rigid vessel	Short to long	No	Yes—gases	After analysis	High	Medium
Gas bag	Short to long	No	Yes—gases	After analysis	High	Close
Evacuated/critical orifice	Medium to long	No	Yes—gases	After analysis	Medium	Distant
Direct-reading portable meters						
Nonspecific	Instantaneous or recorder	NA	Yes	Immediate	High	Slightly distant
Specific	Instantaneous or recorder	NA	No	Immediate	Medium	Slightly distant
Multiple compound	Instantaneous or recorder	Some	Yes	Almost immediate	High	Slightly distant
Mass monitor	Short	Yes	No	Almost immediate	High	Slightly distant
Particle counters	Short	No	No	Almost immediate	High	Slightly distant
Sensor with datalogger	Short or long	No	No	Hours	Medium	Close
Fixed station						
High volume	Medium to long	Yes	Yes—particulate	After analysis	Low	Remote
Horizontal or vertical elutria-tor	Long to short	Yes	Yes—particulate	After analysis	Low	Remote
Installed monitor	Short to long	Some	No	Almost immediate	Low	Remote
Freeze trap	Medium	Yes	Yes—vapors	After analysis	Low	Remote
FTIR	Instantaneous	No	Yes—gases	Immediate	Low	Remote

Table 2.4 (*Continued*)

Method	Specificity	Convenience Rating	Sample Transportability	Recheck of Analysis Possible	Accuracy
Personal sampler/solid sorbent					
Sorption only	High by analysis	High	Good	Elution: yes. Thermal des.: no.	Good
Sorption plus reaction	High by analysis	High	Good	Yes	Good
Personal sampler/filter					
Gross gravimetric	None for weight only—high by analysis	High	Fair	Yes	Good
Respirable gravimetric	High by analysis	Medium	Fair	Yes	Fair
Count	Fair—depends on particle identification	High	Good	Yes	Poor
Combination filter/sorbent	High by analysis	Medium	Good	Yes	Fair
Passive dosimeter	High by analysis	Very high	Good	Yes	Fair
Breathing zone impinger/bubbler					
Analysis	High by analysis	Low	Poor	Yes	Fair
Count	Fair—depends on particle analysis	Low	Poor	Yes	Poor
Dector tubes					
Grab	Medium—some interference	High	No sample	No	Fair
Long period	Medium—some interference	High	No sample	No	Fair
Gas vessels					
Rigid vessel	High by analysis	Low	Fair	Yes	Good
Gas bag	High by analysis	Low	Fair	Yes	Good
Evacuated/critical orifice	High by analysis	Low	Good	Yes	Good
Direct reading portable meters					
Nonspecific	None—total of measured class	High	No sample	No	Good
Specific	Medium—some interference	High	No sample	No	Good
Multiple compound	Medium—frequent overlap	Medium	No sample	No	Fair
Mass monitor	Mass only	High	No sample	Not usually	Fair
Particle counters	Count/size only	High	No sample	No	Fair
Sensor with dataloger	Medium—some interference	High	No sample	No	Good
Fixed station					
High volume	High by analysis	Low	Fair	Yes	Good
Horizontal or vertical elutriator	High by analysis	Low	Fair	Yes	Good
Installed monitor	Medium—may be interferences	High	No sample	No	Good
Freeze trap	High by analysis	Very low	Poor	Yes	Fair
FTIR	Medium—may be interferences	High	No sample	No	Fair

6.6.4 Ability to Measure Mixtures

Most sampling methods provide a sample that can be analyzed for more than one gas or vapor, but usually not for both gases and vapors or particulates.

6.6.5 Time to Result

Certain decisions (vessel entry) must be made immediately while others can wait until after the sample is transferred to a laboratory and analyzed.

6.6.6 Intrusiveness

When the method requires the presence of a person to collect the sample or the wearing of a heavy or awkward sampling apparatus, this intrusion of the sampling system into the work situation may affect worker behavior and exposure.

6.6.7 Proximity to Nose and Mouth

As discussed earlier, locating a sampler inlet even a small distance from a worker's mouth may bias the exposure measurement. Samplers remote from the worker may not be measuring the air inhaled at all.

6.6.8 Specificity

Some methods give only nonspecific information like total weight of all dust particles or concentration of all combustible gases while others provide a sample that can be analyzed for any species or element.

6.6.9 Convenience Rating

These are estimates of the amount of work or difficulty involved in collecting samples.

6.6.10 Sample Transportability

If the sample must be transported to a distant laboratory for analysis, the ability to withstand shock, vibration, storage, and temperature and pressure changes without being altered or destroyed is important.

6.6.11 Recheck of Analysis Possible

Some samples may only be analyzed once while others are in a form such that rechecks, reanalyses at different conditions, or analysis for other substances are possible.

6.6.12 Accuracy

Given all the possibilities for error from sampler calibration, sample collection, transport and analysis, an overall coefficient of variation (CV) of 10 percent is considered good. Some count methods are subject to such counter variability that poor accuracy is usual. Method inaccuracy should not be judged alone but should be seen in combination with the inaccuracy caused by environmental variability, which is usually larger, in making decisions whether a method is sufficiently accurate for a purpose.

Table 2.4 shows that not all sampling strategies are possible since for some strategies the sampling and analytical method with the necessary combination of attributes may not exist. The industrial hygiene technology gaps thus revealed are fruitful areas for future research and development.

REFERENCES

ACGIH (1985). American Conference of Governmental Industrial Hygiene, Particle Size-Selective Sampling in the Workplace, Pub. No. 0830, ACGIH, Cincinnati, Ohio.

ACGIH (1992). American Conference of Governmental Industrial Hygienists, "Threshold Limit Values for Chemical Substances and Physical Agents in the Work Environment with Intended Changes for 1992–1993," ACGIH, P.O. Box 1937, Cincinnati, Ohio, 45201.

Agarual, J. K. and B. Y. H. Liu (1980). *Am. Ind. Hyg. Assoc. J.*, **41**, 191–197.

AIHA (1976). American Industrial Hygiene Association, *Direct Reading Calorimetric Indicator Tubes Manual*, AIHA, Akron, Ohio.

Aitchinson, J. and J. A. C. Brown (1963). *The Lognormal Distribution*, University Press, Cambridge, England.

Anderson, M. E., M. G. MacNaughton, H. J. Clewell, and D. P. Paustenbach (1987). *Am. Ind. Hyg. Assoc. J.*, **48**, 335–343.

Baretta, D. B., R. D. Stewart, and J. E. Mutcheler (1969). *Am. Ind. Hyg. Assoc. J.*, **30**, 537–544.

Bartley, D. L. and G. M. Brewer (1982). *Am. Ind. Hyg. Assoc. J.*, **43**, 520–528.

Baselt, R. C. (1980). *Biological Monitoring Methods for Industrial Chemicals*, Biomedical Publications, Davis, CA.

Bencala, K. E. and J. H. Seinfeld (1976). *Atmos. Environ.*, **10**(11), 941–950.

Berlin, A., R. H. Brown, and K. J. Saunders, Eds. (1987). *Diffusive Sampling, An Alternative Approach to Workplace Air Monitoring*, Royal Society of Chemistry, London.

Berner, B. and E. R. Cooper (1987). *Models of Skin Permeability in Transdermal Delivery of Drugs*, A. F. Kydoniew and B. Berner, Eds., Vol. II, CRC Press, Boca Raton, FL, pp. 41–55.

BMA (1979). *Measurement and Evaluation of Concentrations of Airborne Toxic or Health Hazardous Work-Related Substances*, W. German Federal Ministry of Labor (BMA), TRgA401 Sheet 1.

Breslin, A. J., L. Ong, H. Glauberman, A. C. George, and P. LeClare (1967). *Am. Ind. Hyg. Assoc. J.*, **28**, 56–61.

Brief, R. S. and A. R. Jones (1976). *Am. Ind. Hyg. Assoc. J.*, **37**, 474–478.

Brief, R. S. and R. A. Scala (1975). *Am. Ind. Hyg. Assoc. J.*, **36**, 467–469.

Buringh, E. and R. Lanting (1991). *Am. Ind. Hyg. Assoc. J.*, **52**, 6–13.

Butterworth, R. and J. K. Donoghue (1970). *Health Phys.*, **18**, 319.

Calabrease, E. J. (1977). *Am. Ind. Hyg. Assoc. J.*, **38**, 443–446.

Campbell, E. E. (1976). *Am. Ind. Hyg. Assoc. J.*, **37**, 6, A-4.

CEFIC (1982). Report on *Occupational Exposure Limits and Monitoring Strategy*, European Council of Chemical Manufacturers Association, Brussels.

Chan, T. L., J. B. D'Arey, and R. M. Schreck (1986). *Am. Ind. Hyg. Assoc. J.*, **47**, 411–417.

Chatterjee, B. B., M. K. Williams, J. Walford, and E. King (1969). *Am. Ind. Hyg. Assoc. J.*, **30**, 643–645.

CMA (1989). Chemical Manufacturers Association, *Improving Air Quality: Guidance for Estimating Fugitive Emissions from Equipment*, 2nd ed., CMA, Washington, D.C.

Cohen, B. S., A. E. Chang, N. H. Harley, M. Lippmann (1982). *Am. Ind. Hyg. Assoc. J.*, **43**, 239–243.

Corn, M. (1976). *Am. Ind. Hyg. Assoc. J.*, **37**, 353–356.

Cronen, E. and R. B. Stoughton (1962). *Arch. Dermatol.*, **36**, 265.

Davies, C. N. (1968). *J. Appl. Phys.*, Ser. 2, **1**, 921–932.

Donaldson, H. M. and W. T. Stringer (1976). "Beryllium Sampling Methods," Department of Health, Education and Welfare Publication (NIOSH), 76-201, Cincinnati, Ohio, p. 21.

Douglas, D. D. (1977). *Am. Ind. Hyg. Assoc. J.*, **38**, A-6.

Droz, P. and M. Yu (1990). Biological Monitoring Strategies, in *Exposure Assessment for Occupational Epidemiology and Hazard Control*, S. M. Rappaport and T. J. Smith, Eds., Lewis Publishing Co., Chelsea, MI.

Esmen, N. A. and S. Erdal (1990). *Environ. Health Perspect.*, **88**, 277–286.

Fairchild, C. I. and M. I. Tillery (1977). *Am. Ind. Hyg. Assoc. J.*, **38**, 277–283.

Ferguson, D. M. (1976). *Ann. Occup. Hyg.*, **19**, 275–284.

Fiserova-Bergerova, V., J. T. Pierce, and P. O. Dray (1990). *Am. J. Ind. Med.*, **17**, 617–635.

Flynn, G. L. (1990). Physiochemical Determinants of Skin Absorption, in *Principles of Route-to-Route Extrapolation for Risk Assessment*, T. R. Geritz and C. J. Henry, Eds., Elsevier Science, pp. 93–127.

Fuchs, N. A. (1975). *Atoms, Environ.*, **9**, 697–707.

Fukabori, S. and N. Nakaaki (1976a). *J. Sci. Labour*, **53**, 89–95.

Fukabori, S. and N. Nakaaki (1976b). *J. Sci. Labour*, **52**, 67–81.

Geuskens, R. B. M., M. J. M. Jongen, J. C. Ravensberg, J. Vander Tuin, L. H. Leenheers, and J. F. Vander Wal (1990). *Appl. Occup. Environ. Hyg.*, **6**, 364–369.

Halder, C. A., G. S. Van Gorp, N. S. Hatowm, and T. M. Warne (1986). *Am. Ind. Hyg. Assoc. J.*, **47**, 164–172.

Harper, M. and C. J. Purnell (1987). *Am. Ind. Hyg. Assoc. J.*, **48**, 214–218.

Harris, R. L. (1993). *Guidelines for Collection of Industrial Hygiene Exposure Assessment Data for Epidemiologic Use*, Chemical Manufacturers Association, Washington, D.C.

Hart, C. (1987). *J. Environ. Health*, **49**, 282–287.

Harvey, R. P. (1981). Statistical Aspects and Air Sampling Strategies, in *Detection and Measurement of Hazardous Gases*, C. F. Culles and J. G. Firth, Eds., Heinemann, London–New York, p. 147.

Hawkins, N. C., S. K. Norwood, and J. C. Rock, Eds. (1991). *A Strategy for Occupational Exposure Assessment*, Am. Ind. Hyg. Assoc., Akron, Ohio.

Hayden, J. W., E. G. Comstock, and B. S. Comstock (1976). *Clin. Toxicol.*, **9**, 169–184.

Helper, E. W., N. K. Napier, M. E. D. Hillman, and S. Davidian (1979). *Product/Industry Profile on Art Materials and Selected Craft Materials*, U.S. Consumer Product Safety Commission, CPSC-C-78-0091.

Hickey, J. L. S. and P. C. Reist (1977). *Am. Ind. Hyg. Assoc. J.*, **38**, 613–621.

Hodgson, A. T., J. M. Daisey, and R. A. Grot (1991). *J. Air Waste Mgt. Assoc.*, **41**, 1461–1468.

Horowitz, L. D. (1976). *Am. Ind. Hyg. Assoc. J.*, **37**, 227–233.

HSE (1989). Health and Safety Executive, *Monitoring Strategies for Toxic Substances, Guidance Note*, EH 42, HSE, Bootle, Meseyside, UK.

Jayjock, M. A. (1988). *Am. Ind. Hyg. Assoc. J.*, **49**, 380–385.

Jentysch, D. and D. A. Fraser (1981). *Am. Ind. Hyg. Assoc. J.*, **42**, 810–823.

Johanson, G. and A. Boman (1991). *Br. J. Ind. Med.*, **48**, 788–792.

Kalpasanor, Y. and G. Kurchatora (1976). *J. Air. Pollut. Control Assoc.*, **26**, 981.

Keane, M. J. (1981). *Ann. Am. Conf. Governm. Ind. Hyg.*, **1**, 241–245.

Key, M. M., Ed. (1977). *Occupational Diseases, A Guide to Their Recognition*, Government Printing Office, Washington, D.C.

Kneip, T. and J. V. Crable, Eds. (1988). *Methods for Biological Monitoring*, APHA, New York.

Korobkin, R., A. K. Asbury, A. J. Sohner, and S. L. Nielsen (1975). *Arch. Neurol.*, **32**, 158–162.

Kumagi, S. and I. Matsunaga (1992). *Ann. Occup. Hyg.*, **36**, 131–143.

Larsen, R. I. (1971). A Mathematical Model for Relating Air Quality Measurements to Air Quality Standards, U.S. Environmental Protection Agency, Publication AP-89, Research Triangle Park, NC.

Lauwerys, R. R. (1983). *Industrial Chemical Exposure: Guidelines for Biological Monitoring*, Biomedical Publications, Davis, CA.

Lauwerys, R. R., T. Dath, J. M. Lachapelle, J. P. Buchet, and H. Roels (1978). *J. Occup. Med.*, **20**, 17–20.

LeBlanc, J. V. (1977). *J. Occup. Med.*, **19**, 276–277.

Leichnitz, K. (1976). *Detector Tube Handbook*, 3rd ed., Drägerwerk, Lübeck, Germany.

Leidel, N. A., K. A. Busch, and W. E. Crouse (1975). *Exposure Measurement Action Level and Occupational Environmental Variability*, Government Printing Office, Washington, D.C.

Leidel, N. A., K. A. Busch, and J. R. Lynch (1977). *Occupational Exposure Sampling Strategy Manual*, Department of Health, Education and Welfare, Publication (NIOSH) 77-173, Cincinnati, OH.

Lilienfeld, P. and J. Dulchunos (1972). *J. Am. Ind. Hyg. Assoc.*, **33**, 136–145.

Linch, A. L. and H. V. Pfaff (1971). *Am. Ind. Hyg. Assoc. J.*, **32**, 745–752.

Linch, A. L., E. G. Wiest, and M. D. Carter (1970). *Am. Ind. Hyg. Assoc. J.*, **31**, 170–179.

Lioy, P. J. (1990). *Environ. Sci. Technol.*, **24**, 938–945.

Lipton, S. and J. Lynch (1987). *Health Hazard Control in the Chemical Process Industry*, Wiley, New York.

Lippmann, M. (1988). *Environ. Res.*, **46**, 86–105.

Lynch, J. R. (1970a). ''Air Sampling for Cotton Dust,'' Transactions of the National Conference on Cotton Dust and Health, University of North Carolina, Chapel Hill, NC, p. 33.

Lynch, J. R. (1970b). "Uses and Misuses of Detector Tubes," Transactions of the 32nd Meeting of the American Conference of Governmental and Industrial Hygienists, ACGIH, Cincinnati, OH.

Lynch, J. R. (1976). *Nat. Saf. News*, **113**(5), 67–72.

Malek, R. F., J. M. Daisey, and B. S. Cohen (1986). *Am. Ind. Hyg. Assoc. J.*, **47**, 524–529.

Marbury, M. C. and J. E. Woods (1991). *In Indoor Air Pollution*, J. M. Samet and J. D. Spengler, Eds., Johns Hopkins University Press, Baltimore, MD, pp. 306–332.

Mason, J. W. and H. Dershin (1976). *J. Occup. Med.*, **18**, 603–606.

McCammon, C. S., Jr. and J. W. Woodfin (1977). *Am. Ind. Hyg. Assoc. J.*, **38**, 378–386.

McCann, M. (1979). *Artists Beware*, Watson-Guptill, New York.

McClenny, W. A., J. D. Pleil, G. F. Evans, K. D. Oliver, M. W. Holdren, and W. T. Winberry (1991). *J. Air Waste Manage Assoc.*, **41**, 1308–1318.

McDermott, H. J. and S. E. Killiany (1978). *Am. Ind. Hyg. Assoc. J.*, **39**, 110–117.

Merchant, J. A. (1990). *Environ. Health Perspect.*, **88**, 287–293.

Nagda, N. L., H. E. Rector, and M. D. Koontz (1987). *Guidelines for Monitoring Indoor Air Quality*, Hemisphere Pub., New York.

Nicas, M. and R. C. Spear (1993). *Am. Ind. Hyg. Assoc. J.*, **54**, 211–227.

Nicas, M., B. P. Simmons, and R. C. Spear (1991). *Am. Ind. Hyg. Assoc. J.*, **52**, 553–557.

NIOSH (1991). *Manual of Analytical Methods*, published 1984 with updates through 1991.

Oliver, J. S. (1977). *Lancet*, **1**, 84–86.

OSHA (1970). Occupational Safety and Health Act of 1970, PL 91-596.

OSHA (1982). Use of Personal Sampling Devices During Inspection, 47 Fed. Reg. 55478.

OSHA (1991). *Analytical Methods Manual*, Pub. 1985 with updates through 1991.

OSHA (1992). *Occupational Safety and Health Standards*, 29 CFR 1910.

Ott, W. R. (1989). Human Activity Patterns; A Review of the Literature for Estimating Time Spent Indoors, Outdoors and in Transit in Proceedings of the Research Planning Conference on Human Activity Patterns, USEPA 600, 04.89.004.

Oxley, G. R. (1976). *Ann. Occup. Hyg.*, **19**, 163–167.

Peterson, J. E., R. A. Copeland, and H. P. Hayles (1962). *Am. Ind. Hyg. Assoc. J.*, **23**, 345–352.

Poklis, A. and C. D. Burkett (1977). *Clin. Toxicol.*, **11**, 35–41.

Preat, B. (1987). *Am. Ind. Hyg. Assoc. J.*, **48**, 877–884.

Raiffa, H. (1970). *Decision Analysis: Introductory Lectures on Choices under Uncertainty*, Addison-Wesley Publishing, Reading, MA.

Rappaport, S. M. (1991). *Ann. Occup. Hyg.*, **15**, 61–121.

Rappaport, S. M. and T. J. Smith, Eds. (1991). *Exposure Assessment for Epidemiology and Hazard Control*, Lewis, Chelsea, MI.

Rappaport, S. M., S. Sebrin, and S. Roach (1988). *Appl. Ind. Hyg.*, **3**, 310–315.

Rappaport, S. M., S. Selvin, R. C. Speer, and C. Keil (1981a). *Am. Ind. Hyg. Assoc. J.*, **42**, 831–838.

Rappaport, S. M., S. Selvin, R. C. Speer, and C. Keil (1981b). *ACS Symp. Series*, **149**, 431.

Roach, S. A. (1966). *Am. Ind. Hyg. Assoc. J.*, **27**, 1–12.

Roach, S. A. (1977). *Ann. Occup. Hyg.*, **20**, 65–84.

Roach, S. A., S. Holmes, W. H. A. Beverley, J. L. Bonsall, L. H. Capel, R. D. Hunt, M. Jacobsen, J. G. Morris, W. H. Smither, J. Steel, R. Sykes, and S. J. Silk (1983). *Ann. Occup. Hyg.*, **1**, 1–55.

Rock, J. C. (1982). *Am. Ind. Hyg. Assoc. J.*, **43**, 297–313.

Roy, B. R. (1977). *J. Am. Ind. Hyg. Assoc.*, **38**, 327–332.

Runion, H. E. (1977). *Am. Ind. Hyg. Assoc. J.*, **38**, 391–393.

Ryan, P. B. (1991). *J. Exp. Ana. Environ. Epi.*, **1**, 453–474.

Samit, J. M. and J. D. Spengler (1991). *Indoor Air Pollution*, Johns Hopkins Univ. Press, Baltimore.

Schulte, K. A., D. J. Larsen, R. W. Hornung, and J. V. Crable (1974). "Report on Analytical Methods Used in a Coke Oven Effluent Study," National Institute for Occupational Safety and Health.

Sem, G. J., K. Tsurubayashi, and K. Homma (1977). *Am. Ind. Hyg. Assoc. J.*, **38**, 580–588.

Sherwood, R. J. (1966). *Am. Ind. Hyg. Assoc. J.*, **27**, 98–109.

Spear, R. C., S. Selvin, and M. Francis (1986). *Am. Ind. Hyg. Assoc. J.*, **48**, 365–368.

Sterling, T. D. and D. M. Koboyashi (1977). *Environ. Res.*, **3**, 1–35.

Stern, A. C., Ed. (1976). *Air Pollution*, 3rd ed., Vol. 3, *Measuring Monitoring, and Surveillance of Air Pollution*, Academic Press, New York, p. 799.

Stewart, P. A. and R. F. Herrick (1991). *Appl. Occup. Environ. Hyg.*, **6**, 421–427.

Stewart, P. A., R. F. Herrick, A. Blair, H. Checkoway, P. Dray, L. Fine, L. Fischer, R. Harris, T. Kauppinens, and R. Saracci (1991). *Scand. J. Work Environ. Health*, **17**, 281–285.

Thompson, E. M., H. N. Treaftis, T. F. Tomb, and A. J. Beckert (1977). *Am. Ind. Hyg. Assoc. J.*, **38**, 523–535.

Tuggle, R. M. (1981). *Am. Ind. Hyg. Assoc. J.*, **42**, 493–498.

Tuggle, R. M. (1982). *Am. Ind. Hyg. Assoc. J.*, **43**, 338–346.

Turner, D. (1976). *Ann. Occup. Hyg.*, **19**, 147–152.

USEPA (1988). Protocol for Generating Unit-Specific Emission Estimates for Equipment Leaks of VOC and VHAP, Research Triangle Park, NJ, Pub. N., EPA 450/3-88-010.

USEPA (1990). *Code of Federal Regulations (CFR)*, Title 40 Part 60 Appendix A, Reference Method 21, Determination of Volatile Organic Compound Leaks.

USEPA (1991). Chemical Engineering Branch Manual for the Preparation of Engineering Assessments, PN 3786-64.

USEPA (1992). *Guidelines for Exposure Assessment*, FR 57, 104:22888-22939, May 29, 1992.

USPHS (1977). U.S. Public Health Service, National Institute for Occupational Health and Safety, "The Right to Know: Practical Problems and Policy Issues Arising from Exposures to Hazardous Chemicals and Physical Agents in the Workplace," Department of Health, Education and Welfare, PHS-NIOSH, Cincinnati, OH.

Vincent, J. H. and D. Mark (1981). *Ann. Occup. Hyg.*, **24**, 375–390.

Wallace, L. A., E. D. Pellizzar, T. D. Hartwell, R. Whitmore, H. Zelon, R. Perritt, and L. Sheldon (1988). *Atmos Environ.*, **22**, 2141–2163.

Warriner, R. A. III, A. S. Nies, and W. J. Hayes (1977). *Arch. Environ. Health*, **32**, 203–205.

Weeks, R. W., Jr., B. J. Dean, and S. K. Yasuda (1977). *Occup. Health Saf.*, **46**, 19–23.

Weisenberger, B. L. (1977). *J. Occup. Med.*, **19**, 569–570.

WGD (1991). *Health-Based Recommended Occupational Exposure Limit for Gasoline*, Report WGO 91-318-5, Directorate General of Labour, Den Haag, Netherlands.

Wigle, D. T. (1977). *Arch. Environ. Health*, **32**, 185–190.

Zielhuis, R. L. (1978). *Scand. J. Work Environ. Health*, **4**, 1–18.

Zumwalde, R. K. and J. M. Dement (1977). "Review and Evaluation of Analytical Methods for Environmental Studies of Fibrous Particulate Exposures," Department of Health, Education and Welfare, Publication (NIOSH) 77-204, NIOSH, Cincinnati, Ohio, p. 66.

CHAPTER THREE

Analytical Methods

Peter M. Eller, Ph.D., CIH

1 INTRODUCTION

The accurate determination of toxic substances in personal, breathing-zone air samples is vital to industrial hygiene decision making. The terms *analysis* and *method* are defined to include the taking of the air sample *and* the laboratory- or field-based measurement of quantity of analyte present, as well as all intermediate steps such as calibration and sample preparation. Thus, the determination of an airborne contaminant begins with the selection of the proper sampling device and continues through to the reporting of results. Unless this overall process of *sampling plus measurement* is included in method development and application, important sources of variation may be neglected, and comparison of results and use of data for industrial hygiene decision making will be more difficult (Keith, 1991; Keith et al., 1983).

This chapter summarizes the principles of sampling and measurement methods necessary to achieve accurate industrial hygiene data.

2 DEVELOPMENT AND EVALUATION OF METHODS

Analytical variability may be small with respect to environmental variations in analyte concentration to which workers are exposed (Nicas et al., 1991). The accurate analysis of individual industrial hygiene samples depends on the use of methods that are evaluated and shown to yield results that have small error and bias.

The *quality* of a given analytical result may be defined as its *accuracy*, the degree to which it approaches the actual (''true'') concentration of analyte present in the

Patty's Industrial Hygiene and Toxicology, Third Edition, Volume 3, Part A, Edited by Robert L. Harris, Lewis J. Cralley, and Lester V. Cralley.
ISBN 0-471-53066-2 © 1994 John Wiley & Sons, Inc.

air sample (Dux, 1986). In order to characterize a method fully with respect to accuracy, the method must be evaluated using standard reference materials, realistic sample matrices, and an independent reference method. The following progression in method evaluation is typical:

- Step 1. Draft method from original research or adapted from the literature
- Step 2. Define capabilities of method (potential for quantitative collection from air, acceptable flow rates, sensitivity, selectivity, sample capacity, storage stability before and after sampling)
- Step 3. "Ruggedized" method (major variables identified and controlled) (Youden and Steiner, 1975)
- Step 4. Method calibrated with standard reference materials (Maier, 1991; McKenzie, 1990)
- Step 5. Field-tested method vs. independent reference method
- Step 6. "User-checked" method (critically reviewed using spiked samples by another chemist or laboratory) (Hull, 1984)
- Step 7. Collaboratively tested method
- Step 8. Continuing quality control efforts to ensure that the method as used is in control.

2.1 Method Evaluation Protocols

Protocols for the development of methods for industrial hygiene must contain experiments designed to estimate bias and precision of the analysis of realistic samples. In the Standards Completion Program (SCP) conducted by the Occupational Safety and Health Administration (OSHA) and the National Institute for Occupational Safety and Health (NIOSH), for example, the validation process consisted of analysis both of samples fortified with known quantities of standard materials to estimate recovery and of samples taken from controlled atmospheres generated in the laboratory to estimate overall precision (Busch and Taylor, 1981; Taylor et al., 1977; Gunderson et al., 1980). The coefficients of variation from these two independent sets of data were used to estimate all components of precision in order to produce an overall estimate of the precision of the method. The bias with respect to an independent method was estimated as well and was required to be less than 10 percent. These SCP methods form the basis for most of the analytical methods published by OSHA and NIOSH (OSHA, 1990; NIOSH, 1984). The protocol (Taylor et al., 1977) has been augmented with a test for sample storage stability (Gunderson et al., 1980) and is used for continuing method development efforts at NIOSH.

A separate protocol has been developed for the evaluation of passive monitors (Cassinelli et al., 1987; Bartley et al., 1987). This protocol includes the determination of analytical recovery using spiked monitors. Generated, standard atmospheres are used to determine sampling rate and capacity, reverse diffusion, storage stability, and temperature effects. In addition, a factorial design is used to study the following significant factors and interactions when sampling standard atmospheres:

- Analyte concentration (at 0.1 and 2 times air exposure limit)
- Exposure time (at minimum and maximum recommended sampling times)
- Face velocity (at 10 and 150 cm/sec)
- Relative humidity (at 10% RH and 80% RH)
- Interferences (at 0 and 1 times air standard)
- Orientation (with monitor parallel or perpendicular to airflow)

2.2 Method Classification

The data obtained in these steps may be combined and summarized in a method classification scheme, which describes the degree to which the method has been evaluated. One such scheme was used for methods published by NIOSH, in which the methods were classified as E (proposed), D (operational), C (tentative), B (recommended), or A (accepted) (Crable and Smith, 1975). To move from class E to class D required the successful use of the method on at least 15 field samples. Class C was reserved for methods in general use by other laboratories, but not evaluated for the particular industrial hygiene application. Validation of the method using standard materials, generated samples, and a reference method was required for class B status. Successful field and collaborative testing upgraded the class B method to class A.

2.3 Quality Control

Quality control of the method in day-to-day use is essential for accurate results. Even the most thoroughly evaluated method can produce erroneous results if improperly applied. Each method should have a quality control plan that addresses the unique requirements of the method. An example of the benefits that can accrue from a detailed, diligently applied quality control plan is the improvement in interlaboratory precision of the phase-contrast fiber counting method for asbestos (Abell et al., 1989; Abell and Doemeny, 1991; Schlecht and Shulman, 1986).

A regular program of double-blind quality control samples is invaluable in discovering potential errors in the sampling and measurement phases. These samples should closely match field samples. In most cases "spikes" of standard solution applied to the sampler will suffice. For example, quality control samples of a number of volatile organic compounds on charcoal tubes prepared in this manner have been shown to be stable for several years when stored at $-4°C$ (Eller, P. M., NIOSH–DPSE, Cincinnati, unpublished work). In some cases gas-phase spikes (Lunsford et al., 1990) or sampling of generated, standard atmospheres (Groff and Schlecht, 1991) may be required for the preparation of realistic quality control samples.

Collaborative testing can provide realistic estimates of interlaboratory variability, as does a proficiency testing scheme such as the American Industrial Hygiene Association (AIHA)–NIOSH Proficiency Analytical Testing (PAT) program (Abell and Doemeny, 1991; Groff and Schlecht, 1991). The PAT program does not specify methods to be used by the participating laboratories. In typical, recent results, the

mean relative standard deviations among the designated reference PAT laboratories
were (Groff, J. G., NIOSH–DPSE, Cincinnati, 1991, personal communication):

Lead on filters	3.7%
Cadmium on filters	4.2
Organics on charcoal	4.4
Chromium on filters	6.9
Silica on filters	22.
Asbestos on filters	22.

The PAT samples for organic solvents, silica, and asbestos were prepared by sampling generated atmospheres. These relative standard deviations (RSD) therefore contain components of sample generation variability as well as measurement variability.

Over a recent 10-yr period, the relative standard deviation of interlaboratory asbestos fiber counts on PAT samples decreased by approximately one-third (Abell and Doemeny, 1991; Schlecht and Shulman, 1986); a similar trend was seen with silica analyses (Shulman et al., 1992). Several factors may have contributed to these results: improved individual laboratory performance with increased experience, improved analytical methodology, and attrition of the less-capable laboratories. Laboratories with less experience in the program were significantly more likely to obtain extreme results on a given sample.

3 SAMPLING

Sampling is an integral part of the analysis process. The accuracy of the final result can be no better than the accuracy of the sampling procedure. The sampler must be chosen carefully to achieve the degree of specificity needed (e.g., separation of gas-phase species from particulate interferences) in the sampling steps. The sampler must also have sufficient capacity and adequate working range in order to address time-weighted average or short-term exposure limits, as appropriate.

3.1 Common Errors in Sampling

Aside from concerns about sampling frequency and location, some common errors in sampling are:

- Contamination, yielding a positive bias.
  Remedy: Analyze, and correct for, field blanks.
- Flow rate uncertainty.
  Remedy: Calibrate sampling train before and after sampling; check frequently during sampling.
- Interfering substances in air.
  Remedy: Take corresponding bulk samples.

- Loss of analyte.
  Remedy: Avoid overloading sampler. Ship and store samples cold.
- Low collection efficiency.
  Remedy: Decrease sampling time, temperature, or flow rate. Use fresh collection reagents and samplers.
- Sample too small for precise analysis.
  Remedy: Increase flow rate or sampling time.

3.2 Matching the Sampler to the Properties of the Analyte

For aerosols with no appreciable vapor pressure, glass fiber filters or membrane filters with "pore size" in the range 0.8–5 μm provide excellent collection efficiency in the inhalable size range (Davies, 1968). Vapor pressure becomes an important consideration if the mass of analyte calculated to be present as a gas approaches 10 percent of the total mass collected under the particular sampling conditions at one-half the permissible exposure limit (PEL) or recommended exposure limit (REL). Aerosols with appreciable vapor pressure require the addition of a sorbent for vapor in line downstream from the filter. In some cases equilibrium vapor pressures can be used to predict the need for a sorbent (Perez and Soderholm, 1991); in others, lack of data or anomalous behavior by the analyte may require further experimental work to achieve efficient collection (Hill and Arnold, 1979; Costello et al., 1983).

For gases the sampler must provide sufficient physical or chemical attraction to collect and retain the analyte. In some cases prefilters (to collect interfering particulate matter) must be inert toward the gases that are collected on sorbents downstream. For example, a method for the collection of sulfur dioxide gas relies on an inert prefilter (to separate interfering sulfite and sulfate salts) in front of the active sorbent (Eller, 1989). In another method, for particulate and gaseous fluorides, the cellulose backup pad for the prefilter was found to interfere by adsorbing hydrogen fluoride (HF); the pad was replaced by a nonabsorbing, porous plastic support (Lorberau and Cassinelli, 1990).

3.3 Sampler Capacity

For every sampler an upper limit on the amount of analyte to be sampled must be defined. The sampler must not be overloaded or analyte will be lost. Where a derivatizing reagent is used, the sampler capacity may be a calculated constant.

For solid sorbents the maximum sampler capacity depends on temperature, humidity, flow rate, and the presence of other interfering substances in the air sample. The maximum sampler loading has been defined as two-thirds of the breakthrough capacity (breakthrough capacity is the mass of analyte collected at the point when effluent $\div$ sampled concentration = 0.05) when sampling an atmosphere at twice the exposure limit (Taylor et al., 1977; Gunderson et al., 1980; OSHA, 1990). Other considerations such as relative humidity may be important as well; competition between water molecules and analyte molecules for active sorbent sites will result in reduced capacity for the analyte.

For aerosols, sampling should not result in such heavy deposits on the filter as to cause reductions in flow rate or loss of sample from the filter. One rule of thumb is to attempt to limit aerosol samples to 2 mg total particulate per filter (Morring et al., 1984).

3.4 Sample Flow Rate Considerations

Lower limits of acceptable flow rate for active sampling of gases are determined by the diffusion rate of the analyte (important only below $\approx$ 5 mL/min) (Heitbrink, 1983). Thus, a minimum flow rate of 20 mL/min (the minimum flow that can be conveniently calibrated with currently available field equipment) is appropriate for active sampling methods. Maximum flow rates depend on the efficiency of adsorption, particularly if collection involves derivatization on the sorbent. Typically, sampling of many organic vapors on coconut shell charcoal (physical adsorption) is quantitative at 200 mL/min or less (Taylor et al., 1977; Gunderson et al., 1980; OSHA, 1990), while lower maximum rates apply in the case of derivatization [e.g., formaldehyde (Kennedy, 1989)].

Passive samplers have inherently lower sampling rates than active samplers, but their simplicity and ease of use are attractive. For many gases passive samplers give accurate results (Lewis et al., 1983; Pristas, 1991).

The minimum effective capture velocity for aerosols, with aerodynamic particle size $\leq$ 10 μm, by the 4-mm diameter inlet of the commonly used 37-mm cassette samplers is about 0.5 L/min (Davies, 1968). This leads to acceptable collection efficiency for "respirable" aerosols, but not for "thoracic" or "inhalable" aerosols (ACGIH, 1991). Most sampling with membrane filters is done at 1–3 L/min, but flows up to 14 L/min for such samplers have been suggested (Baron, 1989).

3.5 Working Range

Every method has characteristic operating limits that must be respected in order to obtain accurate results or claimed performance. A working range for a method can be defined as those combinations of air sample volumes and analyte concentrations in air that yield masses of analyte on a single sample that are between the limit of quantitation and some upper limit (sampler capacity or upper limit of linearity for the measuring instrument) (Eller, 1986). Thus, minimum and maximum air sample volumes are defined (at a specified target concentration such as an OSHA PEL):

$$V_{MIN} \text{ (liters)} = \frac{LOQ \text{ (mg)} \times 10^3}{\text{target concentration (mg/m}^3)}$$

$$V_{MAX} \text{ (liters)} = \frac{\text{sampler capacity (mg)} \times 10^3}{\text{target concentration (mg/m}^3)}$$

where LOQ is limit of quantitation. With sample masses below the working range, analytical precision suffers because of proximity to the detection limit. On the other hand, sample volumes smaller than V_{MIN} may be desirable if the concentration of analyte in the atmosphere is much greater than the concentration at which the break-through capacity was measured. With sample volumes greater than V_{MAX}, one runs the risk of breakthrough at these concentrations. At concentrations much less than the target concentration, larger sample volumes may be appropriate if no break-through occurs.

4 MEASUREMENT

The measurement technique must be sufficiently sensitive to quantitate amounts of the analyte in typical industrial hygiene samples and sufficiently selective to detect the analyte in the presence of other substances. Important properties of measurement methods are limit of detection, limit of quantitation, blank values, and selectivity.

4.1 Common Errors in Measurement

Some common sources of error in measurement are:

- Bias (positive or negative) in method.
  Remedy: Calibrate with reference material or reference method; ruggedize to minimize bias.
- Contamination (positive bias).
  Remedy: Analyze, and correct for, media, field, and reagent blanks.
- Desorption efficiency low.
  Remedy: Determine recovery with spiked samplers over the concentration range of the field samples.
- High recovery with standard additions.
  Remedy: Correct for nonlinear calibration curve.
- Interferences (positive or negative bias).
  Remedy: Analyze bulk samples to identify interferents; calibrate by standard additions.
- Matrix effects (positive or negative bias).
  Remedy: Match sample and standard matrices.
- Sample processing losses (negative bias).
  Remedy: Carry standards through same treatment.

4.2 Detection Limit

Detection limit, or limit of detection (LOD), is a decision point used to decide whether to report a significant analyte signal from the sample. It is defined as the mass of analyte that gives a signal $3\sigma_b$ above the mean blank signal, where σ_b is the standard deviation of the blank signal (Keith, 1991; Keith et al., 1983).

The following protocol for estimation of LOD uses the variability of low-level analyte responses, in a matrix approximating that of the samples, as an estimate of

σ_b. The standard deviation of analyte signal is assumed to be constant over the narrow range in step 1 below. If this is not the case, alternate approaches are needed.

1. For each sample set, prepare five or more *low-level* calibration standards, *spiked on sampling media*, to cover the range from less than the expected LOD to *no greater than* 10 times the expected LOD. These shall be separately prepared standards, not simply replicate analyses. A reagent blank, spiked on sampling media, may be used as one of the standards, providing that it produces a positive analytical response.

2. Analyze the low-level calibration standards under the same conditions as for the field samples.

3. Graph the responses of the low-level calibration standards versus mass of analyte. Include all available data for masses that are less than 10 times the expected LOD. Obtain the linear regression equation, $Y = mX + b$, and the predicted responses ($\hat{Y}_i$) at each X (mass of analyte).

4. Calculate the standard error of the regression, $s_y = [\Sigma(\hat{Y}_i - Y_i)^2/(N - 2)]^{1/2}$, where N is the number of data points obtained in step 2 above.

5. Calculate LOD $= 3s_y/m$, where m is the slope of the regression line.

6. Check the calculated LOD against other available data (e.g., availability of mass spectrometric data, background present in the field samples, strip chart recording of peaks, etc.) to make sure that it is a realistic number. If there are doubts as to whether the calculated LOD is realistic, so state in the report.

7. Interpretation: False Positives. In the absence of interferences, we know, with ≈ 99 percent confidence, that an individual sample giving a signal equal to, or greater than, that of the LOD level *does contain* the analyte. That is, the probability of false positives is ≈ 1 percent.

8. Interpretation: False Negatives. The probability of false negatives at the LOD level is 50 percent (i.e., half of samples containing this much analyte will fail to give a detectable signal). More generally, the probability of the analyst detecting the analyte when it is *actually present* varies from < 1 percent when the concentration is $\ll$ LOD, to 50 percent at the LOD, to ≈ 99 percent at 2 $\times$ LOD. If it is necessary for the end user of the data to operate at a lower rate of false negatives than 50 percent, one can say, for example, that "at a level of 2 $\times$ LOD or above, 99 percent of analyte-containing samples have been detected and reported."

4.3 Limit of Quantitation

Limit of quantitation (LOQ) is the lowest mass that can be reported with acceptable precision (Keith, 1991; Keith et al., 1983). Limit of quantitation is the *larger* of: (a) the mass corresponding to the mean blank signal $+ 10\sigma_b$ (i.e., $\pm 30\%$ precision) or (b) the mass above which recovery is ≥ 75 percent. Calculate LOQ $= 3.33 \times$ LOD, *or* LOQ $=$ mass above which recovery is ≥ 75 percent, *whichever is greater*.

4.4   Reporting of Measurements

Sample results below the LOD should be reported as "not detected" (ND). Sample results between the LOD and LOQ should be reported numerically, to two significant figures, and enclosed in parentheses to emphasize the imprecision of the result. The calculated values of LOD should be reported to one significant figure and the LOQ to two significant figures.

4.5   Blank Values

Since the variability of blank values contributes to the limits of detection and quantitation, it is important that blank values be estimated with each sample set analyzed. It is useful to define several types of blanks:

4.5.1   Field Blanks

These are complete collection media, treated exactly like field samples (taken to the field, opened and closed, shipped to the laboratory for analysis) except that no air is pulled through. Laboratory results for the field blanks are reported to the end user of the data (i.e., the industrial hygienist); they should not be subtracted from other sample results by the laboratory. The results reported on field blanks should be interpreted by the person who took the samples to estimate the amount, if any, of contamination that may occur during preparation for sampling, packaging of samples, shipment to the laboratory, or storage before analysis.

4.5.2   Media Blanks

These are unexposed, unopened collection media that the laboratory uses for estimating media background. In some cases the media quality is consistent and known, reducing the need for repeated analysis of media blanks. For example, most commercially available coconut shell charcoal tubes give zero gas chromatographic response for many organic compounds. Media blanks are most important when the sampling device includes reagents (e.g., coated sorbent tubes). In these cases at least six media blanks (from the same lot as the field samples) should be analyzed with each sample set. The laboratory should analyze the media blanks by the same procedure as for the field samples (i.e., using all appropriate reagents and sample treatment steps) and should correct the field sample results for the mean media blank value.

4.5.3   Reagent Blanks

The background arising from contamination or interference due to the reagents used in the sample treatment and measurement steps is usually contained in the media blank values, and therefore taken into account (see Section 4.5.2). Where this is not the case, or where unusually high and variable media blank values are obtained, separate studies should be undertaken to isolate and reduce reagent blank values. An example of such a study is the reduction of an artifact in the determination of particulate cyanides in air, thereby reducing the media blank value (Perkins, et al., 1990).

4.6 Selectivity, Sensitivity, and Qualitative Analysis

In addition to adequate precision and sensitivity, the measurement technique must be selective enough to quantitate the analyte in the presence of potential interferences. In some cases a preliminary qualitative screening process may be applied to delineate compounds of interest for further quantitation.

For example, because of the inherent selectivity of gas chromatography, this technique has been used widely to determine compounds that will elute at column temperatures below about 300°C. First applied to industrial hygiene samples in the 1970s (White et al., 1970; Reckner and Sachdev, 1975; Saalwaechter et al., 1977), gas chromatography has progressed in sensitivity and selectivity. Capillary columns have largely replaced packed columns, with vastly improved resolution of chromatographic peaks. Improved instrument design and operation have made complex separations possible in less than one minute (Ke et al., 1992). In the analysis of complex mixtures such as pesticides, the use of a second capillary column for simultaneous confirmation of the analytes is cost effective (Vargo and Mosesman, 1992). In another variation of two-dimensional chromatography, two different, parallel columns can be timed to use a common detector and to produce two different chromatograms (Gupta and Nikelly, 1991). Coupling of supercritical fluid extraction to the gas chromatograph promises to improve both selectivity and sensitivity (Raymer and Velez, 1991). Improvements in detector technology have been no less dramatic. For example, electron capture detectors for halogen-containing compounds, nitrogen- and phosphorus-specific detectors, and mass-selective detectors are valuable in industrial hygiene analyses.

''Hyphenated'' techniques are another example of the evolution of measurement methods toward improved selectivity. The application of a nondestructive separation technique followed in tandem by another powerful analytical measurement is characteristic of ''hyphenated'' techniques. Thus, if the exit of a gas chromatograph (GC) is coupled to the inlet of a mass spectrometer (MS), the resulting GC–MS finds wide application in the identification of unknown volatile compounds. Some recent examples are:

- A transportable (Jeep-mounted), thermal-desorption GC–MS was used to determine PCBs at a hazardous waste site (Rabbat et al., 1992).
- The combination of ion chromatography and particle beam mass spectrometry has been used to determine several aromatic sulfonic acids (Hsu, 1992).
- Highly sensitive multielement analysis capability results when an inductively coupled plasma is interfaced to a mass spectrometer. The sample solution is introduced to the plasma, which produces ions and sends them to the mass spectrometer for quantitation. With detection limits similar to those obtained by graphite furnace atomic absorption spectrophotometry and greatly reduced

interferences, ICP–MS instruments have been commercialized recently (Tye and Hitchen, 1992).

- A broad range of organic compounds, including aromatic sulfonic acids, chlorinated phenoxy acid herbicides, and polyaromatic hydrocarbons, are amenable to analysis by liquid chromatography–particle beam mass spectrometry (Brown et al., 1991).

Classical techniques remain favored in some cases because of sensitivity or ease of use. For example, the chromotropic acid method for formaldehyde (Altshuller et al., 1962; Kennedy, 1989) is among the most sensitive methods available. Direct-reading, color indicator detector tubes have evolved into versions that easily give direct readout of time-weighted average concentrations. Both active and passive sampling modes of detector tubes are available for some compounds. Certification of detector tubes, conducted by NIOSH for about 12 years and discontinued in 1983 (*U.S. Code of Federal Regulations*, 1973), is now done by the Safety Equipment Institute (Wilcher, 1987).

5 SUMMARY

Determination of the concentration of toxic substances in workplace atmospheres is a challenging application of analytical chemistry. Each step in the process, from selection of the sampling device to sample workup and measurement, contributes to the accuracy of the result. Quality assurance steps should be applied before and during sampling, as well as in the laboratory measurement procedures.

Future improvements in analytical methods for industrial hygiene should include improved sorbent materials to provide better selectivity in sampling, improvements in the sensitivity of measurement methods, and more selective qualitative identification techniques.

REFERENCES

Abell, M. T. and L. J. Doemeny (1991). *Am. Ind. Hyg. Assoc. J.*, **52**, 336–339.

Abell, M. T., S. A. Shulman, and P. A. Baron (1989). *Appl. Ind. Hyg.*, **4**, 273–285.

ACGIH (1991). *Appl. Occup. Environ. Health*, **6**, 817–818.

Altshuller, A. P., L. J. Leng, and A. F. Wartburg (1962). *Int. J. Air Wat. Poll.*, **6**, 381–387.

Baron, P. A. (1989). In *NIOSH Manual of Analytical Methods*, 3rd ed., P. M. Eller, Ed., U.S. Department of Health & Human Services Publ. (NIOSH) 84-100, U.S. Government Printing Office, Washington, D.C., Method 7400 (Fibers).

Bartley, D. L., G. J. Deye, and M. L. Woebkenberg (1987). *Appl. Ind. Hyg.*, **2**, 119–122.

Brown, M. A., R. D. Stephens, and S. Kim (1991). *Trends Anal. Chem.*, **10**, 330–336.

Busch, K. A. and D. G. Taylor (1981). In G. Choudhary, Ed., *Chemical Hazards in the Workplace, ACS Symposium Series No. 149*, American Chemical Society, Washington, D.C., pp. 503–517.

Cassinelli, M. E., R. D. Hull, J. V. Crable, and A. W. Teass (1987). In A. Berlin, R. H. Brown, and K. J. Saunders, Eds., *Diffusion Sampling: An Alternative to Workplace Air Monitoring*, Royal Society of Chemistry, London, pp. 190–202.

Costello, R. J., P. M. Eller, and R. D. Hull (1983). *Am. Ind. Hyg. Assoc. J.*, **44**, 21–28.

Crable, J. V. and R. G. Smith (1975). *Am. Ind. Hyg. Assoc. J.*, **36**, 149–151.

Davies, C. N. (1968). *Brit. J. Appl. Phys.*, *Ser. 2, Vol. 1*, 921–932.

Dux, J. P. (1986). *Handbook of Quality Assurance for the Analytical Chemistry Laboratory*, Van Nostrand Reinhold, New York, p. 2.

Eller, P. M. (1986). *Appl. Ind. Hyg.*, **1**, 91–94.

Eller, P. M. (1989). In *NIOSH Manual of Analytical Methods*, 3rd ed., P. M. Eller, Ed., U.S. Department of Health & Human Services Publ. (NIOSH) 84-100, U.S. Government Printing Office, Washington, D.C., Method 6004 (Sulfur Dioxide).

Groff, J. H. and P. C. Schlecht (1991). *Appl. Occup. Environ. Hyg.*, **6**, 1001–1002.

Gunderson, E. C., C. C. Anderson, R. H. Smith, and L. J. Doemeny (1980). *NIOSH Research Report: Development and Validation of Methods for Sampling and Analysis of Workplace Toxic Substances*, U.S. Department of Health and Human Services Publ. (NIOSH) 80-133.

Gupta, P. K. and J. G. Nikelly (1991). *Anal. Chem.*, **63**, 1264–1270.

Heitbrink, W. A. (1983). *Am. Ind. Hyg. Assoc. J.*, **44**, 453–462.

Hill, R. H., Jr., and J. E. Arnold (1979). *Arch. Environ. Contam. Toxicol.*, **8**, 621–628.

Hsu, J. (1992). *Anal. Chem.*, **64**, 434–443.

Hull, R. D. (1984). In *NIOSH Manual of Analytical Methods*, 3rd ed., P. M. Eller, Ed., U.S. Department of Health & Human Services Publ. (NIOSH) 84-100, U.S. Government Printing Office, Washington, D.C., pp. 29–36.

Ke, H., S. P. Levine, R. F. Mouradian, and R. Berkley (1992). *Am. Ind. Hyg. Assoc. J.*, **53**, 130–137.

Keith, L. H. (1991). *Environmental Sampling and Analysis: A Practical Guide*, Lewis Publishers, Boca Raton, FL, pp. 93–119.

Keith, L. H., R. A. Libby, W. Crummett, J. K. Taylor, J. Deegan, Jr., and G. Wentler (1983). *Anal. Chem.*, **55**, 2210–2218.

Kennedy, E. R. (1989). In *NIOSH Manual of Analytical Methods*, 3rd ed., P. M. Eller, Ed., U.S. Department of Health & Human Services Publ. (NIOSH) 84-100, U.S. Government Printing Office, Washington, D.C., Method 3500 (Formaldehyde).

Lewis, R. G., R. W. Coutant, G. W. Wooten, C. R. McMillin, and J. D. Mulik (1983). *EPA-600/D-83-044*, NTIS PB-83-194357.

Lorberau, C. and M. E. Cassinelli (1990). In *NIOSH Manual of Analytical Methods*, 3rd ed., P. M. Eller, Ed., U.S. Department of Health & Human Services Publ. (NIOSH) 84-100, U.S. Government Printing Office, Washington, D.C., Method 7902 (Fluorides, Aerosol & Gas).

Lunsford, R. A., Y. T. Gagnon, J. Palassis, J. M. Fajen, D. R. Roberts, and P. M. Eller (1990). *Appl. Occup. Environ. Hyg.*, **5**, 310–320.

Maier, E. A. (1991). *Trends Anal. Chem.*, **10**, 340–347.

McKenzie, R. L., Ed. (1990). *Standard Reference Materials Catalog, NIST Special Publication 260*, National Institute of Standards and Technology, Washington, D.C.

Morring, K., J. Clere, and F. Hearl (1984). In *NIOSH Manual of Analytical Methods*, 3rd ed., P. M. Eller, Ed., U.S. Department of Health & Human Services Publ. (NIOSH) 84-100, U.S. Government Printing Office, Washington, D.C., Method 0500 (Nuisance Dust).

Nicas, M., B. P. Simmons, and R. C. Spear (1991). *Am. Ind. Hyg. Assoc. J.*, **52**, 553–557.

NIOSH (1984). *NIOSH Manual of Analytical Methods*, 3rd ed., P. M. Eller, Ed., U.S. Department of Health & Human Services Publ. (NIOSH) 84-100, U.S. Government Printing Office, Washington, D.C.

OSHA (1990). Occupational Safety & Health Administration, *OSHA Analytical Methods Manual*, USDOL OSHA Analytical Laboratory, P.O. Box 65200, Salt Lake City.

Perez, C. and S. C. Soderholm (1991). *Appl. Occup. Environ. Health*, **6**, 859–864.

Perkins, J. B., D. G. Tharr, J. Palassis, D. B. Fannin, and P. M. Eller (1990). *Appl. Occup. Environ. Health*, **5**, 836–837.

Pristas, R. (1991). *Am. Ind. Hyg. Assoc. J.*, **52**, 297–304.

Rabbat, A., Jr., T.-Y. Liu, and B. M. Abraham (1992). *Anal. Chem.*, **64**, 358–364.

Raymer, J. H. and G. R. Velez (1991). *J. Chromatog. Sci.*, **29**, 467–475.

Recker, L. R. and J. Sachdev (1975). *Collaborative Testing of Activated Charcoal Sampling Tubes for Seven Organic Solvents*, U.S. Department of Health, Education, and Welfare Publ. (NIOSH) 75-184.

Saalwaechter, A. T., C. S. McCammon, Jr., C. P. Roper, and K. S. Carlberg (1977). *Am. Ind. Hyg. Assoc. J.*, **38**, 476–486.

Schlecht, P. C. and S. A. Shulman (1986). *Am. Ind. Hyg. Assoc. J.*, **47**, 259–269.

Shulman, S. A., J. H. Groff, and M. T. Abell (1992). *Am. Ind. Hyg. Assoc. J.*, **53**, 49–56.

Taylor, D. G., R. E. Kupel, and J. M. Bryant (1977). *Documentation of the NIOSH Validation Tests*, U.S. Department of Health, Education, and Welfare, Publ. (NIOSH) 77-185.

Tye, C. T. and P. Hitchen (1992). *Am. Environ. Lab.*, **4**, 20–24.

U.S. Code of Federal Regulations Title 42, Ch. I, SubCh. G, Part 84 (May 8, 1973).

Vargo, C. and N. Mosesman (1992). *Am. Environ. Lab.*, **4**, 25–30.

White, L. D., D. G. Taylor, P. A. Mauer, and R. E. Kupel (1970). *Am. Ind. Hyg. Assoc. J.*, **31**, 225–232.

Wilcher, F. E., Jr. (May, 1987). *Occup. Health & Safety*, 81–84.

Youden, W. J. and E. H. Steiner (1975). *Statistical Manual of the Association of Official Analytical Chemists*, AOAC, Washington, D.C.

CHAPTER FOUR

The Emission Inventory

Robert L. Harris, Ph.D.
Earl W. Arp, Jr., Ph.D.

1 INTRODUCTION

An emission inventory for an industrial or commercial enterprise is a compilation of information from which one can calculate or estimate the rates (quantity per unit time) at which pollutants are released to the environment. For purposes of this chapter, only the emissions that contaminate workroom or community air are considered; emissions to surface or ground water, to soil, or to other environmental receptors are treated only as they may, in turn, result directly in emissions to workplace or community air.

An emission inventory may be simple or complex. A rudimentary inventory may consist of source location, date, identification of process, a qualitative listing of materials used, and an index of size (e.g., annual production rate) for the subject enterprise. Such an inventory, along with emission factors generated by studies of similar processes, will permit the making of an estimate of annual emissions. A comprehensive emission inventory, on the other hand, may contain sufficient detail to permit quantification of emissions, including temporal variations, for a number of specific materials from each point of release in a complex industrial process.

An inventory of emissions, along with various other kinds of companion information, discussed later in this chapter, permits the making of estimates of the nature, and sometimes the intensity, of exposures to airborne agents in workplaces or in the community. The level of detail needed and achievable for the inventory depends both on the purposes for which it is to be used and the data sources, or data-generating efforts, that can be utilized.

Patty's Industrial Hygiene and Toxicology, Third Edition, Volume 3, Part A, Edited by Robert L. Harris, Lewis J. Cralley, and Lester V. Cralley.
ISBN 0-471-53066-2 © 1994 John Wiley & Sons, Inc.

2 ELEMENTS OF AN EMISSION INVENTORY

The compilation of emission inventories is a well-established, specifically identifiable, activity in the field of air pollution control. Practices and procedures have been highly developed and description of the technology are available (USEPA, 1973; Hammerle, 1977; Thron, 1986; Engdahl et al., 1986; Berry et al., 1986). Emission inventory has been practiced in industrial hygiene for many years but has not been identified as a categorical work area in this field to the extent that it has been in the field of air pollution control. Although the types of data used and the techniques for obtaining them vary somewhat, the same basic elements appear in emission inventories in both fields of work.

2.1 Identification of Agents

Recognition, evaluation, and control of hazards are the three basic steps in the practice of industrial hygiene and community air pollution control. An emission inventory, regardless of whether it is specifically identified as such, is necessary in all three steps and is particularly important in the first two, recognition and evaluation. It is clear that if a hazard is to be dealt with, it must first be recognized; this recognition, and the identification of an emission, is a rudimentary emission inventory. Evaluation requires more than identification. In situations involving release of chemical agents to the air, evaluation may include obtaining additional information such as quantity, character, and temporal variations of emissions, all of which are part of the emission inventory. The design and implementation of emission control requires detailed information about the emission source that goes beyond the level of detail usually required for an emission inventory.

The federal Toxic Substances Control Act of 1976 (TSCA, 1976), among other things, provides for the collection of information regarding chemical substances manufactured, imported, or produced in the United States. Such information will permit preliminary assessment of potential exposures and possible effects on health and the environment. Compliance with the act requires identification, by process and location, of many chemical agents in industry and commerce. The notification, reporting, and record-keeping provisions of the act and its implementing regulations facilitate the agent identification component of a comprehensive, plantwide emission inventory.

Not all materials handled in industry and commerce are hazardous. According to the American Chemical Society's Chemical Abstract Service, its registry listed more than 11.8 million compounds as of June 1993 (CAS, 1993). The number of registered compounds is growing at a rate of about 7000 per week. Of the millions of compounds that exist, some 61,500 are considered, for purposes of the Toxic Substances Control Act, to be in commercial use as of January 1993 (CAS, 1993). Some toxic dose information, based on experimental animal work or other observations, is available on about 89,000 compounds (NIOSH, 1992). Some of these are relatively nontoxic, others are not in common use, and for many the toxicity information is fragmentary. Probably fewer than 1000 compounds have been identified with occupational health or community air pollution problems sufficiently to permit development of workplace or air quality standards. The identification, by means of an emission inventory, of materials that are released from a process or operation, however, is a fundamental step in the recognition of those that represent a potential hazard in the workplace or the community.

For manufacturing processes, preliminary identification of potentially hazardous agents often can be based on the identification of process raw materials, intermediate and by-product materials, and the process end products. In commercial enterprises the identification of materials that are handled permits the singling out of those that may represent potential hazards. For combustion sources, information on the composition of fuel and the type of combustion equipment used permits qualitative identification of pollutant components.

The evaluation phase of an industrial hygiene or community air pollution problem requires, in addition to identification of the agents of concern, a number of other kinds of information that can be obtained in an emission inventory. Among the most fundamental of these is the identification and description of the site or location at which the contaminant is released to the air.

2.2   Identification of Emission Sites

The most cursory emission inventory may identify the site or location of an emission source only as a particular plant or commercial establishment. Such location information is generally useful only for preliminary surveys of community air pollution or for indicating the need for more thorough workplace exposure evaluation. Any emission inventory use other than agent identification alone requires more specific emissions location information. For example, in air pollution control the identification of specific stacks, vents, and other points of emission is necessary for diffusion modeling and for most impact evaluations and emission regulatory activities. For industrial hygiene purposes the location of a specific workplace, process point, and perhaps even a particular process equipment opening (e.g., a mixer charging port) or work practice (e.g., the handling of shipping bags after use) may be needed. Such location information is vital to the hazard assessment process. It is necessary for identifying the workers subject to exposure from the particular source and for identifying alternatives from which to select the means for control of any hazard caused by the emission.

An emission inventory that identifies both the materials emitted from a process and the specific locations at which these materials are released can serve as the first step, and perhaps the only step necessary, in the evaluation of a hazard and initiation of a control effort.

2.3   Time Factors in Emissions

Time resolution in emission inventories may be yearly, seasonal, monthly, weekly, daily, hourly, or even less than hourly, depending on specific needs and the availability of data. For initial surveys of community air pollution, yearly average emissions by plant site may be satisfactory. At the other extreme, the assessment of emissions to workplaces that involve cyclic or intermittent operations may require

use of time intervals of 15 min or less. For short time or intermittent operations—
for example, the taking of materials samples at process sampling ports—the actual
time interval of emission and the frequency with which the operation occurs should
be recorded in the inventory.

In some cases the interval of record for the quantity of material released is relatively long—for example, a monthly record of solvent use—even though actual release may be cyclic or may occur over short intervals. In such cases the emission
inventory record should contain a sufficiently detailed description of the process or
operation to permit estimation of actual emission intervals and the quantities of materials released during these intervals.

It is important that the emission inventory record include both the date on which
the inventory was done and the calendar interval for which it applies. When a change
in process or operation occurs that materially affects the composition, quantity, or
condition of an emission, the emission inventory record should be updated to reflect
the change. In the absence of any substantial change, the inventory should be revalidated at convenient intervals, perhaps annually, or as may be required by governmental regulation. When an emission inventory record is updated or revalidated,
the old record should be retained; a sequential inventory file over a long period may
be invaluable in future retrospective environmental epidemiologic studies.

3 QUANTIFYING EMISSIONS

Emissions can be quantified either by direct measurement, such as source testing,
or by indirect means. Indirect means include techniques such as process materials
balance or the determination of an index parameter—for example, a production rate—
that can be related empirically to emissions.

3.1 Source Sampling

Source sampling is ordinarily associated with measurement of air pollutant emissions. Under some circumstances the techniques can be applied to industrial hygiene
investigations as well. The techniques of air pollutant source testing have been described in detail by Paulus and Thron (1977) and updated by Thron (1986); the first
(1977) chapter, "Stack Sampling," lists 72 references, the updated chapter (1986)
lists 90. The Environmental Protection Agency (EPA) has published stepwise procedures on source sampling for particulates (USEPA, 1974); this publication contains a number of data recording forms that are useful in sampling not only for
particulates but for other agents as well.

Source sampling ordinarily consists of withdrawing a representative sample from
a contaminant-bearing gas stream in a duct or stack. Analysis of the sample yields
data on concentration of the contaminant in the gas stream. Concentration data,
combined with companion data on gas flow rates in the ducts or stacks, yield values
for contaminant emission rates for gas streams released to the atmosphere.

The critical concern in source sampling is the representativeness of the sample.
Both composition and flow rate of a contaminated gas stream may vary as the pro-

cesses and operations that generate it vary. Thus, representativeness of a source sample depends very much on the representativeness of processes and operations at the time of sampling.

When the contaminant is particulate, the collection of a representative sample requires isokinetic sampling and use of an unbiased sampling traverse pattern. Isokinetic sampling is performed by taking the sample at such a flow rate that the sampled gas stream enters the inlet nozzle of the sampling probe with velocity equal to that which prevails at the specific point in the cross section of the stack or duct from which the sample is being withdrawn. When the velocity of gas at the sampling point in the duct or stack is greater than that in the sampling nozzle, part of the approaching gas stream is deflected around the nozzle. Smaller particles tend to follow the deflected gas stream while larger ones, by virtue of their momentum, tend to continue their trajectories and enter the nozzle; this results in a nonrepresentative overabundance of larger particles in the sample. When the velocity of the gas stream at the sampling point in the duct or stack is lower than the velocity entering the sampling nozzle, the gas stream converges into the nozzle inlet, carrying with it the smaller particles but losing some of the larger ones, which are carried past the nozzle by their momentum; the sample then is nonrepresentative because of a deficiency of larger particles.

The velocity of the gas stream in a duct or stack is not uniform throughout its cross section. For this reason, and because particles are not necessarily uniformly distributed within a duct or stack, a specific traverse pattern is ordinarily used in source sampling and in the measurement of velocity for determination of flow rates. For purposes of a sampling traverse, the cross-sectional area of the duct or stack is divided into equal-sized subareas; sample increments and velocity readings are taken at the centers of these subareas. Sampling time should be the same for each traverse point in a duct or stack, and whenever possible the entire sampling traverse should be accomplished during some interval of time during which the stack or duct flow rate and the composition of emissions do not change. Isokinetic sampling and equal area traverses are discussed in detail, including descriptions of apparatus and calculation methods, in the source sampling references cited earlier (Paulus and Thron, 1977; Thron, 1986; USEPA, 1974).

The sampling of gases and vapors differs from sampling of particulates in that isokinetic sampling is not required unless concentrations differ from place to place in the duct cross section. The collection apparatus and reagents used in a gas sampling train also differ from those used for particulates. Filtration, inertial size classification, and impingement with capture in liquid media are the collection mechanisms used for particulates source sampling; liquid absorption, adsorption on solids, and freeze-out are the methods usually employed for gas and vapor sampling. Information on sampling apparatus, collecting media, and analytic methods for a number of gases and vapors have been tabulated by Paulus and Thron (1977) and updated by Thron (1986).

As mentioned earlier, the techniques of source sampling are most often applied in air pollution emission measurements. They can, in some circumstances, be applied for inplant industrial hygiene purposes as well. When a workplace is served by general dilution ventilation in the exhaust mode, the techniques of source sampling can be applied to the exhausted airstreams to determine the rate at which contaminants are released to the workplace air. The calculation methods described in Section 5.3.1 can be used to estimate emissions when the concentrations of contaminant in the supply and exhaust air, and the ventilation rate are known with the system at equilibrium. In applying the equations to exhaust air streams, the factor K of Eq. (15) has the value 1, and the contaminant generation rate is the sum of the generation rates calculated for all exhaust discharges. The procedure is most applicable when general dilution ventilation in the exhaust mode is the sole, and controlling, ventilation regimen for the workplace, that is, when there is no mechanical local exhaust ventilation. If local exhaust ventilation is used in the workplace, its influence as general dilution ventilation must be taken into account when using this technique to estimate workplace emissions.

Source sampling techniques may also be used to quantify emissions from individual points of release in a workplace. The emissions may be captured using a temporary exhaust ventilation setup, and the amount of material released may be determined by sampling from that exhaust stream. Application of this technique to a single source or emission point in a space that contains several sources of the air contaminant requires either elimination of the influence of the other sources or correction for them. A mechanical arrangement can be provided to supply contaminant-free outside air to the test source. The exhaust stream from the test source will then contain only contaminant from that source and will be unbiased by contaminant from other sources. Alternatively, monitoring of the concentration of contaminant in room air that supplies the source test exhaust system permits correction for other sources; emissions from the test source can be determined by the difference in contaminant concentration in supply and exhaust air of the test system.

A method for measuring emission rates of gases or vapors from point or surface sources using tracer gas has been reported (Antonsson, 1990). Tracer gas is released at a measured rate at the location of the emission source; the emission rate is calculated from the tracer gas emission rate and the correlation of the concentrations of the emitted gas or vapor and the tracer gas measured at the same locations close to the source. As discussed in the paragraph above, correction must be made in the source emission rate if the supply air is contaminated or there are other sources of the emitted gas or vapor in the workplace.

3.2 Materials Balance

In some cases knowledge of processes and operations permits determination of the amount of material released to the air of a workplace without emission measurements. If it is known that a gas or vapor is generated by chemical reaction or otherwise, and is released to workplace air in proportion to the use of a raw material or a production rate, that index of generation can be used to determine the release rate. Examples include the generation of products of combustion by unvented open flames, as is the case with direct fired unit heaters. Here fuel composition and use rate are indices of contaminant emissions. Uses of volatile solvents in which the solvents do not become part of a product, but evaporate completely into the workplace air, are common in industry. Solvent use rate, in such cases, is also an emission rate. When exhaust ventilation is applied to some operations in a workplace and not to others, distinction must be made between that portion of the material that is captured by exhaust ventilation and that released in the occupied workplace; only the portion that is released directly to the workplace air is used to estimate workplace exposures. For estimating community air pollution emission rates, the total amount of volatile material that evaporates into the atmosphere is taken as the emission rate regardless of whether the material is released to workroom air or through exhaust ventilation systems.

The American Petroleum Institute (API) has reported mathematical relationships that describe evaporation losses of petroleum products from tanks during loading and unloading (API, 1959). The materials balance concepts of the API procedure can be applied to estimating vapor emissions from the loading of volatile liquids into vessels that are vented to workroom air. The mass of vapor expelled by displacement when a volatile liquid is transferred into a vessel is

$$M = 1.37 \ VSP_v \frac{mw}{T} \tag{1}$$

where M = mass of vapor expelled (lb)
V = volume of liquid transferred to the vessel (ft^3)
S = fraction of vapor saturation of expelled air
P_v = true vapor pressure of the liquid (atm)
mw = molecular weight of vapor
T = temperature of the tank vapor space (°R)

Except for S, the fraction of vapor saturation, the various parameters of Eq. (1) are ordinarily known or measurable quantities. For splash filling of a vessel that was initially vapor free, or for the refilling of a vessel from which the same liquid has just been withdrawn, the value of S can ordinarily be taken as 1 (API, 1959).

Equation (1) may overestimate the mass of vapor emission if the vessel walls are substantially colder than the volatile liquid. The true vapor pressure depends on the temperature of the liquid; in the case of the cold vessel, however, some vapor may condense on the inner wall surfaces and fail to escape through the vent into the workroom air.

When complete evaporation of a volatile material does not take place, some index other than total use is needed for quantifying emissions. In the simple case, when the amount used and the amount remaining can be determined, the amount released as vapor can be obtained by difference. When this is not possible, more sophisticated means such as exhaust air sampling must be employed.

With complete evaporation of a mixture of volatile materials, the quantity of each component that vaporizes is merely the quantity of that material in the mixture. Partial evaporation of a mixture, however, does not necessarily yield vapor quantity of each component in proportion to the quantity of that component in the liquid mixture. According to Raoult's law, the equilibrium partial pressure of each component of a perfect solution is the product of the vapor pressure of the pure liquid and its mole fraction in the solution:

$$p_n = P_n x_n \tag{2}$$

where p_n = partial pressure of component n
P_n = vapor pressure of pure liquid n
x_n = mole fraction of component n in the liquid mixture

Thus, vapor yielded by partial evaporation from a mixture of volatile materials is richer in the more volatile and leaner in the less volatile components than is the original liquid solution. The use of Raoult's law permits estimation of emission rates of components of a solution when partial evaporation takes place. The composition of the parent solution in each case must be known; values for vapor pressures of pure liquids can be found in chemical handbooks. When a substantial fraction of a liquid mixture evaporates, the change in its composition as the fractions of more volatile components decrease should be taken into account in applying Raoult's law.

Raoult's law should be used with caution in estimating emissions from partial evaporation of mixtures; not all mixtures behave as perfect solutions. Elkins, Comproni, and Pagnotto measured benzene vapor yielded by partial evaporation of mixtures of benzene with various aliphatic hydrocarbons, chlorinated hydrocarbons, and common esters, as well as partial evaporation of naphthas containing benzene (Elkins, et al., 1973). Most measurements for all four types of mixtures showed greater concentrations of benzene vapor in air than were predicted by Raoult's law. Of five tests with naphtha-based rubber cements, one yielded measured values of benzene concentration in air in agreement with calculated values, the other four showed measured benzene concentrations in air to be 3–10 times greater than those calculated using Raoult's law.

Substantial deviation from Raoult's law is not always the case, however, even with benzene. Runion compared measured and calculated concentrations in air of benzene in vapor mixtures yielded by evaporation from a number of motor gasolines and found excellent agreement (Runion, 1975).

In the absence of other more certain means, Raoult's law can be used to estimate emissions generated by partial evaporation of mixtures that approximate ideal solutions or mixtures in which the solution is nearly pure in one component. The applicability of Raoult's law to the mixtures being assessed should be validated, or quantitative measurements of emissions should be done, if accurate emission values are needed.

For dilute solutions the partial pressure of the component present in lower concentration is given by Henry's law, expressed as follows:

$$p_n = H_n x_n \tag{3}$$

where p_n = partial pressure of component n
H_n = Henry's law constant
x_n = mole fraction of component n in the liquid mixture

Henry's law is also applicable to the solubility of a gas in dilute liquid solution, and solubilities of gases in liquid may be expressed in terms of Henry's law constants. These constants and the applicable concentration ranges for valid use of Henry's law can be determined only empirically. Henry's law constants for solutions of some common gases in water and equilibrium data for some gases in a few other solvents may be found in handbooks such as *Chemical Engineers' Handbook* (Perry and Green, 1984).

Application of materials balance concepts for determination of emissions other than those for combustion, chemical reaction, or evaporation of volatile materials, ordinarily requires engineering analysis on a case-by-case basis. A procedure for estimating emission rates from multiple indoor sources has been reported (Franke and Wadden, 1987). The same modeling procedure has been used for estimating emissions from open-top vapor degreasers (Scheff et al., 1992).

3.3 Emission Factors

The need exists for emissions estimates for large numbers of sources in community air pollution studies, but the impracticability of source-by-source emissions tests has led to the development of emission factors. An emission factor is a pollutant emission rate for a particular type of emission source expressed as a quantity of pollutant released per unit of activity of that type source. The unit of activity chosen in each case is one that can be determined and can be related quantitatively to emissions; it may be ton of product, million Btu of heat produced, mile of vehicle travel, or other such index unit. Emission factors represent typical emissions from a class of sources and ordinarily cannot be applied with confidence to individual sources. In the absence of other information, however, emission factors for a particular source type can give useful insights into the character and general levels of emissions from individual sources of that type.

The most reliable emission factors are those based on a combination of emission measurements, process data, and engineering analysis for a large number of sources. Those that do not have a theoretical basis and are derived from only one type of data, or from data from only a few sources, should be used with caution. In some cases sufficient knowledge or information is available to permit development of empirical or analytic relationships between emission rate and some process parameter such as material composition or stream temperature. Such factors are generally the most reliable of all and can even be applied to individual sources with reasonable confidence.

Several thousands of individual air pollutant emission factors for a large number of source types have been tabulated and reported by the EPA's Office of Air and Waste Management (USEPA, 1985); supplements have been published through September 1991. Process descriptions and emission control practices, along with typical collection performance for various types of control, are presented for most of the source types covered. Table 4.1 lists major source types for which emission factors appear in the current publication. In a separate document the EPA has published emission factors for arsenic, asbestos, beryllium, cadmium, manganese, mercury, nickel, and vanadium for processes involving these materials (Anderson, 1973).

Table 4.1 Major Source Types for Which Air Pollution
Emissions Factors Have Been Adopted

1. External combustion sources
   Bituminous (and subbituminous) coal combustion[a,b]
   Anthracite coal combustion[a,b]
   Fuel oil combustion[a]
   Natural gas combustion[a,d]
   Liquified petroleum gas combustion
   Wood waste combustion in boilers[a]
   Lignite combustion[a]
   Bagasse combustion in sugar mills
   Residential fireplaces[d]
   Residential wood stoves[b,c,d]
   Waste oil combustion[b]
2. Solid waste disposal
   Refuse combustion[b,c,d]
   Automobile body incineration
   Conical burners
   Open burning
   Sewage sludge incineration[b,c]
3. Stationary internal combustion sources
   Highway vehicles
   Off highway mobile sources
   Stationary gas turbines for electric utility power plants
   Heavy-duty natural-gas-fired pipeline compressor engines
   Gasoline and deisel industrial engines
   Stationary large bore and dual fuel engines
4. Evaporation loss sources
   Dry cleaning
   Surface coating[b]
   Magnetic tape manufacturing industry[c]
   Surface coating of plastic parts for business machines[c]
   Nonindustrial surface coating[d]
   Storage of organic liquids
   Transportation and marketing of petroleum liquids
   Cutback asphalt, emulsified asphalt, and asphalt cement
   Solvent degreasing
   Waste solvent reclamation
   Tank and drum cleaning
   Graphic arts
   Commercial/consumer solvent use
   Textile fabric printing
   Polyester resin plastics product fabrication[b]
   Waste water collection, treatment and storage[d]
5. Chemical process industry
   Adipic acid
   Synthetic ammonia
   Carbon black
   Charcoal

Table 4.1 (*Continued*)

Chlor-alkali
Explosives
Hydrochloric acid
Hydrofluoric acid
Nitric acid
Paint and varnish
Phosphoric acid
Phthalic anhydride
Polyvinyl chloride and polypropylene[d]
Poly(ethylene terephthalate)[d]
Polystyrene[d]
Printing ink
Soap and detergents[b]
Sodium carbonate[a]
Sulfuric acid
Sulfur recovery
Synthetic fibers (manufacturing)[c]
Synthetic rubber
Terephthalic acid
Lead alkyl
Pharmaceuticals production
Maleic anhydride

6. Food and agricultural industry
Alfalfa dehydrating
Coffee roasting
Cotton ginning
Grain elevators and processing plants[b]
Fermentation
Fish processing
Meat smokehouses
Ammonium nitrate fertilizers
Orchard heaters
Phosphate fertilizers
Ammonium phosphates[d]
Starch manufacturing
Sugar cane processing
Bread baking
Urea
Beef cattle feedlots
Defoliation and harvesting of cotton
Harvesting of grain
Ammonium sulfate

7. Metallurgical industry
Primary aluminum production[a]
Coke production[a]
Primary copper smelting[a]
Ferroalloy production[a]
Iron and steel production[a]

Table 4.1 (*Continued*)

Primary lead smelting[a, c]
Zinc smelting[a]
Secondary aluminum operations[a]
Secondary copper smelting and alloying
Gray iron foundries[a, c]
Secondary lead processing[a]
Secondary magnesium smelting
Steel foundries
Secondary zinc processing
Storage battery production
Lead oxide and pigment production
Miscellaneous lead products
Leadbearing ore crushing and grinding

8. Mineral products industry

Asphaltic concrete plants[a]
Asphalt roofing
Bricks and related clay products[a]
Calcium carbide manufacturing
Castable refactories
Portland cement manufacturing[a, d]
Ceramic clay manufacturing
Clay and fly ash sintering
Coal cleaning
Concrete batching[a]
Glass fiber manufacturing
Frit manufacturing
Glass manufacturing[a]
Gypsum manufacturing
Lime manufacturing[a, b]
Mineral wool manufacturing
Perlite manufacturing
Phosphate rock processing
Construction aggregate processing
Sand and gravel processing[d]
Crushed stone processing[a, b]
Coal conversion
Taconite ore processing[a]
Metallic minerals processing
Western surface coal mining[a, d]

9. Petroleum industry

Petroleum refining
Natural gas processing

10. Wood products industry

Chemical wood pulping[a, c]
Pulpboard
Plywood veneer and layout operations
Woodworking waste collection operations

Table 4.1 (*Continued*)

11. Miscellaneous sources
Wildfires and prescribed burning[b, c, d]
Fugitive dust sources
Unpaved roads[b]
Aggregate handling and storage piles[b]
Industrial paved roads[a, b, c]
Industrial wind erosion[b, c]
Explosives detonation[c]
Wet cooling towers[d]
Industrial flares[d]

[a] USEPA (1985), Supplement A, October 1986.
[b] USEPA (1985), Supplement B, September 1988.
[c] USEPA (1985), Supplement C, September 1990.
[d] USEPA (1985), Supplement D, September 1991.

Emission factors in the EPA tabulation generally apply to identifiable point sources in processes or operations from which pollutants are released to the atmosphere through vents or stacks. As such, they have limited applicability to in-plant industrial hygiene assessments. They do, however, identify some of the air contaminants that are generated by these processes; the same contaminants may represent potential in-plant exposures.

3.4 Fugitive Sources

Contaminant emissions from point sources such as tank vents or transfer points, and from processes and operations that clearly involve release of a process material—for example, release of volatile components in cementing or painting operations—are ordinarily capable of identification and quantification. There are other types of sources not so easily accommodated and sometimes neglected in emission inventories. Such sources, often called fugitive sources, may be intermittent, temporary, or unpredictable; many are unrecognized or ignored in emission inventories. Fugitive sources and the generation of secondary pollutants deserve attention in emission inventories, however, even though all of them may not be capable of quantification or prediction.

Fugitive sources that are of consequence primarily in the field of air pollution, and for which emission factors have been developed, include unpaved roads, aggregate handling and storage piles, heavy construction operations, and paved roads (USEPA, 1985). Other sources that are consequential from the standpoints of both air pollution and industrial hygiene include the following:

1. Urban fires
2. Industrial process fires
3. Materials spills (accidents, equipment failure or malfunction)

4. Sample collection and analysis
5. Process leaks (flanges and piping, valves, packing glands, conveyors, pumps, compressors, tanks and bins, etc.)
6. Relief valves and control device bypasses
7. Maintenance activities (tank and vessel cleaning, filter cleaning, replacing piping, pumps, etc.)
8. Emissions from waste streams and reemissions of collected materials
9. Secondary reactions (nonproduct process reactions, extraprocess reactions)

Urban structural fires are intermittent phenomena predictable only in the aggregate and not as individual events; they are considered to be emission sources primarily in the air pollution sense. An industrial hygiene consequence of urban fires, however, is the exposure to toxic materials of firefighters and others who may be involved in rescue or control activities. A number of toxic atmospheric contaminants are generated by structural fires; these tend to vary from one fire to another. Of concern in all structural fires, however, is emission of carbon monoxide. Burgess et al., (1977) have described exposures of emergency personnel to carbon monoxide emissions in real fire situations; the maximum sustained air concentrations of carbon monoxide to which these persons were exposed in the events in which it was measured was about 2 percent. Materials balance using pyrolysis and oxidation processes offer one means of estimating the quantities of pollutants that may be generated in any particular case.

Estimates of emissions from industrial fires require individual analysis. The quantities and characteristics of combustion products depend on the nature of the materials that burn or are present in the fire, and on the circumstances of combustion (e.g., whether open or confined). Again, consideration of pyrolysis and oxidation phenomena may permit estimation of the nature and quantities of air contaminants generated by any particular event. The likelihood and consequences of accidental fires is an appropriate consideration in industrial emission inventories.

Material spills in manufacturing, transporting, and uses of industrial materials are not infrequent occurrences. Studies of processes and operations with specific attention to spills can reveal the frequencies and magnitudes of spills if they are usual occurrences (Smith, 1972). Such spills can then become part of an emissions inventory. When spills are not a usual occurrence, an emission inventory can do little more than trigger consideration of the possibilities and consequences of such events. Guidance for protection of workers when a spill of a hazardous substance occurs, developed as a joint project of the National Institute for Occupational Safety and Health (NIOSH), the Occupational Safety and Health Administration (OSHA), the U.S. Coast Guard, and the U.S. Environmental Protection Agency (EPA), has been published (Streng et al., 1983).

Process leaks are sometimes of major consequence from the standpoint of industrial hygiene. Control of process leaks, for example, has been a major factor in achieving acceptable working conditions in vinyl chloride monomer and polymerization plants. Process leaks can be identified by inspection or instrumental methods; timely repair may obviate quantifying for an emission inventory. Should quantification be necessary, the techniques of source testing or materials balance may suffice. Quantifying emissions of fugitive dusts has generally been a difficult exercise. A chamber technique for quantifying fugitive dust emissions from handling of granular materials has been reported (Lundgren, 1986).

As acceptable exposure levels become lower, attention to detail in emission evaluation takes on added importance. Routine tasks such as process sample collection may present a potential for excessive exposure when such sampling involves open manhole covers, open sample containers, or unconfined sampling streams. Bell et al. addressed this problem for vinyl chloride sampling and suggested techniques that may find application in other industries, particularly in petrochemical operations (Bell et al., 1975).

Release of materials to community air or to a workplace through process relief valves occurs from time to time. The frequency of operation of these devices and the magnitude of releases, obtained from plant records or other sources, can be used for emission inventory purposes. Recognition of the existence of relief valves or control device bypasses in a process is important in an emission inventory. A common source of emissions in processing plants is leakage from seals of pumps used to transport volatile liquids. It has been reported that emission rates can increase as much as fivefold when a pump stops; the reason for this is not known (Flitney and Nau, 1986). Pumps designed to prevent leakage through shaft seals are now available.

Kletz has stated that many of the foregoing sources of fugitive emissions from plant processes can be controlled with emergency isolation valves (Kletz, 1975). In addition to suggesting a method of control, this author also presents a leakage profile for both an olefin processing plant and an aromatic processing plant that could serve as a basis for developing a fugitive emissions checklist for use at similar operations.

Volatile air contaminants may be released to the atmosphere from process sewers, drainage ditches, and/or collecting ponds. Such releases are of concern from the standpoint of both air pollution and industrial hygiene. Two examples are offered. Consider a case in which a gravity separator is used in an enclosed benzene recovery system to separate the organic and aqueous streams. Water from this separation may be reused for other purposes elsewhere in the plant. Even though benzene is only slightly soluble in water, such water reuse can result in measurable concentrations of benzene in the air at workplaces in the plant where benzene is not used. Consider another case in which alkaline scrubbing water is used to remove fluorides from a process waste gas stream. Discharge of the scrubbing water to a waste pond in which the pH is low permits release to the atmosphere of the collected fluorides. The use of the air pollution control scrubber in such a case only relocates the site of fluoride emission from the process stack to the waste pond. Air sampling and materials balance techniques are means for assessing emissions from waste streams and reemissions of collected materials.

Descriptions of processes and products do not always reveal whether by-products are formed or whether any by-products and their parent chemicals are stable throughout the manufacturing process and in the environment. This issue has been discussed, and an approach to assessing the significance of by-products and secondary reactions have been presented (Sowinski and Suffett, 1977). The likelihood and consequences of hydrolysis, pyrolysis, oxidation, and other reactions are factors in the assessment. The identification of processes and principal materials is the first step in exploring for secondary pollutant emissions in an emission inventory. Literature review and perhaps laboratory exercises may yield indicators of the presence of secondary contaminants; air sampling may be required for validation.

Although it is not always possible to make quantitative, or perhaps even qualitative, assessments of fugitive sources and secondary pollutants, the possibilities of their occurrence and the opportunities for intervention to protect workers and the community merit scrutiny in emission inventories.

4 IN-PLANT EMISSION ESTIMATES FROM RECORDS AND REPORTS

In the absence of in-plant measurements of emissions, estimates derived from secondary sources of information can be used for inventory purposes. In some cases data are available from similar processes, and estimates by analogy may prove to be sufficient for a given purpose. In other cases unique features of a process or operation may make estimates by analogy inappropriate. An alternative to measurement data or analogy makes use of records and reports whose basic purposes were other than for industrial hygiene, but whose content may be extracted, combined, or otherwise manipulated into usable estimate of emissions and potential exposure. Vital information may be recorded in a variety of business documents including engineering, accounting, production, quality control, personnel, and governmental records. Although these records represent a rich source of information for emission inventory purposes, they can serve other purposes in the practice of industrial hygiene and occupational health research as well. Some of these purposes are mentioned from time to time in the following discussions of specific types of records.

4.1 Research and Development Records

Industrial research and development units often issue a variety of reports to producing units as aids to bringing a new product on stream, to ensure a required level of uniformity in a given item produced by different plants, or to make changes or improvements in operating equipment, materials, or techniques. Valuable insight with respect to materials and work practices employed in the past may be gleaned from several of these records. Among such records are the following:

1. **Product Specification.** Final product and intermediate component data included in a product specification often give dimensions, weight, materials of construction, processing aids, processing conditions, tools and fabricating equipment, and special notes.
2. **Standard Operating Procedure (SOP).** This record may exist under a variety of names; SOP, Standard Practice, Uniform Methods, and (Company Name) Manual of Operations appear among the myriad of titles. All, however,

share fairly specific operational instructions, usually in a step-by-step approach by processing sequence. Vital data from this record can include:

a. Fabrication equipment

b. Authorized materials of construction

c. Acceptable processing aids

d. Processing conditions

e. Alternate materials for cyclic operations (e.g., summer vs winter stocks)

f. Change notices affecting materials or techniques

g. Instructions regarding protective equipment

3. **Formulations and Raw Materials.** Detailed data concerning the qualitative aspects of potential exposures can be derived from study of information describing the raw materials employed in a process.

a. Raw material specifications may contain composition data, restrictions on contaminants, and acceptance testing schedules.

b. A listing of authorized vendors can offer clues to potential contaminants based on knowledge of a vendor's source of supply or processing methods.

c. Formulation records frequently contain information on the composition of mixtures as well as the methods, equipment, quantity, and conditions of processing.

4.2 Analytical Laboratory Records

Quality control considerations often dictate that incoming raw materials be monitored for selected chemical and physical properties, that supplies from prospective vendors be subjected to acceptance testing, and that production quality be assayed by sampling at intermediate stages of the process. A single journal entry of analytical results may include a wealth of intelligence applicable to an emissions inventory. Analytical laboratory records include the following:

1. **Raw Material Test Results.** Recorded in either the analyst's journal or on prepared forms will be entries such as:

a. Analytical results vis-à-vis specifications

b. Date of analysis, and perhaps date of receipt

c. Quantity of the shipment

d. Vendor

e. Method of analysis

f. Analyst

2. **Authorized Vendors.** Quality control laboratories often maintain vendor-supplied information regarding purchased materials. Material Safety Data Sheets, product specifications, production methods, and quality control statements all offer bits of useful data.

4.3 Process and Production Records

Records generated by production units relate to such matters as scheduling, quality control, cost control, and to a lesser extent, training exercises. Depending on the product and process, production-type records can be of considerable value in generating an emissions inventory.

1. **Flowcharts.** A detailed diagram of a process is particularly useful in identifying points and potential for release of air contaminants.

2. **Operating Conditions.** Time, temperature, and pressure data are useful in assessing possible release of reaction products, by-products, degradation products, and unreacted raw materials.

3. **Scheduling.** Shift, cyclic, and seasonal variations may influence the emissions of contaminants. Such variations can be ascertained through production scheduling records. Included under this heading are records of batch versus continuous production modes. Gantt chart records offer a particularly attractive systematic source of scheduled activities that normally are both precise and detailed.

4.4 Plant Engineering Records

Virtually all industrial organizations include an engineering component that is responsible for the physical plant and supporting facilities. This responsibility results in a plant archive that is particularly useful in emissions inventory activity and in research that requires reconstruction of past environmental conditions.

1. **Plant Layout.** The plant floor plan, coupled with elevation drawings, affords a visualization of the process, materials flow, and occupied areas, and possibly insights relating to contaminant generation and control. Furthermore, many engineering drawings are cross referenced to other drawings of equipment, emissions control features, and adjacent work areas. Useful characteristics of plant engineering files are:

   a. Processing equipment type, extent, and location appear on layouts.

   b. Department floor space showing geographic boundaries and physical barriers appear in some plates.

   c. Engineering controls, particularly ventilation systems, are depicted in mechanical equipment layouts, which normally show the location, site, type, and rating of air-moving equipment, as well as a reference to the detailed plate covering that system.

   d. Dates of change or modification are often included on drawings, either by direct entry or by reference to a new set of plates.

2. **Equipment Specifications.** Files on processing equipment installed by or under the supervision of plant engineers often include specifications that supply clues to potential problems such as process emissions, sound power rating

and directivity data, or emission control features incorporated in the equipment.

3. **Project Records.** Each major project of construction, modification, or installation normally proceeds as an integral project with one individual assigned as coordinator. Usually this project officer accumulates records that include data on:

   a. Equipment specifications

   b. Performance checks and acceptance test results

   c. Prime contractor and subcontractors

4. **Material Balance.** A detailed accounting of materials, products, and side streams, either measured or theoretical, serves as a basis for a quantitative estimate of emissions.

4.5 Industrial Engineering Records

The practice of industrial engineering includes the analysis of work procedures to improve either work conditions or productivity or both. Especially for the period since World War II, industrial engineering records contain information on work methods and standards that are useful for current emission inventories and for estimating past emissions and exposures that may have resulted from them. Such records include:

1. **Job Descriptions.**

   a. Description numbers, department listings, and simple descriptive job titles can provide the link between the tasks performed and the work history recorded in an individual's personnel file.

   b. Job location within a plant, thus the potential for various exposures, may be ascertained from the geographic area, processing equipment, or department listing contained in a job description.

   c. The listing in a job description of tools, equipment, and processing aids can help establish the nature of emissions and characteristics of potential exposures (e.g., inhalation, skin absorption).

   d. Task evaluations often include judgments concerning such health-related items as work conditions, safety requirements, hazards, skills, effort, and responsibility.

2. **Time and Motion Studies.** Task breakdowns are useful to establish contact with materials of interest, and the fraction of time spent on certain tasks coupled with cycle time can afford an estimate of the extent of exposure.

3. **Process Analysis.** Evaluations of an entire process focus on tasks performed by all the members of a crew rather than those of individual workers. Particularly useful in process analysis reports are listings of fractions of a workday spent at different workstations by members of the operating crew, including break time and sequencing of tasks.

4.6 Accounting Records

Accounting records cover virtually all phases of an operation for both outside reporting and internal control purposes. Major areas of interest for emission inventory purposes include records on incoming materials, internal use of these raw materials, and plant production factors (Anthony, 1970).

1. **Purchasing.** Raw materials for both production and support operations, as well as processing equipment, are ordinarily obtained from outside the company, and an acquisition record is generated upon receipt. Even when raw materials are drawn from captive sources, records of resource depletion may be available.

   a. Listings of materials, quantities, and supplier may be available through the purchasing ledger, often in the accounts payable portion.

   b. Capital equipment is normally amortized by a recognized accounting procedure, the evaluation of which can be employed to determine the date of acquisition. Alternatively, many organizations maintain a property book that contains the same information.

2. **In-plant Issues of Materials**

   a. Data on raw materials consumption often can be developed through accumulation of issues to departments, cost centers, or plants through collected expense or work-in-process accounts.

   b. Material variance accounts may serve as surrogates for material balance data if the latter are unavailable. The absolute variance value between actual and standard material usage may be of marginal value, but changes in this quantity can be important in establishing trends.

   c. Nonproduction supplies are often issued to maintenance and other support departments through a ''general store'' account, with the receipt listing the item and quantities issued.

3. **Inventory.** Inventory control may be maintained through stock records that list material, vendor, date and quantity received, date and quantity of issue, department to which issued, and location and quantity on hand.

4. **Production.** Finished goods and work-in-process production accounts normally include an identification of the work schedule, days operated, and average production level.

4.7 Personnel Records

Personnel records, per se, provide some insights into the nature and locations of contaminant releases applicable to emission inventories. In addition they are vital to other industrial hygiene occupational health research activities. In any study of causal association between conditions of work and health experience, the work experience of each individual is an integral component. Thus, accurate and reliable personnel-type records are vital. Organizational units that normally can develop, or assist in developing, useful records include the personnel, payroll, industrial relations, safety, and medical departments.

1. **Personnel Records.** The file on a particular employee usually lists department and job assignments chronologically and gives a brief description of tasks, limitations, and periods of absence. Earnings records sometimes provide a clue to potential exposure through a listing of special pay or rates for hazardous tasks, and exposure duration or regimens by shift differentials or overtime pay.

2. **Industrial Relations.** Seniority listings and work force distribution charts are useful in establishing the locations of jobs and personnel currently or at various times in the past.

3. **Safety and Medical.** Insurance carrier survey reports, fire and explosion inspections reports, sickness/accident/compensation records, administrative control documents that limit time at a particular task or require worker rotation, along with written SOPs for activities such as entering tanks and vessels and other hazardous tasks, represent a record of emission sites and past working conditions, however subjective, suitable at least to rank-order jobs by exposure potential to certain agents.

4. **Union Records.** Seniority listings and union membership rolls can help to identify the working populations of past years, and records of negotiated agreements between labor and management often contain provisions for work schedules, protective clothing and equipment, provisions for personal hygiene, clothing changes, and other health-related clauses that help to describe conditions of work.

4.8 Government-mandated Records and Reports

The 1980s may be described as the decade of disclosure regarding potential exposure to toxic and hazardous chemical agents. As part of the general "right-to-know" movement, OSHA, EPA, and state agencies promulgated numerous regulations intended to ensure that employees, customers, and general public and government agencies would have the necessary information with which to evaluate potential exposures resulting from chemical releases to the work place and ambient environment. The resultant records and reports provide valuable information to compiling the emission inventory and will sometimes afford insights to potential exposure conditions. Copies of past reports of this type are likely to be filed in the plant or corporate office(s) that deal(s) with environmental affairs and that deal(s) with government regulatory agencies.

Some of the sentinel government requirements are briefly described below along with insights to potential exposure conditions.

4.8.1 OSHA Hazard Communication

OSHA's Hazard Communication Standard (29CFR 1910.1200) (OSHA, 1992) requires that employers inform their employees of potential hazards associated with chemical agents in their places of work. It also requires that manufacturers and distributors of regulated chemicals make hazard information available to their downstream customers. Chemical manufacturers, distributors and processors have achieved compliance through written programs, employee training, container labeling, and Material Safety Data Sheets (MSDS). Firms will generally maintain documentation of all sectors of their program for compliance purposes. As a result of hazard communication requirements being effective in all OSHA-regulated sectors, MSDSs are now ubiquitous in work places across America.

4.8.2   Access to Medical (and Exposure) Records

In 1980 OSHA promulgated its access to medical records rule (29CFR 1910.20), which requires employers to maintain exposure records for 30 years, and to inform employees of their right to review their individual exposure records (OSHA, 1992). Many firms routinely advise workers of monitoring results through workplace postings or personal letter reports. This general rule is applicable to all exposure records not the subject of a specific OSHA requirement.

4.8.3   OSHA Standards Governing Specific Chemicals

Several OSHA standards governing exposure to specific chemical agents include requirements for workplace monitoring and maintaining exposure records beyond the provisions of the general rule identified in the preceding section. Examples of these requirements are found in the standards for benzene (29CFR 1910.1028), lead (29CFR 1910.1025), and formaldehyde (29CFR 1910.1048).

4.8.4   Toxic Substances Control Act

The Toxic Substances Control Act (TSCA, 1976) imposed on chemical manufacturers and distributors a number of new testing, record-keeping, and reporting obligations. Under Section 8(e) of TSCA, anyone observing a new substantial risk to health or the environment resulting from chemical exposure must report this finding to the U.S. EPA within 15 days of discovery. EPA in turn, summarizes the results of TSCA 8(e) reports and makes these summaries available to the general public.

Under the requirements of TSCA Section 8(c), chemical manufacturers and distributors are required to maintain records of significant adverse reactions alleged to be caused by a (specific) chemical substance or mixture whether or not this allegation can be substantiated by the firm. These internal company files are available to the EPA for inspection purposes.

Of particular usefulness in in-plant emission inventories are agent listings prepared for compliance with regulatory requirements under the TSCA (TSCA, 1976). Regulations under TSCA require the listing of any of some 61,000 agents, by CAS number or tradename, which are used in a plant or which may be present in plant processes (Lubs et al., 1982). Industrial hygienists and engineers with knowledge of plant processes can make judgments regarding the potential for release of each agent in these plant-specific lists in each workplace in a plant. The determination may be either qualitative, that is, a yes/no determination that a particular agent is present at a particular workplace, or semiquantitative with ordinal estimates of frequency of release and of magnitude of emissions for each agent and workplace.

Lynch (1991) has described the application of such a system for estimation of workers' exposures in petrochemical plants.

4.8.5 SARA Title III

The Emergency Planning and Community Right to Know Act (EPCRA, 1986), also known as Title III of the Superfund Amendments and Reauthorization Act of 1986 (SARA, 1986) established comprehensive requirements for public reporting of chemical inventories and releases to the environment. Reports are required of facilities operating in the manufacturing sector (SIC Codes 20-39) that employ 10 or more full-time employees and that manufacture, process, or use any listed chemical agent in quantities equal to or greater than threshold quantities established by the federal government. The list of regulated substances in 1992 stands at approximately 300 chemicals, and threshold quantities necessary to trigger reports are 25,000 pounds per year of manufactured or processed chemical agents, and 10,000 pounds of any listed material otherwise used on an annual basis.

SARA reports fall into two basic categories: on-site inventory (Sections 311 and 312) and releases to the environment (Section 313 and Form R). Under the provisions of Section 311, regulated facilities are required to submit either a list of chemical substances stored or an individual MSDS for each regulated chemical. Section 312 requires owners or operators to file inventory reports listing the amounts and locations of regulated chemicals either by hazard class (Tier I report) or by specific chemical agents (Tier II reports). Inventory reports are filed with the Local Emergency Planning Commission (LEPC), the local fire department, and the State Emergency Response Committee (SERC) (40CFR Part 370, 52 FR38344).

Section 313 of SARA requires an annual report estimating quantities of any of the some 300 regulated toxic chemicals released to the environment (40CFR Part 372, 53 FR4500). Regulated toxic chemical release inventory reports must be filed with the SERC in the state wherein the reporting facility is located and the EPA. Individual facility, toxic release inventory (TRI) reports are compiled by EPA and made available to the public for a nominal fee. Under SARA, the list of regulated toxic chemicals and threshold reporting quantities are subject to change from time to time. Beginning in 1992, additional reporting requirements were added to Form R as a result of the Pollution Prevention Act of 1990 (EPA Form 9350-1; revised 5/14/92).

Industrial hygienists may use reports required under SARA Title III for estimating in-plant emissions and potential for workplace exposures in the same ways they may use TSCA reports as described above.

4.8.6 Resource Conservation and Recovery Act

Owners and operators of transportation, storage, and disposal (TSD) facilities that manage hazardous wastes regulated by the Resource Conservation and Recovery Act (RCRA, 1976) incur numerous obligations to maintain records and submit reports detailing materials handled and conditions in effect during periods of operation. Some of the more instructive reports for purposes of the emission inventory include manifest and shipping papers (40CFR 264.71), inspection records (40CFR 264.15), and the biennial report of operations (40CFR 262.4).

4.9 Governmental Records and Reports

Data so broad as to characterize an entire industry, or so specific as to deal with a single chemical compound, are routinely and systematically collected, assembled, and published by various government agencies. Most of these records are available to the enterprising investigator, usually at little or no cost. A few of these record types are mentioned here as illustrations.

1. **Nationwide Statistics.** The U.S. Tariff Commission publishes an annual listing of Production and Sales of Synthetic Organic Chemicals. It includes listings of raw materials, the production chemicals by use category (surface-active agents, etc.), and a directory of manufacturers of each material. Time-series analyses of the entries can indicate the dates of introduction, extent of use, and dates of decline in use of a given material.

2. **Industrywide Data.** Government-directed or controlled operations such as government owned–contractor operated plants compile data for required reports, many of which find their way into the public domain. For example, information on virtually all facets of the synthetic rubber industry during the 1940–1945 period are available through the various reports of the Rubber Reserve Company (Rubber Reserve Company, 1975; Whitby, 1954). Included are flowcharts, raw materials, formulations, plant capacities, and locations.

3. **Specific Products.** On June 15, 1844, Charles Goodyear received U.S. Patent 3633 for a compound consisting of gum elastic, sulfur, and white lead (Babcock, 1966). Thus, the approximate period of time at which rubber compounders could be considered as potentially exposed to white lead as a rubber accelerator is established. More than 5 million U.S. patent numbers have accrued since the mid-nineteenth century, doubtless including many materials, processes, and devices of interest to health investigators.

4.10 Record Location

Within most corporate industrial organizations, records of major additions or changes, as well as events affecting the entire operating units, are likely to be available at a central location, such as the division or corporate offices. Examples of these types of records are overall production levels, descriptions of uniform practices, and product specifications. More specific records, or those dealing with minor plant alterations, are ordinarily available only at the local plant level. Current records are ordinarily found in the departments identified in Sections 4.1 through 4.8 of this chapter; historical records may be found in those departments or in a central plant storage facility. At the plant level project officers often retain personal copies of papers related to their projects; if a plant record has been discarded, such personal records may provide valuable historical data.

4.11 Output from Records

Measures derivable from secondary sources of data are limited by the nature and detail of the record content. In general, the more distant the time of interest, the less complete will be the record, and the lower will be the confidence in its accuracy. When data are sparse and only plantwide data are available, estimated measures, such as gallons per cubic foot of plant building volume per month or pounds used per employee per day, may be the limit of detail for emission estimates, but this may serve for plant-to-plant contrasts. More detailed data such as quantities of materials used in specific departments and descriptions of controls for specific work areas permit more detailed estimates. Material generation rates coupled with general area ventilation rates may enable one to estimate an area average concentration. Such data may be particularly useful for within-plant contrasts or for plant-to-plant studies of particular areas or processes.

Records pertaining to the detailed nature of tasks, such as standard operating procedures or job descriptions, make possible a ranking of jobs—thus a ranking of increments of time in individuals' work histories—by potential for exposure to any particular agent or combination of agents of interest. In this fashion personnel with increments of time in their work histories having similar exposures, for example, homogeneous exposure groups (HEGs) (Hawkins et al., 1991), can be grouped for comparison of health experience with other workers who share different common exposure profiles, that is, who have histories of work in other HEGs.

As an example of the use of secondary information, let us offer an illustration of the type of emissions output that has been accomplished for retrospective research purposes through the use of such sources of information.

In Figure 4.1 the standard operating procedure is the pivotal document that identifies both tasks and materials for a step in manufacturing. A *formulation* record, or series of records, for each material used in that particular manufacturing step identifies each raw material *constituent* of each material. Examination of the raw material *specification* for each constituent then permits identification of *agents* (e.g., coal tar naphtha) that may be emitted in the workplace, often with additional compositional detail (e.g., aromatic content or benzene content). Because formulations, constituents, and specifications change from time to time, the listing of potential emissions thus identified must show inclusive dates on an agent-by-agent basis.

The *tasks* portion of the SOP includes a *job description*, which in turn has a specific *job code*. The job code is listed in the personnel record of each person who has held that particular job. Examination of the personnel records of the plant's current and past workers identifies complete *work history*, including dates and duration of each job code assignment for each *individual* who has held the specific job of interest.

In this way the potential for contact with a particular chemical agent by an identified individual, including dates and duration of the contact, can be developed, even to events that occurred in the distant past.

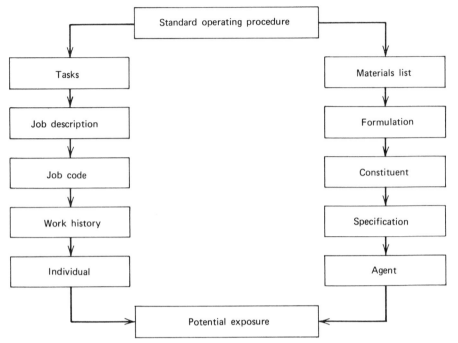

Figure 4.1 Synthesis of potential exposures associated with a manufacturing step for which a standard operating procedure is available.

5   USES OF EMISSION INVENTORIES

Emission inventories are fundamental to the identification, evaluation, and control of industrial hygiene and community air pollution hazards associated with chemical agents. Not all uses for inventory information can be examined here, but a few are discussed.

5.1   Regulatory Requirements

Various governmental regulatory programs in occupational safety and health, air pollution, and toxic substances control have reporting requirements that involve emission inventory information. Regulations regarding the nature and content of these reports, the agents covered, and the types of establishment with reporting obligations change periodically. Persons responsible for industrial or commercial enterprises that involve emissions to community air, or the handling or production of agents that may be covered in occupational health or toxic substances regulatory programs, should keep abreast of the specific requirements imposed by state and federal regulations. Emission inventory programs can then be so designed and operated that the emissions data necessary to satisfy reporting requirements are obtained.

5.2 Community Air Pollution Dispersion Estimates

Data on emissions of a pollutant to community air and companion meteorological data permit estimation and prediction of ground-level concentrations of that pollutant at any desired location in the community. Methods and procedures for manual calculations to estimate the impact of individual sources (Turner, 1969), and a basic computer program for calculating concentration resulting from emissions from large numbers of point and area sources (Busse and Zimmerman, 1973) have been reported. The mathematical relationships and computational techniques are too extensive to be repeated here. Application is best undertaken with the workbook or guide in hand. Suffice it to say that both methods can be used by engineers and industrial hygienists when the necessary data on emissions, source locations, receptor locations, and meteorology are available.

5.3 Workplace Exposure Estimates

Qualitative estimates of the potential for emissions and consequent exposures of workers can be made when agents have been identified and conditions of their use are known. Even the most rudimentary emission inventory—the identification of agent, process, and location of use—permits this kind of assessment. An operating program in which the industrial hygiene department of a large research facility systematically obtains descriptions of all uses of the many hundreds of chemical agents available at the facility has been described (Porter et al., 1977). The system includes a hazard rating for each agent based on its toxicity, flammability, reactivity, and special properties such as may be found in the agent's MSDS. Use information does not identify emissions per se, but retrieval of use information for a particular agent permits rapid subjective assessment of its emissions potential throughout the facility. Quantitative information on emissions, either measured, calculated, or estimated, can be combined with data on workplace ventilation to estimate workroom concentrations of an agent. Such estimates may occasionally be of use in validating sampling data, but are more likely to have value in such applications as assessing growth or decay of concentrations in nonequilibrium situations, in predicting the consequences of new processes or changes in operations, or in estimating retrospectively the levels of a contaminant to which particular groups of workers may have been exposed. The latter use is of particular interest in reconstructing emergency or accident episodes or in retrospective epidemiological studies of worker populations.

5.3.1 Gases and Vapors

Emission rates of gases and vapors are used in the mathematical relationships that yield values for concentrations of contaminants in workroom air. In addition to emission rates, one must know physical dimensions or volume of the workplace and the ventilation rate. The mathematical relationships (Harris, 1960) were derived using the concept of materials balance. The rate of change in the quantity of contaminant in the air of a workplace is described by Eq. (4). All emissions of a contaminant to workroom air comprise a contaminant generation rate G.

$$V \frac{dX}{dt} = G + QX_s - QX \tag{4}$$

Integration of Eq. (4) yields

$$X_t = X_i \exp\left(-\frac{Qt}{V}\right) + \left(X_s + \frac{G}{Q}\right)\left[1 - \exp\left(-\frac{Qt}{V}\right)\right] \tag{5}$$

where V = volume of the ventilated workplace (m³)
  G = rate of contaminant generation as volume per unit time of contaminant gas or vapor (m³/min)
  Q = effective exhaust ventilation rate, volume/unit time of air plus contaminant (ordinarily the volume rate of contaminant, G, can be ignored in Q; furthermore, only with perfect mixing can the actual ventilation rate be used in dilution ventilation calculations, see the following discussion of the factor K) (m³/min)
  t = interval of time elapsed since conditions represented by X_i and $t = 0$ (min)
  X = fraction of contaminant in workplace air (10^{-6} ppm)
  X_i = fraction of contaminant present in workplace air at time $t = 0$
  X_t = fraction of contaminant present in workplace air a time t
  X_s = fraction of contaminant in incoming dilution air

Equation (5) applies only for intervals of time during which G and Q are constant. If either G or Q changes, then a new value of X_i and a new t_0 must be established, that is, X_t at the end of one time interval of constant G and Q becomes X_i for the beginning of the time interval that immediately follows with a new t_0.

At equilibrium, when t is great, Eq. (5) simplifies to

$$X = X_t = X_s + \frac{G}{Q} \tag{6}$$

or with equilibrium and clean air supply:

$$X = X_t = \frac{G}{Q} \tag{7}$$

When the concentration of the contaminant in the workplace air increases with time, the maximum concentration for the time interval t is X_t; for decay situations the maximum concentration for the interval t is X_i. Exposure guides and standards may list both maximum allowable short-term concentrations and allowable average concentrations. The average concentration over time t, also derivable from materials balance, is

$$X_{av} = X_s + \frac{G}{Q} + \left(\frac{V}{Qt}\right)\left(X_i - X_s - \frac{G}{Q}\right)\left[1 - \exp\left(-\frac{Qt}{V}\right)\right] \tag{8}$$

As is the case with Eq. (5), at equilibrium, when t is great, Eq. (8) simplifies to Eq. (6).

Equation (5) may be used to calculate the concentration of a gas or vapor remaining in a vessel or tank at any time after purging ventilation (decay) or contaminant introduction (growth) has begun. It may also be used in the following form to estimate the interval of time required for a particular ventilation rate to cause a particular concentration X_t to be reached:

$$t = \frac{V}{Q} \ln\left(\frac{G + QX_s - QX_i}{G + QX_s - QX_t}\right) \tag{9}$$

In some repetitive operations the generation rate for a contaminant may vary in a cyclical manner, being at one rate for one interval of time and at a different rate for another interval. If, in such a case, the greater generation rate is identified as G, the lower rate as G', and the time intervals t and t' are companion to G and G', respectively, when the cycle has equilibrated so that the workroom concentration pattern has become repetitive, the maximum concentration that occurs during the cycle is represented by

$$X_{max} = \frac{\left(X_s + \frac{G}{Q}\right)\left[1 - \exp\left(\frac{Qt}{V}\right)\right] - \left(X_s + \frac{G'}{Q}\right)\left[1 - \exp\left(\frac{Qt'}{V}\right)\right]}{\exp\left(-\frac{Qt'}{V}\right) - \exp\left(\frac{Qt}{V}\right)} \tag{10}$$

For the same cyclic operation after equilibrium has been reached, the average concentration over a large number of cycles is

$$X_{av} = X_s + \frac{Gt + G't'}{Qt + Qt'} \tag{11}$$

In the simple case when there is no generation during the time interval t', the term G' becomes zero in Eq. (10) and (11).

In 1988, for use in a research proposal, Dr. Michael R. Flynn, Department of Environmental Sciences and Engineering, University of North Carolina at Chapel Hill, derived equations that describe change in concentration with time, and expected inhalation dose, when a volatile solute aerates from a nonvolatile solvent during a batch process when there is no replacement of solute. In such a batch-type process the concentration of solute in solution decays with a half-time characteristic of the solute and the process. The instantaneous generation rate G is proportional to the concentration of solute in the solution, so the value of G will decay with time. The value of G at any time t, after the initiation of such a batch process, is represented by

$$G_t = G_0 \exp{(-K't)} \tag{12}$$

where G_t = the generation rate at time t
       G_0 = the initial generation rate at $t = 0$
       K' = 0.693/half-time of solute concentration in the solution

The substitution of $G_0 \exp{(K't)}$ for G in Eq. (4) and integration yields for air concentration at time t after beginning of a batch process:

$$X_t = X_s + \left(\frac{G_0}{Q - K'V}\right) [\exp{(-K't)}]$$
$$+ \left[X_i - X_s - \left(\frac{G_0}{Q - K'V}\right)\right]\left[\exp{\left(-\frac{Qt}{V}\right)}\right] \tag{13}$$

Equation (13) is applicable over any time interval following the beginning of a batch process with decaying generation rate described by Eq. (12), which has an initial value G_0, and for which Q is constant.

The integration of Eq. (13) over time t, during which the workspace is occupied after beginning of the batch process, yields the concentration–time product (i.e., the inhalation dose for a worker who is present) for decaying generation rate (batch process):

$$Xt = X_s t + \left[\frac{G_0}{K'(Q - K'V)}\right] [1 - \exp{(-K't)}]$$
$$+ \left(\frac{V}{Q}\right)\left[X_i - X_s - \frac{G_0}{Q - K'V}\right]\left[1 - \exp{\left(-\frac{Qt}{V}\right)}\right] \tag{14}$$

In real workroom situations instantaneous perfect mixing does not take place. Usually when dilution ventilation is applied, the contaminant is emitted at locations that are in the occupied portion of a workroom; not all the ventilation air mixes with the contaminant while it is in this occupied space. With known or estimated true values of ventilation rate an empirical factor K may be used in calculation of estimated exposures to account for departures from perfect mixing. The definition of K is

$$K = \frac{Q_{\text{actual}}}{Q} \tag{15}$$

where Q_{actual} = ventilation rate, volume/unit time, as measured or obtained from engineering records (m^3/min)

Q = ventilation rate, volume/unit time, that is effective in determining the concentration at an exposure site, i.e., the value to be used in calculations (m^3/min)

The effective ventilation rate Q is to be used in Eq. (5)–(14) for calculation of exposure estimates. The choice of a value for K is a matter of judgment for an investigator; it is ordinarily in the range of 1–10 (ACGIH, 1992). Calculations using engineering records for ventilation rates, actual emission rates, and a number of concentration measurements of three different contaminants in each of two actual dilution ventilation situations have yielded K values ranging from 2.2 to 5.0 with a mean value of 3.25 (Baker, 1977). In these two dilution ventilation situations, solvent vapors were being emitted in a number of locations in the workrooms at elevations that were approximately the same as breathing zone and sampling elevations. They may be considered to represent generally good mixing of emissions and ventilation air in actual application of dilution ventilation in industrial situations. The observed values are consistent with the lower values of the range of typical K values; the greater the departure from efficient mixing of emissions and ventilation air, the greater would be the value of K. It should be noted that K is not a safety factor; it is an index of departure from perfect mixing. For example, for a dilution ventilation situation in which all dilution air and all emissions pass through a single exhaust fan, the K factor for the discharge air at the fan location has the value 1.

In an excellent analysis of dilution ventilation relationships, Roach (1977) has shown that the variance of emission G is inversely proportional to the length of the time interval t over which emissions take place, and that the variance in concentration X is then inversely proportional to the product of ventilation rate Q and V^2, the square of room volume. Thus, for a given average emission rate, the smaller the workspace and the smaller the ventilation rate, the greater will be the fluctuations in concentration, while the greater the volume of the space and the greater the ventilation rate, the smaller will be the fluctuations.

5.3.2 Particulates

Caution must be exercised in the application of dilution ventilation theory and rationale, as described for gases and vapors, to particulate contaminants that are released to workroom air. Gravitational settling may be an appreciable factor in the decay of air concentrations of particulates. The smaller the particles, the greater is the likelihood that they will remain airborne and the more applicable are dilution ventilation relationships. When emissions of submicrometer sized particles—for example, most components of smoke and metal oxide fumes—are known, concentrations can be approximated by use of the dilution ventilation equations for gases and vapors.

Dusts with a wide range of particle sizes do not necessarily behave in air in the same was as fumes. Estimates of the relative concentrations of respirable dusts may be made with dilution ventilation equations when particulate emissions data are sufficient for quantification of the respirable fraction.

In assessing exposures to particulates, particles larger than those considered respirable should not be ignored. Of these larger particles, those that are inhaled and deposited in the nasopharynx and tracheobronchial system before they can reach the pulmonary spaces of the lungs represent ingestion exposures. Airborne concentrations of such particles generally decay at appreciable rates because of gravitational settling; dilution ventilation relationships do not apply to them.

6 RECORDS RETENTION PROGRAM

Emissions records may be generated for a variety of purposes. Some records are for purposes that can be served in a relatively short time; then the record may be considered a candidate for discarding. Other emissions records may be generated specifically for purposes such as health research, which require long retention. Emissions records made for various short- or long-term purposes may have secondary use in long-term community air pollution or occupational health research. Whether they represent typical operations or unusual events, such records may be invaluable in the future for reconstructing conditions of work to compare with long-term health experience of persons who engaged in that work. Since not all specific agents of future interest are identifiable in advance, it is prudent to retain emissions records for as many agents as possible.

Emissions records, or records from which emissions estimates can be made, such as those identified in Section 4 of this chapter, should be retained for a long time. A single specific age at which a record no longer has value cannot be stated. For research purposes exposure records have value equal to, or greater than, that of medical examination records. Because long latent periods are associated with health effects of some agents, early exposure records may be invaluable in the search for causal associations. Exposure records, or emission records, that identify contact with agents or from which exposure estimates can be made, should be retained long enough to be used in studies of the mortality experiences of the worker populations to which they apply.

The need for long-term retention of environmental records is recognized in the record-keeping provisions of the OSHA regulation of toxic and hazardous substances (OSHA, 1992). Time periods for which employee exposure records must be retained vary with the substance being regulated. For example, records of exposures to asbestos must be retained for at least 20 years, those for vinyl chloride must be retained for not less than 30 years, and those for coke oven emissions must be retained for the duration of employment plus 20 years, or for 40 years, whichever is longer. Emission data are pertinent to these records. Such retention is appropriate for all records of emissions of air pollutants or of chemical agents released in places of work.

The responsibility for retention of emission records for any establishment should be clearly defined. This responsibility may well be placed with the same organizational unit that has responsibility for retention of air monitoring records and/or health experience records. Records that serve as secondary sources of health-related information, such as those identified in Section 4 of this chapter, may be kept in different locations. A checklist of all such records that are needed for emission records purposes should be assembled and kept up to date for each establishment and retained by the unit having responsibility for emissions records. Responsibility for retention of each of these secondary records should be clearly defined, and this responsibility, along with location of each record set, should appear on the records checklist.

REFERENCES

ACGIH (1992). *Industrial Ventilation—A Manual of Recommended Practice*, 21st ed, ACGIH Inc., Cincinnati, OH.

Anderson, D. (1973). *Emission Factors for Trace Substances*, EPA 450-2-73-001, EPA, Research Triangle Park, NC.

Anthony, R. N. (1970). *Management Accounting, Text and Cases*, 4th ed, Irwin, Homewood, IL.

Antonsson, A-B. (1990). *AIHAJ*, **51,** 352–355.

API (1959). American Petroleum Institute, *Evaporative Loss from Tank Cars, Tank Trucks, and Marine Vessels*, API Bulletin 2514, API Washington, D.C.

Babcock, G. D. (1966). *History of the United States Rubber Company, A Case Study in Corporate Management*, Indiana Business Report 39, Graduate School of Business, Indiana University.

Baker, J. C., Jr. (1977). Testing a Model for Predicting Solvent Vapor Concentrations in an Industrial Environment, M.S.E.E. Technical Report, Department of Environmental Sciences and Engineering, University of North Carolina, Chapel Hill, NC.

Bell, Z., J. Lefflen, R. Lynch, and G. Work (1975). *Chem. Eng. Prog.*, **71**(9), 54.

Berry, J. C., D. Beck, R. Crume, D. Crumpler, F. Dimmick, K. C. Hustvedt, W. L. Johnson, L. Keller, R. McDonald, D. Markwordt, M. Massoglia, D. Salman, S. Shedd, J. H. E. Stelling, III, G. Wilkins, and G. Wood (1986). In *Air Pollution*, Vol. 7, 3rd ed., A. C. Stern, Ed., Academic Press, New York, Ch. 9, pp. 395–508.

Burgess, W. A., et al. (1977). *AIHAJ*, **38,** 18–23.

Busse, A. D. and J. B. Zimmerman (1973). *User's Guide for the Climatological Dispersion Model*, EPA-R4-024, EPA, Research Triangle Park, NC.

CAS (1993). Chemical Abstract Service, American Chemical Society, P.O. Box 3012, Columbus, OH 43210.

Elkins, H. B., E. M. Comproni, and L. D. Pagnotto (1973). *AIHAJ*, **24,** 99–101.

Engdahl, R. B., R. E. Barrett, and D. A. Trayser (1986). In *Air Pollution*, Vol. 7, 3rd ed., A. C. Stern, Ed., Academic Press, New York, Ch. 8, pp. 339–393.

EPCRA (1986). Emergency Planning and Community Right to Know Act, see SARA Title III.

Flitney, R. K. and B. S. Nau (1986). *Ann. Occup. Hyg.*, **30,** 241–247.

Franke, J. E. and R. A. Wadden (1987). *Envr. Sci. Technol.*, **21,** 45–51.

Hammerle, J. R. (1977). In *Air Pollution*, Vol. 3, 3rd ed., A. C. Stern, Ed., Academic Press, New York, pp. 717–784.

Harris, R. L. (1960). Industrial Hygiene Engineering Training Course, U.S. Public Health Service, Occupational Health Field Headquarters, Cincinnati, OH.

Hawkins, N. C., S. K. Norwood, and J. C. Rock, Eds. (1991). *A Strategy for Occupational Exposure Assessment*, American Industrial Hygiene Association, Akron, OH.

Kletz, T. (1975). *Chem. Eng. Prog.*, **71,** 9.

Lubs, P. L., E. J. Kerfoot, and J. M. McClellan (1982). *AIHAJ*, **43**, 418–422.

Lundgren, D. A. (1986). *J. Aerosol Sci.*, **17**, 632–634.

Lynch, J. R. (1991). Implementation and Validation of an Exposure Estimating System, Paper 96 presented at the 1991 AIHCE, Salt Lake City, UT.

NIOSH (1992). *Registry of Toxic Effects of Chemical Substances*, on-line reference service, Cincinnati, OH (most recent hard-copy publication, 1987).

OSHA (1992). *Occupational Safety and Health Standards, Subpart Z—Toxic and Hazardous Substances*, 29CFR 1910 (with amendments through June 8, 1992, 57 FR24310), Department of Labor, Washington, D.C.

Paulus, H. J. and R. W. Thron (1977). In *Air Pollution*, Vol. 3, 3rd ed., A. C. Stern, Ed., Academic Press, New York, pp. 525–587.

Perry, R. H. and D. Green, Eds. (1984). *Chemical Engineers Handbook*, 6th ed., McGraw-Hill, New York.

Porter, W. E., E. L. Hunt, and N. E. Bolton (1977). *AIHAJ*, **38**, 51–56.

RCRA (1976). Resource Recovery and Conservation Act, 42 USC 6901, and 40CFR Parts 260–271.

Roach, S. A. (1977). *Ann. Occup. Hyg.*, **20**, 65–84.

Rubber Reserve Company (1975). Report on the Rubber Program 1940–1945, February 24, 1975.

Runion, H. E. (1975). *AIHAJ*, **36**, 338–350.

SARA (1986). Superfund Amendments and Reauthorization Act, Title III, 42 USC 11001 et seq.

Scheff, P. A., R. L. Friedman, J. E. Franke, L. M. Conroy, and R. A. Wadden (1992). *Appl. Occup. Environ. Hyg.*, **7**, 127–134.

Smith, M. (1972). *Investigations of Passenger Car Refueling Losses*, Scott Research Laboratories, Inc., San Bernardino, CA, National Technical Information Service, PB-121 592, Washington, D.C.

Sowinski, E. and I. H. Suffett (1977). *AIHAJ*, **38**, 353–357.

Streng, D. R., W. F. Martin, L. P. Wallace, G. Kleiner, J. Gift, and D. Weitzman (1983). *Hazardous Waste Sites and Hazardous Substances Emergencies*, DHHS (NIOSH) Pub. No. 83-100, PHS CDC, NIOSH, Cincinnati, OH.

Thron, R. W. (1986). In *Air Pollution*, Vol. 7, 3rd ed., A. C. Stern, Ed., Academic Press, New York, Ch. 4, pp. 163–217.

TSCA (1976). Toxic Substances Control Act, Public Law 94-469, 15 USC 2601 et seq.

Turner, D. B. (1969). *Workbook of Atmospheric Dispersion Estimates*, U.S. Public Health Service Publication 999-AP-26, National Air Pollution Control Administration, Cincinnati, OH.

USEPA (1973). U.S. Environmental Protection Agency, *A Guide for Compiling a Comprehensive Emission Inventory*, APTD 1135, EPA, Research Triangle Park, N.C.

USEPA (1974). U.S. Environmental Protection Agency, *Administrative and Technical Aspects of Source Testing for Particulates*, EPA 450/3-74-047, EPA, Research Triangle Park, NC.

USEPA (1985). U.S. Environmental Protection Agency, *Compilation of Air Pollutant Emission Factors*, A.P. 42, 4th. ed. (through Supplement D, 1991), USEPA, Research Triangle Park, NC.

Whitby, G. S., Editor-in-Chief (1954). *Synthetic Rubber*, Wiley, New York.

CHAPTER FIVE

Philosophy and Management of Engineering Control

William A. Burgess, CIH

1 INTRODUCTION

This chapter will cover the engineering methods widely used to minimize adverse effects on health, well-being, comfort, and performance at the workplace and in the community. Major attention will be given to controls for chemical airborne contaminants, although many of the control principles are also applicable to biological and physical hazards. Brief case studies will illustrate the applications of the techniques. The target for control may be a hand tool, a piece of equipment or device, an integrated manufacturing process such as an electroplating line, or a complete manufacturing facility in a dedicated building such as a foundry.

The critical zones of contaminant generation, dispersion, and exposure are shown in Figure 5.1. Ideally the goal is to design each element of the process to eliminate contaminant generation. If it is impossible to achieve this goal and a contaminant is generated, the second defense is to prevent its dispersal in the workplace. Finally, if we fail in that defense and the material released from the operation results in worker exposure, the backup control is collection of the air containing the contaminant by exhaust ventilation. Frequently, the contaminant is removed from the exhaust airstream by air cleaning before returning the air to the workplace or the general environment. In the United States we frequently give little emphasis to the prevention of contaminant generation and release and rely on ventilation as a principal remedy. In this chapter we will confirm the importance of ventilation, but

Patty's Industrial Hygiene and Toxicology, Third Edition, Volume 3, Part A, Edited by Robert L. Harris, Lewis J. Cralley, and Lester V. Cralley.
ISBN 0-471-53066-2 © 1994 John Wiley & Sons, Inc.

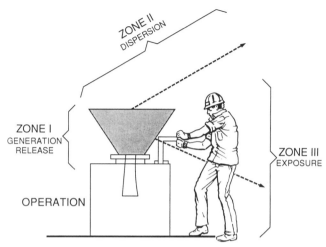

Figure 5.1 Contaminant generation, release, and exposure zones in the workplace.

emphasis will be placed on the primary controls associated with process and material design to minimize generation and release of the contaminant.

The Occupational Safety and Health Administration (OSHA) considers engineering controls as any modification of plant, equipment, processes, or materials to reduce employees' exposure to toxic materials and harmful physical agents. As noted in Table 5.1, authors have used a variety of approaches to classify the various en-

Table 5.1 Classifications of Methods for Control of Air Contaminants

First (1983)	Burgess (this chapter)
Material substitution	Toxic materials
Processing change	Eliminate
Equipment change	Replace
Local exhaust ventilation	Dust control
General (dilution) ventilation	Reduce impurities
Equipment enclosure	Equipment and process
Employee enclosure	Task modification
Gideon et al. (1979)	Facility layout
Control at source	Ventilation
Material substitution	General ventilation
Process or equipment change	Local exhaust ventilation
Isolation	Sherwood and Alsbury (1983)
Local exhaust ventilation	Automation, alternative methods
Work practice	Substitute materials, methods
Control at workplace	Plant layout, enclosures, remote
General exhaust ventilation	Local exhaust ventilation
Housekeeping	Dilution ventilation
Control at worker	Personal protective equipment
Isolation/personal protective equipment	

gineering control techniques. These approaches provide a useful outline for discussion and act as checklists to ensure a comprehensive review. A comprehensive algorithm for control of air contaminants has been designed by Sherwood and Alesbury (1983). Brandt (1947, p. 50) states: "Measures for preventing the inhalation of excessively contaminated air have been discussed by many authors and there are as many classifications of these methods as there are papers on the subject. The principles expounded, however, are always essentially the same." Brandt also reminds the reader that: "The control of an atmospheric health hazard is rarely accomplished by a single measure. It usually involves a combination of measures."

In this chapter engineering controls will include all techniques except personal protective devices and administrative controls, those changes made in the work schedule to reduce the time-weighted average exposure of the worker (Table 5.1).

2 GENERAL GUIDELINES

The strategy to be used in solving a control technology problem varies depending on the setting. As described above, the control problem may involve a specific piece of equipment, a process, or an integrated manufacturing plant. In a common situation faced by many in the 1990s, a shop repairing electric motors must consider a change in solvents in a vapor-phase degreaser. The number of persons participating in the review is small, the decision will be made in a short time, and it will probably rely heavily on vendor information. The range of options presented in this chapter probably will not be thoroughly reviewed. An industrial hygienist will not be involved unless the service shop is a support facility operated by a major company or consultation is available from an insurance carrier, consulting industrial hygienist, or state OSHA consulting program.

A far more complex situation is the scale-up of a chemical process to manufacture a photographic dye. The research to develop the new molecule and synthesize a small quantity for initial tests is completed by a research chemist at a laboratory bench. At this stage little is known about the chemical or its toxicity. However, since gram quantities are handled in a well-controlled laboratory hood, the risk is negligible. The next step involves a scale-up of the operation in a pilot plant facility designed to manufacture kilogram quantities of the material for testing purposes. This testing will provide limited data on the physical properties of the new chemical, its product potential, and toxicity data. If the results are encouraging, production scale-up is considered. A premanufacturing notification must be submitted to the Environmental Protection Agency (EPA) prior to production. This submission, which requires a description of air contaminant control technology and an estimation of worker exposure, involves a comprehensive review of the operation as noted in Table 5.2. It is decided to manufacture the chemical at an existing plant using conventional chemical processing equipment. If the industrial hygienist has been involved in the pilot plant operation, he or she is well placed to participate in a review of the control technology necessary for the full-scale plant. As noted in Table 5.2, the range of talent available to define the engineering controls is extensive and the time frame is extended. Hopefully, step-by-step review to identify the appropriate

Table 5.2 Specialists Participating in Risk Assessment of New Chemical Manufacture

	Administrative	Legal	Sales	Marketing	Chemists	Engineers	Toxicologists	Occupational Hygienists	Environmental Scientists	Statisticians	Risk Analyst	Transportation Systems Analysts	Economists	Information Specialists
Part I: General Information														
A. Manufacturer identification	X	X												
B. Chemical identity					X									X
C. Marketing data			X	X						X			X	
D. *Federal Register* notice	X	X		X	X	X	X	X	X					
E. Schematic flow diagram					X	X		X	X					
Part II: Risk Assessment Data														
A. Test data		X			X		X			X	X	X		
B. Exposure from manufacture														
1. Worker exposure					X	X	X	X			X	X		
2. Environmental release					X	X				X	X	X		
3. Disposal					X	X	X	X	X		X			
4. By-products, etc.					X	X					X			X
5. Transportation		X		X	X						X	X		
C. Exposure from process operations														
1. Worker exposure					X	X	X	X			X	X		
2. Environmental release					X	X				X	X	X		
3. Disposal					X	X	X	X	X		X			
D. Exposure from consumer use	X	X	X	X	X						X			
Part III: Risk Analysis and Optional Data														
A. Risk analysis and optional data	X	X			X		X			X	X			
B. Structure activity relationships					X		X				X			
C. Industrial hygiene	X	X			X	X	X	X			X			
D. Engineered safeguards					X	X	X	X	X		X			
E. Industrial process and use restriction data					X	X								
F. Process chemistry					X	X								
G. Nonrisk factors: Economic and non-economic benefits	X	X	X	X	X	X					X		X	

Source: Arthur D. Little, Inc.

engineering controls will minimize worker exposure. If exposure problems are not anticipated, controls must be retrofitted at significant cost whereas early integration of controls are most easily accepted and economically implemented.

The most complex engineering control package is encountered in the design and construction of a major production facility. The project is complex due to the range of issues that are encountered, however, the design process is started with a clean design sheet and state-of-the-art controls can be included in the design. In the example in Table 5.2 a large number of specialists are included in the review team. If the design is done in-house, the industrial hygienist should be a key participant. If the design work is contracted to a design and construction firm, plant personnel are one step removed from the design process. The company will have a project engineer to interface between the outside team and company engineering group. In this case clear communication must be established by the industrial hygienist to ensure involvement in the design review.

For the industrial hygienist facing his or her first process hazard review, the job may seem awesome. An initial step is to identify your responsibility. Does it include merely health issues at the workplace or are environmental and fire protection problems to be included in your review? Once your responsibility has been defined, the issues of concern for this particular plant should be identified. Mature industries such as metallurgical, glass, and ceramics have been studied, and extensive information is available on standard operations and sources of contamination. As noted in Chapter 4, the EPA has published emission rates for many processes that are useful in process hazard reviews. Another source of information on control technology for mature industries is the series of over 50 technical reports published by the National Institute for Occupational Safety and Health (NIOSH) (see Appendix A).

If the company has similar operations elsewhere, obtain material and process data, industrial hygiene survey data, and air sampling information from that plant to assist in the review. A step-by-step review of the operations is then undertaken. This health and safety review should be as rigorous as the review of percent yield on the process and should produce a detailed process flow diagram showing the locations of contaminant loss in addition to the major physical stresses such as noise, heat, and ergonomics. It is important to also consider the ancillary systems including compressed air, refrigeration, cooling towers, and water treatment. Bulk handling of chemicals and granular minerals frequently presents a major industrial hygiene problem and should be given special attention. Finally, a detailed description of the workers' tasks including required emergency response actions are required since work location and movement have a major effect on contaminant control. A comprehensive checklist for review of environmental and occupational health and safety issues in planning major plant construction is given by Whitehead (1987).

An element that is frequently overlooked in the design of engineering controls is the worker. A review of a specific operation conducted by several workers frequently will identify subtle differences in the way the task is performed. A worker inspired by a wish to reduce lifting, provide "cool down" in the heat, or reduce dust exposure may modify the tasks to accomplish this end. Workers frequently have insight into details of the operation that the design engineer will overlook. The video, real-time monitoring approach discussed in Section 3.2.1 is useful in defining these modifications in tasks.

It is difficult to recommend engineering controls for emerging technologies. Hopefully each industrial group will develop and publish its own engineering controls. Such is the case with advanced composite manufacturing, a technology that has seen expanding applications in recent years in the manufacture of aircraft. An

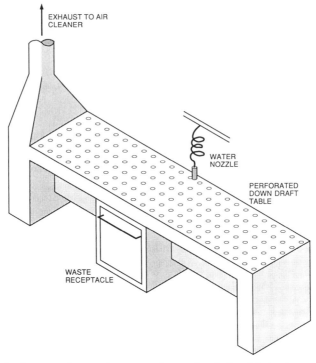

Figure 5.2 Integrated controls on an asbestos insulation fabrication workbench.

association of companies involved in this work has published a volume describing the major hazards and the controls that should be applied (SACMA, 1991).

Frequently the best control cannot be achieved directly on the design board. In the 1970s a number of cases of asbestosis in shipyard workers fabricating asbestos insulating pads for steam propulsion plants prompted the plant to install an integrated workstation incorporating a number of control features. Initial designs of the down-draft work table were fabricated and evaluated by workers resulting in a number of recommendations that were incorporated in the final design improving its acceptance and performance (Fig. 5.2).

3 CONTROL STRATEGY

The general strategy as noted earlier is to eliminate the generation source. If that is not possible, then prevent escape of the contaminant, and if it does escape, collect and remove the contaminant before the worker is exposed. This strategy will be reviewed by considering action on toxic materials, equipment, processes, job tasks, plant layout, and exhaust ventilation.

3.1 Toxic Materials

The use of toxic materials range in complexity from simple wet degreasing to the synthesis of new chemicals, as noted above. In degreasing the exposure is associated with the use of the cleaning agent or handling its waste. In synthesis of a new chemical, exposure may occur to raw materials, intermediates, by-products, the new chemical and waste streams. In minimizing potential worker exposure to these materials, it is necessary to consider the following options.

3.1.1 Eliminate the Toxic Material

This option refers to elimination without replacement. Frequently toxic chemicals are used in a process to improve yield or in a product to improve function or appearance. In review, it may be possible to eliminate the chemical. The pressure to do so may come from potential health effects, the cost of workplace and environmental controls to introduce the toxic chemical into the plant, or it may arise from pressure from the marketplace. An example of the latter resulted from legislation in Switzerland requiring precautionary labeling of batteries containing mercury. The conventional Leclanche dry cell battery utilizes electrodes of manganese dioxide and zinc with an electrolyte of ammonium chloride. A characteristic of the cell is zinc corrosion with subsequent power loss. A mercury compound is added to inhibit this corrosion and resulting performance degradation. At least one manufacturer has determined the performance enhancement from mercury is not necessary, and the mercury compound has been removed from the battery formulation (Ahearn et al., 1991). This change has eliminated exposure to mercury for several hundred workers in this company.

3.1.2 Replace with Alternate Material

This control approach holds the greatest potential for significant reduction in worker exposure to toxic chemicals. Historically, the approach has had success in the United States. One of the most effective steps was the cooperative action taken by the United States Public Health Service, the State of Connecticut Department of Public Health, industry, and labor unions in 1941 to prevent mercurialism by replacing mercuric nitrate with nonmercury compounds in the carroting process in felt hat manufacture.

A recent major action is the virtual banning of asbestos for a wide range of applications. At this writing this action by EPA is being contested in the courts. Another environmental concern is the exposure of children to lead from lead-based paint and potable water supplies containing lead–tin–copper sweat joints. In 1986 the Amendments to the Safe Drinking Water Act effectively banned the use of lead-based solder. The acceptable alternative solders are alloys of tin in the range of 90–95 percent, and the balance is copper, silver, zinc, or antimony (Kireta, 1988). The introduction of the alternative solder will minimize exposure of manufacturing, construction, and maintenance personnel to lead. The banning of mercury in hat manufacture demonstrates an action taken specifically for occupational health reasons; in the second and third examples the action was prompted by environmental concerns but will have a positive impact on the health of all workers.

A total of 10 materials have been banned for specific operations in the United Kingdom (HSE, 1988). An important element of the United Kingdom regulation is the prohibition of silica sand for abrasive blasting. Although there have been attempts to restrict the use of sand for this purpose in the United States, as the most frequently used blasting material it continues to present a major risk to workers.

Greater emphasis will be given to replacement control technology in the next decade. As discussed below, the impetus will continue to result not from the workplace health issue but from the environmental health and ecological impact. Health and safety characteristics including toxicity, smog contribution, ozone depletion, and fire potential must be evaluated. In addition to the review of the performance of the replacement material, the changes in process and product design to satisfactorily change over to the new material must be considered in the final decision.

3.1.2.1 National Toxic Use Reduction Programs.

3.1.2.1 National Toxic Use Reduction Programs. Until the 1980s the EPA emphasis was on pollution control and dealt principally with "end-of-pipe" abatement technology. In 1984 the Hazardous and Solid Waste Amendments to the Resource Conservation and Recovery Act of 1976 redirected environmental quality efforts from the conventional emphasis on waste treatment and disposal to waste minimization. This approach to waste management is characterized by source reduction, recycling, and reuse. In 1986 the Office of Technology Assessment published a report that described the concept of prevention versus control (OTA, 1986). Since these environmental initiatives have great impact on the replacement of toxic chemicals in industry and, therefore worker exposure, it is important that the industrial hygienist understand the environmental management nomenclature (Table 5.3). The Pollution Prevention Act of 1990 consolidated this approach by specifying the following waste management hierarchy as national policy: first source reduction as the most desirable approach, then recycling and reuse, then treatment, and finally disposal as the least desirable approach (U.S. Congress, 1990).

To initiate this strategy, in 1989 the EPA established the Waste Reduction Innovative Technology Evaluation (WRITE) Program. This program funded programs in six states and one county over a 3-year period to (1) provide engineering and

Table 5.3 Environmental Control Approaches

Form of Control	Emphasis	End Rule
Pollution control	"End-of-pipe" control	Does not eliminate pollution, but merely transfers contaminant from one media to another. Incurs an environmental risk from transporting toxic chemicals.
Pollution prevention	Reduction in the production of contaminants	Direct waste reduction, covers all pollutants
Toxic use reduction	Reduction in pollutants generated in manufacturing by changes in plant operation procedures, production processes, materials, or end products Focus on a target list of chemicals	Priority setting permits control of high potential risk chemicals

Source: Adapted from Rossi et al. (1991).

Table 5.4 Examples of Material Replacement from EPA WRITE Program

Operation	Original Material	Replacement
Anodizing aluminum parts	Chromic acid anodizing, hexavalent chromium exposure	Sulfuric acid anodizing, eliminates exposure to chromium
Removal of paint on defective parts	Methylene chloride-based paint stripper	Abrasive blasting with plastic beads, no vapor exposure
Flexograph printing	Application of alcohol-based ink labels	Water-based inks eliminate vapor exposure
Cleaning and deburring small metal parts	Vapor degreasing and alkaline tumbling	Automatic aqueous rotary washer. Eliminates vapor and caustic mist exposure
Cleaning flexographic plates	Solvent cleaning with Stoddard solvent, acetone, toluene, and alcohol	Cleaning with terpenes and aqueous cleaners

Source: Adapted from Harten and Licis (1991).

economically feasible solutions to industry on specific pollution prevention problems, (2) provide performance and cost information on these techniques, and (3) promote early introduction of these pollution prevention programs into commerce and industry. A review of completed and ongoing WRITE programs that demonstrates the importance of replacement technology is shown in Table 5.4 (Harten and Licis, 1991). Although initiated for environmental quality reasons, each of these changes will have a major impact on worker health.

Denmark, one of the many European countries embarking on a national toxic use reduction program, base their work on the simple step model shown in Table 5.5 (Goldschmidt, 1991). The simplicity of this approach suggests that replacement technology can be done on an ad hoc basis; however, Goldschmidt warns that chemical replacement for occupational health reasons is a complex task and requires a systematic review of both the existing conditions and those prevailing after the replacement action. A review of 162 individual replacement actions by the Danish Occupational Health Services confirms that replacement is successful if done in a

Table 5.5 Step Model for the Substitution Process

Step	
1	Problem identification
2	Identification and development of a range of alternatives
3	Identification of consequences of the alternatives
4	Comparison of the alternatives
5	Decision
6	Implementation
7	Evaluation of the result

Source: Goldschmidt (1991)

rational manner following the step model shown in Table 5.5. Among the impressive examples of this program is the replacement with water-based systems of the solvent-based systems in all indoor paints and most outdoor paints.

The principal advances in worker protection in the next decade will probably result from replacement technology accomplished as a result of such worldwide environmental health initiatives.

3.1.2.2 State Toxic Use Reduction Programs.   The recent emphasis on pollution prevention and specifically on toxic use reduction in manufacturing is reflected in recent legislative action in 17 states and by the federal government (Rossi et al., 1991). The Massachusetts Toxic Use Reduction Act of 1989 (Mass. General Laws, 1989) is described as a seminal program in that it does focus on toxic use reduction and excludes off-site recycling, off-site nonproduction unit recycling, transfer, or treatment. Rossi et al. (1992) identifies the five toxic use reduction techniques acceptable under the Massachusetts act as:

In-process Recycling or Reuse.   Recycling, reuse, or extended use of toxics by using equipment or methods that become an integral part of the production unit of concern, including, but not limited to, filtration and other closed-loop methods.

Improved Operations and Maintenance.   Modification or addition to existing equipment or methods including, but not limited to, such techniques as improved housekeeping practices, system adjustments, product and process inspections, or production unit control equipment or methods.

Changes in the Production Process.   Includes both the modernization of production equipment by replacement or upgrade based on the same technology or the introduction of a new production equipment of an entirely new design.

Input Design.   Replacement of a chemical used in production with a chemical of a lower toxicity.

Product Reformulation.   Redesign of the product to produce a product that is nontoxic or less toxic on use, release, or disposal.

Rossi proposes two divergent routes the manufacturing group may take. The route for ease of implementation includes, first, recycling and improved operations and maintenance, then production process changes, and finally replacement of toxic chemicals and product reformulation. However, for effective toxic use reduction the first step is to eliminate or replace the toxic chemical in the process or reformulate the product, then initiate production process changes, and finally recycle and apply operations and maintenance control techniques.

An institute funded by industry assessment has been established at the University of Massachusetts at Lowell with the function to provide a teaching and research facility to encourage aggressive action in toxic use reduction.

3.1.2.3 Industry Programs.   A number of company initiatives have been described that respond to specific legislation or broad environmental concerns. One company program has multiple goals to reduce the use of toxic chemicals, minimize toxic emissions to all environmental media, encourage recycling of chemicals, and reduce all waste (Ahearn et al., 1991). In the design of this program the company followed many of the recommendations of the Office of Technology Assessment report on the reduction of hazardous waste (OTA, 1986). In this company program specific goals for toxic use and waste reduction are set for all production managers, and their annual performance reviews include this issue.

Each chemical used in the company is assigned to one of five Environmental Risk Categories (ERC). The materials in ERC I pose the greatest risk to the environment and ERC II, ERC III, and ERC IV represent progressively less risk. A fifth category includes plastic, steel, paper and other waste identified as rubbish, rubble, and trash. A critical issue in the design of the program was the decision logic for assigning categories. A number of options were considered including ranking chemicals based on workplace exposure guidelines such as threshold limit values (TLVs) or permissible exposure limits (PELs), label signal words (ANSI, 1988), regulatory lists published by OSHA and EPA, and the mathematical weighting of a range of risk factors including toxicity and chemical and physical properties (Karger and Burgess, 1992).

The company chose to assign the ERCs based on a ''wise person'' approach based on a battery of risk factors reflecting the materials' total occupational and environmental impact from release both as an isolated incident such as a spill or the chronic release of small amounts of the chemical to air, water, and soil. The majority of risk factors are toxicity based and chosen to identify significant adverse health and environmental impact. Included in the group of risk factors are acute and chronic toxicity based on animal studies or structural analysis, human case history and epidemiological studies, carcinogenicity, mutagenicity, teratogenicity, and reproductivity effects. A limited number of physical and chemical properties associated with the ''releasability'' of the chemicals were also included in a second group of risk factors. The third group of risk factors reflected broad environmental impacts. The specific criteria for assignment to the five categories are shown in Table 5.6.

This company has a major chemical production facility involved in the manufacture of photographic dyes, which require frequent changes in processes. Ahearn et al. (1991) state that in the design of new processes and products the evaluation of toxic use and waste minimization technology is given equal importance to considerations of cost and performance of the new process or product. The engineering group developing the new processes rigorously review all materials and process alternatives to ensure that the target reductions are met. In this company the ''end of pipe'' abatement strategy is given secondary attention relative to optimization of process and product design for toxic use reduction.

An important part of the program is the goal of a 10 percent reduction in the use of toxic materials per unit of production each year for the first 5 years following the general strategy of elimination of use, replacement with a material of lower risk, reduction in quantity used, and a hierarchal program of waste handling. The goal is to eliminate, replace, or reduce the use of category I and II materials and assure the maximum reuse of categories III and IV materials.

In describing the success of the program the authors (Ahearn et al., 1991) present an example of a process in which a change resulted in elimination of a carcinogenic

Table 5.6 Categories for a Corporate Toxic Use Reduction Program

Categories 1 and II: Use to be reduced
  Human or animal carcinogens, teratogens, or reproductive agents
  Highly toxic chemicals
  Chemicals with human chronic toxicity
  Chemicals for which there are adverse environmental impacts
Category III: Transportation off-site to be reduced via source reduction or recovery and
reuse on site
  Suspected animal carcinogens
  Moderately toxic materials
  Chemicals that cause severe irritation of eyes and respiratory tract at low concentrations
  Chemicals with limited chronic toxicity
  Corrosive chemicals
  Chemicals for which there are environmental considerations
Category IV: Disposal volume to be reduced via waste reduction or reuse following on-site
recycling
  All other chemicals not included in Category V
Category V: Any material such as paper, metal parts, and so on that is identified as rub-
bish, rubble, or trash.

Source: Karger and Burgess (1992).

material. The original process required an aqueous oxidation step using a hexavalent chromium compound; the new process is based on a catalyzed air oxidation step, which does not require the chromium compound (Fig. 5.3). As a result of this change worker exposure to hexavalent chromium was eliminated, and this process change saved the company over $1 million. It is the conclusion of this company that the process that uses the least quantity of toxic chemicals is usually the most economical.

3.1.2.4 Outcome of Replacement Technology. Although the replacement of toxic chemicals may provide impressive gains, the replacement of a target material must be done with great caution. If the replacement technique is to be used successfully, it must be carefully reviewed to define its impact on worker health and environmental impact, product yield, and product quality. It is important that the impact be evaluated in both pilot plant operation and field trials before making the change.

Industrial hygienists practicing from the 1950s to the 1990s devoted more attention to solvent replacement for metal cleaning than any other control problem. It is useful to track the history of such cleaning agents during those years since it shows the complexity of the replacement technology approach and indicates pitfalls that may arise in using this technique. Wolf et al. (1991, p. 1056) traced the major decisions on general degreasing solvents during this period as follows:

> In the 1960s and 1970s in particular, chlorinated solvents were widely substituted for flammable solvents because they better protected workers from flammability. As smog regulations were promulgated in the late 1960s and early 1970s users moved from the photochemically reactive flammable solvents and TCE to the other "exempt" chlori-

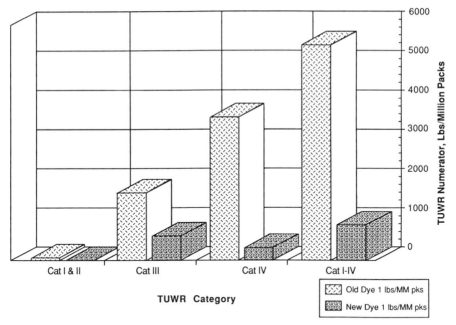

Figure 5.3 Impact of replacement technology on a chemical process resulting in the elimination of hexavalent chromium from the workplace and its impact on toxic use and waste
reduction categories (TUWR). Courtesy of Polaroid Corp.

nated solvents. In the 1980s there was increased scrutiny of the chlorinated solvents.
TCE, PERC and METH were considered undesirable because of their suspect carcinogenicity, and TCE and CFC-113 were being examined for their contribution to strat-
ospheric ozone depletion.

Each change starting in the 1950s, progressing from flammable solvents to chlorinated solvents to nonphotochemical reactive materials to chlorofluorocarbon (CFC)
solvents, introduced a new set of problems. The six major classes of solvents in use
in 1992 or that can be considered for application as general solvents and the critical
characteristics that determine their acceptability are shown in Table 5.7 (Wolf et al.,
1991). The alternatives to the common chlorinated solvents are not encouraging.
The flammable solvents require rigorous in-plant fire protection controls. Both the
flammable and combustible solvents including the terpenes are smog-producing and
therefore are subject to tight federal and local air district regulations. The chlorinated
solvents will continue to see pressure for replacement due to the array of human
toxicity and environmental effects. The increased use of CFCs in the 1970s and
1980s to the present level of 180 thousand metric tons per year is based on performance and reduced chronic toxicity, however, the impact of these chemicals on
ozone depletion has resulted in the ban of CFC-113 worldwide by the year 2000 and
1,1,1-trichloroethane (TCA) by 2002. The hydrochlorofluorocarbons are considered
interim candidates for industrial degreasing but are considered for banning in 2020
and 2040. The hydrofluorocarbons and the other candidate solvents not containing

Table 5.7 Characteristics of Generic Solvent Categories

Generic Solvent Category	ODP[a]	Photo-chemical Reactivity	GWP[b]	Flash Point[c]	Tested for Chronic Toxicity
Flammable solvents	—	Yes	—	F	
Isopropyl alcohol					Rule issued
Mineral Spirits					No
Combustible solvent[d]	—	Yes	—	C	
Terpenes					Limited[e]
DBE					No
NMP					Rule issued
Alkyl Acetates					No
Chlorinated solvents					
TCE	—	Yes	—	—	Yes
PERC	—	Yes[f]	—	—	Yes
METH	—	No	—	—	Yes
TCA	0.1	No	0.02	—	Yes
Chlorofluorocarbons					
(CFCs)		No		—	
CFC-11	1.0		1.0	—	Yes
CFC-113	0.8		1.4	—	Yes
Hydrochlorofluorocarbons					
(HCFCs)		No			
HCFC-123	0.02		0.02	—	In testing
HCFC-141b	0.1–0.18		0.09	—	In testing
HCFC-225	NA		NA	—	In testing
Hydrofluorocarbons					
(HFCs) and Fluorocarbons (FCs)					
Pentafluoropropanol	—	NA	NA	NA	No

Source: Wolf et al. (1991). Reprinted by permission of *Journal of the Air & Waste Management Association.*

[a] ODP: The oxygen depletion potential (ODP) is the potential for ozone depletion of one kilogram of a chemical relative to the potential of one kilogram of CFC-11, which has a defined ozone depletion potential of 1.0.

[b] GWP: The global warming potential (GWP) of a chemical is the potential of one kilogram of the chemical to cause global warming relative to the potential of one kilogram of CFC-11 to cause global warming. CFC-11 has a GWP of 1.0.

[c] F refers to flammable; C refers to combustible.

[d] DBE is dibasic esters; NMP is *N*-methyl-2-pyrrolidone.

[e] One of the terpenes, *d*-limonene, has been tested.

[f] Although PERC is not photochemically reactive, it is not exempt under the Clean Air Act.

NA means not available.

chlorine do not present an ozone depletion risk but may present unacceptable toxicity concerns.

A vigorous research effort is pursued by dozens of companies to develop alternatives to the cleaning agents listed in Table 5.7. It is doubtful that a single agent will be found that will have the broad application of the chlorinated solvents, however progress is being made in non-ozone-depleting, hydrocarbon-based systems.

3.1.3 Dust Control

A number of techniques have been proposed to minimize worker exposure to pneumoconiosis-producing and toxic dusts based on changes in physical form or state of the material. The common techniques to reduce dustiness in this manner are discussed below.

3.1.3.1 Moisture Content of Material. The relationship between moisture content of granular materials and worker exposure to dust is well known. The use of water as a dust suppressant has been of great interest to the mining, mineral processing, and foundry industries. In the 1960s research by the British Cast Iron Research Association demonstrated that silica in air concentrations in foundries were maintained below existing exposure standards if the moisture content of foundry sand was kept above 30 percent (BCIRA, 1977). A sand handling technique that involves direct blending of new moist sand directly from the mixer with shakeout sand to add moisture to cool the shakeout sand and reduce dustiness is described by Schumacher (1978). The author also demonstrated a correlation between results of a simple laboratory dustiness test and in-plant dust exposure. Goodfellow and Smith (1989, p. 180) identify the importance of a dustiness index: ''Dustiness testing can be a useful testing tool in establishing the type and efficiency of dust control required for different materials and materials handling systems. Further field data and verification are required to develop this procedure into a useful tool for practitioners.''

The British Occupational Hygiene Society (BOHS, 1985) has also explored various dustiness estimation methods that may be useful in design of control systems. In chemical processing a solid product is frequently isolated in cake form that is wet with solvent or water. Frequently this material must be dried to ensure product quality and shelf life and to permit packing and shipping. However, if the product is to be used ''in-house,'' one should determine if the drying operation can be omitted permitting direct handling of the wet cake thereby saving money and reducing dustiness.

3.1.3.2 Particle Size. In general, the more extensive the grinding or comminution of a granular material the greater the dust hazard the material will present during transport, handling, and processing. When possible, purchase the most coarse form of the chemical that is suitable for the process. There are production implications that may override considerations of worker exposure. As an example, if the material must be placed in solution the large particle size will slow this process.

3.1.3.3 Dust-controlled Forms. In the last two decades the rubber, pharmaceutical, pigments, and dyestuff industries have given attention to the dustiness of the raw materials they produce and use. Dustiness testing has been extended and significant product design changes have been adopted to minimize dustiness, worker exposure, and product acceptance. In tire manufacture at least a dozen chemicals in granular form are added in small quantities to the batch mix. In an effort to minimize worker exposure to dust from these chemicals, the British Rubber Manufacturing Association has sponsored design of low dusting forms of these common chemicals.

Table 5.8 Dust-controlled Forms of Rubber Chemicals—Comparative Performance

Property	Untreated Powder	Wax or Otherwise Bound				Polymer Bound	
		Soft Paste	Putty	Prills	Pellets/ Granules	Slab	Pellets/ Granules
Active content	5	3	4	5	4	4	4
Convenience of handling	3	1	2	4	5	2	5
Freedom from dust	1	5	5	4	4	5	5
General cleanliness and safety	1	1	2	3	3	5	5
Suitability for automatic weighing	3	1	1	4	4	1	5
Wastage	3	2	4	4	4	5	5
Ease of disposal of containers	3	1	3	3	4	4	4
Identification	—	—	—	—	—	—	—
Mill mixing behavior	3	1	3	3	3	5	4
Internal mixing behavior	5	2	3	5	5	5	5
Dispersion in rubber	5	5	4	3	4	4	4
Total							
	32	22	31	38	40	40	46

Source: Hammond (1980). Reprinted by permission of the British Occupational Hygiene Society.
5 = excellent; 4 = good; 3 = average, 2 = below average; 1 = poor.

The properties of seven dust-controlled forms of rubber chemicals is reviewed by Hammond (1980) in Table 5.8. The author notes that the disadvantages of the most effective approach, coating the chemical with a polymer, include the variability in active chemical content based on bulk weight, the reduced chemical content, and the unsuitability of the polymer in the formulation.

3.1.3.4 Slurry Form. This application has limited application, but in those cases where it can be used it does have great impact on dust concentrations. In the tire industry "master batch" rubber, rubber processed with all chemicals except the vulcanizing agent, is processed from the Banbury to the drop mill where it is "sheeted off" for storage. Until 1960 dry talc or limestone was dusted on the slabs of master batch material to keep it from sticking. This operation resulted in poor housekeeping and a significant exposure to talc. The present technique, adopted in the 1960s, involves dipping the stock in a slurry of talc in water before racking for storage. This simple change resulted in a significant reduction in worker exposure to talc.

3.1.4 Impurities in Production Chemicals

In low concentrations impurities or unreacted chemicals in raw or final product may represent a potential exposure that warrants attention.

3.1.4.1 Residual Monomer in Polymer. In polymer manufacture there is frequently unreacted monomer in the final product. Residual monomer had not been given much attention until the early 1970s when angiosarcoma, a rare liver cancer, noted in workers manufacturing the vinyl chloride polymer, was attributed to the monomer exposure. Investigation showed that the polymer used in subsequent fabricating operations had unreacted monomer present in concentrations as high as 0.4 percent (Braun and Druckman, 1976). This level of contamination prompted concern about the monomer exposure of workers handling the bulk polymer in plastic fabrication operations such as injection molding. As a result of this concern the vinyl chloride manufacturers modified the manufacturing process to reduce the monomer concentration to less than 1 ppm (Berens, 1981) thereby eliminating significant worker exposure. Residual monomer is frequently present in concentrations up to 1 percent in many of the common polymers and may warrant attention. If significant air concentrations are noted when handling the polymer, engineering control is first based on the removal or reduction of the monomer content in the polymer with other controls considered later if this is not adequate.

3.1.4.2 Solvent Impurities. Impurities may pose an unrecognized risk, especially in solvents of high volatility. In the manufacturing of automobile tires, the various rubber components are "layed-up" on a tire building machine. To effect good bonding between the components, the rubber is made tacky by applying a small amount of solvent to the surface with a pad. For several decades this solvent was benzene. The worker exposure, probably in the range of 1–10 ppm, may be responsible for the excess leukemia seen in older tire builders. Starting in the 1950s the industry started to replace benzene with white gasoline. In studies completed in the 1970s by the Harvard School of Public Health Joint Rubber Studies Group, the residual benzene content in white gasoline was 4–7 percent, and air sampling indicated that one-third of the air samples on tire builders exceeded 1 ppm (Treitman, 1976). The PEL for benzene at that time was 10 ppm although it was anticipated that it would be dropped to 1 ppm (this change did occur in 1987). Technical grade chemicals commonly have significant impurities. The level of contamination should be identified, and, if sufficiently high, worker exposure should be evaluated. At that time the necessity for reduction in the impurity level can be determined.

3.2 Equipment and Processes

In Section 3.1 a variety of engineering control options were focused on the choice of materials to minimize the generation and release of airborne contaminants. An equally important step is a review of the various alternatives in the choice of equipment and processes.

In the discussion on dusty materials in Section 3.1.3, techniques to determine the relative index of dustiness were mentioned to assist in the choice of material form and the dust suppression treatment. We do not have such an index for equipment and processes, however, there are a number of operational insights that should be considered in choice of facility. Wolfson (1993) has emphasized the importance of this step in stating that the removal of the dispersal device should be a first step in the engineering control of air contamination.

3.2.1 Diagnostic Air Sampling

Contaminant control cannot be achieved until the significant operational elements of the process that generate and release the contaminant are identified. Occasionally this can be done simply by a critical review of the operations but usually diagnostic air sampling is necessary. The value of short interval, task-oriented air sampling using conventional integrated sampling with subsequent analysis has been clearly stated by Caplan (1985a, p. 625). In describing this approach, illustrated in Table 5.9, Caplan states:

> For the job analyzed in Table 5.9 presumably a single sample would have shown a concentration of 0.21 mg/m³. The task-oriented sampling, however, reveals several interesting things. Column 5 shows that tasks B and C are the major contributors to the day's exposures and that a significant reduction in the concentration at either of those tasks would be adequate to bring the 8 hr exposure well below the TLV. This is true even though task C in itself is below the TLV. In addition, it shows that task F, well above the TLV concentration, is of such short duration that significant improvement in that part of the exposure would not have a large effect on the 8-hour exposure.

The difficulty in this approach is that frequently it is not possible to measure the air concentrations during brief individual tasks and activities due to the low air sampling rate and the limited sensitivity of the analytical procedures. To reveal important generation points in the job, it is necessary to resolve the air concentration profile in a time frame of seconds. The advent of real-time, direct-reading air sampling instruments for particles, gases, and vapors with response times of less one second now permit the investigator to identify these critical contaminant elements in process events and work practices. In the 1980s this air sampling technique saw expanded application with coincidental video taping of the worker during a work

Table 5.9 Task-oriented Air Sampling

Task	GA or BZ[a]	Minutes/ Day	Concentration (mg/m³)	Minutes/Day × Concentration
A. Charge pot	BZ	40	0.12	4.8
B. Unload pot	BZ	80	0.50	40.0
C. General survey	GA	250	0.16	40.0
D. Lab—sample trips	GA	20	0.05	1.0
E. Change room	GA	30	0.08	2.4
F. Pump room—repack	BZ	30	0.32	9.6
G. Lunch room	GA	30	0.07	2.1
		480		99.9

$$\text{TLV} = 0.2 \text{ mg/m}^3 \qquad \text{Wt. avg.} = \frac{99.9}{480} = 0.21 \text{ mg/m}^3$$

Source: Caplan (1985a). Reprinted by permission of John Wiley and Sons from *Patty's Industrial Hygiene & Toxicology*, Cralley and Cralley, Eds., © 1985.

[a]GA = general air; BZ = breathing zone.

cycle. In its most sophisticated form the time-coupled, real-time contaminant concentration at the workers breathing zone is superimposed on the video display permitting the viewer to analyze the data display to identify the specific time and location of release. Control technology is then applied to those tasks or incidents.

The video display may also be used to identify work practices that may either positively or adversely affect worker exposure. In a talc bagging operation reviewed in Vermont in the 1970s one individual consistently had the lowest dust exposure although visual inspection did not reveal any differences in equipment or work practice. If the real-time technique were available, the worker's ''secret'' could have been identified and applied to the other workers. If a ''correct way of doing the job'' can be identified, the video concentration format is an excellent educational tool for workers.

A series of studies by NIOSH investigators describe the application of this technique to a range of in-plant tasks (Gressel et al., 1987; O'Brein et al., 1989). The air sampling instrument is a real-time monitor with a response time much shorter than the period of the shortest worker activity or movement. The instrument is equipped with a data logger with a clock ''locked-in'' or synchronized with the video. The data logger is down-loaded to an IBM-compatible computer, and the data file is analyzed permitting a graphical overlay of air contamination data on the video tape. In Figure 5.4 the video display from a study of a bag dumping operation is recreated in a line drawing for clarity to show the time-coupled concentration on the video screen display (Cooper and Gressel, 1992). Graphical representation of the air sampling data for three jobs with the concentration from the video overlay is shown in Figure 5.5 (Gressel et al., 1988).

The application of this technique to control technology is shown by Gressel et al. (1989) in a study of a chemical weighing and transfer station. The information obtained permitted the investigators to redesign the workstation controls based on a perimeter exhaust hood and an air shower to eliminate eddies induced by the worker's body. Effective control of the dust exposure was obtained with one-third the airflow of the original system. In addition to improved worker protection, the cost savings of the new system resulted in a payback period of 4.5 years.

This diagnostic tool will see expanded application to workstation design in the next decade since it does permit the engineer to identify the specific tasks, equipment function, or worker movement that contribute to worker exposure.

3.2.2 Equipment

3.2.2.1 Ancillary Equipment. The stepwise review of the dispersal potential noted above should include all equipment, not only the major machinery. The importance of this approach is highlighted when one looks at the history of the simple air nozzle in industry. For most of the 1900s the widespread application of the air nozzle operating at line pressure (100 psi) was used to remove chips and cutting oil from machined parts and to dry parts after cleaning with solvent. In the 1970s OSHA required that the operating pressure for these nozzles for cleaning purposes be dropped from 100 to 30 psi. This change had many positive effects including a reduction in noise, eye injuries, solvent mist, and vapor exposures.

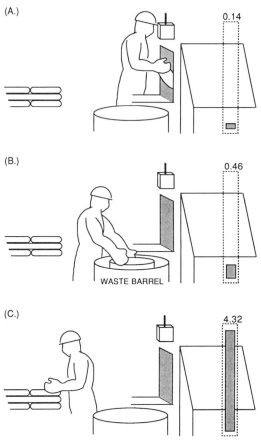

(A.)

0.14

(B.)

0.46

WASTE BARREL

(C.)

4.32

Figure 5.4 Relative dust exposure during bag dumping as determined by video-air sampling technique: (A) Operator slits bag and dumps the granular material into a hopper equipped with an exhaust hood with minor dust exposure. (B) Operator drops the bag into the waste barrel; there is a significant increase in dust concentration. (C) As the operator pushes the bag into the waste barrel, a cloud of dust is released and the relative dust concentration increases to 4.32. [Reprinted with permission from *Applied Occupational and Environmental Hygiene*, **7**(4), 1992.]

Another common piece of equipment, albeit much more complex, is the centrifugal pump. The difficulty of equipment selection is typified by the experience of a chemical engineering team designing a modern chemical plant or refinery having hundreds of centrifugal pumps. A major issue in such plants is the impact of fugitive losses from rotating machinery such as pumps on the workplace and general environment (NIOSH, 1991; BOHS, 1984). The total loss percentage from rotating machinery varies from 8 to 24 percent with overall uncontrolled emissions ranging from 31 to 3231 tons/year for a model chemical plant or a large refinery (Lipton and Lynch, 1987). The authors describe two possible control programs for pump fugitive losses—either a monitoring and maintenance program or engineering control by in-

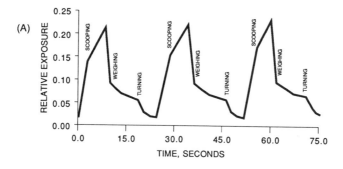

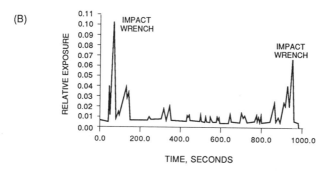

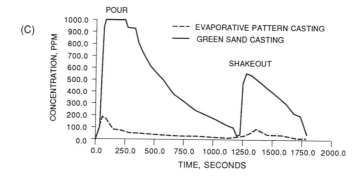

Figure 5.5 Peak dust exposures identified by video–real-time monitoring: (A) Relative exposure versus time for three bags during manual weighout of powders. (B) Relative dust exposure during automotive brake servicing. (C) Carbon monoxide emissions from evaporative pattern casting and green sand processes. (From Gressel et al, 1988. Reproduced by permission of *Applied Industrial Hygiene*).

stallation of dual seal pumps, seal-less pumps, and a closed exhaust hood with air cleaning.

The phase-out of 1,1,1-trichloroethane and trichlorotrifluoroethane solvents in vapor-phase degreasers in the next 10 years has accelerated the search for alternative solvents as described in Section 3.1.2.4. Initial steps to reduce emissions from this common equipment in the 1980s included the use of covers, increased freeboard height, refrigerated condenser fluids, and lower hoist speeds. Although these changes do reduce losses and therefore worker exposure, the survival of this type of equipment rests with the availability of a closed system unit as described by Mertens (1991). Such vapor-tight degreasers are available in Europe and are under development in the United States. In the past, closed system technology has been associated with major processing facilities. In the future individual job shop equipment will utilize this approach to meet critical workplace and environmental constraints.

3.2.2.2 Major Processing Equipment. The complexity of the control technology problem becomes apparent when one moves from choices of ancillary equipment to decisions on major pieces of processing equipment. As an example, a small chemical processing plant for organic synthesis is a multifloor plant with a series of major operations staffed by six chemical technicians. The condensed version we will review (Fig. 5.6) permits discussion of equipment alternatives to minimize the generation and release of air contaminants. The three-step process includes, first, a reaction step to form a solid in suspension, conducted in a pressure vessel. In this case there is no alternative equipment. The reaction must be conducted in a closed pressure vessel equipped with the safety controls shown in Fig. 5.6 to minimize the possibility of a chemical accident. The second step, the isolation of the solid reaction product, can be done by at least three different techniques. The environmental contaminant potential of these processes varies considerably. However, the choice cannot be made on this basis alone, since these techniques do have specific production capabilities that may be the major determinant in the choice. Finally the reaction product is dried and drummed for sale or use in a subsequent operation. Again there are several ways to accomplish the drying. The ranking of these isolation and drying operations based on their environmental impact is shown in Figure 5.6.

The responsibility for the choice of specific equipment in the processing plant will rest with the project design group and will be based on production capacity, ancillary services required, maintenance history, cost and delivery, in addition to health and safety considerations. It is not expected that the industrial hygienist will have complete operational and application data on all chemical unit operations, but their environmental impacts must be understood and made known to the engineering group.

It would be extremely helpful to the industrial hygienist if the individual equipment were assigned an index of contaminant generation/dispersal similar to the index of dustiness or fugitive losses from a pump. This approach is not currently realistic, however, a better characterization of the performance of such equipment will emerge in the next decade.

3.2.3 Processes

A review of the production engineering literature shows a range of processes that can be considered for a manufacturing facility. The choice is usually made based on production output and cost data. Industrial processes designed to accomplish a given task have one other parameter, the ability to contaminate the workplace; this characteristic of the process should be given equal weight to production criteria (Peterson, 1985).

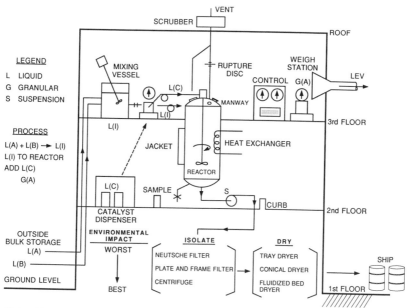

Figure 5.6   Chemical processing facility. Liquid reagents A and B are transferred to a mixing vessel. After mixing, the resulting liquid C is transferred to the reactor by vacuum. Granular chemical X is weighed under local exhaust ventilation (LEV) and dumped into the reactor through a change port also equipped with LEV. A catalyst D is transferred to the reactor by a positive displacement pump. After the reaction is complete, the new chemical is isolated by one of three techniques—Neutsche filter, plate and frame filter, or centrifuge. The resulting wet cake is then dried by one of three methods—tray drier, rotary drier, or fluidized bed drier. The isolation and drying techniques are ranked by environmental impact with the worst process noted first.

The development of technology in a given area such as welding is pushed by production needs not by health concerns. Since 1900 the technology has become more complex and frequently more environmentally challenging. Infrequently, studies are conducted to influence the process design to minimize air contamination. This was the case with Gray and Hewitt (1982) who defined the welding operating parameters that influence fume generation and proposed operational configurations that minimize fume formation rate and the chemical composition of the fume thereby permitting improved efficiency of controls.

An exercise that demonstrates the value of a ranking approach for process environmental impacts is shown in the manufacture of an electronic cabinet (Fig. 5.7). As in the case of the chemical process shown above, the fabrication steps are simplified for ease of discussion. The four enclosure components are fabricated from lightweight aluminum stock using sheet metal shear, brake, and punch. These operations may present a noise hazard but do not generate air contaminants and will not be included in our review. The enclosure can be assembled and painted by a number of techniques; the processes are ranked from worst to best in terms of adverse effect from air contamination.

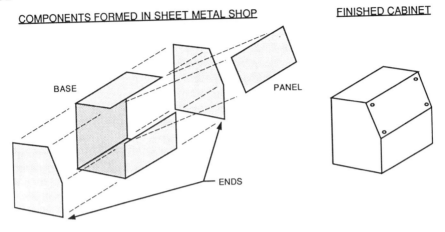

FINISHING OPTIONS

Metal cleaning	Welding	Painting
Wet degreasing with chlorinated HC	Torch welding	Air atomization
Vapor with chlorinated HC	Inert gas shielded arc welding	Airless spraying
Wet degreasing with terpenes	Submerged arc	Hot paint
Water with detergent	Resistance	Electrostatic
		Powder
		Dipping

Figure 5.7 Manufacture of an electronic cabinet. The operational options for cleaning, abrasive blasting, and painting are ranked in terms of environmental acceptance with the worst process noted first.

Another general approach that applies to many processes is the use of containment. This control approach was considered by an engineering control panel sponsored by the Environmental Protection Agency and the following containment ranking was proposed (EPA 1986, R 12).

Standard control measures may often be categorized for each process: a sealed and isolated system where airborne contaminant levels are very low (ppb range for vapors); a substantially closed system where airborne contaminant levels are still relatively low (fraction of ppm or low ppm range for vapors); a semi-closed system which is typical of non-dedicated equipment used in job shop chemical processing facilities; and an open system.

An example of a sealed and isolated system is the modern refinery where such an approach is required both for production reasons and health and safety. A substantially closed system is typified by a semiconductor facility, and a semiclosed system by the chemical processing plant in Figure 5.6. Most common facilities, that is, foundries, electroplating, painting, welding, and machining, are examples of open systems.

The contaminant released as a mist from electrolytic plating operations can be trapped by a layer of plastic chips or a persistent foam blanket on the surface of the bath. Also surfactants may be added to the electrolyte to reduce mist escape. The plate and frame filter in the chemical processing industry is difficult to handle with local exhaust ventilation. Frequently a solvent wash of the cake with a highly toxic material results in high exposure when the filter is broken and the solvent-wet cake is removed. In some cases after the initial wash, the cake can be washed with iso-propanol to strip out the toxic solvent. It is then washed with water so that when the filter is broken and the product is removed exposure is nil.

A series of process changes developed in the WRITE program is shown in Table 5.10.

3.3 Work Task Modification, Automation, and Robotics

It is well known that simple changes in work tasks may have significant impact on job outcome. In the early 1900s, workplace time and motion studies were used to improve productivity. Later a job placement technique devised by Hanman (1968) based on a detailed analysis of the time and effort of the job tasks was effective in reducing on-the-job injuries. More recently, the analytical tools of the ergonomic specialist permits identifying difficult tasks contributing to occupational injuries and illness.

As indicated in Section 3.2.1, the industrial hygienist has techniques to investigate the source of contaminants, the generation mechanism, and to make a semi-quantitative assessment of the generation rate. This approach has tremendous value in analyzing not only the critical generation points on the machine but in viewing the impact of specific worker actions and movement on air concentrations. Preliminary studies reveal that minor changes in work position and movement may offer significant reductions in exposure.

Occasionally the specific modification in work practice is dictated by knowledge

Table 5.10 Examples of Process Change Technology from EPA WRITE Program

Operation	Process Change
Hand mixing of paint	Proportional mixer blends paint at the gun thereby eliminating handling paint and solvent at a mix operation.
Conventional manual air spray painting	Computer-controlled robotic painting with an electrostatic spray to reduce worker exposure to paint mist and solvent vapors
Performance testing of electronic parts with CFC-based cooling system	Installed compressed air cooling system

Source: Adapted from Harten and Licis (1991).

of the mechanisms of generation and release of the contaminant. Such is the case with the flow of granular material at material transfer points. The generation mechanism is the airflow induced by the falling granular material (Anderson, 1964). The induced airflow can be minimized by restricting the open area of the upstream face, reducing the free-fall distance, and reducing the material flow rate as defined by Anderson's equation.

Automation is a general technique that separates the worker from the individual process. In the 1960s and the 1970s this was usually done by simple electromechanical equipment design. A good example of this procedure is the manufacture of asphalt roof shingles. In the 1970s competition resulted in the automation of all parts of the process from the dipping of the stock to the bundling of the package.

Another example of a simple automation process that reduces worker exposure is tire curing. In the early plants the worker lifted the tire out of a curing press at the end of the curing cycle and placed another tire in the mold for curing. During this period he was directly exposed to the emissions released from the press. By 1970 most plants had automated this process. The worker now places the uncured tire on a holding rack in front of the curing press line. When the curing cycle is completed, the tire is ejected from the press to a belt conveyor and the next tire to be cured is transferred from the rack to the press without exposure of the worker.

The advent of robotic techniques has permitted almost all industrial procedures to be candidates for automation. The movement of the worker who buffs rubber boots can now be captured by a robotic system, as shown in Figure 5.8. The ultimate application of robotic techniques is to spray painting. In this case the robotic system can reproduce the movement of a skilled painter. Although ventilation is still required on this job, the worker overseeing the operation is separated from the point of release of the paint mist and solvent.

Robotics may have been a solution to a difficult problem in the 1950s. In a large

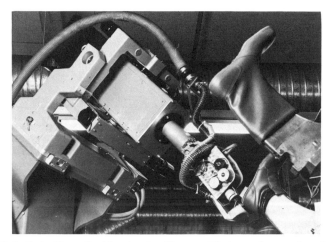

Figure 5.8   Robotic buffing of rubber boots. (*Source:* Photograph reproduced by permission of Matti Koivumaki.)

generator shop the generator coils were preformed of copper bar stock. The coils were then wrapped with insulation tape and painted with an asphalt compound. Protective clothing notwithstanding, the workers had skin exposure to the asphalt, which required aggressive cleaning at the end of the shift. Dermatitis and photosensitivity were frequent occurrences. At that time coil winding experts devoted time and money to developing machine wrapping concepts without success. This type of problem can be solved by robotics in the 1990s.

3.4 Facility Layout

As stated earlier, the most efficient way to do a job is probably the one that impacts the least on the workplace environment. Certainly this is true insofar as overall plant layout is concerned (Caplan, 1985b). In the shop fabricating the enclosure in Figure 5.7 the desired flow of materials is from the incoming truck dock or railroad siding to sheet metal fabrication and then to assembly, finishing, inspection, and finally shipping. In a facility manufacturing pharmaceuticals, the input chemicals are transported to bulk storage and then to chemical processing, packaging, inspection, and shipping. In all manufacturing processes from handling metal to fine chemicals, it is important to minimize the distance the material is moved and the number of times it is picked up and transferred. If this rule is followed worker exposure to air contaminants will be minimized.

3.4.1 Material Transport

This issue is given major attention in industry where large quantities of raw material, intermediates, and final products are handled. Examples are injection molding of children's toys, manufacture of automobile tires, and paint manufacture. In some cases material handling alone defines the plant layout. Frequently bulk storage is in large silos located outside the plant with delivery to the workstations by mechanical or pneumatic conveyors. This is true in the manufacture of plastic tape where tons of PVC granules are used each day (Fig. 5.9). This plant also requires large quantities of solvent delivered from bulk storage by piping to mix tanks and then directly to the coating heads. The intent of this system is to have all material handling to the individual coaters done in closed systems.

In organic synthesis it is common practice to have outside bulk storage of at least a dozen solvents. The solvents are transferred to an inside reaction vessel by piping to a reactor manifold with necessary valving and meters. Granular materials used in small quantities are stored in adjoining storage areas and delivered to the reactor and charged by manual dumping, with a dumping fixture, or occasionally with a transfer lock. When a large number of drums must be dumped, a bulk handling system is used with direct delivery to the reactor vessel.

3.4.2 General Considerations

The location of the process within the facility may influence the worker's exposure. This is true of operations such as foundries where jobs may be easily classed as

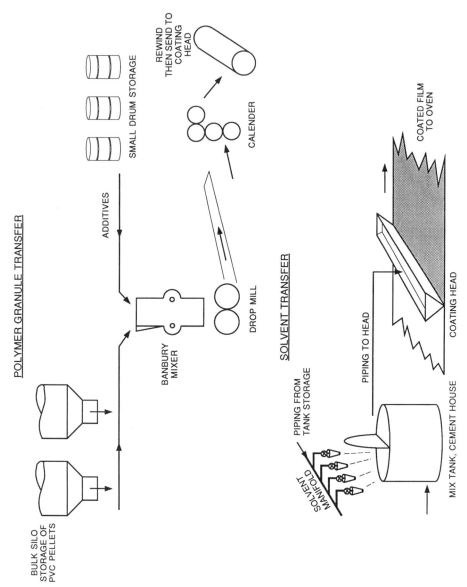

POLYMER GRANULE TRANSFER

BULK SILO
STORAGE OF
PVC PELLETS

ADDITIVES

SMALL DRUM STORAGE

BANBURY
MIXER

DROP MILL

CALENDER

REWIND
THEN SEND TO
COATING
HEAD

SOLVENT TRANSFER

SOLVENT
MANIFOLD

PIPING FROM
TANK STORAGE

PIPING TO HEAD

MIX TANK, CEMENT HOUSE

COATING HEAD

COATED FILM
TO OVEN

Figure 5.9 Manufacture of electrical tape showing the closed bulk transport of granular plastic and the piping of solvents to the mix tank.

clean or dirty and exposures vary greatly. The conventional iron foundry provides a good example of the importance of plant layout. It is common practice to define the optimal plant layout as one with airflow from the cleanest to the dirtiest operations with the exhaust focal point establishing this gradient [Fig. 5.10(A)]. The layout of a similar foundry that does not follow this guideline is shown in Figure 5.10(B). In the latter case the relatively clean molding line is positioned adjacent to the shakeout, an area where the control of airborne foundry dust and thermal degradation products is difficult. This poor layout results in silica and other contaminants released by shakeout moving into the molding area and exposing this work population. If the desired concentration gradient cannot be achieved by ventilation or distance, then structural walls or plastic barriers are a possibility. Although such

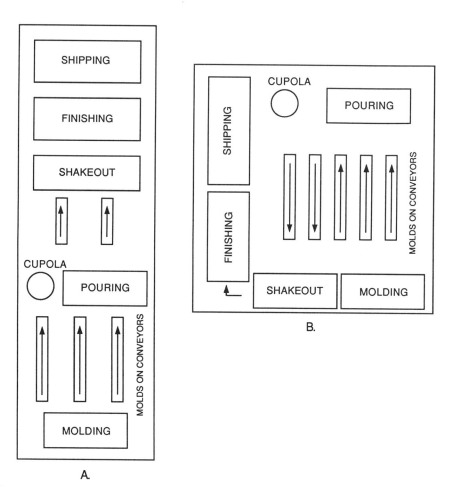

Figure 5.10 Layout of two gray iron foundries. Foundry in (A) is designed for straight process flow and gradual transition from activities that represent limited contamination to those of greater contamination. In foundry (B) the process flow reverses and the dirty shakeout ends up adjacent to the molding line.

barriers define the space, compartmentalizing complicates the design and application of local exhaust ventilation, replacement air, HVAC, and other important services.

3.4.3 Workstations

A foundry in Finland faced problems not only of airborne contaminants but heat, noise, and housekeeping in casting cleaning. The problems were resolved by redesigning the open work space to provide individual workstations designed with integrated services positioned for ease of work (Fig. 5.11). This layout resulted in improved material flow, better housekeeping, reduced air contamination, better

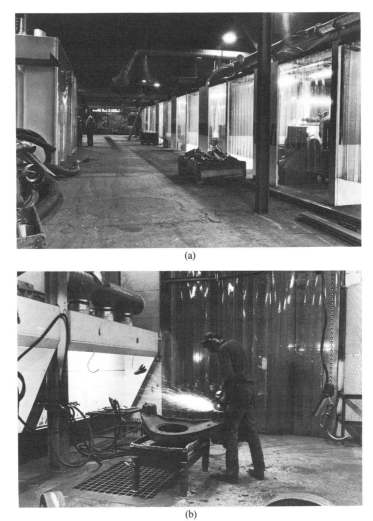

(a)

(b)

Figure 5.11 Well-integrated workstation in a foundry. (A) Overall layout of foundry finishing area; (B) individual workstation with exhaust hood. (*Source:* Photograph reproduced by permission of Matti Koivumaki.)

ergonomics, and improved productivity. The success of such an installation was due, in part, to close employee–employer involvement in the design.

3.4.4 Service and Maintenance

Frequently, close attention is given production area layout as noted in the above example, but rarely is attention given the working environment of the maintenance group. An example of an area where maintenance was not considered was tire curing presses in the plants of the 1970s. The presses were arranged in double rows back to back with limited space between rows for steam, water, and electrical services. While this layout and density represented optimal use of space, repairs of steam and water piping were done in an extremely tight space and controls for the worker were difficult to set up.

The best example of a planned layout that provides proper work space for the trades is noted in the semiconductor industry (Fig. 5.12). In this layout all transfer pumps, distribution lines, and vacuum pumps are located in a service alley separated from the fabrication bay. This design reflects principal concern, not for the comfort of the maintenance worker, but rather to protect the fabrication area against contamination. Notwithstanding, it does result in adequate space for the trades to carry out their work in a safe manner.

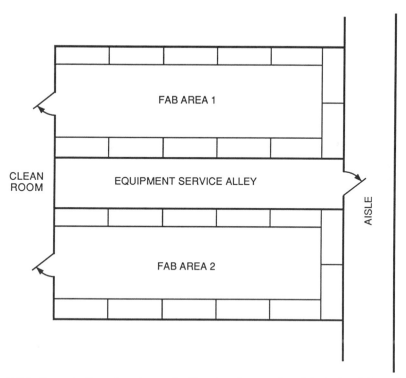

Figure 5.12 Layout of semiconductor facility showing service aisle isolated from, but adjoining, the fabrication bays.

An extension of this concern is the installation of a small field service bench with necessary tools and equipment in plant production areas where maintenance is frequently done. This arrangement permits the maintenance person to do many repairs at the site and not transport equipment back to a main facility with the potential for chemical spills.

3.4.5 Segregation of Operation

It is frequently necessary to segregate or remove an operation from the main production area as an engineering control measure. In the semiconductor and fiber optics industry each time a new facility is designed, the engineers must choose whether to store small amounts of highly toxic gases close to the production tool or store large quantities at a segregated position some distance from the plant to be distributed by double-walled piping to the production tools. This latter segregation technique permits the plant to reduce the number of persons at risk while providing extensive controls at the work site. This option must be weighed against storing small quantities of gas directly at the tool thereby minimizing failures in the transfer systems but requiring frequent change of gas cylinders with the entailed risks.

3.4.6 Isolation of the Worker

The practice of isolating the worker as a control measure is placed under this section since it should be considered in conjunction with layout of the major equipment. This widely used technique is mandated by OSHA coke oven regulations, which requires enclosures with clean, conditioned air. The larry car and pusher cars have controlled environment cab enclosures; the lidsmen have enclosures designed to remove them from the hostile coke oven environment during available rest periods.

In other industries this approach has been chosen to eliminate worker exposure to air contaminants while providing a comfortable working environment. These applications include enclosed booths on a variety of construction equipment, front-end loaders in smelters, crane cabs in metallurgical industries, operator cabs in steel rolling mills, and pouring stations in foundries. Unfortunately off-the-shelf control booths are not available for the range of applications seen in industry, nor have engineering guidelines been published.

3.5 Ventilation

In the introduction to this chapter we stated that ventilation is the third step in the control hierarchy of: (1) do not generate the contaminant; (2) if you do generate a contaminant, do not allow it to be released; and (3) if it is released, collect the contaminant before it reaches the worker. The design goal of industrial ventilation is to protect the worker from airborne contamination in the workplace. To the newcomer this may suggest installing a system that will reduce exposure below the permissible exposure guidelines or an appropriate action level. This is not the case. The professional will design the system to meet the goal of "as low as reasonably practical" (BOHS, 1987).

Within this control approach the effectiveness of the major ventilation techniques is shown in Table 5.11 (BOHS, 1987). There is no agreement on the position of low-volume–high-velocity systems since its effectiveness varies greatly depending on the degree of integration with the tool.

Soule (1991) reviews the application of the two major ventilation control approaches, dilution and local exhaust ventilation (Table 5.12). In most industries dilution is not the primary ventilation control approach for toxic materials. It is accepted that local exhaust ventilation will not provide total capture of contaminant, and dilution ventilation is frequently applied to collect losses from such systems. In addition, it is used for multiple, dispersed, low-toxicity releases.

3.5.1 Design Phase

The specific design methods for both general exhaust ventilation and local exhaust ventilation are presented by Soule (1991) and in other volumes dedicated to ventilation control (Burgess et al., 1989; Burton, 1989; and ACGIH, 1992). An important predesign phase identified by Burton (1989) as problem characterization is frequently given little attention (Table 5.13). This is a topic that can be best addressed

Table 5.11. Ventilation Control Hierarchy

Type of Ventilation[a]	Hood Type	Example
LEV	Total enclosure	Glove box
LEV	Partial enclosure	Laboratory hood
LEV	Low-volume–high-velocity, tool integrated	Portable grinder
LEV	Exterior hood	Welding hood
GEV	Mechanical exhaust	Roof ventilators
GEV	Natural	Wind induced

Source: Adapted from BOHS (1987).
[a]LEV = local exhaust ventilation; GEV = general exhaust (dilution) ventilation.

Table 5.12 Application of Local Exhaust and General Exhaust Ventilation

Local Exhaust Ventilation	General Exhaust Ventilation
Contaminant is toxic.	Contaminant has low order of toxicity.
Workstation is close to contaminant release point.	Contaminants are gases and vapors not particles.
Contaminant generation varies over shift.	Uniform contaminant release rate.
Contaminant generation rate is high with few sources.	Multiple generation sources, widely spaced.
Contaminant source is fixed.	Generation sites not close to breathing zone.
	Plant located in moderate climate.

Source: Adapted from Soule (1991).

Table 5.13 Information Needs for Problem Characterization

Emission Source Behavior
    Location of all emission sources or potential emission sources.
    Which emission sources actually contribute to exposure?
    What is the relative contribution of each source to exposure?
    Characterization of each contributor:
        Chemical composition, temperature, rate of emission, direction of
        emission, initial emission velocity, continuous or intermittent,
        time intervals of emission.
Air Behavior
    Air temperature
    Air movement
    Mixing potential
    Supply and return flow conditions
    Air changes per hour
    Effects of wind speed and direction
    Effects of weather and season
Worker Behavior
    Worker interaction with emission source
    Worker location
    Work practice
    Worker education, training, cooperation

Source: Burton (1989). Reprinted with permission of D. Jeff Burton.

by an industrial hygienist who can provide data on emissions, air patterns, and worker movement and actions.

If the process is new, a video tape of a similar operation with the same unit operations may be available. The best of all worlds would be the availability of a video concentration tape as discussed in Section 3.2.1. If the facility is a duplicate of one in the company, the industrial hygienist should visit the operation with the ventilation designer. Frequently the designer is an outside contractor. In this case it is important that the problem characterization approach be followed and an information package be provided the designer.

The precautions that should be reflected in the design have been discussed in detail elsewhere, but should include worker interface, access for maintenance, and routine testing. Computer-aided manufacturing and design (CAM/CAD) technology now permits precise placement of equipment and ductwork so that ad hoc placement by the installer should be a thing of the past. As discussed earlier, it may be worthwhile to "mock-up" a specific design solution prior to final design and construction. This is especially true when a large number of identical workstations are to be installed as was the case in a shipyard asbestos insulation workroom, as noted in Figure 5.2.

General cautions are appropriate on the use of available design data for control of industrial operations. The ACGIH *Industrial Ventilation Manual* provides the most comprehensive selection of design plates for general industry (ACGIH, 1992). Each of these plates provides four specific design elements: hood geometry, airflow rate, minimum duct velocity, and entry loss. The missing element is the performance of the hood in terms of percent containment. Roach (1981) has recommended such a index as a minimum performance specification for local exhaust hoods.

As noted by Burgess (1993) there has been little consolidation and publication of successful designs by industrial groups. With the exception of the *Steel Mill Ventilation* volume published by the American Iron and Steel Institute in the 1960s (AISI, 1965) and the *Foundry Environment Control* manual in 1972 (AFS, 1972), there is no evidence that companies in the United States wish to share ventilation control technology. Major sources of information on the performance of ventilation systems are the technical reports on engineering control technology published by NIOSH (Appendix A). Many of these reports couple ventilation assessment with measurement of worker exposure.

This discussion indicates that ventilation control designs on standard operations in the mature industries have been published although performance has usually not been reported. It is important to evaluate performance by diagnostic air sampling both to ensure the worker is protected and to prevent overdesign. The latter was shown to be the case in the design for control of a push-pull system for open surface tanks (Sciola, 1993). A mock-up of one tank demonstrated that satisfactory control could be achieved at minimal airflow. Operating at the reduced airflow rate saved $100,000 in installation costs and $263,000 in annual operating costs.

3.5.2 Acceptance and Start-up Testing

It is important that new ventilation systems be inspected and tested on completion of the installation to ensure the system conforms to the design specifications and that worker exposure is kept as low as reasonably possible. Unfortunately it is the author's experience that only 10 percent of the new systems in general industry undergo such scrutiny. The situation is baffling given the importance of contaminant control, the general practice of industry to test other services before acceptance, and the cost of ventilation systems ($10 to $30 per cfm to install and $1 to $4 per cfm per year to operate).

3.5.2.1 Construction Details. The ventilation system design, completed by the plant facilities and engineering group or a consulting firm, is based on certain assumptions and specifications on hardware details such as elbows, entries, expansions, contractions, and so forth. The hood construction has been detailed by the draftsperson. If the hardware elements provided for installation are supplied by a standard manufacturer, the losses will be approximately those used by the designer. If the system is large, it may be worthwhile obtaining samples of the components to test for losses. The hoods warrant special attention. The entry losses on standard hood designs as noted in the ACGIH ventilation manual are quite reliable. However, if the hood design is not typical, it may be difficult to estimate entry loss. In such a case the best solution is to have a single hood fabricated and tested by a laboratory to define the entry loss. This is especially important when a large number of hoods are to be installed. With a sample hood the coefficient of entry can be defined for use in routine hood static suction measurements to calculate airflow. This approach will be discussed in the next section.

An initial physical inspection of the completed system should be conducted to determine if the system is built according to the design. The size and type of fittings should be checked and the hood construction reviewed. The fan type and size should be as specified and the direction of rotation, speed, and current drain should be noted. A detailed inspection sheet should be completed. This type of inspection may seem redundant but in the author's experience it is worthwhile. In one case a mitre elbow was added in a large submain to save space; the single elbow presented high losses that caused the system to malfunction.

The air cleaning component should be inspected to determine if it is installed according to specifications. If ancillary equipment such as pressure sensing or velocity measuring equipment is included in the systems, it should also be inspected and calibrated.

All conventional ductwork and air cleaning equipment should be mounted on the suction side of the fan. If this is not done, toxic contaminants may leak to the occupied space. In one such case an installation handling a volatile and odorous organic chemical with the duct under positive pressure contaminated the workplace. Due to the complexity of the system, the cost of rectifying this problem was $50,000. This problem should have been picked up in the design review. If a duct run must be under positive pressure, special design features must be utilized by the engineer.

3.5.2.2 Airflow. The general format for testing a new local exhaust ventilation system balanced without blast gates is shown in Figure 5.13. The airflow rate through each branch servicing a hood should be evaluated by a pitot-static traverse. The pitot-static tube is a primary standard and, if used with an inclined manometer, it does not require calibration. Care is required in the choice of traverse location and

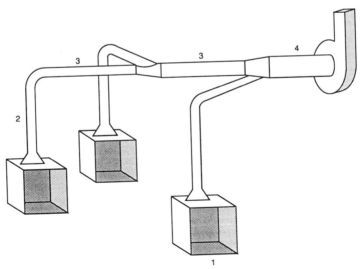

Figure 5.13 Locations for ventilation measurements: (1) At the face of booth-type hoods, (2) just downstream of a hood at a branch location, (3) in the main to define hood exhaust rate by difference, and (4) in the main to define total system airflow.

the device is limited at low velocities. In the last decade electronic manometers with microprocessor-based instruments are available for direct reading of pressure and velocity. When used in conjunction with a small portable recorder, the pitot traverse can be conducted by one person with ease.

One of two critical design velocities will be specified depending on the hood type. A partial enclosure such as a paint spray hood has a design velocity or control velocity at the plane of the hood face. Exterior hoods such as a simple welding hood will have a capture velocity specified at a certain working distance. The control and capture velocities may be calculated from the airflow rate and the hood dimensions. Frequently the actual velocities must be measured to satisfy plant or regulatory requirements. Face velocities at partial enclosures are usually evaluated by one of three direct-reading anemometers: rotating vane, swinging vane, or heated element. These instruments have a wide range of applications, but care should be taken to observe their limitations (Burgess et al., 1989). These simple direct-reading instruments can be used for the required inspections described in Section 3.5.2.

The measurement of hood static suction is also of value in periodic inspections and should be measured at the acceptance tests. The measurement, made 2–4 diameters downstream of the hood, provides a simple method of calculating airflow (Burgess et al., 1989).

The data collected in the acceptance tests are important baseline information for subsequent periodic inspections described in Section 3.5.3.

3.5.2.3 Performance. The term *performance*, as used in this discussion, describes the ability of the local exhaust ventilation to minimize the release of air contaminants from equipment serviced by the system. This measurement can be done by qualitative, semiquantitative, and quantitative methods.

3.5.2.3.1 Qualitative Methods. The use of smoke tracers released at the generation point, if done properly, provides an excellent qualitative method of exploring the performance of the hood. The control boundary can be established with this technique. In addition, it can be an excellent teaching tool for the worker who can view the impact of his own body and actions and that of external disturbances such as drafts from windows, doors, and traffic on the ventilation. Corrosive smoke cannot be used in semiconductor facilities; both dry ice and ultrasonic streams have been evaluated for this purpose. In Europe another qualitative technique, light scatter with back lighting, is widely used to identify the source and generation mode of particle contamination. Exquisite photographs of dust release based on this technique have been used as a design input for ceramic industry dust control.

3.5.2.3.2 Semiquantitative. The local exhaust ventilation system is first assessed to determine if it meets good practice in terms of general design and airflow rate. If that is acceptable, the system is checked qualitatively by smoke and then semiquantitatively using a tracer (Fig. 5.14). A series of release grids are designed that model the actual release area. An oil mist generator operating with corn oil is used to generate a submicron-sized aerosol. The generator is positioned to release the mist deep in the hood so that all mist is collected. The probe of a forward light

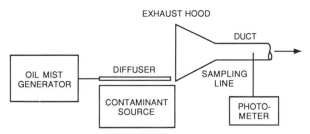

Figure 5.14 Containment fraction for a hood can be determined in the field using a tracer introduced first inside the hood to establish the 100 percent benchmark and then at the normal contaminant release point. [From Ellenbecker et al. (1983). Reproduced with permission of the *Journal of the American Industrial Hygiene Association.*]

scatter photometer samples the duct stream laden with the mist, and the reading is defined as 100 percent containment. The grid is then positioned at the actual work release point and a new reading is taken in the duct. If all the mist is collected, the reading will be 100 percent; if only one-half the oil mist is collected, the reading of containment is 50 percent.

This system and others now available using tracers such as sulfur hexafluoride are semiquantitative. They do not attempt to closely model the specific chemical or its generation rate. However, the approach provides insight into performance of specific systems and permits ''tune-up'' of operating parameters such as airflow rate, hood geometry, baffles, and so forth before the quantitative studies are conducted.

3.5.2.3.3 Quantitative. The containment efficiency of a local exhaust ventilation system is usually impossible to establish in a plant using the actual chemical. To conduct such a test the system must either be modified so that all the contaminant enters the hood to establish 100 percent containment or the fraction lost from the hood must be identified and that fraction added to the material collected by the hood. The fraction lost to the room usually cannot be measured.

In most cases, however, quantitative measurement of performance is made by air sampling on the worker. In this case the demonstrated performance is not specifically of hood containment but rather the ability of the local exhaust ventilation system to achieve control at the breathing zone of the worker. The most realistic test involves the operator using the actual chemical. Occasionally, as in the case of a premanu-facturing notification (PMN) chemical, it may be useful to demonstrate performance of the controls using a substitute material. In one such test a simulant chemical was used for the actual PMN chemical. The simulant was a nontoxic blue dye that was inexpensive, reasonably dusty, and easy to sample and analyze. A later test with the actual chemical confirmed the information obtained with the tracer.

3.5.3 Periodic Testing and Maintenance

The local exhaust ventilation system must be inspected periodically to ensure that the system meets performance and design standards and complies with various reg-ulations. The inspection frequency varies and may range from a monthly to a semiannual interval. The system is inspected visually. The motor-fan components are inspected to determine the condition of the fan blades, direction of rotation, belt tension, guarding, and lubrication. The hoods and ductwork are inspected for corrosion damage, plugging, leaks, and any local modification by the plant. The airflow rate at each station is determined, preferably by the hood static suction method since it is accurate and fast. On semi-enclosing hoods such as paint spray booths the face velocity must be evaluated. Certain regulations may require measurement of capture velocity, that is, the velocity at the contaminant release point. This measurement is difficult to conduct accurately. The measurement of airflow rate to achieve that velocity is the more accurate measurement.

The records management system for the inspection program is of great importance especially for facilities with hundreds of hoods. The need for such a system is described by Stott and Platts (1986, p. 145).

> The recording and manipulation of the necessary data is time-consuming, tedious and prone to inaccuracies. Most is recorded in note books which must be transcribed onto official forms which means that comparisons with previous measurements is often difficult. So at this stage, although a wealth of information is available regarding the extraction systems, the opportunity to analyze the data, initiate maintenance work, or diagnose faults is wasted.

These authors developed a computer-based system to satisfy these needs. The person conducting the field test uses a hand-held data terminal that has been loaded with a program on the ventilation system under test. Each ventilation station is named, the position to be tested is identified, and a space is provided for entry of the test results. In the plant the tester calls up the station and conducts the velocity or pressure test measurement. The data are inserted into the portable data logger. On return to the office the data logger is down loaded to a personal computer. The data are scanned, and those stations where a preset minimum value is not achieved are identified by the computer for follow-up. Additional measurements are requested, and based on these data a diagnosis of the ventilation system problem is identified by the computer for referral to plant engineering.

The importance of well-qualified personnel to design, install, operate, and inspect local exhaust ventilation systems has been discussed in detail (BOHS, 1987; Burton, 1989). It is especially important to develop an in-house training program for personnel who conduct the periodic inspections (Burgess, 1993; Johnson, 1993).

The maintenance problems encountered in system operation and hopefully identified by the acceptance and periodic inspection schedule have been discussed by Burton (1989). The cost of conducting an inspection program has not been published but in evaluating the cost–benefits of such a program, one should recall the installation cost of the system and its operating cost in addition to the important role it plays in worker health. In many locations the required inspection frequency has prompted the installation of direct-reading monitoring equipment such as pressure sensors at a hood static suction location, a pitot-static tube in the hood branch, a heated element sensor or a swinging vane anemometer at the hood face. In a nuclear energy facility an in-line orifice meter is installed to measure flow through a glove box. This mass flow measuring technique has limited application due to energy loss, erosion of the orifice, and contamination.

In discussing the performance of local exhaust ventilation, DallaValle (1952) stated: "Whereas hoods are installed to eliminate a health hazard, tests should be conducted to establish their effectiveness. The ultimate criterion is not the provision of a 'strong' suction but the handling of an air volume which reduces the concentration of the contaminants in question below the MAC level."

4 SUMMARY

In this chapter the advances in control technology in the past decade are reviewed. The ability to identify critical exposure conditions has been enhanced by the coupling of sophisticated, direct-reading air sampling instruments with video taping of the job tasks. This technique permits the investigator to characterize the nature and origin of emissions in great detail with the subsequent application of effective controls. A second major shift during the decade is the emphasis on the reduction in the use of toxic materials with replacement by materials "kinder" to the worker and the environment. Equipment advances during this period include robotic techniques that permit the worker to be separated from the point of greatest exposure. Finally the application of ventilation control is enhanced by improved knowledge of airflow patterns into hoods, new techniques for evaluating hood containment, and management tools for ventilation systems.

REFERENCES

ACGIH (1992). Committee on Industrial Ventilation, *Industrial Ventilation, A Manual of Recommended Practice*, 21 ed., American Conference of Governmental Industrial Hygienists, Cincinnati, OH.

AFS (1972). *Foundry Environmental Control*, American Foundrymen's Society, Des Plaines, IL.

Ahearn, J., H. Fatkin, and W. Schwalm (1991). "Case Study: Polaroid Corporation's Systematic Approach to Waste Minimization," *Pollution Prevent. Rev.*, 1(13), Summer.

AISI (1965). *Steel Mill Ventilation*, American Iron and Steel Institute, New York.

Anderson, D. M. (1964). "Dust Control by Air Induction Technique," *Ind. Med. Surg.*, **34**, 168.

ANSI (1988). "Precautionary Labeling of Hazardous Industrial Chemicals," ANSI Z129.1-1988, American National Standards Institute, New York.

Berens, A. R. (1981). "Vinyl Chloride Monomer in PVC: From Problem to Probe," *Pure Appl. Chem.*, **53**, 365–375.

BCIRA (1977). *Proceedings of the Working Environment in Iron Foundries*, British Cast Iron Research Association, Birmingham, UK.

BOHS (1984). British Occupational Hygiene Society, *Fugitive Emissions of Vapors from Process Equipment*, Technical Guide No. 3, Science Reviews Ltd., Northwood, Middlesex, UK.

BOHS (1985). British Occupational Hygiene Society, *Dustiness Estimation Methods for Dry Materials, Their Uses and Standardization*, Technical Guide No. 4, Science Reviews Ltd., Northwood, Middlesex, UK.

BOHS (1987). British Occupational Hygiene Society, *Controlling Airborne Contaminants in the Workplace*, BOHS Technical Guide No. 7, Science Reviews Ltd., Northwood, Middlesex, UK.

Brandt, A. D. (1947). *Industrial Health Engineering*, Wiley, New York.

Braun, P. and E. Druckman, Eds. (1976). ''Public Health Rounds at the Harvard School of Public Health—Vinyl Chloride: Can the Worker Be Protected,'' *N.E. J. Med.*, **294**, 653–657.

Burgess, W. A. (1993). ''The International Ventilation Symposia and the Practitioner in Industry,'' *Ventilation '91, Proceedings of the Third International Symposium on Ventilation for Contaminant Control*, September 16–20, 1991, ACGIH, Cincinnati, OH.

Burgess, W. A., M. J. Ellenbecker, and R. D. Treitman (1989). *Ventilation for Control of the Work Environment*, Wiley, New York.

Burton, D. J. (1989). *Industrial Ventilation Work Book*, DJBA, Inc., Salt Lake City, UT.

Caplan, K. (1985a). ''Philosophy and Management of Engineering Controls,'' in L. J. Cralley and L. V. Cralley, Eds., *Patty's Industrial Hygiene & Toxicology*, Wiley, New York.

Caplan, K. C. (1985b). ''Building Types,'' in L. V. Cralley, L. J. Cralley, and K. C. Caplan, Eds., *Industrial Hygiene Aspects of Plant Operations*, Vol. 3, McMillan, New York.

Cooper, T. and M. Gressel (1992). ''Real-time Evaluation at a Bag Emptying Operation—A Case Study,'' *Appl. Ind. Hyg.*, **7**(4), 227–230.

DallaValle, J. M. (1952). *Exhaust Hoods*, Industrial Press, New York.

Ellenbecker, M. J., R. J. Gempel, and W. A. Burgess (1983). ''Capture Efficiency of Local Exhaust Ventilation Systems,'' *Am. Ind. Hyg. J.*, **44**, 752–755.

EPA (1986). ''Workshop: Predicting Workplace Exposure to New Chemicals,'' *Appl. Ind. Hyg.*, **1**(3), R-11–R-13.

First, M. W. (1983). ''Engineering Control of Occupational Health Hazards,'' *Am. Ind. Hyg., Assoc. J.*, **44**(9), 621–626.

Gideon, J., E. Kennedy, D. O'Brien, and J. Talty (1979). ''Controlling Occupational Exposures—Principles and Practices,'' National Institute for Occupational Safety and Health, Cincinnati, OH.

Goldschmidt, G. (1991). ''Improvement of the Chemical Working Environment by Substitution of Harmful Substances. An Iterative Model and Tools,'' personal communication.

Goodfellow, H. D. and J. W. Smith (1989). ''Dustiness Testing—A New Design Approach for Dust Control,'' in J. H. Vincent, Ed., *Ventilation '88, Proceedings of the Second International Symposium on Ventilation for Contaminant Control*, Pergamon Press, Oxford, UK.

Gray, C. N. and P. J. Hewitt (1982). ''Control of Particulate Emissions from Electric-Arc Welding by Product Modification,'' *Ann. Occup. Hyg.*, **25**(4), 431–438.

Gressel, M. and T. Fischbach (1989). ''Workstation Design Improvements for the Reduction of Dust Exposures During Weighing of Chemical Powders,'' *Appl. Ind. Hyg.*, **4**(9), 227–233.

Gressel, M., W. A. Heitbrink, J. McGlothlin, and T. Fischbach (1987). ''Real-Time, Integrated, and Ergonomic Analysis During Manual Materials Handling,'' *Appl. Ind. Hyg.*, **2**(3), 108–113.

Gressel, M., W. Heitbrink, J. McGlothlin, and T. Fischbach (1988). "Advantages of Real Time Data Acquisition for Exposure Assessment," *Appl. Ind. Hyg.*, **3**(11), 316–320.

Hammond, C. M. (1980). "Dust Control Concepts in Chemical Handling and Weighing," *Ann. Occup. Hyg.*, **23**(1), 95–109.

Hanman, B. (1968). *Physical Abilities to Fit the Job*, American Mutual Liability Insurance Company, Wakefield, MA.

Harten, T. and I. Licis (1991). "Waste Reduction Technology Evaluations of the U.S. EPA WRITE Program," *J. Air Waste Manage. Assoc.*, **41**(8), 1122–1129.

HSE (1988). Health and Safety Executive, "Control of Substances Hazardous to Health Regulations 1988, Approved Codes of Practice," HMSO, London.

Johnson, G. Q., R. Ostendorf, D. Claugherty, and C. Combs (1993). "Improving Dust Control System Reliability," *Ventilation '91, Proceedings of the Third International Symposium on Ventilation for Contaminant Control*, September 16–20, 1991, ACGIH, Cincinnati, OH.

Karger, E. and W. A. Burgess (1992). Personal communication.

Kireta, A. G. (1988). "Lead Solder Update," *Heat./Pip./Air Cond.*, 119–125.

Lipton, S. and J. Lynch (1987). *Health Hazard Control in the Chemical Process Industry*, Wiley, New York.

Massachusetts General Laws (1989). "Massachusetts Toxics Use Reduction Act," Chapter 211, July 24, 1989.

Mertens, J. A. (1991). "CFCs: In Search of a Clean Solution," *Envir. Prot.*, 25–29.

NIOSH (1991). Technical Report, *Control of Emissions from Seals and Fittings in Chemical Process Industries*, National Institute for Occupational Safety and Health, DHHS (NIOSH) Publication No. 81-118, Cincinnati, OH.

O'Brien, D., T. Fischbach, T. Cooper, W. Todd, M. Gressel, and K. Martinez (1989). "Acquisition and Spreadsheet Analysis of Real Time Dust Exposure Data: A Case Study," *Appl. Ind. Hyg.*, **4**(9), 238–243.

OTA (1986). Office of Technology Assessment, Serious Reduction of Hazardous Waste for Pollution and Industrial Efficiency, Washington, D.C., U.S. Government Printing Office.

Peterson, J. (1985). "Selection and Arrangement of Process Equipment," in L. V. Cralley, L. J. Cralley, and K. C. Caplan, Eds., *Industrial Hygiene Aspects of Plant Operations*, Vol. 3, McMillan, New York.

Roach, S. A. (1981). "On the Role of Turbulent Diffusion in Ventilation," *Ann. Occup. Hyg.*, **24**(1), 105–133.

Rossi, M., K. Geiser, and M. Ellenbecker (1991). "Techniques in Toxic Use Reduction: From Concept to Action," *New Solutions*, **2**(2), 25–32.

SACMA (1991). Suppliers of Advanced Composite Materials Association, "Safe Handling of Advanced Composite Materials," Arlington, VA.

Schumacher, J. S. (1978). "A New Dust Control System for Foundries," *Am. Ind. Hyg. Ass. J.*, **39**(1), 73–78.

Sciola, V. (1993). "The Practical Application of Reduced Flow Push Pull Plating Tank Exhaust Systems," *Ventilation '91, Proceedings of the Third International Symposium for Contaminant Control*, September 16–20, 1991, Cincinnati, OH.

Sherwood, R. J. and R. J. Alsbury (1983). "Occupational Hygiene, Systematic Approach and Strategy of," in L. Parmeggiani, Ed., *Encyclopaedia of Occupational Health and Safety*, International Labor Organization, Geneva.

Soule, R. D. (1991). "Industrial Hygiene Engineering Controls," in G. Clayton and F. E. Clayton, Eds., *Patty's Industrial Hygiene and Toxicology*, Vol. 1, Part B., Wiley, New York.

Stott, M. D. and P. J. Platts (1986). "The Ventdata Ventilation Plant Monitoring and Maintenance System," in H. D. Goodfellow, Ed., *Ventilation '85 Proceedings of the First International Symposium for Contaminant Control*, Elsevier, Amsterdam.

Treitman, R. L. (1976). Personal communication.

U.S. Congress (1990). Pollution Prevention Act of 1990, *Congressional Record*, Section 6606 of the Budget Reconciliation Act, p. 12517, October 26, 1990.

Whitehead, L. W. (1987). "Planning Considerations for Industrial Plants Emphasizing Occupational and Environmental Health and Safety Issues," *Appl. Ind. Hyg.*, 2(2), 79–86.

Wolf, K., A. Yazdani, and P. Yates (1991). "Chlorinated Solvents: Will the Alternatives be Safer?" *J. Air Waste Manage. Assoc.*, **41**(8), 1055–1061.

Wolfson, H. (1993). "Is the Process Fit to Have Ventilation Applied?" in *Proceedings of the Third International Symposium on Ventilation for Contaminant Control*, September 16–20, 1991, Cincinnati, ACGIH.

APPENDIX A: NIOSH CONTROL TECHNOLOGY REPORTS*

RN: 00192317
TI: Engineering Health Hazard Control Technology for Coal Gasification and Liquefaction Processes. Final Report
AU: Anonymous
SO: Division of Physical Sciences and Engineering, NIOSH, U.S. Department of Health and Human Services, Cincinnati, Ohio, Contract No. 210-78-0084, 105 pages, 78 references
PY: 1983

RN: 00092812
TI: Control Technology Assessment: The Secondary Nonferrous Smelting Industry
AU: Burton-DJ; Coleman-RT; Coltharp-WM; Hoover-JR; Vandervort-R
SO: Division of Physical Science and Engineering, NIOSH, Cincinnati, Ohio, NIOSH Contract No. 210-77-0008, 393 pages
PY: 1979

RN: 00092805
TI: An Evaluation of Occupational Health Hazard Control Technology for the Foundry Industry
AU: Scholz-RC
SO: Division of Physical Sciences and Engineering, NIOSH, Cincinnati, Ohio, DHEW (NIOSH) Publication No. 79-114, Contract No. 210-77-0009, 436 pages, 56 references
PY: 1978

Source: Courtesy of National Institute for Occupational Safety and Health.

RN: 00145199
TI: Engineering and Other Health Hazard Controls in Oral Contraceptive Tablet
Making Operations
AU: Anastas-MY
SO: Division of Physical Sciences and Engineering, NIOSH, U.S. Department of
Health and Human Services, Cincinnati, Ohio, 94 pages
PY: 1984

RN: 00080675
TI: Development of an Engineering Control Research and Development Plan for
Carcinogenic Materials
AU: Hickey-JLS; JJ-Kearney
SO: Applied Ecology Department, Research Triangle Institute, Research Triangle
Park, North Carolina NIOSH Contract No. 210-76-0147, 168 pages, 19 references
PY: 1977

RN: 00074602
TI: Engineering Control of Welding Fumes
AU: Astleford-W
SO: Division of Laboratories and Criteria Development, NIOSH, Cincinnati, Ohio,
DHEW (NIOSH) Publication No. 75-115, Contract No. 099-72-0076, 122 pages,
13 references
PY: 1974

RN: 00074249
TI: Engineering Control Research and Development Plan for Carcinogenic Materials
AU: Hickey-J; Kearney-JJ
SO: Division of Physical Sciences and Engineering, NIOSH, Cincinnati, Ohio,
Contract No. 210-76-0147, 167 pages, 19 references
PY: 1977

RN: 00052419
TI: Engineering Control Research Recommendations
AU: Hagoapian-JH; Bastress-EK
SO: Division of Physical Sciences and Engineering, NIOSH, Cincinnati, Ohio,
Contract No. 099-74-0033, 210 pages, 181 references
PY: 1976

RN: 00182292
TI: Control Technology Assessment of Enzyme Fermentation Processes
AU: Martinez-KF; Sheehy-JW; Jones-JH
SO: Division of Physical Sciences and Engineering, NIOSH, U.S. Department of
Health and Human Services, Cincinnati, Ohio, DHHS (NIOSH) Publication No.
88-114, 81 pages, 34 references
PY: 1988

RN: 00177551
TI: Minimizing Worker Exposure during Solid Sampling: A Strategy for Effective Control Technology
AU: Wang-CCK
SO: Division of Physical Sciences and Engineering, NIOSH, U.S. Department of Health and Human Services, 28 pages
PY: 1983

RN: 00177548
TI: A 3-E Quantitative Decision Model of Toxic Substance Control through Control Technology Use in the Industrial Environment
AU: Wang-CCK
SO: NIOSH, U.S. Department of Health and Human Services, Cincinnati, Ohio, 53 pages, 4 references
PY: 1982

RN: 00168700
TI: The Illuminating Engineering Research Institute and Illumination Levels Currently Being Recommended in the United States
AU: Crouch-CL
SO: The Occupational Safety and Health Effects Associated with Reduced Levels of Illumination. Proceedings of a Symposium, July 11–12, 1974, Cincinnati, Ohio, NIOSH, Division of Laboratory and Criteria Development, HEW Publication No. (NIOSH). 75-142, pages 17–27
PY: 1975

RN: 00133474
TI: Control Technology Assessment of the Pesticides Manufacturing and Formulating Industry
AU: Fowler-DP
SO: Division of Physical Sciences and Engineering, NIOSH, Cincinnati, Ohio, Contract No. 210-77-0093, 667 pages
PY: 1980

RN: 00148166
TI: NIOSH Technical Report: Control Technology Assessment: Metal Plating and Cleaning Operations
AU: Sheehy-JW; Mortimer-VD; Jones-JH; Spottswood-SE
SO: NIOSH, U.S. Department of Health and Human Services, Cincinnati, Ohio, Publication No. 85-102, 115 pages, 71 references
PY: 1984

RN: 00144935
TI: A Study of Coal Liquefaction and Gasification Plants: An Industrial Hygiene Assessment, a Control Technology Assessment, and the Development of Sampling and Analytical Techniques. Volume II

AU: Cubit-DA; Tanita-RK
SO: Division of Respiratory Disease Studies, NIOSH, U.S. Department of Health
and Human Services, Morgantown, West Virginia, Contract No. 210-78-0101, 179
pages, 15 references
PY: 1983

RN: 00144934
TI: A Study of Coal Liquefaction and Gasification Plants: An Industrial Hygiene
Assessment, a Control Technology Assessment, and the Development of Sampling
and Analytical Techniques: Volume I
AU: Cubit-DA; Tanita-RK
SO: Division of Respiratory Disease Studies, NIOSH, U.S. Department of Health
and Human Services, Morgantown, West Virginia, Contract No. 210-78-0101, 192
pages, 22 references
PY: 1983

RN: 00136362
TI: Control Technology Assessment of Selected Petroleum Refinery Operations
AU: Emmel-TE; Lee-BB; Simonson-AV
SO: Division of Physical Sciences and Engineering, NIOSH, U.S. Department of
Health and Human Services, Cincinnati, Ohio, NTIS PB83-257-436, Contract No.
210-81-7102, 122 pages
PY: 1983

RN: 00136196
TI: Proceedings of the Second Engineering Control Technology Workshop, June
1981
AU: Konzen-RB
SO: Division of Training and Manpower Development, NIOSH, U.S. Department
of Health and Human Services, Cincinnati, Ohio, NTIS PB-83-112-755, NIOSH
Report No. 80-3794, 138 pages
PY: 1982

RN: 00133178
TI: Principles of Occupational Safety and Health Engineering, Instructor's Guide
AU: Zimmerman-NJ
SO: Division of Training and Manpower Development, NIOSH, U.S. Department
of Health and Human Services, Cincinnati, Ohio, P.O. No. 81-3030, 262 pages, 17
references
PY: 1983

RN: 00135180
TI: Health Hazard Control Technology Assessment of the Silica Flour Milling In-
dustry
AU: Caplan-PE; Reed-LD; Amendola-AA; Cooper-TC

SO: Division of Physical Sciences and Engineering, NIOSH, U.S. Department of Health and Human Services, Cincinnati, Ohio, 60 pages, 18 references
PY: 1982

RN: 00135171
TI: Engineering Noise Control Technology Demonstration for the Furniture Manufacturing Industry
AU: Hart-FD; Stewart-JS
SO: NIOSH, U.S. Department of Health, Education, and Welfare, Grant No. 1-RO1-OH-00953, 123 pages, 8 references
PY: 1982

RN: 00132081
TI: Control Technology Assessment for Chemical Processes Unit Operations
AU: Van-Wagenen-H
SO: NIOSH, Cincinnati, Ohio, Contract No. 210-80-0071, NTIS PB83-187-492, 19 pages
PY: 1983

RN: 00134239
TI: Engineering Control of Occupational Health Hazards in the Foundry Industry. Instructor's Guide
AU: Scholz-RC
SO: NIOSH, Cincinnati, Ohio, NTIS PB82-231-234, 156 pages, 49 references
PY: 1981

RN: 00133884
TI: Demonstrations of Control Technology for Secondary Lead Reprocessing
AU: Burton-DJ; Simonson-AV; Emmel-BB; Hunt-DB
SO: NIOSH, U.S. Department of Health and Human Services, Rockville, Maryland, Contract No. 210-81-7106, 291 pages
PY: 1983

RN: 00132005
TI: Pilot Control Technology Assessment of Chemical Reprocessing and Reclaiming Facilities
AU: Crandell-MS
SO: Engineering Control Technology Branch, NIOSH, Cincinnati, Ohio, NTIS PB83-197-806, 11 pages
PY: 1982

RN: 00130228
TI: Occupational Health Control Technology for the Primary Aluminum Industry
AU: Sheehy-JW

SO: Public Health Service, NIOSH, U.S. Department of Health and Human Services, Cincinnati, Ohio, DHHS Publication No. 83-115, 59 pages, 6 references
PY: 1983

RN: 00130213
TI: Control Technology Assessment in the Pulp and Paper Industry
AU: Schoultz-K; Matthews-R; Yee-J; Haner-H; Overbaugh-J; Turner-S; Kearney-J
SO: NIOSH, Public Health Service, U.S. Department of Health and Human Services, Cincinnati, Ohio, Contract No. 210-79-0008, 974 pages
PY: 1983

RN: 00106373
TI: Mechanical Power Press Safety Engineering Guide, Wilco, Inc., Stillwater, Minnesota
AU: Anonymous
SO: NIOSH, Division of Laboratories and Criteria Development, U.S. Department of H.E.W., Cincinnati, Ohio, 207 pages
PY: 1976

RN: 00123450
TI: Assessment of Engineering Control Monitoring Equipment. Volume II
AU: Anonymous
SO: Enviro Control, Inc., Rockville, Md., NIOSH, Cincinnati, Ohio, 393 pages, 23 references
PY: 1981

RN: 00123377
TI: Control Technology Assessment of Selected Process in the Textile Finishing Industry
AU: Collins-LH
SO: Bendix Launch Support Division, Cocoa Beach, Florida, NIOSH, Cincinnati, Ohio, 227 pages, 44 references
PY: 1978

RN: 00119991
TI: Phase I Report on Control Technology Assessment of Ore Beneficiation
AU: Todd-WF
SO: NIOSH, U.S. Department of Health and Human Services, Cincinnati, Ohio, 82 pages, 80 references
PY: 1980

RN: 00117613
TI: Symposium Proceedings, Control Technology in the Plastics and Resins Industry
AU: Anonymous

SO: NIOSH, U.S. Department of Health and Human Services, Cincinnati, Ohio, 333 pages, 23 references
PY: 1981

RN: 00117359
TI: Proceedings of the Symposium on Occupational Health Hazard Control Technology in the Foundry and Secondary Non-Ferrous Smelting Industries
AU: Anonymous
SO: NIOSH, U.S. Department of Health and Human Services, 401 pages, 45 references
PY: 1981

RN: 00116023
TI: Control Technology Summary Report on the Primary Nonferrous Metals Industry, Vol. IV, Appendix C
AU: Hoover-JR
SO: NIOSH, Center for Disease Control, Public Health Service, U.S. Department of Health, Education, and Welfare, 120 pages
PY: 1978

RN: 00116022
TI: Control Technology Summary Report on the Primary Nonferrous Metals Industry, Vol. 5, Appendix D
AU: Hoover-JR
SO: NIOSH, Center for Disease Control, Public Health Service, U.S. Department of Health, Education, and Welfare, 180 pages
PY: 1978

RN: 00115792
TI: Control Technology for Primary Aluminum Processing
AU: Sheehy-JW
SO: Department of Health and Human Services, Public Health Service, Center for Disease Control, NIOSH, Division of Physical Sciences and Engineering, Cincinnati, Ohio, pages 1–22
PY: 1980

RN: 00115530
TI: An Evaluation of Engineering Control Technology for Spray Painting
AU: O'Brien-DM; Hurley-DE
SO: NIOSH, Center for Disease Control, Public Health Service, U.S. Department of Health and Human Services, 117 pages, 52 references
PY: 1981

RN: 00114224
TI: Control Technology Summary Report on the Primary Nonferrous Metals Industry, Vol. II, Appendix A

AU: Coleman-RT; Hoover-JR
SO: Division of Physical Science and Engineering, NIOSH, 291 pages, 10 references
PY: 1978

RN: 00114223
TI: Control Technology Summary Report on the Primary Nonferrous Metals Industry, Vol. III, Appendix B
AU: Coleman-RT
SO: Division of Physical Science and Engineering, NIOSH, Cincinnati, Ohio, 102 pages
PY: 1978

RN: 00113373
TI: Control Technology Summary Report on the Primary Nonferrous Smelting Industry, Volume 1: Executive Summary
AU: Coleman-RT; Hoover-JR
SO: Division of Physical Science and Engineering, NIOSH, Cincinnati, Ohio, Contract No. 210-77-0008, Radian Corporation, Austin, Texas, 36 pages, 8 references
PY: 1978

RN: 00112367
TI: Control Technology Assessment of Raw Cotton Processing Operations
AU: Anonymous
SO: Envirocontrol Inc., Rockville, Md., NIOSH, Cincinnati, Ohio, Contract No. 210-78-0001, 351 pages
PY: 1980

RN: 00112366
TI: Engineering Control Technology Assessment for the Plastics and Resins Industry
AU: Anonymous
SO: Division of Physical Sciences, NIOSH, U.S. Department of Health, Education and Welfare, Contract No. 210-76-0122, Cincinnati, Ohio, 234 pages, 60 references
PY: 1977

RN: 00102819
TI: Proceedings of the Symposium on Occupational Health Hazard Control Technology In the Foundry and Secondary Non-Ferrous Smelting Industries
AU: Scholz-RC; Leazer-LD
SO: Department of Health, Education, and Welfare, Public Health Service Center for Disease Control, NIOSH, Cincinnati, Ohio, 447 pages, 10 references
PY: 1980

RN: 00094260
TI: Assessment of Selected Control Technology Techniques for Welding Fumes

AU: Van-Wagenen-HD
SO: Division of Physical Sciences and Engineering, NIOSH, Cincinnati, Ohio, NIOSH Publication No. 79-125, 29 pages, 17 references
PY: 1979

RN: 00094228
TI: Control Technology for Worker Exposure to Coke Oven Emissions
AU: Sheehy-JW
SO: Division of Physical Sciences and Engineering, NIOSH, Cincinnati, Ohio, NIOSH Publication No. 80-114, 29 pages, 26 references
PY: 1980

RN: 00092888
TI: Proceedings of NIOSH/University Occupational Health Engineering Control Technology Workshop
AU: Talty-JT
SO: Proceedings of the Workshop on Occupational Health Engineering Control Technology, May 16–17, 1979, Division of Physical Sciences and Engineering, NIOSH, Cincinnati, Ohio, 149 pages
PY: 1979

RN: 00092810
TI: Control Technology Summary Report on the Primary Nonferrous Metals Industry. Volume V. Appendix D: Review of the Testimony Presented at the 1977 OSHA Public Hearing on Sulfur Dioxide
AU: Hoover-JR
SO: Division of Physical Science and Engineering, NIOSH, Cincinnati, Ohio, NIOSH Contract No. 210-77-0008, 183 pages
PY: 1978

RN: 00092809
TI: Control Technology Summary Report on the Primary Nonferrous Metals Industry. Volume IV. Appendix C: Review of the Testimony Presented at the 1977 OSHA Public Hearings on Inorganic Lead
AU: Hoover-JR
SO: Division of Physical Science and Engineering, NIOSH, Cincinnati, Ohio, NIOSH Contract No. 210-77-0008, 123 pages
PY: 1978

RN: 00092808
TI: Control Technology Summary Report on the Primary Nonferrous Metals Industry. Volume III. Appendix B: Review of the Testimony Presented at the 1975/6 OSHA Public Hearing on Inorganic Arsenic
AU: Coleman-RT
SO: Division of Physical Science and Engineering, NIOSH, Cincinnati, Ohio,

NIOSH Contract No. 210-77-0008, 105 pages
PY: 1978

RN: 00092807
TI: Control Technology Summary Report on the Primary Nonferrous Metals Industry. Volume II: Appendix A
AU: Coleman-RT; Hoover-JR
SO: Division of Physical Science and Engineering, NIOSH, Cincinnati, Ohio, NIOSH Contract No. 210-77-0008, 298 pages, 9 references
PY: 1978

The following reports are not yet listed in NIOSHTIC:
Control Technology for Ethylene Oxide Sterilization in Hospitals, V. D. Mortimer and S. L. Kercher, Division of Physical Science and Engineering, NIOSH, Cincinnati, Ohio, Pub No. 89 -120, 167 pages, 82 references, 1989

Control of Asbestos Exposure During Brake Drum Service, J. W. Sheey, T. C. Cooper, D. M. O'Brien, J. D. McGlothlin, and P. A. Froelich, Division of Physical Science and Engineering, NIOSH, Cincinnati, Ohio, Pub No. 89 -121, 69 pages, 48 references, 1989.

CHAPTER SIX

Personal Protective Equipment

Bruce D. Reinert, CIH

1 INTRODUCTION

"I hate respirators!" A respirator researcher made this statement after a meeting of an ANSI Ad Hoc Committee on Respirator Certification. When questioned why someone who had spent a good deal of his career doing respirator research all of a sudden appeared to hate respirators, he was quick to explain. He said he did not like respirators because they were uncomfortable, did not provide absolute protection, required training, were dependent on employees using them properly, degraded a worker's performance, required a maintenance program, were expensive, and so forth. All of these complaints made him wish we could just eliminate the hazards. However, eliminating hazards in the workplace is not always an option, so we look to engineering controls as the next best option, leaving personal protective equipment such as respirators as the last choice.

Engineering controls—all health and safety professionals know that this is the first line of defense and all want them put in place. Unfortunately, it is not a perfect world and there are no perfect engineering controls. They also are expensive and do not always fit the ever-changing workplace. In some cases, engineering controls can, like personal protective equipment, degrade the performance of workers. There is no simple solution to all hazard control situations, and workplaces frequently require a mixture of control strategies to protect employees. Within the mixture of control strategies, personal protective equipment frequently shows up as the method of choice, particularly where the hazards can eventually be controlled by engineering

Patty's Industrial Hygiene and Toxicology, Third Edition, Volume 3, Part A, Edited by Robert L. Harris, Lewis J. Cralley, and Lester V. Cralley.
ISBN 0-471-53066-2 © 1994 John Wiley & Sons, Inc.

methods. For the purpose of this chapter, personal protective equipment includes respirators, hearing protectors, and gloves or suits used for skin protection.

Unfortunately, it is not always the case that personal protective equipment programs are put into place, work themselves through their appropriate lifetime, and then are removed. The personal protective equipment program may become self-preserving. The individuals administering the program do not go back and look at the problem of the hazard itself and redetermine if personal protective equipment continues to be the control method of choice. Once the personal protective equipment program is in place, it tends to stay in place indefinitely. The opportunity is rarely taken to look at whether the situation can now afford a good engineering control strategy or be modified to remove the hazard altogether. Personal protective equipment programs need to be operated with a philosophy that continuously looks at the reason for keeping such programs in place. In looking at this the administrator of the program must go back to the "mission" of health and safety professionals to protect workers. One can never forget that, in almost all instances, use of personal protective equipment exacts a toll from the worker and in some ways may create significant hazards.

In spite of the disadvantages of personal protective equipment, it does play a very important role in protecting workers. When properly used, personal protective equipment can be a good adjunct to the overall health and safety program. A personal protective equipment program will ensure that workers are protected by properly operating equipment and that any adverse impacts on the worker from the equipment are minimized. A good program will cover hazard evaluation, selection, fitting, training, medical evaluation, use, maintenance and program administration.

2 HAZARD EVALUATION

The primary approach to evaluating hazards requires the health and safety professional to thoroughly understand the industrial process involved. Once the process is understood, the best approach can be taken to selecting the appropriate controls. As is frequently mentioned in industrial hygiene textbooks, eliminating the hazard should be the primary objective. Second, one can select engineering and/or administrative controls, and only then use personal protective equipment as a last choice. The health and safety professional must ask a few simple questions before choosing a particular control strategy: Is the hazard a long-term or short-term hazard? How often are the employees at risk? What do the regulations say? Is there more than one type of hazard that would require skin, eye, and respiratory protection? Is noise a concern? There are a significant number of references such as textbooks, Occupational Safety and Health Administration (OSHA) regulations, material safety data sheets, American National Standards Institute (ANSI) standards, and so forth that can be used to help evaluate the hazard.

One concern that is not always looked at by the health and safety professional is the tendency of many individuals to be overly conservative when prescribing personal protective equipment. Frequently an individual will in his or her own mind say "let's err on the safe side." Unfortunately, use of personal protective equipment is not necessarily "on the safe side." Simply putting a respirator on someone does not necessarily protect the individual. One must always ask why we must use personal protective equipment, and therefore what are the hazards involved in using it. A risk–benefit analysis is appropriate in choosing personal protective equipment as well as in determining if any controls should be used at all.

The hazard evaluation must be conducted in a way that identifies all characteristics of the hazard that might influence the selection of the personal protective equipment. All chemical or physical characteristics of the hazard should be reviewed to see how they might affect equipment selection. Workplace conditions that might influence choice of equipment must be reviewed. All such related factors must be identified prior to equipment selection to ensure that the best personal protective equipment is used for the conditions expected.

Particular attention should be paid to combinations of hazards that might render the first choice of personal protective equipment ineffective. A device that may be chosen for a particular hazard may be less than effective when a second hazard is present. For example, the use of butyl rubber gloves is sufficient for skin protection from phenol. However, if toluene is also present, it could swell the matrix of the glove, allowing the phenol to permeate through to the skin. The possible interactions between hazards may require an in-depth analysis before final selection of the personal protective equipment.

3 SELECTION

Function and performance are extremely important to the selection of personal protective equipment. Regulatory requirements are also important since frequently OSHA, National Institute for Occupational Safety and Health (NIOSH), or ANSI regulations/standards will prescribe a particular type of personal protective equipment for a given hazard. It is rare, however, that a guide to selection will take into account the worker's preference. If a worker does not like a particular brand of respirator or ear plug, he or she will not use it. The health and safety professional is faced with the problem of deciding what provides the best protection while taking into account the degradation in performance provided by the personal protective equipment, the comfort to the worker, the interaction of combinations of different types of personal protective equipment such as eye protection and half facepiece respirators, the ease of use of the personal protective equipment, and how it will affect performance of the job, as well as the protection of the worker. Significant impairment in performance may result in a second look at engineering controls or the value of eliminating the hazard altogether. It might be a good opportunity for the health and safety professional to ask himself or herself if use of the personal protective equipment would indeed be something that would provide proper protection or would it be improperly used and therefore the actual protection afforded be reduced.

Does the personal protective equipment really protect the worker? Frequently what is thought of as the best level of protection may not in fact be best in all situations. There is a tendency in respiratory protection to look at self-contained breathing apparatus (SCBAs) as the ultimate. They are used for the most dangerous situations; therefore, how could one go wrong by choosing a SCBA? Unfortunately, SCBAs have limitations like all personal protective equipment. For example, the use of a SCBA in a low atmospheric pressure situation may provide no protection. If the only hazard to the worker is the low partial pressure of oxygen due to the low total atmospheric pressure, the SCBA will be delivering air at that reduced pressure and therefore providing no more oxygen than the ambient air. There is also a hazard associated with the increased weight of the SCBA. One must always ask the right questions. Why are we selecting this device? What does it do to protect the worker? How does it stress the worker? Does it appropriately protect against the hazard involved?

4 FITTING

The fit of personal protective equipment influences the performance of the equipment and the performance of the employee doing the job. Fit is more important to the performance of ear plugs and respirators, for example, than it is with protective clothing, where improper fit is more likely to be detected by the employee. Protective clothing that has a restricted fit can adversely influence the ability of the employee to move about while overly large protective clothing can result in danger if it catches on moving parts of machinery. In short, if workers are restricted by personal protective equipment or if it annoys, irritates, or adversely affects the employee's ability to do a job, then he or she will be reluctant to wear, and in many cases may not wear, the protective equipment. It is therefore important to ensure that fitting programs properly select personal protective equipment that will achieve the chosen goal while being comfortable.

Specific standards such as OSHA 1910.134 (1) and ANSI Z88.2 (2) require fitting programs for respirators. All applicable standards and regulations must be reviewed to assure that the fitting program is the best available to do the job.

The best approach to establishing proper fitting programs is to (1) look at requirements provided by standards; (2) determine the influence that fit has on performance of the equipment, use of the equipment, and performance of the employee; and (3) remember that fit is in itself just that, it does not guarantee good performance in the workplace. The performance of the personal protective equipment in the workplace must be covered by other parts of the overall personal protective equipment program such as training, medical, and so forth discussed below.

5 TRAINING

While training programs must be tailored to the type of personal protective equipment being used, there are a variety of elements that apply in general to all personal protective equipment training programs. These include:

1. The need for the personal protective equipment, why the employee must use it, and the benefits derived from using the personal protective equipment.

2. The nature, extent, and effects of the hazards in the workplace. Workers can adopt the correct attitude toward the program only if they fully understand what the problem is that necessitates the wearing of personal protective equipment.

3. The worker must learn that it is important to let his or her supervisor know of any problems experienced while using the personal protective equipment.

4. The worker must understand why engineering controls are not an effective option for controlling the hazard.

5. When there are a variety of types available, why that particular type of personal protective equipment was selected.

6. An explanation of the operating capabilities and limitations of the personal protective equipment as related to protection of the worker as well as performance of the worker's job.

7. Instructions for inspecting, wearing, maintaining, cleaning, assuring fit, and so forth to determine if the personal protective equipment will operate properly.

8. Explanation of what type of fitting will be done to assure proper selection for the hazard used.

9. Instructions on how to deal with emergencies during use of personal protective equipment. The worker must understand that there are certain situations in which the failure of the equipment necessitates immediate withdrawal from the workplace.

10. The worker must understand the regulations that govern the use of personal protective equipment as well as the workers responsibilities for always using personal protective equipment when the work conditions require it.

6 MEDICAL EVALUATION

The OSHA Safety and Health Standard 29 CFR 1910.134, Respiratory Protection, requires that "Persons should not be assigned to tasks requiring use of respirators unless it has been determined that they are physically able to perform the work and use the equipment. The local physician shall determine what health and physical conditions are pertinent. The respirator user's medical status should be reviewed periodically (for instance, annually)" (1). The description provided here by OSHA can apply equally to all personal protective equipment programs. Some evaluation must be performed to assure the physical capability of the employee to perform work activities while wearing personal protective equipment. The physician must consider a variety of factors including the worker's health, both physical and psychological; the type of personal protective equipment that is going to be used; and the conditions of use. The list of conditions includes environmental conditions (including airborne health hazards), the duration of use, how the equipment may or may not be used in an emergency, and any other special conditions that would influence the effect the personal protective equipment has on the wearer. It is important also that the physician be apprised of any combinations of personal protective equipment that will be used such as protective suits with respirators and gloves, or respirators, goggles, and ear protection.

The medical evaluation should include a medical history, a medical examination, and any diagnostic procedures that can determine the ability of the worker to use the personal protective equipment. A program should be established to periodically review the worker's medical conditions to assure that nothing has occurred that would degrade his or her ability to use personal protective equipment. It is incumbent upon the health and safety professional to provide the physician with sufficient information about conditions of work to allow him or her to adequately evaluate whether or not the employee is physically capable of using the personal protective equipment. To this end, information on workplace monitoring, job description, emergency conditions, and so on should be provided to the physician before a medical evaluation is performed.

A variety of standards such as those proposed by OSHA as well as ANSI standards such as Z88.2, Respiratory Protection Programs (2) and Z88.6, Standard for Respirator Use-Physical Qualifications for Personnel (3) can be reviewed to help establish the appropriate medical evaluation program. The primary function of the program is to assure that the worker can perform his or her job without any adverse effect on himself or herself or co-workers caused by use of the personal protective equipment.

7 USE

The use of the personal protective equipment is probably the most important part of the personal protective equipment program. There are conditions where improper use of personal protective equipment can be worse than no use. This occurs primarily when the employee gains a false sense of security from the personal protective equipment. There are many hazards for which personal protective equipment is used where the employee is unable to determine personally whether or not the environment is hazardous. It is in these situations where an employee can be put in jeopardy if the equipment is not properly used.

The manager of the program should review use of the personal protective equipment. Particular emphasis should be placed on proper wearing of the equipment, user feedback, how the employee determines when maintenance is required, employee inspection of the equipment, and supervisory review of the employee's performance during use of personal protective equipment. The first line supervisor plays the most important role in review of use. This individual is in the best position to determine if the equipment is being used properly and to ensure that the equipment is not having any adverse effect on the employee. A strong training program can best prepare the supervisor for this important role.

8 MAINTENANCE

The maintenance program that is set up for personal protective equipment can significantly affect the performance of respiratory protective equipment or protective suits and/or gloves. Hearing protection maintenance programs are limited if used at all, and the impact of proper maintenance is less severe with hearing protection than it is with other types of personal protective equipment. Maintenance programs for personal protective equipment generally consist of a set schedule of routine maintenance activities that when properly conducted help ensure the protection afforded by this equipment. They also provide an opportunity to check the performance of the equipment through quality checks and assure that the performance has not degraded with use.

Proper maintenance programs include a set schedule for maintenance as well as unscheduled maintenance when significant problems occur. The maintenance schedule must be determined based on the use of the personal protective equipment. For instance, the periodic maintenance of respirators that are used daily should be more frequent than that of respirators that are used only on rare occasions or respiratory protective equipment that is used only for emergency operations. The health and safety professional in the organization must set up the schedule based on best judgment and previous experience with the personal protective equipment to ensure that the proper concerns are addressed in relation to degradation of the equipment during routine or emergency use. It is important that personal protective equipment be refurbished on a scheduled basis even if the equipment has not been used. This is important since the equipment may degrade in storage. In the case of suits or respirators, storage that causes folding or bending could result in degradation of performance. For example, the rubber in a respirator may take a set and result in improper fit when finally used. The manufacturer of the particular type of personal protective equipment can be a good source of information on the proper maintenance schedule.

One of the basic concerns in determining what type of maintenance program to establish is the cost of either maintaining or replacing personal protective equipment after use. The individual responsible for the maintenance program should conduct a cost–benefit analysis based on expected use, cost and frequency of maintenance, cost of stocking appropriate parts for the personal protective equipment, and numbers of equipment in use at the facility. An example of personal protective equipment where replacement is probably less costly than maintenance would be hearing protectors. Most hearing protectors can be replaced at lower cost than they can be repaired. However, in the case of earmuff-type protectors there are some, particularly those that include communication devices, where a maintenance program would be cost effective.

In determining whether personal protective equipment will be reused, it is strongly recommended that a determination be made of the hazards of reuse. An example of hazardous reuse would be a maintenance program where protective gloves are decontaminated and then reused. Frequently, with gloves and protective suits, use of a solvent for removal of material from the surface will cause some contamination to be driven into the matrix of the fabric. This can result in future migration of the hazardous material to the inner surface of the glove or suit and unexpected contamination of the wearer. Before a maintenance program for suits or gloves is established, a very thorough evaluation of the problem of cleaning must be made. No general guidance can be given in this area because of the wide variety of chemicals that are used in occupational situations. Each use of personal protective equipment such as gloves and suits must be evaluated and a determination made on the cost as well as hazard of maintenance and reuse.

One of the determinations that must be made in establishing a maintenance program is whether performance testing is required after maintenance. With certain personal protective equipment such as respirators, performance testing after maintenance is very important. Since respirator maintenance programs require disassembly and reassembly of the respirators, a performance test to assure proper reassembly and performance is important. However, there is no simple performance test for gloves, suits, or hearing protectors. In these areas, the maintenance program itself must undergo strict evaluation to assure that the glove, suit, or hearing protector can perform properly after completion of maintenance.

The impact of failure of the personal protective equipment determines how stringent the maintenance and performance testing program is. A program where respirators are used in environments that are immediately dangerous to life or health must have a significantly more stringent maintenance/performance program than one where respirators are used only for nuisance dust. No standard program can be established for any type of personal protective equipment. As in all other areas of the overall personal protective equipment program, the maintenance program must be tailored to the overall situation encountered at the particular facility.

9 PROGRAM ADMINISTRATION

When establishing a program for personal protective equipment, it is important to keep in mind that while record keeping must be done, it cannot drive the program. The program administrator must assure that the program is directed toward protecting workers. It cannot be emphasized enough that this must be the driving reason for using personal protective equipment. Although regulatory requirements may result in very lengthy and complex record-keeping requirements, the primary function of program administration is to ensure that the personal protective equipment is properly used to protect the worker.

The program administrator must be properly trained and knowledgeable to direct the program. He or she must review the program to ensure compliance with proper standard operating procedures and/or regulatory requirements. It is also important to keep abreast of new technology that may be available to improve the program.

The program administrator must review all aspects of the program as well as the basic question: "Why are we doing this?" Particular emphasis, if possible, should be placed on eliminating the need for personal protective equipment. The program administrator must routinely look at the possibility of changing to environmental controls to protect the worker.

If the program administrator conducts his or her job properly, personal protective equipment will be chosen, used, maintained, and eventually not needed as a means of protecting workers.

REFERENCES

1. Code of Federal Regulations, Title 29, Part 1910, Section 134, revised July 1, 1991.
2. American National Standards Institute, ANSI Z88.2 Practices for Respiratory Protection, (1992).
3. American National Standards Institute, ANSI Z88.6 — 1984 Respirator Use — Physical Qualifications for Personnel, 1984.

CHAPTER SEVEN

Occupational Exposure Limits, Pharmacokinetics, and Unusual Work Schedules

Dennis J. Paustenbach, Ph.D., CIH, DABT

1 INTRODUCTION

Over the past 40 years, many organizations in numerous countries have proposed occupational exposure limits (OEL) for airborne contaminants. The limits or guidelines that have gradually become the most widely accepted both in this country and abroad are those issued annually by the American Conference of Governmental Industrial Hygienists (ACGIH), which are termed threshold limit values (TLVs) (LaNier, 1984; ACGIH, 1992).

The usefulness of establishing exposure limits for potentially harmful agents in the working environment has been demonstrated repeatedly since their inception (Stokinger, 1970; Cook, 1986). It has been claimed that whenever these limits have been implemented in a particular industry, no worker has been shown to have sustained serious adverse effects on health as a result of exposure to TLV concentrations of an industrial chemical (Stokinger, 1981). Although it is arguable whether all the TLVs have provided ample protection against harm, there is little doubt that during the first 50 years of this century, tens of thousands of persons were adversely affected by workplace exposure to chemical agents (Hunter, 1978; OSHA, 1989) and that the adoption of exposure limits like the TLVs significantly reduced the incidence of occupational disease.

Patty's Industrial Hygiene and Toxicology, Third Edition, Volume 3, Part A, Edited by Robert L. Harris, Lewis J. Cralley, and Lester V. Cralley.
ISBN 0-471-53066-2 © 1994 John Wiley & Sons, Inc.

Since this chapter was last written (Paustenbach, 1985), there have been concerns raised by numerous persons regarding the adequacy or health protectiveness of OELs and, in particular, TLVs (Castleman and Ziem, 1988; Ziem and Castleman, 1989; Roach and Rappaport, 1990). The key question raised in these papers was "what percent of the working population is truly protected when exposed to the TLV?"

In the first of their two papers, Castleman and Ziem (1988) claimed that the TLVs were excessively influenced by corporations and, as a result, they suggested that the TLVs lacked objectivity. In addition, they indicated that the scientific documentation for many, if not most, of the TLVs was woefully inadequate. They concluded by suggesting that "an ongoing international effort is needed to develop scientifically based guidelines to replace the TLVs in a climate of openness and without manipulation by vested interests." In a subsequent editorial by Tarlau (1990), it was suggested that each industrial hygienist consider deriving his or her own occupational exposure limits and, where available, to use the OELs suggested by the National Institute of Occupational Safety and Health (NIOSH).

In their second paper, Ziem and Castleman (1989) further discussed their views about the inadequacies of the TLVs. To a large extent, this paper represented an expansion of their 1988 paper. They once again concluded that the TLVs were not derived with sufficient input from physicians and that many TLVs were simply not low enough to protect most workers. They believed that there was more than circumstantial evidence to show that there had been an excessive amount of industrial influence on the TLV committee and that this resulted in TLVs that were not sufficiently low to protect workers.

The response to these two papers by occupational physicians and industrial hygienists was significant (Finklea, 1988; Paustenbach, 1990c). Over the 12 months that followed, more than a dozen letters to the editor were published, and editorials appeared in *The Journal of Occupational Medicine*, *American Journal of Industrial Medicine*, and *The American Industrial Hygiene Association Journal* (AIHAJ). One editorial, written by Tarlau (1990), suggested that industrial hygienists would be better off not relying on the TLVs. This prompted a rather lengthy response by Paustenbach (1990c) wherein he discussed the historical benefits of the TLVs. He suggested that the papers criticizing the TLVs had merit but that the critics were applying the social expectations and scientific standards of 1990 on work that was often performed more than 30–40 years ago.

During 1988–1990, the claims that the TLVs were not well based in science were, to a large extent, subjective or anecdotal. Although Castleman and Ziem (1988) identified inconsistencies in the margin of safety inherent in various TLVs, alleged that companies had undue influence on the TLV committee, and that objective analyses had not been conducted, the significance of these claims with respect to whether workers were sufficiently protected at the TLV remained unclear. This remained the situation until two professors, one from the University of California at Berkeley and the other from England, published a rather lengthy paper that analyzed the scientific basis for a number of the TLVs (Roach and Rappaport, 1990). In this paper they showed that for many of the irritants and systemic toxicants, the TLVs were at or near a concentration where 10–50 percent of the population could be expected to experience some adverse effect. Although for many chemicals the adverse effect might be transient and not very significant, for example, temporary eye, nose, or throat irritation, these authors did offer substantial evidence that for some chemicals there was only a small margin of safety between the TLV concentration and concentrations that had been shown to cause some adverse effect in exposed animals or persons.

Roach and Rappaport summarized their work in this manner:

> Threshold Limit Values (TLVs) represent conditions under which the TLV Committee of the American Conference of Governmental Industrial Hygienists (ACGIH) believes that nearly all workers may be repeatedly exposed without adverse effect. A detailed research was made of the references in the 1976 *Documentation* to data on "industrial experience" and "experimental human studies." The references, sorted for those including both the incidence of adverse effects and the corresponding exposure, yielded 158 paired sets of data. Upon analysis it was found that, where the exposure was at or below the TLV, only a minority of studies showed no adverse effects (11 instances) and the remainder indicated that up to 100% of those exposed had been affected (8 instances of 100%). Although, the TLVs were poorly correlated with the incidence of adverse effects, a surprisingly strong correlation was found between the TLVs and the exposures reported in the corresponding studies cited in the *Documentation*. Upon repeating the search of references to human experience, at or below the TLVs, listed in the more recent 1986 edition of the *Documentation*, a very similar picture has emerged from the 72 sets of clear data which were found. Again, only a minority of studies showed no adverse effects and the TLVs were poorly correlated with the incidence of adverse effect and well correlated with the measured exposure. Finally, a careful analysis revealed that authors conclusions in the references (cited in the 1976 *Documentation*) regarding exposure-response relationships at or below the TLVs were generally found to be at odds with the conclusions of the TLV Committee. These findings suggest that those TLVs which are justified on the basis of "industrial experience" are not based purely upon health considerations. Rather, those TLVs appear to reflect the levels of exposure which were perceived at the time to be achievable in industry. Thus, ACGIH TLVs may represent guides or levels which have been achieved, but they are certainly not thresholds. (Roach and Rappaport, 1990, p. 727)

The authors reported the following as their key findings:

> Three striking results emerged from this work, namely, that the TLVs were poorly correlated with the incidence of adverse effects, that the TLVs were well correlated with the exposure levels which has been reported at the time limits were adopted and that interpretations of exposure-response relationships were inconsistent between the authors of studies cited in the 1976 *Documentation* and the TLV Committee. Taken together these observations suggest that the TLVs could not have been based purely on consideration of health.

> While factors other than health appear to have influenced assignments of particular TLVs, the precise nature of such considerations is a matter of conjecture. However, we note that one interpretation is consistent with the above results, namely, that the TLVs represent levels of exposure which were perceived by the Committee to be realistic and attainable at the time. (Roach and Rappaport, 1990, p. 745)

A number of scientists published comments on the Roach and Rappaport (1990) analysis. One of the more thorough discussion papers was written by the past-chairs of the ACGIH (Adkins et al., 1990). In their letter to the editor, they claimed that the Roach and Rappaport paper was flawed and that it did not assess the validity of the bulk of the TLVs. The essence of their criticism was that the

> . . . conclusions which they draw concerning the protection afforded by TLVs are based on incomplete consideration of all of the data relative to a given substance. The authors present information in their tables as though the effects and exposures are valid and generally accepted by the occupational health community. No single epidemiologic study normally stands by itself. Requirements for inferring a causal relationship between disease and exposure in epidemiological studies are well established and include criteria for temporality, biological gradient with exposure, strength of the association, consistency with other studies, and biological plausibility of the observed effect. Roach and Rappaport present an uncritical analysis of various reports which would lead the uninformed reader to conclude that these criteria have been satisfied. In developing exposure recommendations, the TLV Committee and most other scientific organizations consider all of the relevant data before drawing conclusions. This includes judgments as to the validity and quality of individual studies in addition to the overall weight of the scientific evidence. (Adkins et al., 1990, p. 750)

Another rather lengthy counteranalysis of Ziem and Castleman, which contained a good deal of historical perspective, was written by the ACGIH Board of Directors (1990). In that paper, the Board stated that:

> While some criticisms may be valid, these articles do not fairly present the facts concerning historical development of TLVs nor do they accurately portray procedures followed by the TLV Committee in developing and reviewing TLV recommendations. Both articles contain a substantial number of errors and omissions and freely exercise selective quotation and quotation out of context in an effort to make their points. The section of Ziem and Castleman's article which discusses "Origins of TLVs" is a masterpiece of selective quotation and quotation out of context. This begins with their quoting a statement made by L.T. Fairhall concerning the role of industrial hygienists in setting health standards: "He [industrial hygienist] is in contact with the individuals exposed and therefore soon learns whether the concentrations measured are causing any injury or complaint." The authors use this quote to imply that physicians were excluded from the process of developing exposure guidelines. Taken in context, Fairhall's statement is as follows: "The industrial hygienist is in contact with not one, but a number of plants, using a given toxic substance. He knows, as no one else knows, the actual aerial concentration of contaminant encountered in practice. And he is in contact with the individuals exposed and therefore soon learns whether the concentrations measured are causing any injury or complaint. His judgement and the combined judgment of this entire Conference group is therefore most valuable in helping formulate maximum allowable concentration values." Contrary to Ziem and Castleman's comments, Fairhall advocated a multidisciplinary approach, including physicians, to making exposure recommendations. This has continued to be the operating philosophy of the TLV Committee. The conference in its first ten meetings was chaired by five physicians in six of the ten years. (ACGIH Board of Director, 1990, p. 343)

Although the merit of the Roach and Rappaport analyses, or for that matter, those of Ziem and Castlemen, will be debated for a number of years, it is clear that the rigor and documentation by which TLVs, and other OELs, will be set will probably never be as it was between 1945 and 1990. That is, that the rationale, as well as the degree of "risk" inherent in a TLV will be more explicitly described in the documentation for each TLV and that the definition of "virtually safe" or "insignificant risk" will change as the values of society change (Paustenbach et al., 1990b).

As a result of these papers, the 1990 Professional Conference on Industrial Hygiene (PCIH), held in Vancouver, was devoted to analyzing and discussing occupational exposure limits and health risk assessment. The speakers were experts in the fields of toxicology, risk assessment, industrial hygiene, and occupational medicine. Although a consensus was not reached, it was clear to most persons at the meeting that a larger fraction of the working population may well need to be protected by future TLVs. As a result of this on-going evaluation of TLV's and the criteria by which occupational exposure criteria limits will be set, as well as society's increasing expectations that risks be controlled to very low levels, this will be one of the top issues in the 1990s.

Between 1989 and 1992, a significant fraction of the members of the TLV Committee for Chemical Substances changed. Dr. John Doull assumed the chairmanship from Dr. Mastramateo and, by 1992, the philosophical objectives of the TLV committee had clearly moved toward protecting a larger fraction of the working population than had been previously considered necessary (ACGIH, 1992). For example, the TLV committee began to consider the results of risk assessment models to predict the likelihood of health effects (Alavanja et al., 1990; Notice of Intended Changes, 1990). Although they made it clear that the results would not dictate their decision, fairly complex explanations of the mechanism by which chemicals "may" exert their adverse effects began to be considered by the committee. Further, the "new" TLV committee had a greater sensitivity to the public's expectation that the risks to which workers should be exposed had to be much less than had previously been considered acceptable (Infante, 1992).

During 1991 and 1992, the TLV committee began to reevaluate the meaning of the term "intended to protect all workers." Roach and Rappaport's paper seemed to provide enough information to suggest that for the past 50 years the TLVs had been set at concentrations that industry could generally achieve yet were believed to be low enough to protect the majority (as much as 60–90 percent) of the exposed population from the primary adverse health effects. The "new" TLV committee acted swiftly to change the criterion of acceptability. For example, for formaldehyde, where perhaps as much as half of the general population might observe some degree of eye irritation at the previous TLV of 2.0 ppm, the TLV proposed in 1989 (0.3 ppm as a ceiling) attempted to protect virtually everyone (ACGIH, 1992; Paustenbach et al., 1994).

As recently as October, 1993, the TLV committee had not yet decided what percent of the working population the TLVs should attempt to protect. Although not specifically stated, it is clear that the committee has wanted to protect as much as 95 percent of the exposed population for chemicals like formaldehyde (ACGIH, 1992b). If such a criterion were to be adopted in the future, and if one considers the interindividual differences in susceptibility, it is probable that new TLVs for irritants may need to be as much as 10- to 50-fold less than their 1990 values to achieve this goal. Likewise, for the carcinogens, future TLVs will probably be much different than those established over the past 10 years. Currently, most TLVs pose a theoretical cancer risk of 1 in 10 to 1 in 1000 (Alavanja et al., 1990). However, as evidenced in the proposed TLV for benzene, which has a theoretical cancer risk of about 5 in 1000 (compared to 5 in 100 with the current TLV), future TLVs will be much lower than those of previous years. (Infante, 1992)

The degree of reduction in TLVs or other OELs that is likely to occur will vary depending on the type of adverse health effect to be prevented, for example, central nervous system depression, acute toxicity, odor, irritation, developmental effects, etc. It is also unclear to what degree the TLV committee will rely on various predictive toxicity models or the risk criteria they will adopt in the 1990s. Nonetheless, irrespective of the toxic effect, it is clear that most OEL's, including the TLV's, will be lower in the coming years due to higher societal expectations regarding the level of risk deemed acceptable for the workforce.

1.1 The "New" Philosophy (Two Case Studies)

The change in how future TLVs will probably be established in the coming years is illustrated by evaluating two chemicals for which guidelines were proposed in 1989 and 1990: formaldehyde and benzene, respectively. Since the latter is a known human carcinogen and the former is an irritant (as well as an animal carcinogen at high concentrations), they serve as good examples of the new expectations of the industrial hygiene community and, specifically, the ACGIH TLV committee.

1.2 Formaldehyde Case Study

Formaldehyde is a commercially significant chemical. It is an irritant and animal carcinogen and has historically been known to have a TLV that contains a low margin of safety against eye irritation. The 1984 TLV was 2.0 ppm as a short-term exposure limit (STEL), and in 1986 it was reduced to 1 ppm time-weighted average (TWA). In 1989, the committee proposed a TLV of 0.3 ppm as a ceiling value (CV). The 0.3 ppm CV was considered adequate to protect virtually everyone from its irritant effects, as well as to virtually eliminate the cancer hazard. Because this proposal was considered by some to be unjustifiably restrictive, legal action against the ACGIH was threatened by a trade association. As a result, formaldehyde became one of the first TLVs to receive attention and scrutiny similar to that associated with federal regulatory rule making. For example, by the middle of 1992, the documentation for the proposed TLV for formaldehyde was 90 double-spaced pages in length (24 pages in print) and contained more than 150 references. In contrast, the 1984 documentation of the TLV for formaldehyde was only 5 pages (in print) and only about 30 references were cited (ACGIH, 1992b).

The economic importance of a TLV–CV of 0.3 ppm was sufficiently important that a panel of experts on formaldehyde and occupational exposure limits was assembled by the Industrial Health Foundation (through funding from the Formaldehyde Institute). Their goal was to carefully review all available data and to suggest an OEL to the TLV subcommittee. The panel evaluated all the references discussed by the TLV committee, as well as others, and eventually concluded that a concentration of 0.3 ppm as an 8-hr TWA should be protective of as much as 99 percent of the potentially exposed population. They also concluded that a ceiling value was not warranted (Paustenbach et al., 1994).

This author has estimated that perhaps as much as 4000–6000 hr (2–3 person-years) of some of Americas most knowledgeable health scientists were expended during 1989–1992 trying to identify the most appropriate TLV for formaldehyde; a significant commitment of effort. The point is that whatever TLV is adopted, it will have received a great deal more effort than that which has normally been required to set perhaps 20 other TLVs. If such an effort is invested in many other chemicals, the TLV process will be slowed dramatically. This may well eliminate one of the reasons why the ACGIH has continued to recommend exposure limits even after the formation of the Occupational Safety and Health Administration (OSHA) and NIOSH. Specifically, one of the key attributes of the TLV's has been the speed with which the ACGIH can set them. Up to now, any shortcomings due to using an accelerated process were deemed acceptable to most practicing hygienists since it was considered more important to help protect large numbers of workers by setting many guidelines rather than agonize over setting precise ones.

1.3 Benzene Case Study

The experience with benzene is not unlike that of formaldehyde. Benzene is a very important industrial chemical used in large volumes in the United States. In July 1990, the ACGIH published a ''Notice of Intended Change—Benzene (1990),'' whereby it proposed a revision of the TLV for benzene from an atmospheric concentration of 10 ppm to 0.1 ppm as a TWA with a skin notation and the designation as an A1 carcinogen (confirmed human carcinogen).

The proposed TLV of 0.1 ppm benzene was based on (1) the results of quantitative risk assessments of leukemia, with special emphasis on the Rinsky et al. (1987) assessment using the NIOSH case-control data; (2) direct inspection of observational data pertaining to benzene exposure levels associated with leukemia cases/deaths from the Dow Chemical Company cohort mortality study; and (3) benzene exposure levels associated with chromosomal breakage in epidemiologic and toxicologic studies (Infante, 1992). The theoretical cancer risk associated with the proposed benzene TLV was either 1 in 10,000 or 1 in 1000 for persons exposed to 0.1 ppm for 8 hr/day, 5 day/week, for 40 years. The difference is due to the exposure assessment and risk model one chooses to accept (Crump and Allen, 1984; Paustenbach et al., 1992; Crump, 1993).

After weighing all the available data, the TLV committee chose in June, 1992, to extend the period for comment. During 1992–1994, the TLV committee will wrestle with the issue of ''What theoretical cancer risk is acceptable and how certain do our risk estimates have to be?'' Depending on the results of those deliberations, a different proposed TLV for benzene may surface.

The debates surrounding the TLVs for benzene and formaldehyde focus on issues that are not unique to those chemicals. Our knowledge of the quantitative hazard posed by many chemicals will continue to be lacking, if not very incomplete, for at least another 30 years. Consequently, a good deal of professional judgment and a reliance on untested models will be necessary. For both of the aforementioned chemicals, a relatively large number of epidemiology studies, many volumes of animal data, and nearly 70 years of use in industry were available. In both cases, more or better data would not necessarily lessen the uncertainty in our estimate of risk at certain doses. The issue at hand is ''how much risk is acceptable when persons are aware of the magnitude of the danger and are compensated (to some degree) for allowing themselves to be exposed?'' A number of recent papers and books address this issue of voluntary vs involuntary risk in significant detail (Wildavsky, 1988; Rodricks et al., 1987; Paustenbach, 1990b; Travis et al., 1987; Alvanja et al., 1990; Travis and Hester, 1990c).

During the 1990s, it can be expected that the philosophical objectives of the TLVs will change. The method by which they will be established, and the values selected, will probably become more like than unlike those used to set environmental limits. It can be anticipated that for many of the systemic toxicants, the TLVs could drop by a factor of 5 or 10 or greater. For some of the irritants, reductions of 10- to 50-fold can be expected. It is possible that the TLVs for some carcinogens could be reduced by a factor of as much as 50–500. To measure such concentrations using personal monitoring devices or to design engineering controls that will prevent such exposures in the workplace will present challenges that industrial hygienists have not frequently been asked to meet. Almost certainly, these issues will reshape the practice of industrial hygiene as we enter the twenty-first century.

1.4 Objectives of This Chapter

The purpose of this chapter is twofold: (a) to understand the principles upon which OELs have been established and (b) to convey enough information on the pharmacokinetics and toxicological aspects of chemicals to allow hygienists to know how to adjust OELs for unusual work schedules.

2 HISTORY AND METHODOLOGY FOR ESTABLISHING OCCUPATIONAL EXPOSURE LIMITS

2.1 History of Exposure Limits

The contribution of OELs to the prevention or minimization of disease is now widely accepted, but for many years such limits did not exist and even when they did, they were often not observed (Cook, 1945; Smyth, 1956; Stokinger, 1981; LaNier, 1984). It was, of course, well understood as long ago as the fifteenth century that airborne dusts and chemicals could bring about illness and injury, but the concentrations and lengths of exposure at which this might be expected to occur were unclear (Ramazinni, 1700).

As reported by Baetjer (1980, p. 774), "early in this century when Dr. Alice Hamilton began her distinguished career in occupational disease, no air samples and no standards were available to her, nor indeed were they necessary. Simple observation of the working conditions and the illness and deaths of the workers readily proved that harmful exposures existed. Soon however, the need for determining standards for safe exposure became obvious."

Cook has reported that the earliest efforts to set an OEL were directed to carbon monoxide, the toxic gas to which more persons are occupationally exposed than to any other. The work of Max Gruber at the Hygienic Institute at Munich was published in 1883. The paper described exposing 2 hens and 12 rabbits to known concentrations of carbon monoxide for up to 47 hr over 3 days. He stated that "the boundary of injurious action of carbon monoxide lies at a concentration on all probability of 500 parts per million, but certainly (not less than) 200 parts per million." In arriving at this conclusion, Gruber had also exposed himself. He reported no symptoms or uncomfortable sensations after 3 hr on each of two consecutive days at concentrations of 210 and 240 ppm (Cook, 1986).

According to Cook (1986), the earliest and most extensive series of animal experimentations on exposure limits were those conducted by K. B. Lehmann and others under his direction at the same Hygienic Institute where Gruber had done his work with carbon monoxide. The first publication in the series entitled "Experimental Studies on the Effect of Technically and Hygienically Important Gases and Vapors on the Organism" was a report on ammonia and hydrogen chloride gas that ran to 126 pages in Volume 5 of *Archiv für Hygiene* (Lehmann, 1886). This series of reports on animal experimentation with a large number of chemical substances by Lehmann and associates continued through Part 35 in Volume 83 (1914), followed by a final comprehensive paper of 137 pages on chlorinated hydrocarbons in Volume 116 (1936) of the German *Archiv*.

Kobert (1912) published one of the earlier tables of acute exposure limits. Concentrations for 20 substances were listed under the headings: (1) rapidly fatal to man and animals, (2) dangerous in 0.5 to 1 hr, (3) 0.5 to 1 hr without serious disturbances, and (3) only minimal symptoms observed (Cook, 1986). In his paper on "Interpretations of Permissible Limits," Schrenk (1947) notes that the "values for hydrochloric acid, hydrogen cyanide, ammonia, chlorine and bromine as given under the heading 'only minimal symptoms after several hours' in the foregoing Kobert paper agree with values as usually accepted in present-day tables of MACs for reported exposures." However, values for some of the more toxic organic solvents, such as benzene, carbon tetrachloride, and carbon disulfide, far exceeded those currently in use (Cook, 1986).

One of the first tables of exposure limits to originate in the United States was that published by the U.S. Bureau of Mines (Fieldner et al., 1921). Although its title does not so indicate, the 33 substances listed are those encountered in workplaces. Cook (1986) also noted that most of the exposure limits through the 1930s, except for dusts, were based on rather short animal experimentation. A notable exception was the study of chronic benzene exposure by Leonard Greenburg of the U.S. Public Health Service conducted under the direction of a committee of the National Safety

Council (NSC, 1926). He ultimately recommended an acceptable exposure limit based on long-term animal experimentation.

According to Cook (1986), for dust exposures, permissible limits established before 1920 were based on exposures of workers in the South African gold mines where the dust from drilling operations was high in crystalline free silica. The effects of the dust exposure were followed by periodic chest x-ray examination, and the dust concentrations were monitored by an instrument known as the konimeter, which collected a nearly instantaneous sample. In 1916, on correlating these two sets of findings, an exposure limit of 8.5 million particles per cubic foot of air (mppcf) for the dust with an 80–90 percent quartz content was set (Report of Miners, 1916). Later, the level was lowered to 5 mppcf. Cook (1986) also reported that, in the United States, standards for dust, also based on exposure of workers, were recommended by Higgins et al. following a study at the southwestern Missouri zinc and lead mines in 1917. The initial level established for high-quartz dusts was 10 mppcf, appreciably higher than was established by later dust studies conducted by the U.S. Public Health Service. In 1930, the USSR Ministry of Labor issued a decree that included the first actual approval of workplace maximum allowable concentrations for USSR with a list of 12 industrial toxic substances.

The most comprehensive list of occupational exposure limits up to 1926 was that for 27 substances which was published in Volume 2 of International Critical Tables (Sayers, 1927). Sayers and Dalle Valle (1935) published a table giving the physiological response to five levels of concentrations of 37 substances. The first four refer to acute effects, but the fifth is for the maximum allowable concentration for prolonged exposure. About this time, Lehmann and Flury (1938) and Bowditch et al. (1940) published papers that presented tables with a single recommended value for repeated exposures to each substance.

As noted by Cook (1986), many of the exposure limits developed by Lehmann were included in the Henderson and Haggard monograph (1943) initially published in 1927 and a little later in Flury and Zernik's *Schadilche Gase* (1931). According to Cook (1986), this book acted as the bible on effects of injurious gases, vapors, and dusts in industrial exposures until Volume II of *Toxicology of Patty's Industrial Hygiene and Toxicology* (1949).

Henschler (1984) noted that the first experimentally based report on the existence of a dose that does not cause harm was that of Flury and Heubner (1919) demonstrating that exposure to an acceptable level of hydrogen cyanide, for no matter how long a period of time, will not cause a distinct toxic effect (Cook, 1986). Over the years 1930–1939, in the United States, investigations providing information on concentrations of toxic substances involving occupational exposures were conducted at the Bureau of Mines Experimental Station in Pittsburgh. A series of 13 papers under the general title of *Acute Response of Guinea Pigs to Vapors of Some New Commercial Organic Compounds* was published by Sayers, Yant, Schreck, Patty, and others in *Public Health Reports* of the U.S. Public Health Service during this period.

According to Baetjer (1980), the first list of standards for chemical exposures in industry were called maximum allowable concentrations (MAC), and these were prepared in 1939 and 1940. They represented a consensus of opinion of the American Standard Association and a number of industrial hygienists who had formed the

ACGIH in 1938. These "suggested standards" were published in 1943 by James Sterner.

A committee of the ACGIH met in early 1940 to begin the task of identifying safe levels of exposure to workplace chemicals. They began by assembling all the data that they could locate that would relate the degree of exposure to a toxicant to the likelihood of producing an adverse effect (Stokinger, 1981; LaNier, 1984). This task, as might be expected, was a formidable one. After much painstaking research and labor-intensive documentation, the first set of values were released in 1941 by this committee, which was composed of Warren Cook, Manfred Boditch (reportedly America's first hygienist employed by industry), William Fredrick, Philip Drinker, Lawrence Fairhall, and Alan Dooley (Stokinger, 1981).

In 1941 a committee, designated as Z-37, of the American National Standards Institute (ANSI), then known as the American Standards Association, developed its first standard—carbon monoxide—with an acceptable value of 100 ppm. The ANSI issued separate bulletins through 1974 including exposure standards for 33 toxic dusts and gases. In 1983, the American Industrial Hygiene Association (AIHA) established a Committee on Occupational Health Standards to work through the procedures of ANSI for the development of additional consensus standards (Cook, 1986).

At the Fifth Annual Meeting of the ACGIH in 1942, the newly appointed Subcommittee on Threshold Limits presented in its report a table of 63 toxic substances with the "maximum allowable concentrations of atmospheric contaminants" from lists furnished by the various state industrial hygiene units. The report contains the statement: "The table is not to be construed as recommended safe concentrations. The material is presented without comment" (Cook, 1986).

In 1945 a list of 132 industrial atmospheric contaminants with MACs was published by Cook (1945). The most important part of the publication was the inclusion of references on the original investigations leading to the values. The table included the then current values for the six states, California, Connecticut, Massachusetts, New York, Oregon, and Utah, values presented as a guide for occupational disease control by the U.S. Public Health Service and 11 standards of the ANSI. In addition, the author included a list of MACs that appeared best supported by the references based on original investigations. Approximate round number equivalents of the parts per million values were included with the author's list as milligrams per cubic meter (Cook, 1986).

At the 1946 Eighth Annual Meeting of ACGIH, the Subcommittee on Threshold Limits presented their second report with the values of 131 gases, vapors, dusts, fumes, mists, and 13 mineral dusts. As stated in the report, the values were "compiled from the list reported by the subcommittee in 1942, from the list published by Warren Cook in *Industrial Medicine*, **14,** 936–946 (1945), and from published values of the Z-37 Committee of the American Standards Association. "The committee emphasized that the list of M.A.C. values is presented . . . with the definite understanding that it be subject to annual revision."

At the Ninth International Congress on Industrial Medicine in London in 1948 and also at the Fifteenth International Congress on Occupational Health in Vienna in 1966, papers were presented on MACs. The first of these by Drinker and Cook presented the idea that, since the boundaries of MAC values were not sharp lines of demarcation, it would be of practical value for the understanding of relative hazards that one of six ranges of values, rather than a single value, be proposed for each chemical: less than 0.1 ppm, 0.1–2 ppm, 2–20 ppm, 20–100 ppm, 100–500 ppm, and 500–2500 ppm. Examples were given for each range. It was noted that the applications of these zones could involve a danger that is to be recognized and avoided. The accepted MAC for a substance may be close to the lower limit of a zone, such as toluene in zone V (its MAC at the time was 200 ppm). It is not to be assumed that average exposures to toluene could be permitted up to 500 ppm on the basis that the zone includes concentrations from 100 to 500 ppm. The zoning scheme was for classification of information and not for justification of excessive exposures (Cook, 1986).

The overall impact of these efforts to develop quantitative limits to protect humans from the adverse effects of workplace air contaminants and physical agents could not have been anticipated by the early TLV committees. To their credit, even though toxicology was then only a fledgling science, their approach to setting limits has generally been shown to be correct even by today's standards. For this reason many of the techniques for setting limits established by this committee are still in use today (Cook, 1945; Smyth, 1956; LaNier, 1984; Stokinger, 1970; WHO, 1977; Dourson and Stara, 1983). The principles they used to set OELs were similar to those used to identify safe doses of food additives and pharmaceuticals (Lehmann and Fitzhugh, 1954).

From the perspective of the hygienist, engineer, and businessperson, there have been many benefits of setting OELs. The establishment of limits, by their very nature, implies that at some concentration or dose, exposure to a toxicant can be expected to pose no significant risk of harm to exposed persons. By incorporating this fundamental toxicological principle into the realm of business management, the practice of industrial hygiene has been able to make significant strides.

The key to the success of limits has not been only that they were established on solid scientific principles; rather, the setting of any goal gives a sense of purpose and direction to industrial hygiene, occupational, or medical programs which prior to the TLVs had been difficult to evaluate. The setting of goals, such as controlling workplace concentrations below a TLV or permissible exposure limit (PEL), sets an objective that can then be mutually pursued by the occupational health team, engineers, and management. By introducing the concept of ''safe level of exposure'' and by establishing a type of ''management by objectives,'' occupational health programs can establish and pursue a clear list of priorities (Paustenbach, 1990c).

2.2 The Fundamentals behind the TLVs

The ACGIH TLVs and most other OELs are limits that refer to airborne concentrations of substances and represent conditions under which ''it is believed that nearly all workers may be repeatedly exposed day-after-day without adverse health effects'' (ACGIH, 1992). It is important to recognize that unlike some exposure limits for ambient air pollutants, contaminated water, or food additives set by other professional groups or regulatory agencies, exposure to the TLV will not necessarily prevent discomfort or injury for everyone who is exposed (Adkins et al., 1990). The ACGIH recognized long ago that because of the wide range in individual suscepti- bility, a small percentage of workers may experience discomfort from some sub- stances at concentrations at or below the threshold limit and that a smaller percentage may be affected more seriously by aggravation of a preexisting condition or by de- velopment of an occupational illness (Cooper, 1973; Omenn, 1982; ACGIH, 1992). This is clearly stated in the introduction to the ACGIH's annual booklet ''Threshold Limit Values for Chemical Substances and Physical Agents and Biological Exposure Indices'' (ACGIH, 1992).

This limitation, although perhaps less than ideal, has been considered a practical one since airborne concentrations so low as to protect hypersusceptibles have tra- ditionally been judged infeasible due to either engineering or economic limitations. This shortcoming in the TLVs has, until the past 4 years, not been considered a serious one. If society wishes to significantly reduce workplace exposures, by using engineering controls and personal protective equipment, it is now possible to do so. In light of the dramatic improvements of the past 10 years in our analytical capa- bilities, personal monitoring/sampling devices, biological monitoring techniques, and the use of robots as a plausible engineering control, we are now technologically able to consider more stringent occupational exposure limits. Whether we are willing to invest the financial resources needed to control workplace exposure to much lower levels or whether it makes good economic sense to do so is not yet known (U.S. EPA SAB, 1990).

Threshold limit values, like most other OELs used in other countries, are based on the best available information from industrial experience, experimental human and animal studies, and, when possible, from a combination of the three (Smith and Olishifski, 1988; ACGIH, 1992). The rationale for each of the values differs from substance to substance. For example, protection against impairment of health may be a guiding factor for some, whereas reasonable freedom from irritation, narcosis, nuisance, or other forms of stress may form the basis for others. The age and com- pleteness of the information available for establishing most occupational exposure limits also varies from substance to substance; consequently, the precision of each particular TLV is subject to variation. The most recent TLV and its documentation should always be consulted in order to evaluate the quality of the data upon which that value was set.

The background information and rationale for each TLV is published periodically in the *Documentation of the Threshold Limit Values* (ACGIH, 1989). Some type of documentation is occasionally available for OELs set in other countries. The ratio- nale or documentation for a particular OEL should always be consulted before in- terpreting or adjusting an exposure limit, as well as the specific data considered in establishing it (ACGIH, 1992).

Even though all of the publications that contain OELs emphasize that they were intended for use only in establishing safe levels of exposure for persons in the work- place, they have been used at times in other situations. It is for this reason that all exposure limits should be interpreted and applied only by someone knowledgeable of industrial hygiene and toxicology. The TLV committee did not intend that they be used, or modified for use:

1. As a relative index of hazard or toxicity
2. In the evaluations of community air pollution
3. Estimating the hazards of continuous, uninterrupted exposures or other extended work periods
4. As proof or disproof of an existing disease or physical condition
5. For adoption by countries whose working conditions differ from those of the United States (ACGIH, 1992)

The TLV committee and other groups that set OELs warn that these values should not be ''directly used'' or directly extrapolated to predict safe levels of exposure for other exposure settings. However, if one understands the scientific rationale for the guideline and the appropriate approaches for extrapolating data, they can be used to predict acceptable levels of exposure for many different kinds of exposure scenarios and work schedules (Hickey and Reist, 1979; NASA, 1991).

2.3 Philosophy of Exposure Limits

Threshold limit values were originally prepared to serve only for the use of industrial hygienists who could exercise their own judgment in their application. They were not to be used for legal purposes (Baetjer, 1980). However, in 1968 the Walsh–Healey Public Contract Act incorporated the 1968 TLV list, which covered about 400 chemicals. When the OSHA was passed, it used national consensus standards to establish federal standards. The only consensus standards available were the 20 standards the American Standard Association (now called the American National Standard Association) has promulgated. However, OSHA could adopt the 1968 TLV list because it was already in a federal law. OSHA could not indicate that these standards were TLVs and cannot update these values in line with the yearly revision of the TLV list.

An understanding of the philosophy used in setting occupational exposure limits is critical to the practice of industrial hygiene and toxicology. Exposure limits for workplace air contaminants are based on the premise that, although all chemical substances are toxic at some concentration when experienced for a specific period of time, a concentration (e.g., dose) does exist for all substances at which no injurious effect should result no matter how often the exposure is repeated. A similar premise applies to substances whose effects are limited to irritation, narcosis, nuisance, or other forms of stress (Stokinger, 1981; ACGIH, 1992).

This philosophy thus differs from that applied to physical agents such as ionizing radiation, and some chemical carcinogens, since it is possible that there may be no threshold or no dose at which zero risk would be expected (Stokinger, 1981). Even though many would say that this position is too conservative in light of our lack of understanding of the mechanisms by which cancer occurs, there are some data on some genotoxic chemicals that support this theory (Bailer et al., 1988; Travis and Hester, 1990). On the other hand, many scientists believe that a threshold dose exists for those chemicals that are carcinogenic in animals but that act through a nongenotoxic (sometimes called epigenetic) mechanism (Watanabe et al., 1980; Stott et al., 1981; Butterworth and Slaga, 1987; Reitz et al., 1990a). Still others maintain that a practical threshold exists for even genotoxic chemicals, although they agree that the threshold may be at extremely low doses (Seiler, 1977; Stott et al., 1981; Wilkinson, 1988; Bus and Gibson, 1985). With this in mind, some occupational exposure limits proposed by regulatory agencies in the early 1980s were set at levels that, although not completely without risk (safe), were no greater than classic occupational hazards such as electrocution, falls, and so forth. Because the workplace mortality risk is about 1 in 1000, this has been the rationale for selecting this theoretical cancer risk for setting some OELs (Rodricks et al., 1987; Travis et al., 1987).

2.4 Exposure Limits Other Than TLVs

The philosophical underpinnings for the various occupational exposure limits vary between the organizations and countries that develop them. For example, in the United States at least six groups recommend exposure limits for the workplace. These include the TLVs of the ACGIH, the recommended exposure limits (RELs) suggested by NIOSH of the U.S. Department of Health and Human Services, the workplace environment exposure limits (WEEL) developed by the AIHA, standards for workplace air contaminants suggested by the Z37 Committee of the ANSI, the proposed workplace guides of the American Public Health Association (APHA, 1991), and lastly, recommendations that have been made by local, state, or regional government. In addition to these recommendations or guidelines, PELs, which are regulations that must be met in the workplace because they are law, have been promulgated in the U.S. by the Department of Labor and are enforced by the OSHA (OSHA, 1989a,b).

Outside the United States, as many as 50 other countries or groups of countries have established OELs (Stokinger, 1970; WHO, 1977; Cook, 1986). Many, if not most, of these limits are nearly or exactly the same as the ACGIH TLVs developed in the United States (Letavet, 1961; Magnuson, 1964; Cook, 1986). In some cases, such as in the East European countries, including former Soviet-bloc countries, and in Japan, the limits can be dramatically different than those used in the United States. Differences among various limits recommended by other countries can be due to a number of factors:

1. Difference in the philosophical objective of the limit and the untoward effects they are meant to minimize or eliminate
2. Difference in the predominant age and sex of the workers
3. Duration of the average workweek
4. Economic state of affairs in that country
5. Lack of enforcement (therefore acting simply as a guide)

For example, limits established in what is now the organization of Russian States, are often based on a premise that they will protect "everyone, rather than nearly everyone, from any (rather than most) toxic or undesirable effects of exposure" (Elkins, 1961; Stokinger, 1970; WHO, 1977; Magnuson, 1964; Stokinger, 1963; Lyublina, 1962; LaNier, 1984).

The former USSR also established many of its limits with the goal of eliminating any possibility for even reversible effects, such as those involving subtle changes in behavioral response, irritation, or discomfort. The philosophical differences between limits set in the former USSR and in the United States have been discussed by Letavet (1961, p. 142), a Russian toxicologist, who stated that:

> The method of conditioned reflexes, provided it is used with due care and patience, is highly sensitive and therefore it is a highly valuable method for the determination of threshold concentrations of toxic substances.

> At times disagreement is voiced with Soviet MAC's for toxic substances, and the argument is that these standards are founded on a method which is excessively sensitive, namely the method of conditioned reflexes. Unfortunately, science suffers not a surplus of excessively sensitive methods, but their lack. This is particularly true with regard to medicine and biology.

> Although the methods of examination of the higher nervous activity are very sensitive, they cannot be considered to always be the most sensitive indicator of an adverse response and to enable us always to discover the harmful after-effects of being exposed to a poison at the earliest time.

Such subclinical and fully reversible responses to workplace exposures have, thus far, been considered too restrictive to be useful in the United States and in most other countries. In fact, due to the economic and engineering difficulties in achieving such low levels of air contaminants in the workplace, there is little indication that these limits have actually been achieved in countries that have set them. Instead, the limits appear to serve more as idealized goals rather than limits that manufacturers are legally bound or morally committed to achieve (Elkins, 1961; Stokinger, 1963).

Occupational health professionals who attempt to determine which of the more than 40 different exposure limits for a chemical is appropriate must understand the underlying rationale and documentation upon which the limit for any particular country was established. A number of fairly thorough discussions of the history and philosophical basis upon which OELs have been set have been published and these should be reviewed (see supplemental reference list). A comprehensive listing of the various OELs used throughout the world can be found in two references that are not always well known among practitioners of industrial hygiene. The first is *Occupational Exposure Limits for Airborne Toxic Substances*, 3rd ed., published by International Labour Office of the World Health Organization (WHO, 1990), and the other is *Occupational Exposure Limits—Worldwide* published by the American Industrial Hygiene Association (Cook, 1986).

2.5 Basis for Current Exposure Limits

Occupational exposure limits established both in the United States and elsewhere are derived from a wide number of sources. As shown in Table 7.1, the 1968 TLVs (those adopted by OSHA in 1970 as federal regulations) were based largely on human experience. This may come as a surprise to many hygienists who have recently

Table 7.1 Distribution of Procedures Used to Develop ACGIH TLVs for 414 Substances Through 1968[a]

Procedure	Number	Percent Total
Industrial (human) experience	157	38
Human volunteer experiments	45	11
Animal, inhalation—chronic	83	20
Animal, inhalation—acute	8	2
Animal, oral—chronic	18	4.5
Animal, oral—acute	2	0.5
Analogy	101	24

Source: Stokinger (1970).

[a]Exclusive of inert particulates and vapors.

entered the profession since it indicates that, in most cases, the setting of an exposure limit has been after it has been found to have toxic, irritational, or otherwise undesirable effects on humans. As might be anticipated, many of the more recent exposure limits for systemic toxins, especially those internal limits set by manufacturers, have been based primarily on toxicology tests conducted on animals, which is in contrast with waiting for observations of adverse effects in exposed workers (Paustenbach and Langner, 1986). Of course, even as far back as 1940, animal tests were acknowledged by the TLV committee to be very valuable, and they do, in fact, constitute the second most common source of information upon which these guidelines have been established (Stokinger, 1970).

Several approaches for deriving occupational exposure limits from animal data have been proposed and put into use over the past 40 years. The approach used by the TLV committee and others is not markedly different from that which has been used by the U.S. Food and Drug Administration (FDA) in establishing acceptable daily intakes (ADI) for food additives. An understanding of the FDA approach to setting exposure limits for food additives and contaminants can provide good insight to industrial hygienists who are involved in interpreting occupational exposure limits and should be reviewed (Rodricks and Taylor, 1983; Dourson and Stara, 1983).

Various approaches for setting both environmental and occupational exposure limits from animal data have been discussed elsewhere (Calabrese, 1983). Discussions of methodological approaches that can be used to establish workplace exposure limits based exclusively on animal data have also been presented (Weil, 1972; WHO, 1977; Calabrese, 1978; Zielhuis and van der Kreek, 1979a,b; Calabrese, 1983; Dourson and Stara, 1983; Finley et al., 1993). Although these approaches have some degree of uncertainty, they seem to be much better than a qualitative extrapolation of animal test results to humans.

The criteria used to develop the TLVs may be classified into four groups: morphologic, functional, biochemical, and miscellaneous (nuisance, cosmetic) (Table 7.2). As noted, approximately 50 percent of the 1968 TLVs have been derived from the human data, and approximately 30 percent are derived from animal data (Table 7.1). Of the human data, most are derived from the effects noted in workers who

Table 7.2 Classification of Criteria for ACGIH TLVs Applicable to Humans and Animals

Morphologic	Functional	Biochemical	Miscellaneou
Applied Criteria			
Systems of organs affected—lung, liver, kidney, blood, skin, eye, bone, CNS, endocrines, exocrines	Changes in organ function—lung, liver, kidney, etc. Irritation Mucous membranes Epithelial linings Eye Skin Narcosis Odor	Changes in amounts biochemical constituents including hematologic Changes in enzyme activity Immunochemical allergic sensitization	Nuisance Visibility Cosmetic Comfort Esthetic (Analogy)
Carcinogenesis Roentgenographic changes			
Potentially Useful Criteria			
Altered reproduction Body—weight changes Organ/body weight changes Food consumption	Behavioral changes Higher nervous functions Conditioned and unconditioned reflexes—learning Audible and visual responses Endocrine glands Exocrine glands	Changes in isoenzyme patterns Radiomimetic effects Teratogenesis Mutagenesis	

Source: Stokinger (1970).

were occupationally exposed to the substance for many years. Consequently, most of the existing TLVs have been based on the results of workplace monitoring, compiled with qualitative and quantitative observations of the human response (Stokinger, 1970; Park and Snee, 1983). In more recent times, TLVs for new compounds have been based primarily on the results of animal studies rather than human experience.

The rationale on which most of the existing TLVs have been established are shown in Table 7.3. It is noteworthy that in 1968 only about 50 percent of the TLVs were set to prevent systemic toxic effects. Roughly 40 percent were based on irritation and about 5 percent were intended to prevent cancer.

In January 1989, OSHA amended its existing Air Contaminants Standard § 1910.1000 including Tables Z-1, Z-2, and Z-3. This amendment made 212 PELs more protective, set new PELs for 164 substances not regulated by OSHA, and maintained other PELs unchanged. Changes included revision of some PELs, inclusion of STELs to complement 8-hr TWA limits, establishment of skin designation, and addition of ceiling limits as appropriate.

OSHA reviewed health, risk, and feasibility evidence for all 428 substances for which changes to the 1988 PEL were considered. In each instance where a revised or new PEL was adopted, OSHA determined that the new limits substantially reduced the significant risk of material health impairment among American workers,

Table 7.3 Distribution of Criteria Used to Develop ACGIH TLVs for 414 Substances Through 1968[a]

Criteria[b]	Number	Percent	Criteria	Number	Percent[a]
Organ or organ system affected	201	49	Biochemical changes	8	2
Irritation	165	40	Fever	2	0.5
Narcosis	21	5	Visual changes (halo)	2	0.5
Odor	9	2	Visibility	2	0.5
Organ function changes	8	2	Taste	1	0.25
Allergic sensitivity	6	1.5	Roentgenographic changes	1	0.25
Cancer	6	1.5	Cosmetic effect	1	0.25

Source: Stokinger (1970)

[a] Exclusive of inert particulates and vapors.

[b] Number of times a criterion was used of total number of substances examined × 100, rounded to nearest 0.25 percent. Total percentages exceed 100 because more than one criterion formed the basis of the TLV of some substances.

Table 7.4 OSHA's Estimate of the Number of United States Workers Potentially at Risk of Experiencing Adverse Effects Due to Exposure to Workplace Chemicals by Type of Effect[a]

Adverse Health Effect	No. of Workers Potentially Exposed to Substances Associated with Effect, Minimum Estimate	No. of Workers Potentially Exposed to Substances Associated with Effect, Maximum Estimate	No. of Workers Exposed Above Final Limits for Substances, Minimum Estimate	No. of Workers Exposed Above Final Limits for Substances, Maximum Estimate
Physical irritant effects	3,375,472	3,889,261	222,191	222,191
Odor effects	519,318	521,938	3,597	3,597
Systemic toxicity	4,305,578	5,038,573	457,104	490,282
Mucous membrane irritation	10,730,691	14,906,090	789,461	1,141,133
Metabolic interferences	4,015,702	4,205,530	1,233,413	1,241,564
Liver/kidney disease	3,292,933	3,806,226	536,945	546,429
Ocular disturbances	2,482,449	2,569,950	83,272	110,560
Respiratory disease	4,231,235	4,782,280	1,405,501	1,568,579
Cardiovascular disease	166,077	166,868	44,403	44,403
Neuropathy	2,212,358	2,463,583	379,974	401,576
Narcosis	6,966,025	10,520,982	941,472	1,073,717
Cancer	1,712,799	1,851,342	465,013	528,650
Allergic sensitization	2,545,551	2,648,973	305,955	305,955

Source: OSHA, 1989a.

[a] OSHA noted that double counting of employees simultaneously exposed to more than one substance in different adverse health effects categories prevents the summation of workers exposed to all adverse health effects using this table.

Table 7.5 Example Page from OSHA (1989) PEL Preamble which Presents the Rationale for the PEL for Individual Substances

H.S. Number	CAS Substance Name	Primary Basis Number	Preamble for Limits	Section
1001	Acetaldehyde	75–07–0	Sensory Irritation	VI.C.3
1002	Acetic acid	64–19–7	Sensory irritation	VI.C.3
1003	Acetic anhydride	108–24–7	Analogy	VI.C.12
1004	Acetone	67–64–1	Sensory irritation	VI.C.12
1005	Acetonitrile	75–05–8	Systemic toxicity	VI.C.8
1006	Acetylsalicylic acid (asprin)	50–78–2	Systemic toxicity	VI.C.8
1007	Acrolein	107–02–8	Sensory Irritation	VI.C.3
1008	Acrylamide	79–06–1	Cancer	VI.C.12
1009	Acrylic acid	79–10–7	Analogy	VI.C.12
1010	Allyl alcohol	107–18–6	Sensory irritation	VI.C.3
1011	Allyl chloride	107–05–1	Liver and kidney effects	VI.C.4
1020	Amitole (3-amino-1,2,4-triazole)	61–82–5	Cancer	VI.C.15
1021	Ammonia	7664–41–7	Sensory irritation	VI.C.3
1022	Ammonium chlorine (fume)	12125–02–9	Sensory irritation	VI.C.3
1024	Ammonium sulfamate (ammate)	7773–06–0	Physical irritation	VI.C.10
1025	Aniline	62-53-3	Biochemical/metabolic effects	VI.C.13
1028	Asphalt fumes	8052–42–4	Cancer	VI–C–15
1029	Atrazene	1912–24–9	NOAELs	VI–C–9
1031	Barium sulfate	7727–43–7	Physical irritation	VI.C.10
1032	Benomyl	17804–35–2	Physical irritation	VI.C.10
1033	Beryllium and compounds	7440–41–7	Cancer	VI.C.15
1034	Bismuth telluride (SE-doped)	1304–82–1	Respiratory effects	VI.C.6
1035	Bismuth telluride (undoped)	1304–82–1	Physical irritation	VI.C.10
1036	Borates, tetra, sodium (anhydrous)	1330–43–4	Sensory irritation	VI.C.3
1037	Borates, tetra, sodium (decahydrate)	1303–94–4	Sensory irritation	VI.C.3
1038	Borates, tetra, sodium (pentahydrate)	12179–04–3	Sensory irritation	VI-C-3

Source: OSHA, 1989a. [NOTE: The Courts ruled in 1992 that these PELs were to be revoked and that the prior list be used.]

and that the new limits were technologically and economically feasible. The revised standards were believed to provide additional occupational health protection to 4.5 million workers at an annual cost of approximately $150 per employee protected (Table 7.4). This cost represented a fraction of 1 percent of sales for all affected sectors (OSHA, 1989). As a result of a court decision in 1992, the 1989 PELs were overturned and the prior PELs remained in place.

As illustrated in Table 7.5, OSHA divided the approximate 700 chemicals into 14 categories (under the preamble for the limits) according to their primary adverse health effect. Although no tabulation of the percentage of the total fit into a particular category, the percentage of chemicals in each cateogry remains about the same as in 1968.

2.6   Setting OELs for Irritants

Unlike the period 1940–1975, when OELs for irritants had to be based on the human experience, we now have a fairly good ability to predict safe levels of exposure using models. Two kinds of models are available. One is based on tests that consider the response of rats and/or mice (Kane and Alarie, 1977; Alarie, 1981; Abraham et al., 1990; Nielsen, 1991) to irritants; the second is based on the chemical properties of the chemical (Leung and Paustenbach, 1988). It is recommended that these be used to set preliminary OELs for chemicals for which appropriate data are available.

In one of their more comprehensive papers, Alarie and Nielsen (1982) described the capacity of their model to predict safe exposure limits for the benzenes and alkylbenzenes.

The potency of the alkylbenzenes increased with chain length. However, the ratio of equipotent concentration/saturated vapor concentration for these alkybenzenes varied very little. Knowing the vapor pressure enabled prediction of sensory irritation potency up to a chain length of C_6. As previously suggested [Y. Alarie (1981), *Food Cosmet. Toxicol.*, **19**, 623–626], RD_{50} values multiplied by 0.03 gave values for toluene, ethylbenzene, isopropylbenzene, and p-terri-butyltoluene in close agreement with established threshold limit value (TLV) for industrial exposure. No established TLV values exist for the other alkylbenzenes investigated. Using 0.03 RD_{50} acceptable TLVs for n-hexylbenzene would be 50, 20, 20, 10, and 5 ppm, respectively. Minimal or no pulmonary irritation was observed with these alkylbenzenes. Benzene was inactive as a sensory irritant or pulmonary irritant up to 8500 ppm. Benzene in particular, and to some extent toluene, stimulated the respiratory rate. The effect was maximum approximately 10 to 20 min after onset of exposure. A model for the sensory irritating action of alkylbenzenes is proposed on the basis of their physical interaction with a receptor protein in a lipid layer.

This work was expanded over the next 10 years to address a variety of irritants including ketones, alcohols, alkanes, disulfides, and styrene like chemicals (Nielsen, 1991; Schaper, 1993).

An approach for using the pK_a to set preliminary OELs for organic acids and bases was recently developed and validated (Leung and Paustenbach, 1988). It was shown that the OELs for these chemicals correlated very well ($r \geq 0.80$) with pK_a (Figs. 7.1 and 7.2). These authors suggested that for organic acids and bases for which no OEL has been established, the following equations could be used to set a preliminary occupational exposure limit.

For organic acids:

$$\log OEL \ (\mu mol/m^3) = 0.43 \ pK_a + 0.53$$

For organic bases:

$$OEL \ (\mu mol/m^3) = -200 \ pK_a + 2453$$

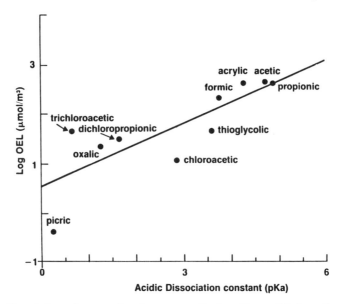

Figure 7.1 Correlation of occupational exposure limits with equilibrium dissociation constants of organic acids. The correlation coefficient $r = 0.80$. The regression equation is log OEL (μmol/m^3) = 0.43 pK$_a$ + 0.53. (From Leung and Paustenbach, 1988a.)

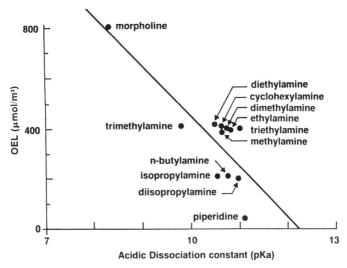

Figure 7.2 Correlation of occupational exposure limits with equilibrium dissociation constants of organic bases. The correlation coefficient $r = -0.81$. The regression equation is OEL (μmol/m^3) = -200 pK$_a$ + 2453. (From Leung and Paustenbach, 1988a.)

Table 7.6 Occupational Exposure Limits for Selected High Volume Organic Acids and Bases Recommended by a Mathematical Formula Based on the Disassociation Constant[a]

Acid	mg/m³	ppm	Base	mg/m³	ppm
Acrylic	16	5	Allylamine	29	12
Butyric	35	10	Dialylamine	58	15
Caproic	49	10	Dibutylamine	43	8
Crotonic	30	8.5	Isobutylamine	21	7
Hepatanoic	55	10	Propylamine	21	7
Isobutyric	35	10	Trialylamine	109	20
Isocaproic	49	10			
Isovaleric	42	10			
Methacrylic	30	8.5			
Pentenoic	32	7.8			
Propiolic	1.5	0.5			
Valeric	42	10			

Source: Leung and Paustenbach, 1988a.

[a]Exposure limits were calculated by using the equations: acid: $\log \text{OEL} \ (\mu\text{mol/m}^3) = 0.43 \ \text{pK}_a + 0.53$; base: $\text{OEL} \ (\mu\text{mol/m}^3) = -200 \ \text{pK}_a + 2453$.

Table 7.6 presents a few suggested OELs that were calculated using these formulae. During the 5 years since this paper was published, many hygienists who have used this approach have reported that it yields acceptable results.

2.7 Limits for Chemical Carcinogens

The impetus to have the TLV committee develop a classification of occupational carcinogens arose in 1970, when it felt that the lists published by numerous agencies and different groups claiming that a large number of substances were occupational carcinogens was getting out of hand. Substances of purely laboratory curiosity, such as acetylaminofluorene and dimethylaminobenzene, which were found to be tumorigenic in animals, were classed along with known human carcinogens of high potency, such as bis-chloromethyl ether (BCME). In short, no distinction was made between an animal tumorigen and a human carcinogen. Union leaders, workers, and the public would often become worried equally about each one: a situation that the ACGIH believed to benefit no one (Stokinger, 1977).

The TLV committee felt that the finding of a substance to be tumorigenic, often in a half-dead mouse or rat due to intolerable doses, as was the case for chloroform and trichloroethylene, was not ipso facto evidence that it will be carcinogenic in man under most conditions found in the workplace. It is for this reason that the ACGIH Chemical Substance TLV Committee, as early as 1972 made a clear distinction between animal and human carcinogens.

By setting exposure limits, the ACGIH, as well as its sister organizations throughout the world, has acknowledged that chemical carcinogens are likely to have a threshold, or at least a "practical threshold." This position, simply stated is that at some level of exposure it would not be expected to cause a significant cancer risk (e.g., a risk no greater than 1 in 1000 or 10,000 workers).

In 1977, Herbert Stokinger, then chairman of the ACGIH TLV committee, summarized the past philosophy of the ACGIH with respect to carcinogen TLVs:

Experience and research findings still support the contention that TLVs make sense for carcinogens. First and foremost, the TLV Committee recognizes practical thresholds for chemical carcinogens in the workplace, and secondly, for those substances with a designated threshold, that the risk of cancer from a worker's occupation is negligible, provided exposure is below the stipulated limit. There is no evidence to date that cancer will develop from exposure during a working lifetime below the limit for any of those substances.

Where did the TLV Committee get the idea that thresholds exist for carcinogens? We have been asked "Where is the evidence?" . . . Well, the Committee thinks it has such evidence, and here it is.

It takes three forms:

1. Evidence from epidemiologic studies of industrial plant experience, and from well-designed carcinogenic studies in animals,

2. Indisputable biochemical, pharmacokinetic, and toxicologic evidence demonstrating inherent, built-in anticarcinogenic processes in our bodies.

3. Accumulated biochemical knowledge makes the threshold concept the only plausible concept. (Stokinger, 1977, p. 56)

The TLV committee in its recent assessments has kept pace with our increased understanding of the hazards posed by exposure to chemical carcinogens. For example, beginning in 1985, they began to consider not only the results of cancer (low-dose) models but also in vitro data, case reports, studies of the mechanisms of action, and other information on carcinogens. Due to the variability in risk estimates between the various models, and their inability to account for protective biological processes, the TLV committee has been, and continues to be, reluctant to place too much emphasis on them. As a result, the TLV committee has not embraced the Environmental Protection Agency (EPA) approach to setting acceptable levels of exposure to carcinogens.

In the ACGIH TLV booklet, it is noted that (ACGIH, 1992):

The TLV Committee considers information from the following kinds of studies to be indicators of a substance's potential to be a carcinogen in humans; epidemiology studies, toxicology studies, and, to a lesser extent, case histories. Scientific debate over the existence of biological thresholds for carcinogens is unlikely to be resolved in the near future. Because of the long latent period for many carcinogens, and for ethical reasons, it is often impossible to base timely risk-management decisions on results from human studies.

In order to recognize the qualitative difference in research results, two categories of carcinogens are designated in the booklet: A1—Confirmed Human Carcinogens, and A2—Suspected Human Carcinogens.

Exposure to carcinogens must be kept to a minimum. Workers exposed to A1 carcinogens without a TLV should be properly equipped to eliminate to the fullest extent possible all exposure to the carcinogen. For A1 carcinogens with a TLV and for A2 carcinogens, worker exposure by all routes should be carefully controlled to levels as low as reasonably achievable (ALARA) below the TLV.

Appendix A of the Annual TLV Booklet (p. 45) contains the listing of carcinogens and the category into which they were placed (ACGIH, 1992). The Chemical Substances TLV Committee shares the public's concern over chemicals or industrial processes that cause or contribute to the increased risk of cancer in workers. The goal of the committee has been to synthesize the available information in a manner that will be useful to practicing industrial hygienists, without overburdening them with needless details. The committee has noted, however, that their current practice does not make adequate allowance for degrees of uncertainty regarding results from both human and animal studies. During 1989–1991, the TLV committee reviewed current methods of classification used by other groups and developed a new procedure (Alavanja et al., 1990). The current categories for occupational carcinogens are (ACGIH, 1992):

A1—Confirmed Human Carcinogen: The agent is carcinogenic to humans based on the findings of epidemiologic studies of, or convincing clinical evidence in, exposed humans.

A2—Suspected Human Carcinogen: The agent is carcinogenic in experimental animals at dose levels and by routes of administration that are considered relevant to worker exposure. Available epidemiologic studies are conflicting, controversial, or inadequate to confirm an increased risk of cancer in exposed humans.

A3—Animal Carcinogen: The agent is carcinogenic in experimental animals at relatively high dose or by routes of administration that are not considered relevant to worker exposure. Available epidemiologic studies do not confirm an increased risk of cancer in exposed humans. Available evidence suggests that the agent is not likely to cause cancer in humans except under uncommon or unlikely exposure situations.

A4—Not Classifiable as a Human Carcinogen: There are inadequate (or no) data at the present time on which to classify the agent in terms of it carcinogenicity in humans and/or animals.

A5—Not Suspected as a Human Carcinogen: The agent is not suspected to be a human carcinogen on the basis of properly conducted epidemiologic studies in humans. Those studies have sufficiently long follow-up, reliable exposure histories, and adequate statistical power to conclude that exposure to the agent does not convey a significant risk of cancer to humans.

It can be anticipated that as the TLV committee continues to evaluate and pass judgment on the various chemical carcinogens, they will identify risk assessment methods in which they have confidence. At that time, they may well base their recommended TLVs, in part, on model-derived cancer risks (Alavanja et al., 1990).

It is noteworthy that in their recent deliberations on benzene and formaldehyde, the TLV committee considered all relevant information, including low-dose extrapolation models, physiologically based pharmacokinetic models (PB-PK), biologically based cancer models (MVK), and other information before they made their recommendation.

2.8 The Role of Dose–Response Models

Even though the TLV committee, as well as many other groups that recommend exposure limits, may believe that there is likely to be a threshold for carcinogens at very low doses, another equally credible school of thought is that there is little or no evidence for the existence of thresholds for chemicals that are genotoxic (Bailer, et al., 1988; Travis and Hester, 1990). In an attempt to take into account the philosophical postulate that chemical carcinogens do not have a threshold even though a no observed effect level (NOEL) is often observed in an animal experiment, modeling approaches have been developed (Crump et al., 1976; Paustenbach, et al., 1990a; Paustenbach, 1990b).

The rationale for a modeling approach is that it is impossible to conduct toxicity studies at doses near those measured in the environment because the number of animals necessary to elicit a response at these doses would be too great (Crump et al., 1976; Perera, 1984). Consequently, results of animal studies conducted at high doses are extrapolated by models to those levels found in the workplace or the environment. By the early 1980s, mathematical modeling approaches for evaluating the risks of exposure to carcinogens became popular among various regulatory agencies. Even though the limits recommended by these models have rarely been the sole factor on which regulatory limits have been established, they have significantly influenced the values selected (Rodricks et al., 1987; Travis et al., 1987; Finley et al., 1993).

The most popular models for low-dose extrapolation are the one-hit, multistage, Weibull, multihit, Logit, and Probit. Some are presented in Figure 7.3. Since it is usually presumed for genotoxic carcinogens that at any dose, no matter how small, a response could occur in a sufficiently large population, an arbitrary risk level is usually selected (i.e., 1 in 1000 to 1 in 1,000,000) as presenting an insignificant or de minimus level of risk. By identifying these de minimus levels as virtually safe levels, regulatory agencies do not give the impression that there is an absolutely safe exposure or that there is a threshold below which no response would be expected. Having calculated an exposure level or virtually safe dose (VSD), regulatory agencies can begin to make judgments about the biologic and economic feasibility and reasonableness of establishing a standard based on mathematical modeling. Often the use of models to help assess risks of exposure to carcinogens has been erroneously called ''risk assessment'' (Paustenbach, 1989). In practice, however, modeling is only one part of the risk assessment process (Park and Snee, 1983). A true risk assessment to determine safe levels of occupational exposure requires exhaustive analysis of all of the information obtained from studies of mutagenicity, acute toxicity, subchronic toxicity, chronic studies in animals, and metabolism data as well as human epidemiology data before a limit is recommended (Paustenbach, 1990b).

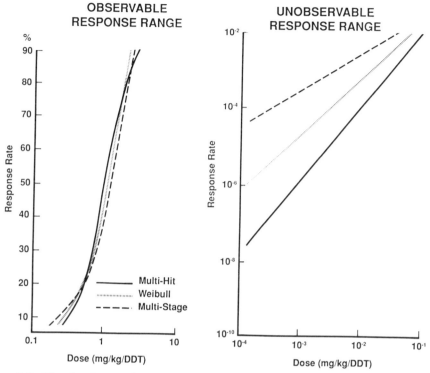

Figure 7.3 The fit of most dose–response models to data in the range tested in animal studies is generally similar. However, because of the differences in the assumptions on which the equations are based, the risk estimates at low doses can vary dramatically between the different models. (From Paustenbach et al., 1990b.)

At this time, the use of quantitative risk modeling can be useful in the overall process of setting occupational exposure limits, but because of the dozens of short-comings associated with them, especially their inability to consider complex biological events that surely must occur at low dose levels, they are not used as the sole basis for deriving OELs. Nonetheless, it has been proposed that the TLV and other OEL committees at least consider the results of this model when selecting a recommended guideline (Infante, 1992).

Several papers have compared the plausible cancer risk for workers exposed to the TLVs using models with risks often deemed acceptable by the EPA and FDA (Alavanja et al., 1990; Rodricks, et al., 1987; Paustenbach, 1990b). The results are shown in Table 7.7. The occupational risk issue is even more interesting when one considers the estimated steady-state tissue level following chronic exposure to the TLV versus that due to background exposure to the same chemicals in our diet (Table 7.8) (Leung and Paustenbach, 1988b).

The principal reason for the wide disparity between the EPA and the CS–TLV committee can be explained primarily by the underlying philosophical principles governing the two organizations rather than the technical differences between the

Table 7.7 Model Derived Estimates of Lifetime Risks of Death from Cancer per 1000 Exposed Persons Associated with Occupational Exposure at Pre-1986 and Post-1987 OSHA Permissible Exposure Limits (PELS) for Selected Substances

Substance	Cases/1000 at Previous PEL	Cases/1000 at Revised PEL
Inorganic arsenic	148–767	8
Ethylene oxide	63–109	1–2
Ethylene dibromide (proposal)	70–110	0.2–6
Benzene (proposal)	44–152	5–16
Acrylonitrile	390	39
Dibromochloropropane (DBCP)	—	2
Asbestos	64	6.7

Source: Table reprinted from Rodricks et al. (1987); reprinted with permission of *Regulatory Toxicology and Pharmacology.*

two scientific methods. The CS–TLV committee is governed by the precept that "threshold limit values refer to airborne concentrations of substances and represent conditions under which it is believed that nearly all workers may be repeatedly exposed day after day without adverse effect" (ACGIH, 1992). In contrast, when the EPA must promulgate regulations under the Clean Air Act they are compelled to be more conservative. For example, the Clean Air Act states that primary air standards must protect the public health with an adequate margin of safety. The requirement for an "adequate margin of safety" is intended both to account for inconclusive

Table 7.8 Estimated Steady-State Adipose Tissue Concentration of Chemicals Following Chronic Exposure at the OEL Compared With the Levels Due to Background Exposure Alone

Chemical	OEL	$t_{1/2}$ (yr)	Adipose Tissue Level Background[a]	Exposed[b]	E/B[c]
DDT[d]	1 mg/m^3	1.5	6 ppm	480 ppm	80
Dieldrin	0.25 mg/m^3	1	0.29 ppm	80 ppm	276
PCB[e]	1 mg/m^3	2.5	1 ppm	800 ppm	800
TCDD[f]	0.2 ng/m^g	8	7 ppt	180 ppt	26

Source: Leung and Paustenbach (1988b).

[a] Background levels refer to those in nonoccupationally exposed general population.

[b] The levels in persons occupationally exposed to the OEL are calculated with the equation presented in Calabrese (1978).

[c] E/B = ratio of predicted steady-state adipose tissue level in persons occupationally exposed at the current TLV versus that measured in persons exposed to background levels.

[d] DDT-dichlorodiphenyl trichloroethane.

[e] PCB-polychlorinated biphenyl.

[f] TCDD-2,3,7,8-tetrachlorodibenzo-*p*-dioxin.

[g] OEL is the value suggested by Leung and Paustenbach (1988).

scientific and technical information and to provide a reasonable degree of protection against hazards that research has not yet identified.

Recognizing that imposing zero emission for some substances could produce a heavy economic burden on society, EPA has addressed the problem by proposing that best available technology (BAT) be used to control carcinogens. If BAT controls leave an unreasonable residual risk, further controls will be considered (Alavanja et al., 1990). The use of models that are conservative and adoptions of a 1 in 100,000 or 1 in 1,000,000 risk criterion have been justified by EPA because of a strong desire to protect virtually everyone in the public including the aged, young, and infirmed and to account for the fact that they can be continually exposed for 70 years.

2.9 Shortcomings in Extrapolating Animal Data

Part of the difficulty in extrapolating from animals to humans is the inability to predict species differences. Species differences in the uptake, metabolism, and elimination as well as in the sensitivity of the animal to chemicals are well documented (Williams, 1974). In many cases, even when physiological differences are accounted for mathematically (Andersen, 1981), the extrapolation of animal data to human is not always predictable. Therefore, some arbitrary safety factor is usually used that allows for the possibility that humans will be more sensitive to the substance than the species tested.

There is another difficulty that is often not recognized in extrapolating toxicity data from animal tests to occupational standards; animal tests are conducted under different conditions than occupational exposures. For example, inhalation toxicity tests are often limited to a few weeks while exposures in the workplace could last for as many as 40 years. Another difference in testing is that animals are often exposed for only 4–6 hr a day. While these short exposure periods are convenient for laboratory personnel who must load and unload inhalation chambers or manually dose the animals, the results of these tests are used to set TLVs for humans exposed as many as 6–12 hr per day. Another shortcoming is that the physical activity of animals during an inhalation toxicity test is minimal since test animals are usually confined to small cages that restrict movement. The result is that the rat, for example, will have a ventilation rate as much as two- to fivefold lower than the normal rate. The impact of exertion or strenuous work in the uptake and elimination of airborne toxicants has been studied by a number of persons (Astrand and Gamberale, 1978; Astrand, 1975; Droz and Fernandez, 1977; Zenz and Berg, 1970; Astrand et al., 1972). Lastly, by dosing during the sleep phase of the test animal, the potential effects of the circadian rhythm on the toxic response can occur. The applicability of these results versus those that would be obtained during a night exposure schedule continue to be of interest to investigators. For some chemicals, these effects can be dramatic (Craft, 1970). Another shortcoming of the toxicity testing is that the concentration of the test substance in the inhalation chamber is maintained at a constant level. In contrast, 94 percent of the TLVs are not constant exposure limits but are TWA limits (Stokinger, 1970). This means that the workplace concentration of these substances could deviate above the TLV as long as the average concentration was at, or below, the TLV. During the 1970s and a portion of the 1980s, a maximum excursion level was recommended to prevent excessively high exposures for short periods of time, and these excursions were based on the magnitude of the TLV. The excursion factors recommended up until 1976 were: 3-fold for TLVs from 0 to 1 ppm, 2-fold for TLVs from 1 to 10 ppm, 1.5-fold for TLVs from 10 to 100 ppm, 1.25-fold for TLVs from 100 to 1000 ppm.

Because the general formula approach to adjusting TLVs for short periods had shortcomings, the ACGIH TLV committee in 1976 began adopting short-term exposure limits STELs for each of the chemicals. Because the first set of STELs was established in a generic manner, the ACGIH TLV committee decided in 1984 that these 15-min limits lacked strong scientific merit. As a result, they began the task of dropping those for which they thought there was inadequate data to establish such a limit. They continue to consider and evaluate pharmacokinetic approaches for setting new STELs. There continues to be interest in identifying a method that can predict the lesser or greater hazard posed by exposure to fluctuating air concentrations of chemical vapors. It has been shown that for some chemicals, more severe adverse effects may occur when the animals are exposed to intermittent doses than when exposed to the same dose (ppm-hr) delivered using a constant concentration (van Stee et al., 1982). This may be particularly important for some carcinogens. Historically, it has been held that the average daily dose is what dictates the cancer risk.

Fortunately, the availability of desk-top computers and physiologically–based pharmacokinetic (PB-PK) models should soon allow toxicologists to estimate the impact of pulsed doses and different routes of exposure on the severity of toxic effect. These models allow us to adjust for the numerous physiological factors (Andersen et al., 1987a; Leung, 1991). For example, it has been well established that biological response to inhaled substances is a function not only of concentration and time but also of uptake characteristics. Some substances well tolerated under TLV conditions may become hazardous when exposure is continuous, while other substances may be tolerated well at TLV levels even for continuous exposure. This parameter can be accounted for using PB-PK models. This can be important since the use of nonnormal work schedules by industry can be anticipated to become significant, thus justifying the use of computer models to adjust OELs. Third, the variation of concentrations for some chemicals in workroom air may be more hazardous than exposure to a constant level of an air contaminant. It is well known that variable concentrations are the rule rather than the exception in most workplaces.

3 CORPORATE OCCUPATIONAL EXPOSURE LIMITS

It has been claimed that the implementation of OELs has been instrumental for the near elimination of serious occupation disease in the Western world. Although exposure limits or guides for most large-volume chemicals have been established, the majority of the 10,000 chemical routinely used in industry do not have them. As a result, many firms have chosen to establish internal or corporate limits to protect their employees as well as the persons who purchase those chemicals (Paustenbach and Langner, 1986).

The process by which internal limits are set is generally initiated by the manufacturing divisions, although the Corporate Health, Safety and Environmental Affairs Department may also initiate the process. A panel of toxicologists, industrial hygienists, physicians, and epidemiologists usually gather the scientific data and make the technical assessment much like the ACGIH TLV committee. The data considered by the group are similar to that considered by the TLV committee (Table 7.9). Their deliberations often are reviewed by an oversight group, which integrates the scientific input with information provided by the business, law, regulatory, and other groups to establish, as appropriate, internal exposure levels, including in some instances, maximum exposure levels and short-term exposure levels (Paustenbach and Langner, 1986).

The oversight group often includes individuals from occupational health, industrial hygiene, toxicology, product safety, medical, law, and the principally affected

Table 7.9 Data often Used in Developing an Occupational Exposure Limit

Physical properties
  Lipid solubility
  Water solubility
  Vapor pressure
  Odor threshold
Acute toxicity data
  Oral toxicity, LD_{50}
  Dermal toxicity, LD_{50}
  Dermal and eye irritation
  Inhalation toxicity, LC_{50}
Subacute and subchronic data (oral, dermal, or inhalation)
  14 day, NOEL
  90 day, NOEL
  6 month, NOEL
Other data
  Developmental (teratology and embyotoxicity)
  Mutagenicity (Ames test, drosophilia, etc.)
  Fertility
  Reproductive (3-generation)
  Reversability study
  Dermal absorption test
  Pharmacokinetics
  Cancer bioassay (2 yr)
Epidemologic data
  Morbidity
  Mortality
  Base reports
Industrial hygiene exposure data
  Area samples
  Personal samples

Source: Paustenbach and Langner (1986).

operating or product group. Finally, the rationale for the internal exposure level is documented. Figure 7.4 illustrates the process for setting internal exposure levels typical of those used by numerous firms.

Several philosophical underpinnings should be accepted by firms who set these limits. Foremost is the concept that guidelines are needed whenever employees are being exposed. Second, firms should document the rationale for establishing their guides. Third, if adequate toxicology data are not available, it would still be valuable to set tentative exposure limits for a chemical based on exposure levels that have been measured and found to be acceptable. Since scientific data are generally lacking for establishing ''exact'' exposure levels, this approach is reasonable and more prudent then simply waiting until adverse affects are observed.

Most firms who have established OELs believe that the management of occupational health concerns requires criteria much like a manufacturing group needs quality control criteria. Without these limits as guides, operations managers would not know when conditions are unhealthy, when personnel need to be protected and, if monitoring is performed, how the results should be interpreted.

Nearly all the firms who set OELs have found that one of the most difficult and controversial aspects is the legal ramifications. For example, lawyers have noted that if a company develops internal standards on their chemicals or if they choose to adopt values for a chemical that are more conservative than a regulatory agency, the firm had best plan to comply with them. On the other hand, most lawyers agree that perhaps an equal legal exposure exists with those firms who know a great deal about the potential hazards of a chemical yet do not set internal limits. Admittedly, such a scenario puts manufacturers between a rock and a hard place. For example, some firms may feel that the workmen's compensation immunity does not encourage them to set internal limits on their own chemicals. It is, however, worth bearing in mind that as the manufacturer of a chemical they could be sued by someone else's employee who, if injured, could claim that they did not supply enough data.

4 SHIFTWORK AND UNUSUAL WORK SCHEDULES

The concern about the adverse health effects of night shift work has existed for over 100 years. In 1860, for example, some scientists were worried that bakers might be at increased risk of physical and emotional illness because they always worked at night and, subsequently, some effort was made to try to regulate their work hours and their working conditions (Rentos and Shepard, 1976). Since then, it has been shown that some persons who have been very productive while working standard 8-hr/day, 40-hr/week daytime schedules can become especially fatigued, unhappy, less productive, and perhaps even more susceptible to the effects of chemical agents and physical agents after they are placed on shift work (see supplemental references). Even though a number of studies have been conducted, the degree to which shift work effects a worker's capabilities, longevity, mortality, and overall well-being is still not fully understood.

So-called unusual work shifts and work schedules have been implemented in a number of industries in an attempt to eliminate or at least reduce, some of the prob-

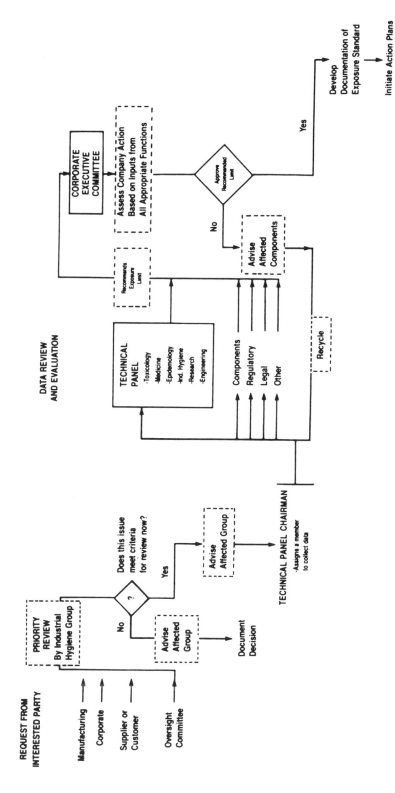

Figure 7.4 This figure illustrates the organization and methods often used to set corporate occupational exposure limits (see Paustenbach and Langner, 1986).

lems caused by normal shift work that require three work shifts per day. These unusual or "other-than-normal" shifts have been termed odd, novel, extended, extraordinary, compressed, nonnormal, nonroutine, prolonged, exceptional, nonstandard, unusual, peculiar, weird, and nontraditional (Brief and Scala, 1975; Paustenbach, 1985). In 1981 the AIHA established a committee to address the potential occupational health aspects of shift work and the need to adjust exposure limits for persons who work schedules that were markedly different than the "normal" workweek, which consists of five consecutive 8-hr daylight workdays followed by two days off. At the committee's first meeting, it was agreed that the term *unusual work shift* should be used, for sake of consistency, to describe these other-than-normal shifts.

In general, most unusual work schedules will involve workdays markedly longer than 8 hr in duration, however, because many persons are regularly exposed to high concentrations of xenobiotics for very short periods during shifts and, because TWA occupational exposure limits were not necessarily intended for use during short exposures, these exposure periods were also classified as "unusual" by the committee. The assessment of the health aspects of unusual shifts is complex since some of these schedules require the worker to alternate between night and day work every few days (rapid rotation). These schedules have been called rapidly rotating, fast, or simply rapid-roto shifts (Brandt, 1969; Yoder and Botzum, 1971; Wilson and Rose, 1978; Botzum and Lucas, 1980; Vokac et al., 1981; Knauth et al., 1983; Smolensky et al., 1985).

In order for the health professional to ensure protection of workers who are exposed to airborne chemicals during unusually long or unusually short work periods, he or she should be familiar with the toxicology and pharmacokinetics of the chemical of interest as well as understand the rationale for its occupational exposure limit, since most limits were established only for exposure during a normal 8-hr/day, 5-day/workweek schedule.

4.1 Background on Shift Work

Traditionally, a fixed work schedule is established by the employer and consists of five 8-hr days each week starting at 8 a.m. and ending at 5 p.m. Initially, the need to operate certain manufacturing processes 24 hr/day served as the impetus for using 3 shifts of workers to cover the 24-hr workday. Later, shift work was implemented because the economics of having idle equipment 70 percent of the week was prohibitive. Of course, the use of three shifts per day allows uninterrupted production throughout the year so that a process never has to shut down. Many continuous process operations, such as those found in oil refineries, chemical plants, steel and aluminum mills, pharmaceutical manufacturing, glass plants, and paper mills, cannot be shut down without causing serious production and financial losses so they require round-the-clock manning (Brief and Scala, 1975; Wilson and Rose, 1978).

In many professions adherence to the 40-hr/week schedule is unusual. For example, craftspeople and equipment repair persons frequently work beyond 8 hr/day (overtime) to accommodate routine equipment repair, fill in for absent workers, or deal with seasonal fluctuations in demand (Wilson and Rose, 1978). For some persons, such as those in the railroad industry and the military, a workday of 16–24 hr is not uncommon. In certain industries, very complicated work schedules that have involved both short and long periods of work have been implemented for a wide number of reasons. The advantages and disadvantages of alternative (unusual) work schedules, as well as the number of persons involved in various types of work schedules have been discussed (USBLS, 1979; Nollen and Martin, 1978; Nollen, 1981, 1982).

Dozens of alternatives to traditional work schedules have been developed and are now used in a number of industries. As observed during the recession of the 1980s and early 1990s, the length of time worked by a person was often altered by the use of part-time employees and many workweeks were shortened or lengthened due to wide fluctuations in the demand for a given product. Overall, the high cost of manufacturing equipment, increased foreign competition, and the inability to halt certain chemical and physical processes after they have begun has forced industry to make shift work and modified work schedules a permanent part of manufacturing (Wilson and Rose, 1978).

The shift worker employed in a 24-hr/day operation, usually on a rotating, 8-hr schedule, faces many disruptive events that arise due to his or her work schedule. Overall, the primary complaint of these workers surrounds the unsatisfactory impact shift work has on their social lives (Thiis-Evensen, 1958; Yoder and Botzum, 1971; Hedges, 1975; Dunham and Hawk, 1977; Fottler, 1977; Wilson and Rose, 1978; Botzum and Lucas, 1980). When they are home, most other persons are asleep, at work, or at school. In general, these shift workers have only one full weekend per month during which they are not at work. As a result, shift workers and members of their family are frequently disappointed with the quantity and quality of time that they spend together (Taylor, 1969; Wheeler et al., 1972; Kenny, 1974; Wilson and Rose, 1978; Botzum and Lucas, 1980; Landry, 1981).

Researchers who have studied the effects of shift work on human health have noted that shift workers may have trouble sleeping, often feel fatigued, often feel overly tired during days off work, and are often chronically irritable (Taylor, 1969; Smolensky et al., 1978; Smolensky, 1980; Landry, 1981; Smolensky et al., 1985; Smolensky, 1993). Other undesirable effects of shift work that are more easily measured have included constipation, gastritis, gastroduodenal ulcers, peptic ulcers, high absenteeism, and lessened productivity (Bjerner et al., 1955, 1964; Colquhoun et al., 1968a,b, 1969; Colquhoun, 1971; Johnson et al., 1981). As a consequence, numerous kinds of unusual work schedules have been developed and implemented by companies who are seeking alternatives to standard shift work. Throughout this chapter, standard shift work is always defined as a workweek of 8 hr/day and 5 days that occurs during daylight hours, usually between 8 a.m. to 5 p.m., and is followed by 2 days off work.

4.2 Unusual Work Schedules

One kind of work schedule classified as unusual is the type involving work periods longer than 8 hr and varying (compressing) the number of days worked per week (e.g., a 12-hr/day, 3-day workweek). Another type of unusual work schedule that involves a series of brief exposures to a chemical or physical agent during a given work schedule (e.g., a schedule where a person is exposed to a chemical for 30 min, 5 times per day with 1 hr between exposures). Another type of unusual schedule is that involving the "critical case" wherein persons are continuously exposed to an air contaminant (e.g., spacecraft, submarine).

Compressed workweeks are a type of unusual work schedule that has been used primarily in nonmanufacturing settings. It refers to full-time employment (virtually 40 hr/week) that is accomplished in less than 5 days/week. Many compressed schedules are currently in use, but the most common are (a) 4-day workweeks with 10-hr days; (b) 3-day workweeks with 12-hr days; (c) 4.5-day workweeks with four 9-hr days and one 4-hr day (usually Friday); and (d) the 5/4, 9 plan of alternating 5-day and 4-day workweeks of 9-hr days (Nollen and Martin, 1978; Nollen, 1981, 1982).

Over the past two decades, unusual work shifts and schedules including some form of the compressed workweek have been implemented in many manufacturing facilities (Brief and Scala, 1975; Hickey and Reist, 1977). Examples of the types of schedules and one type of industry that has used them include four 10-hr workdays per week (chemical); a 6-week cycle of three 12-hr workdays for 3 weeks followed by four 12-hr workdays for 3 weeks (pharmaceutical); a 6-hr per day, 6-day workweek (rubber); a 56/21 schedule involving 56 continuous days of work of 8 hr per day followed by 21 days off (petroleum); a 14/7 schedule involving 14 continuous days of work of 8–12 hr per day followed by 7 days off (petroleum); a 3/4 schedule involving only three 12-hr workdays in one week followed by a week of four 12-hr workdays (pharmaceutical); four 12-hr workdays followed by three 10-hr workdays, followed by five 8-hr workdays then 4 days off; five 8-hr workdays, followed by two 12-hr days, then 5 more 8-hr workdays followed by 5 days off (petrochemical); a 2/3 schedule involving 18 hr of work for 2 days then 3 days off (military); and numerous other variations of these (Brief and Scala, 1975; Brandt, 1969; Wilson and Rose, 1978). Of all workers, those on unusual schedules represent only about 5 percent of the working population (Nollen, 1982). Of this number, only about 50,000–200,000 Americans who work unusual schedules are employed in industries where there is routine exposure to significant levels of airborne chemicals. In Canada, the percentage of chemical workers on unusual schedules is thought to be even greater than in most other countries.

5   CHEMICAL PHARMACOKINETICS

Pharmacokinetics is defined as the study of the rate processes of absorption, distribution, metabolism, and excretion of drugs and toxicants in intact animals (Gibaldi and Perrier, 1975). Many of the approaches or models that will be described for adjusting exposure limits for both a series of short and very long or continuous exposure periods are based on the pharmacokinetics of the inhaled toxicant.

Pharmacokinetic modeling is the science of describing, in mathematical terms, the time course of drug and metabolite concentrations in body fluids and tissues (O'Reilly, 1972; Withey, 1979; O'Flaherty, 1987). Most often, the modeling of a chemical's behavior is based on the blood concentration of a contaminant versus the time course of its concentration in the blood following exposure. It is generally assumed that blood is in dynamic equilibrium with all tissues and fluids of the body; consequently, it is the biologic medium of choice for monitoring. In some cases it may be necessary or useful to measure the output of its parent chemical or drug and the metabolites in the urine, breath, and feces and then model this profile mathematically.

A pharmacokinetic model is regarded as being accurate if it correctly predicts the concentration of a substance or its metabolite in the blood, urine, breath, or feces for as long as these substances can be measured analytically. The model may be tested by comparing its predictions with experimental data after exposure to the drug by intravenous, oral, or inhalation administration. The effects of various dosing regimens can also be modeled. Usually, recovery of +95 percent of the administered dose(s) as unchanged drug and metabolites in urine, breath, and feces, and agreement between the recovered amounts and the pharmacokinetic model is considered confirming evidence that the model accounts for the fate of the absorbed chemical or drug (Dittert, 1977).

The principles of industrial pharmacokinetics (i.e., absorption, distribution, metabolism, and excretion) are discussed here with emphasis on those principles necessary to understand the various models for adjusting occupational exposure limits for unusual periods of exposure.

5.1 Absorption of Chemicals

For workers, the two predominant routes of entry for industrial chemicals are absorption through the respiratory tract and the skin. Absorption that takes place through the lungs can be roughly assessed through analysis of the air that is breathed. The results can be interpreted by comparison with an exposure limit. Absorption via the skin, on the other hand, evades this method of measuring exposure. Uptake by the skin cannot be directly accounted for except by biological monitoring. It is acknowledged that methods for analyzing most chemicals in various excretory pathways or fluids have not been developed for most chemicals; therefore, when it is anticipated that a significant portion of the daily dose may be due to skin absorption, it can be assumed that a proportionally greater amount will be absorbed by persons who work on shifts longer than 8 hr/day. Indirect methods such as those used in health risk assessment are another approach to estimating dermal uptake (Paustenbach et al., 1992).

Absorption of organic vapors may take place both in the upper and in the lower parts of the respiratory tract. Two basic processes may be observed here depending on the physical properties of the toxic substance. If the toxic substance is present in air in the form of an aerosol, absorption will often be preceded by deposition of the substance in the upper respiratory tract. When very small particles are present (less than 10 μm diameter), the deposition will be in the alveolar region (Doull et al., 1990).

The basic absorption mechanism for most industrial chemicals is gas diffusion, and the factor that predicts the efficiency of this process is the partition coefficient between air and blood. Values for the coefficient between air and blood have recently been estimated for only a few substances (Gargas et al., 1989). The coefficients for partition between air and water are available for a far greater number of compounds. Usually, these have been obtained while working out sampling procedures involving absorption of a gaseous substance in water. By knowing the oil/water partition coefficient, one can often predict the relative efficiency of absorption through the skin and the lung. In addition, the partition coefficient and degree of solubility in water of a substance can give some insight as to the degree of distribution among the various tissues and the likely rate of elimination. For chemicals that have not been studied pharmacokinetically, these physical properties (especially the partition coefficient) can be very useful for predicting the chemical's behavior during unusually long exposures (Piotrowski, 1977).

The respective values of the air/water partition coefficients for organic compounds may vary by several orders of magnitude: from about 10 for carbon disulfide, through 10^{-3} for acetone, acrylonitrile, and nitrobenzene, to 10^{-5} for aniline and toluidine. The degree of pulmonary absorption (i.e., retention of vapors in the lungs) increases with a decreasing partition coefficient for air/blood (water); however, variation of the retention is much less than the variation of the respective partition coefficients. For example, lung retention of aniline vapor (partition coefficient 10^{-5}) is about 90 percent; nitrobenzene (partition coefficient 10^{-3}) up to 80 percent; benzene (partition coefficient 10^{-1}) up to about 50–75 percent; and finally, carbon disulfide (partition coefficient 10^{0}) up to about 40 percent (Piotrowski, 1977).

5.2 Concept of Half-Life

Most biologic processes follow first-order kinetics. In other words, the rate of elimination or metabolism is known to be a function of the concentration of all reactive species and, where there is only one of these, the reaction is termed first-order and the rate is directly proportional to the concentration:

$$\text{Rate} = dC/dt = kC \tag{1}$$

This is the differential form of a first-order reaction and is of little practical use since dC/dt, although it can be found from the tangent of the time–concentration curve, is very difficult to measure precisely.

The integrated form of Eq. (1) is more useful:

$$C_t = C_0 e^{-kt} \tag{2}$$

C_0 is the initial concentration and is sometimes designated a; it is not unusual to find C_t, the concentration at time t, designated $a - x$, where x is the amount that has disappeared at the time t. Thus Eq. (2) sometimes appears as

$$\ln \frac{a}{a - x} = kt \tag{3}$$

By definition, the half-life ($t_{1/2}$) for any first-order kinetic process is the time taken for the original amount or concentration to be reduced by one-half. So from Eq. (3), if $C_t = \frac{1}{2}C_0$ at time $t_{1/2}$, then

$$\ln \frac{C_0}{C_0/2} = kt_{1/2}$$

$$\ln 2 = kt_{1/2}$$

$$t_{1/2} = \frac{\ln 2}{k} = \frac{2.303 \times 0.3010}{k} \qquad \text{(e.g., min, hr)} \qquad (4)$$

Note that $t_{1/2}$ is independent of concentration for a first-order process and that the larger the half-life, the slower the elimination or reaction rate, that is, the smaller the rate coefficient k. The fundamental concepts are illustrated in Figure 7.5.

When describing the elimination of a chemical from humans or experimental animals, it is often convenient to use the term *biologic half-life*. The biologic half-life of a chemical is the time needed to eliminate 50 percent of the absorbed material, either as the parent compound or one of its metabolites. As will be shown, the biologic half-life is the most important criterion for assessing whether a TLV should be adjusted for either very short or very long durations of exposure. The second key factor is the type of adverse effect or hazard posed by the chemical. The biologic half-life and the exact exposure regimen (time exposed and recovery time) are the two factors that determine the degree of adjustment needed to provide predicted equal protection.

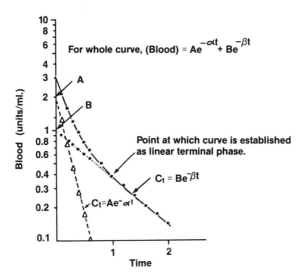

Figure 7.5 Semilogarithmic plot of blood concentration versus time after intravenous administration illustrating biphasic elimination (two compartments) and the use of ''curve stripping'' to resolve the A intercept.

5.3 Concept of Steady State

The concept of steady state can be a difficult one for persons to comprehend. It is easy to visualize how persons who are exposed to a chemical with a short half-life will rapidly eliminate the chemical so that the blood and tissue levels of the chemical return to zero before returning to work the next day (e.g., carbon disulfide). The other extreme is represented by chemicals such as the 2,3,7,8 tetrachloro-dibenzo-*p*-dioxin (TCDD), which have extremely long half-lives (whole body bio-logic half-life = 7 to 11 years). Here, the body burden never returns to zero before reexposure occurs so each successive dose adds to the existing burden. These phenomena are shown in Figure 7.6.

The goal of the modeling approach to adjusting occupational limits, however, is to identify a dose that ensures that the daily peak body burden or weekly peak body burden does not exceed during unusual work schedules that which occurs during a normal 8-hr/day, 5-day/week shift. The potential for the body burden to exceed

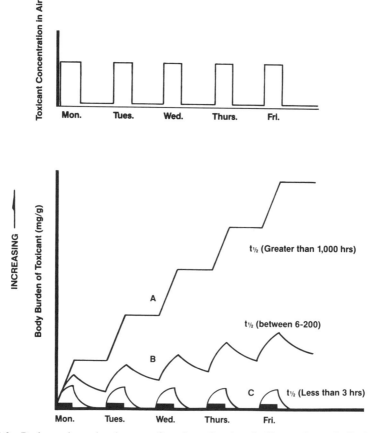

Figure 7.6 Body or tissue burden as a function of a chemical's uptake and elimination rate (i.e., biologic half life).

normal levels during unusually long work periods exists whenever the biologic half-life for the chemical in man is in the range of 3–200 hr (see Fig. 7.7). The key point to understand here is that for any chemical, a steady-state plasma or tissue level will eventually be achieved following regular exposure to any work schedule and even during continuous exposure. At first glance, most persons might think that with continuous exposure, the body burdens will continuously increase as long as exposure is maintained. For any chemical, a steady-state blood or tissue level of the contaminant will be achieved when the rate of elimination of the drug equals the rate of absorption (Notari, 1985; Gibaldi and Perrier, 1975.) (See Figure 7.8.) A rule-of-thumb is that steady-state body burdens occur when exposure occurs for a period greater than five biologic half-lives. For example, for TCDD, the steady-state concentrations in adipose tissue occurs after about 55–60 years (5 × 11 yr).

During continuous inhalation exposure to volatile workplace gases and vapors, the concentration of the chemical in the blood increases toward an equilibrium between absorption, on the one hand, and metabolism and elimination, on the other. This is accompanied by a decreasing retention of the absorbed gases and vapors during each breath. This decrease of retention during the early periods of continuous inhalation exposure may be observed in practice for compounds whose air/blood (water) partition coefficient is of the order of 10^{-3} or greater. Decreasing retention is characteristic for carbon disulfide, tri- and tetrachloroethylene, benzene, toluene, nitrobenzene, and other chemicals (carbon disulfide is illustrated in Fig. 7.9). If

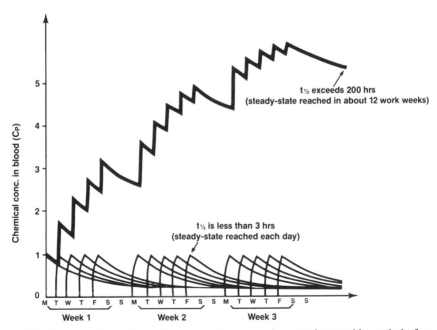

Figure 7.7 The principle of graphic summation assuming regular weekly periods free of exposure (From: Piotrowski, 1971a). Illustrates that biologic half-life determines the degree of day-to-day accumulation and the time to steady-state.

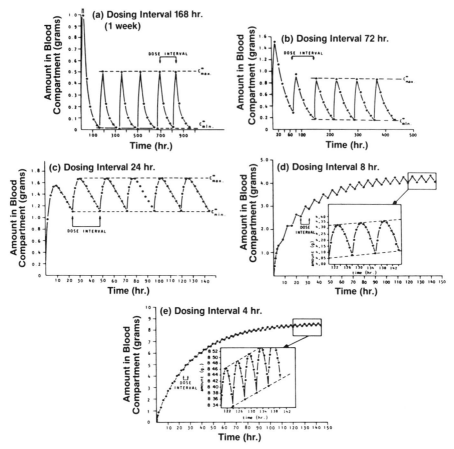

Figure 7.8 Multiple-dosing curves arising from one-compartment pharmacokinetic model with first-order absorption and elimination. These curves illustrate that the steady-state blood level and the time to reach steady state are dependent on the period of time between exposures and the biologic half-life of the substance. [Reprinted with permission from Withey (1979).]

metabolism and distribution are rapid, this phenomenon is less pronounced and, for example, has not been observed for aniline (low partition coefficient) or for styrene (Piotrowski, 1977). In the case of styrene, analogous to acrylonitrile, the explanation for lack of time-dependent decrease of retention seems to lie not so much in the magnitude of the physical partition coefficient as in the metabolism of the chemical.

The overall or average retention (R) of an organic vapor has been studied on human volunteers in chamber-type experiments, where it can be determined for a particular exposure period directly from the ratio of concentrations of the chemical in the inhaled and expired air:

$$R = \frac{C_i - C_e}{C_i} \tag{5}$$

C_i and C_e denote the concentrations in inhaled and exhaled air, respectively.

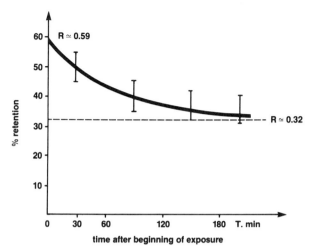

Figure 7.9 Retention of carbon disulfide vapors by the respiratory tract with continuous exposure. (Based on work of Jakubowski, 1966.)

It is often useful for the industrial hygienist to know how to calculate an individual's anticipated daily uptake (dose) of a toxicant. One of the key determinants or elements in these calculations is the ventilation rate. In chamber-type experiments, in which volunteers are exposed in a sitting position and not subject to additional physical effort, the ventilation rate is usually of the order of 0.3–0.4 and 0.4–0.5 m^3/hr, in females and males, respectively (Piotrowski, 1977). Differences are due to the differences in body weights and heart rates. It can be assumed that for people engaged in light work, corresponding to a slow walk, the rate is at least doubled. Specific ventilation rates have been determined for hundreds of tasks and should be used when applicable (Durnin and Passmore, 1967).

In calculating the amount of a substance absorbed through the pulmonary tract during industrial work activities, various authors have used different average ventilation rates as typical for workers performing light work: from about 0.8 m^3/hr to about 1.25 m^3/hr. The latter figure (10 m^3 per 8-hr working shift) is most often used by industrial health professionals in the United States since it is almost always higher than actual levels and, therefore, conservative.

Taking the above factors into account, the total amount of a substance absorbed through the respiratory tract over a period of exposure would be the product of air concentration (C), duration of exposure (T), ventilation rate (V), and average fractional retention rate (R) for the time of exposure:

$$C \ (\text{mg}/\text{m}^3) \times T \ (\text{hr}) \times V \ (\text{m}^3/\text{hr}) \times R = \text{absorbed dose} \qquad (6)$$

This simple formula is useful for estimating the acceptability of an air concentration of a toxicant in the workplace if the person is to be exposed for periods markedly longer than 8 hr/day or 40 hr/week (Leung and Paustenbach, 1988b). Even if the industrial hygienist knows little else about the chemical's biologic or physical properties, he or she can limit the absorbed dose during the longer workday to that expected for a normal workday by calculating the air concentration for the nonnormal condition, which would yield the same daily dose allowed for the normal schedule.

6 PHARMACOKINETIC MODELS

The dynamic behavior of toxic substances can be described in terms of mathematical compartments in which a compartment represents all of the organs, tissues, and cells for which the rates of uptake and subsequent clearance of a toxicant are sufficiently similar to preclude kinetic resolution. This simplification of the body introduces few errors in the final analysis and is much more realistic than attempting to describe the fate of a chemical in each major tissue (e.g., liver, heart, lung, etc.). The mathematical description of how a chemical behaves in a living organism requires a model.

In general, most models divide the body into from one to five compartments. Fiserova-Bergerova et al. (1980) have suggested an approach to compartment categorization based on perfusion, ability to metabolize the inhaled substance, and the solubility of the substance in the tissue. Lung tissue, functional residual air, and arterial blood form the central compartment LG (lung group) in which pulmonary uptake and clearance take place. The partial pressure of inhaled vapor equilibrates with four peripheral compartments. Vessel-rich tissues form two peripheral compartments: BR (blood rich) compartment includes brain, which lacks capability to metabolize most xenobiotics, and is treated as a separate compartment because of its biological importance and the toxic effect of many vapors and gases on the central nervous system. VRG (vessel-rich group) compartment includes tissues with sites of vapor metabolism such as liver, kidney, glands, heart, and tissues of the gastrointestinal tract. Muscles and skin form compartment MG (muscle group), and adipose tissue and white marrow form compartment FG (fat group). The FG compartment is treated separately, since the dumping of lipid-soluble vapors in this compartment has a smoothing effect on concentration variation in other tissues, and these variations caused by changes in exposure concentrations, minute ventilation, and exposure duration would be more dramatic if not for the buffering effect of adipose tissue.

An example of a complex pharmacokinetic model that describes the possible fate of an inhaled substance is depicted in Figure 7.10. An illustration of the possible fate of a common solvent like tetrachloroethylene in the various imaginary compartments is shown in Figures 7.11 and 7.12. The key point illustrated here is that fat- or lipid-soluble chemicals will quickly reach equilibrium in the highly perfused tissues, and they will also be removed quickly following cessation of exposure. However, the less perfused tissues, which are usually high in lipid, do not reach saturation quickly and, more importantly, they take a great deal of time to reach background levels (Paustenbach, et al., 1988b).

6.1 One-Compartment Model

The simplest pharmacokinetic model is the one-compartment open model illustrated in Figure 7.13. Here, the body is viewed as a single homogeneous box with a fixed

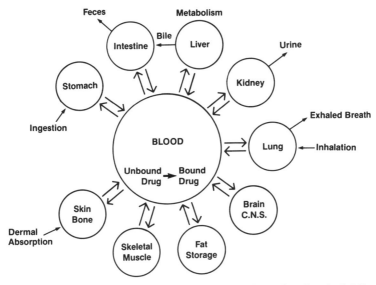

Figure 7.10 A diagram illustrating the potential distribution of a chemical following exposure. The differences in the rates of absorption or elimination among these tissues is the reason for using multicompartment pharmacokinetic models.

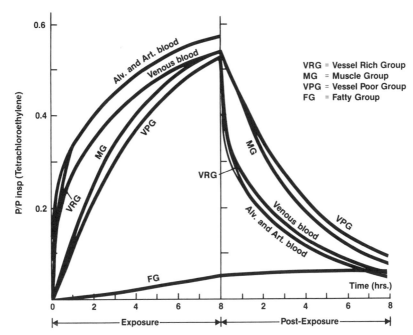

Figure 7.11 Predicted partial pressure of tetrachloroethylene in alveolar air, mixed venous blood, and tissue groups during and after 8 hr of exposure to a constant air concentration of tetrachloroethylene. Partial pressures in alveoli, blood, and tissues (P) are expressed as a fraction of the constant partial pressure in ambient air during exposure (P_{insp}). (From: Guberan and Fernandez, 1974).

235

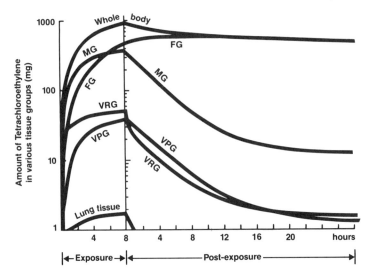

Figure 7.12 Predicted distribution of tetrachloroethylene to the tissue groups during and after 8 hr exposure to 100 ppm. In contrast with many other volatile chemicals, due to its relatively high lipid solubility, tetrachloroethylene demonstrates some persistence in fatty tissues. (From: Guberan and Fernandez, 1974).

volume. A drug or chemical entering the systemic circulation by any route is instantaneously distributed throughout the body. Although the absolute concentrations in all body tissues and fluids are not identical, it is assumed that they all rise and fall in parallel as drug is added and eliminated.

Overall elimination of drug from the body by urinary excretion and/or metabolism usually obeys first-order kinetics; that is, the rate of elimination is proportional to drug concentration in blood or plasma, Cp, as shown by the following equation:

$$\frac{dCp}{dt} = -k_{el}Cp \tag{7}$$

Rearrangement and integration of Eq. (7) gives

$$\log Cp = -\frac{k_{el}(t)}{2.3} + \log Cp^0$$

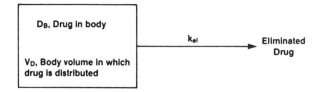

Figure 7.13 One-compartment open pharmacokinetic model.

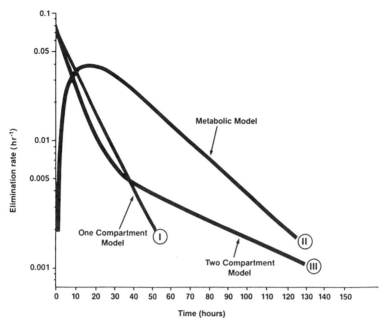

Figure 7.14 Three basic types of elimination curves following a single instantaneous exposure (IV) at times equals zero. (From Piotrowski, 1977.)

which says that following an intravenous dose, a semilog plot of plasma concentrations versus time will be a straight line with a slope of $-k_{el}/2.3$ and an initial plasma concentration of Cp^0.

The best way to determine whether a one-compartment model is appropriate for a given drug is to plot the intravenous (IV) plasma concentrations against time on semilog paper and see if the plot is a straight line (see Fig. 7.14). This approach permits determination of the two key parameters of the model: k_{el} the overall elimination rate constant, and V_D, the apparent volume of distribution of the drug in the body.

6.2 Two-Compartment Model

Often, it is clear from the blood plasma curves that there is a slower distribution to some tissues. In this case the kinetic model that will allow the interpretation of the observed experimental data is a two-compartment model, illustrated in Figure 7.15. In the two-compartment model the body is essentially reduced to an accessible compartment, the blood, and a second less accessible and diffuse compartment, the tissues. In the physiological sense, as soon as the molecules in the administered dose mix with the blood, they will be rapidly carried to all parts of the body and brought into intimate contact (by perfusion) with organs, tissues, fat depots, and even bone. Some tissues, like the liver and the kidney, are very well perfused; others, such as muscle and fat, may be poorly perfused. Molecules that are carried to the perfused

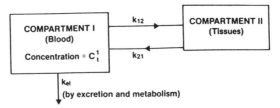

Figure 7.15 Schematic of the two-compartment pharmacokinetic model.

organs and tissues will then diffuse from the blood across cellular membranes and a dynamic equilibrium will usually be established (Fig. 7.16) (O'Flaherty et al., 1982; Withey, 1979).

The distribution of a chemical as described by a two-compartment model is readily apparent in a plot of blood versus time following exposure. A blood concentration–time curve following an intravenous bolus administration, which exhibits the characteristics of distribution, is shown in Figure 7.17. An inspection of this kind of plot reveals an initial rapid (steep slope) nonlinear portion followed after a time (which depends on the nature and characteristics of the administered substance) by a slower linear portion. This linear portion, sometimes referred to as the terminal phase, has a definite slope from which a rate coefficient can be evaluated. If the rate

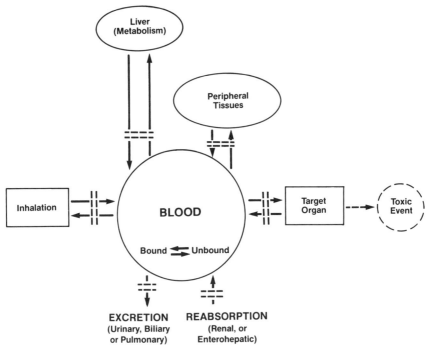

Figure 7.16 The physiologic model for toxic action proposed by Withey (1979) modified to emphasize the fate of inhaled substances. The dashed lines represent diffusion barriers.

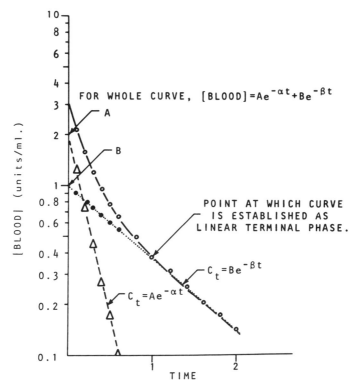

Figure 7.17 Semilogarithmic plot of blood concentration against time after intravenous administration illustrating biphasic elimination (two compartments) and the use of "curve stripping" to resolve the A intercept.

coefficient for this linear terminal phase is defined as β, and the intercept of the line extrapolated back to the ordinate axis is B, then this part of the relationship can be described by

$$C_t^1 = Be^{-\beta t} \tag{8}$$

From the initial curved portion of the elimination curve, the rate at which the central compartment releases the toxicant is the slope of the feathered line. This can be determined easily from the method of residuals (Gibaldi and Perrier, 1975). Having determined A, the intercept of the feathered line, and α, the slope, the whole experimentally observed curve can be described by the biexponential equation

$$C_t^1 = Ae^{-\alpha t} + Be^{-\beta t} \tag{9}$$

where C_t^1 is the concentration of toxicant in blood at any time following exposure.

It is essential to note that in this case $Ae^{-\alpha t}$; that is, the initial phase of the curve, dominated by the rate coefficient α, is essentially over and completed when $e^{-\alpha t}$ approaches 0 and thereafter the curve is described by $C_t = Be^{-\beta t}$.

The individual rate coefficients α and β, obtained empirically from the plotted data, and the intercepts A and B can then be used to determine the individual rate coefficients of the model (Withey, 1979):

$$k_{21} = \frac{A\alpha - B\beta}{\alpha - \beta}$$

$$k_e = \frac{\alpha\beta}{k_{21}}$$

$$k_{12} = \alpha + \beta - k_{21} - k_e \tag{10}$$

where k_{21} = rate transfer coefficient for the movement of molecules from compartment 2 to compartment 1

k_{12} = rate transfer coefficient for the movement of two molecules from compartment 1 to compartment 2

k_e = rate transfer coefficient for the elimination of the chemical from the system

Curve-fitting computer programs are available which require only rough estimates of the four parameters, that is, the slopes α and β and the intercepts A and B, to allow more precise calculation of the individual rate coefficients from the animal or human data.

To consider the body as one or two compartments involves considerable simplification of reality. To account for each of the 10 compartments shown in Figure 7.10 would involve considerable experimental and mathematical difficulty and could conceivably require 9 different equations to resolve the chemical's behavior. A more complex 3-compartment model, as shown in Figure 7.18, is usually the most rigorous model used for data analysis. Due to the minimal practical benefits of performing such rigorous mathematical analysis, combined with the comparatively less quantitative basis by which most occupational exposure limits are established, toxicologists should use Occam's razor in that the simplest model describing the available data is always the best one!

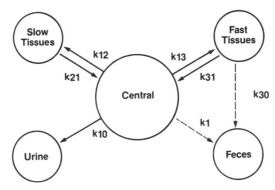

Figure 7.18 Schematic illustrating a three-compartment open model.

The one-compartment model (which is the one used in all pharmacokinetic models recommended for adjusting exposure limits) is consistent with the data in a great many cases (Dittert, 1977), but in some others it provides a poor fit (Riegelman et al., 1968). For the purpose of adjusting exposure limits, when a particular chemical appears to exhibit the characteristics of a multicompartmental distribution, it is suggested that it be forcibly resolved into no more than two compartments, and preferably one. This can be accomplished through the use of several computer programs currently available (Sedman and Wagner, 1976).

6.3 Accumulation of Chemicals in the Body

Cumulation of a substance in the body is a process wherein the concentration of a particular substance increases following repeated or continuous exposure. When using biologic monitoring to evaluate occupational exposure, cumulation is likely when the concentration of a substance in the analyzed media (urine, feces, blood, or expired air) increase with each day or week of exposure. This phenomenon has been observed even during normal work schedules below the TLV (Lehnert et al., 1978) and, therefore, exposure during unusually long work periods might be expected to increase the cumulation. In particular cases, steady increases in the concentration of the inhaled contaminant in adipose tissue may be indicative of any potential for adverse effects of chronic exposure.

Cumulation, if it occurs, results from a slow turnover of the substance. Thus it may take place under conditions of every kinetic model, provided the elimination rate constant is low (long biologic half-time). From theoretical considerations, it can be expected that the highest value of the rate constant (single compartment model) at which cumulation might occur would not exceed 0.1 hr^{-1} (Mason and Dershin, 1976). The tendency of a chemical to cumulate in the body with repeated exposure can be due to several factors. For substances that are eliminated mainly unaltered in the breath, the cause of slow turnover may be the low air/blood partition coefficient or deposition in poorly perfused compartments such as fat; for substances excreted by the kidney, a low clearance may result from poor glomerular filtration, or intensive tubular reabsorption, or both. In practical situations, two other events may be of significance, namely slow biotransformation (when excretion takes place in metabolized form) and deposition in adipose tissue. These factors can act in combination, as has been shown for nitrobenzene and DDT (Piotrowski, 1977).

The likelihood that a chemical will cumulate in the body or a key tissue is dependent on the overall rate at which a chemical is absorbed, metabolized, and excreted. The time required for the concentration of the parent chemical or its metabolite to be reduced by one-half in a particular medium (e.g., blood) or a tissue (e.g., fat), is called the biologic half-life. It must be emphasized that the biologic half-life varies for each substance and is dependent on the species studied and maybe influenced by the route of exposure. As a result, the estimation of the biologic half-life of a substance in humans based on animal data usually requires considerable effort. As shown in Figure 7.19, chemicals with moderate half-lives (greater than 3 hr and less than 200 hr) are likely to show some degree of day-to-day cumulation during the workweek even at exposures at or near the TLV (Hickey and Reist, 1977).

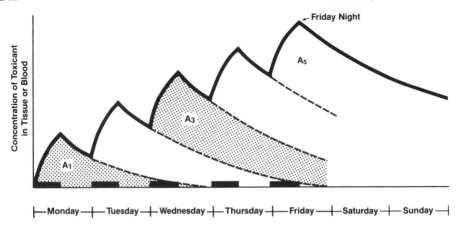

Figure 7.19 Day-to-day increase of toxicant concentration in tissue or blood following exposures of 8 hr/day, 5 day/week at TLV levels. For this type of cumulation to occur, the half-life of the air contaminant or its metabolite would be about 6 → 200 hrs., as shown in Figures 7.6 and 7.7.

Cumulation of a chemical during an exposure regimen (workweek or year) is not necessarily detrimental as long as the peak or steady-state tissue levels do not reach levels that are above the threshold concentration for such endpoints as cytotoxicity, behavioral toxicity, and so forth. This weekly increase in body burden in workers periodically exposed has been demonstrated for some chemicals (Lehnert et al., 1978).

One of the primary concerns with unusual work schedules is the possibility that day-to-day increases in the toxicant concentration at the site of action might occur during unusually long work shifts to a much greater degree than that which occurs during standard 8 hr/day schedules. This potential problem is illustrated in Figure 7.20 where it is shown that a chemical with a 24-hr biologic half-life would clearly provide a greater peak body burden following 4 days of exposure on a 10-hr/day schedule than that observed in the workers exposed 8 hr/day for 5 days unless the chemical concentration in air was not lowered. The point here is that even when the number of hours of exposure per week are the same, for some chemicals, the peak body burden may be different when the work schedules vary. In this figure, C_T represents the concentration in the blood following a certain period of exposure and C_s is the concentration in the blood at the saturation level attained following continuous exposure to that same air concentration. In this figure t represents a schedule involving five 8-hr workdays per week and t' represents a workweek consisting of four workdays: a 10 hr/day, 8 hr/day, 12 hr/day, and another 10 hr/day followed by 3 days off work. Even though this unusual work schedule involves 40 hr of exposure to 50 ppm of a chemical vapor, exposed workers are likely to have greater peak body burdens than persons who are also exposed for 40 hr per week but only 8 hr per day.

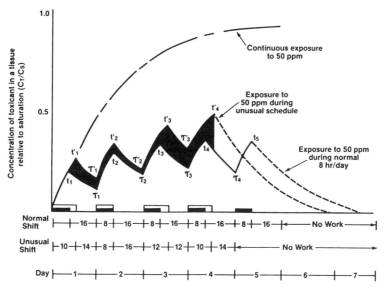

Figure 7.20 The accumulation of a substance during irregular periods of intermittent exposure that could occur during unusual work shifts and overtime. This plot illustrates that even when the total weekly dose (40 hr × 50 ppm) is unchanged, the peak body burden for two different shift schedules may not be equivalent. To help ensure that the peak tissue concentration for the unusual exposure schedule does not exceed the presumably "safe" level of the normal schedule, the air concentration of toxicant in the workplace may have to be reduced. K_{elim} has a value of 0.03 hr^{-1} in this illustration. (Courtesy of Dr. J. Walter Mason.)

6.4 Nonlinear Pharmacokinetic Models

All the models discussed so far have been based on the assumption that all pharmacokinetic pathways can be described by linear differential equations with constant coefficients and first-order rate constants. This may not always be true and it is important to recognize that in some cases nonlinear behavior may occur. This can have a great deal of effect on the estimation of risks associated with exposure (Gehring and Blau, 1977; Withey, 1979; Ramsey and Gehring, 1980; Hoel, et al., 1983). So, in general, the concentration of an air contaminant at or near the occupational exposure limit will not be sufficient to saturate the blood or bring about nonlinear behavior (Fiserova-Bergerova and Holaday, 1979). However, this could occur if exposures were continuous or at high concentrations for short periods, even though the acceptable 8-hr TWA concentration was not exceeded. This phenomenon must be understood by the hygienist and toxicologist when interpreting some oncogenicity studies wherein a maximum tolerated dose (MTD) must be administered to at least one group of animals. In certain cases it is only when saturation occurs that an oncogenetic response will occur and, in these cases, the minimal likelihood that Michaelis-Menten kinetics (see Fig. 7.21) will exist in exposed workers must be considered (Reitz, et al., 1989; Reitz, et al., 1990a).

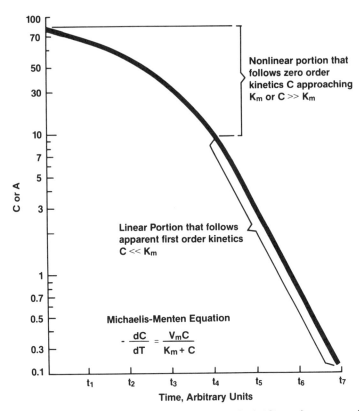

Figure 7.21 Simulated plasma concentration of a chemical (C) or the amount in the body (A) as a function of time for a chemical that displays dose-dependent or nonlinear pharmacokinetics described by the Michaelis-Menten equation: dc/dT is the change in concentration with time, V_m is the maximum rate of the process, and K_m is the Michaelis constant. (From: Gehring and Young, 1979.)

A dose-dependent change in pharmacokinetic behavior may arise from several causes. At saturation, the capacity of some biological system in the overall process can become overloaded. Some examples of chemicals that exhibit dose-dependent pharmacokinetic at concentrations well in excess of the TLV include vinyl chloride, vinylidene chloride, methylene chloride, methanol, benzene, and 2,4,5-T. Dose-dependent effects are of special interest to the toxicologist since some toxic effects are associated only with large or excessive doses of drugs. Alcohol pharmacokinetics is one example of dose-dependent kinetics. As is frequently the case, the dose dependence is due to metabolic saturation of the particular enzyme; in this case alcohol dehydrogenase (Piotrowski, 1977). When this enzyme system is saturated, the person will be exposed to higher levels of unmetabolized alcohol via the blood than would normally occur at lower levels of alcohol ingestion.

In many of the cases cited in the literature, the observed dose dependence has been attributed to changes in the metabolic reaction involved. The metabolism of

Table 7.10 Possible Causes of Dose-Dependent Pharmacokinetics

Metabolic saturation	a. Self-inhibition by excess drugs
	b. Insufficient metabolic enzyme
	c. Competition for coenzyme or cosubstrate
Excretory saturation	a. Competition for tubular secretion or re-sorption mechanism
	b. Changed drug distribution
Changed drug distribution	a. Protein and tissue binding effect
	b. Overflow into new volumes of distribution

Source: Withey (1979).

the drug may be due to an enzymatic reaction and thus would behave according to the classical Michaelis-Menten model of enzyme kinetics (Fig. 7.21). Such a system involves reaction of drug as substrate with enzyme to produce a metabolite via an intermediate enzyme–drug complex. At low drug concentrations, the reaction is overall first-order with a rate constant, $k_m = V_m/K_m$. At high drug concentrations, the capacity of the limited amount of enzyme present for reaction is exceeded and the reaction rate becomes constant (V_m). Thus, the production of metabolite becomes a constant rate or zero-order process (Bus and Gibson, 1985).

The transformation from first-order to zero-order reaction rate is not an abrupt jump but a graded transition. During this transition the kinetics may remain apparently first-order but the elimination constant changes. Thus, both types of dose-dependent effects mentioned earlier may be explicable by a model of the Michaelis-Menten type. For those chemicals for which the rationale for the TLV or PEL is based on effects that occur only when the metabolic process is overloaded or saturated, such as vinyl chloride's carcinogenicity, it is suggested that no adjustment to the TLV is probably needed to protect workers on long shifts as long as the average weekly exposure throughout the month is about 40 hr. This suggestion is based on the observation that the occupational exposure limits are set at air concentrations far below those at which saturation and dose-dependent kinetics are likely to occur (Fiserova-Bergerova and Holaday, 1979; Bus and Gibson, 1985). Some common causes of dose-dependent pharmacokinetics are presented in Table 7.10 (Withey, 1979).

7 WHY ADJUST EXPOSURE LIMITS FOR UNUSUAL WORK SCHEDULES

It has been speculated by a number of researchers that some workers may be at increased risk of injury during unusual work schedules. In particular, industrial hygienists and occupational physicians have been concerned that unusual work schedules and the potential effects on the circadian rhythm might eventually compromise a worker's capacity to cope with exposure to airborne toxicants. In some cases, it is clear that adjustments to the TLV or any other limit are necessary to provide workers on long workdays the same degree of protection given those persons on normal 8-hr workdays (Brief and Scala, 1975). Establishing acceptable limits of exposure for unusual work schedules is a difficult task since, at best, each person's susceptibility to stressors is dependent on many factors that are unique to that individual (ACGIH, 1982; Omenn, 1982; Castlemen and Ziem, 1988).

Because tens of thousands of persons are exposed to airborne substances for unusually short or long periods of time, several organizations have set limits for these situations. Pennsylvania promulgated short-term exposure limits soon after the OSHA regulations took effect in 1971 (Commonwealth of Pennsylvania, 1971). Brief and Scala (1975) and OSHA (1979) have developed simple, easily applied models aimed at adapting exposure limits to unusual situations. The National Institute for Occupational Safety and Health has applied its recommended standard for chloroform (NIOSH, 1975) and benzene (NIOSH, 1974) to both 8 and 10-hr work shifts.

The National Research Council has described the need to adjust exposure limits for long-term continuous exposure, such as might be encountered in space exploration (NASA, 1991). ACGIH has incorporated short-term exposure limits into its TLV list. Numerous pharmacokineticists and other researchers (Roach, 1966, 1977, 1978; Mason and Dershin, 1976; Hickey, 1977; Hickey and Reist, 1977; Andersen et al., 1987a) have proposed complex models for adjusting limits and a few toxicologists have begun to investigate the differences in toxicologic response between administration of the same daily dose over particularly short and long periods of exposure. These will be discussed in detail later in the chapter.

Two questions regarding the occupational health aspects of the 12-hr schedules surfaced shortly after its commencement. One of the initial concerns of occupational physicians and industrial hygienists was whether the recommended limit for exposure to noise in the workplace, 85 dBA for 8 hr, was sufficiently low to protect workers on the 12-hr shift. There was speculation that regulatory agencies such as OSHA might arbitrarily lower the standard by 3–5 dBA for workers on these shifts. Second, since most federal standards for occupational exposure to airborne toxicants are based a work schedule of 8 hr per day, 5 days per week, there was concern that a marked decrease in the concentration of air contaminants in the workplace might be necessary to protect workers. The concern that the existing exposure limits might not protect those on unusual work schedules was legitimate since both the noise and air contaminant limits were based on the results of either animal testing, which was conducted for 6.0 hr or less per day, or the epidemiological experience of workers who were exposed during normal, 8-hr/day work shifts.

7.1 Limits for Exposure to Noise During Unusual Shifts

Currently, ACGIH recommends that exposures to sound be limited to 85 dBA for periods of 8 hr per day and to no more than 80 dBA if exposed 16 hr per day (ACGIH, 1992). The setting of a special guideline for those persons working 12-hr shifts has not been specifically addressed by the ACGIH, but one can infer that some noise level between 80 and 85 dBA would be appropriate. The regulatory agency responsible for setting standards that protect worker health, OSHA, proposed in 1974 a legal limit for occupational exposure to noise of 85 dBA time-weighted average. This proposed standard was to apply to all persons who worked 8-hr/day or longer, but the total exposure could not exceed 50 hr per week.

As recently as 1992, it was the opinion of several experts in the field of noise-induced hearing loss that routine exposure to a 12-hr/day, 3- or 4-day/week work schedule probably does not require a special noise limit as long as the workers are exposed throughout the year an average of no more than 40 hr per week. This seems reasonable since there is evidence that injury due to exposure to certain physical agents such as noise or radiation is solely a function of the intensity and duration of exposure. For most physical agents, the risk is dictated by the total dose over a month, year, or lifetime rather than each day or week, as long as daily threshold for acute effects are excepted. These should not be markedly affected by slight alterations in the dosage regimens. In part, this is because the potential problems of cumulative effects due to the biologic half-life is of little or no concern for most physical agents.

7.2 Exposure Limits for Air Contaminants During Unusual Schedules

In the preface of the ACGIH publication *Threshold Limit Values For Chemical Substances and Physical Agents and Biological Exposure Indices in the Work Environment with Intended Changes for 1992-93*, it states that the TLVs refer to airborne concentrations of substances and represent conditions under which it is believed that nearly all workers may be repeatedly exposed day after day without adverse effect. The values refer to a TWA concentration for a normal 8-hr workday and a 40-hr workweek (ACGIH, 1992). Implicit in this evaluation is an assumption that a balance exists between the accumulation of a contaminant in the body while exposed at work, and the elimination of the contaminant while away from work (during which time there is presumably no exposure). Because of these assumptions, it is appropriate to consider whether and to what extent the TLVs and other limits require modification for unusual work schedules.

It is readily apparent to a toxicologist or pharmacokineticist that some modification of the TLVs may be necessary if these limits are to provide an equivalent amount of protection to persons working unusual shifts. For example, a 12-hr work shift involves a period of daily exposure that is 50 percent greater than that of the standard 8-hr workday and the period of recovery before reexposure is shortened from 16 to 12 hr (Brief and Scala, 1975). For certain systemic toxins having half-lives between 5 and 500 hr, it can be envisioned that shifts longer than 8 hr/day would present a correspondingly greater hazard than that incurred during normal workweeks if the exposure (i.e., dose) were not reduced by some albeit small amount.

The very basis of occupational exposure limits for inhaled toxicants is that there is a maximum concentration of the air contaminant in the workplace to which persons can safely be exposed and suffer no adverse effect. Consequently, it can be envisioned that there is a corresponding maximum body burden at which an adverse effect is not likely. TLVs represent conditions under which it is believed that nearly all workers may be exposed day after day without adverse effect, so exposure to a substance at its TLV should result in some maximum burden of the substance in a particular target organ of the body (Hickey, 1977). Therefore, it seems logical that TLVs for certain chemicals should be modified if persons are exposed more than 8 hr/day so as to prevent this maximum body burden from being exceeded (see Fig. 7.22).

Since the rationales for the more than 700 ACGIH TLVs vary, identifying adjusted TLVs for some, if not most, of the air contaminant guidelines is not a trivial task. Several factors should be considered. For example, the need to lower the average exposure limit for certain industrial chemicals would be especially important in those situations where the safety factor in the TLV is small, toxicity data are limited, the toxic effect is serious, accumulation of the chemical is possible following several days of repeated exposure, or where there is the possibility for extreme variability in worker response to a toxicant.

The procedures or models for modifying occupational exposure limits that have been proposed may appear cumbersome or complicated. It is for this reason that this chapter has been written. Furthermore, because many occupational health professions are interested in how to make the best use of exposure limits and since most modification procedures are based on pharmacokinetic principles, a basic discussion of this topic follows.

8 MODELS FOR ADJUSTING OCCUPATIONAL EXPOSURE LIMITS

Several researchers have proposed mathematical formulas or models for adjusting occupational exposure limits (PELs, TLVs, etc.) for use during unusual work schedules, and these have received a good deal of interest in the industrial and regulatory arenas (Brief and Scala, 1975; Mason and Dershin, 1976; Hickey, 1977; Hickey and Reist, 1977; Roach, 1977, 1978; Hickey and Reist, 1979; Hickey, 1980; Hickey, 1983; Andersen et al., 1987a). Although OSHA has not officially promulgated specific exposure limits applicable to unusual work shifts, it has published guidelines for use by OSHA compliance officers for adjusting exposure limits (OSHA, 1989a,b; OSHA, 1990). These generally apply to work shifts longer than 8 hr/day.

Under the General Duty Clause of OSHA, Section 5A1, employers that have employees who, during long shifts, are exposed to workplace air concentrations in excess of adjusted exposure limits are citeable for noncompliance. All of the models that an OSHA compliance officer might use in determining the acceptability of an air contaminant level are presented and discussed in this chapter. Each of the models has advantages and disadvantages. The shortcomings in determining modified exposure limits for unusual schedules have been discussed (Calabrese, 1977, 1983). It should be noted that OSHA has, thus far, cited few (if any) firms that have unusually long shifts and are not in compliance with the limits suggested by these models.

8.1 Brief and Scala Model

In the early 1970s, due to the increasingly large number of workers who had begun working unusual schedules, the Exxon Corporation began investigating approaches to modifying the occupational exposure limits for their employees on 12-hr shifts. In 1975, the first recommendations for modifying TLVs and PELs were published by Brief and Scala (1975), wherein they suggested that TLVs and PELs should be modified for individuals exposed to chemicals during novel or unusual work schedules.

They called attention to the fact that, for example, in a 12-hr workday the period of exposure to toxicants was 50 percent greater than in the 8-hr workday, and that the period of recovery between exposures was shortened by 25 percent, or from 16 to 12 hr. Brief and Scala noted that repeated exposure during longer workdays might, in some cases, stress the detoxication mechanisms to a point that accumulation of a toxicant might occur in target tissues, and that alternate pathways of metabolism might be initiated. It has generally been held that due to the margin of safety in the TLV (albeit small), there was little potential for frank toxicity to occur due to unusually long work schedules.

Brief and Scala's approach was simple but important since it emphasized that unless worker exposure to systemic toxicants was lowered, the daily dose would be greater, and due to the lesser time for recovery between exposures, peak tissue levels might be higher during unusual shifts than during normal shifts. This concept is illustrated in Figure 7.22. Their formulas [Eqs. (11) and (12)] for adjusting limits are intended to ensure that this will not occur during novel work shifts:

$$\text{TLV reduction factor (RF)} = \frac{8}{h} \times \frac{24h}{16} \tag{11}$$

where h is hours worked per day.

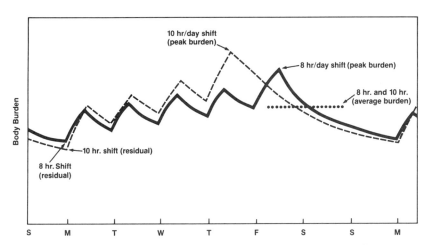

Figure 7.22 Comparison of the peak, average, and residual body burdens of an air contaminant following exposure during a standard (8-hr/day) and unusual (10-hr/day) workweek. In this case the weekly average body burdens are the same for both schedules since each involved 40 hr/week. The residual (Monday morning) body burden of the 8-hr shift worker, however, is greater than the 10-hr shift worker and the peak body burden of the person who worked the 10-hr shift is higher than the 8-hr worker. Based on Hickey (1977).

For a 7-day workweek, they suggested that the formula be driven by the 40-hr exposure period; consequently, they developed Eq. (12), which accounts for both the period of exposure and period of recovery:

$$\text{TLV reduction factor (RF)} = \frac{40}{h} \times \frac{168h}{128} \qquad (12)$$

where h = hours exposed per week.

One advantage of this formula is that the biologic half-life of the chemical and the mechanism of action are not needed in order to calculate a modified TLV. Such a simplification has shortcomings since the reduction factor for a given work schedule is the same for all chemicals even though the biologic half-lives of different chemicals varies widely. Consequently, this formula overestimates the degree to which the limit should be lowered.

Brief and Scala (1975, p. 469) were cautious in describing the strength of their proposal and offered the following guidelines for its use. They should be considered when applying this model and also the others:

1. Where the TLV is based on systemic effect (acute or chronic), the TLV reduction factor will be applied and the reduced TLV will be considered as a time-weighted average (TWA). Acute responses are viewed as falling into two categories: (a) rapid with immediate onset and (b) manifest with time during a single exposure. The former are guarded by the C notation and the latter are presumed time and concentration dependent, and hence, are amenable to the modifications proposed. Number of days worked per week is not considered, except for a 7-day workweek discussed later.

2. Excursion factors for TWA limits (Appendix D of the 1974 TLV publication) will be reduced according to the following equation:

$$EF = (EF_8 - 1)RF + 1 \qquad (13)$$

where EF = desired excursion factor

    EF_8 = value in Appendix D for 8-hr TWA

    RF = TLV reduction factor

3. Special case of 7-day workweek. Determine the TLV Reduction Factor based on exposure hours per week and exposure-free hours per week.

4. When the novel work schedule involves 24-hour continuous exposure, such as in a submarine, spacecraft or other totally enclosed environment designed for living and working, the TLV reduction technique cannot be used. In such cases, the 90-day continuous exposure limits of the National Academy of Science should be considered, where applicable limits apply.

5. The techniques are not applicable to work schedules less than seven to eight hours per day or $\le$ 40 hours per week.

Brief and Scala (1975) also correctly noted that:

> The RF value should be applied (a) to TLV's expressed as time-weighted average with respect to the mean and permissible excursion and (b) to TLV's which have a C (ceiling) notation except where the C notation is based solely on sensory irritation. In this case the irritation response threshold is not likely to be altered downward by an increase in number of hours worked and modification of the TLV is not needed.

In short, the Brief and Scala formula is dependent solely on the number of hours worked per day and the period of time between exposures. For example, for any systematic toxin, this approach recommends that persons who are employed on a 12-hr/day, 3- or 4-day workweek should not be exposed to air concentrations of a toxicant greater than one-half that of workers who work on an 8-hr/day, 5-day schedule.

In their publication, Brief and Scala acknowledged the importance of a chemical's biologic half-life when adjusting exposure limits, but because they believed this information was rarely available, they were comfortable with their proposal. They noted that a reduction in an occupational exposure limit is probably not necessary for chemicals whose primary untoward effect is irritation since the threshold for irritation response is not likely to be altered downward by an increase in the number of hours worked each day (Brief and Scala, 1975), that is, irritation is concentration rather than time dependent. Although this appears to be a reasonable assumption, some researchers believe that it may not be the case for all chemicals since duration of exposure could possibly be a factor in causing irritation in susceptible individuals who are not otherwise irritated during normal 8-hr/day exposure periods (Alarie and Nielsen, 1982; Schaper, 1993).

Illustrative Example 1 (Brief and Scala Model). Refinery operators often work a 6-week schedule of three 12-hr workdays for 3 weeks, followed by four 12-hr workdays for 3 weeks. What is the adjusted TLV for methanol (assume TLV = 200 ppm) for these workers? Note that the weekly average exposure is only slightly greater than that of a normal work schedule.

Solution

$$\text{RF} = \frac{8}{12} \times \frac{24 - 12}{16} = 0.5$$

$$\text{Adjusted } TLV = \text{RF} \times \text{TLV}$$

$$= 0.5 \times 200 \text{ ppm}$$

$$= 100 \text{ ppm}$$

Note: The TLV reduction factor of 0.5 applies to the 12-hr workday, whether exposure is for 3, 4, or 5 days per week.

Illustrative Example 2 (Brief and Scala Model). What is the modified TLV for tetrachloroethylene (assume TLV = 10 ppm) for a 10-hr/day, 4-day/week work schedule if the biologic half-life in humans is 144 hr?

Solution

$$RF = \frac{8}{10} \times \frac{24 - 10}{16} = 0.7$$

New TLV = $0.7 \times 10 = 7$ ppm.

Note: This model and the one used by OSHA *do not consider* the pharmacokinetics (biologic half-life) of the chemical when deriving a modified TLV. Other models to be discussed later *do* take this information into account.

Illustrative Example 3 (Brief and Scala Model). In an 8-hr day, 7-day workweek situation, such as the 56/21 schedule, persons work 56 continuous days followed by 21 days off. What is the recommended TLV for H_2S (assume TLV = 10 ppm) for this special case of a 7-day workweek? Assume that the biologic half-life in humans for H_2S is about 2 hr and the rationale for the limit is the prevention of irritation and systemic effects.

Solution.   Exposure hr per week = $8 \times 7 = 56$ hr
                Exposure-free hr per week = $(24 \times 7) - 56 = 112$ hr

$$RF = \frac{40}{56} \times \frac{112}{128} = 0.625$$

Adjusted TLV = RF $\times$ 10 ppm = 6 ppm.

Illustrative Example 4 (Brief and Scala Model). Ammonia has a TLV of 25 ppm and is an upper respiratory tract irritant. What is the modified TLV for this chemical for a work schedule of 14 hr/day for 3 day/week?

Solution. Since the rationale for the limit for ammonia is the prevention of irritation, *no* adjustment (lowering) of the limit is needed.

8.2 OSHA Model

Most toxicologists believe that, in general, the intensity of a toxic response is a function of the concentration that reaches the site of action (Fiserova-Bergerova et al., 1980; Andersen et al., 1987b). This principle is simplistic and, while it may not apply to irritants, sensitizers, or carcinogens, it is clearly true for the systemic toxics. This assumption is the basis for the OSHA model for modifying PELs for unusual shifts (OSHA, 1979). The originators of the model assumed that for chemicals that cause an acute response, if the daily uptake (concentration × time) during a long workday was limited to the amount that would be absorbed during a standard workday, then the same degree of protection would be given to workers on the longer shifts. For chemicals with cumulative effects (i.e., those with a long half-life), the adjustment model was based on the dose imparted through exposure during the normal workweek (40 hr) rather than the normal workday (8 hr).

OSHA recognized that the rationale for the occupational exposure limits for the various chemicals was based on different types of toxic effects. After OSHA adopted the same 500 TLVs of 1968 as PELs (29 CFR 1910.1000), and again did the same in 1989, they placed each of the chemicals into one of six ''Work Schedule'' categories (Table 7.11) to assure that an appropriate adjustment model would be used by their hygienists. As can be seen in Examples 5, 6, and 7, the degree to which an exposure limit is to be adjusted, if at all, is based to a large degree on the work schedule category (and type of adverse effect) in which a chemical is placed (Table 7.12).

The rationale that OSHA and its expert consultants used when categorizing the various chemicals was based on the primary type of health effect to be prevented, biologic half-life (if known), and the rationale for the limit. The categories include (a) ceiling limit, (b) prevention of irritation, (c) technological feasibility limitations, (d) acute toxicity, (e) cumulative toxicity, as well as (f) acute and cumulative toxicity. A review of the degree of adjustment required for each type of chemical can be found in Table 7.12. Table 7.13 contains names of the categories of the different health effects into which the 700 chemicals with PELs were placed. Table 7.14 presents some examples of how chemicals are classified according to primary adverse health effects. The complete list can be found in the *Federal Register* (OSHA, 1989a).

Irrespective of the model that will be used to make the adjustments, including the pharmacokinetic models to be discussed, the table in the *Federal Register* (OSHA, 1989a) should be consulted before the hygienist begins the task of modifying an exposure limit. The use of this table, combined with some professional judgment, will prevent hygienists from requiring control measures when they are unnecessarily restrictive as well as minimize the risk of injury or discomfort from overexposure during an unusual exposure schedule. For example, as noted by OSHA, substances in Category 1A, 1B, and 1C do not require adjustment during long shifts due to the rationale for those limits (Table 7.12). An unusual schedule can involve exposures as short as 15 min or as long as 24 hr/day (continuous).

As discussed briefly, OSHA has proposed two simple equations for adjusting occupational health limits. These equations are offered to their compliance officers as an alternative to the more complex models of Brief and Scala (1975) or Hickey and Reist (1977). The first equation, which appears in Chapter 13 of the *OSHA Field Manual* (OSHA, 1979), is to be used with chemicals whose primary hazard is acute injury (Category 2, Table 7.11). In these cases, the objective is to modify the limit for the unusually long shift to a level that would produce a dose (mg) that would be *no greater* than that obtained during 8 hr of exposure at the PEL. Examples of chemicals with exclusively acute effects include carbon monoxide or phosphine. The

Table 7.11 Work Schedule Categories Listed by OSHA

Category 1A. Ceiling Limit Standard. Substances in this category (e.g., butylamine) have ceiling limit standards that were intended never to be exceeded at any time, and so, are independent of the length of frequency of work shifts. The ceiling PELs for substances in this category should not be adjusted.

Category 1B. Standards Preventing Mild Irritation. Substances in thie category have a PEL designed primarily to prevent acute irritation or discomfort (e.g., cyclopentadiene). There are essentially no known cumulative effects resulting from exposures for extended periods of time at concentration levels near the PEL. The PELs for substances in this category should not be adjusted.

Category 1C. Standards Limited by Technologic Feasibility. The PELs of substances assigned to this category have been set either by technologic feasibility (e.g., vinyl chloride) or good hygiene practices (e.g., methyl acetylene). These factors are independent of the length or frequency of work shifts. The PELs for substances in this category should not be adjusted.

Category 2. Acute Toxicity Standards.
a. The substances in this category have PELs that prevent excessive accumulation of the substance in the body during 8 hr of exposure in any given day (e.g., carbon monoxide).
b. The following equation determines a level that ensures that employees exposed more than 8 hr/day will not receive a dosage (i.e., length of exposure × concentration) in excess of that intended by the standard.

$$\text{Equivalent PEL} = \text{8-hr PEL} \times \frac{8 \text{ hr}}{\text{hours of exposure in 1 day}} \quad \text{(daily adjustment)}$$

c. The industrial hygienist should normally conduct sampling for the entire shift minus no more than 1 hr for equipment setup and retrieval (e.g., at least 9 hr of a 10-hr shift). In situations where an employee works multiple shifts in a day (e.g., two 7-hr shifts), and the industrial hygienist can document sufficient cause to expect exposure concentrations to be similar during the other shifts, the sampling should be done during only one shift.

Category 3. Cumulative Toxicity Standards

a. Substances assigned to this category present cumulative hazards (e.g., lead, mercury, etc.). The PELs for these substances are designed to prevent excessive accumulation in the body resulting from many days or even years of exposure.
b. The following equation ensures that workers exposed more than 40 hr/week will not receive a dosage in excess of that intended by the standard.

$$\text{Equivalent PEL} = \text{8-hr PEL} \times \frac{40 \text{ hr}}{\text{hours of exposure in 1 week}} \quad \text{(weekly adjustment)}$$

c. It is the responsibility of the industrial hygienist to conduct sufficient sampling to document exposure levels for the entire week when evaluating conditions on the basis of this equivalent PEL. For most operations, the industrial hygienist will be able to sample during one shift only and then document sufficient cause to predict exposure concentrations during the other shifts.

Category 4. Acute and Cumulative Toxicity Standards. Substances in this category may present both an acute and a cumulative hazard. For this reason, the PELs of these substances should be adjusted by either equation; i.e., whichever provides the greatest protection.

Table 7.11 (*Continued*)

Refined Adjustment Equations for Specific Standards. The adjustment equation presented for Categories 2 and 3 reflect an oversimplification of the actual accumulation and removal of a toxic agent from the body. Additional research, however, is needed in order to apply more complex equations to estimate resulting body burden and health risk due to prolonged exposure periods.

Source: OSHA (1979).

following equation is recommended by OSHA for calculating an adjustment limit (equivalent PEL) for these types of chemicals:

$$\text{Equivalent PEL} = 8\text{-hr PEL} \times \frac{8 \text{ hr}}{\text{hours of exposure per day}} \tag{14}$$

The other formula recommended by OSHA applies to chemicals for which the PEL is intended to prevent the cumulative effects of repeated exposure. For example, PCBs, PBBs, mercury, lead, and DDT are considered cumulative toxins because repeated exposure is usually required to cause an adverse effect, and the overall biologic half-life is clearly in excess of 10 hr. The goal of PELs in this category is to prevent excessive cumulation in the body following many days or even years of exposure. Chemicals whose rationale is based on cumulative toxic effects are placed in Category 3 in Table 7.11. Accordingly, Eq. (15) is offered to OSHA compliance officers as a viable approach for calculating a modified limit for chemicals whose half-life would suggest that not all of the chemical will be eliminated before returning to work the following day. Its intent is to ensure that workers exposed more than 40 hr/week will not eventually develop a body burden of that

Table 7.12 Prolonged Work Schedule Categories

Category[a]	Classification	Adjustment Criteria
1A	Ceiling standard	None
1B	Irritants	None
1C	Technologic limitations	None
2	Acute toxicants	Exposed—8 hr/day
3	Cumulative toxicants	Exposed—40 hr/week
4	Both acute & cumulative	Exposed—8 hr/day and/or Exposed—40 hr/week

Source: OSHA (1979).

Note: The health effects and classification of violation sections of this chapter have been reviewed by a panel of toxicologists and industrial hygienists from NIOSH using policy guidelines established by OSHA.

[a] This column indicates the code designation for prolonged work schedules that may require an adjustment to the PEL.

7.13 OSHA Substance Toxicity Table (Rationale for Placing a Chemical into One of the Categories Listed in the OSHA *Officers Field Manual*)

Health Code Number	Health Effect
1	Cancer—Currently regulated by OSHA as carcinogens; chiefly work practice standards
2	Chronic (cumulative) toxicity—Suspect carcinogen or mutagen
3	Chronic (cumulative) toxicity—Long-term organ toxicity other than nervous, respiratory, hematologic or reproductive
4	Acute toxicity—Short-term high hazards effects
5	Reproductive hazards—Fertility impairment or teratogenesis
6	Nervous system disturbances—Cholinesterase inhibition
7	Nervous system disturbances—Nervous system effects other than narcosis
8	Nervous system disturbances—Narcosis
9	Respiratory effects other than irritation—Respiratory sensitization (asthma)
10	Respiratory effects other than irritation—Cumulative lung damage
11	Respiratory effects—Acute lung damage/edema
12	Hematologic (blood) disturbances—Anemias
13	Hematologic (blood) disturbances—Methemoglobinemia
14	Irritation—eye, nose, throat, skin—Marked
15	Irritation—eye, nose, throat, skin—Moderate
16	Irritation—eye, nose, throat, skin—Mild
17	Asphyxiants, anoxiants
18	Explosive, flammable, safety (no adverse effects encountered when good housekeeping practices are followed)
19	Generally low-risk health effects—Nuisance particulates, vapors or gases
20	Generally low-risk health effects—Odor

Source: OSHA (1979, 1989b).

Table 7.14 Example of OSHA Classification of Chemicals by Primary Adverse Health Effect (OSHA, 1989)

Substance	Health Code No.	Health Effects	Work Category
Abate	6	Cholinesterase inhibition	3
Acetaldehyde	14	Marked irritation—eye, nose, throat, skin	1B
Acetic acid	14	Marked irritation—eye, nose, throat, skin	1B
Acetic anhydride	14	Marked irritation—eye, nose, throat, skin	1B
Acetone	16, 8	Mild irritation—eye, nose, throat/narcosis	1B
Acetonitrile	16, 4	Mild irritation—eye, nose, throat/acute toxicity (cyanosis)	4
2-Acetylaminofluorene	1	Cancer	1C

Table 7.14 (*Continued*)

Substance	Health Code No.	Health Effects	Work Category
Acetylene	18, 17	Explosive/simple asphyxiation	1C
Acetylene tetrabromide	3, 10	Cumulative liver and lung damage	4
Acrolein	14	Marked irritation—eye, nose, throat, lungs, skin	1B
Acrylamide—skin	7, 3	Polyneuropathy, dermatitis/skin, eye irritation	4
Acrylonitrile—skin	2, 5	Suspect carcinogen	4
Aldrin—skin	2, 3	Suspect carcinogen/cumulative liver damage	4
Allyl alcohol—skin	4, 14	Eye damage/marked irritation—eye, nose, throat, bronchi, skin	1B
Allyl chloride	3, 14	Liver damage/Marked irritation—eye, nose, throat	4
Allyl glycidyl ether (AGE)—skin	14	Contact skin allergy/Marked irritation—eye, nose, throat, bronchi, skin	1B
Allyl propyl disulfide	14	Marked irritation—eye, nose, throat	1B
Aluminum oxide	18, 19	Nuisance particulate	1C
4-Aminodiphenyl—skin	1	Cancer	1C
2-Aminopyridine	4, 7	CNS stimulation/headache/increased blood pressure	4
Ammonia	11, 14	Marked irritation—eye, nose, throat, bronchi, lungs	1B
Ammonium chloride (fume)	16	Mild irritation—eye, nose, throat	1B
Catechol pyrocatechol	14, 3	Eye and skin irritation/kidney damage	4
Cellulose (paper fiber)	19	Nuisance particulate	1C
Cesium hydroxide	15	Moderate irritation—eye, nose, throat, skin	1B
Chlordane—skin	3, 2	Cumulative liver damage/suspect carcinogen	4
Chlorinated camphene (toxaphene)—skin	3	Cumulative liver damage	3
Chlorinated diphenyl oxide	3	Cumulative liver damage/dermatitis	3
Chlorine	11, 14	Lung injury/marked irritation—eye, nose, throat, bronchi	1B
Chlorine dioxide	11, 14	Lung injury/marked irritation—eye, nose, throat, bronchi	1B
Chlorine trifluoride	11, 14	Marked irritation—eye, nose, throat, bronchi, lungs	1A

Table 7.14 (*Continued*)

Substance	Health Code No.	Health Effects	Work Category
Chloroacetaldehyde	14	Marked irritation—eye, nose, throat, lungs, skin	1A
alpha-Chloroacetophenone (phenacyl chloride)	14	Marked irritation—eye, nose, throat, bronchi, lungs, skin	1B
Chlorobenzene (monochloro-benze)	3, 8	Cumulative systemic-toxicity/narcosis	4
o-Chlorobenzylidene malonitrile—skin	14	Marked irritation—eye, nose, throat, skin	1B
2-Chloro-1,3-butadiene	(See Chloroprene)		
Chlorobromomethane	3, 8	Cumulative liver damage/narcosis	3
Chlorodifluoromethane (F-22)	18	Good housekeeping practice	1C
Chlorodiphenyl (42% Cl)—skin	2, 3	Suspect carcinogen/Chloracne/cumulative liver damage	4
Chlorodiphenyl (54% Cl)—skin	2, 3	Suspect carcinogen/chloracne/cumulative liver damage	4
Chloroform (trichloromethane)	2, 3, 8	Suspect carcinogen/cumulative liver and kidney damage/narcosis	4
bis-Chloromethyl ether	1	Cancer (lung)	1C
1-Chloro-1-nitropropane	15	Moderate irritation—eye, nose, throat, skin	1B
Chloropicrin	14, 11	Marked irritation—eye, nose, throat, bronchi, lungs, skin	1B
Chloroprene—(2-chloro-1,3 butadiene)—skin	5, 3, 2	Reproductive hazard/systemic toxicity/suspect mutagen	4
Chloropyrifos (Dursban[R])—skin	6	Cholinesterase inhibition	4
o-Chlorostyrene	3	Cumulative liver, kidney damage	3
o-Chlorotoluene—skin	2, 15	Mild irritant—eye, skin	1B
2-Chloro-6-trichloromethyl pyridine (N-Serve[R])	18	Good housekeeping practice	1C
Chromates, certain insoluble forms (as Cr)	2, 10, 3	Suspect carcinogen/cumulative lung damage/dermatitis	4
Chromic acid & chromates	2, 10, 3	Suspect carcinogen/cumulative lung damage/nasal preforation, ulceration	4
Chromium, soluble chromic, chromous salts (as Cr)	10, 3	Cumulative lung damage/dermatitis	3
Clopidol (Coyden[R])	18	Good housekeeping practice	1C
Coal dust	10	Pneumoconiosis	3
Coal tar pitch volatiles	2, 10	Suspect carinogen/cumulative lung changes	4

Table 7.14 (*Continued*)

Substance	Health Code No.	Health Effects	Work Category
Cobalt, metal, fume & dust (as Co)	9, 10, 3	Asthma/cumulative lung changes/dermatitis	3
Coke oven emissions	1, 3	Cancer—lungs, bladder, kidney/skin sensitization	1C
Copper dusts & mists (as Cu)	16	Mild irritation—eye, nose, throat, skin	4
Copper fume (as Cu)	15, 11	Moderate irritation—eye, nose, throat, lung	4
Corundum (Al_2O_3)	19	Nuisance particulate	1C
Cotton dust (raw)	9, 10	Asthma/cumulative lung damage (bysinosis)	4
Crag[R] herbicide	3	Cumulative liver damage	3
Cresol (all isomers)—skin	14, 4, 3	Marked irritation—eye, skin/acute toxicity (CNS), liver and kidney damage	1B
Cristobalite	10	Pneumoconiosis	3
Crotonaldehyde	14	Marked irritation—eye, nose, throat, lungs	1B
Crufomate[R]	6	Cholinesterase inhibition	3
Lindane—skin	7, 3, 2	Cumulative CNS and liver damage/suspect carcinogen	4
Lithium hydride	14, 11, 7	Marked irritation—eye, nose, throat, skin/lung damage/CNS effects	1B
LPG (liquified petroleum gas)	18, 17, 8	Explosive/asphyxiant/narcosis	2
Magnesite	19	Nuisance particulate/accumulation in lungs	1C
Magnesium oxide fume	11	Lung effects (fume fever)	2
Malathion—skin	6	Cholinesterase inhibition	3
Maleic anhydride	14, 9, 2	Marked irritation—eye, nose, throat, lungs (edema), skin/asthma	2
Manganese & compounds (as Mn)	7, 10	Cumulative CNS damage/lung damage	1A
Manganese cyclopentadienyl tricarbonyl (as Mn)—skin	4, 7, 3	Acute CNS and blood effects/cumulative kidney damage	4
Marble	19	Nuisance particulate/accumulation in lungs	1C
Mercury, (organo) alkyl compounds, (as Hg)—skin	7, 3, 14	Acute and cumulative CNS damage/marked skin irritation	4
Mercury, inorganic (as Hg)—skin	7, 3, 2	Acute and cumulative CNS damage/gastrointestinal effects/gingivitis/suspect carcinogen	4

Table 7.14 (*Continued*)

Substance	Health Code No.	Health Effects	Work Category
Mesityl oxide	16	Mild irritation—eye, nose, throat	1B
Methane	18, 17	Explosive/simple	1C
Methanethiol	(See Methyl mercaptan)		
Methomyl (Lannate[R])—skin	6	Cholinesterase inhibition	3
Methoxychlor	3	Cumulative kidney damage	3
Methyl acetate	16, 8, 7	Mild irritation—nose, throat, lungs/narcosis/CNS effects	4
Methyl acetylene (propyne)	18, 8	Explosive/narcosis	1C
Methyl acetylene—propadiene mix (MAPP)	18	Flammable	1C
Methyl acrylate—skin	14, 4, 3	Marked irritation—eye, nose, throat, skin/acute lung damage	1B
Tetrahydrofuran	15, 8	Moderate irritation—eye, nose, throat/narcosis	2
Tetramethyl lead (as Pb)—skin	3, 7, 4	Cumulative liver, CNS and kidney damage/acute CNS effects	3
Tetramethyl succinonitrile—skin	4	Acute systemic toxicity (CNS)—headache, nausea, convulsions	4
Tetranitromethane	14, 4, 3	Marked irritation—eye, nose, throat/acute CNS and lung effects (edema)/cumulative systemic damage	4
Tetryl (2,4,6-Trinitrophenyl Methylnitramine)—skin	3	Contact dermatitis, skin sensitization/cumulative systemic toxicity	3
Thallium (soluble compounds)—skin (as Tl)	3	Cumulative systemic toxicity	3
4,4'-Thiobis (6-tert-butyl-*m*-cresol)	19	Apparent low toxicity	1C
Thiram[R]	4, 5	Acute systemic toxicity (antabuselike effects)/suspect teratogen	4
Tin (Inorganic compounds, except oxide) (as Sn)	4, 3	Acute and chronic systemic toxicity	4
Tin (Organic compounds) (as Sn)	14, 3	Marked irritation—skin/cumulative systemic toxicity	3
Tin oxide	10	Pneumoconiosis (stannosis)	3
Titanium dioxide	19	Nuisance particulate (accumulation in lungs)	1C
Toluene—skin	15, 8	Moderate irritation—eye, nose, throat/narcosis	2

Table 7.14 (*Continued*)

Substance	Health Code No.	Health Effects	Work Category
Toluene-2,4-diisocyanate (TDI)	9, 14, 3	Asthma/marked irritation—eye, nose, throat, bronchi, lungs/dermatitis	4
o-Toluidine—skin	13, 4, 2	Methemoglobinemia/acute systemic effects/suspect carcinogen	4
Toxphene	(See Chlorinated camphene)		
Tributyl phosphate	15, 7	Moderate irritation—nose, throat, lungs/headache	4
1,1,2-Trichloroethane—skin	3, 8	Cumulative liver damage/narcosis	4
Trichloroethylene	8, 3, 2	Narcosis/cumulative systemic toxic effects/suspect carcinogen	4

Source: OSHA (1989).

substance in excess of persons who work on normal 8-hr/day, 40-hr/week schedules.

$$\text{Equivalent PEL} = 8\text{-hr PEL} \times \frac{40 \text{ hr}}{\text{hours of exposure in one week}} \quad (15)$$

The specific approaches that an OSHA compliance officer can use to evaluate a workplace using unusual work shifts is described in detail in the *OSHA Field Operations Manual* (OSHA, 1979, 1989b). The OSHA models, although less rigorous than the pharmacokinetic models that will be discussed, have certain advantages due to their simplicity, e.g., they do account for the kind of toxic effect to be avoided, require no pharmacokinetic data, and tend to be more conservative than the pharmacokinetic models.

It is interesting that in the first OSHA occupational health regulation that discusses the long workday (unusual shifts), they prohibited the use of its own adjustment scheme to establish acceptable levels of exposure (Hickey, 1983). Because regulatory agencies must make decisions based on political, social, and scientific information, and, especially because they must survive legal scrutiny, it is important that the industrial hygienist, toxicologist, or physician understand that regulatory guidelines have to consider each and every exposure condition. Consequently, professionals should take the time to become familiar with the rationales for the TLV, PEL, AIHA WEEL, European MAC, and any other occupational exposure limit before adjusting it for unusually short or long periods of exposure. In this respect, Tables 7.11 to 7.14, as well as the various books that document the rationales for the various limits, are important aids.

Illustrative Example 5 (OSHA Model). An occupational exposure limit of 1 $\mu g/m^3$ has been suggested by NIOSH for polychlorinated biphenyls (PCBs). Studies of exposed workers indicate that the biologic half-life of PCBs is as long as several years. What adjustment to the occupational exposure limit might be suggested by NIOSH for workers on the standard 12-hr work shift involving 4 days of work per week if they adopted the simple OSHA formulas?

Solution

$$\text{Recommended limit} = 8 \text{ hr PEL} \times \frac{8 \text{ hr}}{12 \text{ hr}}$$

$$\text{Recommended limit} = 1 \ \mu g/m^3 \times 0.667 = 0.667 \ \mu g/m^3$$

Note: Since PCBs (chlorodiphenyl) are listed as both cumulative and acute toxicants (Category 4) in Table 7.14, Eq. (14) rather than (15) should be used since it yields the more conservative results.

Illustrative Example 6 (OSHA Model). Many industries such as boat manufacturing are seasonal in their workload. During the months of January, February, March, and April, the builders of boats work 5 days per week, 14 hr per day and could be exposed to concentrations of toluene diisocyanate (TDI) at the TLV of 0.005 ppm. What occupational exposure limit is recommended for TDI for a person who works 14 hr per day for 5 days per week but only works 8 weeks per year?

Solution. No adjustment is made.

Note: TDI is categorized in Table 7.14 as a Category 1A chemical (i.e., one that has a ceiling limit). Substances in this category have limits that should never be exceeded and consequently the limits are independent of the length or frequency of exposure. Exposure limits for chemical irritants such as these are currently thought not to require adjustment. Until more is known about human response to irritants during unusually long durations of exposure, the physician, nurse, and hygienist should make note of the employee tolerance to the presence of irritants at levels at or near the TLV. Eventually, human experience will tell us whether irritation is a time-dependent phenomenon and whether the response varies with the different chemicals.

Illustrative Example 7 (OSHA Model). The permissible exposure limit for elemental mercury is 50 $\mu g/m^3$. It has a half-life in humans in excess of several days. What adjustment to the limit would be recommended by OSHA for workers on a shift involving 4 days at 12 hr/day followed by 3 days of vacation, then 3 days of 12 hr/day followed by 4 days off?

Solution

$$\text{Equivalent PEL} = 8 \text{ hr PEL} \times \frac{40 \text{ hr}}{48 \text{ hr}}$$

$$= 50 \ \mu\text{g}/\text{m}^3 \times 0.833 = 40 \ \mu\text{g}/\text{m}^3$$

Note: Since elemental mercury is classified as a cumulative toxin (Category 3) according to Table 7.14, Eq. (15) should be used. The 48-hr workweek was used in this example since it yields a more conservative adjustment than the 36-hr workweek. It could be argued that a smaller adjustment factor based on the average number of hours worked every 2 weeks more accurately reflects the exposure [i.e., $(48 + 36)/2 = 42$] since the chronic effects of mercury are due to many weeks or years of excess exposure. A more detailed discussion of how to deal with toxins that tend to accumulate can be found in Hickey (1983).

One major drawback to the simple formulas suggested by Brief and Scala (1975) as well as OSHA is that they are conservative (i.e., they suggest a modified TLV or PEL that is lower than that predicted by more accurate pharmacokinetic models). This occurs because they *do not* take into account (quantitatively) the toxicant's overall biologic half-life (i.e., metabolism and elimination). As will be shown later in this chapter, pharmacokinetic models usually recommend a smaller degree of reduction in the air contaminant limit and, therefore, the achievement of the adjusted limit should be less costly yet still provide adequate protection.

8.3 Iuliucci Model

Robert Iuliucci of Sun Chemical proposed a formula (1982) for adjusting limits for long workdays that is similar to Brief and Scala's except that it accounts for the number of days worked each week as well as the number of hours worked each day. It is mentioned only for the sake of completeness since it poses no particular advantages over Brief and Scala's approach and because it is limited only to the schedule Iuliucci described (12-hr/day, 4-day/week schedule). For the work schedule, Iuliucci recommended the following equation for modifying the TLV:

$$\text{TLV}_x = \text{TLV}_s \times \frac{8 \text{ hr worked}}{12 \text{ hr worked}} \times \frac{12 \text{ hr recovered}}{36 \text{ hr recovered}} \times \frac{4 \text{ day workweek}}{5 \text{ day workweek}} \quad (16)$$

In addition to this formula, he recommended that exposure to carcinogens during 12-hr shifts always be reduced by 50 percent. Although this may be prudent, it would seem to be unnecessarily strict based on pharmacokinetic considerations.

8.4 Pharmacokinetic Models

Pharmacokinetic models for adjusting occupational limits have been proposed by several researchers (Mason and Dershin, 1976; Hickey, 1977, 1980, 1983; Hickey and Reist, 1977, 1979; Roach, 1966, 1977, 1978; Veng-Pederson et al., 1987; Andersen et al., 1987a). These models acknowledge that the maximum body burden arising from a particular work schedule is a function of the biological half-life of the substance. Pharmacokinetic models, like the other models, generate a correction factor based on the pharmacokinetic behavior of the substance as well as the number of hours worked each day and week, and this is applied to the standard limit in order to determine a modified limit. Unlike the OSHA, Brief and Scala, and Iuliucci approaches, by accounting for a chemical's pharmacokinetics, these models can also identify those exposure schedules where a reduction in the limit is not necessary.

The rationale for a pharmacokinetic approach to modifying limits is that during exposure to the TLV for a normal workweek, the body burden rises and falls by amounts governed by the biological half-time of the substance (Figs. 7.6 and 7.7). A general formula provides a modified limit for exposure during unusual work shifts so that the peak body burden accumulated during the unusual schedule is no greater than the body burden accumulated during the normal schedule. This is the goal of all of the pharmacokinetic models that have, thus far, been developed (Hickey and Reist, 1977).

It is worthwhile to note that the maximum body burden arising from continuous uniform exposure under the standard 8-hr/day work schedule always occurs at the end of the last work shift before the 2-day weekend. On the other hand, the maximum body burden under an extraordinary work schedule may not occur at the end of the last shift of that schedule (see Example 15). This is especially true when the duration and spacing of work shifts that precede the last shift differ markedly from the standard week. Because unusual work schedules can be based on a 2-week, 3-week, 4-week, or even 11-week cycle and the work shift may be 10, 12, 16, or even 24 hr in duration, no generalization regarding the time of peak body burden can be offered. The time of peak tissue burden for unusual schedules must therefore be calculated for each specific schedule and chemical.

8.4.1 Mason and Dershin Model

Mason and Dershin (1976) were the first to propose a pharmacokinetic model for adjusting exposure limits for unusual exposure schedules. Apparently, due to the manner in which the model was presented, their publication did not receive the attention and use that later researchers enjoyed. In spite of this, their approach is entirely accurate. Like the other pharmacokinetic models to follow, it accounted for the biologic half-life of the chemical and the number of hours of exposure per day and per week.

The pharmacokinetic model they developed accounted for a number of factors known to influence the rate of accumulation of a chemical. These factors included the toxicant concentration to which the individual is exposed, the physiochemical form of the material, the rate of metabolism and excretion, as well as the distribution of the material in the body following absorption. The major drawback of the model is that it assumes the body acts as a single compartment. The one-compartment approach (discussed previously) assumes the chemical to be uniformly distributed throughout blood and aqueous body fluids without significant storage in specific tissues except where such tissues may constitute the rate limiting step. In Mason

and Dershin's model, the overall respiratory exchange, metabolism, and renal excretion were accounted for by using a single effective clearance constant, k.

In their study they noted that for a simple single compartment model, Ruzic (1970)
had shown that the overall rate of change in accumulation can be expressed as

$$\frac{d[A]}{dt} = k_i^*[M] - k_c[A] \tag{17}$$

where $[M]$ = concentration of the contaminant in the environment (alveolar
        spaces) mg/L
   $[A]$ = concentration in the compartment, mg/L
   k_i^* = effective rate constant for uptake, hr^{-1}
$k_c[A]$ = overall clearance constant, hr^{-1}

Note: The effective rate constants may also include other factors: for example,
changes in vital capacity, minute volume, membrane permeability, or absorption
from other sources as the cutaneous absorption and loss of carbon disulfide.

In a cyclic pattern of exposure and recovery, Eq (17) has a general solution following the final period of recovery of

$$[A](t_m) = \frac{k_i^*}{k_c}[M][e^{-k_c(t_m - t_{m-1})} - e^{-k_c(t_m - t_{m-2})}$$

$$+ e^{-k_c(t_m - t_{m-3})} \cdots - e^{-k_c(t_m - t_{m-i})} \qquad \text{for } t_{(m-i)} \geq 0 \tag{18}$$

where k^*, k, and M are held constant and
   $t_{(m)}$ = time elapsed from the onset of the initial exposure, hours, including re-
        covery following the last exposure,
$t_{(m-1)}$ = time elapsed from the onset of exposure through the completion of the ith
        phase of uptake or recovery (loss)

and the initial body burden is assumed to be negligible.
Similarly, Eq. (18) may be solved for the body burden obtained at the close of
the last exposure by substituting $t_{(m-1)}$ for $t_{(m)}$ so

$$[A]_{(t_{m-1})} = \frac{k_i^*}{k_c}[M][1 - e^{-k_c(t_{m-1} - t_{m-2})}$$

$$+ e^{-k_c(t_{m-1} - t_{m-3})} \cdots - e^{-k_c(t_{m-1} - t_{m-i})}] \tag{19}$$

As noted by the authors, in intermittent exposure and recovery, an upper limit to
accumulation should be achieved for many of the typical, lipid-soluble solvents,
within 5 days, provided that the periods of recovery and rate of excretion are sufficiently large. The fraction of the saturation value (from continuous exposure) attained at the close of a series of cycles of exposure and recovery varies with the

clearance coefficient k_c and the pattern and duration of exposure according to the function

$$[e^{-k_c(t_m - t_{m-1})} - e^{-k_c(t_m - t_{m-2})} + e^{-k_c(t_m - t_{m-3})} \cdots - e^{-k_c(t_m - t_{m-i})}] \qquad (20)$$

which is dependent on both the concentration of exposure and final equilibrium position. Using Henderson and Haggard's (1943) definition of the equilibrium distribution coefficient,

$$D = \frac{C}{C_1} \qquad (21)$$

where $\quad D$ = distribution coefficient, dimensionless
$\qquad C$ = concentration of contaminant in the fluid phase (mg/L)
$\qquad C_1$ = concentration of contaminant in the vapor phase of the alveolar air (mg/l)

and solving Eq. (21) at equilibrium,

$$\frac{[A]}{[M]} \approx \frac{k_i^*}{k_c} \approx D \qquad (22)$$

This completes the data requirements for calculation of an expected body burden due to a series of intermittent exposures.

The authors acknowledged the limitations inherent in the assumption regarding the use of a one-compartment model but noted that in most cases this limitation had little practical significance with respect to adjusting exposure limits. Other kineticists, although not all of them, would probably agree with this observation (Dittert, 1977). In addition, Mason and Dershin supported the prior recommendations of Brief and Scala wherein a modeling approach should not be used to adjust limits whose goal is to minimize the likelihood of irritation, sensitization, or a carcinogenic response. Illustrative Examples 8, 9, and 10 demonstrate the use and general applicability of the model.

Illustrative Example 8 (Mason and Dershin Model). Several workers exposed to methanol in a printing operation complained that their exposure left them dizzy and with optic neuritis at the end of work on Wednesday, 56 hr after reporting to work Monday morning. If the distribution coefficient for methanol at saturation is 1700 to 1 for body water over the concentration in alveolar air; and the concentration in workroom air is 350 ppm (TLV = 200 ppm), what concentration would a worker have obtained in blood at (1) the end of the last exposure and (2) the time at which he or she would report to work on Thursday morning?

A schematic diagram illustrating the behavior of methanol of this situation is shown in Figure 7.23.

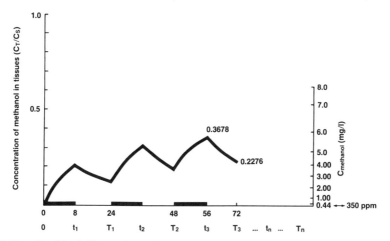

Figure 7.23 Graphical illustration of the behavior of methanol following the exposure schedule described in Example 8. Day-to-day increase in body burden were predicted by pharmacokinetic models. Exposure begins at $t = 0$, the 1st period ends at T_1, followed by 1st period of recovery (beginning at t_1), which ends at T_1; and repeats through t_n if the last point (T) is during exposure, and t_n if in recovery. (Courtesy of Dr. J. Walter Mason.)

Part I. What is the saturation fraction at the close of the last shift, 56 hr after initial exposure? Shifts are 8 hr in length and are separated by 16 hr without exposure (recovery).

A.

$$C_T/C_S = [1 - e^{-k(T)} + e^{-k(T-t_1)} - e^{-k(T-T_1)} + e^{-k(T-t_n)} - e^{-k(T-\tau_n)}]$$

where $t_1 = 8$ hr, $\tau_1 = 24$ hr, $t_2 = 32$ hr, $\tau_2 = 48$ hr, $t_3 = 56$ hr $= T$ (which is quitting time on Wednesday), k for methanol (MeOH) $= 0.03$ hr^{-1},

$$C_T/C_S = [1 - e^{-k(56)} + e^{-k(56-8)} - e^{-k(56-24)} + e^{-k(56-32)} - e^{-k(56-48)}]$$

$$C_{56}/C_S = [1 - e^{-0.03(56)} + e^{-0.03(48)} - e^{-0.03(8)}]$$

$$C_{56}/C_S = 0.3678$$

B. To calculate the tissue concentration (whole body) at saturation:

$$C_S = DC_{MeOH} = 1700 \times 0.44 \text{ mg/l} = 747 \text{ mg/l (body water) @ } 37°C$$

where D is the distribution coefficient for MeOH (body fluids/alveolar air) and C_{MeOH} the ambient methanol concentration

$$C_{56} = 747 \text{ mg/l} \times 0.3678 = 275 \text{ mg/l}$$

Note: If this was the only exposure to MeOH, exposure to $C_{8,5}/C_{Uns} \times TLV_{8,5}$ would not allow an increase in the tissue concentration over that experienced under the 8-hr day, 5-day week. This provides the general solution for adjusting standards to unusual shifts as

$$TLV_{unusual} = C_{5,8}/C_{Uns} \times TLV_{5,8}$$

where

$$C_{5,8}/C_{Uns} = F$$

Then,

$$TLV_{unusual} = F \times TLV_{5,8}$$

Part II. What is the saturation fraction at the end of the final period of recovery 72 hr after the onset of the first exposure? Shifts are 8 hr as before.

A.

$$C_T/C_S = [-e^{-k(T)} + e^{-k(T-t_1)} - e^{-k(T-\tau_1)}$$
$$+ e^{-k(T-t_2)} - e^{-k(T-\tau_2)} + e^{-k(T-t_n)}]$$

where $T = 72$, $t_1 = 8$, $\tau_1 = 24$, $t_2 = 32$, $\tau_2 = 48$, $t_3 = 56$, $\tau_3 = 72 = T$ (which is the starting time for work on Thursday), k for MeOH $= 0.03$ hr^{-1},

$$C_T/C_S = [-e^{-k(72)} + e^{-k(72-8)} - e^{-k(72-24)}$$
$$+ e^{-k(72-32)} - e^{-k(72-48)} + e^{-k(72-56)}]$$
$$C_{72}/C_S = [0 - e^{-0.03(72)} + e^{-0.03(64)} \cdots + e^{-0.03(16)}]$$
$$C_{72}/C_S = 0.2276$$

B. The tissue concentration is then found by

$$C_S = DC_{MeOH} = 747 \text{ mg}/1 \text{ (body water) @ } 37°C,$$
$$C_{72} = 747 \text{ mg}/1 \times 0.2276 = 170 \text{ mg}/1$$

Illustrative Example 9 (Mason and Dershin Model)

Part I. Workers are exposed to methanol for 8 hr/day for 5 days/week. What is the body burden of those workers at the end of the fifth day knowing that the distribution coefficient (blood/air) is 1700, the effective clearance constant, k_c, is $0.03h^{-1}$ ($t_{1/2} = 24$ hr) and the concentration to which they are exposed (1993 TLV) is 0.26 mg/1 (200 ppm)?

Solution. Setting $1700 = D$, $k_c = 0.03$ hr^{-1}, and the periods of exposure and recovery at 8 and 16 hr, respectively; for which $t_m = 5(24) = 120$, $t_{(m-1)} = 120 - 16 = 104$, $t_{(m-2)} = 104 - 8 = 96$ and so forth; the body burden remaining at the close of recovery on the fifth day may be calculated as a function of the concentration of exposure.

$$[A]_{(tm)} = 1700[M][e^{-0.03(16)} - e^{-0.03(24)} + e^{-0.03(40)} - e^{-0.03(48)}$$

$$+ e^{-0.03(64)} - e^{-0.03(72)} \cdots]$$

$$[A]_{(tm)} = (1700M)(0.249) = 420M$$

Similarly, if the concentration at the end of the last exposure is of interest t_m is set equal to $t_{(m-1)}$ and

$$[A]_{(tm-1)} = 1700[M][1 - e^{-0.03(104-96)} + e^{-0.03(104-80)} \cdots - e^{-0.03(104-72)}]$$

$$A_{(tm-1)} = 1700M(0.401) = 680M \text{ (peak burden during 8 hr/day schedule)}$$

Part II. Having calculated the body burden (peak) at the end of 5 days of exposure during a normal 8-hr workday, what modified TLV would be recommended for a 14-hr workday and 4-day workweek?

Solution. If the modified exposure limit is chosen so that the tissue concentration attained at the close of the last work phase under standard conditions is equal to the accumulation allowed under the novel shift arrangement:

$$[A]_{(tm-1)n} = [A]_{(104)5}$$

where $[A]_{(tm-1)n}$ = concentration that would be obtained at the close of the final exposure period in a novel shift arrangement
$[A]_{(104)5}$ = concentration obtained after a 5-day workweek under standard conditions

The accumulation at the close of the last exposure may be obtained directly with Eq. (22) or from tables constructed with the exponential term of the same equation. The final form is then reduced to

$$[M]_n = \frac{[A]_{(104)5}}{D} \times \frac{1}{\text{saturation fraction for } (t_{m-1})_n}$$

where $[M]_n$ = alveolar concentration of the contaminant resulting in a body burden equal to that attained by exposure under standard conditions
TLV_n = $[M]_n$ adjusted to ambient conditions

Therefore

$[A]_{96-56}$ = body burden of toxicant following 4 days of exposure during 14 hr/day schedule (i.e., 10hr/day or recovery)

By substitution

$$[A]_{96} = 1700M[1 - e^{-0.03(96-82)} + e^{-0.03(96-72)} - e^{-0.03(96-58)} + e^{-0.03(96-48)} \cdots]$$

$$[A]_{96} = (1700M)[1 - 0.657 + 0.487 - 0.320 + 0.237 - 0.156 + 0.115 - 0.75]$$

$$[A]_{96} = (1700M)0.630 = 1070M \text{ (peak burden during 14-hr/day schedule)}$$

To calculate the modified TLV for the 4-day, 14 hr/day schedule:

$$F[A]_{(tm-1)n} = [A]_{(104)5}$$

$$F = \frac{A_{(120)}}{A_{(96)}} = \frac{680M}{1070M} = 0.636$$

Adjusted TLV = 0.636 (8 hr TLV) = 0.636 (0.26 mg/l) = 0.165 mg/l

Conclusion. The concentration of airborne methanol for the 14-hr/day schedule should be reduced from 0.260 to 0.165 mg/L (37% lower) in order to have the same peak body burden as that noted during the standard workweek.

Figure 7.24 shows how a series of curves can be generated for a particular chemical which would permit the rapid identification of an adjustment factor for a number of schedules. The factor suggested by Mason and Dershin's formula is compared to that recommended by Brief and Scala (dotted line) in this figure.

Illustrative Example 10 (Mason and Dershin). In the oil producing regions of Canada and the North Sea, work schedules can become very complex. Calculate a modified exposure limit for a 2-week work schedule where persons will be exposed to cyclohexane (assume biologic half-life in humans of 23 hr) for four 10-hr workdays, followed by 4 days off, then four 12-hr workdays followed by 2 days off, to complete 2 calendar weeks, which involved 88 hr of exposure.

Equation A

$$[A]_{(tm)} = \frac{k_i^*}{k_c} [M] [e^{-k_c(tm-tm-1)} - e^{-k_c(tm-tm-2)} + e^{-k_c(tm-tm-3)} \cdots - e^{-k_c(tm-tm-i)}]$$

Equation B

$$[A]_{(tm-1)} = \frac{k^*}{k_c} [M] [1 - e^{-k_c(tm-1-tm-2)} + e^{-k_c(tm-1-tm-3)} = e^{-k_c(tm-1-tm-i)}]$$

$$\text{for } t_{(m-i)} \geq 0$$

where k_i^*, k_c and $[M]$ are held constant and

t_m = time elapsed since initial onset of exposure (t = 0) through the last phase of recovery = 336 hr

t_{m-1} = time elapsed to the point at which work ends = 276 hr

This schedule and the day-to-day increase in blood plasma concentration are shown in Figure 7.25.

The solid line was calculated by iteration [as a check against Eq. (A)]. But, the point at 252 and 336 hr were obtained via Eq. (B) and (A), respectively. Data points for the iteration are:

$A(t)/A_{sat}{}^a$	t, hr	$A(t)/A_{sat}{}^a$	t, hr	$A(t)/A_{sat}{}^a$	t, hr
0	0	0.3132	96	0.3178	240
0.2592	10	0.1525	120	0.5240	252
0.1703	24	0.0742	144	0.3656	264
0.3854	34	0.0361	168	0.5574	276
0.2532	48	0.0176	192	0.3889	288
0.4468	58	0.3146	204	0.4893	312
0.2935	72	0.2195	216	0.0921	336
0.4766	82	0.4555	228		

^{a}This is the "saturation fraction."

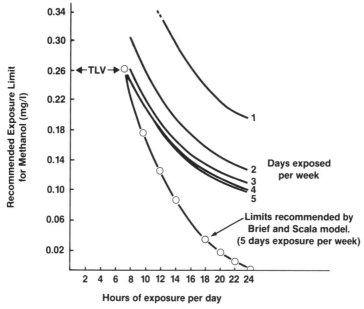

Figure 7.24 Modified exposure limits for methanol for various work schedules as determined by Mason and Dershin's formula. The dotted line illustrates the limits recommended by the Brief and Scala models. (From Mason and Dershin, 1976.)

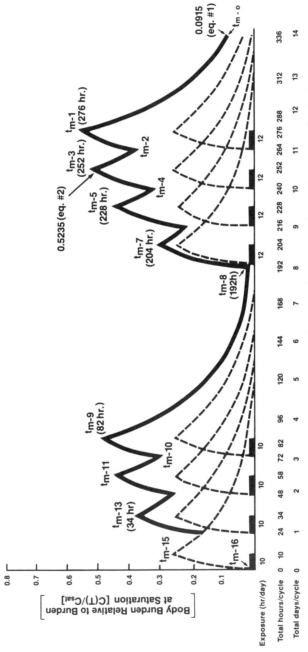

Figure 7.25 The likely body burden of cyclohexane in persons who work an unusual schedule involving four 10-hr days followed by four 12-hr days as described in Example 10. The dashed line represents the behavior of the chemical during each day of exposure and the dark lines represent the overall behavior of the chemical due to repeated exposure.

The construction of a table is the easiest way to solve Eq. (A) and (B). For $A(t_m)$ use Eq. (A) $[e^{-k\Delta t} - e^{-k\Delta t} \cdots]$. For $A(t_{m-1})$ use Eq. (6) $[1 - e^{-k\Delta t} \cdots]$.

t_m	t_{m-n}	Δt	$-k$	$-k\,\Delta t$	$e^{-k\Delta t}$
$336 - 276_1$	$=$	60	-0.03	-1.80	$+0.1653$
$336 - 264_2$	$=$	72	-0.03	-2.16	-0.1153
$336 - 252_3$	$=$	84	-0.03	-2.52	$+0.0805$
$336 - 240$	$=$	96	-0.03	-2.88	-0.0561
$336 - 228$	$=$	108	-0.03	-3.24	$+0.039200$
$336 - 216$	$=$	120	-0.03	-3.60	-0.027300
$336 - 204$	$=$	132	-0.03	-3.96	$+0.019100$
$336 - 192$	$=$	144	-0.03	-4.32	-0.013300
$336 - 82$	$=$	254	-0.03	-7.62	$+0.0005$
$336 - 72$	$=$	264	-0.03	-7.92	-0.0004
$336 - 58$	$=$	278	-0.03	-8.34	$+0.0024$
$336 - 48$	$=$	288	-0.03	-8.64	-0.000177
$336 - 34$	$=$	302	-0.03	-9.06	$+0.000116$
$336 - 24$	$=$	312	-0.03	-9.36	-0.000086
$336 - 10_{15}$	$=$	326	-0.03	-9.78	$+0.000057$
$336 - 0_{16}$	$=$	336	-0.03	-10.08	-0.000042
				$\Sigma e^{-k\Delta t} =$	0.0915

A diagram comparing the behavior of the body (blood) levels for the work schedule described in this example with that of a standard schedule is shown in Figure 7.25. Each day's exposure or "daily additions" are represented by the dashed lines at the bottom of the plot. The solid line, which leads up to point t, represents the sum of the dashed line values for the same point. Piotrowski and others have used this approach to illustrate the principle of summation (Lehnert et al., 1978).

Examples 8 through 10 illustrate one of the advantages of the Mason and Dershin model in that some persons feel that it is more flexible than the other models for calculating a modified exposure limit for complex work schedules. Specifically, it is useful whenever there is no fixed number of hours worked each day or a fixed number of days worked per week. In short, the best aspect of this approach is that the period in the cycles need not be of equal duration or number.

8.4.2 Hickey and Reist Model

In 1977 Hickey and Reist published a paper describing a general formula approach to modifying exposure limits that was equivalent to that of Mason and Dershin. The benefits of their work were manifold. They confirmed the soundness of the previous model but, equally importantly, they also validated it to some extent by comparing the results with published biological data. In addition, they proposed broader uses of the pharmacokinetic approach to modifying limits and presented a number of graphs that could be used to adjust exposure limits for a wide number of exposure schedules. The graphs were based on (a) the biologic half-life of the material, (b) hours worked each day, and (c) hours worked per week.

Over the next 3 years they published studies that illustrated how their model could be used to set limits for persons on overtime (Hickey and Reist, 1979) and for seasonal workers (Hickey, 1980). Hickey's treatment of the topic of adjusting exposure limits is quite thorough, and his publications are primarily responsible for most of the interest and research activity in this area.

As discussed, it is clear that for any schedule, the degree of toxicant accumulation in tissue is a function of the biologic half-life of the substance. Figure 7.26 illustrates how a toxicant might behave in a biologic system or a tissue following repeated exposure to a particular average air concentration during a typical work schedule. Note that the peak body burden, rather than the average (Ba) or residual body burden (Br), is the parameter of interest. The biologic half-life not only dictates the level to which a chemical accumulates with repeated exposure, it dictates the time at which steady state will be reached for any given exposure regimen (normal, unusual, or continuous). For example, for moderately volatile substances (e.g., solvents) that have half-lives in the range of 12–60 hr, and for most work schedules, the steady-state tissue burden will be reached after approximately 2–6 weeks of repeated exposure. For most volatile chemicals (low-molecular-weight solvents) with shorter half-lives, the steady-state blood levels will be reached after about 2–4 workdays. Under conditions of continuous uniform exposure, most chemicals will be within 10 percent of the steady-state levels following about 4 times the biologic half-life of the chemical, and after 7 half-lives it will be within 1 percent of the plateau (steady-state) levels (Gibaldi and Perrier, 1975).

Several indices of body or tissue burden could have been chosen as the basis for predicting equal protection for any two different exposure regimens. These indices are the peak, residual, and average body or tissue burden of a substance. Figure 7.24 illustrates these three potential criteria from which one must choose in order to

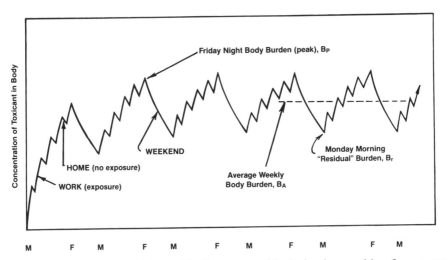

Figure 7.26 Illustration of the weekly fluctuation of body burden resulting from occupational exposure to an inhaled substance. Peak (B_P), residual (B_r), and average body burdens (B_A) are shown. (From Hickey and Reist, 1977.)

build a mathematical model. As in Mason and Dershin's model. Hickey and Reist selected the peak body burden as the criterion since it is a more conservative approach to predicting the occurrence of a toxic effect than either the average or residual tissue concentrations (Amdur, 1973). A thorough discussion of the rationale for selecting the peak burden rather than the residual or average for building the models can be found in Hickey's dissertation (Hickey, 1977).

Other choices for predicting safety are problematic. For most chemicals, the residual body burden goes to virtual zero for most chemicals after a weekend away from exposure. Consequently, modeling to control this criterion would not prevent excessive peak burdens. The use of the average burden reduces the model to Haber's law (David et al., 1981). This, of course, would allow high tissue burdens to occur for long periods even though the TWA burden might be acceptable. Peak burden, therefore, is the best criterion, however, it may not be appropriate when the goal of an exposure limit is to avoid a carcinogenic hazard. In these cases, control of the average weekly or daily exposure (at the TLV or PEL) should generally be adequate to prevent any significant risk.

The Hickey and Reist model can, like Mason and Dershin's approach, be used to determine a special exposure limit for workers on extraordinary schedules, which will prevent peak tissue or body burdens from being greater than that observed during standard shifts. This special limit is expressed as a decimal adjustment factor that, when multiplied by the appropriate exposure limit, would yield the "modified" limit. It is worthwhile to note that all of the researchers have been careful to note that they *did not assert that currently prescribed or recommended occupational health limits are safe, but only that the special limit that can be predicted from their models should yield "equal protection" during a special exposure situation*! Examples 11 through 15 illustrate the use of the Hickey and Reist model.

One limitation of the models of Hickey and Reist (Hickey, 1977; Hickey and Reist, 1977), Mason and Dershin (1976), and Roach (1978) is that they assume that the body acts as one compartment. Although this simplification may not pose many shortcomings for the task of adjusting occupational exposure limits, it is well known that many, if not most, chemicals do not exhibit one-compartment behavior (Riegelman et al., 1968; Withey and Collins, 1980). This is not surprising since after a substance (particulate, gaseous, or vapor) is inhaled in air, it is taken up by the body, distributed, perhaps metabolized, and excreted by complex processes. Even though these processes can now be modeled quite well through the use of complex mathematical models, the one-compartment model has been used by these scientists since, in most cases, even this simple approach will yield results similar to those that incorporate more complex approaches.

As discussed by Hickey, the one-compartment model is the simplest one and it assumes that the body is a homogeneous mass, comparable to a room or compartment containing a clean fluid such as air. More of the air, bearing a contaminant, enters and flows continuously through the compartment, mixing en route with the air therein in a process analogous to inhalation. If contaminated air continues to enter, the contaminant concentration in the compartment increases until it reaches equilibrium with that of the incoming air; that is, as much contaminant is leaving as entering (see Fig. 7.27). When the contaminated air supply is replaced with clean

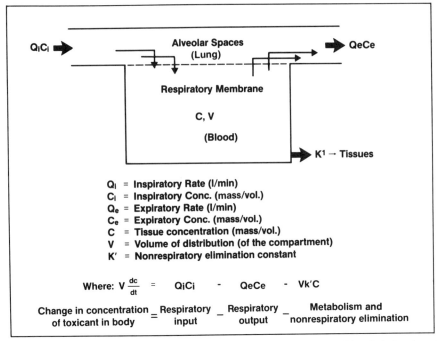

Figure 7.27 Diagram illustrating the driving forces for the pharmacokinetic behavior of an inhaled air contaminant. This simple description is the basis of the models developed by Mason and Dershin (1976), as well as, Hickey and Reist (1977).

air, the process is reversed, and the contaminant concentration in the compartment decreases exponentially (Hickey and Reist, 1977). An analogy can be shown using fluid in tanks, as shown in Figure 7.28. The mass transfer phenomena are described by the following equations:

For uptake:

$$B_t = CWK(1 - e^{-kt}) + B_0(e^{-kt}) \tag{23}$$

For excretion:

$$B_r = B_t(e^{-kt_r}) \tag{24}$$

where B_t = body burden of substance at time t (mass)
  B_0 = initial body burden of substance at time zero (mass)
  B_r = residual body burden of substance at time t (mass)
  C = substance concentration in air (mass/volume)
  K = ratio of the substance's equilibrium solubility in the body to that in air, or "partition coefficient" (dimensionless)
  W = volume of body

k = uptake and excretion rate of substance in the body, equal to L/WK, in which L is the flow rate of air to the body time (time, hr^{-1})

t = time of exposure to substance in air (hr)

t_r = time since cessation of exposure to substance in air (hr).

It should be noted that k may also be expressed in terms of half-life or half-time of the substance in the body, $T_{1/2}$, where $k = (\ln 2)/T_{1/2}$.

Hickey and Reist have noted that while the predictive capability of the pharmacokinetic models is limited by the shortcomings caused by simplification to a one-compartment system, there are practical circumstances that minimize these drawbacks (Hickey, 1977). First, many of the body tissues that are important targets for inhaled substances (''critical tissues'') are highly perfused (Ruzic, 1970; Guberan and Fernandez, 1974; Gibaldi and Perrier, 1975; Notari, 1985; Handy and Schindler, 1976; Piotrowski, 1977; Fiserova-Bergerova and Holaday, 1979; Andersen et al., 1987b), and the concentration of the contaminant in these tissues may follow that of the arterial blood closely, thus in effect, becoming part of the lung–arterial blood compartment. The opposite case occurs when the buildup of contaminant in the critical or target tissue is extremely slow compared to the buildup in the rest of the body. In such a case, the remainder of the body, or more specifically the arterial blood, may be assumed to reach saturation relatively quickly and remain at a virtually constant concentration. In effect, the body (except for the critical tissue) becomes part of the ambient environment, and the critical tissue becomes the one-compartment body (Hickey and Reist, 1977).

Figure 7.25 illustrates how the body takes up and excretes an inhaled air contaminant as described according to Eq. (23) and (24), where exposure to contaminated air occurs during working hours and clean air is inhaled during nonworking hours. The body takes up the contaminant according to rate k during periods of exposure, and during nonworking hours the body excretes the contaminant according to negative rate $-k$ (Hickey and Reist, 1977). The rate constant k is assumed to remain

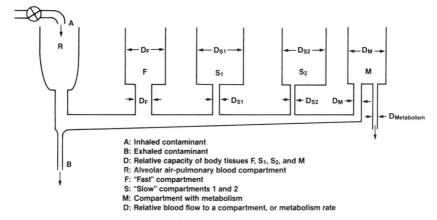

A: Inhaled contaminant
B: Exhaled contaminant
D: Relative capacity of body tissues F, S₁, S₂, and M
R: Alveolar air-pulmonary blood compartment
F: "Fast" compartment
S: "Slow" compartments 1 and 2
M: Compartment with metabolism
D: Relative blood flow to a compartment, or metabolism rate

Figure 7.28 Simulation of body uptake of an air contaminant using fluid in tanks as analogy. [Reprinted from Hickey (1977) with permission.]

unchanged for each chemical regardless of the duration of exposure or whether there are repeated exposures. Small changes in k have been reported following repeated exposure to unusual shifts, but these are not usually large enough to justify mathematical correction. It should be remembered that for a given exposure schedule and any k, the body will eventually reach some equilibrium level with the contaminated air after continuous or repeated exposures (see Fig. 7.8).

The variation in body burden upon exposure to an air contaminant for five workdays per week for a period long enough to reach equilibrium (steady state) is also illustrated in Figure 7.24. Equilibrium implied that the "Monday morning" body burden (B_r) remains the same from week to week for a given exposure schedule. Each schedule also has a characteristic Friday afternoon peak body burden (B_p), and average body burden (B_a). This is illustrated in Figure 7.21 where two different exposure schedules and the resulting body burdens are described for a chemical with a moderately long half-life.

In spite of the fact that nearly all volatile chemicals will demonstrate some degree of two-and three-compartment behavior, the one-compartment assumption is generally satisfactory (if the half-life has been properly calculated). Dittert (1977) has noted that in many, if not most situations, simplification to one-compartment behavior poses a minimal source for error in most calculations.

In their publications, Hickey and Reist (1977) described the derivation and use of the following equation for adjusting limits:

$$(1 - e^{-kT_s}) \left[1 - \exp\left(-kt_n\right) + \exp\left(-k \sum_{i=n-1}^{n} t_i\right) \right.$$

$$\left. - \cdots + \cdots - \exp\left(-k \sum_{i=1}^{n} t_i\right) \right] n$$

$$(1 - e^{-kl_n}) \left[1 - \exp\left(-kt_s\right) + \exp\left(-k \sum_{j=s-1}^{s} t_j\right) \right.$$

$$\left. - \cdots + \cdots - \exp\left(-k \sum_{j=1}^{s} t_j\right) \right] s \qquad (25)$$

in which the t values represent duration of sequential work and rest periods in cycle T for normal (n) and special (s) exposure schedules. The authors noted that in their model, the use of ratios causes many of the imponderable and unknown terms to cancel, leaving only the special work schedule, which will be known, and the substance half-life (or uptake/excretion rate), which may or may not be known.

This equation can be used to determine a modified TLV or PEL for any exposure schedule since it accounts for the number of hours worked per day, days worked per week, time between exposures, and biologic half-life of the toxicant.

Where the special or extraordinary work cycle uses normal days and weeks, Eq. (25) can be simplified to the following form:

$$F_p = \frac{(1 - e^{-8k})(1 - e^{-120k})}{(1 - e^{-hk})(1 - e^{-24dhk})} \tag{26}$$

in which, using hours as the time unit,

F_p = TLV or PEL reduction factor
k = uptake and excretion rate of the substance in the body (biologic half-life)
h = length of special daily work shift
d = number of workdays per "workweek" in the special schedule

The general Eq. (25) for regular repetitive schedules simplifies to

$$F_p = \frac{[1 - e^{-kt_{1n}}][1 - e^{-k(t_{1n} + t_{2n})n}][1 - e^{-kT_s}][1 - e^{-k(t_{1s} + t_{2s})}]}{[1 - e^{-kt_{1s}}][1 - e^{-k(t_{1s} + t_{2s})m}][1 - e^{-kT_n}][1 - e^{-k(t_{1n} + t_{2n})}]} \tag{27}$$

where t_{1n} = length of normal daily work shift (8 hr)
t_{2n} = length of normal daily nonexposure periods (16 hr)
$t_{1n} + t_{2n}$ = length of normal day (24 hr)
T_n = length of normal week (168 hr)
n = number of workdays per normal week (5),
t_{1s} = length of special "daily" work shift (hr)
t_{2s} = length of special nonexposure periods between shifts (hr)
$t_{1s} + t_{2s}$ = length of basic work cycle, analogous to the "day" (hr)
T_s = length of periodic work cycle, analogous to the "day" (hr)
m = number of work "days" per work "week" in the special schedule

The model may be used to predict the permissible level and duration of exposure necessary to avoid exceeding the normal peak body burden during intrashift, short, high-level exposures. The model does this by establishing excursion limits that will provide equal protection for these situations. Equation (28) is used to do this:

$$F_p = \frac{1 - e^{-kt_n}}{1 - e^{-kt_e}} \tag{28}$$

where $k = \ln 2 / T_{1/2}$
t_e = exposure time (hr)
t_n = normal shift length (8 hr)

Hickey and Reist have noted that for substances with short biologic half-lives, less than 3 hr, no adjustment needs to be applied for workers on most extraordinary work shifts since there is no opportunity for accumulation. In Figure 7.29, the normal workweek is compared to workweeks from one to seven 8-h days. It can be

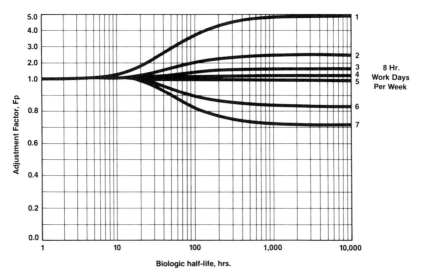

Figure 7.29 Adjustment factor (F_P) as a function of substance half-life ($t_{1/2}$) for various workweeks. (From Hickey and Reist, 1977.)

seen that exposure limits may not be increased, even if exposure is for only one day per week, unless the substance half-life is greater than 6 hr. Similarly, limits need not be decreased for 6- or 7-day workweeks involving exposures of 8 hr/day unless the substance half-life is greater than about 16 hr. For substances with very long half-lives, those in excess of 40 hr, F_p is simply proportional to the number of hours worked per week, as compared to 40 hr.

In an effort to simplify the process of adjusting limits for unusual shifts, Hickey and Reist developed a number of graphs, shown in Figures 7.29–7.31 and 7.38–7.42. Many health professionals have found these to be very useful when estimating safe levels of exposure for chemicals for which they have little or no pharmacokinetic data. In these graphs, the adjustment factor, F_p is usually plotted as a function of substance half-life, $t_{1/2}$ for a particular work schedule(s). For example, Figure 7.30 shows the difference in the occupational exposure limit between a normal workweek and a workweek of four 10-hr days, a workweek of three 12-hr days, and a single 40-hr shift per week.

As is shown in Figure 7.30, for substances with very short half-lives, one hour or less, no correction is needed since the peak body burden is reached very quickly and is the same for a normal workweek as for any longer schedule. Therefore, if B_p is chosen as the predictor of equal protection, no reduction in OSHA limits is necessary for longer-than-normal work shifts as long as the weekly exposure is less than 40 hr.

For substances with very long half-lives in the body, the adjustment factor is proportional merely to the number of hours exposed, not the daily or weekly exposure schedule. Thus, all 40-hr weeks have a special exposure limit for such sub-

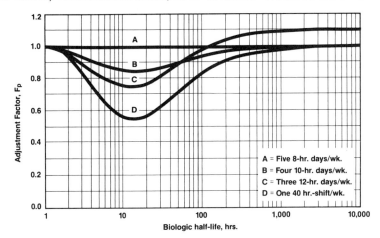

Figure 7.30 Adjustment factor (F_P) as a function of substance half-life ($t_{1/2}$) for various exposure regimens (shift schedules). (From Hickey and Reist, 1977.)

stances equal to the normal limit, or an F_p of unity. Since three 12-hr days total only 36 hr per week, F_p for that schedule is 40/36, or 1.1, for a substance with a very long half-life.

Hickey offered sound advice when he noted that one need not resort to the conservative approaches of Brief and Scala, OSHA, or Iuliucci when the biologic half-life of the substance is not known. By assuming that the chemical has a half-life that would cause the greatest degree of day-to-day accumulation for that particular work schedule, the worst-case F_p can be calculated for any exposure schedule. Some of these worst-case values of F_p for selected schedules are shown in Figure 7.30. For example, since F_p varies as a function of the half-life, the worst-case condition is 0.84 for four 10-hr days, 0.75 for three 12-hr days, and 0.54 for the single 40-hr shift. Where the half-life is not known, the worst-case F_p can be used. Consequently, the pharmacokinetic models can be used to accurately protect workers on any schedule even when the pharmacokinetic behavior of the specific chemical is not known.

Other curves may be generated from Eq. (25) or (27) to compare any two schedules. In Figure 7.31, the correction factor recommended for continuous exposures for several periods of time from one to 1024 days is presented. A rest period equal to three times the exposure periods is needed before reexposure should occur. Again, for substances with very short half-lives, no adjustment to exposure limits is necessary when B_p is the criterion.

Illustrative Example 11 (Hickey and Reist Model). In cities where commuting distances are a burden to the worker, one of the more frequent work schedules is the four day, 10-hr/day "compressed workweek." Assuming that this workweek is used in the textile industry and that persons are routinely exposed to aniline at the PEL of 5 ppm, what adjusted occupational exposure limit would be recommended? Assume that aniline has an overall (beta phase) half-life of about 2 hr in humans.

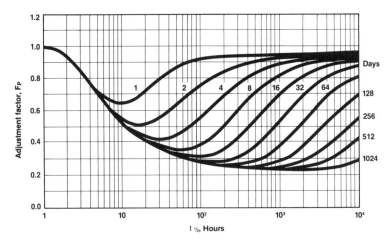

Figure 7.31 Adjustment factor (F_P) as a function of substance half-life ($t_{1/2}$) for continuous exposure schedules (days). (From Hickey and Reist, 1977.)

Solution. Since the workweeks are equal and both schedules have all workdays consecutive, the simplified form of the general Eq. (26) can be used.

$$F = \frac{C_s}{C_n} = \frac{(1 - e^{-8k})(1 - e^{-120k})}{(1 - e^{-10k})(1 - e^{-96k})} \quad \text{for} \quad t_{1/2} = 2 \text{ hr}$$

$$k = \ln 2/2 \text{ hr} = 0.347 \text{ hr}^{-1}$$

$$F = 0.9677 \equiv 0.97$$

$$C_s = C_n F = 5 \times 0.97 = 4.85 \approx 5.0 \text{ ppm}$$

Note: No change is needed for chemicals with a half-life this short unless exposure is for 24 hr/day for several days.

Illustrative Examples 12 (Hickey and Reist Model). Some persons in the petrochemical industry will routinely work 12-hr shifts for 5 consecutive days before having a 4-day period of no work; then they return for three 12-hr days again to be followed by 4 days off. This 5/4, 3/4, 12-hr schedule requires some adjustment of the normal exposure limit for certain systemic toxins if the peak body burden for this unusual shift is not to exceed the normal shift.

Based on Piotrowski's work (1977), the overall half-life (B phase) in the human for trichlorethylene and its metabolites is about 9 hr. What modifications of the 1983 TLV of 50 ppm for trichloroethylene would be recommended for the workers on this new shift schedule during their first week of 5 consecutive 12-hr days of work?

Solution. Begin by determining the body burden for a normal week (B_{pn})

For

$$t_{1/2} = 9 \text{ hr}, \quad k = 0.077$$

$$B_{p_n} = C_n WK \frac{(1 - e^{-8k})(1 - e^{-120k})}{(1 - e^{-168k})(1 - e^{-24k})}$$

$$B_{p_n} = 0.546 C_n WK$$

We do not want the first-week special body burden, B_{p_s} (after switching schedules) to exceed B_{p_n} (normal burden), so

$$B_{p_s}(\text{week 1}) = C_s WK(1 - e^{-12k})$$
$$\cdot (1 \ + e^{-124k} + e^{-48k} + e^{-72k} + e^{-96k}) + B_{p_n}(e^{-178k})$$

	↑	↑	↑	↑	↑	↑
	day	day	day	day	day	residual left
	5	4	3	2	1	from last week
						of old schedule

We set $B_{p_s} = B_{p_n}$ because we want the body burden from the "special" schedule (B_{p_s}) to be equal to the body burden for the normal schedule (B_{p_n}).

$$B_{p_n} - B_{p_n}(e^{-178k}) = B_{p_n}(1 - e^{-178k}) = 0.546\, C_n WK(1 - e^{-178k})$$

$$0.546\, C_n WK(1 - e^{-178k}) = C_s WK(1 - e^{-12k})(1 + e^{-96k})(1 + e^{-96k} + e^{-72k}$$
$$+ e^{-48k} + e^{-24k})$$

$$\frac{C_s}{C_n} = \frac{\text{TLV special}}{\text{TLV normal}}$$

$$= \frac{0.546\,(1 - e^{-178k})}{(1 - e^{-12k})(1 + e^{-96k} + e^{-72k} + e^{-48k} + e^{-24k})}$$

$$= \frac{0.546}{0.716} = 0.76$$

$$C_s = C_n \times 0.76$$

$$C_n = 50 \text{ ppm trichloroethylene}$$

The recommended TLV for trichloroethylene for a 12-hr/day, 5-day schedule = 38 ppm.

Note: Due to the short half-life and 40-hr/week schedule, a short cut approach would yield same result:

$$F = \frac{1 - e^{-8k}}{1 - e^{-12k}} = 0.76$$

Illustrative Example 13 (Hickey and Reist Model). In an 8-hr day, 7-day workweek situation, such as the 56/21 or 14/7 schedules, what should the TLV for H_2S be (assume TLV = 100 ppm)? This is the special case of a 7-day workweek. Biologic half-life in humans is 2 hr, but the rationale for the standard is based on systemic effects and irritation.

Solution. There is no need to reduce limits to prevent excess irritation, but for system effects it should be. For 14 days on and 7 days off:

$$F_p = \frac{(1 - e^{-8k})(1 - e^{-120k})/(1 - e^{-168k})(1 - e^{-24k})}{(1 - e^{-8k})(1 - e^{-(14 \times 24)k})/(1 - e^{-(3 \times 168)k})(1 - e^{-24k})}$$

where   total week = 21 days or 168×3 hr
workweek = 14 days or 14×24 hr

$$F_p = \frac{(1 - e^{1-120k})/(1 - e^{-168k})}{(1 - e^{-336k})/(1 - e^{-504k})} = 1$$

With a $T_{1/2}$ of 2 hr, virtually all of the chemical is lost during the 16 hr of recovery each day, so there is no need to lower the TLV. Also, even though the average hours worked per week is 37.3, the TLV may not be raised by 40/37.3 or by 1.07X. For a 56/21 schedule,

$$F_p = \frac{(1 - e^{-120k})/(1 - e^{-168k})}{(1 - e^{-(56 \times 24)k})/(1 - e^{-(7 \times 168k)})} = 1$$

Likewise, with 56/21 there is no need to reduce TLV, even through the average hr/week is 40.7.

8.4.3 Roach Model

Roach (1978) also proposed a mathematical model for use during extraordinary work shifts. His model, although developed independently, was virtually identical to that proposed by Mason and Dershin, as well as Hickey and Reist. His general equation is shown below:

$$R = \frac{(1 - e^{-8a})(1 - e^{-120a})(1 - e^{-la})}{(1 - e^{-24a})(1 - e^{-168a})(1 - e^{-ma})\Sigma e^{-na}} \tag{29}$$

In this formula the shifts included are those in one complete work cycle prior to the shift end in question and

l = total number of hours for a complete work cycle
m = number of hours duration of the work cycle

n = number of hours from a prior work shift end to the shift end in question

e = the exponent of natural logarithms, 2.718

a = $\log 2/t_{1/2} = 0.693/t_{1/2}$

$t_{1/2}$ = biologic half-time in hours

The minimum value of this ratio, R_{min}, is the value of R obtained for the particular work shift in the cycle in which the maximum body burden occurs.

Persons who wish to use this model are referred to the original article (Roach, 1978). Since it is functionally the same as the prior models, no examples of its use are provided. Table 7.15, however, was developed by Roach and can serve as a useful guide for quickly approximating the modified exposure limit for a number of types of unusual shifts where the biologic half-life ($t_{1/2}$) is known. Roach has shown that for any given work schedule, no matter how complex, a generalized graphical solution that yields the adjustment factor formula for any chemical can be developed. To illustrate this point, Roach developed Figure 7.32 for the particular complex work schedule shown.

Like previous writers, Roach noted that the limits for substances which (a) have very short biologic half-lives, (b) irritants, or (c) are carcinogens may require no alteration when the standard work schedule is altered. Roach suggested that the TLV for substances that have a very long biological half-time, such as mineral dusts, should be modified in proportion to the average hours worked per week, which has also been the recommendation of OSHA, Mason and Dershin (1976), and Hickey and Reist (1977). With such substances, the duration of any practical work cycle is short in comparison with their biological half-time, therefore it is appropriate that the limit would be unaltered so long as persons only worked an average of 40 hr per week. Roach suggested that as an additional precaution appropriate medical surveillance to detect any adverse effects would be advisable if the work schedule is such that R_{min} is less than 0.5 or greater than 2.0.

Table 7.15 Examples of Exposure Limit Adjustment Factors for a Variety of Unusual Work Shifts Based on the Chemical's Biologic Half-Life in Humans[a]

Work Shifts/Week	Hour/Work Shift	R_{min} when $t_{1/2}$ is			
		1 hr	10 hr	100 hr	1000 hr
4	10	1.00	0.85	0.94	0.99
5	9	1.00	0.92	0.89	0.89
5	10	1.00	0.85	0.81	0.80
5	12	1.00	0.75	0.68	0.67
6	6	1.01	1.26	1.18	1.12
6	8	1.00	1.00	0.89	0.84
6	10	1.00	0.85	0.72	0.67
7	6	1.01	1.26	1.09	0.97
7	8	1.00	1.00	0.82	0.73

[a]Calculated by Roach (1978) using a pharmacokinetic model.

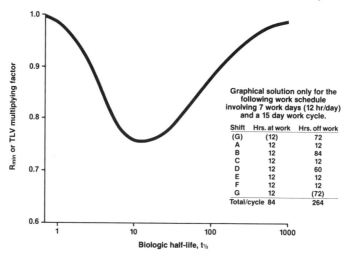

The graph includes the following table:

Graphical solution only for the following work schedule involving 7 work days (12 hr/day) and a 15 day work cycle.

Shift	Hrs. at work	Hrs. off work
(G)	(12)	72
A	12	12
B	12	84
C	12	12
D	12	60
E	12	12
F	12	12
G	12	(72)
Total/cycle	84	264

Figure 7.32 Graph showing the adjustment factor for any chemical to which workers are exposed during this specific work schedule. This approach can be quickly used by industrial hygienists who must set limits for dozens of chemicals for a given shift schedule. (From Roach, 1978.)

8.4.4 Veng-Pedersen Model

As mentioned previously, one of the shortcomings of the existing models for adjusting exposure limits is that a one-compartment model is used to determine the adjustment factor. A model by Veng-Pedersen (1984) has been proposed that takes into account any number of compartments exhibited by a chemical. The equations for adjusting the TLV are based on a "linear systems" approach. The merits of a linear systems approach have been discussed by several authors (Thran, 1974; Veng-Pedersen, 1977; Cutler, 1978). At this time, the usefulness of the Veng-Pedersen proposal is limited since a true "working" formula for adjusting exposure limits was fully developed. The model, however, does have the capability of predicting, based on animal data, whether or not a chemical will accumulate with repeated exposures if the standard kinetic parameters are known (Veng-Pedersen, 1984). The approach has on one occasion been used to evaluate the results of animal toxicologic studies of 12-hr/day exposures (Veng-Pedersen et al., 1987). The benefits of the PB–PK approach outweigh the possible benefits of the linear systems model and, therefore, it likely will not be used in the future.

8.5 Determining a Chemical's Biologic Half-life

Effective half-lives of chemicals are difficult to precisely determine because of the complex manner in which many behave in the body. As discussed previously, the body can be envisioned to be made up of many compartments, and these different parts can take up and excrete substances at different rates. Consequently, each com-

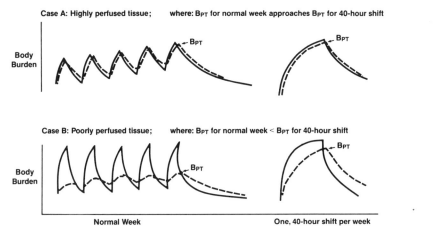

Figure 7.33 Variation in tissue uptake as a function of exposure schedule and blood flow (B_{PT} is peak tissue concentration of toxicant). (From Hickey, 1977.)

partment has a different half-life for each substance. The difficulties in determining the half-life in a tissue such as the liver is the reason why overall or apparent half-lives that represent clearance of contaminants from the blood are used. Figure 7.33 illustrates how the degree of perfusion of various tissues and the exposure schedule will influence the peak concentration of toxicant that will be achieved, as well as the biologic half-life of the chemical in that tissue.

Effective half-lives for human uptake and excretion have been determined for many substances. Roach (1977) and Hickey (1977) have listed the half-lives of several industrial chemicals that they compiled from various sources. Table 7.16 presents a slightly more comprehensive list of substances for which half-lives have been determined in humans. Where available, half-lives from human studies should be applied to the models, and where the half-life is not known, the worst-case approach described by Hickey and Reist (Commonwealth of Pennsylvania, 1971) is convenient. Whenever both the α and β phase half-lives are known, the β phase half-life should be used in the various models since the terminal or β phase half-life conservatively describes the time course of elimination of the chemical. There are numerous differences and pitfalls in accurately determining biologic half-life, and these have been reviewed by Gibaldi and Weintraub (1971).

8.6 Comparing the Various Models

The different models that modify exposure limits without consideration of pharmacokinetic behavior are those of OSHA (1979) and Brief and Scala (1975). The Brief and Scala model equations are:

$$\text{RF} = \frac{8}{h} \times \frac{24 - h}{16} \quad \text{on a daily basis} \tag{30}$$

$$RF = \frac{40}{H} \times \frac{168 - H}{128} \quad \text{on a weekly basis} \tag{31}$$

$$EF = [(EF_8 - 1) \times RF + 1] \quad \text{for excursions} \tag{32}$$

where RF = reduction factor to be applied to the TLV or OSHA limit
   h = hours worked per day
   H = hours worked per week

Table 7.16 Estimated Half-Lives of Various Chemicals or Their Metabolites in Humans

Substance	Compartment (Media Collected)	Half-Life	Reference
Acetone	Overall (blood)	3 hr	Piotrowski (1977)
Aniline	Overall (urine)	2.9	Piotrowski (1977)
Benzene	Overall (blood)	3.0 hr	Piotrowski (1977)
			Nomiyama and Nomiyama (1974)
			Hunter and Blair (1972)
Benzidine	Overall (urine)	5.3 hr	Piotrowski (1977)
Carbon monoxide	Overall (breath)	1.5 hr	Peterson (1962)
Carbon disulfide	Overall (breath)	0.9 hr	Piotrowski (1977)
Carbon tetrachloride	Fast (breath)	20 min	Stewart et al. (1961a)
	Slow (breath)	3.0 hr	
Dichlorofluoromethane	Overall (blood)	9.4 min	Adir et al. (1975)
Dimethyl formamide	Overall (urine)	3.0 hr	Krivanek et al. (1978)
Ethyl acetate	Overall (breath)	2.0 hr	Nomiyama and Nomiyama (1974)
Ethyl alcohol	Overall (breath)	1.5 hr	Nomiyama and Nomiyama (1974)
Ethyl benzene	Overall (urine)	5.0 hr	Piotrowski (1977)
Hexane	Overall (breath)	3.0 hr	Nomiyama and Nomiyama (1974)
Methanol	Overall (urine)	7.0	Piotrowski (1977)
Methylene chloride	Overall (blood)	2.4 hr	DiVincenzo et al. (1972)
Nitrobenzene	Overall (urine)	86.0 hr	Piotrowski (1977)
Phenol	Overall (urine)	3.4 hr	Piotrowski (1971b)
p-Nitrophenol	Overall (urine)	1.0 hr	Piotrowski (1977)
Styrene	Overall (urine)	8.0 hr	Ikeda et al. (1974)
			Ramsey and Young (1978)
			Stewart et al. (1968)
Tetrachloroethylene	Overall (breath)	70 hr	Stewart et al. (1970a, 1961b)
1,1,1-Trichloroethane	Overall (urine)	8.7 hr	Monster (1979)
Trichloroethylene	Fast (breath)	30 min	Stewart et al. (1970b)
	Slow (breath)	24 hr	Kimmerle and Eben (1973)
Trichlorofluoroethane	Overall (blood)	16 min	Adir et al. (1975)
Toluene	Fast (urine)	4 hr	Ogata et al. (1970)
	Slow (urine)	12 hr	Carlsson (1982)
Xylene	Overall (urine)	3.8 hr	Ogata et al. (1970)

EF = adjusted excursion factor
EF = normal excursion factor

Note: ACGIH abandoned the use of excursion factors in 1976. OSHA never adopted them.

Illustrative Example 14 (Hickey and Reist Model). The NIOSH recommended occupational exposure limit for PCB is 1 $\mu g/m^3$. In humans it has been found that the biologic half-life of PCBs is roughly 2.5 years. What adjustments to the occupational exposure limit would you recommend for workers on the standard 12-hr shift involving 4 days of work followed by 3 days of vacation, then 3 days of work followed by 4 days off, etc.

Solution

$$F_P = \frac{(1 - e^{-8k})(1 - e^{-120k})/(1 - e^{-168k})(1 - e^{-24k})}{(1 - e^{-12k} + e^{-24k} - e^{-36k} + e^{-48k} \cdots e^{-228k})/(1 - e^{-336k})}$$

$$= \frac{0.2381}{0.2501} = 0.9524$$

As pointed out earlier, as $t_{1/2}$ gets very large,

$$F_P = \frac{40}{\text{hr/wk special schedule}}$$

$$= \frac{40}{84/2} = 0.9525$$

Adjusted limit = TLV_s = 0.95 × 1 $\mu g/m^3$ (essentially no change is needed).

Illustrative Example 15 (Hickey and Reist Model). In Canada, unusual work shifts have become more commonplace than in the United States. In one industry, the unions and management decided that a combination of the 8-hr/day and 12-hr/day work schedule best fit their needs. What modified occupational exposure limit would be indicated for benzene (proposed TLV of 1.0 ppm) if the limit imposed in the plant were the most rigorous one to which they must adhere during the month (i.e., 12 days of repeated exposure)?

The exact schedule used in this industry involved 5 days of exposure for 8 hr/day followed by 2 days of 12 hr/day then 5 days of exposure for 8 hr/day shift followed by 6 days off work. The schedule then repeats itself so that workers average only 40 hr per week each 4-week cycle. Figure 7.34 illustrates the qualitative behavior of benzene in the body during this exposure schedule.

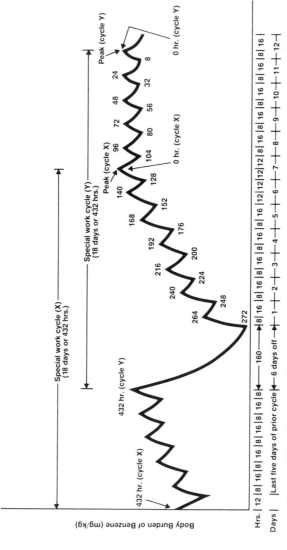

Figure 7.34 Graphical illustration of the likely fluctuations of the body burden of benzene (or its metabolites) following repeated exposure to the complex 8-hr/day and 12-hr/day work schedule described in Example 16.

Solution. Two cycles must be examined to determine which gives the lower F_p:

a. Cycle of 12 on and 6 off.
b. Cycle of 5 on, 6 off and 7 on, with 6th and 7th days having 12-hr shifts.
c. Benzene $T_{1/2} = 10$ hr, $k = \ln 2/10$

$$F_p = \frac{(1 - e^{-8k})(1 - e^{-120k})/(1 - e^{-168k})(1 - e^{-24k})}{(1 - e^{-8k} + e^{-24k} - e^{-32k} + e^{-48k} \cdots e^{-272k})/(1 - e^{-432k})}$$

$$= \frac{0.52502}{0.52524} = 1.0$$

Note: The 12 on, 6 off cycle, gives an F_p of unity.

Why? The residual levels from the two extra exposures from the previous Saturday and Sunday add only $0.0002CWK$ to the $0.52502CWK$ from the normal exposure burdens. To confirm, check the burden at the end of the 12-hr Sunday shift.

$$F_p = \frac{0.52502 \leftarrow \text{normal week}}{(1 - e^{-12k} + e^{-24k} - e^{-36k} + e^{-52k} \cdots - e^{-420k})/(1 - e^{-432k})}$$

$$= \frac{0.525}{0.686} = 0.765$$

$\text{TLV}_s = 0.765 \times 1 \text{ ppm} = 0.765 \text{ ppm}$

The peak body burden occurs at the end of the second 12-hr shift. The moral is that the time of peak burden must be chosen correctly. Otherwise, as could have occurred here, the incorrect factor would be applied to the exposure limit. If it is not obvious, more than one peak time must be tested. (This example problem was developed by Dr. John Hickey.)

As noted by Hickey (1977), when applying the reduction factor to determine allowable exposure limits for nonnormal schedules, Brief and Scala give separate weightings to increased exposure time and decreased recovery time between exposures. Their model incorporated the concept of the "excursion factor" and reduces the TLV by a factor proportional to the decrease in the allowable exposure limit. The Brief and Scala model was not intended to be applied to shorter-than-normal exposures, only to longer-than-normal daily or weekly exposures. It cannot be applied to continuous exposures, as it devolves to zero at this point. Their model does not take into account substances with very short half-lives, but it is evident that rapidity of toxic response was considered in its development.

By comparison, the OSHA model [Eqs. (14) and (15)], if applied to other than 8-hr shifts, could be recast in its simplest form using the Brief and Scala symbols:

$$\text{RF} = \frac{8}{h} \tag{33}$$

This represents the OSHA model as it would be used to determine limits for a substance (with no peak or ceiling limits) if a single uniform exposure occurred for h hours. For example, if a person were exposed only for 4 hr of an 8-hr shift, the adjustment factor would be 2; that is, a concentration of up to twice the TWA limit would meet OSHA regulations. Assuming application of OSHA regulations to 10 hr shifts, the adjustment factor for a 10-hr exposure would be 0.8; that is, the OSHA 8-hr limit would be reduced to 0.8 of its value to meet regulations.

Neither of the OSHA or Brief and Scala models accounts for the uptake rate; the reduction (or adjustment) factors derived from these models can be compared only to the worst-case adjustment factor for the model under development. It must also be assumed that OSHA regulations would apply to exposures longer than 8-hr. With these qualifications, several comparisons are made in Table 7.17.

It is readily apparent that in every case, the pharmacokinetic models call for less reduction in OEL for longer-than-normal exposure periods and for more reduction in short-term exposure limits than those required by the OSHA model. As noted by Hickey (1977), this is a direct reflection of the pharmacokinetic model's use of peak body burden rather than average burden as a criterion to predict equal protection and of the fact that it takes into account the uptake and excretion rate of chemicals.

Even considering the limitations of the pharmacokinetic models, the adjustment factors derived by the model reflect more realistically the necessary protection from exposure than does the OSHA model, and to be a further extension of the concept of the Brief and Scala model (Brief and Scala, 1975; Hickey, 1977).

The various mathematical models proposed each have advantages and disadvantages. For chemicals with acute or chronic toxicity, the OSHA model restricts the daily dose or the weekly dose, respectively, during an unusual work shift to the same amount as that obtained during a standard 8-hr shift. It does not acknowledge the lesser recovery period or biologic half-life of the compound. The Brief and Scala

Table 7.17 Comparison of Adjustment Factors Derived from Different Models

Condition	Worst Case Pharmacokinetic Models, F_p	Brief & Scala, RF	OSHA Current Practice	OSHA Adjustment Factor for Longer Shifts
Five 8-hr days per week	1	1	1	1
Four 10-hr days per week	0.84	0.7	1	0.8
Three 12-hr days per week	0.75	0.5	1	0.67
Five 16-hr days biweekly	0.56	0.25	1	0.5
Alternating weeks of three and four 12-hr days	0.72	0.5	1	0.67
Four hrs of exposure daily[a]	1	[b]	2	[b]
Two hours of exposure daily[a]	1	[b]	4	[b]
One-half hour exposure daily[a]	1	[b]	16	[b]

Source: This chart was developed by Hickey (1977).

[a] For substances with only TWA limits.

[b] Method not applicable.

model does not permit the daily dose of the toxicant under a novel work shift to be greater than that for a standard shift and it also accounts for the lessened time for elimination, however, it does not consider biologic half-life. Consequently, the Brief and Scala model is the most conservative of all of the models. In contrast, the Mason and Dershin, Hickey and Reist, and Roach models account for biologic half-life, increased daily dose, lessened recovery between exposures as well as the weekly dose. As a result these pharmacokinetic models yield less conservative results and they are presumed to be more accurate. Table 7.18 contains some of the general guidelines regarding the adjustment of occupational exposure limits for persons who work unusually long work shifts. Example 16 further illustrates some of the differences in the results of the various models.

Illustrative Example 16 (Comparing the Models). Assuming that 1,1,2-trichloro-ethane has a biologic half-life of 16 hr in humans, what modified TLV or PEL would be appropriate for persons who wished to work a 3-day, 12-hr/day workweek? Note that the dose for the unusual workweek (360 ppm-hr) would be less than for the normal 8-hr/day, 5-day workweek (500 ppm). The present PEL and TLV for 1,1,2-trichloroethane is 10 ppm.

OSHA model: Modified PEL $= 10.0 \text{ ppm} \times \dfrac{8 \text{ hr}}{12 \text{ hr worked}/\text{day}}$

Modified PEL $= 6.66$ ppm

Brief and
Scala model: Modified TLV $= \dfrac{8 \text{ hr}}{12 \text{ hr}} \times \dfrac{24 - 12}{16} \times 10.0 \text{ ppm}$

Modified TLV $= 5.0$ ppm

Hickey and
Reist model: Modified TLV $= 10.0 \text{ ppm} \times \dfrac{(1 - e^{-8k})(1 - e^{-120k})}{(1 - e^{-t_1 k})(1 - e^{-t_2 k})}$

Modified TLV $= 10.0 \text{ ppm} \times \dfrac{(1 - e^{-8(0.04)})(1 - e^{-120(0.04)})}{(1 - e^{12(0.04)})(1 - e^{72(0.04)})}$

Modified TLV $= 7.5$ ppm

Note:

$$k = \frac{\ln 2}{t_{1/2}} = \frac{0.693}{16} = 0.04$$

$t_1 =$ hr worked per day on unusual schedule

$t_2 = 24 \times$ days worked per week on unusual schedule

It is apparent from Examples 11 and 12 that the various models can recommend markedly different limits of exposure. In all cases, the pharmacokinetic approach

Table 7.18 Rules of Thumb for Adjusting Occupational Exposure Limits for Persons Working Unusual Shifts

1. Where the goal of the occupational exposure limit is to minimize the likelihood of a systemic effect, the concentration of toxicant to which persons can be exposed should be less than the TLV if they work more than 8 hr/day or more than 40 hr/week and the chemical has a half-life between 4 and 400 hr.
2. Exposure limits whose goals are to avoid excessive irritation or odor will, in general, not require modification to protect persons working unusual work shifts.
3. Adjustments to TLVs or PELs are not generally necessary for unusual work shifts if the biological half-life of the toxicant is less than 3 hr or greater than 400 hr.
4. The biologic half-life of a chemical in humans can often be estimated by extrapolation from animal data.
5. The four most widely accepted approaches to modifying exposure limits will recommend adjustment factors that will vary. In order of conservatism, the Brief and Scala model will recommend the lowest limit and the kinetic models will recommend the highest.

Brief and Scala > OSHA > ACGIH > Pharmacokinetic

6. Whenever the biologic half-life is unknown, a "safe" level can be estimated by assuming that the chemical has a biologic half-life of about 20 hr. (Note: This will generally yield the most conservative adjustment factor for typical 8-, 10-, 12-, and 14-hr workdays.)

recommends a less strict TWA limit than that generated by models that do not consider the biologic half-life of the chemical.

8.7 A Generalized Approach to the Use of Pharmacokinetic Models

As discussed, four or five different researchers have proposed pharmacokinetic models for adjusting limits and since they are based on the same assumptions, the results will essentially be the same. Roach (1978) and Hickey and Reist (1977) have presented charts that can be used to quickly determine the adjustment factor for many common shifts. However, for all other situations, the industrial hygienist must begin with the basic equations in order to calculate a modified limit.

Because most persons have found Hickey and Reist's approach and their publications to be most easily understood, the following generalized scheme for determining the adjustment factor for any schedule will be based on their equations. The limits derived from their model will be virtually identical to those obtained by use of the other models. Where the nonnormal work exposure schedule does not fit a curve derived by Hickey and Reist, one of several equations may be used.

1. For any regular weekly schedule, Eq. (27).
2. For a sporadic schedule, Eq. (25).
3. For an excursion in a normal shift, Eq. (28).
4. Where continuously rising exposure is expected, Eq. (34), Figure 7.35.

$$F_p = \frac{(1 - e^{kt_e})4k}{kt_e - (1 - e^{-kt_e})} \qquad (34)$$

5. For discrete variations in exposure levels, Eq. (35).

$$F_p = \frac{1 - e^{-kt}}{f_c(1 - e^{-kt_c}) + f_b(1 - e^{-kt_b})(e^{-kt_c}) + f_a(1 - e^{-kt_a})(e^{-kt_b})(e^{-kt_c})} \qquad (35)$$

where t = total exposure time period
f_i = air concentration of substance as fraction of special exposure level (concentrations f_a, f_b, f_c)
t_i = time of exposure at f_i (periods t_a, t_b, t_c)
k = clearance factor, $\ln 2 / t_{1/2}$

As noted by Hickey (1977), this equation has a drawback in that F_p is calculated on the assumption that peak body burden occurs at the end of a shift. If the actual peak occurs within a shift, F_p must be determined on that basis. Thus, to use Eq. (35) correctly, it must be known or calculated in advance at which exposure level the peak body burden will occur. If exposure levels do not decrease during a shift, the peak burden may be predicted to occur at shift's end. Where levels of contaminant decrease during the shift, F_p may be determined for the end of each discrete period and the lowest one applied.

It should be noted that different exposure regimens can affect how the body accumulates a chemical during a particular work shift. To illustrate how the pharmacokinetic models can predict variations in body burden due to unusual exposure periods, Hickey (1977) has offered the following examples (17 and 18).

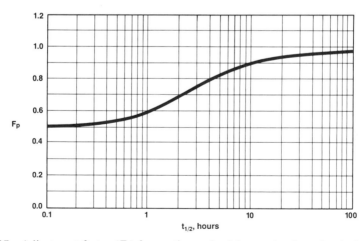

Figure 7.35 Adjustment factor (F_p) for continuously rising contaminant levels during work shift. (From Hickey, 1977.)

Illustrative Example 17 (Evaluating Short-Term Exposures). Assume that three
work schedules involve exposure to trichloroethylene (TCE). In situation 1, the
worker is exposed to the TWA limit for 8 hr per day (normal). In situation 2, the
TCE concentration rises linearly from 0 to 200 ppm during the shift. In situation 3,
the workweek is the same, but the worker is exposed to a worst-case situation: a
discretely rising TCE concentration with peaks of 300 ppm for 5 min every 2 hr.
These situations are depicted in Figure 7.36.

What can be said about compliance to the various OSHA limits? Would one
expect the peak body burdens to vary with the different schedules even though the
absorbed dose (ppm-hr) is the same for all three? Is the peak body burden for any
of these short-term exposure schedules likely to exceed the peak body burden observed during continuous 8-hr exposure to 100 ppm?

Note: TCE has an OSHA PEL of 100 ppm (TWA), a ceiling limit of 200 ppm,
and an OSHA peak limit of 300 ppm of 5 min every 2 hr. TCE has a fast compartment biologic half-life of 15 min and a slow compartment biologic half-life of 7.6
hr. (This example was developed by Dr. John Hickey.)

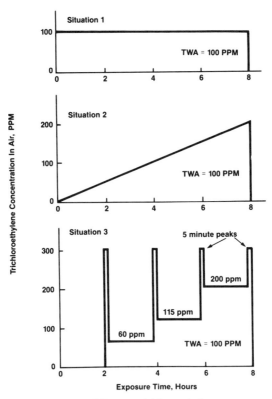

Figure 7.36 Comparison of three different trichloroethylene exposure situations, wherein
the 8-hr TWA concentration is always 100 ppm, but the resulting peak body burden could
vary between the exposure schedules. Based on Example 17. (From Hickey, 1977.)

Solution

PART I. In each case, exposures are within OSHA limits, since all have TWA averages of 100 ppm and none of the short-term exposures exceed an OSHA limit for short exposure periods.

PART II. Using Eq. (34) for situation 2 and Eq. (35) for situation 3, the F_p values for these two exposure regimens can be calculated. F_p is the adjustment applied to the TWA limit in a special exposure situation that will result in a predicted peak body burden equal to that resulting from a normal exposure at the TWA limit. When the adjustment is not made, and C_s is left equal to C_n, then R_p can be determined. R_p is the predicted ratio of special and normal body burdens (B_{ps}/B_{pn}), and is the reciprocal of the predicted F_p.

In the situations at hand, both F_p and R_p values are predicted for situations 2 and 3 relative to normal situation 1. This is shown in Table 7.19 for both the fast compartment of TCE ($t_{1/2}$ = about 15 min) and its slow compartment ($t_{1/2}$ = 7.6 hr). Results are interpreted as follows:

If the slow compartment is the critical tissue, the predicted peak tissue burdens will not exceed the peak burden for normal exposure if the OSHA TWA limit (C_n) is reduced to $0.89C_n$ (or 89 ppm) in situation 2, and to $0.86C_n$ (or 86 ppm) in situation 3. Stated in terms of R_p, if the concentration is not reduced but is left at C_n (100 ppm), the predicted peak body burden will be 1.12 times greater than desired in situation 2 and 1.16 times greater in situation 3. It is clear that departures from normal exposure have only a small effect on the accumulated burden in a compartment with a low uptake rate (long half-life) for TCE. Also, the additional short peak exposures in situation 3 have little further effect on peak burden over that resulting from situation 2.

PART III. If the fast compartment is the critical tissue group, the OSHA TWA limit must be reduced from 100 to 56 ppm for situation 2 and to 48 ppm for situation 3,

Table 7.19 Adjustment Factors (F_p) and the Predicted Ratio of Body Burdens (R_p) for the Three Exposure Schedules Described in Example 17[a]

Predictive Index	Compartment[b]	Exposure Situation[c]		
		Situation 1	Situation 2	Situation 3
F_p	Slow	1	0.89	0.86
	Fast	1	0.56	0.48
R_p	Slow	1	1.12	1.16
	Fast	1	1.8	2.1

[a]This example readily illustrates how the manner in which one is exposed to a particular airborne toxicant can affect the peak body burden even though the total amount of toxicant absorbed each workday remains the same for all three situations.

[b]Fast compartment $t_{1/2}$ = 15 min; slow compartment $t_{1/2}$ = 7.6 hr.

[c]Situations 2 and 3 are compared to situation 1 (normal).

if the predicted peak compartment burden is not to exceed that resulting from normal exposure. Failure to reduce the concentration will result in a peak burden of 1.8 times greater than normal in situation 2 and 2.1 times greater than normal in situation 3. Note that in the fast compartment, the brief peaks in situation 3 add considerably (about 16%) to the burden accumulated in situation 2. Part of this increase is due to the longer exposure at the "ceiling" limit of 200 ppm in Situation 3.

This example illustrates how various exposure regimens can result in widely different body accumulations of a substance, and how the model predicts adjustment factors.

Illustrative Example 18 (Predicting Peak Body Burden). To illustrate how pharmacokinetic models can predict equal peak body burdens for different shift lengths, let us assume two work situations with exposure to TCE. In situation 1, workers are exposed for five 8-hr days per week at the OSHA TWA limit of 100 ppm, and in situation 2, workers are exposed to the same level for four 10-hr days per week (illustrated in Figure 7.22).

What adjustment factor (F_p) must be applied to the normal limit (100 ppm) so that the exposure in situation 2 will result in a body burden no greater than that resulting from exposure in situation 1?

Solution

PART I. Using Eq. (26) or Figure 7.30, it can be determined that for the fast compartment $(t_{1/2} = 15$ min$)$, $F_p = 1$, and for the slow compartment $(t_{1/2} = 7.6$ hr$)$, $F_p = 0.87$. Thus, if the fast compartment is critical, no reduction in the OSHA limit is required in situation 2 to avoid an increase in fast-compartment accumulation. Since the peak burden is reached in the fast compartment in a few hours, the peak tissue concentration should be no greater with 10 hr exposure than with 8 hr.

PART II. If the slow compartment is critical, the OSHA limit must be reduced to 0.87 of the normal limit, or 87 ppm, to avoid a higher predicted peak in situation 2 than in situation 1. The prudent course would be to make the reduction where there is doubt as to which compartment is critical. If the $t_{1/2}$ values are unknown or in doubt, a worst-case F_p of 0.84, or 84 ppm, can be determined from Figure 7.30.

As shown in the previous examples, the use of the pharmacokinetic approach has a good deal of flexibility in that it can predict modified exposure limits for both short and long periods of exposure.

8.8 Special Application to STELs

Hickey (1977) has discussed the use of the pharmacokinetic approach to setting acceptable limits for very short periods of exposure. Figure 7.37 illustrates the model's predicted F_p values for exposure to only a single excursion daily, as a function of substance half-life and the short-term exposure time. Hickey (1977) has suggested an approach to determining the effect on F_p of the spacing of multiple excursions during a shift.

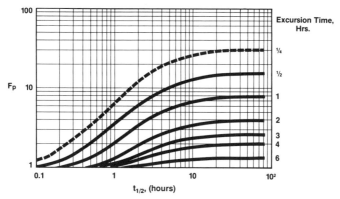

Figure 7.37 Adjustment factor (F_p) as a function of substance half-life ($t_{1/2}$) for three excursion schedules. Based on exposure schedule shown in Fig. 7.41. (From Hickey, 1977.)

Example 17 illustrates how F_p can be calculated for short exposures spaced at intervals other than 60 min. From this example, it is readily seen (Fig. 7.38) that depending on the chemical's half-life, the spacing of the exposures can influence the recommended short term limit.

Illustrative Example 19 (Accounting for sporadic peak exposures). To illustrate how the time between high exposures (above limit) can influence the degree of adjustment, solve F_p for the following two situations.

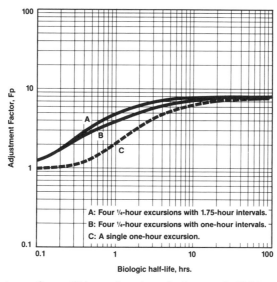

Figure 7.38 Adjustment factor (F_p) as a function of substance half-life ($T_{1/2}$) and excursion time (hours). Dotted line is used to express the high level of uncertainty involved in the prediction. (From Hickey, 1977.)

CASE A. Assume that a person works in a foundry and is exposed to carbon monoxide for only 15 min, 4 times per day when he opens an oven. There is a 1-hr interval between the times he opens the oven (shown in Figure 7.39, Case A). What adjustment factor (F_p) would be suggested according to the Hickey and Reist model?

CASE B. Assume that the worker who opens the ovens can space the times between exposures at 1.75 hr rather than 1.0 hr (Fig. 7.39 Case B). What F_p is needed to show the level of protection as when the exposures last 1 hr?

Answer. As shown in Figure 7.38, F_p for Case A and Case B will vary with the time between short periods of exposure even though the total dose (ppm-hr) remains the same day. In the case of carbon monoxide, which has a biologic half-life in humans of 3.5 hr, the F_p for the 4 excursions with a 1-hr interval is 5.7. When the rest interval is 1.75 hr, the F_p is 6.7.

Case A of Figure 7.39 illustrates air concentrations of a substance over an 8-hr shift, using a 1-hour respite short-term exposure schedule, as compared to normal exposure. Similarly, Case B of Figure 7.39 depicts four 15-min excursions, but spaced equally over the entire shift, with 1.75-hr rest periods between exposures.

Hickey (1977) has noted that generally, as shown in Figure 7.38, it makes little difference in resultant peak body burden whether the excursions are separated by 1-

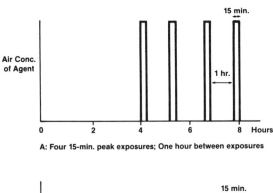

A: Four 15-min. peak exposures; One hour between exposures

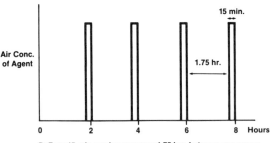

B: Four 15-min. peak exposures; 1.75 hrs. between exposures

Figure 7.39 Comparison of two short-term exposure schedules wherein the peak and TWA concentration(s) are the same but the peak body burden could in some cases be different. Consequently, a lower short-term exposure limit might be needed. Based on Example 19. (From Hickey, 1977.)

hr or by 1.75-hr intervals of nonexposure. The difference increases markedly, however, as the rest intervals between excursions are diminished, culminating in a single 1-hr exposure (four 15-min excursions with zero rest time between). This situation is shown in the dotted curve in Figure 7.38 for contrast, and is the same as the 1-hr excursion curve in Figure 7.37. Note also that the model's predictions reach a limiting STEL of 8.0, thus satisfying the ACGIH statement that the TLV–TWA may not be exceeded. *The curves in Figure 7.38 indicate once again that for long half-life substances, the exposure schedule is of little consequence when establishing STELs.*

Hickey (1977) pointed out that the pharmacokinetic models predict equal protection at somewhat higher STELs than many of those recommended by the ACGIH. This is because many of the ACGIH STEL values continue to be based generally on the excursion limits. By contrast, the model bases STELs solely on first-order uptake rates.

The pharmacokinetic model may be used to predict the permissible level and duration of exposure necessary to avoid exceeding the normal peak burden during short, high-level exposures. The model does this by establishing excursion limits that will predict equal protection for these situations. Assuming exposure is limited to a *single daily excursion* of duration t_e, Eqs. (25) and (28) reduce to

$$F_p = \frac{1 - e^{-k(8)}}{1 - e^{-kt_e}} \tag{36}$$

Adjustment factors derived from Eq. (36) have been plotted in Figure 7.37 as a function of substance half-life and excursion time. To illustrate, if a worker is exposed for one 30-min period during an 7-hr shift to a substance with a half-life in the body of 4 hr ($k = 0.17$), the adjustment factor would be determined as follows:

$$F_p = \frac{1 - e^{-(0.17)8}}{1 - e^{-(0.17)1/2}} = 9.0$$

The model would thus predict a limit of 9 times the OSHA TWA limit. This value may be read from Figure 7.37. If the OSHA cumulative exposure formula (OSHA, 1979) were used to determine a limit, the limit would be 16 times the OSHA TWA limit, assuming no ceiling or peak limits for the substance. In Figure 7.37, the 15-min curve is dotted because the one-compartment model is not precise for very short exposure times.

8.9 Other Applications of the Pharmacokinetic Modeling Approach

Thus far, the use of modeling has been restricted to only inhaled gases and vapors. Hickey and Reist (1977) have suggested that limits for particulates, reactive gases, and vapors can also be adjusted using their approach. For these substances and other situations, peak body burden B_p is still deemed to be the last criterion on which to develop a scheme to provide equal protection for the unusual work shift. The following sections describe these other situations and are based almost exclusively on the work of Hickey (1977) and Mason and Hughes (1985). The usefulness of kinetic modeling for very short durations, for seasonal workers, for exposures off-the-job, and for carcinogens are also discussed.

For many chemicals, only the results of animal studies will be available. These data can be quite useful for estimating the biologic half-life of specific chemicals in humans. It is, however, always inappropriate to assume that the half-life of a chemical obtained in a mouse, rat, hamster, dog, rabbit, or even monkey will describe its likely fate in humans. Unfortunately, animal data have often been directly used to adjust exposure limits for unusual work schedules. In most cases, the biologic half-life of an industrial chemical in an animal will be much less than in humans.

Quantitative approaches for extrapolating animal metabolism and excretion data to humans has been developed and could be used in the existing approaches (Leung, 1991). As shown in Table 7.20, it is readily apparent why the biologic half-life of a chemical in the smaller animals can be much less than in humans. Exceptions to this rule-of-thumb occur when humans cannot metabolize the parent compound (but the animal can). Fortuitously, this will rarely occur. In general, the rate of metabolism or elimination is a function of the alveolar ventilation rate and the cardiac rate. Table 7.20 illustrates the difference in the ventilation rate and the expected rate of clearance of hexane from four species including humans (Andersen, 1981).

Lastly, there are three additional caveats that need to be recognized regarding the determination of a biologic half-life for a substance. First, the half-life in the urine, breath, or feces is not the same as that for blood. Second, unless only one of these routes of elimination is predominant, none may correlate to the blood. Therefore, when available, the blood plasma half-life is the one that should be used. Third, the biologic half life of a particular chemical in animals and humans can vary with repeated exposure. That repeated exposure can vary the biologic half-life was noted in the work of Paustenbach et al. (1986b) that addressed the effects of repeated

Table 7.20 Interspecies Scaling of Hexane Clearance Based on Alveolar Ventilation Rates

Species	Body Weight (g)	V_{alv}[a] $(l/hr/kg)$	Expected CL_{hexane}[b,c] $(l/hr/kg)$	Ratio (Species/Human)
Mouse	30	35.9	11.8	6.9
Rat	250	21.1	7.4	4.1
Rabbit	3,000	11.4	3.8	2.2
Human	70,000	5.2	1.7	1.0

Source: M. E. Andersen (1981).

[a] $V_{alv} = [0.084 \ l/hr \ (b.wt.)^{0.75}]/(body \ wt.)$.

[b] For these clearance calculations, it is assumed that S_b for n-hexane is about equal from species to species and that E_t at low inhaled concentrations is 0.25 in all species.

[c] Rate of metabolism is CL times C_{inh}. Relative ratio is given by the ratio of CL in animals/CL in humans.

exposure of rats for periods of 11.5 hr/day and in the work of O'Flaherty et al. (1982) who noted the progressive changes in the half-life of lead in exposed workers with increasing years of exposure. The many potential pitfalls in determining the biologic half-life of a chemical have been discussed elsewhere.

8.9.1 Particulates

8.9.1.1 Chemical This entire section on particulates and reactive gases was published in Hickey and Reist (1977). In that study, they noted that the modeling approach for particulates is likely to be similar to vapors. Some researchers have suggested that more data needs to be gathered to validate the reasonableness of their model for particulates. Because it seemed appropriate given our state of knowledge, Hickey and Reist asserted that the one-compartment model as applied to particulates is likely to be analogous to inert gases and vapors with one exception; deposition of particulates is presumed to occur in the body at some linear rate proportional to air concentration. In their derivation, Hickey (1977) assumed that the clearance of particulates is presumed to conform to a first-order exponential.

For uptake:

$$B_t = (CLf/k)(1 - e^{-kt}) + B_0(e^{-kt}) \tag{37}$$

For clearance:

$$B_r = B_t(e^{-kt_r})$$

where L = flow rate of air to the body (volume/time)
$\quad\quad\quad f$ = fraction of particulates deposited
$\quad\quad\quad k$ = clearance rate of deposited particulates ($k = \ln 2/t_{1/2}$)

Note: The other symbols are as defined in Eqs. (23) and (24).

As is the case for inert gases, when F_p is determined, L, f, and the k in the denominator cancel, leaving F_p for particulates identical to that for inert gases. The mechanisms are different, but the model, being a ratio, requires only that $L, f,$ and k be the same for normal as for special schedules.

The retention of inhaled particulates has traditionally been considered to vary directly with their concentration in air. The ICRP Task Group on Lung Dynamics has shown that the processes are independent of the air concentration of particulates, whether deposition is by inertia, gravity, or diffusion (NCRP, 1959).

Retention varies significantly with other factors, however. The simple expression Lf in Eq. (37) masks a complex combination of variables including inhalation rate, which in turn affects air velocity in respiratory passages; particulate size, size distribution, density, and shape; respiratory frequency; and breathing habits, such as depth of breathing and mouth or nose breathing. The model does not consider any of these factors in predicting equal peak burdens, as they are canceled out by the assumption that they do not change with exposure time (shift length). In spite of these shortcomings, the assumption of a linear deposition rate appears valid.

As noted by Hickey (1977), the assumption of first-order clearance is tenuous. However, the lung clearance rate is generally thought to vary with the magnitude of lung burden, although there is no general agreement on this point. It is thus not definite that clearance follows a single exponential, although half-lives for particulate clearance have been published. Because of the uncertainty of the half-life or half-lives for body clearance of any particulate substance, Hickey and Reist have suggested that the use of the worst case would seem to be prudent in using the model for adjusting particulate limits to predict equal protection.

When the model is applied to short exposures to particulates at high concentrations, there is the implicit assumption that the predicted allowable higher concentration limit is not so high as to overwhelm the deposition or clearance mechanisms of the body. This can occur at very high concentrations, and application of the model to particulates is limited to this extent.

8.9.1.2 Microbial Aerosols. Particulate aerosols may contain viable microorganisms and the use of the models for this hazard have been discussed by Hickey (1977). These aerosols behave physically as any other airborne particulates until deposition in the host. There they may exhibit the unique characteristic of being able to multiply, either at the deposition site or some secondary site in the host. The resultant adverse effect may be an infection.

The number of viable organisms that must reach the host in order to initiate an infection depends on many factors, including host susceptibility, deposition site, and organism virulence. However, if one presumes that such a number exists, and that this number is analogous to peak body burden, the model may be applied to viable particulates.

There are no TLVs or OSHA limits or any other occupational exposure limits for specific microorganisms or viable particulates in air. There are recommended limits applicable to particular locations, such as hospital areas. However, these have not been correlated with, nor are they claimed to be based on, infectious dose. As Hickey (1977) noted, in the absence of such limits, the determination of F_p (or F_a or F_r) for microorganisms in air becomes academic, as there is no limit to adjust. However, the potential for application of the model exists and awaits the development of relevant limits.

8.9.2 Reactive Gases and Vapors

8.9.2.1 Systemic Poisons. Two types of reactive substances are discussed in relation to the model: those that are both metabolized and excreted through respiration and those that are only metabolized. The equations of substances that are both metabolized and expired have been derived as follows:

For uptake:

$$B_t = CWK[1 - e^{-(k_1 + k_2)t}][k_1/(k_1 + k_2)] + B_0[e^{-(k_1 + k_2)t}]$$

For excretion:

$$B_r = B_t[e^{-(k_1 + k_2)t_r}]$$ (38)

where k_1 = uptake and excretion rate by respiration
k_2 = rate by metabolism

The other symbols are as described in Eqs. (36) and (37).

Again, when F_p is calculated, the additional factor, $k_1/(k_1 + k_2)$, cancels, and F_p is identical to that for inert gases and vapors, except that the effective half-life is described by $k_1 + k_2 = \ln 2/t_{1/2}$. Use of the prepared curves, such as Figure 7.30, would require knowledge of the combined effective half-life of a substance or use of the worst case, as before. For substances that are only metabolized and not exhaled through the lungs, Eq. (37) applies.

8.9.2.2 *Local Irritants and Allergens.* As noted by Brief and Scala, Mason and Dershin, Hickey and Reist, and Roach, the models *do not* appear amenable to deriving adjustment factors for exposure to primary irritants or allergens. The actions of these substances appear to be based on such a small local compartment, as contrasted to the entire body, that the predicted equal protection approach may not be applicable. It is also likely that B_p is unsuitable as a criterion for predicting response. By the same token, however, a prolonged exposure, beyond 8 hr, might not require any reduction in exposure limits. Fiserova-Bergerova (1972) has suggested that a predictive model could be derived for irritants.

8.9.3 Radioactive Material

The mechanism of uptake and excretion of radioactive substances has been studied thoroughly by many researchers and its discussion is outside the scope of this chapter. However, the fate of inhaled radioactive gases and particulates has been modeled thoroughly and maximum allowable concentrations in air for them have been published, based in a large part on effective half-life. One marked difference here is that for many radioactive substances, the actual dose to which one is exposed is much easier to calculate and measure biologically, thus making the modeling and validation much more straightforward.

8.9.4 Mixtures

Hickey (1977) has discussed mixtures and he noted that OSHA regulations state that exposure limits for mixtures of substances in air (except for some dusts) shall be such that

$$C_1/L_1 + C_2/L_2 + \cdots + C_n/L_n \leq 1$$ (39)

in which C_i is the concentration of a substance in air and L_i is its OSHA limit. The equation presumes strictly an additive effect of inhaled substances.

The model does not take into account potentiation or the possible synergistic effects of mixtures. However, if it is assumed that additive effects exist in proportion to peak body burdens, the model may be modified to accommodate mixtures. Instead of equating peak body burden for a special schedule $(B_{p/s})$ to that for a normal one $(B_{p/a})$, as in Eq. (25), the model would set

$$B_{p/s} = B_{p/n}(C_i/L_i) \tag{40}$$

and the adjustment factor for any substance in a mixture, $F_{p/m}$, would be its F_p acting alone reduced by whatever factor is needed to meet the limits of Eq. (39) or

$$F_{p/m} = F_p(C_i/L_i) \tag{41}$$

In practice, F_p would merely be calculated from Eqs. (25) through (27) or read from graphs, but applied as an adjustment to the reduced limit, C_i, as determined for the OSHA formula [Eq. (15)], rather than to the normal OSHA limit. Example 20 is provided to illustrate Hickey's conceptual approach to adjusting limits for exposure to mixtures.

Illustrative Example 20. (Exposure to Mixtures). The pharmaceutical industry was the first to pioneer the use of the 12-hr work shift. In many firms the typical work-week involves 4 days. In many firms the typical workweek involves 4 days, 12 hr/day, then 3 days off followed by 3 days of 12 hr/day followed by 4 days off. Every 2 weeks, everyone will have worked about 40 hr/week.

Acknowledging that persons are often exposed to more than one chemical at a time, what modification would be recommended for exposure to isoamyl alcohol (TLV = 100 ppm) and carbon tetrachloride (TLV = 5 ppm) for this type of shift?

It is assumed that the biologic half-life in humans for isoamyl alcohol is 12 hr and for carbon tetrachloride it is about 5 hr. Note that each is a systemic toxin and that the intent of the TLV for isoamyl alcohol is to prevent CNS depression and liver toxicity, while the intent of the carbon tetrachloride standard is primarily to prevent liver toxicity.

Solution

Part A. For Both isoamyl alcohol and carbon tetrachloride, the following approach could be used:

$$F = \frac{(1 - e^{-8k})(1 - e^{-120k})/(1 - e^{-168k})(1 - e^{-24k})}{[1 - e^{-12k} + e^{-24k} - e^{-36k} + e^{-48k} - e^{-60k} + e^{-72k} - e^{-84k} + e^{-192k} - e^{-204k} + e^{-216k} - e^{-228k} + e^{-240k} - e^{-252k}/(1 - e^{-336k})]} \tag{42}$$

Part B. For CCl_4, $t_{1/2} = 5$ hr; so $k = 0.139$ hr^{-1}. By substitution,

$$F = \frac{0.695}{0.841} = 0.83$$

Therefore, $TLV_s = TLV_N (0.83) = 5 \times 0.83 = 4.2$ ppm for CCl_4.

Part C. Using the same rationale, the modified TLV for isoamyl acetate is found using $k = \ln 2/12$ hr, or $k = 0.058$ hr^{-1}. By substitution,

$$F = \frac{0.493}{0.664} = 0.74$$

Therefore, special TLV = 74 ppm for isoamyl alcohol.

Part D. This does not consider the potential additive or synergistic effect of simultaneous exposures to both chemicals. If it is desired to apply the ACGIH approach for additive effects of exposure to both chemicals, then one could apply their approach for assessing mixtures. The modified TLVs for this situation would be

$$\frac{C_1}{4.2} + \frac{C_2}{74} \leq 1.0$$

and the concentration of one or both chemicals should be reduced until the equation is less than, or equal to, unity.

8.10   A Physiologically Based Pharmocokinetic (PB–PK) Approach to Adjusting OELs

A sophisticated and more accurate approach than those previously proposed involves the use of a PB–PK model for determining adjustment factors for unusual exposure schedules. The PB–PK model requires data on the blood–air and tissue–blood partition coefficients, the rate of metabolism of the chemical, organ volumes, organ blood flows, and ventilation rates in humans. Andersen et al. (1987b) illustrated the use of the approach on two industrially important chemicals—styrene and methylene chloride. Their analysis suggested that when pharmacokinetic data are not available, a simple inverse formula may be sufficient for adjustment in most instances; this makes application of complex kinetic models unnecessary. Pharmacokinetic approaches alone should not be relied on for exposure periods greater than 16 hr/day or less than 4 hr/day because the mechanisms of toxicity for some chemicals may vary for very short or very long term exposure. For these altered schedules, biological information on recovery, rest periods and mechanisms of toxicity should be considered before any adjustment should be attempted.

Central to the development of an appropriate physiological model for nonconventional work shifts is the need to quantify the degree of risk associated with 8 hr exposure to the TLV so that the calculated limit for the long schedule poses an equivalent risk. For chemicals that possess a particular kind of toxic effect, it is a relatively straightforward task to adjust the exposure limits (Table 7.21). For most irritant gases, a ceiling TLV (C-TLV) has been identified that avoids air concentrations that produce direct irritant effects on the lungs or mucous membranes. In these

Table 7.21 Risk Indexes for Various Classes of Chemicals

Class of Toxicant	Appropriate Risk Index
Irritant gases	Maximum air concentration
Cholinesterase inhibitor	Blood and acetylcholine esterase
Heavy metal	Total daily dose
Genotoxic carcinogen	Monthly, annual or lifetime dose
Industrial solvents	Depends on mechanism of toxicity

Source: Andersen et al. (1987a).

cases, the parameter most likely to dictate the intensity of the adverse response, which will be defined as the risk index, is the maximum concentration in air. Consequently, for irritants, the ceiling value is the TLV, irrespective of the duration of the work shift.

At the other extreme are chemicals with long biological half-lives (greater than 400 hr) where the total accumulated body burden is the parameter that will determine the severity of the response (i.e., the risk index) (Andersen et al., 1987b). For the persistent chemicals, many of which are heavy metals or very high molecular weight lipophilic organics, there is a daily or weekly limit implicit in the TLV and that amount (dose) becomes the basis by which the daily limit can be established. In these cases, adjustments to exposure limits can be based on a simple ratio and proportion approach. For example, if 1 mg/m^3 is the 8-hr TLV, then 0.67 mg/m^3 would be the 12-hr exposure limit.

At least half the chemicals for which TLVs have been established have systemic toxic effects (i.e., at tissues other than at the site of entry) and possess biologic half-lives that are not dissimilar to the duration of exposure in most workplaces (e.g., 6–12 hr). For these chemicals something must be known about the mechanism of toxicity before the most appropriate risk index and adjusted occupational limits can be calculated. For irritants, as shown by the dotted line in Figure 7.32, the concentration that causes discomfort will be the exposure limit regardless of exposure duration. For cumulative chemicals, such as PCBs or lead, shown in Figure 7.32 as the solid line, the relationship is a hyperbola where the TLV is related inversely to exposure duration. Most TLVs for industrial chemicals, however, will lie somewhere between lines *A* and *B*. The objective of any approach to adjusting limits, including the PB–PK approach, is to avoid recommendations that would require unnecessarily expensive controls, yet the limits ensure a safe level of exposure (Table 7.16). Potential indexing factors for adjusting OELs are presented in Table 7.22.

A physiological approach for examining the kinetic behavior of inhaled vapors and gases that are nonirritating to the respiratory tract has been developed (Andersen et al., 1987a). In this description (Fig. 7.40), the body is lumped tissue groups corresponding to the:

1. Highly perfused organs, excluding the liver
2. Muscle and skin
3. Fat

Table 7.22 Potential Indexing Factors for
Adjusting Occupations Exposure Limits for
Unusual Work Schedules

Airborne concentration (exposure correlate)
Airborne concentration × time
Blood concentration (body burden)
Blood concentration × time
Peak metabolite concentration
Peak metabolite concentration × time
Target tissue concentration
Target tissue concentration × time

Source: Andersen et al. (1987a).

4. Organs with high capacity to metabolize the inhaled chemical. The physiological parameters of the metabolizing tissue groups are essentially those for the liver.

To describe the metabolism and fate of a chemical in humans or any living organism using a model, basic biological and physiological data on the species are used (Figure 7.40). The concentration of the inhaled contaminant in venous blood leaving each tissue can be determined by the tissue–blood partition coefficient. Blood flows and organ volumes are set consistent with literature values for these parameters. Organ partition coefficients and metabolic constraints for each chemical are determined by simple laboratory experimentation (Gargas et al., 1989).

An essential element of the PB–PK models is a determination of the solubility of the test vapors in various biological fluids and tissues. For vapors, solubility is quantified by determining appropriate partition coefficients. Partition coefficients relate the relative amount of material in the liquid and gaseous phases at equilibrium. A blood–air partition coefficient of 10 means that there is 10 times as much substance in a unit volume on blood as in a corresponding unit volume of air at equilibrium. Partition coefficients can be determined for blood and tissues by a vial equilibration technique in which small amounts of test chemical vapors are added to the head space above the biological samples (Gargas et al., 1989). After equilibration, the head space is sampled for test chemical. The partition coefficient is determined by a calculation based on the difference between the amount in the test vial and that in a control vial. Tissue–blood partition coefficients are determined by dividing the tissue–air by the blood–air values.

These various constraints are generally used in the four mass-balance differential equations that describe the time-dependent changes of tissue concentration in each of the compartments (Paustenbach, et al., 1988). Mixed venous blood concentration is determined as the weighted sum of effluent blood concentration from each tissue group, and the arterial concentration is determined from inhaled air concentration and venous blood concentration on the assumption that arterial blood leaving the lung is equilibrated with end alveolar air. Specifically, in the models built by Andersen et al., (1987a), cardiac output was assigned a value of 5.64 l blood/hr for a

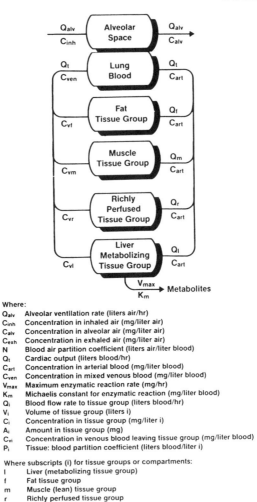

Where:
Q_{alv}	Alveolar ventilation rate (liters air/hr)
C_{inh}	Concentration in inhaled air (mg/liter air)
C_{alv}	Concentration in alveolar air (mg/liter air)
C_{exh}	Concentration in exhaled air (mg/liter air)
N	Blood air partition coefficient (liters air/liter blood)
Q_t	Cardiac output (liters blood/hr)
C_{art}	Concentration in arterial blood (mg/liter blood)
C_{ven}	Concentration in mixed venous blood (mg/liter blood)
V_{max}	Maximum enzymatic reaction rate (mg/hr)
K_m	Michaelis constant for enzymatic reaction (mg/liter blood)
Q_i	Blood flow rate to tissue group (liters blood/hr)
V_i	Volume of tissue group (liters i)
C_i	Concentration in tissue group (mg/liter i)
A_i	Amount in tissue group (mg)
C_{vi}	Concentration in venous blood leaving tissue group (mg/liter blood)
P_i	Tissue: blood partition coefficient (liters blood/liter i)

Where subscripts (i) for tissue groups or compartments:
l	Liver (metabolizing tissue group)
f	Fat tissue group
m	Muscle (lean) tissue group
r	Richly perfused tissue group

Figure 7.40 Illustration of how the body is described in physiologic pharmacokinetic modeling. (From Ramsey and Andersen, 1984.)

0.30-kg rat, and to maintain a ventilation–perfusion ratio of 0.8, alveolar ventilation was set equal to 4.50 L air/hr. The liver, fat, muscle, and richly perfused tissue groups in the rat were assigned volumes, respectively, equal to 4, 7, 75, and 5 percent of body weight. Blood flow distribution to the liver, fat muscle, and richly perfused tissue groups was, respectively, 25, 9, 12, and 54 percent of cardiac output.

As discussed by Andersen et al. (1987a), to adjust the TLV for most gases and volatile liquids, toxicity will usually be related to the area-under-the-blood-curve (AUC) rather than to peak blood concentration (Figure 7.41). For the sake of simplicity, a simple inverse relationship (Fig. 7.41, line *B*) might be an acceptable way of adjusting many TLVs, since the risk index calculated when an AUC is used will be nearly identical to that derived from the inverse relationship. To adjust for shifts

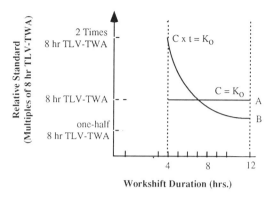

Figure 7.41 Strategies for adjusting 8-hr TLV–TWA to work shifts of shorter or longer duration. For irritants where a given airborne concentration has an effect, there would be no difference in proposed TLV regardless of work shift duration (solid line *A* between dotted lines). For chemicals that may accumulate with repeated exposure, the TLV is related inversely to the work shift duration (solid line *B*). It is not immediately obvious how adjustments would be made for most industrial vapors other than they should lie somewhere between lines *A* and *B*.

of 4–16 hr/day, the PB–PK approach is recommended whenever possible, or to use the direct ratio/proportion approach (Fig. 7.33, line *B*) if there is an insufficient amount of biologic data to develop a PB–PK model. For exposures of less than 4 hr or greater than 16 hr, the use of PB–PK modeling would be fairly difficult since information would be necessary on recovery and the mechanism of toxicity. An example of how a PB–PK model was used to adjust the OEL for styrene for unusual schedules is presented in Figure 7.42.

9 ADJUSTING LIMITS FOR CARCINOGENS

Unlike other toxic effects caused by xenobiotic substances, the carcinogenic response due to exposure to genotoxic agents is not currently believed to have a threshold (Peto, 1978; Bailer et al., 1988). Although this hypothesis cannot be proven or disproved for most industrial chemicals since exposure to very low doses of the less potent carcinogens would require an exposed human population of enormous size before a response would be observed, this theory must be considered when setting and modifying exposure limits. The adjustment of exposure limits for carcinogenic materials has been addressed by Mason and Hughes (1985) and their work is the basis for the following discussion.

9.1 Rationale

In general, toxic substances have been shown to exert their effect in proportion to the concentration in the body or within specific tissues. With most of these substances, it is common to find an all or none toxic response above a critical or thresh-

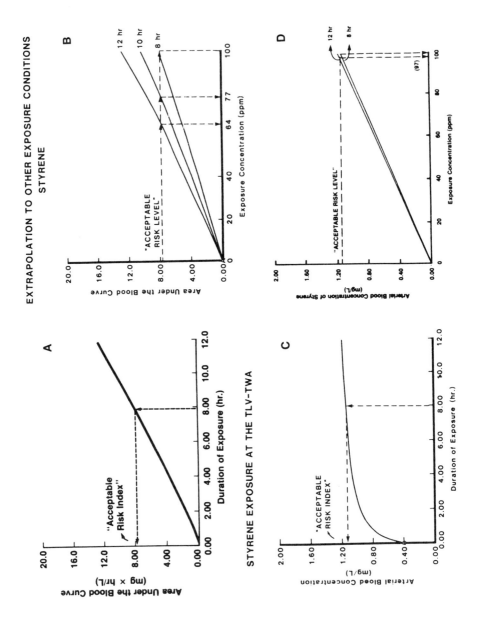

EXTRAPOLATION TO OTHER EXPOSURE CONDITIONS
STYRENE

STYRENE EXPOSURE AT THE TLV-TWA

312

Figure 7.42 Shift adjustments using a physiological pharmacokinetic approach. Panel A: After the decision is made about the appropriate measure of tissue dose, the human PB–PK model is exercised at the TLV–TWA and the acceptable risk index determined for the 8-hr exposure. In this case you find 8 hr on the X axis and read off the accumulated area under the blood curve. Panel B: The curve in panel A is obtained by running a kinetic model at a single concentration. The curves in panel B are different. In this case the models are run for many concentrations with a specified exposure duration. The output of these so-called repetitive runs are used to construct a composite curve relating target tissue dose to various exposure concentrations. Three curves were generated for exposure durations of 8, 10, and 12 hr. In order to calculate the adjusted TLV, the risk index from panel A is used on the Y axis and a line drawn parallel to the X axis. The shift-adjusted TLV–TWA then is determined from the intersection with the particular work shift curves. Panel C: Shift adjustments based on peak blood concentrations. The PB–PK model is run to estimate the time course of styrene in blood and the value for the 8-hr exposure (i.e., the acceptable risk level) is read off the curve. Panel D: Adjusting the TLV to a nonstandard, a 12-hr shift. The model is exercised for exposures of 8 or 12 hr for a variety of exposure concentrations and end exposure peak blood styrene concentration plotted *vs.* exposure concentration. The new 12-hr TLV–TWA is determined as in panel B by taking the risk level from the Y axis and drawing the line parallel to the X axis to its intersection with the 12-hr curve. This is the new acceptable exposure level. (From Anderson et al., 1987a.)

old concentration (Reitz, et al., 1989; Reitz, et al., 1990a,b). In other words, at higher concentrations there will be a correlation between an increase in the response and increasing concentrations in tissue, and below this threshold a given effect will not be observed. For toxicants that act in this fashion, the peak concentration in tissue is generally thought to be the most important parameter to predict a toxic response (Amdur, 1973; Andersen et al., 1987b). This group includes systemically acting substances such as chemical asphyxiants, narcotics and anesthetics, hemolytic agents, and probably some carcinogens or co-carcinogens. It *does not* include irritants that act with absorption, *nor does* it include allergens for which the severity of the response, once triggered, appears to be independent of the severity of exposure.

The threshold principle *may* apply to a small group of chemicals (e.g., carcinogens and mutagens) in which the biological response appears to result from chance molecular interactions that are independent of a tissue threshold. Response to these substances is in some ways similar to the response observed following exposure to low levels of ionizing radiation (Butler, 1979; Saffioti, 1980). For the most part, this latter group of substances act by forming covalent bonds with the genetic material of cells, causing an alteration of the genetic code, which in the case of somatic tissue may lead to cancer (Harvey, 1982; Williams and Weisburger, 1990). With many of these substances, the production of a diseased state appears to take place in separate phases. These begin with the initial chemical reaction or hits at a sensitive target (initiation) but end in a complicated series of interactions involving cell transformation, survival, and replication. Once the process of proliferation is initiated, progression to a diseased state may be independent of the concentration of the initiating substances, but may also be affected by the presence of promoting substances. The promoter may affect genetic repair, cell regulation, or the immune system and subsequently the survival of a transformed cell line (Butterworth and Slaga, 1987). As a result, the overall pattern of dose–response, especially at very low doses, is unclear.

Some chemicals can cause cancer in animals yet have no apparent genotoxicity or lack the ability to initiate cell transformation. These chemicals, which can be promoters, are often called epigenetic or nongenotoxic carcinogens. Although the exact mechanism of action is unclear for these substances, they apparently act in a manner much different from that of initiators (Gehring and Blau, 1977; Reitz et al., 1979; Schumann et al., 1980; Watanabe et al., 1980; Williams, 1980, 1981; Squire, 1981; Williams and Weisburger, 1990; Ames, 1987). Since many experts believe that a threshold exists for nongenotoxic carcinogens, it has been suggested that any approach to adjusting the TLV for these substances should be similar to that used for systemic toxins (i.e., the pharmacokinetic approach) (Butterworth and Slaga, 1987; Leung, et al., 1988; Paustenbach, 1990b).

9.2 Method for Adjusting Exposure Limits for Carcinogens

The approach suggested by Mason and Hughes assumes that the most sensitive or critical step in the carcinogenic process is that of initiation, since it is predicated to result from the chance interaction of a single molecule of the substance, or its metabolic derivatives with an appropriate molecule within a tissue that is generally presumed to be DNA. It then follows that the chance of such an event occurring among a fixed population of receptors will be determined by the concentration of the substance that is available, and the time that it is available, that is,

$$P = f(d, t) \tag{43}$$

where P is the probability of initiation occurring and d and t are, respectively, the tissue concentration (dose) of the substance and duration of the concentration in tissue (Jacobi, 1980). In slightly different terms, the "effective dose" is the integral of the body burden:

$$D_{\text{eff}} = \int_{t_1}^{t_2} C \, dt \tag{44}$$

where t_1–t_2 is the span of interest, C the concentration in tissue, and D_{eff} is the effective dose associated with a given level of response in the exposed population (Butler, 1979). The assumption that it is the long-term average dose that is more important than day-to-day peak concentrations is supported in a study by Bolt et al. (1981).

According to Mason and Hughes (1985), if an acceptable level of response and a concomitant, albeit probably low, dose has been set for a substance to which exposure occurs over a 40-hr (5-day, 8-hr/day) workweek, that standard could be extrapolated to unusual shift work schedules by limiting the effective dose in the unusual shift to that predicted at the exposure limit for the 5-day, 40-hr workweek. Unlike the case of nonstochastic substances for which the peak concentration is of concern, the "dose" attributable to a series of integrated body burdens will be the simple sum of the contributions form the individual shifts. Using the indefinite integral, the effective lifetime contribution to the body burden (C_t) from any one exposure lasting time t_0 is thus

$$(C_t) = (k_i^*(M)/k_0)t_0 \tag{45}$$

where k^* and k_0 are effective mass transfer constants for the substances in humans and (M) its concentration in the environment (Mason and Dershin, 1976).

Mathematically, the total dose over a series of exposure periods simply becomes the sum of the individual shift contributions, regardless of whether the work schedule is unusual or normal. Obviously, this influences the adjustment process since all exposure cycles of the same duration and concentration yield identical doses. Consequently, the authors have suggested that the TLV under unusual shift conditions should be

$$\text{TLV}_{\text{(unusual)}} = \text{TLV}_{\text{(std)}} \times \frac{\text{total exposure time (std shift sequence)}}{\text{total exposure time (unusual shift sequence)}} \tag{46}$$

Graphical depiction of the solution to this equation is presented in Figure 7.43.

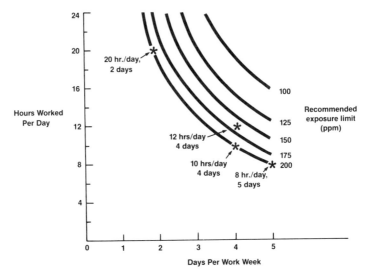

Figure 7.43 Graphical approach to adjusting exposure limits for genotoxic carcinogens based on the method of Mason and Hughes (1985). Curves for adjusting limits for these substances are based on limiting the time integral of the tissue concentration for the work schedule rather than limiting the peak tissue concentration. Consequently, the three 40-hr workweeks illustrated have the same allowable concentrations. Exposure for longer periods, e.g., in four 12-hr shifts, requires lowering the degree of exposure. (Courtesy of Dr. J. Walter Mason.)

The mechanics of shift arrangement for carcinogens then become a factor only if the novel work schedule results in a longer (or shorter) workweek, work month, work year, or working lifetime. This is quite different from the case of nonstochastic agents for which the novel TLV is determined by iteration of the exponential function $(1 - e^{-k_0 t_0})$ to obtain the peak body burdens, which would result in the respective series (Mason and Dershin, 1976). Consequently, biological kinetics become important only in making comparisons between substances and not in the extrapolation process.

In spite of its possible shortcomings and our lack of biological understanding of the carcinogenic processes, this approach seems reasonable for any chemical that has been shown to be positive in animal tests and also has demonstrated genotoxic potential.

9.3 Potential Shortcoming of the Proposed Approach for Adjusting TLVs for Carcinogens

As discussed by Mason and Hughes (1985), the adjustment of TLVs for substances that produce a response in proportion to the time integral of the body burden is simple to accomplish. However, the extent to which integrated body burdens adequately describe biological response for chemical carcinogens in humans is at present unknown (Rupp et al., 1978; Osteman-Golkar and Ehrenberg, 1983; Moolgavkar et al., 1988). For some electrophilic substances, for example, ethylene oxide, this approach appears to be reasonable (Ehrenberg et al., 1974), while for others, such as benz(a)pyrene, it may not be appropriate because of the apparent role that enzyme induction, caused by repeated exposure, may play in modifying the biologic response. Also, for some chemicals, there may be an inverse relationship between dose and the onset of an observable response (time to response), and because this is not considered in the Mason and Hughes model, a different approach would be necessary.

10 ADJUSTING OCCUPATIONAL EXPOSURE LIMITS FOR MOONLIGHTING, OVERTIME, AND ENVIRONMENTAL EXPOSURES

In many cases, persons who work unusual shifts have much free time away from work. For example, these schedules often allow persons to farm 20–30 hr/wk. Free time for these shifts usually includes as few as 3 full days off work each week or as many as 6 continuous full days off work every 2 weeks. These long periods of time give persons an opportunity to work second jobs. In fact, in studies of 12-hr shift workers, it was reported that many persons hold a part-time job along with their regular job so as to gainfully occupy these large blocks of free time (Wilson and Rose, 1978).

The adjustment of exposure limits for persons on long shifts has been a topic of interest since 1975, but if these persons are also exposed during their off hours, additional adjustment of the OEL may be needed. The approach to estimate the adjustment factor is the same for correcting limits to account for overtime as well as for environmental exposures. The following approach was developed and has been discussed by Hickey and Reist (1979).

10.1 Adjustment of Limits for Overtime and Moonlighting

As before, the one-compartment biological model is used in this approach. The one-compartment model predicts that no adjustment is necessary to exposure limits for substances with very long (over 1000 hr) or very short (less than 1 hr) half-lives, but that adjustment is necessary for substances with intermediate half-lives (usually 6–100 hr).

For regular moonlighting or overtime on a 5-day/week basis, Eq. (27) devolves to

$$F_p = \frac{1 - e^{-8k}}{1 - e^{-t_s k}} \tag{47}$$

in which it is the daily exposure in hours. Adjustment factors derived from Eq. (47) for this situation are shown in Figure 7.44, which was developed by Hickey and Reist (1979). Similar equations and curves can be developed from Eq. (27) for any

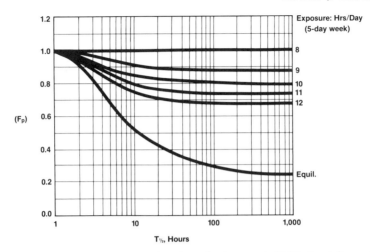

Figure 7.44 Adjustment factor (F_p) as function of substance half-life ($t_{1/2}$) for various daily exposures. The bottom line (equil.) represents the recommended adjustment factors for continuous exposure (24 hr/day). (From Hickey and Reist, 1979.)

schedule. For regular weekend moonlighting or overtime, working six or seven 8-hr days, Eq. (47) becomes

$$F_p = \frac{1 - e^{-120k}}{1 - e^{-124km}} \tag{48}$$

where m is the number of workdays per week. F_p values from Eq. (48) are shown in Figure 7.45.

Hickey has noted that the model is more complex when dealing with irregular or unplanned overtime or moonlighting added to an otherwise normal schedule. For example, suppose an employer decides Friday afternoon that workers must work overtime that same day or on Saturday. Recalling that in this model workers are normally presumed to have accumulated their allowable peak body burden of a substance by Friday afternoon, how does one adjust the exposure limit to prevent a predicted excess body burden accumulation?

If work (and presumed exposure) is to continue past normal quitting time Friday, the limit should be adjusted so that the body burden becomes no greater than the peak that would occur as a result of five 8-hr/day exposures per week at the TLV (PEL), which is the usual Friday-P.M. peak. In this case, Eq. (27) devolves to an equilibrium situation

$$F_p = \frac{(1 - e^{-8k})(1 - e^{-120k})}{(1 - e^{-168k})(1 - e^{-24k})} \tag{49}$$

For substances with short half-lives, no adjustment to exposure level is needed (the body is already at equilibrium), whereas for substances with long half-lives, the

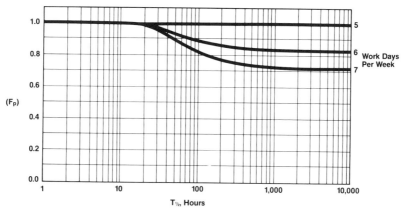

Figure 7.45 Adjustment factor (F_p) as a function of substance half-life ($t_{1/2}$) for workweeks of 5, 6, and 7 days. (From Hickey and Reist, 1979.)

level should be reduced to 40/168 of normal. The problem does not end there because the worker has lost part of his or her weekend recovery time, and next week's exposure must be lowered to compensate. This involves complex manipulation of Eq. (27). Other irregular overtime and moonlighting situations must also be modeled individually to determine an appropriate F_p value. Since there are infinite variations, only the foregoing examples are given here.

The point to be emphasized is that these exposures do add to the body burden and should be taken into account in setting exposure limits. An illustrative example of the mathematical approach is shown in Example 21. Figure 7.46 and others have been developed by Hickey (Hickey and Reist, 1979) for use during second shifts and are useful.

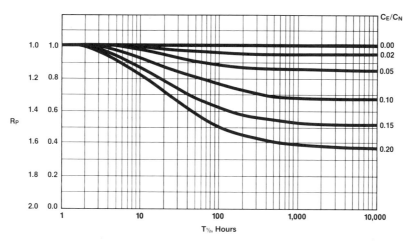

Figure 7.46 Predicted effect of off-the-job exposures to contaminants on adjustment factor (F_p) and peak body burden (R_p). (From Hickey and Reist, 1979.)

Illustrative Example 21 (Calculating an Adjustment Factor with Consideration Given to Overtime and Moonlighting). Many persons work two jobs. If an employee were self-employed as a furniture stripper in his off hours, what modification to the normal daily TWA exposure limit would be necessary for methylene chloride given the following information?

A person works 10 hr/day, 4 days/week at a paint plant and is exposed to methylene chloride. He usually strips furniture as a second job for only 2 hr on the days he also works at the factory, but he usually strips furniture 8 hr/day during 2 of the 3 days off work each week. He is not exposed to methylene chloride on Sunday. The biologic half-life for methylene chloride in humans is 6 hr (assume TLV = 100 ppm).

Solution. The model considers only exposure time, not total work period, so F_p is based on four 2-hr daily exposures followed by two 8-hr daily exposures per week.

$$F_p = \frac{(1 - e^{-8k})(1 - e^{-120k})/(1 - e^{-24k})}{\begin{array}{c}1 - e^{-8k} + e^{-24k} - e^{-32k} + e^{-54k} - e^{-56k} + e^{-78k}\\ - e^{-80k} + e^{-102k} - e^{-104k} + e^{-126k} - e^{-148k}\end{array}}$$

where $k = \ln 2/6$ hr $= 0.1155$.

$$F_p = \frac{0.6434}{0.6413} = 1.0, \qquad \text{TLV}_{\text{special}} = 1 \times 100 = 100 \text{ ppm}$$

The shorter exposure week (32 versus 40 hr) would allow an increase in exposure limit except that the $t_{1/2}$ is short (6 hr), the daily weekend exposure is the governing factor. Since weekend exposure is 8 hr/day, as in the normal schedule, no increase in TLV is predicted. This is illustrated in Figure 7.29.

Note that in the absence of weekend exposure, the model predicts an allowable increase in regular job exposure:

$$F_p = \frac{(1 - e^{-8k})(1 - e^{-120k})}{(1 - e^{-2k})(1 - e^{-96k})} = 2.92$$

The modified limit for 2 hr/day, 4 days/week exposures is 2.92 × 100 or 292 ppm.

10.2 Simultaneous Occupational and Environmental Exposures

As noted by Hickey, regulatory limits and recommended guidelines often assume zero off-the-job exposure to a substance. This is not always the case. The models discussed derive an expression for F_p, given a situation in which a person is exposed to an exposure limit of a substance during normal working hours, and to its environmental limit for the remainder of the time. The Hickey and Reist model was used to determine, first, how much the worker's peak body burden would be increased over that acquired without any off-the-job exposure, and second, how much the on-job limits should be reduced so that the peak body burden would be no higher with both on-job and off-job exposure than it would be with normal on-job exposure and zero off-job exposure (Hickey and Reist, 1979).

From Eq. (27) a value of F_p is found, representing, as before, the adjustment needed to the occupational limit to avoid a predicted higher-than-normal body burden accumulation because of the off-job exposure. If, on the other hand, it is assumed that no adjustment is made to the normal occupational limit, the relative increase in body burden due to the additional off-job exposure can be determined from the ratio of body burdens accumulated from normal and dual exposure. This approach is discussed in more detail in the original article (Hickey and Reist, 1979).

It is clear (Fig. 7.46) that for agents with relatively long half-lives and with relatively high environmental limits (as compared to the normal exposure limits, C_n), the environmental exposure adds significantly to the body burden of a substance if no adjustment is made to the occupational limit. Under these conditions, significant reductions in occupational limits are necessary to avoid any increase in body burden as a result of the additional environmental exposure.

As noted by Hickey and Reist (1979), the Environmental Protection Agency environmental limits for SO_2 and NO_2 are less than one percent of the OSHA occupational exposure limits, so dual exposure makes little difference to body burden accumulation. However, the EPA limit for carbon monoxide of 10 mg/m^3 (9 ppm), an 8-hr limit but in effect a ceiling limit, is 18 percent of the OSHA limit. This has some effect on F_p and R_p. For example, as shown in Figure 40, if a worker is exposed to 50 ppm CO on the job and 9 ppm the remainder of the time, his predicted peak body burden ($t_{1/2}$ in humans for carbon monoxide = 3–4 hr) with off-the-job exposure would be 1.06 times his on-the-job burden. If this environmental exposure occurs, the on-the-job exposure should be reduced to 0.94 of normal to avoid the predicted excessive peak body burden.

Illustrative Example 22 (Hickey and Reist Approach for Combined Environmental and Occupational Exposure). In certain regions of the country, the ambient concentration of carbon monoxide averages 9.0 ppm, which is the current EPA limit for environmental exposure. Many persons are occupationally exposed to carbon monoxide at concentrations at or near the TLV (25 ppm in 1993). What modified occupational exposure limit would be suggested if a person worked 12 hr/day for 4 days each week if one also wanted to take into consideration the background concentration to which the person would be exposed when away from his job? Carbon monoxide has a biologic half-life of about 4 hr in humans.

Solution. Our objective is to find the allowable CO concentration at work (C_s) that will not result in a $B_{p/s}$ greater than $B_{p/n}$. In determining C_s, you account for both the work-related CO and the background CO. Of course, you can reduce only the work-related contribution of CO since the environmental levels are fixed. Consequently, the additional CO concentration during work is $C_s - C_e$.

If there were no additional environmental exposure, the TLV adjustment required by the special schedule would be

$$B_{p(\text{normal})} = C_N WK \, (f\colon t_n, k)$$

where

$$f\colon t_{n,k} = \frac{(1 - e^{-8k})(1 - e^{-120k})}{(1 - e^{-168k})(1 - e^{-24k})}$$

$$B_{p(\text{special})} = (C_s - C_e)WK \, (f\colon t_s, k) + C_e WK \, (f\colon t_e, k)$$

and

$$F_p = \frac{C_{\text{special}}}{C_{\text{normal}}}$$

The additional reduction from environmental exposure is

$$F_p = \frac{(f\colon t_n, k)}{(f\colon t_s, k)} - \frac{C_e}{C_n}\left[\frac{1}{f\colon t_p, k} - 1.0\right]$$

where $t_{1/2} = 4$ hr

$$F_p = \frac{(1 - e^{-8k})(1 - e^{-120k})/(1 - e^{-168k})(1 - e^{-24k})}{(1 - e^{-120k})(1 - e^{-96k})/(1 - e^{-168k})(1 - e^{-24k})}$$

$$- \frac{9}{50}\frac{1}{(1 - e^{-12k})(1 - e^{-96k})/(1 - e^{-168k})(1 - e^{-24k})}$$

Combined $F_p = 0.8571 - 0.0225 = 0.835$.

Modified TLV for the combined exposures: TLV $= 0.83 \times 25$ ppm $= 20.8$ ppm.

Since 9 ppm of carbon monoxide is already present in the ambient air, the work environment should contain no more than 11.8 ppm in order to maintain a peak body burden for this special situation ($B_{p/s}$) at the same level as a person occupationally exposed to 50 ppm ($B_{p/n}$).

11 ADJUSTMENT OF OCCUPATIONAL EXPOSURE LIMITS FOR SEASONAL OCCUPATIONS

Seasonal occupations are particularly associated with agriculture and related activities, including fertilizer and pesticide manufacture and use, seed treatment and distribution, cotton ginning, and food canning, as well as construction and innumerable other pursuits. Occupational Safety and Health Administration (OSHA) permissible exposure limits have not taken seasonal exposure patterns directly into account. In response to the lack of information in this area, Hickey (1980) developed an approach to adjusting TLVs or PELs to suit particular seasonal exposures. The following discussion is based on his suggestions.

A one-compartment model was used to determine a "special" exposure limit that would predict equal protection to a worker in some special exposure situation that the TLVs (or PELs) provide in a "normal" exposure situation. This special limit is expressed as an adjustment factor (F_p), which is the ratio of the exposure limit for any special exposure schedule to the normal exposure limit (TLV or PEL). That is, the normal exposure limits times F_p equals the special exposure limit, or

$$\text{(TLV normal)}(F_p) = \text{(TLV special)} \tag{50}$$

The subscript p indicates that the adjustment factor is based on "peak" body burden of a substance as the criterion for equal protection.

F_p is a function of the work schedule and the biological half-life $(T_{1/2})$ of a substance in the body. The general equation for F_p as modified to apply to seasonal exposure is

$$F_p = \frac{(1 - e^{-8k})(1 - e^{-120k})(1 - e^{-8400k})}{(1 - e^{kH})(1 - e^{-24kD})(1 - e^{-168kW})} \tag{51}$$

where H = length of daily work shift in the special schedule (hr/day)
   D = length of the special workweek (days/week)
   W = number of weeks in the special work season (weeks/year)
   $-k$ = excretion rate of a particular substance in the body ($k = \ln 2/t_{1/2}$)
   $t_{1/2}$ = biological half-life of the substance in the body

In this general equation, the special work schedule factors in the denominator are balanced against the "normal" work schedule factors in the numerator. These are a workday of 8 hr, 5 work days per week (120 hr), and fifty 168-hr weeks per year (8400 hr). This schedule assumes a 2-week vacation annually.

11.1   Application of the Hickey and Reist Model for Seasonal Shift Work

Hickey has offered the following example to illustrate the use of his model for seasonal shift work.

Illustrative Example 23 (Adjusting for Seasonal Work): Take, for example, a seasonal job in which workers are exposed to a substance six 16-hr days/week for a 17-week season. Values of $H = 16$, $D = 6$, and $W = 17$ may be substituted in Eq. (51) and F_p found for any k value.

In this example, F_p goes from a value of 1.0 (no adjustment to TLVs) for exposure to substances with very short half-lives (high k values) to a low point of 0.43 for substances with a 400-hr half-life, and then rises to 1.2 for substances with very long half-lives (low k values) (Hickey, 1980).

This phenomenon is explained as follows. If $t_{1/2} = 1$ hr, the peak body burden is reached in 5 or 6 hr exposure and gets no higher even if exposure continues for 16 hr/day. As the $t_{1/2}$ gets longer, the abnormally long workday and workweek require a reduction in the TLV to avoid a predicted higher-than-normal body burden. For a substance with a $t_{1/2}$ of 400 hr, F_p is 0.43, or nearly as low as the ratio of the normal 40-hr to the special 96-hr workweek $(40/96 = 0.42)$.

To this point, there has been no effect on F_p from the short season. For substances with a half-life longer than 400 hr, as shown in Figure 7.47, F_p increases (curve B) to 1.2 at $t_{1/2} = 11$ years. This value approaches the ratio of the normal (2000 hr) to the special (1632) work year $(2000/1632 = 1.22)$. F_p can exceed unity (i.e., TLVs can be adjusted upward) for substances with very long half-lives, because with such substances, the total intake/year is more important than whether the intake takes place over a 17- or a 50-week period. Since the seasonal work year has fewer work hours than a normal work year, the total intake and thus the predicted peak body burden will be no more than for a normal year even if the exposure level is increased by 20 or 22 percent.

Let us now consider the adjustment factor if the workday and workweek are normal and only the work year varies. In this case, Eq. (51) reduces to

$$F_p = \frac{1 - e^{-8400k}}{1 - e^{-168kW}} \qquad (52)$$

F_p values from Eq. (52) are plotted in Figure 7.48 against $t_{1/2}$ for work seasons from 10 to 50 weeks. It can be seen that no ''credit'' (i.e., increase in TLV) can be taken for a short season unless the $t_{1/2}$ of the substance exceeds 30 hr (many particulate substances do have much longer half-lives). For substances with very long half-lives, the F_p again becomes the ratio of hours worked per normal year to hours worked per special season.

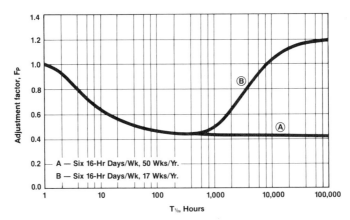

Figure 7.47 Adjustment factor (F_p) as a function of substance half-life $(t_{1/2})$ for the two seasonal exposure schedules indicated. (From Hickey, 1980.)

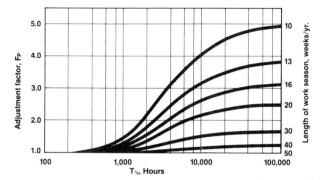

Figure 7.48 Adjustment factor (F_p) as a function of substance half-life ($t_{1/2}$) for various work seasons. (From Hickey, 1980.)

To avoid calculating each schedule separately, a quick approximation may be used to determine whether F_p is significantly different from 1.0 for a particular agent and work season. One may arbitrarily state that the effect of the shorter season on F_p is not worth considering if it permits raising the TLV by no more than 5 percent, or

$$1/(1 - e^{-168kW}) < 1.05 \tag{53}$$

Solving this equation for $t_{1/2}$ ($k = \ln 2/t_{1/2}$):

$$t_{1/2} \text{ in hours} < 38\ W \tag{54}$$

That is, if the substance half-life in hours is less than 38 times the number of weeks (W) in the work season, the short season does not materially change the predicted peak body burden from that expected in a normal work year. Stated another way, the TLV may not be raised by virtue of the shorter work year unless the substance half-life in hours is greater than $38W$. Example 23 illustrates the use of this approach.

Illustrative Example 24 (Approach to Adjusting Limits for Seasonal Occupations). Seasonal jobs are common in many industries. Assuming that persons are exposed to styrene 16 hr/day for 6 days/week for an 18-week season, what modified TLV would be suggested if the biologic half-life in humans were 6 hr? (assumed TLV = 50 ppm.)

Solution
Part A. Since $t_{1/2}$ in hours $< 38 \times$ (18 week), or < 684 hr, the exposure limit may not be raised by virtue of a shorter work *year*.
Part B. Must it be lowered because of the longer workday?

$$F = \frac{(1 - e^{-8k})(1 - e^{-120k})/(1 - e^{-168k})(1 - e^{-24k})}{(1 - e^{-16k})(1 - e^{-144k})/(1 - e^{-168k})(1 - e^{-24k})}$$

$$= \frac{(1 - e^{-8k})(1 - e^{-120k})}{(1 - e^{-16k})(1 - e^{-144k})}$$

$$= \frac{(0.6031)(1)}{(0.8425)(1)} = 0.72$$

This equation shows that only the length of the workday is important in determining TLV adjustment in this particular case.

This adjustment is the same as that predicted by Figure 7.47 for a 16 hr/day and a chemical with a 6-hr biologic half-life.

12 BIOLOGIC STUDIES

Very few studies have investigated the potential effects of unusual exposure regimens on the severity of toxic response. It appears that the health risks that might occur if an exposure limit were not lowered during an unusual work shift will, in all likelihood, be too subtle to be measured quantitatively in most animal studies. There is, however, some evidence that, in general, exposures that are intermittent or unusually long are likely to potentiate the response (Coffin et al., 1977; van Stee et al., 1982). On a theoretical basis, measurable but perhaps clinically insignificant changes in the rates and routes of elimination of inhaled substances can be expected during long work shifts (Dittert, 1977; Fiserova-Bergerova and Holaday, 1979; O'Flaherty et al., 1982; Paustenbach et al., 1986a,b; Andersen et al., 1987a).

Only a few biologic monitoring studies (Dixon et al., 1984) have thus far been conducted to demonstrate whether a significant difference in effects between a normal and unusual schedule is likely to occur for any chemical at levels near its TLV. *It will, in all likelihood, be many years before clinical studies in humans, or toxicologic studies in animals will show whether there is the possibility of increased health risk for some chemicals if exposure limits are not lowered for most unusual workshifts.*

The rationale for such adjustments are therefore, in part, philosophical and involves a judgment about theoretical risk. Our knowledge of the pharmacokinetics of chemicals clearly tells us that if the airborne concentrations of a chemical are not lower during extra long periods of exposure, the peak levels of the toxicant in tissue *will be higher* during that period than would occur during "normal" 8-hr/day work schedules. Consequently, if health professionals believe that persons on both schedules should be at an equivalent level of risk, they can use one of the models available for determining the concentrations at which parity can be expected. We can, however, understand something about the magnitude of the hazard from the few studies that have been conducted.

Lehnert et al. (1978) conducted a cross-sectional epidemiology study of workers on normal 8-hr/day shifts that involved the analysis of 6126 biological samples.

These persons were exposed to either trichloroethylene, benzene, or toluene during their workday, and trichloroacetic acid, phenol, or hippuric acid, respectively, were measured in their urine as an indicator of exposure as well as body burden. They found that even during normal 8-hr/day schedules, exposures at or near the TLVs of these chemicals caused the concentration of the metabolites phenol and trichloroacetic acid in the urine to be higher on Friday afternoon than Monday afternoon. Lehnert's et al. data suggest that there may be some degree of accumulation, although of no apparent toxicological significance, occurring in these workers during normal schedules. From this, it might be expected that for those chemicals where day-to-day accumulation normally occurs, accumulation might be exaggerated when the exposure period is longer and the recovery period shorter.

At least one biological monitoring study was specifically conducted in an effort to determine whether persons exposed to dimethylformamide (DMF) for 12 hr/day accumulated the substance to a greater degree than persons who were exposed during a normal workweek (Dixon et al., 1984). The conditions of the study were ideal in that a baseline set of urinary excretion data were collected on 80 employees who had been working the 8-hr/day shift for at least several months, and this was compared to the urinary data of the same group after they were placed on the 12-hr/day shift (only day shift). Consequently, through use of the paired-T test, the differences in the elimination for each person as well as the group could be determined. Their results indicate that *no difference* was observed in the concentration or the quantity of the urinary metabolites after they were placed on the 12-hr/day, 4-day/week shift. It should be noted that the implications of this study, like most other single studies, cannot be generally extrapolated to other chemicals because exposures to DMF averaged about 2 ppm, which is markedly below the TLV of 10 ppm. Second, the biologic half-life of DMF in humans is about 4-hr, therefore, it would not be expected to accumulate from day to day.

Few pharmacokinetic and toxicological studies have been conducted in animals exposed for 12 hr/day with the intent of determining any differences due to the longer exposure period. It appears that MacGregor conducted one of the first toxicity studies comparing the response due to 12 and 8 hr of exposure (MacGregor, 1973). She determined uptake factors for hexafluoroacetone using rats and predicted uptakes for periods of 6, 8, and 12 hr. MacGregor then developed a first-order model for determining uptake of inert gases in which she determined uptake constants for carbon monoxide using human volunteers. These predictions were compared to the data of Petersen (1962) and good agreement was reported between the observed and predicted values of CO uptake. The Hickey and Reist model predictions of peak CO burdens have been compared to the predictions of the MacGregor model (Hickey, 1977). In Hickey's validation procedure, exposure to 50 ppm CO was assumed for five 8-hr days per week, three 12-hr days per week, and 24 hr/day, 7 days per week. The comparisons were made in terms of F_p. The predictions of both models agreed quite closely. Hickey has noted that MacGregor also developed a predictive model for reactive vapors but that it has not been validated (Hickey, 1977).

Paustenbach et al. (1986b) conducted an experiment to evaluate differences in toxicity, distribution and pharmacokinetics due to exposures of 8 and 12 hr/day. One group of rats was exposed for 8 hr/day for 10 of 12 consecutive days (simulating 2 weeks on standard work schedule) and another for 11.5 hr/day for 7 of 12 consecutive days (simulated 12-hr/day schedule). Thus, each group received essentially the same dose (ppm-hr) of toxicant (carbon tetrachloride) during the 2-week test period. The results showed that the 11.5-hr/day exposure schedule produced minor changes in the distribution and concentration of CCl_4 in various tissues as compared to rats exposed 8 hr/day. There was no significant difference in hepatotoxicity between the groups following each week of exposure as measured by histopathology. However, exposure to the 11.5-hr/day dosing regimen consistently produced significantly higher levels of serum sorbitol dehydrogenase (SDH), an enzyme that indicates liver damage, than exposure to the 8-hr/day schedule (Table 7.23) (Paustenbach et al., 1986a).

It is noteworthy that the rates and routes of elimination were measurably different for the two schedules (Figs. 7.49 and 7.50) (Paustenbach, et al., 1986b). Following 2 weeks of exposure to the 8-hr/day schedule, ^{14}C activity in the breath and feces comprised 52 and 41 percent of the total ^{14}C excreted. Following 2 weeks of exposure to the simulated 12-hr/day work schedule, the values were 32 and 62 percent indicating that the longer work shift altered both the rate and route of elimination of CCl_4. It was found that 97–98 percent of the ^{14}C activity in the expired air was $^{14}CCl_4$. The elimination of $^{14}CCl_4$ and $^{14}CO_2$ in the breath followed a two-compartment, first-order pharmacokinetic model ($r^2 = 0.98$). For rats exposed 8 hr/day, the average half-life for elimination of $^{14}CCl_4$ in the breath for the fast (α) and slow (β) phases for the 2-week schedule averaged 85 and 435 min, respectively. For rats exposed 11.5 hr/day, the average half-lives for the α and β phases over the 2 weeks averaged about 95 and 590 min, respectively. Differences in the rate of elimination of $^{14}CO_2$ and ^{14}C activity in the urine and feces were also observed (Table 7.24).

The results of Paustenbach et al. (1986a,b) suggest that even if the weekly dose (ppm-hr) is held constant, subtle changes in dosage regimen, like those involving unusual (12 hr/day) work schedules, will have an effect on distribution, the degree of toxic response (Table 7.23), the pharmacokinetics of elimination (Fig. 7.50) and, perhaps, the metabolism of the half-life is greater than 4–6 hr. This study also showed that the effects of long exposures is most important for the lipophilic chemicals. For example, the markedly longer elimination half-lives of $^{14}CO_2$ and $^{14}CCl_4$ observed in the groups exposed to the simulated 12-hr/day work schedule compared to the groups exposed 8 hr/day indicates that the four additional hours of daily exposure places a greater percentage of the absorbed dose to poorly perfused lipid depots.

When a PB–PK model was developed to describe these data, the relevance to humans was quickly understood (Paustenbach et al., 1988). As shown in Figure 7.51 a and b , it is clear that rats exposed to 5 ppm do not reach the same peak blood levels after repeated exposure as humans exposed to 5 ppm. Humans accumulated CCl_4 with repeated exposure while rats did not. The difference between the species is enhanced when exposed for 12 hr/day. This is, of course, the critical concern regarding unusually long work schedules; do workers achieve significantly higher body burdens or peak blood levels? This study was a good illustration of how the PB–PK model can successfully scale-up rat data to predict the human response.

A comparative toxicity study involving 8- and 12-hr/day exposures was conducted by Kim and Carlson (1986a) for dichloromethane (methylene chloride). In

Table 7.23 Statistical Comparison of Serum Sorbitol Dehydrogenase (SDH) Activity in Selected Groups of Rats Following Exposure to 100 ppm of Carbon Tetrachloride (CCl$_4$) for Either 8 Hr/Day or 11.5 Hr/Day under a Number of Different Dosage Regimens[a]

Treatment Groups Compared	Dosage Regimen	SDH Activity Mean ± SE (IU/mL)
1	1 day, 8 hr	7.0 ± 1.5[b]
2	1 day, 11.5 hr	14.8 ± 3.7
3	2 days, 8 hr/day	11.5 ± 2.2
4	2 days, 11.5 hr/day	18.3 ± 4.0
5	3 days, 8 hr/day	21.0 ± 3.2
6	3 days, 11.5 hr/day	29.0 ± 6.2
7	5 days, 8 hr/day	22.5 ± 2.7[b]
8	4 days, 11.5 hr/day	68.3 ± 9.3
9	5 days, 8 hr/day[c]	20.3 ± 4.4[b]
10	4 days, 11.5 hr/day[d]	12.3 ± 0.6
11	5 + 5 days, 8 hr/day	39.0 ± 8.8[b]
12	4 + 3 days, 11.5 hr/day	65.0 ± 16.2
13	5 + 5 days, 8 hr/day[e]	4.9 ± 1.1[b]
14	4 + 3 days, 11.5 hr/day[f]	14.3 ± 1.3
7	5 days, 8 hr/day	22.5 ± 2.7[b]
11	5 + 5 days, 8 hr/day	39.0 ± 8.8
8	4 days, 11.5 hr/day	68.3 ± 9.3
12	4 + 3 days, 11.5 hr/day	65.0 ± 16.2
9	5 days, 8 hr/day[c]	20.3 ± 4.4[b]
13	5 + 5 days, 8 hr/day[e]	4.9 ± 1.1
10	4 days, 11.5 hr/day[d]	12.3 ± 0.6[b]
14	4 + 3 days, 11.5 hr/day[f]	14.3 ± 1.3
12	4 + 3 days, 11.5 hr/day	65.0 ± 16.2[b]
15	4 + 4 days, 11.5 hr/day	110.0 ± 26.7
12	4 + 3 days, 11.5 hr/day	65.0 ± 16.2[b]
16	4 + 5 days, 11.5 hr/day	102.3 ± 14.3

Source: Paustenbach et al. (1986a).

[a] SDH activity was determined immediately after exposure except where indicated. Values shown are the mean of four rats per group.
[b] Significant difference in SDH activity between groups ($p < 0.05$).
[c] SDH activity was determined 64 hr after exposure.
[d] SDH activity was determined 84 hr after exposure.
[e] SDH activity was determined 64 hr after exposure.
[f] SDH activity was determined 108 hr after exposure.

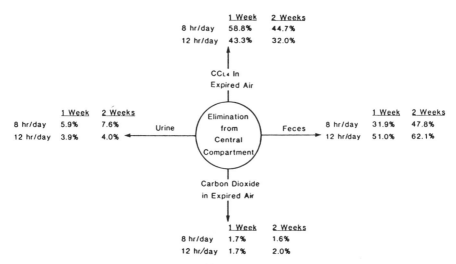

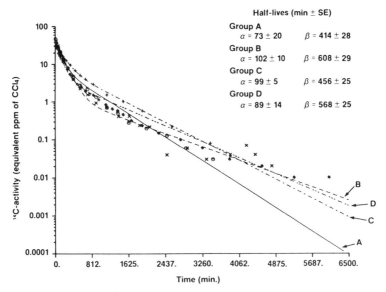

Figure 7.49 Differences in excretion of ^{14}C activity from rats following one and two weeks of exposure to 100 ppm of carbon tetrachloride during an 8-hr/day and 11.5-hr/day exposure schedule. (From Paustenbach et al., 1986b.)

Figure 7.50 Elimination of ^{14}C activity (98% CCl_4) in the expired air of four groups of rats (4 per group) exposed to 100 ppm of carbon tetrachloride. Two groups were exposed for 8 hr/day for either 5 or 7 days (A) or 10 of 14 days (C). The other were exposed for 11.5 hr/day for either 4 or 7 days (B) or 7 of 10 days (D) and have markedly longer half-lives than those exposed 8 hr/day. (From Paustenbach et al., 1986b.)

Table 7.24 Half-lives for Elimination of ^{14}C Activity in Rats Following Inhalation Exposure to 100 ppm of $^{14}CCl_4$ Following Several Different Dosing Regimens[a]

Dosing Regimen	$^{14}CCl_4$ in Expired Air		^{14}C Activity in Urine	$^{14}CO_2$ in Expired Air		^{14}C Activity in Feces
	$t_{1/2}$ (α) ± SE	$t_{1/2}$ (β) ± SE	$t_{1/2}$ (β) ± SE	$t_{1/2}$ (α) ± SE	$t_{1/2}$ (β) ± SE	$t_{1/2}$ (β) ± SE
8 hr/day, 5 days	73 ± 20[b]	414 ± 28[b,c]	1344 ± 149	123 ± 17[b-d]	1017 ± 95[c,d]	4900 ± 2100
11.5 hr/day, 4 days	102 ± 10	608 ± 29[c,d]	963 ± 107	221 ± 24[c,d]	1209 ± 124[c,d]	4300 ± 1400
8 hr/day, 5 + 5 days	96 ± 4	455 ± 24[c]	1066 ± 250	305 ± 33[c]	829 ± 81[c]	3700 ± 800[c]
11.5-hr/day, 4 + 3 days	89 ± 14	568 ± 25	944 ± 208	455 ± 42	1824 ± 175	6700 ± 1400

Source: Paustenbach et al. (1986b).

[a]Results are an average of data obtained from four rats that made up each group exposed to a particular dosage regimen. Half-lives are expressed in minutes.

[b]Indicates that this pharmacokinetic parameter is statistically different ($p < 0.05$) from that calculated for the group exposed to the 11.5-hr/day, 4-day dosing regimen.

[c]Indicates that this pharmacokinetic parameter is statistically different ($p < 0.05$) from that calculated for the group exposed to the 8-hr/day, 5 + 5-day dosing regimen.

[d]Indicates that this pharamacokinetic parameter is statistically different ($p < 0.05$) from that calculated for the group exposed to the 11.5-hr/day, 4 + 3-day dosing regimen.

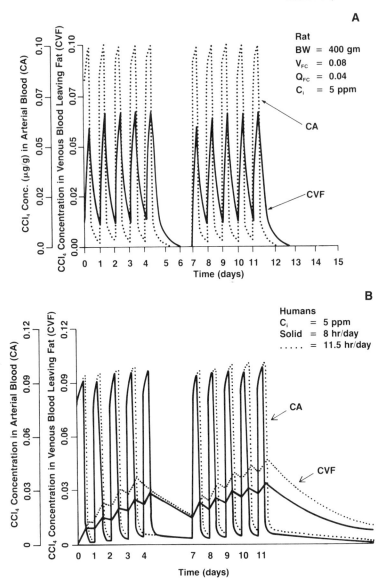

Figure 7.51 Comparison of the model-predicted concentrations of CCl$_4$ in arterial blood (CA) with those likely to be observed in venous fluid leaving the fat (CVF) for humans exposed to 5 ppm CCl$_4$. It is clear that there are significant differences between the rat and human even following just 8 hr of exposure at the 1993 TLV (5 ppm). As shown here, a PB–PK model can easily account for the pharmacokinetic differences between test animals and humans. (From Paustenbach et al., 1988.)

this study, carboxyhemoglobin (COHb) formation and elimination in rats and mice exposed to an 8 hr/day, 5 day workweek or a 12 hr/day, 4 day simulated workweek at dichloromethane (DCM) concentrations of 200, 500, or 1000 ppm were compared. They showed that the effect of the 12-hr exposure period was insignificant as determined by the COHb level after first day's exposure, immediately prior to the second day's exposure, after the last workday's exposure and 2 or 3 days after the last exposure. They also measured the half-lives of COHb and DCM in blood. The relatively short half-lives of COHb and DCM in these two species indicated that neither COHb nor DCM would be present for prolonged periods after DCM exposure ceased. Treatment with SKF-525A did not affect the half-life of DCM, suggesting that DCM was rapidly exhaled. Even after correcting for physiological differences between the mice and rat and human, this study indicated that for compounds like DCM, with half-lives less than 4 hr in man, and where there are readily reversible biological effects, no increased toxicity would be expected in persons who were exposed 12-hr/day. Another related study of the same authors (Kian and Carlson, 1986b) evaluated aniline.

The Haskell Laboratories of DuPont Chemical Company have evaluated several chemicals to determine the influence of exposure duration on toxicological response (Pastoor and Burgess, 1983). In their study of aniline vapor, they exposed adult male rats to either 10, 30, or 90 ppm aniline vapors for either 3, 6, or 12 hr per day for 2 weeks. Daily indices of toxicological response included body weight and methemoglobin measurements. After the final exposure, red blood cell counts as well as spleen and liver weights were measured and these were examined histopathologically. Their results showed that aniline-induced hemolysis and consequent splenic enlargement and deposition of hemosiderin is slightly related to exposure duration, and strongly related to exposure concentration. Aniline-induced methemoglobin formation, however, was not related to exposure duration, but was linearly correlated with exposure concentration. Their results suggest that aniline concentration, rather than duration of exposure, predominantly influences toxicological response.

Although there is not a large number of careful studies that have evaluated the biologic response of humans exposed to unusual work schedules, our knowledge of the pharmacokinetics of chemicals is usually sufficient to allow toxicologists to make good predictions of the effects of extra long or continuous periods of exposure. The best method for approaching these problems is to use PB–PK models.

13 UNCERTAINTIES IN PREDICTING TOXICOLOGICAL RESPONSE

It has been noted that any model for adjusting exposure limits will have a number of limitations because models, by definition, are based on several assumptions. These limitations have been reviewed by Calabrese (1977) and Hickey (1977). As has been noted, one key assumption is that the pharmacokinetic models consider the body to function as one homogeneous compartment. The second, and possibly more important, limitation of the modeling approach is the toxicological assumption that neither repeated exposure, the length of the exposure, or the type of shift schedule (e.g., rapid rotation) will alter the way humans absorb, metabolize, and eliminate the substance. In many cases, this limitation will not be significant.

The major possible shortcoming of the modeling approach is the error involved in assuming that lengthening the exposure does not change the way the body handles the chemical. In an effort to predict the effects of changes in toxicity due exclusively to changes in dosage regimen (such as that involved in 12-hr shifts), toxicologists have frequently used Haber's law as a rule-of-thumb. Haber's law claims that the dose is of central interest rather than the time over which that dose is administered. A thorough study of the limitation of Haber's law was published by David et al. (1981), and they showed that the severity of liver toxicity was markedly influenced by the time period over which a given dose (ppm-hr) was administered. Although the conclusions are similar to those obtained by other researchers (Kazmina, 1976) who have studied the limitations of extrapolating data from short-term inhalation exposures, this study is unique in that four different dosing regimens were used. Although the heavy emphasis on the total dose, rather than dose per unit time, has not been relied upon for nearly 40 years, industrial hygienists must all too often evaluate risks based on the amount of toxicant taken up per day rather than consider the pharmacokinetics. For example, this principle is used when hygienists extrapolate the results of standard 4- to 6-hr inhalation tests to estimate the adverse effects of exposure to 12-hr work schedules.

The affect of repeated exposure and long periods of exposure on a chemical's pharmacokinetic behavior are not accounted for in the models. This phenomena could have some affect on the toxicity of the material if the shift schedule were markedly different than an 8-hr/day, 5-day week. For example, MacGregor (1973) has found that the rate of metabolism and elimination during 12-hr/day exposure periods is measurably different than that observed during 8-hr/day exposure periods. (O'Flaherty et al., 1982; Paustenbach et al., 1986b) also showed that measurable differences in the rates and routes of elimination in the feces, breath, and urine between two different exposure schedules can occur.

There are several reasons why alterations in excretion rates and routes with repeated exposure have not been frequently reported. Colburn and Matthews (1979) have noted that unless all of the inhalation exposures involve radiolabeled material, rather than using labeled material only during the first and last weeks, the 'last in–first out' phenomena may take place. When this occurs, potential effects on distribution, metabolism, and elimination due to repeated exposure may not be detected because nonlabeled chemical may not be uniformly distributed or excreted with the labeled material. In short, the first dose may be equilibrated more deeply in the fat than the second or subsequent doses.

The observation that repeated exposure to unusual work schedules may affect the distribution and excretion of some chemicals will not be limited to carbon tetrachloride, carbon monoxide, and 2-HFA. For example, similar effects would be expected for industrial solvents such as cyclopropane, cyclohexane, 1,1,1-trichloroethane, and perchloroethylene, which are similar to classic anesthetics (such as halothane) in that they are quite lipid soluble and not appreciably metabolized. For example, Fiserova-Bergerova and Holaday (1979) have reported that the half times of uptake for halothane for the vessel-rich group (VRG), muscle group (MG), and fatty groups (FG) are in the magnitude of 2 min, 30 min, and 20 hr, respectively. They noted that most clinical anesthesia lasts 1–4 hr and that by the end of anesthesia the partial pressures of anesthetic agents in tissues of the FG compartment are far from equilibrium and that a steady–state has not been reached in that compartment. The result is a redistribution of vapor in the body after the offset of anesthesia. While clearance of the VRG and MG compartments starts instantly, the FG compartment continues uptake until the partial pressure in arterial blood declines to the partial pressure in the FG compartment. Consequently, desaturation curves are very much affected by the duration of exposure.

Another potential shortcoming of the modeling approach is the assumption that exposure to the toxicant does not inhibit or induce the microsomal enzyme system responsible for its metabolism (Wisniewska-Knypl et al., 1975). If this occurs, subsequent doses of the substance will alter the rate of toxication or detoxification. Along the same lines, models assume that one or more metabolic pathways will not be saturated. Although saturation is not very likely for exposure at or near the TLV, the likelihood that the metabolism of a compound will remain constant following repeated dosing is less likely. The induction or inhibition of its own metabolism by previous exposure has been demonstrated for nitrobenzene, acetone, and a few other xenobiotics (Wisniewska-Knypl et al., 1975).

All of the aforementioned factors, as well as those involving the likely affects of the circadian rhythm on toxic response, chronopharmacology and chronokinetics, make the modeling approach to the setting of modified exposure limits a crude approximation of the likely biologic processes that probably take place during unusual shifts. However, even though the current models may not account for the dozens of biologic phenomena that may be occurring, the available information clearly suggests that they are adequate to adjust OELs. Even though the existing biological data seem to suggest that, at levels near the TLV the increased risk of injury will usually not be appreciable, in light of the relative ease of adjustment, the procedure seems to be a worthwhile exercise.

It cannot be overstressed that modification of occupational exposure limits for situations other than 8-hr/day, 40-hr workweeks requires a clear understanding of the rationale for a particular limit. Blind use of any of the modeling approaches to modifying limits for either very short or very long periods of exposure can lead to either a lack of protection for those workers on unusual shifts or, as is more likely the case, a good deal of overprotection. Overprotection, although perhaps prudent, is not an optimal use of the limited resources allotted for minimizing occupational disease and, in some cases, could bring about undue economic hardships for both the employer and the employee.

REFERENCES

Abraham, M. H., G. S. Whiting, Y. Alarie et al., (1990). "Hydrogen Bonding 12. A New QSAR for Upper Respiratory Tract Irritation by Airborne Chemicals in Mice," *Quant. Struct. Relat.*, **9**, 6–10.

ACGIH (1990). Board of Directors, "Threshold Limit Values: A More Balanced Appraisal," *Appl. Occup. Environ. Hyg.*, **5**(6), 340–344.

ACGIH (1991a). *Documentation of The Threshold Limit Values and Biological Exposure*

Indices, Sixth Edition, American Conference of Governmental Industrial Hygienist, P.O. Box 1937, Cincinnati, OH 45201.

ACGIH (1992a). *Threshold Limit Values for Chemical Substances and Physical Agents and Biological Exposure Indices for 1992–1993*, American Conference of Governmental Industrial Hygienists, P.O. Box 1937, Cincinnati, OH 45201.

ACGIH (1991b). "Formaldehyde, in *Documentation of The Threshold Limit Values and Biological. Exposure Indices for 1991–1992*. American Conference of Governmental Industrial Hygienists. Cincinnati, Ohio 45201, pp. 664–688.

Adir, J. et al. (1975). "Pharmacokinetics of Fluorocarbon 11 and 12 in dogs and Humans," *J. Clin. Pharmcol.*, **15**, 760–770.

Adkins, L. E. et al. (1990). Letter to the Editor, *Appl. Occup. Environ. Hyg.*, **5**(11), 748–750.

Alarie, Y. (1981). "Dose Response Analysis in Animal Studies: Prediction of Human Responses," *Environ. Health Perspect.*, **42**, 9–13.

Alarie, Y. and G. D. Nielsen (1982). "Sensory Irritation, Pulmonary Irritation, and Respiratory Stimulation by Airborne Benzene and Alkylbenzenes: Prediction of Safe Industrial Exposure Levels and Correlation with their Thermodynamic Properties," *Toxicol. Appl. Pharmacol.*, **65**, 459–477.

Alavanja, M. C. R., C. Brown, R. Spirtas, and M. Gomez (1990). "Risk Assessment of Carcinogens: A Comparison of the ACGIH and the EPA," *Appl. Occup. Environ. Hyg.*, **5**(8), 510–517.

Amdur, M. O. (1973). "Industrial Toxicology," in *The Industrial Environment—Its Evaluation and Control*, National Institute for Occupational Safety and Health, Rockville, MD, Chapter 12.

Ames, B. N. (1987). "Six Common Errors Relating to Environmental Pollution," *Regul. Toxicol. Pharm.*, **7**, 379–346.

Andersen, M. E. (1981). "Pharmacokinetics of Inhaled Gases and Vapors," *Neurobehavioral Toxicol. Tertol.*, **3**, 383–389.

Andersen, M. E., M. G. MacNaughton, H. J. Clewell, and D. J. Paustenbach (1987a). "Adjusting Exposure Limits for Long and Short Exposure Periods Using a Physiological Pharmacokinetic Model," *Am. Ind. Hyg. Assoc. J.*, **48**, 335–343.

Andersen, M. E., H. J. Clewell, M. L. Gargas, F. A. Smith, and R. H. Reitz (1987b). "Physiologically Based Pharmacokinetics and the Risk Assessment Process for Methylene Chloride," *Toxicol. Appl. Pharmacol.*, **87**, 185–205.

APHA (1991). *Health Based Exposure Limits and Lowest National Occupational Exposure Limits*, Draft #5, November 6, 1991, American Public Health Association (APHA), Washington, D.C.

Astrand, I. (1975). "Uptake of Solvents in the Blood and Tissues of Man—A Review," *Scand. J. Work Environ. Health*, **1**, 199–218.

Astrand, I. and F. Gamberale (1978). "Effects on Humans of Solvents in the Inspiratory Air: A Method of Estimation of Uptake," *Environ. Res.*, **15**, 1–4.

Astrand, I., H. Ehrner-Sanuel, A. Kilbom, and P. Ovrum (1972). "Toluene Exposure. I. Concentration in Alveolar Air and Blood at Rest and During Exercise," *Work. Environ. Health*, **9**, 119–130.

Baetjer, A. M. (1980). "The Early Days of Industrial Hygiene. Their Contribution to Current Problems," *Am. Ind. Hyg. Assoc. J.*, **41**, 773–777.

Bailer, J. C., E. A. C. Crouch, R. Shaikh, and D. Spiegelman (1988). "One-Hit Models of Carcinogenesis: Conservative or Not?" *Risk Anal.*, **8**, 485–490.

Bjerner, B., A. Holm, and A. Swensson (1955). "Diurnal Variation in Mental Performance: A Study of 3-Shift Workers," *Br. J. Ind. Med.*, **12**, 103–110.

Bjerner, B., A. Holm, and A. Swensson (1964). "Studies on Night and Shiftwork," in *Shiftwork and Health*, A. Aanonsen, Ed., Scandinavian University Books, Oslo, Norway.

Bolt, H. M., J. G. Filser, and A. Buchter (1981). "Inhalation Pharmacokinetics Based on Gas Uptake Studies, III: A Pharmacokinetic Assessment in Man of Peak Concentrations of Vinyl Chloride," *Arch. Toxicol.*, **48**, 213–228.

Botzum, G. D. and R. L. Lucas (1980). "Slide Shift Evaluation—A Practical Look at Rapid Rotation Theory," *Proceedings of Human Factors Society*, pp. 207–211.

Bowditch, M., D. K. Drinker, P. Drinker, H. H. Haggard, and A. Hamilton (1940). "Code for Safe Concentrations of Certain Common Toxic Substances Used in Industry," *J. Ind. Hyg. Tox.*, **22**, 251–260.

Brandt, A. (1969). "On the Influence of Various Shift Systems on the Health of the Workers," *XVI Int. Congr. Occup. Health, Tokyo*, pp. 106.

Braun, W. H., P. J. Gehring, and J. D. Young (1979). "Application of Pharmacokinetic Principles in Practice," *American Chemical Society* (Sept.):110 (Abstracts).

Brief, R. S. and R. A. Scala (1975). "Occupational Exposure Limits for Novel Work Schedules," *Am. Ind. Hyg. Assoc. J.*, **36**, 467–471.

Bus, J. S. and J. E. Gibson (1985). "Body Defense Mechanisms to Toxicant Exposure," in *Patty's Industrial Hygiene and Toxicology*, L. J. Cralley and L. V. Cralley, Eds., 2nd eds., Vol. 3B, Wiley, New York.

Butler, G. C. (1979). "Estimation of Doses and Integrated Doses," in *Principles of Ecotoxicology*, G. C. Butler, Ed., Scientific Committee on Problems of the Environment (SCOPE), Wiley, New York.

Butterworth, B. E. and T. Slaga (1987). *Nongenotoxic Mechanisms in Carcinogenesis: Banbury Report 25*, Cold Spring Harbor Laboratory, Cold Spring Harbor, New York.

Calabrese, E. J. (1977). "Further Comments on Novel Schedule TLVs," *Am. Ind. Hyg. Assoc. J.*, **38**, 443–446.

Calabrese, E. J. (1978). *Methodological Approaches to Deriving Environmental and Occupational Health Standards*, Wiley, New York.

Calabrese, E. J. (1983). *Principles of Animal Extrapolation*, Wiley, New York.

Carlsson, A. (1982). "Exposure to Toluene: Uptake, Distribution, and Elimination in Man," *Scand. J. Work Environ. Health*, **8**, 43–56.

Castleman, B. I. and G. E. Ziem (1988). "Corporate Influence on Threshold Limit Values," *Am. J. Ind. Med.*, **13**, 531–559.

Coffin, D. L., D. E. Gardner, G. I. Sidorenko, and M. A. Pinigin (1977). "Role of Time as a Factor in the Toxicity of Chemical Compounds in Intermittent and Continuous Exposures. Part II. Effects of Intermittent Exposure," *J. Toxicol. Environ. Health*, **3**, 821–828.

Colburn, W. A. and H. B. Matthews (1979). "Pharmacokinetics in the Interpretation of Chronic Toxicity Tests: The Last-In, First-Our Phenomenon," *Toxicol. Appl. Pharm.*, **48**, 387–395.

Colquhoun, W. P. (1971). "Circadian Variations in Mental Efficiency," in *Biological Rhythms and Human Performance*, W. P. Colquhoun, Ed., Academic Press, London and New York.

Colquhoun, W. P., M. J. F. Blake, and R. S. Edwards (1968a). "Experimental Studies of Shiftwork. I: A Comparison of 'Rotating and Stabilized' 4-Hour Shift System," *Ergonomics,* **11,** 437–447.

Colquhoun, W. P., M. J. F. Blake, and R. S. Edwards (1968b). "Experimental Studies of Shiftwork. II: Stabilized 8-Hour Shift Systems," *Ergonomics,* **11,** 527–537.

Colquhoun, W. P., M. J. F. Blake, and R. S. Edwards (1969). "Experimental Studies of Shiftwork. III: Stabilized 12-Hour Shift Systems," *Ergonomics,* **12,** 865–875.

Commonwealth of Pennsylvania (November 1, 1971). Threshold Limit Values and Short-term Limits. Title 25, Part 1, Subpart D, Article IV, Chapter 201, Subchapter A, Threshold Limits, Rules and Regulations, 1 Pa. B. 1985.

Cook, W. A. (1945). "Maximum Allowable Concentrations of Industrial Contaminants," *Ind. Med.,* **14**(11), 936–946.

Cook, W. A. (1986). *Occupational Exposure Limits—Worldwide,* Am. Ind. Hyg. Assoc., Akron, OH.

Cooper, W. C. (1973). "Indicators of Susceptibility to Industrial Chemicals," *J. Occ. Med.,* **15**(4), 355–359.

Craft, B. F. (1970). "The Effects of Phase Shifting on the Chronotoxicity of Carbon Tetrachloride in the Rat," *Ph.D. Dissertation,* University of Michigan, Ann Arbor.

Crump, K. S. and B. C. Allen (1984). "Quantitative Estimates of Risk of Leukemia from Occupational Exposure to Benzene," OSHA Docket H-059B, Exh. 152. Information Relevant to Setting a PEL for Benzene. Washington, D.C.

Crump, K. S., D. G. Hoel, G. H. Langley, and R. Peto (1976). "Carcinogenic Processes and Their Implications for Low Dose Risk Assessment," *Cancer Res.,* **36,** 2973–2979.

Cutler, D. J. (1978). "Linear System Analysis in Pharmacokinetics," *J. Pharmacokin. Biopharm.,* **6,** 265–282.

David, A., E. Frantik, R. Holvsa, and O. Novakova (1981). "Role of Time and Concentration on Carbon Tetrachloride Toxicity in Rats," *Inter. Arch. Occup. Environ. Health,* **48,** 49–60.

Dittert, L. W. (1977). "Pharmacokinetic Prediction of Tissue Residues," *J. Tox. and Environ. Health,* **2,** 735–756.

DiVincenzo, G. D., F. J. Yanno, and B. D. Astill (1972). "Human and Canine Exposure to Methylene Chloride Vapor," *Am. Ind. Hyg. Assoc. J.,* **33,** 125–135.

Dixon, S. W., G. J. Graepel, D. L. Leser, and L. F. Percival (1984). "Effect of a Change from an 8-Hr to a 12-Hr Shift on the Levels of DMF Metabolites in the Urine," A Report by Haskell Labs, Dupont Corp, Wilmington, DE.

Doull, J., C. Klaasen, and M. Amdur (1990). *Toxicology: The Basic Science of Poisons,* 4th ed., McMillan, New York.

Dourson, M. L. and J. F. Stara (1983). "Regulatory History and Experimental Support of Uncertainty (Safety) Factors," *Regulatory Toxicol. Pharm.,* **3,** 224–238.

Droz, P. O. and J. G. Fernandez (1977). "Effect of Physical Workload on Retention and Metabolism of Inhaled Organic Solvents—A Comparative Theoretical Approach and Its Applications with Regards to Exposure Monitoring," *Inter. Arch. Occup. Environ. Health,* **38,** 231–240.

Dunham, R. B. and D. I. Hawk (1977). "The Four-Day/Forty-Hour week: Who Wants It?" *Academy of Management J.,* **20,** 4–6.

Durnin, J. V. G. and R. Passmore (1967). *Energy, Work, and Leisure,* Heinemann Educational Books, LTD, London.

Ehrenberg, L., K. D. Hieschke, S. Osteman-Golkar, and I. Wennberg (1974). "Evaluation of Genetic Risks of Alkylating Agents: Tissue Doses in the Mouse from Air Contaminated with Ethylene Oxide," *Mutation Res.*, **24**, 83–103.

Elkins, H. B. (1961). "Maximum Acceptable Concentrations, a Comparison in Russia and the United States," *AMA Arch. Environ. Health*, **2**, 45–50.

Fieldner, A. C., S. H. Katz, and S. P. Kenney (1921). "Gas Masks for Gases Met in Fighting Fires," *USA Bureau of Mines Bulletin 248*, Pittsburgh, PA.

Finklea, J. A. (1988). "Threshold Limit Values: A Timely Look," *Am. J. Ind. Med.*, **14**, 211–212.

Finley, B. F., D. Meyer, and D. J. Paustenbach (1993). "A Recommended RFC for CR(III) and CR(VI)," *Reg. Toxicol. Pharmacol.*, **16**, 161–176.

Fiserova-Bergerova, V. (1972). "Simulation of Uptake, Distribution, Metabolism, and Excretion of Lipid Soluble Solvents in Man," *Aerospace Medical Research Laboratory Report*. No. AMRL-TR-72-130 (Paper No. 4), Wright-Patterson Air Force Base, OH.

Fiserova-Bergerova, J. and D. A. Holaday (1979). "Uptake and Clearance of Inhalation Anesthetics in Man," *Drug Metab. Rev.*, **9**(1), 43–60.

Fiserova-Bergerova, V., J. Vlach, and J. C. Cassady (1980). "Predictable Individual Differences in Uptake and Excretion of Gases and Lipid Soluble Vapors Simulation Study," *Br. J. Ind. Med.*, **37**, 42–49.

Flury, F. and W. Heubner (1919). "Uber Wirkung und Eingiftung eingeatmeter Blausaure," *Biochem. Z.*, **95**, 249–256.

Flury, F. and F. Zernik (1931). *Schadliche Gase, Dampfe, Nebel, Fauch-und Staubarten*, Julius Springer Verlag, Berlin.

Fottler, M. D. (December, 1977). "Employee Acceptance of Four Day Workweek," *Acad. Mangt. J.*, **20**, 100.

Gargas, M. L., R. J. Burgess, D. E. Voisard, G. H. Cason, and M. E. Andersen (1989). "Partition Coefficients of Low-Molecular-Weight Volatile Chemicals in Various Liquids and Tissues," *Toxicol. Appl. Pharmacol.*, **98**(1), 87–99.

Gehring, P. J. and G. E. Blau (1977). "Mechanisms of Carcinogenesis: Dose Response," *J. Environ. Path. Tox.*, **1**, 163–179.

Gibaldi, M. and D. Perrier (1975). *Pharmacokinetics*, Marcel Dekker, New York.

Gibaldi, M. and H. Weintraub (1971). "Some Considerations as to the Determination and Significance of Biologic Half-life," *J. Pharm. Sci.*, **60**, 624–626.

Guberan, E. and J. Fernandez (1974). "Control of Industrial Exposure to Tetrachloroethylene by Measuring Alveolar Concentrations: Theoretical Approach Using a Mathematical Model," *Br. J. Ind. Med.*, **31**, 159–167.

Handy, R. and A. Schindler (1976). Estimation of Permissible Concentrations of Pollutants for Continuous Exposure, U.S. Environmental Protection Agency, Pub. No. EPA-600/2-76-155, Washington, D.C.

Harvey, R. G. (1982). "Polycyclic Hydrocarbons and Cancer," *Am. Scientist*, **70**, 386–393.

Hedges, J. N. (April, 1975). "How Many Days Make a Workweek?" *Monthly Labor Review*, **98**.

Henderson, Y. and H. H. Haggard (1943). *Noxious Gases and the Principles of Respiration Influencing their Action*, Reinhold Publishing, New York.

Henschler, D. (1984). "Exposure Limits: History, Philosophy, Future Developments," *Ann. Occup. Hyg.*, **28**, 79–92.

Hickey, J. L. S. (1977). Application of Occupational Exposure Limits to Unusual Work Schedules and Excursions, Ph.D. Dissertation, University of North Carolina at Chapel Hill, Chapel Hill, NC.

Hickey, J. L. S. (1980). "Adjustment of Occupational Exposure Limits for Seasonal Occupations," *Am. Ind. Hyg. Assoc. J.*, **41**, 261–263.

Hickey, J. L. S. (1983). "The 'TWAP' in the Lead Standard," *Am. Ind. Hyg. Assoc. J.*, **44**(4), 310–311.

Hickey, J. L. S. and P. C. Reist (1977). "Application of Occupational Exposure Limits to Unusual Work Schedules," *Am. Ind. Hyg. Assoc. J.*, **38**, 613–621.

Hickey, J. L. S. and P. C. Reist (1979). "Adjusting Occupational Exposure Limits for Moonlighting, Overtime, and Environmental Exposures," *Am. Ind. Hyg. Assoc. J.*, **40**, 727–734.

Hoel, D. G., N. L. Kaplan, and M. W. Anderson (1983). "Implication of Nonlinear Kinetics on Risk Estimation in Carcinogenesis," *Science*, **219**, 1032–1037.

Hunter, D. (1978). *Occupational Diseases*, 6th ed., Little Brown, Boston, MA.

Hunter, C. G. and D. Blair (1972). "Benzene: Pharmacokinetic Studies in Man," *Arch. Occup. Hyg.*, **15**, 193–199.

Ikeda, M., T. Immamura, M. Hayashi, T. Tabuchi, and I. Hara (1974). "Biological Half-Life of Styrene in Human Subjects," *Int. Arch. Arbeitsmed.*, **32**, 93–100.

Infante, P. F. (1992). "Benzene and Leukemia: The 0.1 ppm ACGIH Proposed Threshold Limit Value for Benzene," *Appl. Occup. Environ. Hyg.*, **7**(4), 253–262.

Iuliucci, R. L. (1982). "12-Hour TLV's," *Pollution Eng.*, November, 25–27.

Jacobi, W. (1980). "Basic Concepts of Radiation Protection," *J. Ecotox. Environ. Safety*, **4**(4), 434–443.

Jakubowski, M. (1966). "Cutaneous Sense Organs of Fishes the Lateral-Line Organs in Some *Cobitidae Cobitis-Taenia Misgurnus-Fossilis Nemachilus-Barbatulus*," *Acta Biol. Cracov. Ser. Zool.*, **9**(1), 71–80.

Johnson, L. C., D. I. Tepas, W. P. Colquhoun, and M. J. Colligan, Eds. (1981). *The Twenty Four Hour Workday: Proceedings of a Symposium on Variations in Work Sleep Schedules*, U.S. Dept. of Health and Human Services, Cincinnati, OH.

Kane, L. E. and Y. Alarie (1977). "Sensory Irritation to Formaldehyde and Acrolein During Single and Repeated Exposures in Mills," *Am. Ind. Hyg. Assoc. J.*, **38**, 509–522.

Kazmina, N. P. (1976). "Study of the Adaptation Processes of the Liver to Monotonous and Intermittent Exposures to Carbon Tetrachloride" (in Russian), *Gig. Tr. Prof. Zabol.*, **3**, 39–45.

Kenny, M. (1974). "Public Employee Attitudes Toward the Four-Day Workweek," *Publ. Personnel Management*, **3**, 100–110 (March–April).

Kim, Y. and G. P. Carlson (1986a). "The Effect of an Unusual Workshift on Chemical Toxicity: I. Studies on the Exposure of Rats and Mice to Dichloromethane," *Fund. Appl. Toxicol.*, **6**, 162–171.

Kim, Y. C. and G. P. Carlson (1986b). "The Effect of an Unusual Workshift on Chemical Toxicity: II. Studies on the Exposure of Rats to Aniline," *Fund. Appl. Toxicol.*, **7**, 144–152.

Kimmerle, G. and A. Eben (1973). "Metabolism, Excretion and Toxicology of Trichloroethylene After Inhalation. II. Experimental Human Exposure," *Arch. Toxicol.*, **30**, 127–138.

Knauth, P., B. Eichhorn, I. Lowenthal, K. H. Gartner, and J. Rutenfranz (1983). "Reduction of Nightwork by Re-designing of Shift-Rotas,'' *Inter. Arch. Occup. Environ. Health,* **51,** 371–379.

Kobert, R. (1912). ''The Smallest Amounts of Noxious Industrial Gases which are Toxic and the Amounts which May Perhaps Be Endured,'' in *Compendium of Practical Toxicology,* 5th ed., Stuttgart, p. 45.

Krivanek, N. et al. (1978). ''Monmethylformamide Levels in Human Urine after Repetitive Exposure to Dimethylformamide Vapor,'' *J. Occup. Med.,* **20,** 179–187.

Landry, R. F. (1981). ''Off-Beat Rhythms and Biological Variables,'' *Occup. Health and Safety,* **50,** 40–43.

LaNier, M. E. (1984). *Threshold Limit Values: Discussion and 35 Year Index with Recommendations (TLVs: 1946–81),* American Conference of Governmental Industrial Hygienists, Cincinnati, OH.

Lehmann, A. and O. G. Fitzhugh (1954). ''100-fold Margin of Safety,'' *Q. Bull-Assoc. Food Drug Off.,* **18,** 33–35.

Lehmann, K. B. (1886). ''Experimentelle Studien uber den Einfluss Technisch und Hygienisch Wichtiger Gase und Dampfe auf Organismus: Ammoniak und Salzsauregas,'' *Arch. Hyg.,* **5,** 1–12.

Lehmann, K. B. and F. Flury (1938). *Toxikologie und Hygiene der technischen Losungsmittel,* Julius Springer Verlag, Berlin.

Lehmann, K. B. and L. Schmidt-Kehl (1936). ''Die 13 Wichtigsten Chlorkohlenwasserstoffe der Fettreihe vom Standpunkt der Gewerbehygiene,'' *Arch. Hyg. Barkt.,* **116,** 131–268.

Lehnert, G., R. D. Ladendorf, and D. Szadkowski (1978). ''The Relevance of the Accumulation of Organic Solvents for Organization of Screening Tests in Occupational Medicine. Results of Toxicological Analyses of More than 6000 Samples,'' *Inter. Arch. Occup. Environ. Health,* **41,** 95–102.

Letavet, A. A. (1961). ''Scientific Principles for the Establishment of the Maximum Allowable Concentrations of Toxic Substances in the USSR,'' in *Proceedings of the 13th Annual Congress on Occupational Health July 25–29, 1960,* Book Craftsmen Assoc., New York.

Leung, H. W. and D. J. Paustenbach (1988a). ''Setting Occupational Exposure Limits for Irritant Organic Acids and Bases based on Their Equilibrium Dissociation Constants,'' *Appl. Ind. Hyg.,* **3,** 115–118.

Leung, H. W. and D. J. Paustenbach (1988b). ''Application of Pharmacokinetics to Derive Biological Exposure Indexes from Threshold Limit Values,'' *Am. Ind. Hyg. Assoc. J.,* **49**(9), 445–450.

Leung, H. W. and D. J. Paustenbach (1990). ''Organic Acids and Bases: A Review of Toxicological Studies,'' *Am. J. Ind. Med.,* **18,** 717–735.

Leung, H. W. (1991). ''Development and Utilization of PB-PK Models for Toxicological Applications,'' *J. Toxicol. Environ. Health;* **32,** 247–267.

Leung, H. W., F. J. Murray, and D. J. Paustenbach (1988). ''A Proposed Occupational Exposure Limit for 2,3,7,8-TCDD,'' *Am. Ind. Hyg. Assn. J.,* **49**(9), 466–474.

Lyublina, E. I. (1962). ''Some Methods Used in Establishing the Maximum Allowable Concentrations,'' *MAC of Toxic Substances in Industry, IUPAC,* p. 109–112.

MacGregor, J. A. (1973). *Application of Pharmacokinetics to Occupational Health Problems,* Ph.D. Dissertation, University of California at San Francisco.

Magnuson, H. L. (1964). ''Industrial Toxicology in the Soviet Union Theoretical and Applied,'' *Am. Ind. Hyg. Assoc. J.,* **25,** 185–190.

Mason, J. W. and H. Dershin (1976). ''Limits to Occupational Exposure in Chemical Environments under Novel Work Schedules,'' *J. Occup. Med.,* **18,** 603–607.

Mason, J. W. and J. Hughes (1985). A Proposed Approach to Adjusting TLVs for Carcinogenic Chemicals (unpublished report), Univ. of Alabama, Birmingham.

Monster, A. C. (1979). "Difference in Uptake, Elimination and Metabolism in Exposure to Trichlorethylene, 1,1,1-trichlorethane and Tetrachloroethylene," *Int. Arch. Occup. Environ. Health* **42**, 311–317.

Moolgavkar, S. H., A. Dwangi, and D. J. Venson (1988). "A Stochastic Two-Stage Model for Cancer Risk Assessment: The Hazard Function and the Probability of Tumor," *Risk Anal.*, **8**, 383–392.

NASA (1991). *Spacecraft Maximum Allowable Concentrations for Formaldehyde*, National Aeronautics and Space Administration (NASA), Johnson Space Center, Houston, TX.

NCRP (1959). Maximum Permissible Body Burdens and Maximum Permissible Concentrations of Radionuclides in Air and in Water for Occupational Exposure, *NBS Handbook No. 69*, U.S. Government Printing Office, Washington, D.C.

NIOSH (1974). *Criteria for a Recommended Standard—Occupational Exposure to Benzene*, National Institute for Occupational Safety and Health, HEW Publication (NIOSH) 74-137, Rockville, MD.

NIOSH (1975). *Criteria for a Recommended Standard—Occupational Exposure to Chloroform*, National Institute for Occupational Safety and Health, HEW Publication (NIOSH) 75-114, Rockville, MD.

NSC (1926). Final Report of the Committee of the Chemical and Rubber Sector on Benzene, National Safety Council (NSC), National Bureau of Casualty and Surety Underwriters, May.

Nielsen, G. D. (1991). "Mechanisms of Activation of the Sensory Irritant Receptor by Airborne Chemicals," *Crit. Reviews in Toxicol.*, **21**, 183–208.

Nollen, S. D. (1981). "The Compressed Workweek: Is It Worth the Effort?" *Ind. Eng.*, 58–63.

Nollen, S. D. (1982). "Work Schedules," in *Handbook of Industrial Engineering*, G. Salvendy, Ed., Wiley-Interscience, New York.

Nollen, S. D. and V. H. Martin (1978). "Alternative Work Schedules. Part 3: The Compressed Workweek," *AMACOM*, New York.

Nomiyama, K. and H. Nomiyama (1974). "Respiratory Retention, Uptake and Excretion of Organic Solvents in Man: Benzene, Toluene n-hexane, Ethyl Acetate and Ethyl Alcohol," *Int. Arch. Arbeits Med.*, **32**, 75–83.

Notari, R. E. (1985). *Biopharmaceutics and Pharmacokinetics, 3rd ed.*, Marcel Dekker, New York.

Notice of Intended Changes—Benzene (1990). *Appl. Occup. Environ. Hyg.*, **5**(7), 453–463.

O'Flaherty, E. J., P. B. Hammond, and S. I. Lerner (1982). "Dependence of Apparent Blood Lead Half-Life on the Length of Previous Lead Exposure in Humans," *Fund. Appl. Toxicol.*, **2**, 49–54.

O'Flaherty, E. J. (1987). *Pharmacokinetics of Chemicals*, Wiley, New York.

Ogata, M., K. Tomokuni, and Y. Takatsuka (1970). "Urinary Excretion of Hippuric Acid and m- or p-methylhippuric Acid in the Urine of Persons Exposed to Vapors of Toluene and m- or p-xylene as a Test of Exposure," *Br. J. Ind. Med.*, **27**, 43–50.

Omenn, G. S. (1982). "Predictive Identification of Hypersusceptible Individuals," *J. Occup. Med.*, **24**, 369–374.

O'Reilly, W. J. (1972). "Pharmacokinetic in Drug Metabolism and Toxicology," *Canadian J. Pharm. Sci.*, **7**, 66–77.

OSHA (1979). *OSHA Compliance Officers Field Manual*, Occupational Safety and Health Administration, Department of Labor, Washington, D.C.

OSHA (1989a). Air Contaminants: Final Rule, Occupational Safety and Health Administration, *Federal Register*, **54**, 2332–2983.

OSHA (1989b). *The Field Operations Manual*, June 15, Washington, D.C.

OSHA (1990). *The Occupation Safety and Health Administration Technical Manual (OTM)*, Feb. 5, Washington, D.C.

Osteman-Golkar, S. and L. Ehrenberg (1983). "Dosimetry of Electrophilic Compounds by Means of Hemoglobin Alkylation," *Ann. Rev. Public Health* **4**, 317–402.

Park, C. and R. Snee (1983). "Quantitative Risk Assessment: State of the Art for Carcinogenesis," *Fund. Appl. Toxicol.*, **3**, 320–333.

Pastoor, T. P. and B. A. Burgess (1983). "Effect of Concentration and Duration of Exposure on the Inhalation Toxicity of Aniline for Periods of 3, 6, 9 and 12 Hours," presented at the 1983 Joint Conference on Occupational Health.

Paustenbach, D. J. (1985). "Occupational Exposure Limits, Pharmacokinetics, and Unusual Work Schedules," in *Patty's Industrial Hygiene and Toxicology*, Vol. 3A, *The Work Environment*, L. J. Cralley and L. V. Cralley, Ed., Wiley, New York, pp. 111–277.

Paustenbach, D. J. (1989). "Health Risk Assessments: Opportunities and Pitfalls," *Columbia J. Environ. Law*, **14**(2), 379–410.

Paustenbach, D. J. (1990a). *What Does the Risk Assessment Process Tell Us About the TLVs?* presented at the 1990 Joint Conference on Industrial Hygiene, Vancouver, B.C. (Oct. 24th).

Paustenbach, D. J. (1990b). "Health Risk Assessment and the Practice of Industrial Hygiene," *Am. Ind. Hyg. Assoc. J.*, **51**, 339–351.

Paustenbach, D. J. (1990c). "Occupational Exposure Limits: Their Critical Role in Preventative Medicine and Risk Management" (editorial), *Am. Ind. Hyg. Assoc. J.*, **51**, A332–A336.

Paustenbach, D. J., Y. Alarie, T. Kulle, N. Schachter, R. Smith, J. Swenberg, H. Witshi, and S. B. Horowitz (1994). "Setting an Occupational Exposure Limit for Formaldehyde: Conclusions of an Expert Panel," *Reg. Toxicol. Pharm.* (in review).

Paustenbach, D. J. and R. R. Langner (1986). "Setting Corporate Exposure Limits: State of the Art," *Am. Ind. Hyg. Assoc. J.*, **47**, 809–818.

Paustenbach, D. J., G. P. Carlson, J. E. Christian, and G. S. Born (1986a). "The Effect of the 11.5 hr/day Exposure Schedule on the Distribution and Toxicity of Inhaled Carbon Tetrachloride in the Rat," *Fund. Appl. Toxicol.*, **6**, 472–483.

Paustenbach, D. J., G. P. Carlson, J. E. Christian, and G. S. Born (1986b). "A Comparative Study of the Pharmacokinetics of Carbon Tetrachloride in the Rat Following Repeated Inhalation Exposure of 8 hr/day and 11.5 hr/day," *Fund Appl. Toxicol.*, **6**, 484–497.

Paustenbach, D. J., H. J. Clewell, M. L. Gargas, and M. E. Andersen (1988). "A Physiologically Based Pharmacokinetic Model for Inhaled Carbon Tetrachloride," *Toxicol. Appl. Pharmacol.*, **96**, 191–211.

Paustenbach, D. J., J. D. Jernigan, B. L. Finley, S. R. Ripple, and R. E. Keenan (1990). "The Current Practice of Health Risk Assessment: Potential Impact and Standards for Toxic Air Contaminants," *JAPCA* **40**(12), 1620–1630.

Paustenbach, D. J., P. Price, W. Ollison, J. D. Jernigan, R. D. Bass, and D. Peterson (1992). "A Reevaluation of Benzene Exposure for the Pliofilm (Rubber Worker) Cohort (1936–1976)," *J. Toxicol. Environ. Health*, **36**, 177–231.

Perera, F. (1984). "The Genotoxic/Epigeneric Distinction: Relevance to Cancer Policy," *Environ. Res.*, **34**, 175–180.

Perera, F. and P. Boffetta (1988). "Perspectives on Comparing Risks of Environmental Carcinogens," *J. Natl. Cancer Inst.*, **80**, 1282–1293.

Petersen, J. E. (1962). "Absorption and Elimination of Carbon Monoxide by Inactive Young Men," *Arch. Environ. Health*, **21**, 165–171.

Peto, R. (1978). "Carcinogenic Effects of Chronic Exposure to Very Low Levels of Toxic Substances," *Environ. Health Perspect.*, **22**, 155–159.

Piotrowski, J. (1971a). *The Application of Metabolic and Excretion Kinetics to the Problems of Industrial Toxicology*, National Library of Medicine, U.S. Government Printing Office, Washington, D.C.

Piotrowski, J. K. (1971b). "Evaluation of Exposure to Phenol: Absorption of Phenol Vapors in the Lungs and Through the Skin and Excretion in Urine," *Br. J. Ind. Med.*, **28**, 172–178.

Piotrowski, J. K. (1977). *Exposure Tests for Organic Compounds in Industrial Toxicology*, Department of Health, Education and Welfare (NIOSH), Cincinnati, OH, pp. 77–144.

Ramazinni, B. (1700). *De Morbis Atrificum Diatriba* (Diseases of Workers) (translated by W. C. Wright in 1940), University of Chicago Press, Chicago.

Ramsey, J. C. and M. E. Andersen (1984). "A Physiologically-Based Description of the Inhalation Pharmacokinetics of Styrene in Rats and Humans," *Toxicol. Appl. Pharm.*, **73**, 159–175.

Ramsey, J. C. and P. J. Gehring (1980). "Application of Pharmacokinetic Principles in Practice," *Fed. Proc.*, **39**, 60–65.

Reitz, R. H., A. M. Schumann, P. G. Watanabe, T. F. Quast, and P. J. Gehring (1979). "Experimental Approach for Evaluating Genetic and Epigenetic Contributions to Chemical Carcinogenesis," *Proc. Am. Assoc. Cancer Res.*, **20**, 266–281.

Reitz, R. H., G. S. Smith, M. E. Andersen, H. J. Clewell, III, and M. L. Gargas (1989). "Use of Physiological Pharmacokinetics in Cancer Risk Assessments: A Study of Methylene Chloride," Chapter 5 in *The Risk Assessment of Environmental and Human Health Hazards: A Textbook of Case Studies*, D. Paustenbach, Ed., Wiley, New York, pp. 238–265.

Reitz, R. H., A. L. Mendrala, R. A. Corley, J. F. Quast, M. L. Gargas, M. E. Andersen, D. Staats, and R. B. Connolly (1990a). "Estimating the Risk of Liver Cancer Associated with Human Exposures to Chloroform Using Physiologically Based Pharmacokinetics Modeling," *Toxicol. Appl. Pharamacol.*, **105**, 443–459.

Reitz, R. H., P. S. McCroskey, C. N. Park, M. E. Andersen, and M. L. Gargas (1990b). "Development of a Physiologically Based Pharmacokinetic Model for Risk Assessment With 1,4-dioxane." *Toxicol. Appl. Pharmacol.*, **105**, 37–54.

Rentos, P. G. and R. D. Shepard, Ed. (1976). *Shift Work and Health—A Symposium*, U.S. HEW PHS, National Institute Occupational Safety and Health, Washington, D.C.

Report of Miners' (1916). Phthisis Prevention Committee, Johannesburg, Union of South Africa.

Riegelman, S., J. C. Loo, and M. Rowland (1968). "Shortcomings in Pharmacokinetic Analysis by Conceiving the Body to Exhibit Properties of a Single Compartment," *J. Pharm. Sci.*, **57**, 117–125.

Rinsky, R. A., A. B. Smith, R. Hornug, R. G. Filloon, R. J. Young, H. A. Okun, and P. J. Landrigan (1987). "Benzene and Leukemia: An Epidemiologic Risk Assessment," *N. Engl. J. Med.*, **316**(17), 1044–1050.

Roach, S. A. (1966). "A More Rational Basis for Air Sampling Programs," *Am. Ind. Hyg. Assoc. J.*, **27**, 1–19.

Roach, S. A. (1977). "A Most Rational Basis for Air Sampling Programs," *Ann. Occup. Hyg.*, **20**, 65–84.

Roach, S. A. (1978). "Threshold Limit Values for Extraordinary Work Schedules," *Am. Ind. Hyg. Assoc. J.*, **39**, 345–364.

Roach, S. A. and S. M. Rappaport (1990). "But They Are Not Thresholds: A Critical Analysis of the Documentation of Threshold Limit Values," *Am. J. Ind. Med.*, **17**, 727–753.

Rodricks, J. and M. R. Taylor (1983). "Application of Risk Assessment to Food Safety Decision Making," *Reg. Toxicol. Pharm.*, **3**, 275–307.

Rodricks, J. V., S. Brett, and G. Wrenn (1987). "Significant Risk Decisions in Federal Regulatory Agencies," *J. Reg. Toxicol. Pharmacol.*, **7**, 307–320.

Rodricks, J. V. (1989). "Origins of Risk Assessment in Food-Safety Decision-Making," *J. Amer. Coll. Toxicol.*, **7**, 379–391.

Rupp, E. M., D. C. Parzychk, R. S. Booth, R. J. Ravidon, and B. L. Whitfield (1978). "Composite Hazard Index for Assessing Limiting Exposures to Environmental Pollutants: Application through a Case Study," *Environ. Sci. Technol.*, **12**(7), 802–807.

Ruzic, A. (1970). "Pharmacokinetic Modelling of Various Theoretical Systems," *J. Pharm. Sci.*, **11**, 110–150.

Saffioti, U. (1980). "Identification and Definition of Chemical Carcinogens: Review of Criteria and Research Needs," *J. Toxicol. Environ. Health*, **6**(5), 1029–1058.

Sayers, R. R. (1927). *Toxicology of Gases and Vapors. International Critical Tables of Numerical Data, Physics, Chemistry and Toxicology*, Vol. 2, McGraw-Hill, New York, pp. 318–321.

Sayers, R. R. and J. M. DalleValle (1935). "Prevention of Occupational Diseases Other than Those That Are Caused by Toxic Dust," *Mech. Eng.*, **57**, 230–234.

Schaper, M. (1993). "Development of a Database for Sensory Irritants and Its Use in Establishing Occupation Exposure Limits," *Am. Ind. Hyg. Assoc. J.*, **54**(9), 488–544.

Schrenk, H. H. (1947). "Interpretation of Permissible Limits," *Am. Ind. Hyg. Assoc. O.*, **8**, 55–60.

Schumann, A. M., J. F. Quast, and P. G. Watanabe (1980). "The Pharmacokinetics and Macromolecular Interactions of Perchloroethylene in Mice and Rats as Related to Oncogenicity," *Toxicol. Appl. Pharm.*, **55**, 207–219.

Sedman, A. J. and J. G. Wagner (1976). "CSTRIP, A Fortran IV Computer Program for Obtaining Initial Poly Exponential Parameter Estimates," *J. Pharm. Sci.*, **65**, 1006–1020.

Seiler, J. P. (1977). "Apparent and Real Thresholds: A Study of Two Mutagens," in *Progress in Genetic Toxicology*, D. Scott, B. A. Bridges, and F. H. Sobels, Eds., Elsevier Biomedical Press, New York.

Smith, R. G. and J. B. Olishifski (1988). "Industrial Toxicology," in *Fundamentals of Industrial Hygiene*, J. Olishifski, Ed., National Safety Council, Chicago, pp. 354–386.

Smolensky, M. H. (1980). "Human Biological Rhythms and Their Pertinence to Shift Work and Occupational Health," *Chronobiologia*, **7**, 378–390.

Smolensky, M. H., J. A. Jovonovich, G. M. Kyle, and B. Hsi (1978). "Chrontoxicity in Rodents Challenged with Propranolol HCL (Inderal®) *Chronopharmacology*," in *Proceedings of the Satellite Symposium of 7th Intercongress of Pharmacology*, Reinberg and Smolensky, Eds., Paris.

Smolensky, M. H., D. J. Paustenbach, and L. E. Schering (1985). "Biological Rhythms, Shiftwork and Occupational Health," in *Patty's Industrial Hygiene and Toxicology*, 2nd ed., Vol. 3b, *Biological Responses*, L. J. Cralley and L. V. Cralley, Eds., Wiley, New York, pp. 175–312.

Smyth, H. F. (1956). "Improved Communication; Hygienic Standard for Daily Inhalation," *Am. Ind. Hyg. Assoc. Qtrly.*, **17**, 129–185.

Squire, R. A. (1981). "Ranking Animal Carcinogens: A Proposed Regulatory Approach," *Science*, **214**, 877–880.

Stewart, R. D., H. H. Gay, D. S. Erley, C. L. Hake, and J. E. Peterson (1961a). "Human Exposure to Carbon Tetrachloride Vapor-Relationship of Expired Air Concentration to Exposure and Toxicity," *J. Occup. Med.*, **3**, 586–590.

Stewart, R. D., A. Arbon, H. H. Gay, D. S. Erley, C. L. Hake, and A. W. Schaffer (1961b). "Human Exposure to Tetrachloroethylene Vapor," *Arch. Environ. Health*, **2**, 516–522.

Stewart, R. D., H. C. Dodd, E. D. Baretta, and A. W. Schaffer (1968). "Human Exposure to Styrene Vapor," *Arch. Environ. Health*, **16**, 656–662.

Stewart, R. D., E. D. Baretta, H. C. Dodd, and T. R. Torkelson (1970a). "Experimental Human Exposure to Tetrachloroethylene," *Arch. Environ. Health*, **20**, 224–229.

Stewart, R. D., H. C. Dodd, H. H. Gay, and D. S. Erley (1970b). "Experimental Human Exposure to Trichloroethylene," *Arch. Environ. Health*, **20**, 64–71.

Stokinger, H. E. (1963). "International Threshold Limits Values," *Am. Ind. Hyg. Assoc. J.*, **24**, 469–484.

Stokinger, H. E. (1970). "Criteria and Procedures for Assessing the Toxic Responses to Industrial Chemicals," in *Permissible Levels of Toxic Substances in the Working Environment*, International Labor Office, World Health Organization, Geneva.

Stokinger, H. E. (1977). "The Case for Carcinogen TLV's Continues Strong," *Occup. Health and Safety*, **46**(March–April), 54–58.

Stokinger, H. E. (1981). "Threshold Limit Values: Part I," in *Dangerous Properties of Industrial Materials Report*, May–June, pp. 8–13.

Stott, W. T., R. H. Reitz, A. M. Schumann, and P. G. Watanabe (1981). "Genetic and Nongenetic Events in Neoplasia," *Food. Cosmet. Tox.*, **19**, 567–576.

Tarlau, E. S. (1990). "Industrial Hygiene with No Limits. A Guest Editorial," *Am. Ind. Hyg. Assoc. J.*, **51**, A9–A10.

Taylor, P. J. (1969). *The Problems of Shift Work*, Proceedings of an International Symposium on Night and Shiftwork, Oslo, Sweden.

Thiis-Evensen, E. (1958). "Shift Work and Health," *Ind. Med. Surg.*, **27**, 493–513.

Thran, C. D. (1974). "Linearity and Superposition in Pharmacokinetics," *Pharmacol. Rev.*, **26**, 3–31.

Travis, C. C. and H. S. Hattemer-Frey (1988). "Determining an Acceptable Level of Risk," *Environ. Sci. Technol.*, **22**, 873–876.

Travis, C. C., R. K. White, and R. C. Ward. 1990a. "Interspecies Extrapolation and Pharmacokinetics," *J. Theor. Biol.*, 142: 285–304.

Travis, C. C., J. L. Qullen, and A. D. Arms, 1990b. "Pharmacokinetics of Benzene," *Toxicol. Appl. Pharmacol.*, 102, 400–420.

Travis, C. C. and S. T. Hester (1990c). "Background Exposure to Chemicals: What is the risk?" *Risk Anal.*, **10**, 463–466.

Travis, C. C., S. A. Richter, E. A. Crouch, R. Wilson, and E. Wilson (1987). "Cancer Risk Management: A Review of 132 Federal Regulatory Decisions," *Environ. Science Technol.*, **21**(5), 415–420.

USBLS (1979). U.S. Bureau of Labor Statistics, *News, Number of Days in the Workweek*, press release, March.

U.S. E.P.A. (1990). *Reducing Risk: A Comparative Assessment of Environmental Problems*, EPA Science Advisory Board, Washington, D.C.

van Stee, E. W., G. A. Boorman, M. P. Moorman, and R. A. Sloane (1982). ''Time-varying Concentration Profile as a Determinant of the Inhalation Toxicity of Carbon Tetrachloride,'' *J. Toxicol. Environ. Health*, **10**, 785–795.

Veng-Pedersen, P. (1977). ''Curve Fitting and Modeling in Pharmacokinetics and Some Practical Experiences with NONLIN and New Program FUNFIT,'' *J. Pharmacokin. Biopharm.*, **5**, 513–531.

Veng-Pedersen, P. (1984). ''Pulmonary Absorption and Excretion of Compounds in the Gas Phase. A Theoretical Pharmacokinetic and Toxicokinetic Analysis,'' *J. Pharm. Sci.*, **230**, 101–106.

Veng-Pedersen, P., D. J. Paustenbach, G. P. Carlson, and L. Suarez (1987). ''A Linear Systems Approach to Analyzing the Pharmacokinetics of Carbon Tetrachloride in the Rat Following Exposure to an 8-hour/day and 12-hour/day Simulated Workweek,'' *Arch. Toxicol.*, **60**, 355–364.

Vokac, Z., P. Magnus, E. Jebens, and N. Gundersen (1981). ''Apparent Phase-Shifts of Circadian Rhythms (Masking Effects) During Rapid Shift Rotation,'' *Int. Arch. Occup. Environ. Health*, **49**, 53–65.

Watanabe, P. G., J. D. Young, and P. J. Gehring (1977). ''The Importance of Non-Linear (Dose-Dependent) Pharmacokinetics in Hazard Assessment,'' *J. Environ. Path. Toxicol.*, **1**, 147–159.

Watanabe, P. G., R. H. Reitz, A. M. Schumann, M. J. McKenna, and P. J. Gehring (1980). ''Implications of the Mechanisms of Tumorigenicity for Risk Assessment,'' in *The Scientific Basis of Toxicity Assessment*, M. Witschi, Ed., Elsevier/North-Holland Press, Amsterdam, pp. 69–88.

Weil, C. S. (1972). ''Statistics Versus Safety Factors and Scientific Judgment in the Evaluation of Safety for Man,'' *Toxicol. Appl. Pharmicol.*, **21**, 454–463.

Wheeler, K., R. Gurman, and D. Tarnowieski (1972). ''The Four-Day Week,'' *AMACOM*, New York.

Wildavsky, A. (1988). Searching for Safety, Transaction Pub., Oxford.

Wilkinson, C. F. (1988). ''Being More Realistic about Chemical Carcinogenesis,'' *Environ. Sci. Technol.*, **9**, 843–848.

Williams, G. M. (1980). ''Classification of Genotoxic and epigenetic Hepatocarcinogens Using Liver Culture Assays,'' *Ann. NY Acad. Sci.*, **349**, 273–282.

Williams, G. M., J. H. Weisburger, and D. Brusick (1981). ''The Role of Genetic Toxicology in a Scheme of Systematic Carcinogen Testing,'' *The Pesticide Chemist and Modern Toxicology*, ACS, Washington, D.C., March 2, pp. 57–87.

Williams, G. M. and J. H. Weisburger (1990). ''Chemical Carcinogens,'' in *Cassarets and Doull's Toxicology: The Basic Science of Poisons*, (4th ed.). Macmillan, New York, Ch. 5. C. D. Klaassen, M. O. Amour and J. Doull, ed.

Williams, R. T. (1974). ''Inter-species Variations in the Metabolism of Xenobiotics,'' *Biochem. Soc. Trans.*, **2**, 359–377.

Wilson, J. T. and K. M. Rose (1978). *The Twelve-Hour Shift in the Petroleum and Chemical Industries of the United States and Canada: A Study of Current Experience*, Wharton Business School, University of Pennsylvania, Philadelphia, PA.

Wisniewska-Knyil, J. M., J. K. Jablonska, and J. K. Piotrowski (1975). "Effect of Repeated Aniline, Nitrobenzene, and Benzene on Liver Microsomal Metabolism in the Rat," *Brit. J. Ind. Med.*, **32**, 42–48.

Withey, J. R. (1979). "Pharmacokinetic Principles," in *First International Congress of Toxicology*, G. Plaa, Ed., Academic Press, New York.

Withey, J. R. and B. T. Collins (1980). "Chlorinated Aliphatic Hydrocarbons Used in the Foods Industry: The Comparative Pharmacokinetics of Methylene Chloride, 1,2-Dichloroethane, Chloroform, and Trichloroethylene after I.V. Administration in the Rat," *J. Environ. Pathol. Toxicol.*, **3**, 313–332.

WHO (1977). *Methods Used in Establishing Permissible Levels in Occupational Exposure to Harmful Agents*, Tech. Report 601, International Labor Office, World Health Organization, Geneva.

WHO (1990). *Occupational Exposure Limits for Airborne Toxic Substances*, 3rd ed., Occ. Safety and Health Series, No. 37, International Labor Office, World Health Organization, Geneva.

Yoder, T. A. and G. D. Botzum (1971). *The Long-Day Short-Week in Shift Work—A Human Factors Study*, 16th Annual Meeting of the Human Factors Society, Indianapolis, IN.

Zeckhauser, R. and W. K. Viscusi (1990). "Risk within Reason," *Science*, **248**, 559–564.

Zenz, C. (1990). *Occupational Medicine: Principles and Practical Applications.* Year Book Medical Pub., New York.

Zenz, C. and B. A. Berg (1970). "Influence of submaximal Work on Solvent Uptake," *J. Occup. Med.*, **12**, 367–369.

Zielhuis, R. L. (1974). "Permissible Limits for Occupational Exposure to Toxic Agents: A Discussion of Differences in Approach Between US and USSR," *Int. Arch. Arbeitsmed.*, **33**, 1.

Zielhuis, R. L. and F. van Der Kreek (1979a). "Calculations of a Safety Factor in Setting Health Based Permissible Levels for Occupational Exposure. A Proposal I," *Int. Arch. Occup. Environ. Health*, **42**, 191–201.

Zielhuis, R. L. and F. W. van Der Kreek (1979b). "Calculations of a Safety Factor in Setting Health Based Permissible Levels for Occupational Exposure. A Proposal II. Comparison of Extrapolate and Published Permissible Levels," *Int. Arch. Occup. Environ. Health*, **42**, 203–215.

Ziem, G. E. and B. I. Castleman (1989). "Threshold Limit Values: Historical Perspective and Current Practice," *J. Occup. Med.*, **13**, 910–918.

CHAPTER EIGHT

Interpreting Levels of Exposures to Chemical Agents

Stephen M. Rappaport, Ph.D.

1 INTRODUCTION

The last two decades witnessed great advances in the technology of measuring exposures to toxic agents in the workplace (see Chapter 2). It is now a relatively simple matter to assay the levels of airborne chemicals, among various members of a particular group of workers, by personal monitoring over the full work shift. Unfortunately, the interpretation of levels of exposure has not achieved the same degree of sophistication. Indeed, decisions still tend to rely upon one-to-one comparisons of the highest measured air concentrations with occupational exposure limits (OELs), much as they did 20 years ago (see, e.g., Leidel et al., 1977; Ulfvarson, 1977; Rock, 1981, 1982; Rappaport, 1984). If the highest air concentration is less than the OEL, then exposure is acceptable (''in compliance'') and vice versa.

This traditional approach, where a minimal number of measurements is compared directly with an OEL, will be termed *compliance testing*. In some countries, such as the United States, compliance testing can result in legal sanctions against the employer when evidence indicates that employees are ''overexposed.'' However, even in situations where the legal basis for monitoring is unclear, the practice is still common because of a strong historical precedence and given the seductive simplicity that is appealing to the occupational hygienist. Thus, compliance testing should be viewed as a general mechanism for assessing exposure that transcends the legal milieu in which air sampling is conducted.

Patty's Industrial Hygiene and Toxicology, Third Edition, Volume 3, Part A, Edited by Robert L. Harris, Lewis J. Cralley, and Lester V. Cralley.
ISBN 0-471-53066-2 © 1994 John Wiley & Sons, Inc.

The decisions arising from compliance testing would be reasonable if the air concentration experienced by a worker or group was constant. Unfortunately, exposures vary to such an extent that a person found to be exposed at, say, half of the OEL on one day might be exposed at twice the OEL on the next and one individual in a group might be exposed, on the average, to 10 times that of another. This is illustrated in Figure 8.1 using data obtained from four different groups of workers exposed, respectively, to inorganic lead (group 1), benzene (groups 2 and 3), and styrene (group 4). The characteristics and sources of the four data sets are summarized in Table 8-1. Since each point in Figure 8.1 represents the exposure received by a given worker over a single work shift, it is obvious that exposures varied considerably within individuals from day to day and, in some cases, between individuals in the same job group. It should also be noted that each data set documents exposures at or above the operative permissible exposure limit (PEL) of the U.S. Occupational Safety and Health administration (OSHA).

This variability, and the lack of guidance currently available for dealing with it, greatly complicates the assessment of occupational exposure. The industrial hygienist can be at a loss to decide not only how many measurements to make and when and where to make them, but also how to interpret exposure levels relative to OELs. Such fundamental problems have motivated movement away from compliance testing and toward statistical methods where the number of measurements as well as the timing and duration are selected a priori to allow appropriate inferences to be drawn. Such methods and the underlying testing structures are often referred to as *sampling strategies*.

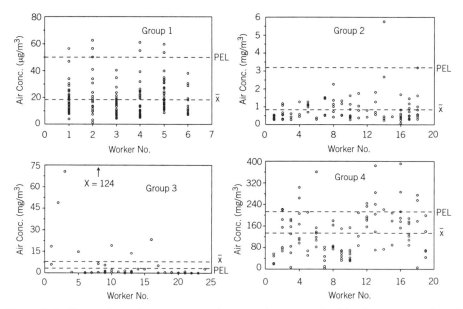

Figure 8.1 Scatterplots of exposures to air contaminants received by four groups of workers. (Groups are described in Table 8.1). Each point represents an 8-hr TWA air concentration. Legend: $\bar{X}$ is the mean exposure, PEL is the permissible exposure limit.

Table 8.1 Descriptions of Four Groups of Workers Exposed to Air Contaminants

Group	Contaminant	Industry	Job	N	k	Reference
1	Inorganic lead	Alkyl-lead manufacturing	Unknown	177	6	Cope et al. (1979)
2	Benzene	Petroleum refining	Refining operator	90	18	Spear et al. (1987)
3	Benzene	Petroleum refining	Transfer/ movement op.	48	24	Spear et al. (1987)
4	Styrene	Boat manufacturing	Sprayer/ laminator	103	19	Unpublished data

Legend: N is the number of measurements; k is the number of workers in the sample.

Because exposures must be evaluated in the face of great variability, statistical methods are central to proper assessment procedures. Likewise, since monitoring is motivated by, and must ultimately relate to, exposure limits, the impact of variability of exposure on the interpretation of OELs is important. The purpose of this chapter is to integrate statistical methods and the philosophical bases for exposure limits within a conceptual framework for interpreting exposure data. The methods and interpretations of assessment practices will be illustrated with the four data sets represented in Figure 8.1. Thus, an Appendix, which provides a complete listing of these data, is included for those who wish to work through the various examples given in the text.

2 STATISTICAL ASPECTS

The object of a prospective sampling program is to allow inferences to be drawn concerning the degree of exposure of the individual or group over periods of months or years or, in the narrower context mentioned earlier, to determine whether exposures are acceptable relative to particular OELs. This section focuses on the statistical issues related to characterizing distributions of exposure.

2.1 Stationarity

The concept of stationarity will not be dealt with in detail. It implies that the statistical descriptors of the underlying process that give rise to exposure, namely, the mean, variance, and autocorrelation functions, do not change over the time period of interest. Thus, if stationarity prevails, one is on relatively firm ground in using the estimated exposure distribution to draw inferences about the situation over the period sampled, but not in using these data to forecast the future or to reconstruct past conditions. Thus, hygienists should view exposure assessment as an ongoing activity in which inferences about the levels of airborne chemicals are made periodically. Bias can be minimized by collecting measurements throughout a sampling period that covers the full range of operations, activities, and environmental conditions that would be expected to influence air levels. Coenen (1971) suggested, for example, that sampling extend over a full year so that seasonal patterns both in the weather and in the production schedule might be identified.

2.2 The Lognormal Model

The sample sizes obtained from most industrial surveys are so small that only by adopting a model is it possible to draw inferences about exposure. The report of Leidel et al. (1975), for example, examined the exposure distributions of 59 job groups based on data compiled between 1971 and 1974 by the National Institute for Occupational Safety and Health (NIOSH) in the United States; sample sizes ranged from 3 to 40 with a median value of 5.

When the need for a parametric model for occupational exposures was recognized in the 1950s and 1960s, attention quickly focused upon the two-parameter lognormal distribution (which will subsequently be referred to as simply a lognormal distribution). There were probably several reasons for this including the following: convenience (a simple transformation allows the normal distribution to be used), prior applications in related areas of urban air pollution and particle size analysis, and empirically the good fit noted by early investigators. However, given the widespread application of the lognormal distribution to occupational exposures today, it is surprising that there have been few studies where the assumption of lognormality was investigated in any way.

The first application of the lognormal model to occupational exposure appears to have been reported by Oldham (1953) and Roach (1950) who noted that the distribution of 779 randomly collected dust measurements gathered from the breathing zones of Welsh coal miners was well characterized by a lognormal distribution. (Each measurement was of 3-min duration.) Juda and Budzinski (1964) analyzed 13 sets of dust data collected in various industries and came to the same conclusion (sample sizes ranged between 19 and 211 observations with a median value of 149). They tested for goodness of fit to the lognormal distribution by applying the Komogorov–Smirnov test to the log-transformed data; the hypothesis of lognormality was rejected for only one of the 13 data sets at a significance level of 0.05. Hines and Spear (1984) reported that lognormality could not be rejected for a distribution of 30 short-term exposures to ethylene oxide in hospitals, and Kromhout et al. (1987) made the same observation for exposures to various agents measured in five factories with sample sizes between 43 and 205. Buringh and Lanting (1991) reported that lognormality was not rejected for 77 percent of 145 data sets of 3 measurements and for 93 percent of 48 data sets containing 6 or more measurements. Likewise Waters et al. (1990) could not reject the null hypothesis for 15 of the 23 occupational data sets (sample sizes ranged between 5 and 40).

Gale (1965, 1967), Hounam (1965), Sherwood (1966), and Langmead (1970) plotted air concentrations of radionuclides on log-probability paper to gain some insight into how well (or poorly) a set of data fitted the lognormal model. This technique has since been applied to nonradioactive airborne chemicals (e.g., Coenen, 1966; Sherwood, 1971; Jones and Brief, 1971; Brief and Jones, 1976; Leichnitz, 1980) most extensively by Kumagai et al. (1989). In commenting on this graphing procedure Aitchison and Brown (1957, p. 32) noted that the method "can hardly be regarded as a rigorous statistical test of lognormality it nevertheless provides a quick method of judging whether the population may feasibly be lognormal."

This small collection of published papers provides empirical evidence that occupational exposures are often quasi-lognormally distributed. There are also theoretical grounds for presuming that the variation in air concentration with time arises from the multiplicative interaction of a series of random variables including the source, the ventilation, and the mobility of the worker (Aitchison and Brown, 1957; Esmen and Hammad, 1977). Koch (1966, 1969) argued in the context of biological processes (which are also quasi-lognormal) that such interactions tend to produce distributions that are right-skewed and approximately lognormal, regardless of the distributions of the individual variables.

Taken together, this combination of empirical and theoretical arguments justifies application of the lognormal model to those occupational data sets that were collected without obvious sources of bias and where air concentrations were small relative to a limiting physical or physiochemical process (such as vaporization relative to a saturation vapor pressure). Whenever possible, it is prudent, nonetheless, to check the fit of data to the model graphically and/or by goodness-of-fit tests. When gross deviations from lognormality are observed, the results of parametric analyses should be interpreted with caution.

2.3 Means, Variances, and Probabilities

Suppose that the exposures of a group of workers has been measured a number of times and were of uniform duration (e.g., 8 hr). If the measurements were collected so that they could be considered representative of the group and of the exposure situation (e.g., with due regard for randomness and independence in relation to individuals and operating conditions, and over time), then it makes sense to use the data to estimate the statistical parameters of the distribution.

2.3.1 Parameters from a Lognormal Distribution

Let $X_{ij} = X_{1,1}, X_{1,2}, \cdots, X_{1,n_1}, \cdots, X_{k,n_k}$ represent a sample of $N = \Sigma_i^k n_i = n_1 + n_2 + \cdots + n_k$ measurements obtained from a group of k workers in a group. It is assumed that this sample was derived from a distribution with mean μ_x and variance σ_x^2 (sometimes referred to as the "arithmetic" mean and variance). If the distribution is lognormal, then exposure can also be characterized by the normal distribution, which is comprised of the logarithms of the air concentrations. Let $Y_{ij} = \ln (X_{ij})$ represent a sample of log-transformed air concentrations. The distribution of Y_{ij} is normal with mean μ and variance σ^2. Figure 8.2 illustrates the relationship between the lognormal distribution of air concentrations and the normal distribution of log-transformed concentrations. Note that the geometric mean, $\mu_g = \exp (\mu)$ and geometric standard deviation $\sigma_g = \exp (\sigma)$ of the lognormal distribution of X_{ij} are related to the corresponding parameters of the normal distribution. Thus, there are three sets of parameters, any of which can be used to define a given lognormal distribution of exposures.

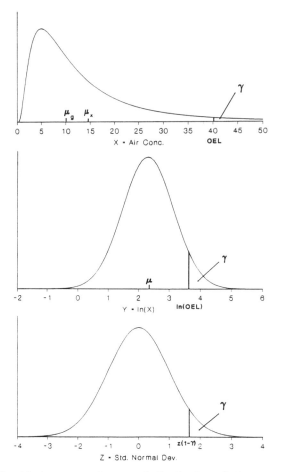

Figure 8.2 Relationship between a lognormal distribution of air concentrations (X) (top) and the resulting normal distributions of the log-transformed concentration (Y) (middle) and the standard normal deviate (Z) (bottom). The relationships are defined in Eq. (6) of the text.

2.3.2 Estimation of Parameters

The parameters of the lognormal distribution, μ_x and σ_x^2, can be directly estimated from the data as the first two moments of the distribution of exposure concentrations, which are, respectively,

$$\overline{X} = \frac{1}{N} \sum_{i=1}^{k} \sum_{j=1}^{n_i} (X_{ij}) \quad \text{and} \quad S_x^2 = \left(\frac{1}{N-1}\right) \sum_{i=1}^{k} \sum_{j=1}^{n_i} (X_{ij} - \overline{X})^2$$

Note that $\overline{X}$ and S_x^2 are unbiased estimates of μ_x and σ_x^2, regardless of the underlying distribution function. The parameters μ and σ, from the normal distribution, are estimated in an analogous manner as the first two moments of the log-transformed air concentrations, designated as $\overline{Y}$ and S_y^2, respectively, as follows:

$$\overline{Y} = \frac{1}{N} \sum_{i=1}^{k} \sum_{j=1}^{n_i} (Y_{ij}) \quad \text{and} \quad S_y^2 = \left(\frac{1}{N-1}\right) \sum_{i=1}^{k} \sum_{j=1}^{n_i} (Y_{ij} - \overline{Y})^2$$

The geometric mean and geometric standard deviation of the lognormal distribution are estimated as $\overline{X}_g = \exp(\overline{Y})$ and $S_g = \exp(S_y)$, respectively.

Finally, the mean μ_x and variance σ_x^2 from the lognormal distribution can be estimated from $\overline{Y}$ and S_y^2 by using the maximum-likelihood formulas:

$$\overline{X}_m = \exp\left(\overline{Y} + \frac{0.5\,(S_y^2)(N-1)}{N}\right)$$

$$S_m^2 = \exp\left(2\overline{Y} + \frac{S_y^2\,(N-1)}{N}\right) \exp\left(\frac{S_y^2\,(N-1)}{N}\right)$$

When samples are large ($n > 50$) the maximum-likelihood estimates of the mean and variance of a lognormal distribution, although slightly biased, have certain optimal properties and should be used in preference to the estimates $\overline{X}$ and S_x^2. If the sample is small ($n < 20$), however, the direct estimate $\overline{X}$ is more precise than the maximum-likelihood estimate $\overline{X}_m$ and may be preferred for certain applications (Selvin and Rappaport, 1989; Chang and Nelson, 1991). Although correction factors are available to remove the biases in the maximum-likelihood estimates (Aitchison and Brown, 1957), the corrections are small and can usually be ignored for $n > 20$ (Selvin and Rappaport, 1989).

In summary, the following symbols are employed to represent the six parameters, which can be used to define a lognormal distribution, and their estimates:

Mean	Standard Deviation
$\mu_x(\overline{X}_m \text{ or } \overline{X})$	$\sigma_x(S_m \text{ or } S_x)$
$\mu_g(\overline{X}_g)$	$\sigma_g(S_g)$
$\mu(\overline{Y})$	$\sigma(S_y)$

2.3.3 Converting Among Parameters

Each set of parameters can be derived from any other set, and Table 8.2 gives the useful relationships that can be used to convert from one to another (Aitchison and Brown, 1957; Leidel et al., 1977). The most important of these is the relationship between the mean of the lognormal distribution, μ_x, and the mean and variance of the normal distribution, μ and σ^2:

$$\mu_x = \exp(\mu + 0.5\sigma^2) \tag{1}$$

Equation (1) indicates that the mean μ_x is always greater than the geometric mean $\mu_g = \exp(\mu)$ (Fig. 8.2) and that the difference between the two increases with σ^2.

Table 8.2 Formulas for Converting among Parameters of a Lognormal Distribution[a]

Given	To Obtain	Use
μ	$\mu_g =$	$\exp(\mu)$
μ_x, σ_x	$\mu_g =$	$\mu_x^2/\sqrt{\mu_x^2 + \sigma_x^2}$
σ	$\sigma_g =$	$\exp(\sigma)$
μ_x, σ_x	$\sigma_g =$	$\exp[\sqrt{\ln[1 + (\sigma_x^2/\mu_x^2)]}]$
μ, σ	$\mu_x =$	$\exp[\mu + (\frac{1}{2})\sigma^2]$
μ_g, σ	$\mu_x =$	$\mu_g \exp[(\frac{1}{2})\sigma^2]$
μ, σ	$\sigma_x =$	$\sqrt{[\exp(2\mu + \sigma^2)][\exp(\sigma^2) - 1]}$
μ_g, σ	$\sigma_x =$	$\sqrt{\mu_g^2[\exp(\sigma^2)][\exp(\sigma^2) - 1]}$
μ_g	$\mu =$	$\ln(\mu_g)$
μ_x, σ	$\mu =$	$\ln\mu_x - \frac{1}{2}(\sigma^2)$
σ_g	$\sigma =$	$\ln\sigma_g$
μ_x, σ_x	$\sigma =$	$\sqrt{\ln[1 + (\sigma_x^2/\mu_x^2)]}$

[a] For legend see List of Symbols.

It is important to differentiate between these two parameters because it is μ_x and *not* μ_g that determines an individual's cumulative exposure over time. Another useful formula in Table 8.2 is $\sigma^2 = \ln(1 + \sigma_x^2/\mu_x^2)$, which allows the variance of the normal distribution to be related to the square of the coefficient of variation of exposure, $CV_x = \sigma_x/\mu_x$, obtained from the lognormal distribution.

Table 8.2 may also be used with the estimated parameters $\overline{Y}$ and S_y^2 or $\overline{X}_m$ and S_m^2 for manipulating samples of data. However, in applying Table 8.2, direct estimates of the parameters (i.e., $\overline{X}$ and S_x^2) of the lognormal distribution should generally not be employed.

2.3.4 Variation Within and Between Workers

Since sampling is usually conducted to assess exposures received by groups of workers employed in the same job, it is necessary to apply methods to differentiate the variability associated with the job or process from that associated with the individual worker and his or her tasks or practices. Since the first type of variation should affect all workers in a particular job more-or-less equally, it is designated as the *within-person* component of variance. Conversely, the latter type of variation is particular to the individuals in a job and, therefore, reflects consistent differences in exposure between the workers; this is designated as the *between-person* (within job) component of variance.

Assume that a group of workers is exposed to air concentrations that are lognormally distributed with mean μ_x and variance σ_x^2. This distribution, representing all exposures experienced by all workers on all days, will be referred to as the *total distribution*. Each scatter plot, shown in Figure 8.1, embodies one sample from the total distribution of exposures received by each of the various groups. Every individual in a group has his or her personal distribution of day-to-day exposures, which can differ significantly from that of the group as a whole; some evidence of such behavior can be seen in the scatter plot of group 3 (exposed to benzene) where workers 8 and 12, for example, appear to have very different overall exposures.

2.3.4.1 Accounting for the Variances.

Since the underlying distribution of exposures to chemical agents is assumed to be lognormal, a convenient way to account for the components of variability is in terms of the variance of the normally distributed variable, Y_{ij}. Assuming that the well-known one-way random-effects model is appropriate for evaluating occupational exposures (for a general discussion see Searle, 1971; for specific applications, see Kromhout et al., 1987; Brunerkreff et al., 1987; and Heederik et al., 1991), then exposure can be expressed as

$$Y_{ij} = \mu + A_i + \epsilon_{ij} \text{ for } i = 1, 2, \cdots, k \text{ and } j = 1, 2, \cdots, n_i \qquad (2)$$

where A_i represents the random deviation in the mean (log-transformed) exposure (designated μ_i) between the ith worker and that of the group as a whole (designated μ) and ϵ_{ij} represents random deviations about μ_i associated with differences in exposure experienced by the ith person from day to day. Under the random-effects model it is assumed that both A_i and ϵ_{ij} are normally distributed with a mean of zero and a variance of either σ_B^2 or σ_W^2, respectively. Thus, it is implicit in this model that the within-person component of variance σ_W^2 is homogeneous across all members of the group and that the mean exposure of the ith worker is, therefore, $\mu_{x,i} = \exp(\mu_i + 0.5\sigma_W^2)$. It follows from Eq. (2) that μ_i is normally distributed with mean μ and variance σ_B^2 and also that $\ln(\mu_{x,i})$ is normally distributed with mean $\exp(\mu + 0.5\sigma_W^2)$ and variance σ_B^2. Thus, the distributions of $\ln(\mu_{x,i})$ and μ_i across a group of workers should both be normal and should both have the same variance (σ_B^2) under this model.

The parameters σ_W^2 and σ_B^2 can be estimated by applying analysis-of-variance (ANOVA) techniques (Searle, 1971). The ANOVA estimator of σ_W^2 is given by

$$_W S_y^2 = \text{MSW}_y = \frac{\sum\limits_{i=1}^{k} \sum\limits_{j=1}^{n_i} (Y_{ij} - \overline{Y}_i)^2}{(N - k)} \qquad (3)$$

where $\overline{Y}_i = (1/n_i) \sum_i^{n_i} Y_i$ represents the estimated mean of the ith worker's log-transformed exposures. This is the mean-squared (error) "within" (MSW_y) given in the ANOVA table. Likewise, the ANOVA estimator of σ_B^2 is given by

$$_B S_y^2 = \frac{(k - 1)(\text{MSB}_y - {_W}S_y^2)}{\left[N - \sum\limits_i^{k} n_i^2 / N \right]} \qquad (4)$$

where $\text{MSB}_y = [\sum_i^k n_i(\overline{Y}_i - \overline{Y})^2]/(k - 1)$ is the mean-squared (error) "between." Since the "total" variance σ^2 is defined as the sum of σ_W^2 and σ_B^2, it can be estimated as the sum of the estimates of the within- and between-worker variances, that is,

$$_TS_y^2 = {_W}S_y^2 + {_B}S_y^2 \tag{5}$$

which should be very nearly equal to the direct estimate, S_y^2, defined earlier.

To illustrate the partitioning of exposure variability into the within- and between-person components of variance, consider the four data sets illustrated in Figure 8.1 and summarized in the Appendix. One-way analyses of variance were performed on the variable Y_{ij} in each case and the mean-squared errors MSW_y and MSB_y computed. For group 1, where six workers were exposed to inorganic lead, $N = 177$, $k = 6$, $\Sigma_i^k n_i^2/N = 31.588$, $MSB_y = 1.0567$, and $MSW_y = 0.3774$. Thus, the estimated variance components are as follows:

$$_WS_y^2 = MSW_y = 0.3774 \quad \text{[from Eq. (3)]}$$

$$_BS_y^2 = (5)(1.0567 - 0.3774)/(177 - 31.588) = 0.0234 \quad \text{[from Eq. (4)]}$$

$$_TS_y^2 = 0.3774 + 0.0234 = 0.4008 \quad \text{[from Eq. (5)]}$$

Table 8.3 summarizes the results of ANOVA for all four of the groups.

2.3.4.2 Implications of the Between-Person Distribution.

Determination of the between-person component of variance allows attention to focus upon the distribution of exposures across a group of individuals in a particular job. Since this (between-person) distribution is central to the proper classification and assessment of exposure, a great deal of insight can be derived from applications of the ANOVA techniques illustrated above. For example, in the case of the lead-exposed workers in group 1, only $_BS_y^2/{_T}S_y^2 = (0.0234/0.3961) = 5.8$ percent of the total variance resulted from systematic differences between workers in the group; thus, all individuals in this group received exposures that were very similar overall. This result can be compared with those from the other groups, summarized in Table 8.3, where the proportion of the total variance attributed to differences between workers was 7.1 percent (group 2), 44.1 percent (group 3), and 36.2 percent (group 4), respec-

Table 8.3 Variance Components of the Log-transformed Exposures Received by Four Groups of Workers (Groups Described in Table 8.1)

Group	N	k	$_WS_y^2$	$_BS_y^2$	$_TS_y^2$	$_BS_y^2/{_T}S_y^2$
1	177	6	0.377	0.023	0.400	0.058
2	90	18	0.466	0.033	0.499	0.071
3	48	24	3.012	2.377	5.389	0.441
4	103	19	0.513	0.292	0.805	0.362

Legend: N is the Number of the measurements; k is the number of workers in the sample; $_WS_y^2$ and $_BS_y^2$ are the estimated within- and between-worker components of variance; $_TS_y^2$ is the estimated total variance.

tively. Obviously, there were differences among the groups regarding the uniformity of exposure; this will be discussed further in Section 5.3.

Additional insight into the nature of exposure across a group can be obtained by testing the validity of the lognormal distribution as a model for exposure among individuals in a given group. Log-probability plots are shown in Figure 8.3 for the four groups of workers; in each case, $\overline{Y}_i$ is plotted versus the corresponding cumulative probability [equal to $i/(k + 1)$ where i refers here to the rank of $\overline{Y}_i$ from low to high] from the between-person distribution. The slope of each graph is related to the estimated standard deviation of $\overline{Y}_i$, which is designated $S_{\overline{Y}_i}$. Since $S_{\overline{Y}_i}^2 = [_BS_y^2 + (_wS_y^2/n')]$ can only be determined with equal precision across the sample of workers when the number of measurements per person, n', is constant (i.e., for the "balanced" case), balanced subsamples were obtained prior to plotting the data. This was done by selecting a value of n' for each group, then excluding individuals with $n_i < n'$, and randomly sampling individuals with $n_i > n'$ (without replacement) to obtain n' measurements per person. It is clear from Figure 8.3 that each of the distributions of $\overline{Y}_i$ fit the normal model reasonably well. This conclusion is supported by the results, shown in Table 8.4, of applying a test for goodness of fit (the Shapiro–Wilk statistic, Shapiro and Wilk, 1965). In all four cases the hypothesis of a lognormal distribution was not rejected at a significance level of 0.05.

The fact that the between-worker distributions of the four groups were approximately lognormal allows the likelihood of "overexposure" relative to an OEL to be extended from the context of isolated days to individual workers. For this reason Rappaport (1991a) defined such lognormally distributed groups as *monomorphic groups* and dealt with them under probabilistic models relating cumulative exposure and risk. In subsequent sections of this chapter, similar analyses will be extended to the four monomorphic groups considered here.

2.3.5 *Probabilities of Exceeding Particular Levels*

The parameters of a lognormal distribution can be used to calculate various likelihoods, such as, the probability that exposure exceeds an OEL. As Figure 8.2 shows, the probability that a random exposure X_{ij} from the lognormal distribution would exceed the OEL is equal to the probability that the corresponding normally distributed value Y_{ij} would exceed ln (OEL). It also follows that this likelihood is equal to the probability that a standard normal deviate, Z, would exceed the value $z_{(1-\gamma)} = [\ln (\text{OEL}) - \mu]/\sigma$, which is associated with the γ-level percentile, $\gamma = P\{X_{ij} > \text{OEL}\}$ (see Fig. 8.2); that is,

$$\gamma = P\{X_{ij} > \text{OEL}\} = P\{Y_{ij} > \ln (\text{OEL})\} = P\left\{Z > \frac{\ln (\text{OEL}) - \mu}{\sigma} = z_{(1-\gamma)}\right\}$$

(6)

Thus, γ represents the likelihood that a typical worker from a group would be exposed during one interval (e.g., one work shift) to an air concentration above the OEL. The same reasoning can be extended to consider the probability that a random

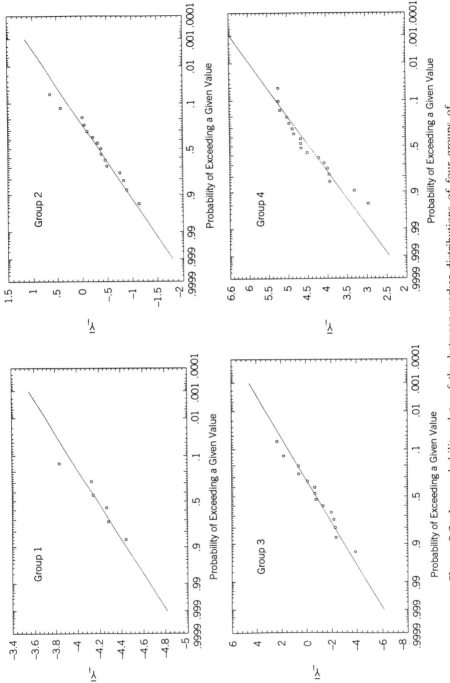

Figure 8.3 Log-probability plots of the between-worker distributions of four groups of workers. (Groups are described in Table 8.1). Each point represents the estimated mean exposure ($\overline{Y}_i$) [with units of $\ln(\text{mg}/\text{m}^3)$] of a worker obtained from a balanced subsample of each data set which is described in Table 8.4.

Table 8.4 Application of Shapiro-Wilk Test of Normality to the Log-transformed Exposures Received by Four Groups of Workers (Groups Described in Table 8.1) (Applied to $\overline{Y}_i$ for balanced subsamples of data)

Group	k'	n'	Shapiro-Wilk Statistic	Reject H_0? (p value)
1	6	15	0.944	No (0.710)
2	15	3	0.973	No (0.860)
3	14	2	0.973	No (0.876)
4	18	3	0.918	No (0.142)

Legend: k' is the number of workers in the balanced subsample; n' is the number of measurements per worker in the balanced subsample; H_0 is the null hypothesis stating that there is no significant lack of fit.

individual from a monomorphic group would be exposed, on the average, above the OEL. This likelihood, designated θ, is given by

$$\theta = P\{\mu_{x,i} > \text{OEL}\} = P\left\{Z > \frac{\ln(\text{OEL}) - (\mu + 0.5\sigma_W^2)}{\sigma_B} = z_{(1-\theta)}\right\} \quad (7)$$

Another important consequence of the lognormal model for occupational exposure is the fact that the mean μ_x and the variance σ_x^2 are not independent as is the case with the normal distribution, but rather, are related. One result of this lack of independence is the fact that μ_x and both γ and θ are also related (Rappaport et al., 1988a). These relationships are easily illustrated by combining Eqs. (1) and (6) or (7) and solving for OEL/μ_x, so that

$$\frac{\text{OEL}}{\mu_x} = \exp\left[0.5z_{(1-\gamma)}^2 - 0.5(z_{(1-\gamma)} - \sigma)^2\right] \quad (8)$$

or

$$\frac{\text{OEL}}{\mu_x} = \exp\left[0.5z_{(1-\theta)}^2 - 0.5(z_{(1-\theta)} - \sigma_B)^2\right] \quad (9)$$

These expressions allow the mean exposure to be related to the likelihoods of overexposure, over either days or individuals, at a given value of either σ or σ_B. Unfortunately, because these variances are seldom known a priori, it is rarely possible to exploit Eqs. (8) and (9) prospectively in sampling strategies. However, since OEL/μ_x is maximal when $\sigma = z_{(1-\gamma)}$ or when $\sigma_B = z_{(1-\theta)}$, there is a corresponding maximum in the probability of overexposure that will be designated either as $\gamma_{\max}$ or $\theta_{\max}$, respectively. It has been shown by Rappaport et al. (1988a) that

$$\frac{\text{OEL}}{\mu_x} = \exp\left(0.5z_{(1-\gamma_{\max})}^2\right) \quad (10)$$

or

$$\frac{\text{OEL}}{\mu_x} = \exp{(0.5z^2_{(1-\theta_{max})})} \qquad (11)$$

Thus, at a given mean exposure, there exists a maximum probability that the OEL will be exceeded, either over days or over individuals, regardless of the variance of the exposure distribution. Table 8.5 lists several examples of values of γ_{max} and θ_{max} (between 1 and 40%) that correspond to given values of (OEL/μ_x). The table shows, for instance, that when $\text{OEL}/\mu_x = 3.9$, γ_{max} and θ_{max} equal 0.05; in other words, if the mean is less than about one-quarter of the OEL, then not more than 5 percent of the exposures (over either days or individuals) are expected to exceed the OEL *regardless of the variability of exposure.* Various implications of this property of the lognormal distribution have been discussed (Rappaport et al., 1988a; Rappaport, 1991a) and will be explored further in this chapter.

2.4 Autocorrelated Exposure Series

It has been shown that the distribution of exposures received by a worker can be summarized by the mean and variance. However, these parameters do not provide any information concerning the correlation of exposures measured at different times. This is given by a third characteristic of the distribution, the autocorrelation function, $\rho(h)$, which defines the relationship between air concentrations separated by h intervals of time, where h is referred to as the lag.

Let $\{X_t\}$ for $t = 1, 2, \cdots, n$ represent a discrete series of air concentrations measured at regular intervals, for example, 8-hr time-weighted average (TWA) air

Table 8.5 Maximum Probability with which Exposures from a Lognormal Distribution with Mean μ_x Can Exceed the OEL

γ_{max} or θ_{max}	z	OEL/μ_x
0.01	2.33	15.0
0.02	2.05	8.2
0.03	1.88	5.8
0.04	1.75	4.6
0.05	1.64	3.9
0.10	1.28	2.3
0.20	0.842	1.4
0.40	0.253	1.03

Legend: γ_{max} is the probability of overexposure during one interval; θ_{max} is the probability of overexposure of an individual worker's mean exposure; z is a value from the standard-normal distribution associated with γ_{max} or θ_{max}.

concentrations measured for a single worker each day for n days. If $\{X_t\}$ is stationary, then the autocorrelation function is defined as the autocovariance coefficient at lag h divided by the variance; that is,

$$\rho(h) = \frac{E[(X_t - \mu_x)(X_{t+h} - \mu_x)]}{\sigma_x^2} \tag{12}$$

where $E[X_t]$ refers to the expected value of the variable X_t. Thus, the autocorrelation function represents the proportion of σ_x^2 that can be attributed to the covariance of sequential values; for a perfectly correlated process $X_t = X_{t+h}$ and $\rho(h) = 1$ for all lags while a purely random series has $\rho(h) = 0$ for all lags (Rappaport and Spear, 1988; Francis et al., 1989; Buringh and Lanting, 1991).

2.4.1 Variance of Autocorrelated Exposures

If exposures are highly autocorrelated, then samples collected during brief surveys may not be sufficiently independent to allow valid statistical inferences to be drawn about the underlying population of exposures. Consider, for example, the estimation of the mean exposure on the basis of a series of n consecutive measurements. Spear et al. (1986) showed that the variance of the mean, $\sigma_{\bar{X}}^2$, can be related to the variance of the underlying distribution of exposures, σ_x^2, according to the following relationship:

$$\sigma_{\bar{X}}^2 = \left(\frac{\sigma_x^2}{n}\right)\left[1 + \left(\frac{2}{n}\right)\sum_{h=1}^{n-1}(n-h)\,\rho(h)\right] \tag{13}$$

Equation (13) indicates not only the variance of the mean will be less than that of the individual exposures but also that the reduction in variability associated with averaging depends on the autocorrelation function $\rho(h)$. If $\{X_t\}$ were a purely random process, then $\rho(h) = 0$ for all lags and $\sigma_{\bar{X}}^2 = \sigma_x^2/n$. However, for all other situations one needs to know the autocorrelation function in order to estimate the variance of $\bar{X}$.

2.4.2 Model for Autocorrelation

Unfortunately, a reasonable estimate of $\rho(h)$ over the first 10–12 lags requires the measurement of at least 50 sequential exposures (Box and Jenkins, 1976). Since industrial hygienists rarely collect such large numbers of measurements, little information is currently available to allow autocorrelation functions to be analyzed. This paucity of time-series data has motivated investigators to rely upon models to gain insight into the potential importance of autocorrelation on the estimation of exposure. The most popular of these has been the first-order autoregressive process [AR(1) process], which depicts the current exposure as a weighted fraction of the previous exposure plus a random input (Coenen, 1976; Roach, 1977; Koizumi et al., 1980; Spear et al., 1986; Preat, 1987; Rappaport and Spear, 1988; Francis et al., 1989).

If it is assumed that the log-transformed exposures $\{Y_t\}$ are adequately described by an AR(1) process, and are normally distributed, then, as shown by Male (1982) and illustrated in this context by Francis et al. (1989) the autocorrelation function of X_t is given by,

$$\rho(h) = \frac{\exp (\sigma^2 w_y^h) - 1}{\exp (\sigma^2) - 1} \tag{14}$$

where w_y^h is the autocorrelation coefficient at lag h of $\{Y_t\}$. Table 8.6 uses calculations employing Eqs. (13) and (14) to illustrate the impact of autocorrelation on the variance of the mean of five consecutive measurements. The three values of $w_y = 0.1$, 0.5, and 0.8 represent series with low, medium, and high levels of autocorrelation, and the three values of $\sigma_g = 1.6$, 2.0, and 3.5 represent low, medium, and high levels of exposure variability. The calculations indicate that if exposures are highly autocorrelated, the variance of the mean of five consecutive measurements will be much greater than that of the corresponding mean from an uncorrelated process; for example, for $w_y = 0.8$ the variance of the mean should be, depending on the value of σ_g, between 3.0 and 3.5 times that expected when $w_y = 0$.

Using similar calculations, Francis et al. (1989) showed that when $\sigma_g = 1.6$ and $w_y = 0.8$, about 30 percent of the estimated means of $n = 5$ would lie outside the 95 percent confidence interval of the mean that would occur if the series were purely random. In other words, if exposures are highly autocorrelated ($w_y = 0.8$), there would be about one chance in three that an estimated mean from five consecutive measurements would be significantly different from one based upon five uncorrelated

Table 8.6 Effect of Autocorrelation on the Variance of the Mean of 5 Consecutive Measurements (σ_x^2) for Different Levels of Variability and Autocorrelation[a]

		Autocorrelation Coefficient $\rho(h)$[b]				
σ_g	w_y	$h = 1$	2	3	4	$\sigma_{\bar{x}}^2 / \sigma_x^{2c}$
1.60	0.1	0.0903	0.0089	0.0009	0.0001	0.231
	0.5	0.4721	0.2295	0.1131	0.0562	0.429
	0.8	0.7818	0.6140	0.4841	0.3828	0.706
2.00	0.1	0.0798	0.0078	0.0008	0.0001	0.228
	0.5	0.4402	0.2069	0.1004	0.0494	0.410
	0.8	0.7598	0.5837	0.4522	0.3526	0.683
3.50	0.1	0.0447	0.0042	0.0004	0.0000	0.215
	0.5	0.3133	0.1263	0.0570	0.0271	0.342
	0.8	0.6598	0.4549	0.3243	0.2371	0.591

[a] For legend see List of Symbols.
[b] Calculated according to Eq. (12).
[c] Calculated according to Eq. (13).

values. So, the question logically arises as to the extent to which occupational exposures are autocorrelated.

2.4.3 Autocorrelation of Actual Exposures

In the context of day-to-day exposure, there are only anecdotal suggestions that significant autocorrelation may exist. Esmen (1979) noted a strong correlation between production rate and exposure, particularly for batch operations; Ulfvarson (1983) commented upon seasonal variations and multiyear trends in exposure; Buringh and Lanting (1991) observed that the variances of data sets with $3 \leq n \leq 6$, obtained from a variety of industries, were smaller when all measurements were collected within a single week than otherwise; and Kumagai et al. (1989) noted that estimates of σ_g were significantly smaller when measured on a given individual over two consecutive days than otherwise. However, in the only study to investigate the question directly, Francis et al. (1989) could find relatively little evidence of autocorrelation in three industrial data sets (17 workers) of sequential shift-long air concentrations extending from 36 to 730 days. For example, their analysis of the three workers from group 1 with the most complete exposure data (workers 1, 3, and 4) had first-lag autocorrelation coefficients, respectively, of 0.03, -0.04, and 0.05, none of which was statistically significant; the largest autocorrelation coefficient observed for a worker in any data was 0.54. If the results of Francis et al. (1989) are typical of the degrees of autocorrelation observed from day to day in most workplaces, then it may well be possible to estimate the distribution of shift-long measurements of exposure on the basis of discrete campaigns of a few days duration. However, this suggestion will require confirmation.

Very little work has been performed concerning the levels of autocorrelation of intrashift variations in exposure. Coenen (1971, 1976) noted that dust concentrations and vinyl-chloride concentrations measured continuously at fixed locations in manufacturing facilities were both highly autocorrelated. Roach (1977), Spear et al. (1986) and Rappaport and Spear (1988) used air exchange rates to estimate the levels of short-term autocorrelation that might exist in air environments where mass transport of the contaminant is governed by turbulent diffusion. In order to simplify the mathematics, they assumed that $\{X_t\}$ (rather than $\{Y_t\}$) is described by an AR(1) process and is normally distributed. In this context, $\rho(h) = w^h = e^{-bh\tau}$, where w^h represents the autocorrelation coefficient of $\{X_t\}$ at lag h, and b represents the air exchange rate. When σ_g is less than about 3, then values of $\rho(h) = w^h$ are approximately equal to those obtained from Eq. (14) and the values of $w_y = 0.1, 0.5,$ and 0.8 are roughly equivalent to air exchange rates of $b = 10, 3,$ and 1 hr^{-1}, respectively, when $\tau = 0.25$ hr. Table 8.6, therefore, suggests that periods of hours can be required between measurements to obtain unbiased estimates of means and variances if the air exchange rate is small. Thus, the industrial hygienist who wishes to predict the frequencies of brief excursions to short-term-exposure limits (STELs) should be wary of doing so on the basis of serial measurements (see Sections 6.4.3 and 6.5.4). Recent advances in the development of personal monitors that measure short-term exposures and that store the data over an entire shift should allow actual autocorrelation functions to be investigated.

2.5 Effect of Averaging Time on Exposure Distributions

Since the durations of measurements are sometimes different, the question logically arises as to the impact of averaging time on the exposure distribution. Are the mean and variance of the distribution of 15-min averages, for example, the same as those of 8-hr averages when sampling the same environment? Experience and common sense suggest that this is not the case and that the variance, in particular, decreases with increasing averaging time (LeClare et al., 1969; Coenen, 1971). Several investigators have treated this question mathematically (Roach, 1966, 1977; Coenen, 1976; Spear et al., 1986; Preat, 1987) and have shown that the mean value of the exposure distribution μ is constant and independent of averaging time if $\{X_t\}$ is stationary. However, the variance σ_x^2 depends on the autocorrelation function of the instantaneous concentrations and, under certain circumstances, does change with averaging time according to Eq. (13) where, in this context, $\sigma_{\bar{X}}^2$ is the variance of the long-term averages, σ_x^2 is the variance of the short-term exposures, and n is the number of short-term intervals in the long-term average (e.g., $n = 32$ for 15-min intervals during an 8-hr work shift).

As shown earlier, if the distribution of exposure is assumed to be lognormal, then the mean and variance define the distribution completely. However, it should be noted that although μ_x is independent of the averaging time of the measurements, the geometric mean μ_g decreases with averaging time as σ_g increases. This indicates that exposure distributions change with averaging time even when all measurements come from the same environment. Thus, an exposure distribution must be considered as the distribution of air concentrations *all collected over the same averaging time*. Minor differences in the duration of measurement are unlikely to have much effect on the estimated parameters of the exposure distribution. However, potential problems can be avoided by adhering to a uniform duration for all measurements to be used in a particular application.

The relationship given in Eq. (13) suggests that substantial reductions in the variance are possible by extending the averaging time beyond one shift. If the amount of autocorrelation observed from shift to shift is small, then $\rho(h)$ approaches zero for all lags and $\sigma_{\bar{X}}^2 = (\sigma_x^2/n)$. Thus, the variance of the estimated mean based on week-long measurements (5 shifts) would be one-fifth of that based on measurement of single shifts. So if the goal is to obtain the most precise information about the long-term mean exposure with the minimum amount of measurement, multishift sampling should be considered. Technological developments in the area of passive monitoring for gaseous pollutants make this a viable alternative for future assessments.

3 EXPOSURE LIMITS

Since most prospective assessments of exposure have been motivated by the need to comply with OELs, one cannot consider the development of sampling strategies without discussing exposure limits. Unfortunately, the groups that have developed OELs in various countries have tended to ignore the long-term variation in exposure that occurs both over time and across the population. This is perhaps understandable because organizations that set standards are usually governmental bodies that always keep in mind the requirement that the limits be enforceable. A "one-shift" standard is enforceable because it can be tested by an inspector during a one-shift visit. Indeed, one could argue that it is not for the standard setter to tell the industrial hygienist how to maintain conditions "in compliance." Yet this unwillingness of standard-setting bodies to address the problems posed by exposure variability has encouraged industrial hygienists to do likewise with the consequence that long-term testing has been largely ignored.

The following analysis considers the implications of exposure variability upon the interpretation of exposure limits used in the United States as an example. Since the official limits used in many countries share the salient features of the American OELs, it is hoped that the analysis will be useful to those hygienists and standard-setting authorities who reside outside the United States. It should also be borne in mind that since formal OELs have been developed for only about 600 of the approximately 70,000 chemical substances currently used in industry, some companies have developed their own internal OELs, which are applied in much the same manner as official limits (Paustenbach and Langner, 1986).

3.1 Threshold Limit Values

Prior to passage of the Occupational Health Act of 1970 (OSH Act, 1970) limits of airborne exposure in the United States generally reflected consensus standards developed by various groups, most notably the threshold limit values (TLVs) of the American Conference of Governmental Industrial Hygienists (ACGIH). The history and development of TLVs are well documented and will not be reiterated here (see, e.g., Lanier, 1984; Paustenbach, 1985; Ulfvarson, 1987). Suffice it to say that, over the last four decades, the TLVs have had a profound influence on the practice of occupational hygiene everywhere in the world. OSHA, several states within the United States, and many other countries have adopted some or all of the TLVs as their official limits.

3.1.1 Long-Term versus Short-Term TLVs

The ACGIH defines TLVs according to criteria for monitoring over both the long term and the short term. The long-term limit, referred to as the TLV–TWA (time-weighted average), is defined as "the time-weighted average concentration for a normal 8-hour workday and 40-hour workweek" (ACGIH, 1991, p. 3). Here the phrase, "normal 8-hour workday and 40-hour workweek," emphasizes that TLV–TWAs represent long-term averages. The short-term exposure limit or TLV–STEL, on the other hand, is defined as an air concentration averaged over 15 min, which represents

the concentration to which workers can be exposed continuously for a short period of time without suffering from (1) irritation, (2) chronic or irreversible tissue damage, or (3) narcosis . . . , and provided that the daily TLV–TWA is not exceeded. It is not a separate independent exposure limit, rather it supplements the . . . TWA limit where there are recognized acute effects from a substance whose toxic effects are primarily of a chronic nature. (ACGIH, 1991, p. 3)

Two of the three criteria, that is, irritation and narcosis, are consistent with the accepted perceptions of STELs in that they refer to transient effects that might arise even though the primary effects are chronic in nature and the 8-hr TWA concentration is at or below the TLV–TWA (Henschler, 1984; Ulfvarson, 1987; Zielhuis et al., 1988). Likewise, the fact that irreversible tissue damage, such as neural death associated with anoxia, might be associated with periods of intense exposure, provides a clear rationale for STELs. However, the notion that chronic damage might depend on the rate of exposure at a given TLV–TWA is more controversial since it presumes that transmission of the transient to the target tissues is both rapid and efficient, and that the relationship between the tissue burden and damage is nonlinear (curving upward) (Rappaport, 1985; Rappaport, 1991a,b).

A third type of limit designated the TLV-C represents a "ceiling" value that should not be exceeded even instantaneously. A review of the 1991 list of TLVs suggests that ceiling limits are generally applied to substances that produce *only* acute effects or irritation.

3.1.2   Health Basis of TLVs

The ACGIH states that its limits represent "airborne concentrations . . . under which it is believed that nearly all workers may repeatedly be exposed day after day without adverse health effects" (ACGIH, 1991, p. 2). This definition of TLVs as levels that protect nearly all workers from "adverse health effects" implies that these limits are based primarily on health considerations. For many agents the paucity of exposure–response data available to, or reported by, the ACGIH has caused this interpretation to be questioned (Henschler, 1984; Zielhuis, 1988; Halton, 1988; Roach and Rappaport, 1990). The documentation supporting a particular TLV is typically a relatively brief review and listing of reference material and contains no formal analysis of exposure–response relationships (ACGIH, 1986). Some TLVs are based on animal experiments, others on reports of human experiences both in the workplace and in controlled experiments with volunteers. After examination of the original references cited in the documentation of those 1976 and 1986 TLVs, which had been based upon human experience, Roach and Rappaport (1990) reported that the risks inherent in certain TLVs were surprisingly large; on the average, between 14 and 17 percent of individuals exposed at, or in some cases below, these TLVs were adversely affected.

Some investigators pointed out correlations between the TLVs and certain physiochemical constants, such as solubility (Sato, 1987), vapor pressure (Ulfvarson, 1987), and acid dissociation constant (Leung and Paustenbach, 1988). Roach and Rappaport (1990) observed that the TLVs were correlated with the levels of exposure reported in the studies cited by the TLV committee in the documentation of its limits. They speculated, therefore, that the TLVs (at least in 1976 and 1986) may have taken into account the committee's perception that more stringent limits were unrealistic given the apparent ability of industry at the time to achieve particular levels of control. Castleman and Ziem (1988) went a step further by suggesting that the TLVs resulted from the direct influence of industry consultants upon the deliberations of the committee. But, regardless of the exact mechanism by which the limits have been set, it is clear that the TLVs have not always been set at levels that would protect nearly all workers from the health consequences of exposure.

3.1.3 Recognition of Exposure Variability

Although the TLV committee has acknowledged short-term fluctuations in air concentrations and has used this variation to justify STELs (of 15-min duration) and ceiling limits, the variability of shift-long exposures has not been addressed (Lanier, 1984; Ulfvarson, 1987; ACGIH, 1991). This neglect of variability of shift-long exposures is difficult to justify today when personal sampling equipment is readily available, but it undoubtedly conditioned much of the thinking concerning assessment of exposure in the past. That is, the TLVs were intended for and used by industrial hygienists as reference points in a world where the available data were extremely limited. Because measurements were few and could not easily be related to actual exposures, shift-long area concentrations were equated with long-term exposures of groups of workers in the vicinity. This practice, and the fact that TLVs historically have included little if any margins of safety, led health professionals to conclude that any measurement above the TLV pointed to a hazardous situation.

The above analysis suggests that compliance testing in the United States resulted primarily from two factors: (1) the inability to measure the individual exposures of large numbers of workers; and (2) the interpretation of TLVs as concentrations not to be exceeded over any full shift. The technological leap from stationary and hand-held samplers to personal sampling equipment in the 1960s (Sherwood and Greenhalgh, 1960) made it possible for industrial hygienists to measure shift-long exposures directly and, therefore, to recognize the importance of variability over shifts and between individuals. Unfortunately, the introduction of OSHA standards in 1971–72 appears to have codified compliance testing in the United States just at the time when it was possible for more sophisticated approaches to evolve.

3.2 OSHA Standards

The OSH Act of 1970 established OSHA as the official standard-setting body in the United States (OSH Act, 1970). The permissible exposure limits (PELs) established by OSHA for chemical agents fall into two categories: those originally adopted as existing standards in 1971, under a provision of the OSH Act, and those subsequently developed as new standards (OSHA, 1987a). Those adopted as existing standards included PELs for about 400 chemicals, most of which had originally been issued as TLVs by the ACGIH in 1968. OSHA recently updated its existing standards to about 600 PELs by adopting what is essentially the entire list of 1987/88 ACGIH TLVs. New OSHA standards specified exposure limits for 12 substances. As shown in Table 8.7, the new standards often include a second exposure criterion

Table 8.7 Exposure Limits Set by OSHA with New Standards[a]

Substance	PEL	AL	STEL
Asbestos	0.2 fiber/cm^3	0.1 fiber/cm^3	
Vinyl chloride	1.0 ppm	0.5 ppm	5.0 ppm
Arsenic	10 μg/m^3	5 μg/m^3	
Lead	50 μg/m^3	30 μg/m^3	
Coke oven emissions	150 μg/m^3		
Cotton dust	200 μg/m^3		
DBCP[b]	1.0 ppb		
Acrylonitrile	2.0 ppm	1.0 ppm	10.0 ppm
Ethylene oxide	1.0 ppm	0.5 ppm	
Benzene	1.0 ppm	0.5 ppm	5.0 ppm
Formaldehyde	1.0 ppm	0.5 ppm	2.0 ppm

[a] PEL is the permissible exposure limit; AL is the action level; STEL is the short-term exposure limit.
[b] 1,2-Dibromo-3-chloropropane.

called an Action Level (AL, an exposure concentration of roughly one-half of the PEL, which triggers certain requirements), and sometimes also a STEL.

3.2.1 Risk and Feasibility

Court decisions over the last two decades have laid down that OSHA may issue a new standard after demonstrating that workers are at "significant risk" of adverse effect at the existing exposure limit. Significant risk has been interpreted by OSHA in situations involving carcinogens, as an individual lifetime risk of at least 1 per 1000 (OSHA, 1987b). Having demonstrated a significant risk at an existing limit, OSHA established a new PEL that represented the lowest level thought to be economically and technologically feasible in that segment of industry that was considered least able to control exposures (OSHA, 1987b). If the residual risk at this new PEL still exceeded 1 per 1000, then the door is left open for OSHA to institute further reductions in the future.

This process, which OSHA has used since 1983 to set its new standards, is illustrated in Figure 8.4. Since the toxic agent in each case has been a carcinogen, OSHA has employed linear cancer models to demonstrate significant risk and to estimate the risk attributable to cumulative exposure at the new PEL over a working lifetime of 45 years. This latter risk (associated with exposure at the new PEL) is referred to here as the *lowest feasible risk* since it is presumed that at least one segment of industry cannot reduce average exposures below the new PEL. In most cases thus far the lowest feasible risk has also exceeded 1 per 1000. Table 8.8 lists selected PELs promulgated since 1983 and the associated levels of risk. It should be noted that both toxicological and epidemiological data were used in the risk assessments and that quite large risks were inherent in the old PELs (originally adopted as 1968 TLVs for arsenic and ethylene oxide or the comparable ANSI standard for benzene). This explicit estimation of risk and feasibility differentiates PELs from TLVs and apparently from most OELs promulgated elsewhere as well (Henschler, 1984; Zielhuis, 1988).

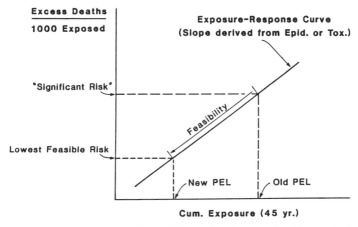

Figure 8.4 Procedure used by OSHA to develop new health standards.

3.2.2 Interpretation of PELs

Since OSHA equates risk with cumulative exposure over 45 years and considers feasibility in the context of average conditions, each new PEL represents the *average* exposure received by the individual worker over 45 years (in at least one segment of the industry). Thus, the philosophy used to establish a new PEL is generally consistent with the idea that the risk born by the ith worker is proportional to $\mu_{x,i}$ over a fixed period of time (e.g., 45 years). OSHA also noted that some segments of the industry affected by the benzene standard could consistently maintain exposures below the new PEL and, if so, that the workers would experience proportionally lower risks (OSHA, 1987b); here again, individual risk as defined by the linear model, would be proportional to $\mu_{x,i}$ over 45 years. These two fact suggest that new PELs represent the mean exposures received by individual workers in that segment of industry that is considered least able to control exposures.

Notwithstanding the fact that the standard-setting process used by OSHA equates risk with cumulative exposure, the official definition maintains the status of the PEL

Table 8.8 Determinations of Individual Risk Associated with New OSHA Standards

	Arsenic	Ethylene Oxide	Benzene
Old PEL	50 $\mu g/m^3$	50 ppm	10 ppm
New PEL	1 $\mu g/m^3$	1 ppm	1 ppm
Sig. Risks[a]	148–425	63–109	37–186
L.F. Risk[a]	2.2–2.9	1.2–2.3	4–22
Source	Epidem.	Toxicol.	Epidem.
Reference	OSHA (1983)	OSHA (1984)	OSHA (1987b)

[a] Significant risk and lowest feasible risk given as 95% confidence intervals of expected number of excess deaths per 1000 workers exposed at the PEL for 45 years.

as a level not to be exceeded for any shift-long measurement. For example, the new standard for benzene specifies the PEL in terms of the onus laid on the employer, as follows (OSHA, 1987b):

> The employer shall assure that no employee is exposed to an airborne concentration of benzene in excess of . . . 1 ppm as an 8-hour time-weighted average. (OSHA, 1987b)

While this statement can reasonably be interpreted to mean that no employee should be exposed to more than 1 ppm of benzene *on the average*, such an interpretation has thus far been rejected (OSHA, 1987b). OSHA similarly specified a STEL for benzene, stating that

> The employer shall assure that no employee is exposed to an airborne concentration of benzene in excess of . . . 5 ppm as averaged over any 15 minute period. (OSHA, 1987b)

Some of this rigidity in interpretation of OELs is undoubtedly a throwback to earlier days when exposure data were rare because measurements were so difficult to make. Today, however, it hard to reconcile a requirement for absolute compliance with the availability of personal-sampling equipment.

3.2.3   From TLVs to PELs

This analysis of OELs in the United States indicates that the process of transition from TLVs to PELs was uneven (Rappaport, 1984). On the one hand, techniques for the setting of new limits became more sophisticated, as is shown by the explicit recognition of individual risk and of the need to ensure feasibility, two factors that, in the context of chronic disease, depend primarily on the mean exposure received by the individual worker over time. Yet, on the other hand, the methods offered to the industrial hygienist for determining compliance with OELs still rely upon interpretation of the limits as levels not to be exceeded over any full shift. As will be seen, this separation of standard setting from a consistent interpretation of "compliance" greatly hinders the long-term assessment of exposures.

3.3   Working Limits

If statistical methods are to be adopted for assessing long-term exposures in the workplace, it is essential that industrial hygienists become aware of the underlying premises and shortcomings of the OELs that they employ. As suggested above, some limits, including many TLVs, contain little or no margins of safety. Likewise, new OSHA standards, which are based on the feasibility of achieving a level in that segment of industry that is least able to control exposure, are not sufficiently protective in other sectors where exposures can be consistently better controlled. Thus, it may be necessary to adopt working limits that incorporate safety factors to allow for uncertainties in the underlying risk to health and for the need to reduce air concentrations to the lowest levels feasible in all segments of industry.

Each safety factor should be chosen on the basis of a reasoned judgment concerning the biological effects, the quality of data supporting the OEL, and the circumstances under which exposures arise (Zielhuis and van der Kreek, 1979a,b; Dourson and Stara, 1983). Large companies in the United States, for example, are capable of maintaining exposures at most facilities well below a new PEL and could choose a working limit of, say, a half to a fourth of the PEL on the basis of feasibility alone. Many have adopted corporate exposure guidelines that do this. More marginal facilities, which might be unable to achieve the desired levels of control, should strive for incremental reductions of exposure over time. In such situations it might initially be necessary to test relative to OEL = PEL. Then annual testing could be performed, with concurrent adoption of controls, until the target level, for example, OEL = PEL/4, is achieved.

4 ESTIMATION OF EXPOSURE

The methods required for collection of exposure data and estimation of the parameters of the exposure distribution(s) are relatively straightforward insofar as personal measurements are obtained by some means of random sampling. The need for random samples cannot be overemphasized since biases introduced by nonrandom collection procedures (e.g., sampling only "worst cases" or during discreet campaigns of a few days) not only can invalidate inferential testing but also can lead to interpretations of the data in general that are of doubtful validity (Ulfvarson, 1983). This is not to say that only purely random sampling needs to be performed over individuals and time, but rather that randomization should be included as far as is practical given the many constraints involved in a particular situation. Any basic statistical text should provide adequate guidance concerning accepted procedures for random sampling, with and without stratification according to factors such as level of exposure, facilities, and time, so that the means of reducing bias from the collection of exposure data is within the grasp of the industrial hygienist.

Since budgetary constraints always loom large in a sampling program, every effort should be expended to increase sample sizes at a given cost. For example, the use of passive monitors for gaseous contaminants can easily increase the number of measurements collected during a survey by 10–fold over pump-and-collector systems. Likewise, it is often possible for measurement to be performed by technicians or by on-site personnel rather than by the industrial hygienist who might only be able to visit the site every 1–5 years. Also, it is generally wise to monitor exposures for a whole shift rather than for short periods since this allows greater numbers of workers to be included in the sample. If short-term sampling is necessary, then the industrial hygienist should be wary of evaluating only a single exposure under the assumption that it represents a worst case; indeed, given the many sources of short-term variability in the workplace, it is advisable to investigate the population of worst cases by suitable schemes of random sampling. Finally, any compromise between sample size and precision of monitoring should come down heavily on the side of larger numbers of measurements since environmental variability is almost always much greater than the error of measurement. In fact, there is a critical need for inexpensive monitoring procedures of modest precision (e.g., $CV_e \leq 50\%$) that require a minimum of calibration and interpretation and that can be used by individuals without professional qualifications.

5 CLASSIFICATION OF THE POPULATION

Suppose that all of the workers in a group are "uniformly exposed," that is, the mean exposure for each individual is the same and, therefore, the same as the mean for the entire group. In this case the only effective interventions are those that lower exposure for everyone, such as changes in tasks, in the process, or in ventilation.

If workers are not uniformly exposed, on the other hand, then some are exposed on the average to higher concentrations than are others because of differences in their tasks, locations, or individual work practices. This is, of course, why industrial hygienists attempt to identify highly exposed individuals and to ascertain whether their exposures are in the acceptable range thereby placing an upper bound on the exposure for the entire group (Roach et al., 1967; Leidel et al., 1977). Unfortunately, this practice of investigating only high-risk workers presumes that such individuals can be identified by inspection; this is not always the case. Moreover it provides little information concerning the population as a whole and can lead to dubious groupings for workers at low and intermediate levels, so that it can obscure important exposure–response relationships if questions of possible health effects should arise at some time in the future.

In order to deal with the issue of uniformity of exposure, the sampling strategy should allow levels of exposure to be classified *across* the population at risk by establishing groups that possess particular characteristics of exposure. Two procedures, which will be described, respectively, as the "observational" and the "sampling" approaches, are available.

5.1 Classification by Observation

The observational approach requires that workers' exposures be classified qualitatively on the basis of common environmental variables related to the process, tasks, job title, and so forth; measurements are then performed to provide quantitative estimates of exposure within each group. Thus, classification precedes the estimation of exposure. Corn and Esmen (1979) attributed the approach to Woitowitz et al., (1970), in the context of retrospective assessments. Subsequent refinements (Corn and Esmen, 1979; Esmen, 1984) resulted in prospective assignments of workers into common groups, referred to as *exposure zones*, for subsequent sampling. Corn and Esmen devised sophisticated schemes that included both occupational titles and a generalized mechanism for categorizing tasks, processes and workplaces.

Since groups are assigned on the basis of inspection and analysis of records and occupational titles, the observational approach requires a great deal of the industrial hygienist's time and professional judgment in order to sort out successfully the many environmental variables that come into play. In fact, it minimizes the number of personal measurements by assigning the same mean exposure to all individuals within a zone. Esmen (1984) even went so far as to warn the industrial hygienist against revising a priori classifications on the basis of measurement data since "it is highly unlikely that a readjustment of zones predicated on results will improve the outcome." It is important, therefore, that the group be uniformly exposed; otherwise misclassification will occur because groups that are apparently distinct, in fact, have overlapping exposure distributions.

5.2 Classification by Sampling

The sampling approach requires multiple measurements of representative workers' exposures to determine the within-worker and between-worker components of variance. Although formal structures for this approach have only recently been described (Spear et al., 1987; Kromhout et al., 1987; Boleij et al., 1987; Rappaport et al., 1988b; Rappaport, 1991a,b), the essential elements were recognized by Oldham and Roach (1952) in an investigation of dust exposures in Welsh coal mines more than 40 years ago. Since workers are assigned to groups on the basis of measurements, the estimation of exposure precedes classification and, because the hygienists' time is devoted primarily to data collection and analysis rather than to inspection, the amount of data is maximized.

5.3 Definitions of Groups

To add structure to the classification of groups, the definition of a *monomorphic group*, alluded to earlier, will again be employed to designate a collection of individuals whose mean exposures can be adequately described by a single lognormal distribution (Rappaport, 1991a). On this basis the four groups of workers introduced in Figures 8.1 and 8.3 constitute monomorphic groups. Since the between-worker distribution is the key to assessing exposures of groups of workers, the concept of a monomorphic group is particularly relevant to the sampling strategy.

The definition of a *uniformly exposed* group will now be defined quantitatively as a monomorphic group in which 95 percent of the individual mean exposures lie within a factor of 2. This definition implies that the ratio of the 97.5th percentile to the 2.5th percentile, denoted by $_BR_{.95}$, is not greater than 2. Thus, $_BR_{.95} = [\exp(\mu + 1.96\sigma_B)/\exp(\mu - 1.96\sigma_B)] = \exp[3.92\sigma_B]$ and a uniformly exposed group would be a monomorphic group in which $_B\sigma_g \leq 1.2$. This definition of a uniformly exposed group is admittedly arbitrary, and some might argue that a factor of 2 is too restrictive. Other factors could be used if desired. For example, a grouping such as that recommended in the United Kingdom (HSE, 1984), where individual (mean) exposures lie within a range of one-half to two times the group mean, would correspond to a monomorphic group with $_BR_{.95} \leq 4$ and $_B\sigma_g \leq 1.4$. Returning to the four groups of exposures described earlier, the factors that would contain 95 percent of the populations of mean exposures are estimated in Table 8.9, where $_B\hat{R}_{.95} = \exp(3.92_BS_y)$ is the estimate of $_BR_{.95}$ in each case. The calculations indicate that groups 1 and 2 where $_B\hat{R}_{.95} = 1.7$ and 2.0, respectively, would be uniformly exposed as defined above; however, groups 3 and 4 with $_B\hat{R}_{.95} = 392$ and 8.3, respectively, would not be uniformly exposed. These results are qualitatively consistent with the

Table 8.9 Between-Person Variability for Four Groups of Workers (Groups Defined in Table 8.1)

Group	k	$_BS_g$	$_B\hat{R}_{.95}$	Uniformly exposed
1	6	1.16	1.8	Yes
2	18	1.20	2.0	Yes
3	24	4.67	421	No
4	19	1.72	8.3	No

Legend: k is the number of workers in the sample; $_BS_g$ is the between-person geometric standard deviation; $_B\hat{R}_{.95}$ is the estimated factor that would contain 95% of the mean exposures from the between-person distribution.

ratios of the between-person to total variance, given in Table 8.3, which indicated much higher proportions of the between-person variance to the total variance for groups 3 and 4 compared to groups 1 and 2.

5.4 Observation versus Sampling

Returning now to the merits of the observational and sampling approaches for assigning workers into groups, consideration of either the estimated factor $_B\hat{R}_{.95}$ or $_BS_g$ as a measure for uniform exposure offers considerable insight. If $_B\hat{R}_{.95}$ or $_BS_g$ is small, then exposure is governed primarily by the process or by environmental conditions shared by all individuals. In these cases, the observational approach should provide valid assessments of exposure. If, on the other hand, $_B\hat{R}_{.95}$ or $_BS_g$ is large, it is likely that individual tasks or practices are important determinants of exposure; such situations might overwhelm the industrial hygienist's observational skills leading to improper classifications.

Since two of the four groups evaluated herein were uniformly exposed (see Table 8.9), it is clear that the observational approach can establish uniformly exposed groups. However, since the other two groups were not uniformly exposed, it is also apparent that the observational approach cannot assure uniform exposure across a group. If these results are indicative of those that would be observed from a larger cross section of workplaces, and preliminary analyses suggest that they are (Rappaport, 1991a,b), then one must question the ability of the industrial hygienist to assign groups solely on the basis of observation. Thus, the actual partitioning of exposure variability by the sampling approach may be not only more efficient (because of reduced professional time) but also less equivocal in many situations. A hybrid scheme that employs a relatively crude observational component followed by a thorough plan for random sampling may well provide the optimal approach to classification of groups.

6 TESTING RELATIVE TO LIMITS

Some alternatives will now be considered for determining whether exposures are acceptable relative to limits. Of the possible tests that might be applied for this purpose, only the three most prominent examples will be considered here, namely, compliance testing, testing the probability that the OEL will be exceeded (i.e., the probability of "overexposure"), and testing the mean exposure relative to the OEL. Although the analysis will focus primarily upon measurements over the whole shift, evaluations relative to STELs will also be mentioned.

6.1 Criteria for Evaluating Strategies

Since each test approaches the problem from a different perspective, objective criteria are needed to evaluate the relative strengths and weaknesses. An ideal test would have the following characteristics:

1. It would provide for rigorous evaluation in the sense that it would test a specific hypothesis, relating some characteristic of the exposure distribution with the appropriate OEL. To discourage the practice of declaring situations acceptable by default, it should be assumed that exposures are unacceptable till proved otherwise (Roach et al., 1967; Coenen and Riediger, 1978; Selvin et al., 1987; Rappaport and Selvin, 1989).

2. It would encourage monitoring. Data should be required to declare exposure acceptable; the amount of data should be greatest in marginal situations.

3. It would be powerful, identifying acceptable exposures with accessible numbers of measurements. In this context, a sample size of $n = 50$ is offered as a practical maximum for an initial determination.

4. It would lead to decisions that relate to the degree of hazard. In other words, the likelihood of declaring exposures acceptable should be directly related to the "risk" of adverse effect.

6.2 Outcome versus Risk

In order to relate compliance with the risk of disease, it is necessary to know what characteristic(s) of the exposure distribution conditions the risk borne by the individual worker and the group. Regarding acute exposures (which can lead to either death or serious effects), it seems clear that risk is primarily related to the likelihood that very large air concentrations will be encountered. Thus, it is theoretically possible to assess exposures to acutely toxic substances by focusing upon the extreme right tail of the distribution (Rappaport et al., 1981b). However, it is naive to suppose that one can accurately predict the likelihood of exposures at levels that are 100 or even 1000 times greater than the mean exposure without collecting very large numbers of measurements and without making tenuous assumptions regarding the underlying distribution of exposures (Rappaport, 1988; 1991a). Thus, it seems unwise to consider sampling strategies in terms of personal monitoring of individuals when the agent in question produces only acute effects. Rather it appears more appropriate, in these circumstances, to continuously monitor either the source of exposure or the air in the vicinity of the worker and to implement practices that will protect the individual from accidental releases of these contaminants. It follows from this that ceiling limits, which relate to substances producing acute effects exclusively, can be implemented rigorously *only* through continuous monitoring of air contaminants over time scales of seconds to minutes.

When considering the risks of chronic disease, it seems unlikely that risk would be unduly influenced by transient exposures of one shift or less but rather should be related to the long-term exposures received from year to year (Rappaport, 1991a,b). Indeed, such reasoning is embodied in the development of new standards by OSHA where it is implicit that the risk of disease is strictly proportional to the cumulative exposure over 45 years (see Fig. 8.4); in this context an individual's risk can be stated simply as

$$\mathrm{Pr}_{i,t} = C\mu_{x,i}t \tag{15}$$

where $\mathrm{Pr}_{i,t}$ represents individual risk after t years of exposure (which is 45 years in this case) and C is a constant of proportionality. If all individuals within a group are equally susceptible to the disease, then it follows from Eq. (15) that the mean risk of the group as a whole, after t years of exposure, should be proportional to the mean exposure μ_x, that is,

$$\mu_{\mathrm{Pr},t} = C\mu_x t \tag{16}$$

This notion that the risk of chronic disease is proportional to either μ_x times t (as assumed in OSHA's new standards) or to the weighted sum of mean exposures averaged every year or so (if either latency or the serial order of exposures is important) is central to the logical development of sampling strategies and will be used in subsequent sections of this chapter to relate outcomes of the various tests with risk.

6.3 Testing Compliance

The usual test for compliance involves direct comparison of the largest of a few measured exposures with an OEL, which has been interpreted as an air concentration not to be exceeded. As noted earlier, this approach is embodied in the wording of OSHA standards in the United States and undoubtedly in other official limits as well. The essential elements of the test, as well as particular applications to OSHA standards, have been analyzed (Leidel et al., 1977; Tuggle, 1981; Rock, 1981; 1982; Rappaport et al., 1981a,b; 1984).

6.3.1 The Probability of Compliance

Two criteria are used to determine the outcome of the test, namely, the probability that a given exposure would exceed the limit, γ, and the number of measurements. The probabilities of compliance $P(\mathrm{Comp.})$ and of noncompliance $P(\mathrm{Noncomp.})$ are easily obtained from the following expressions (Rappaport, 1984):

$$P(\mathrm{Comp.}) = (1 - \gamma)^N \tag{17}$$

$$P(\mathrm{Noncomp.}) = 1 - P(\mathrm{Comp.}) \tag{18}$$

and are illustrated in Figure 8.5 for sample sizes between 1 and 20. Although the OEL would generally be a shift-long limit, these relationships can also be used to

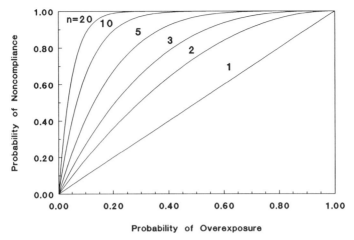

Figure 8.5 Relationship between the probability of overexposure (γ_{OEL}) and the probability of noncompliance. From Eq. (18).

evaluate compliance of short-term exposures relative to STELs. If an action level (AL) is also operative, additional rules and monitoring requirements essentially require that the probability of exceeding the AL (usually half of the PEL) be used in the above expressions. This arises from increases in sample size that are required by OSHA whenever an initial measurement is found to be above the AL but less than the PEL (Rappaport, 1984).

Since $0 < \gamma < 1$, Eqs. (17) and (18) make explicit what is known intuitively, that any population of exposures can be declared out of compliance given a large enough sample. Indeed, Figure 8.5 shows that sample size is often the most important single determinant of compliance. Since N is totally unrelated to the health hazard, this illustrates a weakness of traditional practices. It also follows from Eq. (17) that compliance can be maximized merely by minimizing the number of measurements [i.e., $P(\text{Comp.}) = 1$ for $N = 0$].

To illustrate the importance of sample size on compliance testing, consider again the data presented previously for the four groups of workers. In order to test compliance of these groups, the estimated probability of overexposure, designated $\hat{\gamma}$, is needed. Since it is assumed that the distributions of exposures are lognormal, then it is reasonable to estimate γ by substituting the estimates of μ and σ ($\overline{Y}$ and $_T S_y$) into Eq. (6) along with the operative OEL, either a PEL or an AL. For example, using the new OSHA standard for inorganic lead, which includes an AL of 0.03 mg/m³, this probability is estimated to be

$$\hat{\gamma}_{AL} = P\left\{ Z > \frac{\ln(\text{AL}) - \overline{Y}}{S_y} = z_{(1 - \hat{\gamma})} \right\}$$

$$= P\left\{ Z > \left[\frac{\ln(0.03) - (-4.179)}{0.630} \right) = 1.065 \right] \right\}$$

$$= 0.143$$

The estimated values of γ are summarized in Table 8.10 for all four groups and their respective OELs. The probabilities of noncompliance can easily be estimated by substituting values of $\hat{\gamma}$ into Eqs. (17) and (18). For example, for group 1, $P(\text{Noncomp.}) \simeq 0.14, 0.36, 0.53$, and 0.95 for $N = 1, 3, 5$ and 20, respectively. So an assessment with five measurements would more likely than not turn up a value above the AL; then, additional monitoring would eventually lead to a measurement above the PEL. A decision as to whether exposures of members of group 1 are "in compliance" would, therefore, be reduced to the following incredible statement: If an initial survey included fewer than five measurements, then compliance should be the outcome and if it included five or more, noncompliance is more probable. In the case of group 2 $[\hat{\gamma}(\text{AL}) = 0.102]$, 7 measurements would be required for noncompliance to be more likely; for group 3 $[\hat{\gamma}(\text{AL}) = 0.385]$, two measurements would be required; and, finally, for group 4 $[\hat{\gamma}(\text{PEL}) = 0.203]$, three measurements.

6.3.2 Compliance versus Risk

Since "compliance" is determined by the probability that a measurement will exceed the OEL at a given sample size [according to Eq. (17)], and since "risk" is assumed here to be related to the mean exposure received by the individual or the group [according to Eqs. (15) and (16)], then the relationship between compliance and risk is at best confused (Rappaport, 1984; Spear and Selvin, 1989). In order to investigate the relationship, consider two groups of workers (designated a and b that have equal probabilities of overexposure, that is, where $\gamma_a = \gamma_b = \gamma$. Let RR = Risk (a)/Risk (b) represent the ratio of risks borne by two groups. If the distributions of exposures are lognormal, then RR can be related to the probability of overexposure as follows,

$$\text{RR} = \mu_{\text{Pr},a}/\mu_{\text{Pr},b} = \mu_{x,a}/\mu_{x,b} = \exp\left[(\sigma_b - \sigma_a)(z_{(1-\gamma)} - 0.5(\sigma_a + \sigma_b))\right] \quad (19)$$

Furthermore, let group b represent a reference group that receives a distribution of exposures with minimal variability, both within and between persons. Based on preliminary analyses of a variety of industrial data sets, a group where $\sigma_b = \ln(1.2)$

Table 8.10 Probabilities of Overexposure for Four Groups of Workers Relative to OSHA Standards (Groups Defined in Table 8.1)

Group	PEL (mg/m^3)	AL (mg/m^3)	$\overline{Y}$	$_TS_y^2$	$\hat{\gamma}_{\text{PEL}}$	$\hat{\gamma}_{\text{AL}}$	$\hat{\theta}$ (PEL)
1	0.05	0.03	−4.179	0.401	0.030	0.143	<0.001
2	3.2	1.6	−0.429	0.499	0.012	0.102	<0.001
3	3.2	1.6	−0.221	5.389	0.276	0.383	0.532
4	213	—	4.614	0.805	0.203	—	0.182

Legend: $\overline{Y}$ is the estimated mean of the log-transformed exposures; $_TS_y^2$ is the estimated total variance of the log-transformed exposures; $\hat{\gamma}_{\text{PEL}}$ and $\hat{\gamma}_{\text{AL}}$ are the probabilities that exposure during a random work shift would exceed the respective OEL; $\hat{\theta}_{\text{PEL}}$ is the probability that a random worker from the group would have a mean exposure above the PEL.

= 0.182 appears to represent the minimum level of variability that is encountered in practice (Rappaport, 1991b). Since exposure of this reference group (b) is very nearly constant at the OEL, any increase in variability of exposure received by the group in question (indicated by $\sigma_a > 0.182$) will be reflected by a change in the RR at a given value of γ.

Table 8.11 illustrates the relationship shown in Eq. (19) between the RR and σ_a at various values of γ. Each row in Table 8.11 represents differences in the RR that are associated with increasing variability in the exposure distribution at a given probability of compliance. [Note: $P(\text{Comp.}) = (1 - \gamma)^{1/N}$]. The table shows that the nature of this relationship depends on the value of γ. If γ is small, the risk borne by the reference group is large relative to the group in question and the RR *decreases* with increasing variance (σ_a); for example, when $\gamma = 0.01$, a group with highly variable exposure ($\sigma_a = 1.386$) would be at about 16 percent of the risk of the reference group ($\sigma_b = 0.182$). However, when γ is large (i.e., greater than about 0.4), the situation is reversed and the RR *increases* with increasing variance; for example, when $\gamma = 0.80$ a group with $\sigma_a = 1.386$ would have a RR of 7.08 indicating a risk that is 7-fold greater than that of the reference group. (This example also illustrates that if $N = 1$, there would be one chance in five that a group with a mean exposure of seven times the OEL would be declared in compliance). Indeed, the risk ratios associated with the most variable exposures in Table 10.11 ($\sigma_a = 1.86$) cover a 45-fold range between $0.01 < \gamma < 0.80$!

The examples show that the risks of groups of workers can be very different even though compliance testing would lead to the same decision regarding the exposures. The implications of the relationship between compliance and *individual* risk depend on how σ^2 is partitioned into the within-and between-worker variances. If the group is uniformly exposed, then each individual has the same mean exposure, that is, $\mu_{x,1} = \mu_{x,2} = \cdots = \mu_{x,k} = \mu_x$, and individual risk and group risk become the same. Then there would be no clear relationship between compliance and individual risk just as was shown in Table 10.11 for the group as a whole.

Table 8.11 Risk Ratios (RR)[a] of Pairs of Distributions with the Same Probability of Overexposure (γ)

	Risk Ratio (RR)[b]		
γ	$\sigma_a = 0.693$	$\sigma_a = 1.099$	$\sigma_a = 1.386$
0.01	0.378	0.213	0.156
0.02	0.437	0.275	0.218
0.05	0.538	0.398	0.355
0.10	0.648	0.556	0.549
0.20	0.813	0.832	0.933
0.40	1.100	1.426	1.895
0.80	1.932	3.880	7.082

[a]Risks are given relative to a reference group with $\sigma_b = 0.182$.
[b]Calculated according to Eq. (19).

However, if the group is not uniformly exposed, then some individuals are at greater risk than others because they are exposed, on the average, to higher levels of air contaminants. In this context the dispersion of risk across the group is related to σ_B. Indeed, when σ_B approaches σ, then θ becomes large and, under some circumstances, can even exceed γ, indicating that the proportion of individual mean exposures in excess of the OEL is greater than that of daily exposures (Spear and Selvin, 1989). As shown in Table 8.10, this appears to be the situation with group 3. If a decision were based on a single measurement in such a case, as might happen during an OSHA inspection, then a determination of noncompliance would suggest a greater-than allowable mean exposure (of some individuals) and the associated risk. As a simple expedient, therefore, compliance testing might be useful to the official inspector who cannot measure more than a trivial number of exposures. Unfortunately, as sample sizes increase to values that would reasonably be required to define the exposure distribution, P(Noncomp.) inevitably becomes large regardless of the magnitude of the mean exposure (Fig. 8.5); at this point compliance testing breaks down as a useful tool.

6.3.3 Summary

Several shortcomings make compliance testing a poor choice for evaluating long-term exposures. First, hypotheses can neither be postulated nor tested since the sample size is not controlled. Second, compliance is not necessarily related to risk because distributions with very different mean exposures can have the same probability of compliance. Third, monitoring is discouraged since the probability of compliance decreases as the number of measurements increases. Finally, the test artificially increases the importance of measurement error since decisions hinge on comparison of a single (largest) value with the OEL (Leidel, et al., 1975; Leidel and Busch, 1985). Thus, even though this error term is usually a trivial component of total variation [e.g., if $CV_x = 1.0$ ($\sigma^2 = 0.693$) the typical precision of measurement, where $CV_e \leq 0.10$ ($\sigma^2 \leq 0.00995$) (Leidel et al., 1977), only contributes $[0.00995/(0.693 + 0.00995)] = 0.014 = 1.4\%$ to the total variance], it might appear important when the largest air concentration is only slightly greater than the OEL.

6.4 Testing the Probability of Overexposure

A more sophisticated testing structure emerges by defining an acceptable exposure distribution as one in which the probability of overexposure $\gamma < P$, where P represents some small probability, usually 0.05 (Ulfvarson, 1977; Rappaport et al., 1981b; Tuggle, 1982; Leidel and Busch, 1985; Selvin et al., 1987; Matsunaga et al., 1989). While this approach retains γ as a decision-making element, and can, therefore, be regarded as an outgrowth of compliance testing, rigorous tests of hypotheses can be conducted via one-sided tolerance limits (Selvin et al., 1987). For example, the following testing structure can be used for prospective assessments.

$$H_0: \gamma = P \text{ (unacceptable)}$$

$$H_1: \gamma < P \text{ (acceptable)}$$

where H_0 and H_1 symbolize the *null hypothesis* and the *alternative hypothesis*, respectively. If one is able to use data to reject the null hypothesis, that γ is equal to P (and implicitly greater than P), in favor of the alternative, that γ is less than P, then exposure can be declared acceptable at a given level of significance.

The test statistic is the parametric one-sided tolerance limit, defined as

$$T_u = \overline{Y} + KS_y \tag{20}$$

where K is the tolerance factor chosen such that there is probability α that a lognormal distribution with $\gamma = P$ would be falsely declared acceptable (Selvin, et al., 1987). If $T_u < \ln$ (OEL), then H_0 is rejected at an α level of significance and exposure is declared acceptable.

6.4.1 Sample Size Requirements

The test of γ, when conducted with parametric one-sided tolerance limits, encourages monitoring since the power of the test increases with sample size (in this context the power of the test is the probability of declaring exposures acceptable when $\gamma < P$); this is a direct consequence of the hypothesis-testing structure. Unfortunately, the power of the technique is limited. The approximate sample size (actually an underestimate of N) required to provide a power of $(1 - \beta)$ is given by

$$N \simeq \frac{[z_{(1-\alpha)} + z_{(1-\beta)}]^2}{[z_{(1-P)} + z_\gamma]^2} \tag{21}$$

where the four z terms represent values from the standard normal distribution associated with probabilities $(1 - \alpha)$, $(1 - \beta)$, $(1 - P)$, and γ (Selvin and Rappaport, 1988).

Table 8.12 lists approximate sample size requirements derived from Eq. (21) for $P = 0.05$, $\alpha = 0.05$, and various values of γ and $(1 - \beta)$. It is clear that large sample sizes $(N > 50)$ are required to achieve a 90–95 percent level of power for values of $\gamma > 0.01$. For example, the exposures of group 1 with $\hat{\gamma}$(PEL) $= 0.03$ would require 156 measurements to assure a decision of acceptable exposure with a

Table 8.12 Approximate Sample Size Requirements of the Test of Overexposure for $P = \alpha = 0.05^a$

γ	Sample Size N^b	
	$(1 - \beta) = 0.90$	$(1 - \beta) = 0.95$
0.01	19	24
0.02	53	66
0.03	156	196
0.04	778	982

[a] For legend see List of Symbols.
[b] Calculated according to Eq. (21).

probability of 90 percent. Therefore, if $P = 0.05$, this test can be viewed as practical only in situations where $\gamma < 0.01$ as is the case for group 2; then, about 24 measurements would be required to declare exposure acceptable with a power of 90 percent. Groups 3 and 4, with estimated values of γ of 0.275 and 0.203, respectively, are clearly greater than $P = 0.05$ and could not be declared acceptable with a probability greater than α.

It should be noted that nonparametric tolerance limits can also be used to test γ (Leidel et al., 1977) by comparing the largest measurement to the OEL when the sample size has been chosen a priori on the basis of the desired values of α and P, that is,

$$N = \ln{(\alpha)}/\ln{(1 - P)} \tag{22}$$

If all measurements are less than the OEL, then exposure would be declared acceptable. Thus, the probability of declaring exposure acceptable $P(\text{Acc.})$ with nonparametric tolerance limits is analogous to $P(\text{Comp.})$ and can be calculated from Eq. (17) for given values of γ and N. A few calculations employing Eqs. (17), (18), and (22) quickly demonstrate that requirements for large sample sizes and a lack of power essentially preclude nonparametric tolerance limits from practical consideration. For example, from Eq. (22) it is seen that 59 measurements are required to test H_0: $\gamma = P = 0.05$ at a significance level of 0.05. By employing Eq. (18) with $(N) = 59$, it becomes obvious that even when $\gamma \ll 0.05$, the probability of declaring acceptable exposures to be unacceptable would still be large.

6.4.2 Outcome versus Risk

The test statistic T_u [from Eq. (20)] combines the mean and the variance of the distribution into a single quantity. Thus, all distributions that share a common value of γ (the probability of overexposure) are treated the same regardless of the particular combinations of means and variances that are represented (Selvin et al., 1987). This can again lead to difficulties in relating outcome with risk depending on how exposure variability is partitioned across the group.

If the group is uniformly exposed, then γ relates to the frequency with which days exceed the OEL and a distribution where $\mu_x > \text{OEL}$ cannot be declared acceptable with probability greater than α. This is easily illustrated by referring to Eqs. (8) and (10) where it can be shown that, under the null hypothesis where $\ln{(\text{OEL})} = \mu + z_{(1 - P)}\sigma$, the maximum value that the mean exposure from a lognormal distribution can assume is, $\mu_{x,\max} = 0.5z^2_{(1 - P)}(\text{OEL})$; thus, when $P = 0.05$, $\mu_x < 0.26(\text{OEL})$ regardless of the value of σ. This shows that application of the test of γ to a uniformly exposed group is always conservative from the health perspective. Even so, as shown in Table 8.11, lognormal distributions with different values of μ_x (and the associated levels of risk) can share the same value of γ. For example, when $\gamma = 0.01$ and $\sigma_a = 1.386$, RR $= 0.156$ indicating that group a is at about 16 percent of the risk of group b where $\sigma_b = 0.182$.

If the test of overexposure is applied to a monomorphic group with a large between-worker variance, then it relates directly to the probability θ that an individual mean exposure would exceed the OEL. In this context, risk and outcome are closely related since large values of θ indicate large probabilities that some individuals in the group have mean exposures that exceed the OEL. While such applications of the test to the between-worker distribution appear appropriate, none have thus far been reported.

6.4.3 Application to STELs

Since frequencies with which exposure limits are exceeded are either explicitly or implicitly built into STELs (Rappaport et al., 1988a), there is a rationale for applying the test of γ to short-term exposures. For example, the STELs recommended by the ACGIH refer to 15-min air concentrations that should not be exceeded more than 4 times in an 8-hr shift (ACGIH, 1991). In this case, a null hypothesis of H_0: $\gamma = P = 4/32 = 0.125$ can be tested against the alternative, H_1: $\gamma < 0.125$. Table 8.13 presents approximate sample size requirements to perform such determinations at a significance level of 0.05 and with 90 or 95 percent power. In this situation all measurements would be of 15-min duration.

Caution must be exercised concerning the timing of short-term measurements since exposures are likely to have significant autocorrelation within shifts, particularly in continuous, indoor operations. Since such correlation of serial values would disrupt the assumption of independence, it can be necessary to collect short-term data over several days to achieve the desired sample sizes.

6.4.4 Summary

The test of the probability of overexposure provides a mixed bag to the industrial hygienist. On the one hand, it allows acceptable exposure to be defined in a manner that can be rigorously tested and provides sufficient data to investigate the exposure distribution. Likewise, the probability of overexposure is related to the risk received by the population inasmuch as overexposure is interpreted relative to the between-worker distribution, that is by θ rather than to γ as is typically suggested. But, on

Table 8.13 Approximate Sample Size Requirements of the Test of Overexposure Relative to a STEL for $P = 0.125$ and $\alpha = 0.05^a$

	Sample Size n^b	
γ (STEL)	$(1 - \beta) = 0.90$	$(1 - \beta) = 0.95$
0.01	7	8
0.02	11	14
0.04	24	30
0.06	53	66
0.08	132	167
0.10	492	622

[a] For legend see List of Symbols.
[b] Calculated according to Eq. (21).

the other hand, the inherent lack of power leads to large type II errors in many (if not most) situations of real interest.

6.5   Testing the Mean Exposure

The preceding analysis indicates that a difficulty both with compliance testing and with the test of overexposure is the uncoupling of the assessment of exposure from the evaluation of hazard in some circumstances. One way to circumvent this problem is to make inferences concerning the mean exposure μ_x received by the uniformly exposed group during each period of investigation.

The notion that exposure should be evaluated in terms of the mean air concentration is certainly not new. It was argued in Great Britain in the early 1950s that, because of the cumulative effect of coal dust on respiratory function, exposure to coal dust should be assessed in terms of the mean exposure received over time (Oldham and Roach, 1952; Roach, 1953; Long, 1953; Wright, 1953). During the subsequent 10–15 years, attention shifted to statistical methods that might be applied to estimate long-term averages when the underlying distribution of exposures is lognormal (Tomlinson, 1957; Juda and Budzinski, 1964, 1967; Coenen, 1966, 1971) and to the consideration of exposures to radioactive particles, for which cumulative exposure is also important (e.g., Sherwood, 1966; Langmead, 1971). More recent work has focused on the appropriateness of the mean exposure for determining acceptable conditions in general (Rock, 1981, 1982; Rappaport, 1985, 1988; Rappaport et al., 1988a) and upon statistical methods for testing the means of lognormally distributed exposures relative to limits (Coenen and Riediger, 1978; Galbas, 1979; Rappaport and Selvin, 1987; Evans and Hawkins, 1988). Strategies for evaluating mean exposures relative to OELs have been applied to underground mines in the United States (Corn, 1985; Corn et al., 1985) and for monitoring long-term exposures to hazardous substances in the Federal Republic of Germany (Heidermans et al., 1980; Riediger, 1986).

6.5.1   Hypothesis Testing

The following method can be used for testing hypotheses concerning means from a lognormal distribution (Rappaport and Selvin, 1987),

$$H_0: \mu_x = \text{OEL (unacceptable)}$$

$$H_1: \mu_x < \text{OEL (acceptable)}$$

A test statistic,

$$T_m = (\bar{X}_m - \text{OEL})/S_{\bar{X}_m}$$

is constructed where $S_{\bar{X}_m}$ is the standard error of $\bar{X}_m$ under the condition stated in H_0 where $\mu_x = \text{OEL}$, which is given by

$$S_{\bar{X}_m} = [\mu_x^2(S_y^2 + 0.5S_y^4)/(N - 2)]^{1/2}$$

The statistic T_m has approximately a t distribution with $N - 1$ degrees of freedom. If $T_m < t_{\alpha, N-1}$, then exposures are declared acceptable at an α level of significance.

6.5.2 Sample Size Requirements

The approximate sample size required for this test of the mean exposure from a lognormal distribution is given by the following expression (Rappaport and Selvin, 1987):

$$N \simeq \frac{(z_{(1-\alpha)} + z_{(1-\beta)})^2 (\sigma^2 + 0.5\sigma^4)}{[1 - (\mu_x / \text{OEL})]^2} \tag{23}$$

Table 8.14 lists the sample sizes needed to test the mean exposure against the OEL at a significance level of 0.05 and with 90 percent power. Depending on the variability, it should be possible to apply the test to situations in which μ_x is between a tenth and a half of the OEL with fewer than 50 observations. This can be illustrated by referring again to the exposures received by the four groups of workers. Assuming in each case that $\mu_x < \text{PEL}/2$ and could, therefore, be assessed with modest numbers of measurements. If it is also assumed that $\sigma \simeq {}_TS_y$ then, from Eq. (23), it can be estimated that 12 measurements would be required to reject H_0 at $\alpha = 0.05$ and $(1 - \beta) = 0.90$ both for group 1 ($\overline{X}_m / \text{PEL} = 0.372$, ${}_TS_y = 0.629$) and for group 2 ($\overline{X} / \text{PEL} = 0.260$, ${}_TS_y = 0.706$). In the case of group 3, where $\overline{X}_m / \text{PEL} = 3.51$, it would not be possible to reject H_0 with the probability greater than α. Finally, 143 measurements would be required from group 4 in order to reject H_0 with a power of 90 percent ($\overline{X}_m / \text{PEL} = 0.705$, ${}_TS_y = 0.897$). This last example illustrates that in marginal situations where μ_x approaches the OEL, the monitoring requirements needed to demonstrate acceptable exposure can be so great that the employer could well consider it preferable to institute controls.

6.5.3 Outcome versus Risk

Since an individual's long-term risk of disease is assumed to be related to his or her mean exposure over the period that the distribution remains stationary, the relationship between the outcome of a test and individual risk again depends on how vari-

Table 8.14 Approximate Sample Size Requirements of the Test of the Mean Exposure for $\alpha = 0.05$ and $(1 - \beta) = 0.90$[a]

	Sample Size n[b]				
μ_x / OEL	$\sigma_g = 1.5$	2.0	2.5	3.0	3.5
0.10	2	6	13	21	30
0.25	3	10	19	30	43
0.50	7	21	41	67	96
0.75	25	82	164	266	384

[a] For legend see List of Symbols.
[b] Calculated according to Eq. (23).

ability in exposure is distributed within the group. If the group is uniformly exposed (as is the case of groups 1 and 2), then all individuals share the same mean exposure, and no individual should have a probability greater than α that his or her mean exposure would exceed the OEL.

If, however, the group is not uniformly exposed, then some individuals have larger probabilities of overexposure than others, even if it has been shown that μ_x < OEL for the group as a whole. One mechanism for dealing with this would be to divide the population into successively smaller subgroups until uniform exposures are achieved (HSE, 1984). Then each subgroup could be tested relative to the OEL.

An alternative method takes advantage of the definition of a monomorphic group as a collection of individuals whose mean exposures comprise a lognormal distribution. If a monomorphic group were considered to be acceptable only following demonstration that μ_x < OEL/U, where U is a factor that is introduced a priori to reflect the uncertainty in exposure of any member of the group, then from Eq. (11), it follows that no more than θ_{max} of individuals in the group are expected to have mean exposures above the OEL (Rappaport et al., 1988a). Since θ_{max} is related to U, for example, $\theta_{max} = 0.05$ for $U = 4$, selection of U is an important matter. One could establish a schedule of values of U for various categories of jobs and/or tasks represented by the population at risk.

6.5.4 Application to STELs

Since the mean value from a lognormal distribution is independent of the averaging time of measurements, it is theoretically possible to evaluate the frequency with which short-term exposures exceed a STEL based solely upon knowledge of μ_x (Rappaport et al., 1988a). For example, if it can be demonstrated that μ_x < STEL/4, then no more than 5 percent of exposures are expected to exceed the STEL regardless of the variance of the distribution [Eq. (10)]. And because μ_x can be estimated with measurements covering a full shift (or longer), it should be possible to evaluate excursions over short periods without the need for short-term data. In fact, because autocorrelation should generally decrease with increasing averaging time, long-term measurements are more likely to be independent than short-term measurements and can arguably allow more valid inferences to be drawn about the frequencies of "peak" exposures.

The idea that long-term monitoring data can be used to evaluate short-term exposures is potentially extremely useful. Rappaport et al. (1988a) explored this notion with toluene diisocyanate exposure data obtained from 41 workers employed in 7 facilities that manufactured polyurethane foams. It was demonstrated in each case that where μ_x (or $\overline{X}_m$) was less than 0.005 ppm (the TLV–TWA), then fewer than 5 percent of 15-min exposures exceeded the TLV–STEL of 0.02 ppm = 4 (TLV–TWA), as predicted by theory.

It is, therefore, implicit in any pair of long- and short-term limits that, insofar as exposures are lognormally distributed and μ_x is less than the long-term limit (TWA, e.g., OSHA PEL), the maximum frequency with which the STEL can be exceeded is

$$\gamma_{STEL} = P\{Z > [(2) \ln (STEL/TWA)]^{1/2} = z_{(1-\gamma)}\} \qquad (24)$$

This implicit value of γ_{STEL} can be smaller than that which is explicitly allowed by the standard-setting group. For example, if $\mu_x < $ TLV–TWA $= 50$ ppm for carbon monoxide, then the maximum probability that the STEL $= 400$ ppm would be exceeded is $\gamma_{STEL} = P\{Z > [(2) \ln (400/50)]^{1/2} = 2.04 > z_{(1-\gamma)}\} = 2\%$, provided that exposures are lognormally distributed. Yet, the ACGIH explicitly allows 12.5 percent of 15-min exposures to exceed the STEL (ACGIH, 1991); this clearly would not be possible, regardless of the variance of the exposure distribution. Groups that set OELs should explore this contradiction, which arises from lognormally distributed exposures in establishing pairs of exposure limits for a given agent.

6.5.5 Summary

The test of the mean exposure relative to the OEL or some fraction thereof as discussed in Section 6.5 of this chapter, possesses the features outlined above for an ideal strategy. The approach allows rigorous analysis; monitoring is encouraged; there should be a clear correlation between the mean exposure and the chronic hazard for most toxicants (over at least some time scale in the range of years); the statistical structure provides sufficient power for routine use; and novel applications of the test make it possible, at least in theory, to evaluate excursions above STELs. If made part of a plan that includes randomness and independence in sampling and that requires multiple measurements of representative workers, then this approach should satisfy the needs of prospective exposure assessment in most cases.

7 CONCLUSIONS

This chapter has applied statistical methods to evaluate occupational exposures and to test various measures of exposure relative to OELs. In doing so it has become clear that the great variability in exposure that occurs both within and between individuals performing the same job vastly complicates the problem and calls into question the traditional practices that are inherent in compliance testing. The following general conclusions should be considered as starting points in the development of more sophisticated approaches that might be accepted in the future.

7.1 Exploiting the Lognormal Model

A variety of statistical issues come into play in designing sampling strategies. First, since sample sizes are generally limited, it is often necessary to assume a particular distribution function in order to make statistical inferences. Based on both theoretical grounds and empirical observation, it appears reasonable to assume a two-parameter lognormal model, provided that a uniform averaging time is adhered to and that data are inspected for obvious deviations from lognormality. If the lognormal model is used, it is important to differentiate between the mean exposure, which is a predictor of cumulative dose, and the geometric mean exposure, which has no physiological significance.

An interesting feature of the lognormal distribution concerns the relationship between the mean of the distribution and the probability of overexposure. For any family of lognormal distributions that share a common mean value, there exists a maximum probability that the OEL will be exceeded. If, for example, the mean is less that one-fourth of the OEL, than fewer than 5 percent of exposures from that distribution can exceed the OEL, and so on. This property can be used to great advantage in the development of efficient sampling strategies. It is theoretically possible, for example, to place an upper limit on the frequency of excursions of a STEL solely on the basis of the mean exposure, which has been estimated with long-term measurements.

7.2 Stationary Behavior and Independence

Most statistical testing relies upon the assumption that the distribution does not change with time and upon the independence of the measurements used to make inferences. While the former assumption might be reasonable over some period of time, common sense suggests that the distribution will eventually change owing to changes in process and equipment and in the work force. Thus, it is prudent to view assessment of exposure as an ongoing activity with periodic tests of the air environment. Likewise, industrial hygienists should be reluctant to base judgments entirely upon data collected during brief campaigns (regarding either intra- or intershift sampling) owing to the possibility of significant autocorrelation of exposures. If exposures are highly autocorrelated, then data obtained from a few serial measurements can lead to imprecise estimates of the parameters of the distribution. Although time-series data are limited at present, it appears that intrashift exposures are more likely to be highly autocorrelated than are those measured from day to day.

7.3 Addressing Biological Issues

The rationale for developing biologically relevant strategies lies in the relationship between a worker's exposure distribution and his or her risk of adverse effect. The risk of acute effects is related to the small probability of encountering very high concentrations. Yet, because exposures associated with rare events are extremely difficult to characterize, assessment of acutely toxic agents should focus upon monitoring and control of the source rather than upon the measurement of exposure per se. Thus, sampling strategies should be considered primarily in the context of long-term exposures that give rise to chronic effects. In this case it appears reasonable to assume that individual risk will be primarily related to the mean exposure received by a worker over relatively long time scales (years). The mean air concentration, or the series of mean levels measured every year or so, should, therefore, be regarded as a relevant predictor of individual risk.

7.4 Assessment Relative to Occupational Exposure Limits

Since the risk of chronic disease can be assumed to depend upon the mean exposure received over periods of years, then it becomes clear that monitoring efforts should be devoted primarily to the routine evaluation of long-term mean exposures rather than to monitoring "peak" exposures. This notion is consistent with the interpretation of exposure limits for chronic toxicants as levels that relate to long-term averages, as is implicitly recognized by OSHA in setting its PELs. However, because OELs adopt compromises between minimizing risk, on the one hand, and the feasibility of achieving certain levels, on the other, these limits may not be as protective as required in certain segments of industry. In such cases, working limits, which are more protective than official OELs, become essential to the strategic plan. It must also be borne in mind that, although OSHA sets PELs that equate risk with long-term mean exposures, OSHA has thus far interpreted its PELs as absolute levels not to be exceeded during any given work shift. This lack of consistency creates practical problems that will undoubtedly requires some time to resolve.

7.5   Classifying the Population

The dispersion of risk across the population should be related to the distribution of individual mean exposures. If the variance of this between-worker distribution is large, then the range of individual risks would also be large and vice versa. Thus, a preliminary step in the exposure assessment involves classifying workers into groups according to the level of exposure. Unfortunately, the traditional method for accomplishing this, based on job title and direct observation, is useful only when the group is uniformly exposed, that is, when all individuals' mean exposures lie within a narrow range. Preliminary evidence indicates, however, that most groups are not uniformly exposed. It therefore appears necessary to move away from strictly observational schemes and toward sampling as the basis for assigning groups. This can be accomplished by random-sampling designs in which multiple measurements are collected from representative workers. Then the total variation in exposure can be partitioned into within- and between-worker components to provide a framework for classification. A particularly attractive way of doing this involves assignment of monomorphic groups, that is, collections of individuals whose mean exposures comprise discrete lognormal distributions. Then, the variance of the between-worker distribution can be used as an objective measure of uniformity of exposure across the group.

7.6   Methods for Testing Exposure

Three approaches were evaluated for testing exposures relative to limits: compliance testing, a test of overexposure, and a test of the mean exposure. Several criteria were used for considering the utility of each approach, namely, provisions for rigorous evaluation, correlations between outcome and hazard, incentives for monitoring, and the sample size requirements. Compliance testing was found to fall short on all counts. Since the probability of compliance is maximized by minimizing the amount of data collected, sample sizes remain small and decisions become highly capricious. Furthermore, lognormal distributions representing several fold different mean values can be equally likely to be declared in compliance, depending on the amount of shift-to-shift variation.

The other two approaches for testing are based on the parameters of the exposure distribution and thus encourage monitoring and allow for statistical hypothesis testing. When focused upon the between-worker distribution, both techniques provide consistent bases for evaluating long-term exposures. Yet, regarding the issues of statistical power, correlation between outcome and hazard and overall efficiency, the test for evaluating the mean exposure appears to be superior to that based on over-exposures and presents a more compelling rationale for making decisions.

8 LIST OF SYMBOLS

b	Air exchange rate in the breathing zone, air changes/hr
CV_x	Coefficient of variation of X_{ij}
CV_e	Coefficient of variation of measurement error
h	Lag of $\rho(h)$, i.e., the fixed number of intervals between observations in a time series
k	Number of individuals in a group exposed to X_{ij}
k'	Number of individuals in a balanced subgroup of workers
K	Tolerance factor for a one-sided tolerance limit
n, N	Sample size, i.e., the number of observations (measurements or individuals) in a sample.
n'	Number of observations per individual in a balanced subgroup of workers
P	Allowable frequency of excursions above an exposure limit
$P(\text{Acc.})$	Probability of declaring exposures acceptable with nonparametric one-sided tolerance limits
$P(\text{Comp.})$	Probability of compliance; the likelihood that all measurements from a sample of given size would be less than the OEL
$P(\text{Noncomp.})$	Probability of noncompliance; the likelihood that at least one measurement from a sample of given size would exceed the OEL
$P\{Z > z\}$	Statement indicating the probability that the standard-normal variate Z exceeds the value z
Pr	Risk of disease arising from exposure to X_{ij}, which is borne by an individual in a group (after t years of exposure)
$_B R_{.95}$	Ratio of the 97.5th and 2.5th percentiles of the lognormally distributed mean exposures of a group of workers (between-worker distribution); this is equivalent to a factor containing 95 percent of the individual mean exposures derived from a lognormal distribution (monomorphic group)
RR	Risk ratio; the ratio of risks borne by two populations (designated a and b) at a given probability of overexposure, γ
S_x	Direct estimate of the standard deviation of X_{ij}
S_m	Maximum-likelihood estimate of the standard deviation of X_{ij} when exposures arise from a lognormal distribution
S_g	Estimated geometric standard deviation of X_{ij}
$_W S_g, _B S_g$	Estimated geometric standard deviations of within- and between worker distributions of X_{ij}
S_y	Estimated standard deviation of Y_{ij}

$_wS_y$, $_BS_y$, $_TS_y$	Estimated standard deviations of within-worker, between-worker, and total-distributions of Y_{ij}
$S_{\bar{X}m}$	Standard error of $\bar{X}_m$ under the null hypothesis for the test of the mean exposure test when $\mu_x = $ OEL
t	Time, duration of exposure
T_m	Test statistic for determining whether the mean exposure from a lognormal distribution exceeds the OEL (test of the mean exposure)
T_u	Test statistic for determining whether less than P percent of exposures from a lognormal distribution exceed the OEL (test of overexposures)
U	Uncertainty factor in exposure
w^h	Autocorrelation coefficient of (X_t) at lag h
w_y^h	Autocorrelation coefficient of (Y_t) at lag h
X_{ij}	The exposure concentration received by the ith worker during the jth interval
$\{X_t\}$	Series of exposures received by an individual or group over time
$\bar{X}$	Direct estimate of the mean of X_{ij}
$\bar{X}_g$	Estimated geometric mean of X_{ij}
$\bar{X}_m$	Maximum-likelihood estimation of the mean of X_{ij} when exposures arise from a lognormal distribution
Y_{ij}	Natural logarithm of the exposure concentration, $Y_{ij} = \ln (X_{ij})$
$\bar{Y}$	Estimated mean of Y_{ij}
$\{Y_t\}$	Series of log-transformed exposures; $\{Y_t\} = \{\ln (X_t)\}$.
Z	A Standard normal deviate with a mean of zero and variance of one
$z_{(1-p)}$	Value from the standard normal distribution associated with probability $(1 - p)$
α	Significance of the test of a null hypothesis H_0 against the alternative H_1; probability of a type I error
β	Probability of a type II error; $(1 - \beta)$ is the power of a test of significance
γ	Probability of overexposure; the probability that an individual will be exposed above the OEL during a random interval
θ	Probability that an individual mean exposure in a monomorphic group will be above the OEL
μ_x	Mean of X_{ij}
μ_g	Geometric mean of X_{ij}
μ	Mean of Y_{ij}
μ_{Pr}	Mean risk of a population exposed to X_{ij}
$\rho(h)$	Autocorrelation function of (X_t)
σ_x	Standard deviation of X_{ij}
σ_g	Geometric standard deviation of X_{ij}
$_w\sigma_g$, $_B\sigma_g$	Geometric standard deviations of within- and between-worker distributions of X_{ij}
$\sigma_{\bar{X}}^2$	Variance of $\bar{X}$ for a given sample size

σ	Standard deviation of Y_{ij}
σ_a, σ_b	Standard deviation of Y_{ij} for two groups of workers (designated a and b) whose exposure distributions have the same probability of overexposure γ
σ_W, σ_B	Standard deviations of within- and between-worker distributions of Y_{ij}
τ	Interval between observations in a time series

9 GLOSSARY OF SPECIAL TERMS

Autocorrelation   The correlation between sequential values of a time series.

Between-worker distribution   The distribution of individual mean exposure concentrations across a group of workers during a particular period of time.

Campaign sampling   The collection of exposure data during short surveys of a few days time.

Compliance   A condition in which all measured exposures are less than the operative OEL.

Feasibility   The ability of industry to achieve particular levels of control in the workplaces that harbor air contaminants; explicitly defined by OSHA in the United States in terms of both economic and technologic factors.

Individual risk   The risk borne by a worker at a particular time following exposure to $\{X_t\}$.

Lognormal model   Application of the lognormal distribution function to describe the statistical properties of populations of exposures or burdens.

Lowest feasible risk   The lifetime risk of an individual worker when exposed at a new OSHA PEL for 45 years.

Monomorphic group   A group of workers whose individual mean exposures comprise a single lognormal distribution.

Observational approach   A method of grouping workers into exposure categories by inspection and by the use of occupational titles, locations, and tasks.

Peak exposure   A transient high exposure extending over a period of one shift or less.

Power   The probability of rejecting the null hypothesis when it is false; from the employer's perspective this is the probability of declaring exposure acceptable when it is in fact acceptable.

Representative workers   Individuals chosen at random from a particular group of workers.

Sampling approach   A method of grouping workers based on the results of random measurements of exposures received by representative workers.

Sampling strategy   A comprehensive plan for assessing exposures that define the monitoring requirements and the mechanism by which decisions are made.

Significant risk   A nontrivial risk borne by a worker at the end of a working lifetime when exposed to a toxic substance at a particular air concentration; explicitly defined by OSHA in the United States as a risk of 1 excess death per 1000 workers exposed at the PEL for 45 years.

Stationarity   The condition where the statistical descriptors of a time series (i.e., the mean, variance, and autocorrelation functions) do not change over time.

Time series   A sequence of values of a random variable separated by a fixed interval of time (e.g., exposures measured every 15 min or every 24 hr).

Type I error   The probability of rejecting the null hypothesis when it is true; from the employer's perspective this is the probability of declaring exposure acceptable when it is unacceptable.

Type II error   The probability of accepting the null hypothesis when it is false; from the employer's perspective this is the probability of declaring exposure unacceptable when it is in fact acceptable.

Uniformly exposed group   A group in which all individuals have the same mean exposure; a monomorphic group for which $_BR_{.95}$ is two or less.

Within-worker distribution   The distribution of exposure concentrations received by an individual worker over time.

APPENDIX

Exposure data for the four groups of workers described in Table 8.1 of the text. Air concentrations are given in $\mu g/m^3$ for group 1 and mg/m^3 for groups 2–4.

Group	Worker	Conc.	Group	Worker	Conc.
1	1	10.40	1	1	20.20
1	1	11.40	1	1	15.90
1	1	17.10	1	1	9.00
1	1	28.90	1	1	9.50
1	1	12.40	1	1	19.10
1	1	10.00	1	1	9.00
1	1	24.60	1	1	25.70
1	1	21.50	1	1	46.90
1	1	11.30	1	1	7.90
1	1	15.00	1	1	22.20
1	1	30.40	1	2	30.80
1	1	25.40	1	2	1.30
1	1	34.00	1	2	23.80
1	1	3.90	1	2	10.10
1	1	14.40	1	2	11.40
1	1	18.60	1	2	7.50
1	1	56.40	1	2	50.10
1	1	25.60	1	2	18.00
1	1	10.00	1	2	56.50
1	1	21.90	1	2	9.70
1	1	12.90	1	2	13.70
1	1	8.60	1	2	40.90
1	1	19.60	1	2	20.80
1	1	13.00	1	2	62.60
1	1	10.20	1	2	19.20
1	1	19.50	1	2	21.00

Group	Worker	Conc.	Group	Worker	Conc.
1	2	3.20	1	4	5.80
1	2	17.70	1	4	6.80
1	2	13.10	1	4	6.90
1	2	34.00	1	4	5.00
1	2	16.70	1	4	20.00
1	2	13.50	1	4	19.60
1	2	10.10	1	4	13.20
1	3	9.70	1	4	9.40
1	3	5.10	1	4	9.90
1	3	15.20	1	4	7.90
1	3	34.80	1	4	22.50
1	3	8.70	1	4	22.60
1	3	14.70	1	4	24.60
1	3	13.30	1	4	20.10
1	3	6.60	1	4	39.30
1	3	5.20	1	4	24.40
1	3	7.70	1	4	14.00
1	3	7.70	1	4	10.60
1	3	23.70	1	4	31.90
1	3	18.80	1	4	6.30
1	3	21.40	1	4	10.10
1	3	7.10	1	4	6.90
1	3	17.80	1	4	21.30
1	3	10.20	1	4	10.30
1	3	16.50	1	4	10.80
1	3	11.20	1	4	21.50
1	3	16.00	1	4	55.00
1	3	16.80	1	4	19.90
1	3	17.70	1	4	7.60
1	3	7.70	1	5	8.80
1	3	19.70	1	5	20.80
1	3	8.80	1	5	18.90
1	3	9.20	1	5	59.70
1	3	9.30	1	5	35.00
1	3	4.30	1	5	23.20
1	3	4.90	1	5	15.70
1	3	8.50	1	5	53.50
1	3	14.50	1	5	24.40
1	3	40.40	1	5	11.50
1	3	9.20	1	5	15.70
1	3	27.80	1	5	31.40
1	4	25.10	1	5	33.50
1	4	6.10	1	5	26.90
1	4	19.90	1	5	39.90
1	4	61.10	1	5	11.30
1	4	8.10	1	5	26.90
1	4	13.80	1	5	11.80
1	4	26.50	1	5	14.70

Group	Worker	Conc.	Group	Worker	Conc.
1	5	12.40	2	5	1.09
1	5	13.10	2	5	1.15
1	5	33.10	2	5	0.70
1	5	11.00	2	5	1.28
1	5	27.50	2	6	0.38
1	5	18.60	2	6	0.29
1	5	15.00	2	7	0.54
1	5	8.90	2	7	1.47
1	5	18.10	2	7	0.51
1	5	16.70	2	7	1.53
1	5	16.70	2	8	2.27
1	5	31.00	2	8	0.73
1	5	22.90	2	8	1.37
1	5	22.90	2	8	1.15
1	6	12.00	2	8	0.42
1	6	10.90	2	8	1.02
1	6	8.60	2	9	0.51
1	6	38.20	2	9	0.51
1	6	7.30	2	9	1.18
1	6	13.80	2	9	0.54
1	6	10.80	2	9	1.34
1	6	10.30	2	10	1.15
1	6	8.10	2	10	1.63
1	6	12.00	2	10	0.51
1	6	8.10	2	10	0.86
1	6	19.40	2	10	0.38
1	6	29.00	2	10	0.27
1	6	18.30	2	10	0.42
1	6	30.60	2	10	0.45
2	1	0.48	2	11	1.05
2	1	0.32	2	11	1.76
2	1	0.38	2	11	0.80
2	1	0.27	2	11	0.38
2	1	0.54	2	12	0.80
2	2	0.54	2	12	0.42
2	2	1.09	2	13	1.82
2	2	1.18	2	13	1.28
2	2	0.67	2	13	0.57
2	2	0.32	2	13	0.70
2	2	0.28	2	14	2.68
2	2	0.57	2	14	0.25
2	3	0.42	2	14	0.35
2	3	0.28	2	14	5.75
2	3	0.61	2	14	0.45
2	4	1.28	2	15	0.45
2	4	0.57	2	16	0.61
2	4	0.45	2	16	1.69
2	5	0.99	2	16	0.83

Group	Worker	Conc.	Group	Worker	Conc.
2	16	0.51	3	12	0.13
2	16	0.25	3	13	0.26
2	16	0.10	3	13	0.61
2	16	0.42	3	13	0.22
2	16	0.09	3	13	13.74
2	16	0.35	3	13	1.31
2	17	1.18	3	14	2.75
2	17	0.35	3	15	2.72
2	17	1.47	3	16	23.32
2	17	0.73	3	17	0.16
2	17	0.93	3	17	5.11
2	17	0.29	3	18	0.42
2	17	0.51	3	19	0.32
2	17	0.48	3	19	0.57
2	17	0.30	3	19	0.38
2	18	1.63	3	19	0.32
2	18	3.19	3	20	0.16
2	18	0.57	3	20	0.06
2	18	1.02	3	21	0.01
2	18	0.83	3	21	0.96
2	18	0.31	3	22	0.03
2	18	0.54	3	22	0.01
2	18	0.45	3	23	0.03
3	1	18.53	3	24	2.84
3	1	5.85	4	1	19.39
3	2	48.87	4	1	18.73
3	3	70.60	4	1	21.56
3	4	0.51	4	1	56.20
3	4	0.51	4	1	48.00
3	5	14.79	4	2	66.87
3	6	0.03	4	2	75.08
3	6	0.29	4	2	81.14
3	7	0.22	4	2	184.21
3	7	0.32	4	2	220.50
3	8	6.48	4	2	222.22
3	8	0.51	4	2	155.30
3	8	123.61	4	3	60.52
3	9	0.89	4	3	131.06
3	9	1.34	4	3	185.05
3	9	0.19	4	3	84.66
3	9	5.43	4	3	178.88
3	10	19.23	4	3	6.64
3	10	1.92	4	3	157.30
3	10	0.51	4	4	206.25
3	11	0.80	4	4	303.68
3	12	1.47	4	4	166.63
3	12	0.01	4	4	108.61

Group	Worker	Conc.	Group	Worker	Conc.
4	4	96.78	4	11	209.56
4	4	114.87	4	12	157.58
4	4	264.47	4	12	233.61
4	5	67.86	4	12	181.40
4	5	50.66	4	12	244.71
4	5	210.81	4	12	188.01
4	6	107.50	4	12	194.63
4	6	118.46	4	13	241.17
4	6	133.33	4	13	384.00
4	6	82.65	4	13	139.03
4	6	138.27	4	13	204.40
4	6	153.63	4	13	284.59
4	6	361.13	4	13	71.01
4	7	2.78	4	14	74.20
4	7	10.15	4	14	221.67
4	7	31.64	4	15	89.93
4	7	47.05	4	15	179.08
4	7	72.61	4	15	208.47
4	7	82.00	4	16	189.83
4	8	74.42	4	16	213.78
4	8	179.75	4	16	153.32
4	8	86.23	4	16	176.79
4	8	82.29	4	16	391.33
4	8	163.53	4	16	287.04
4	9	32.08	4	17	108.77
4	9	58.54	4	17	188.63
4	9	39.52	4	17	89.47
4	9	49.02	4	17	123.10
4	9	66.61	4	17	175.80
4	9	69.23	4	17	131.26
4	9	150.06	4	17	169.31
4	10	30.60	4	18	178.20
4	10	56.00	4	18	6.32
4	10	35.09	4	18	256.08
4	10	42.45	4	18	275.59
4	10	70.52	4	19	69.57
4	10	64.07	4	19	43.57
4	10	154.14	4	19	140.49
4	11	122.41	4	19	66.71
4	11	87.96	4	19	199.22

Acknowledgment

The author gratefully acknowledges the advice of L. Kupper regarding the ANOVA model and the assistance of H. Kromhout and W. Braun in performing the goodness-of-fit tests.

REFERENCES

ACGIH (1986). *Documentation of the Threshold Limit Values and Biological Exposure Indices*, 5th ed., American Conference of Governmental Industrial Hygienists, Cincinnati, OH.

ACGIH (1991). *Threshold Limit Values and Biological Exposure Indices for 1991–1992*, American Conference of Governmental Industrial Hygienists, Cincinnati, OH.

Aitchison, J. and J. A. C. Brown (1957). *The Lognormal Distribution*, Cambridge University Press, London.

Boleij, J., D. Heederik, and H. Kromhout (1987). *Karakterisering van Bloot-stelling aan Chemische Stoffen in de Werkomgeving* (Characterization of exposure to chemical substances in the working environment), Purdoc Wageningen, the Netherlands, in Dutch.

Box, G. E. P. and G. M. Jenkins (1976). *Time Series Analysis: Forecasting and Control*, Holden-Day, San Francisco.

Brief, R. S. and A. R. Jones (1976). *Am. Ind. Hyg. Assoc. J.*, **37**, 474–478.

Brunekreef, B., D. Noy, and P. Clausing (1987). *Am. J. Epidemiol.*, **125,** 892–898.

Buringh, E. and R. Lanting (1991). *Am. Ind. Hyg. Assoc. J.*, **52,** 6–13.

Castleman, B. I. and G. E. Ziem (1988). *Am. J. Ind. Med.*, **13,** 531–559.

Chang, K. H. and P. I. Nelson (1991). *Am. Ind. Hyg. Assoc. J.*, **51,** A-813–814.

Coenen, W. (1966). *Staub Reinhalt. Luft*, **26,** 39–45 (English translation).

Coenen, W. (1971). *Staub. Reinhalt. Luft*, **31,** 16–23 (English translation).

Coenen, W. (1976). *Staub Reinhalt. Luft*, **36,** 240–248, in German.

Coenen, W. and G. Riediger (1978). *Staub Reinhalt. Luft*, **38,** 402–409, in German.

Cope, R., B. Panacamo, W. E. Rinehart, and G. L. Ter Haar (1979). *Am. Ind. Hyg. Assoc. J.*, **40,** 372–379.

Corn, M (1985). *Scand. J. Work Environ. Health*, **11,** 173–180.

Corn, M. and N. A. Esmen (1979). *Am. Ind. Hyg. Assoc. J.*, **40,** 47–57.

Corn, M., P. Breysse, T. Hall, G. Chen, T. Risby, and D. L. Swift (1985). *Am. Ind. Hyg. Assoc. J.*, **46,** 4–8.

Dourson, M. L. and J. F. Stara (1983). *Reg. Tox. Pharcacol.*, **3,** 224–238.

Esmen, N. (1979). *Am. Ind. Hyg. Assoc. J.*, **40,** 58–65.

Esmen, N. (1984). "On Estimation of Occupational Health Risks", in, *Occupational and Industrial Hygiene: Concepts and Methods*, N. A. Esmen and M. A. Mehlman, Eds., Princeton Scientific Publishers, Princeton, NJ.

Esmen, N. and Y. Hammad (1977). *J. Environ. Sci. Health*, **A12,** 29–41.

Evans, J. S. and N. C. Hawkins (1988). *Am. Ind. Hyg. Assoc. J.*, **49,** 512–515.

Francis, M., S. Selvin, R. C. Spear, and S. M. Rappaport (1989). *Am Ind. Hyg. Assoc. J.*, **50,** 37–43.

Galbas, H. G. (1979). *Staub Reinhalt. Luft*, **39,** 463–467, in German.

Gale, H. J. (1965). *The Lognormal Distribution and Some Examples of Its Application in the Field of Radiation Protection*, AERE, Harwell, Berkshire, UK.

Gale, H. J. (1967). *Ann. Occup. Hyg.*, **10,** 39–45.

Halton, D. H. (1988). *Reg. Tox. Pharmacol.*, **8,** 343–355.

Heederik, D., J. S. M. Boleij, H. Kromhout, and T. Smid (1991). *Appl. Occup. Environ. Hyg.*, **6,** 458–464.

Heidermanns, G., G. Kuhnen, and G. Riediger (1980). *Staub Reinhalt. Luft*, **40**, 367–373, in German.

Henschler, D. (1984). *Ann. Occup. Hyg.*, **28**, 79–92.

Hines, C. J. and R. C. Spear (1984). *Am. Ind. Hyg. Assoc. J.*, **45**, 44–47.

Hounam, R. F. (1965). *An Application of the Log-Normal Distribution to Some Air Sampling Results and Recommendations on the Interpretation of Air Sampling Data*, AERE, Harwell, Berkshire, UK.

HSE (1984). *Monitoring Strategies For Toxic Substances*, Guidance Note EH 42, Health and Safety Executive, HMSO, UK.

Jones, A. R. and R. S. Brief (1971). *Am. Ind. Hyg. Assoc. J.*, **32**, 610–613.

Juda, J. and K. Budzinski (1964). *Staub Reinhalt. Luft*, **24**, 283–287, in German.

Juda, J. and K. Budzinski (1967). *Staub Reinhalt Luft*, **27**, 12–16 (English translation).

Koch, A. L. (1966). *J. Theoret. Biol.*, **12**, 276–290.

Koch, A. L. (1969). *J. Theoret. Biol.*, **23**, 251–268.

Koizumi, A., T. Sekiguchi, M. Konno, M. Ikeda (1980). *Am. Ind. Hyg. Assoc. J.*, **41**, 693–699.

Kromhout, H. Y. Oostendorp, D. Heederik, and J. Boleij (1987). *Am. J. Ind. Med.*, **12**, 551–562.

Kumagai, S., I. Matsunaga, K. Sugimoto, Y. Kysaka, and T. Shirokawa (1989). *Jpn. J. Ind. Hlth.*, **31**, 216–226, in Japanese.

Langmead, W. A. (1971). *Inhaled Part. Vapors*, **2**, 983–995.

Lanier, M. E., Ed. (1984). *Threshold Limit Values—Discussion and Thirty-Five Year Index with Recommendations*, ACGIH, Cincinnati, OH.

LeClare, P. C., A. J. Breslin, and L. D. Y. Ong (1969). *Am. Ind. Hyg. Assoc. J.*, **30**, 386–393.

Leichnitz, K. (1980). *Staub Reinhalt. Luft.*, **40**, 241–243, in German.

Leidel, N. A. and K. A. Busch (1985). "Statistical Design and Data Analysis Requirements," in *Patty's Industrial Hygiene and Toxicology*, 2nd ed., Vol. 3A, L. J. Cralley and L. V. Cralley, Eds., Wiley, New York, pp. 395–507.

Leidel, N. A., K. A. Busch, and W. E. Crouse (1975). *Exposure Measurement Action Level and Occupational Environmental Variability*, USDHEW NIOSH publ. No. 76-131, USGPO, Washington, D.C.

Leidel, N., K. Busch, and J. Lynch (1977). *Occupational Exposure Sampling Strategy Manual*, USDHEW, NIOSH Publ No. 77-173, National Institute for Occupational Safety and Health, Cincinnati, OH.

Leung, H. W. and D. Paustenbach (1988). *Appl. Ind. Hyg.*, **3**, 115–118.

Long, W. M. (1953). *Br. J. Ind. Med.*, **10**, 241–244.

Male, L. M. (1982). *Atmos. Environ.*, **16**, 2247–2252.

Matsunaga, I., K. Kumagai, and K. Sugimoto (1989). *Jpn. J. Ind. Hlth.*, **31**, 227–234, in Japanese.

Oldham, P. (1953). *Br. J. Ind. Med.*, **10**, 227–234.

Oldham, P. and S. A. Roach (1952). *Br. J. Ind. Med.*, **9**, 112–119.

OSHA (1983). "Supplemental Statement of Reasons for the Final Inorganic Arsenic Standard," *Fed. Reg.*, **48**, 1864–1903.

OSHA (1984). "Occupational Exposure to Ethylene Oxide," *Fed. Reg.*, **49**, 25734–25809.

OSHA (1987a). *OSHA Safety and Health Standards*, 29 Code of Federal Regulations 1900–1910, Subpart Z, Sect. 1000–1047, U.S. GPO, Washington, D.C.

OSHA (1987b). "Occupational Exposure to Benzene," *Fed. Reg.*, **52**, 34460–34579.

OSH Act (1970). Occupational Safety and Health Act of 1970, PL-91-596, S.2193, Dec. 29.

Paustenbach, D. (1985). "Occupational Exposure Limits, Pharmacokinetics, and Unusual Work Schedules," in *Patty's Industrial Hygiene and Toxicology, Vol. III, Theory and Rationale of Industrial Hygiene Practice*, 2nd ed. 3A, L. J. Cralley and L. V. Cralley, Eds., Wiley, New York, pp. 111–278.

Paustenbach, D. and R. Langner (1986). *Am. Ind. Hyg. Assoc. J.*, **47**, 809–818.

Preat, B. (1987). *Am Ind. Hyg. Assoc. J.*, **48**, 877–884.

Rappaport, S. M. (1984). *Am. J. Ind. Med.*, **6**, 291–30.

Rappaport, S. M. (1985). *Ann. Occup. Hyg.*, **29**, 201–214.

Rappaport, S. M. (1988). "Biological Consideration for Designing Sampling Strategies," in *Advances in Air Sampling*, W. John, Ed., Lewis Publishers, Chelsea, MI, pp. 337–352.

Rappaport, S. M. (1991a). *Ann. Occup. Hyg.*, **35**, 61–121.

Rappaport, S. M. (1991b). *Appl. Occup. Environ. Hyg.*, **6**, 448–457.

Rappaport, S. M. and S. Selvin (1987). *Am. Ind. Hyg. Assoc. J.*, **48**, 374–379.

Rappaport, S. M. and R. C. Spear (1988). *Ann. Occup. Hyg.*, **32**, 21–33.

Rappaport, S. M., S. Selvin, R. C. Spear, and C. Keil (1981a). "An evaluation of Statistical Schemes for Air Sampling," in *Measurement and Control of Chemical Hazards in the Workplace Environment*, G. Choudhary, Ed., ACS Symposium Series 149, American Chemical Society, Washington, D. C., pp. 431–455.

Rappaport, S. M., S. Selvin, R. C. Spear, and C. Keil (1981b). *Am. Ind. Hyg. Assoc. J.*, **42**, 831–838.

Rappaport, S. M., S. Selvin, and S. Roach (1988a). *Appl. Ind. Hyg.*, **3**, 310–315.

Rappaport, S. M., R. C. Spear, and S. Selvin (1988b). *Ann. Occup. Hyg.*, **32** (Suppl. 1), 529–537.

Riediger, G. (1986). *Staub Reinhalt. Luft*, **46**, 182–186, in German.

Roach, S. A. (1953). *Br. J. Ind. Med*, **10**, 220–226.

Roach, S. A. (1959). *Br. J. Ind. Med.*, **16**, 104–122.

Roach, S. A. (1977). *Ann. Occup. Hyg.*, **20**, 65–84.

Roach, S. A. and S. M. Rappaport (1990). *Am. J. Ind. Med.*, **17**, 727–753.

Roach, S. A., E. J. Baier, H. E. Ayer, and R. L. Harris (1967). *Am. Ind. Hyg. Assoc. J.*, **28**, 543–553.

Rock, J. C. (1981). "The NIOSH Action Level: A Closer Look," in *Chemical Hazards in the Workplace*, G. Choudhary, Ed., American Chemical Society, Washington, D.C., pp. 472–489.

Rock, J. (1982). *Am. Ind. Hyg. Assoc. J.*, **43**, 297–313.

Sato, A. (1987). *Scand. J. Wk. Environ. Hlth.*, **13**, 81–93.

Searle, S. R. (1971). *Linear Models*, Wiley, New York.

Selvin, S. and S. M. Rappaport (1988). University of California, School of Public Health, Berkeley, California, unpublished observation.

Selvin, S. and S. M. Rappaport (1989). *Am. Ind. Hyg. Assoc. J.*, **50**, 627–630.

Selvin, S., S. M. Rappaport, R. C. Spear, J. Schulman, and M. Francis (1987). *Am. Ind. Hyg. Assoc. J.*, **48**, 89–93.

Shapiro, S. S. and M. B. Wilk (1965). *Biometrika*, **52**, 591–611.

Sherwood, R. J. (1966). *Am. Ind. Hyg. Assoc. J.*, **27**, 98–109.

Sherwood, R. J. (1971). *Am. Ind. Hyg. Assoc. J.*, **32**, 840–846.

Sherwood, R. J. and D. M. S. Greenhalgh (1960). *Ann. Occup. Hyg.*, **2**, 127–132.

Spear, R. C. and S. Selvin (1989). *Risk Anal.*, **9**, 579–586.

Spear, R. C., S. Selvin, and M. Francis (1986). *Am. Ind. Hyg. Assoc. J.*, **47**, 365–368.

Spear, R. C., S. Selvin, J. Schulman and M. Francis (1987). *Appl. Ind. Hyg.*, **2**, 155–163.

Tomlinson, R. C. (1957). *Appl. Statist.*, **6**, 198–207.

Tuggle, R. M. (1981). *Am. Ind. Hyg. Assoc. J.*, **42**, 493–498.

Tuggle, R. M. (1982). *Am. Ind. Hyg. Assoc. J.*, **43**, 338–346.

Ulfvarson, U. (1977). *Scand. J. Work, Environ. Health*, **3**, 109–115.

Ulfvarson, U. (1983). *Int. Arch. Occup. Environ. Health*, **52**, 285–300.

Ulfvarson, U. (1987). *Scand. J. Work Environ. Health*, **13**, 389–398.

Waters, M. A., S. Selvin., and S. M. Rappaport (1991). *Am. Ind. Hyg. Assoc. J.*, **52**, 493–502.

Woitowitz, H. J., G. Schacke, and R. Woitowitz (1970). *Staub Reinhalt. Luft*, **30**, 15–18 (English translation).

Wright, B. M. (1953). *Br. J. Ind. Med.*, **10**, 235–240.

Zielhuis, R. L. (1988). "Occupational Exposure Limits for Chemical Agents", in, *Occupational Medicine—Principles and Practical Applications* 2nd ed., C. Zenz, Ed., Year Book Medical Publishers, Chicago, pp. 491–502.

Zielhuis, R. L. and F. W. van der Kreek (1979a). *Int. Arch. Occup. Environ. Health*, **42**, 191–201.

Zielhuis, R. L. and F. W. van der Kreek (1979b). *Int. Arch. Occup. Environ. Health*, **42**, 203–215.

Zielhuis, R. L., P. C. Noordam, H. Roelfzema, and A. A. E. Wibowo (1988). *Int. Arch. Occup. Environ. Health*, **61**, 207–211.

CHAPTER NINE

Data Automation

Richard L. Patnoe, Ph.D. and Robert L. Fischoff, Ph.D.

1 THE NEED FOR DATA AUTOMATION

This chapter presents the basic concepts of electronic data processing (EDP) systems and describes the steps necessary to plan, design, and implement an effective computer application. A specific example of a monitoring system for industrial chemicals is given. Industrial health applications are also discussed and references are provided to assist those interested in pursuing additional aspects of data automation. A data processing glossary is included for reference.

The purpose of the Occupational Safety and Health Act (OSHA) is to ensure, as far as possible, safe and healthful working conditions for every industrial employee in the nation by providing mandatory occupational safety and health standards. These standards oblige the employer to maintain records of employee exposures, to give employees access to the records, to allow employees the opportunity to observe monitoring or measuring being conducted, to notify the employees of excessive exposures, and to inform them of corrective action being taken. The government is also allowed access to all of the records. In some cases, record retention requirements are 30 or more years.

In addition to recording and reporting requirements, OSHA standards prescribe the training of the employee, suitable protective equipment, control procedures, type and frequency of medical examinations, and posted warnings to ensure employee awareness of hazards, symptoms, emergency treatment, and safe use conditions. Thus the responsibilities of industrial hygiene management can become exceedingly complex. In protecting the health of employees the industrial hygienist must rec-

Patty's Industrial Hygiene and Toxicology, Third Edition, Volume 3, Part A, Edited by Robert L. Harris, Lewis J. Cralley, and Lester V. Cralley.
ISBN 0-471-53066-2 © 1994 John Wiley & Sons, Inc.

ognize potential health hazards, have them evaluated, assure that controls are in place, initiate exposure monitoring procedures, enter and delete employees from the record-keeping system, and comply with changing government requirements. Besides industrial hygienists, other people or functional areas that are affected by the OSHA regulations needing hazard exposure evaluations and reports are line management, employees, safety, transportation, medical, development engineering, manufacturing engineering, facilities engineering, chemical control and disposal, purchasing, shipping and receiving, personnel, laboratory, and similar corporate functions. Because of the workload of record keeping, data analysis, and reporting, many industrial hygienists are turning to electronic data processing. An EDP system can greatly increase the productivity and quality of most information collecting, storage, and retrieval operations, while providing more timely and economical data analyses and reports. Automating a large industrial hygiene program is a complex task requiring careful planning and evaluation if costly mistakes are to be avoided.

The emphasis of industrial hygiene has gradually changed from discovering job-related causes of ill health to monitoring and controlling potentially harmful work environment situations before they result in injury to workers or the public. Associated with this modification has been a significant change in industrial hygiene methodology, namely, an increasing requirement for data collection, record keeping, statistical analysis, and reporting. Added to this professional responsibility for data management are the requirements of the OSHA for record keeping, reporting, and tracking a growing list of harmful or suspected substances used in modern industrial processes.

Discussions with industrial hygiene professionals about the demands of their jobs reveals that many are spending a significant amount of time arranging, refining, and manipulating data. Of the time spent working with data, many people estimate that 80 percent of the time is related to the above tasks and that 20 percent or less is concerned with interpretation of the data and in adding value to the results. This is a tragic waste of effort since the 80 percent discussed above are activities that can very easily, speedily, and conveniently be handled by automation. Computers and computing technology does a fantastic job of manipulating, collecting, and sorting of data. When industrial hygienists are working on these aspects of the problem, their talents and value to the enterprise are not being fully utilized. The industrial hygienist truly becomes valuable when efforts are spent in interpretation of data, deciding what additional data is needed, and presenting this data. The use of data automation frees up industrial hygienists to do the advanced and technical aspects of their jobs. Everyone is interested in making the maximum contribution to their job/employer and increasing their worth to the enterprise. Certainly, the employer shares that interest. Thus, the quest becomes one of how to optimize effectiveness. Automating data management can contribute to this. This chapter discusses how automating data can help to increase this value.

1.1 The Mind Set

Many industrial hygienists pride themselves on the use of state-of-the-art instruments, expert technical advice, and a thorough engagement with the precision and accuracy of their work. Additionally, in planning their work, they are singularly devoted to enhancement of internal (to the industrial hygienist) department work flow. To improve efficiency, efforts are directed toward gathering more and more accurate data. When participating in professional conferences, the sessions attended are related to those that will enhance professional credentials and technical expertise. When assessing capital needs, the individual is most likely to budget for and spend money on a new tool or sampling instrument. If the tool is a software tool, the purpose is to enhance the internal productivity of the industrial hygienist. This type of industrial hygienist we could call *introverted* since the focus is on the industrial hygiene discipline and interdepartmental work.

A second type of industrial hygienist is concerned with how the data is used and conveyed as information to the management, workers, or customers. The central focus is on the end user of the data. For example, how are the workers interpreting the message, how is management using the information, does management understand the information, and what is the value of the information to the users and to the enterprise? The emphasis is on how this information effects the economic health, reputation or viability of the enterprise. Since the focus is outside the industrial hygiene department we may call this industrial hygienist *extroverted*.

For illustration of this we might consider the airborne monitoring results gathered for a particular chemical. The introverted industrial hygienist would focus on gathering more frequent samples, the statistical treatment of the samples, and a means to more quickly gather the samples. In addition to these things, the extroverted hygienist is concerned with how management uses the information, how much gathering the information costs, whether fewer samples could be used, whether a less expensive analytical method could be used, and how any sample exceeding the standards impacts personnel relations, how this would appear if found in the media, the impact of a high reading on the morale and performance of the person sampled, and how they can help the manager of the area get his or her job done.

These two types of industrial hygienists were described to get to a point of discussion about what is needed in data automation. If the intent is to specifically improve the processes internal to the industrial hygienist department, then we should set out to do this. This very well could mean doing electronic logging of samples, and electronic data system for instrument calibration, a means of direct data acquisition from an instrument, or automating the hearing testing laboratory. On the other hand, if one is considering how to best reach the people on the plant floor or get information from the plant floor, other things need to be considered. Often, plant floor systems can be used to give production figures, area samples, inventory of chemicals, and occasionally usage conditions such as ventilation flow rates or ambient temperatures.

By focusing on the business aspects of the enterprise the industrial hygienist contributes to the overall enterprise and to the industrial hygiene arena. The individual industrial hygienist's contribution is increased by considering the flow of data in and out of their area or department.

1.2 Legal Requirements for Data

The purpose of OSHA is to ensure, as far as possible, safe and healthful working conditions for every industrial employee in the nation by providing mandatory occupational safety and health standards. These standards oblige the employer to maintain records of employee exposures, to give employees access to the records, to allow employees the opportunity to observe monitoring or measuring being conducted, to notify the employees of excessive exposures, and to inform them of corrective action being taken. The government is also allowed access to all of the records. In some cases, record retention requirements are 30 or more years. In addition to recording and reporting requirements, OSHA standards prescribe the training of the employee, suitable protective equipment, control procedures, type and frequency of medical exams, and posted warnings to ensure employee awareness of hazards, symptoms, emergency treatment, and safe use conditions. Thus the responsibilities of industrial hygiene management become exceedingly complex. In protecting the health of employees, the industrial hygienist must recognize potential health hazards, have them evaluated, assume that controls are in place, initiate exposure monitoring procedures, enter and delete employees from the recording-keeping system, and comply with changing government requirements.

2 ELECTRONIC DATA PROCESSING CONCEPTS

2.1 Computer Equipment (Hardware)

Electronic data processing is the handling of data by an electronic computer and associated devices (peripherals) such as printers, scanners, FAX machines, and so forth in a planned sequence of operations to produce a desired result. The many types of EDP systems range in size from relatively simple desktop units to complex systems that fill several large rooms with interconnected devices. But regardless of the information to be processed or the complexity of equipment used, all EDP involves four basic functions:

1. Entering the source data into the system (input)
2. Storing the data in addressable locations (storage)
3. Processing the data in an orderly manner within the system (processing)
4. Providing the resulting information in a usable form (output)

These functions are performed by an input device, a storage unit, a central processing unit, and an output device (Fig. 9.1).

2.1.1 Central Processing Unit

The central processing unit (CPU) is the main body or brain of a computer. The CPU is the controlling center of the entire EDP system. It is divided into two parts as shown in Fig. 9.1: the control section and the arithmetic/logical unit. The control section directs and coordinates all computer system functions. It is like a traffic cop that schedules and initiates the operation of input and output devices, arithmetic/ logical unit tasks, and the movement of data from and to storage. The arithmetic/ logical unit performs such operations as addition, subtraction, multiplication, divi-

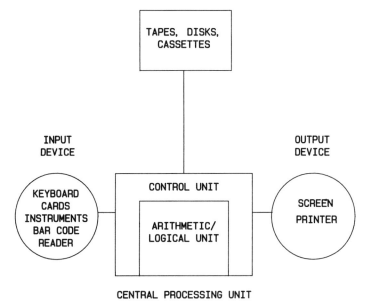

Figure 9.1 Central organization of a computing system.

sion, shifting, moving, comparing, and storing. It also has a capability to test various conditions encountered during processing and to take action accordingly.

2.1.2 Input Devices

Input devices read or sense coded data and make this information available to the computer. Data for input can come from a keyboard, another computer, an instrument, punch cards, a scanner, FAX machine, and so forth. Regardless of the origin of the data, it must be in a readable form by the receiving computer. The need for input readable by the receiving machine is much like you being able to call one of the Arabic countries on the telephone, but unless you speak Arabic (or they speak your language), although the connection is made, you cannot understand each other.

2.1.3 Output Devices

Output devices send information from the computer onto video display terminals, magnetic tape, disk, or drums. They may print information on paper, generate signals from transmissions over teleprocessing networks, produce graphic displays or microfilm images, or take other specialized forms.

Frequently, the same physical device, such as a tape drive, is used for both input and output operations. Thus input and output (I/O) functions are generally treated together. The number and type of I/O devices that may be connected directly to a CPU depends on the design of the system and its application. Note that the functions of I/O devices and auxiliary storage units may overlap—thus a tape drive or disk file may be used both for I/O operations and for data storage.

2.2 Computer Programs (Software)

2.2.1 Operating or Control Programs

To make possible the teleprocessing networks and the orderly operation of many types of I/O devices that may be on-line with a computer, control programs have been developed. Control programs, also known as monitor programs or supervisory programs, act as traffic directors for all the application programs (which solve a problem or carry out an operation or process data), then relinquish control of the computer to the control program. The control program may be constructed to allow the computer to handle random inquiries from remote terminals, to switch from one problem program within the computer to another, to control external equipment, or to do whatever the application requests.

Operating systems are the control programs that tell a computer how to function. There are a number of commercial names for the operating systems for computers. For the PC there is Microsoft Disk Operating System, MS/DOS® (Microsoft Corporation) or Operating System 2, OS/2®, or MacIntosh Disk Operating System, MAC/DOS® (Apple Computer Systems). Generally associated with the workstations is UNIX® (American Telephone and Telegraph), AIX® (IBM Corporation), ULTRIX® (Digital Equipment Corporation), and HP__UX® (Hewlett-Packard Company), and Virtual System, VS® (Wang Computer Corporation). The larger IBM mainframes are Virtual Memory, VM®, and Multiple Virtual Memory System, MVS®. Although this list is not exclusive, it includes the vendors and names of the systems most commonly used. The importance to the industrial hygienist is in understanding that a program written for one operating system will not function in another operating system. Thus, when selecting an application, the hygienist must consider the operating system that is used at his or her work site. Selecting a program that has the same operating system commonly employed at the work site can increase the effectiveness of a particular application. When a program that runs on one operating system is converted to run on another operating system, the act of doing this is called *porting*.

2.2.2 Application Programs

Examples of application programs may be laser inventory, recording results of audiometric testing, results of air monitoring, and so forth. A more complete list is found in Appendix B. Each EDP system is designed to perform a specific number and type of operations. It is directed to perform each operation by an instruction. The instruction defines a basic operation to be performed and identifies the display device, or mechanism, needed to carry out the operation. The entire series of instructions required to complete a given procedure is known as an application (or problem) program.

The possible variations of a stored program afford the EDP system almost unlimited flexibility. A computer can be applied to a great number of different procedures simply by reading in, or loading, the proper program into storage. Any of the standard input devices can be used for this purpose because instructions can be coded into machine language just as data can.

The stored program is accessible to the computer, giving it the ability to alter the program in response to conditions encountered during an operation. Consequently the program selects alternatives within the framework of the anticipated conditions.

2.3   Some Important Considerations

2.3.1   Storage

Storage is like an electronic filing cabinet, completely indexed and instantaneously accessible to the computer. All data must be placed in storage before it can be processed by the computer. Each element of storage has a specific location, called an address, so that the stored data can be located by the computer as needed.

The size or capacity of storage determines the amount of information that can be held in the system at any one time. In some computers, storage capacity is measured in millions of digits or characters (bytes) that provide space to retain entire files of information. In other systems storage is smaller, and data are held only while being processed. Consequently the capacity and design of storage affect the method in which data are handled by the system.

The amount of computer storage available has increased dramatically and the cost per unit continues to drop. Storage is normally measured as kilobytes (10^3 bytes), megabytes (10^6 bytes), or gigabytes (10^9 bytes) which have been shortened to "meg" or "gig."

Rapid improvements and decreased costs have been made by using magnetic storage, where the information is stored on mylar tape coated with a magnetic media, a hard drive, an aluminum alloy disk coated with magnetic media. Or more recently, optical storage, where the metallic disk is coated with a material that is sensitive to laser light, allowing greater amounts of storage. With these refined technologies the amount of storage (from an industrial hygienists perspective) is essentially limitless. Thus, the primary area of concern becomes the frequency and the convenience of access to the stored data rather than the amount of stored data. To illustrate one means of obtaining large amounts of storage, the data may be transferred to a magnetic tape, then the magnetic tape stored in a cabinet. In order to retrieve the data, the user than must get the tape from the cabinet, load it on a tape drive, issue the command for retrieval, and wait for the data. Unless the tapes are well cataloged, this may not be a trivial exercise.

2.4   Cost to Automate

The cost to put in place a data management system varies considerably. A *simplistic view* is to consider only the cost of the software. This is a mistake! This single item may be a small part of the overall cost of implementation. As a base, the hardware costs, software costs (both operating system and application), integration services to connect the application program to other programs, personnel time including both the time to select and specify as well as to train to use, and travel to visit other sites constitute the system cost.

There are numerous trade-offs that will occur in this process. The cost of a solution may be cheaper if it is a PC solution, but it may not have the functionality, capacity, or speed of a more sophisticated solution. And it is only usable by one person at a time. It is tempting to buy little pieces of several different functions from different vendors. Then the systems would not operate in the same way, training costs go up and the utility of the solution suffers. A third example is keyboard entry of material safety data sheets. If the number of sheets is small, keyboard entry may be cheaper than the use of image, but accuracy and speed of entry is lower.

The costs of software alone can be almost nothing if some programming is copied out of a book to do simple things like calculating exposure limits. Some examples of this are found in *Computers in Health and Safety* by the American Industrial Hygiene Association. There is also the opportunity to get these already on a diskette by paying a small fee, a bit of commerce known as *shareware*. Single-application-only software typically costs from $500–$5000 if purchased from a vendor. Some vendors also offer a time share function where you can put your data on their hardware and then pay a charge for the amount of processing and the duration of the connect time. The connection is made through a modem installed in the computer and connected to the vendor over a telephone line.

2.5 Size of a System

Since it is assumed that the reader is familiar with the personal computer, it is well to discuss some of the other computing solutions available. Computer analysts break down the types of computers into four classes: personal computer, workstation, midrange system, and mainframe. There is considerable overlap in price and function between these classes. Also, these different computer sizes are linked together and processing of data occurs on all of the systems, the advantages of each size of system being optimized. In general as you move from PC to mainframe, the power, speed, and capability increase along with the total price for a complete system. These computing systems can be single user/single task, single user/multiple task, multiple user/multiple tasks. In general the PC using MS/DOS® is a single user/single function system. This means that only one person can be using the one function at a time. IBM's OS/2®, or Operating System 2, is a single-user environment but a multiple-tasking environment. With OS/2® on a single PC, the PC may be collecting data from an audiometric measuring device while at the same time printing out a chart of the data. In the workstation environment, usually associated with reduced instruction set computing (RISC), or with a large mainframe, many users may be connected to a system performing multiple tasks at the same time such as accounting, word processing, database searching, graphics processing, industrial hygiene applications, and process control.

3 STEPS IN EDP APPLICATION SELECTION DEVELOPMENT

In designing and implementing any new computer application, there are logical steps of analysis, planning, development, testing, and installation that must be carried out. A means of addressing some of the problems is now described.

3.1 Getting Help

In the same way that a person from the factory floor looks to the industrial hygienist to provide specialized expertise, the industrial hygienist should look to the computer specialist to provide special skills and insight. Too frequently, the industrial hygienist begins data automation by seeking out software providers. Although this step will provide some valuable input into what functions are available and some approximate costs, it will NOT provide the necessary internal foundation to make wise investments in data processing.

One of the first steps necessary is to get someone with the necessary systems background to aid with the decision. Before making a decision on the size of a computing system, what database design is optimal, what are the data transfer and networking requirements, it is well to consult an expert. Most industrial hygienists do not possess the requisite background to successfully answer all the questions, particularly in an efficient and timely manner. The first place to start is with a systems analyst. The job of the systems analyst is to work with the end user in defining requirements and system selection. All large corporations and most locations and divisions of corporations will have someone with either this title or with these skills and duties. Far too many industrial hygienists, in an effort to go it alone, overlook the first step—seek an expert. In all probability, the industrial hygienist can use the extra help also since they are extremely busy.

If a company systems analyst is not available, all of the large computer manufacturing companies have staff available for this purpose. The service of a systems analyst is also available through companies whose primary business is systems integration. The specialists are available by consulting a telephone directory or asking for advice from professional associates.

Once the specialist is on board, he or she will be able to guide the industrial hygienist through the entire process of defining requirements, software selection, providing a justification, giving opinion of provider expertise, providing some reality testing, and assurances that the decisions are the correct ones.

Deciding what, how, when, costs, timing, and so forth for automation are big decisions. It is well worth the time and effort in planning the approach to the problem with all the expertise available.

3.2 What Is Currently Being Used at Your Location?

Operating systems are the control programs that tell a computer how to function. The importance to the industrial hygienist is in understanding that a program written for one operating system will not function in another operating system. Thus, when selecting an application, the industrial hygienist must consider the operating system that is used at the work site. Selecting a program that has the same operating system commonly employed at the work site will increase the effectiveness of a particular application and maximize efficiency of the department.

3.3 Defining and Sizing the Scope of the Automation

When a new computer application is desired, the first task is to *clearly* define the problem. A report should be prepared that states clearly and concisely just what is to be performed by the proposed system and the scope (size) of the effort. This report describes the problem in terms of subject, scope, objectives, and recommendations. This report will establish the basis for communicating needs to the system analyst, to management, and will form the foundation for a proposal for purchase or coding of a program.

3.4 Analyzing the Problem

The analysis phase of a design effort is not an isolated step. The system analyst, who views a new computer application in terms of its scope and objectives, works closely with the users to determine their needs, the information in use in the present system, and the information needed in the new system. The analyst must also consider what equipment would be the most cost-effective for the necessary functions of the new system, determine the basic functional specifications, approximate price, duration of project, manpower requirements, and contractual performance requirements.

3.5 System Selection

Some broad policy level decisions need to be made as a necessary first step. First, the decision must be made as to how many persons the system is to serve. If the answer is a simple application such as recording the building and type of laser at a location for use by one industrial hygienist, the task is straightforward. If this is the only requirement, there are number of very simple databases that could accommodate the needs, and a simple PC can accommodate this task. Frequently, the need is much greater, even for an individual with a simple concept. As in the example above regarding lasers, the industrial hygienist would like a clock or calendar function so if a new laser is registered, the data and time can automatically be noted. If the industrial hygienist wishes to sort by laser type, another variable is needed. Tracking of the users and their education or retraining could be another addition to the system as could the location, last date of inspection, and so forth. Further, it is desirable to have the education record added to the personnel file, so some integration of the data is useful. Another consideration is whether other hygienists or associated personnel such as physicians need to access the information during the normal course of their work or during absences of the industrial hygienist. Thus, it is advantageous if others have knowledge of the way the data management system works so that access can be quick, easy, and reliable. As this simple example illustrates, the problem grows quickly.

The primary question facing the industrial hygienist is: ''Do I need an automated system for my data?'' It depends, is usually the answer. Although this is not a very satisfying answer, it does represent the truth. There are many conditions to be reviewed. Some of the questions are summarized below:

- Quantity of data to be managed?
- How will the data be used? Federal or state reports? Management reports?

- How will the data be accessed? Terminal? Batch reports? Other applications?
- Who needs the data besides the end user? Corporate? Medical? Division?
- How often, when, and under what circumstances do I need the data?

If the decision to automate is affirmative, then several other questions need to be answered:

- Which applications will be automated and in what priority?
- Do the applications stand alone or do they require data from another source such as material safety data sheets?
- Is the project going to be combined with another such as the automation of environmental, medical, or safety data.
- How is the investment justified?
- Is a customized system needed? Semicustomized? Off the shelf?

It is advisable to make every effort in answering all of these questions correctly and thinking through how the data management system will be used. This will save much time and labor later when it is time to bring the system into production.

3.6 Combining Needs with Other Areas

Each person wishing to automate their data needs has to consider many things besides functionality. To the industrial hygienist, the paramount issue may be air sampling data, to the environmental engineer it may be EPA's Superfund Amendment and Reauthorization Act report (also known as SARA, Title 3, form R, or SARA 313), or to the safety engineer, it may be accident data tracking. One legitimate question seems to be whether to combine the project(s) for each of the above four organizations or to go it alone. Experience has shown that compromises are hard to reach since each party wants to have his or her functions done first or has some special need that the others do not have.

How do you resolve this problem? One means is to have a structured discussion format with moderator and analyst. In the study of information science this method is known by various names but the words joint application design (JAD) or joint application requirements (JAR) are often used to describe the techniques. The JAD or JAR has five fundamental pieces: (a) A top-down system design process with an executive sponsor who can resolve issues if not resolvable at a lower level; (b) participants work together in a group session generally of about 8–15 people; (c) synergy is developed between management, users, and information systems; (d) the participants jointly produce requirements and (if a JAD) design specifications; (e) the project is facilitated by experienced people, typically one facilitator and one analyst. The outcome of the effort is a prioritized list of requirements for the project and a design if carried through to this point. Appendix G gives a more detailed description of a JAD.

3.7 Justification of the Investment

Cost justification for automation is a difficult but necessary task. There are three fundamental ways that savings can be accomplished by the use of computerized data management system: (1) the reduction of manpower used to perform specific tasks such as the preparation of reports, summaries, and planning documents; (2) better analysis of data for accurate assessment of trends, quicker analysis of data, more complete analysis such as with the use of statistical tools and patterns analysis; and (3) rapid transmission of this data to those who use the data such as production managers, corporate functions, and so forth.

Regardless of the business model used, most justifications will include reduction in clerical time, shifting of responsibilities from more expensive personnel such as industrial hygienists to clericals or data entry personnel, ability to respond quickly to requests for information such as an OSHA visitation, making the data available to others either in the department, within a division, to an overseas affiliate, or to the corporate offices. Although hard to accurately quantify the benefit, the ability to respond quickly and accurately to questions posed by production floor managers, is certainly valuable. With a computerized database, the industrial hygienists can respond to these requests quickly and accurately, can prepare a presentation for employees in a reasonable time, and then go on to do other things. Better use of existing data, such as on statistical comparisons, epidemiology, and so forth are greatly aided by computerization.

3.7.1 Breaking Down the Costs

The costs of a system may be broken down as follows:

- Costs to purchase
    Hardware
    Software
    Personnel

- Costs to produce
    Initial costs for the functions sought
    Systems analysis costs
    Programming costs
    Testing costs
    Training costs
    Implementation costs
    Documentation costs

- Costs to operate
    Day-to-day costs
    Repair costs (hardware and software)
    Maintenance costs (hardware and software)

3.7.2 Spreading the Costs

Prudence is needed in estimating the costs. For example, a particular application may require a graphically enabled monitor. It is more expensive than a monitor with less capability, but since a monitor is being purchased for some other reason, only the incrementally larger purchase price should be used. If a workstation is included in the purchase order for this application, is this the only application for which the workstation will be used? The same logic holds for larger hardware items such as an upgrade to the size of the mainframe because of additional load. It is tempting to add this upgrade cost onto the price of this one application even though there are likely many other applications running on the same system. The rhetorical question is: "is it the first application put onto the computer system causing the upgrade or the most recently added application?" Larger firms will have a methodology for coping with this seemingly intractable problem. Unfortunately, too often the justification model calls for putting all of the costs onto the most recent project (yours). There is no easy solution to this problem, but recognition that this justification methodology may be used will help to prepare your business case. You must become familiar with the business case used in justification of investment in EDP. Ask the information systems personnel for model proposals.

There are similar problems to consider when purchasing software. There is a whole range of tools that are used by various applications. They may be graphical enablers (those software packages that allow you to use a graphic system), imaging systems that may be used for many applications in addition to the industrial hygiene applications, specific workstation software that may compress and decompress images for storing or viewing, statistical packages, instrument connects, and many more. These special tools packages are usually called by an application program when needed. Some are built into a particular software application but others are called when needed. A good example of the latter is a printer driver program. Most applications need to print out results, thus there is no point in the printer driver being in every application, so it becomes one that is called up by the application as needed. This saves valuable memory space and is more efficient then writing the driver for each application. Also, if the printer is changed, you do not want to change each application to accommodate the new printer.

If a software package is a commercial off-the-shelf system (COTS), the base cost of the application package needs to be determined. Different vendors of software use different pricing structures. Some are based on the size of the CPU processor, others use as a basis the number of users, still others price by the number of employees at the location, and still others use a fixed cost. Regardless of the software providers pricing structure, consideration needs to be given as to whether and where there is training, is the training billed separately or a part of the initial cost, is it on location or at the vendor's location, and so forth. Also, is there any customization included in the cost? The *base cost* of the system may NOT be the ultimate cost if customization is required. If this is a customized package, we must consider the internal costs of a systems analyst, programmer, and so forth. These are costs to the company or site. Sometimes these costs are charged directly to the internal customer (you, the industrial hygienist), sometimes they are fixed fee, and sometimes they are covered in a site overhead-account. Each alternative has a different impact on your budget. Understand this impact before committing!

3.7.3 Common Look and Feel (Usability and Familiarity)

If a software package is purchased, it is still likely that there is a need for some systems analysis time and some customization of the software. The screen design (this is the way a computer screen appears) typically replicates some form that is in current use. Significant decisions need to be made at this point since a computer screen has a different number of lines than an $8\frac{1}{2}$ by 11-in sheet of paper (there are large screens that can accommodate the whole page, but this will increase costs). The users of the system will be much more accepting if it looks familiar. Thus, if the function key (or pF key), pF = 12 is usually used for filing, it is wise to use this same key on your program. If the help key is pF = 1, then this ought to be the one used in your program.

In addition to functional characteristics, buy or build decisions, there are some general considerations that are of great importance. The first of these is the usability of the system. Regardless of the power of the system, if it is difficult to learn and use, chances are pretty good that you will not be satisfied with it. If the decision is made to use several smaller packages, then a number of problems are incurred. One of the real advantages of a common system for a company is the ability for all to use it easily and readily. Another way of saying the same thing is to minimize training costs. If a common system is used, normally one supplied by one provider or vendor, the following advantages will occur. They are:

Minimized training costs
Minimized development or customization costs
Purchasing leverage
Ease of integration with other applications
Remote locations with equal access

The authors would be remiss if they did not discuss a down side to the use of a common system. At the time of this writing, the authors are unaware of any single vendor that has all the functionality that is needed for environmental, safety, and industrial hygiene.

3.8 Buy or Build

The most critical question to be answered is: "What are the requirements for data automation?" If this question is not answered, then it is foolish to even consider the buy or build decision. The hygienist must first determine exactly the need that is being filled. Once this determination is done, then one can begin shopping for what is available. Unfortunately, many people start out by looking at software packages rather than analyzing their own data processing needs. This is truly the cart in front of the horse. While looking around for ideas is certainly an acceptable way to gain insights into what is needed, the base decision comes back to the industrial hygienists' needs.

The list of choices available in making the buy or build decision is as follows:

1. Buy a software package and integrate it into an existing system in the company.
2. Buy a hardware and software package that can be integrated into another system being developed in house.
3. Buy a hardware and software package that will meet the complete needs of the project with no major modification.
4. Buy a hardware and software package that meets part of the needs (50 percent) of the project and contract with the vendor to develop the remaining requirements.
5. Buy all hardware and operating software and contract with a vendor to develop the application software according to your functional and technical requirements.
6. Buy all hardware and operating software and develop the application software completely in house.

None of these options is exclusive. Normally, some additional hardware must be purchased, some additional software must be purchased. Additional hardware is usually needed if facsimile devices are used, scan images of material safety data sheets are used, more people are using computers, and so forth. Even if application software is purchased, there may be some additional report generation features, scanning software, graphics programs, and so forth that are needed. Because different means and ways of doing business are used, the methodology will change. For example, if the calibration of the instruments is part of the software, there is no need for calibration for the equipment.

In the seventies and early eighties, there was a dearth of software programs available for the industrial hygienist. Today, with the popularity of workstations, midrange systems, and mainframes, the industrial hygienist has much more from which to choose. The Center for Environment and Energy Management in Washington, D.C., has put together a list of advantages and disadvantages of buying or building the needed software. Table 9.1 summarizes these.

A few additional comments are appropriate in making the by or build decision (2):

- Cost effectiveness is a major factor in the decision process. But remember that if the system is not used by the right people for whatever reason, the cost per user rises proportionately. The lowest total price is not necessarily the right choice. The cost should be prorated over the entire lifetime of the product, the number of users, the costs to run the system (computer people like to say the number of cycles that are burned), and the cost maintenance as well as convenience of use.

Table 9.1 Build or Buy: Items That May Influence Your Decision

	Pro	Con
Build	Tailored. Good fit. May be able to market the result.	Unique. Support? Cost is substantial ($1.5M+). Timing uncertain. Training on use is less thorough. Loss of a key programmer can be very damaging. Maintenance charges will be substantially higher over time. More industrial hygiene professional staff time will be needed during definition phase and implementation phase.
Buy	State of art. Bite-sized chunks. Many user's groups. Can be online quickly. Likely much cheaper.	Not comprehensive. Upgrades difficult with some.

- The buy or build decision process does not reduce your need to assess what it is you want to do with the information. It does not allow you to skip any of the important phases of the project. A feasibility study (a pilot project is recommended) and a functional system design are still required to determine your project requirements.

- The buy or build decision will carry over and greatly impact future project tasks such as testing, installation, documentation, training, implementation, and maintenance requirements.

- The buy or build decision should take into account how soon you need the product. Build will usually take longer to get to production than when buying a system ready to go.

- What are your long-term ownership requirements and future maintenance requirements? Do you want to be able to do the work yourself?

- Do you want to retain full ownership and possibly market the product at a later date? If a marketing organization is not geared to this product and customer set, funded for the effort, and imbued with the skills, prospects for success are dismal. Regardless, the business case should be built on your needs and any fortunate outcomes considered to be gravy.

3.9 Selecting an Area for Automation (A-Pilot Effort?)

There are numerous application areas that could be selected for automation. These include air sampling data, training, lasers, and so forth. The reader is referred to Appendix A for a more complete list of the possibilities. By using a weighting system for the applications some initial assessment can be made. However, caution is needed in this selection since an item with lower priority may be a stand-alone application and can be done easily, thus the payback is quick and the solution simple. If a simple application is selected first, experience is gained in the automation process that may be applied to future automation efforts.

3.10   Implementation

Too often, industrial hygienists believe that the job is finished when a selection process has taken place or a functional specification has been written. Nothing could be further from the truth. Now the *real work* begins. The implementation phase is absolutely critical to success of the project. No matter how careful requirements are gathered, no matter how much probing and examining is done in the selection or programming of the software, no matter how sophisticated one or two users are, the project is going to have a very rocky start if the *implementation plan is flawed*, under staffed, or poorly organized. It is of paramount importance to get off to a good start, experience some early successes, and get the user community friendly toward the project. There are bound to be some detractors among the users since some will not want to change, some have a vested interest in the old system, some are threatened, and some do not see anything in it for them. Thus, the project organizer has two aspects of the project to address, the technical and organization plus the public relations and communicative aspects; both are addressed below.

3.10.1   Organizational and Technical Project Aspects for the Users

Although the following list is not complete, it outlines the major tasks that must be completed in order to be successful. The outline is for use by an end user or industrial hygienist/manager. Your systems analyst will have a different list of actions to do although there is a relationship between them.

- Define current environment
    Hardware configuration
    Networking
    Hardware available
- Define new hardware requirements
    Hardware, printers, monitors, cables, computers
    System requirements, additional burden to network or mainframe
    Data storage requirements, how much data storage
    Define transaction rate, optimum, mean, low
- Define new software requirements
    New systems software or version of the software
    New workstation or PC software required
    New application software
- Define test project/environment
    Develop test plan
    Define test time frame
    Define beginning/end of test
- Order/receive products
    What is minimum needed for test?
    Lease/buy/borrow?
    Configure

- Define personnel requirements for test
  Who will test?
  Is the test environment representative of working environment?
  What will be left out/included?
  What is success? What concludes the test period? Measurable results?
- Preparation for full installation
  Roll out plan
  Define personnel requirements
  Define training plan
  Create user's manual
  Establish "hotline" or Question & Answer board or electronic bulletin board
- Install full system
  Acquire remaining hardware items
  Acquire remaining software items
  Assign personnel
  Develop security plan— who has access, when?
  Develop project schedule
  Develop transition plan—critical decisions must be made about when and what data is migrated to the new system. Do you start with day 0 and put only new data in the system? Do you migrate only last year's data? Two years? All data? What subsets of the data?
  How do you manage when elements of the old system are still in place and you are bringing up a new system?
- Exploitation of the system
  The new system has capabilities that did not exist before—how do I discover and take advantage of this?
  What changes could be made to take even greater advantage of the system?
  Generation of standard reports
    What are these reports?
    What is the periodicity?
    How do we distribute the reports?
  Plan a celebration.
    Recognition
    Summarization

4 AN INSTRUMENT-CONNECTED SYSTEM FOR MONITORING

An example of an automated air monitoring system is the Industrial Centralized Air Monitoring System (ICAMS) manufactured by Perkin-Elmer, Applied Science Operation in Pomona, California. The ICAMS® is a mass-spectrometer-based centralized ambient air monitor. The system provides broad facility coverage for environmental and industrial hygiene monitoring. It is a double-focusing, magnetic sector spectrometer. Magnetic sector mass spectrometers are well known for their stability over extended periods of time. ICAMS can accommodate up to 25 compounds and up to 50 sampling points that can be located as far as 1000 feet away.

All system operations are controlled by Data System (DS), which is an IBM PS/2® 50Z desktop computer. Spectral data is acquired by the DS and is converted into concentration data. The concentration data is displayed along with the identification of the remote location from which the sample was drawn. Sample sequencing, compound to be monitored, and alarm activation are customized by the user through the DS.

The DS functions as a display to report gas concentrations, as a data storage device for storing Investigative Scan data, a means of entering system configuration data, and as a controller that acquires data from the ICAMS® and processes it into concentration results. A remote terminal interface (RS232-com1) is included with the Data Station.

The printer provided with the DS automatically produces a hard copy report of alarms and allows a printout of any screen on the CRT display. The ICAMS® operating software is a menu-driven program. It is designed to interact with the user by displaying a menu of selections on the Data Station screen. A remote terminal can be connected to the Data Station so that ICAMS can be controlled from a remote location. The user has the ability to monitor system activity remotely as well as change accessible parameters.

Off-line analysis of the data gathered by the ICAMS® system can be accomplished by replacing the Remote Terminal with a PS/2® model 50Z personal computer referred to as the Remote Analysis Station (RAS). The RAS emulates a VT220 terminal as well as being capable performing off-line analysis of Investigative Scan data as well as concentration data via optional software packages.

Along with acting as a remote terminal, the RAS serves as archiving center. It provides data sorting, compression, conversion and record archiving of concentration data, Investigative Scans, and diagnostic messages from up to 99 ICAMS units. The user selects which data will be downloaded to the RAS's hard disk drive and compressed for future retrieval. The data can be retrieved at a later date and used to do trend analysis, long-term worker exposure, and real-time alarm tracking. The RAS may also be used to transfer data to a host computer. Parameters, such as date, site, and data type, are defined by the user for the data files to be retrieved. The RAS provides an efficient means for data storage and retrieval. Different application software packages allow for easy manipulation of the data obtained for the RAS.

Library Search is an application software package that resides in the ICAMS® Remote Analysis Terminal. Library Search enables the user to quickly identify unknown chemical compounds not found in the ICAMS® pre-programmed monitoring list. An Investigative Scan is retrieved from the RAS and run through the library Search Program. The spectral patterns of the Investigative Scan is compared to more than 30,000 spectral patterns of various chemicals that are part of the Library Search program. The reference library is derived from the National Standards Library (NBS). The method used in identifying the unknown is the Probability Based Matching routine. A ''reverse search'' technique is used that compares each mass spectrum in the mass spectral library to the investigative scan of an air sample containing the unknown compounds. The user has the flexibility to define libraries for specific compounds of interest.

The Trend Application software is another applications package that resides in the ICAMS® Remote Analysis Station. The Trend software enables the ICAMS® user to generate reports of concentration values for user-specified time intervals. Via the RAS, concentration data is collected from the ICAMS® and stored in the Trend Historical Databases. Compounds, ports, and the time interval for data collection is defined by the operator using the Trend Analysis setup screen. The operator has the ability to set up several different time periods for data collection on one ICAMS. The Trend Reports generated by the Trend Application software present the data in the form of time-weighted averages. Low, high, and average concentration may be displayed. Several different formats are available for report presentation. The user can specify Trend Reports to be generated by compound, by location, or by trends over time. The Trend Reports are useful for evaluating chemical trends for a facility as well as closely monitoring worker exposure levels. Average concentration readings, standard evaluations as well as the low and high readings can be displayed.

The ICAMS® and the Remote Analysis Station are current examples of automated monitoring systems. Data collection and handling is automated through the use of IBM PS/2® computers. The ICAMS® system collects data and the RAS allows for easy access and data manipulation.

5   CD-ROM Technology

One of the most pressing problems faced by industrial hygienists is finding information quickly, accurately, and as effortlessly as possible. One of the exciting new technologies is a memory disk capable of holding large amounts of data that can be accessed quickly. This technology is now described.

5.1   Description of CD-ROM

CD-ROM (compact disk—read-only memory) is a relatively new technology that allows information to be stored at very high density for rapid retrieval. Data are stored on a clear plastic disc with a metallic underlayer, which is manufactured in a process identical to that of an audio disc sold in music stores. This information is arranged in a continuous spiral track, starting at the center of the disk and spreading outward. *Disc* is the standard convention for spelling the CD-ROM media, while *disk* refers to magnetic storage media such as floppy disks or the hard disk drive of a personal computer.

Very large amounts of information can be put on a single CD-ROM disc. Currently the 120-mm (4.72-in.) diameter disc, which is the most popular size, can contain up to 660 megabytes of data. This is equivalent to over 1800 360-kilobyte floppy disks, or a 7-story high stack of $8\frac{1}{2} \times 11$-in. double-spaced typed pages. With data compression techniques the total can be significantly higher. The upper limit of information storage capacity on a CD-ROM disc is largely determined by limitations of the reading hardware rather than the disc itself.

Information is stored on a CD-ROM disc by a process of burning depressions, called pits, into the track with a laser beam. The pits may be of varying length, and they alternate with the raised, unburned areas, which are called lands. The pits and lands are covered with a reflective coating.

The mass publishing of information on CD-ROM is accomplished by first burning a master disc, followed by duplication of the master into the desired number of copies. While production of the master disc is quite expensive, on the order of $1000, duplication of copies is generally possible at a cost of $2 per disc or less. In the duplicated disks, the pits and lands are contained inside a clear overlayer of plastic, and are thus protected from degradation by the external environment. These discs are mass replicated by a stamping process.

When the disc is used in a special player called a CD-ROM drive, the information is read by another laser beam that follows the spiral track of the pits and lands. When the laser encounters a pit, the beam is reflected back to the reader in focus. In other words, the height of the pit is at a critical focal length for the lens of the laser, such that the reflected light from the pit is still in focus when it returns to the detector.

A photodetector cell measures the intensity of the reflected light, and the time interval of each episode of scattering and reflection is monitored and translated into electrical signals of "off" and "on," which are interpreted as the 0's and 1's of binary machine code. The binary code is related to the original digital input. The duration of a pit or land is equivalent to a 0, and the *transition* between a pit and a land is the 1.

Because two consecutive transitions and hence two adjacent 1's cannot occur, a convention has been developed that translates the 8-bit word of machine code into a 14-bit scheme called eight-to-fourteen modulation (EFM). To increase the accuracy of reading, three additional spacer bits, called *merge bits*, are used between the 14-bit words. Thus the 8-bit word in electronic media becomes a 17-bit word on CD-ROM.

In addition to the data stored on the CD-ROM disc, some form of *retrieval software* is necessary to access data. Retrieval software identifies the locations of all the data on the disc, and also defines the formats of the menus and screens that appear on the monitor of the computer when the CD-ROM system is used, and the searching scheme for accessing the data. The retrieval software can be in the form of separate floppy disks, or can be built into the CD-ROM itself.

Information is usually transferred from files in a mainframe computer to tape, and the information on the tape is transferred to CD-ROM by a complex process of premastering and mastering, which may be performed by a service company. The process of making a master CD-ROM disc from tape can be quite expensive, in the range of several thousand dollars. Once the master disc is made, multiple copies of the master disc can be produced at a much lower cost per disc. Thus information can be put on CD-ROM by an individual company for its own internal use or by an electronic publishing company for commercial distribution. A list of various service companies, publishers, and other resources for CD-ROM is provided in Appendix A.

5.2 Related Technologies

Other developments in related optical technology make the storage of information more direct. One such development is the WORM (write one–read many) optical drive. This machine allows the user to enter data directly into an optical disc. The information then becomes permanently archived on the disc and can be accessed in a similar manner to mass-produced CD-ROM discs. Another technology is called DRAW (direct read after write). Both WORM and DRAW provide direct means for the generator of the data to write directly onto the disc, thus bypassing the expensive mastering step. These technologies are still relatively rare, however, and are not directly compatible with CD-ROM.

Videodiscs, which use an analog recording technique, are related to CD-ROM, and record visual images and sound. While the main market for video discs has been in the entertainment industry, they offer much potential for health and safety training. Videodiscs can be programmed with interactive software, to provide customized training and testing capabilities.

5.3 CD-ROM Hardware and software requirements

CD-ROM discs require a special player called a CD-ROM drive, which is somewhat different from the audio CD players is music systems. There are numerous manufacturers of the CD-ROM drives, but the most widely used models are made by SONY, Hitachi, Toshiba, and LMSI (formerly Philips). There are various ways to accommodate more than one disc when several different CD-ROM systems are installed on the same computer, or when a single system is on more than one disc. Discs can be changed manually by the user, just as with a floppy disk drive on a personal computer (PC). Alternatively, certain models of single CD-ROM drives can be "daisy-chained" (connected in series). There are also newer "juke box" CD-ROM drives that can accommodate more than one disc in a single player. The price of single-unit CD-ROM players has dropped considerably over the past few years, and they are generally now available for $400 to $1000. CD-ROM drives usually have a lifetime of several years with average use.

A freestanding CD-ROM drive may be connected by cable to a special card installed in the CPU of a personal computer. There are also CD-ROM drives that can be built into the PC unit. Besides the CD-ROM drive, a personal computer is required. There are CD-ROM drives and computer applications available for most major brands of PCs including IBM, Macintosh, and the NEC PC9801. The vast majority of CD-ROM information systems are made for IBM and compatible PCs, however. Some portable PCs can accommodate built-in CD-ROM drives.

The configuration of the personal computer is usually not critical for running the CD-ROM systems. Most CD-ROM products will be adequately displayed in CGA, EGA, or VGA monitors. For systems involving information in graphical format, a VGA monitor with a graphics adapter is recommended. A hard disk is required for some CD-ROM products, but CD-ROM systems typically do not require much storage space on the hard disk. The retrieval software may occupy only a few megabytes on the hard disk, but the information continues to live in the CD-ROM. Generally at least 512K of RAM (memory) is required. CD-ROM products will support various kinds of printers.

Most CD-ROM products also require a recent version of Microsoft Extensions software. The Microsoft Extensions program configures the PC to recognize the CD-ROM drive, which is usually designated as the D drive, and allows the programs on the PC to communicate with the CD-ROM drive. Sometimes the required Microsoft Extensions software is provided with the CD-ROM drive.

Occasionally there may be software incompatibilities between different CD-ROM products when they are installed on the same computer. The best CD-ROM publishers have 24-hr technical support groups that can guide the customer step-by-step through the intricacies of diagnosing and correcting these incompatibilities. Usually the problem can be corrected by editing one or more files on the hard disk.

5.4 CD-ROM Delivery Systems

In general, information from a CD-ROM can be distributed in the same ways as any other electronic information. Besides the freestanding PC for a single user, some CD-ROM products can run on a LAN (local area network) for multi-user access. The CD-ROM systems published by Micromedex, Inc., can support both the Novell and Token Ring networks. The theoretical upper limit to the number of simultaneous CD-ROM users on a network is a complex function of the hardware and network software. For a LAN, up to eight users can simultaneously access a CD-ROM without degradation of performance.

Micromedex has also developed proprietary MAS (multiple application server) systems in which the CD-ROM resides on a PC and the PC is connected to a mainframe. By means of these MAS systems, the users can combine the advantages of a mainframe network with those of PC-based systems: Very large networks can be supported, and the process of loading new product upgrades is greatly simplified compared to a mainframe.

The development of portable PCs has allowed CD-ROM information systems to be truly portable as well. At the present time, the HazMat response trucks of the Fairfax County, Virginia, Fire Department have portable computers on the trucks with CD-ROM databases installed. Thus the HazMat response team can have full and rapid access to critical information at the scene of incidents involving hazardous materials.

Portable or laptop PCs can also be used with modems and cellular phones in the field to call a host base where the CD-ROM system resides. This true remote access capability requires additional software programs, including communications software and a data-linking program such as Carbon Copy. Remote access to health and safety data is especially helpful in high-risk field operations, such as logging, construction, petroleum exploration and drilling, mining, and pipeline maritime operations.

5.5 Suitability of Information for CD-ROM

CD-ROM is especially useful when a large amount of information needs to be accessed or stored, when the information does not change rapidly, and when rapid access to any or all of the information is required. All of these conditions need to be fulfilled in order to make CD-ROM storage the most cost-effective technology in comparison to paper, microfiche, or magnetic media.

Some examples of information that are well-suited for CD-ROM in the health and safety field are: access and storing of industrial hygiene, medical and safety surveillance records, analytical laboratory methods, collections of reference visible, infrared, and ultraviolet spectrophotometric, nuclear magnetic resonance, and mass spectral patterns, and analytical laboratory data; computer-based training for hazard communication and other OSHA requirements; collections of material safety data sheets (MSDS); bibliographic databases containing references and abstracts of published studies; compiled databases containing reviews and summaries of health, safety, and environmental data; recommendations for medical treatment of occupational and environmental overexposures and for HazMat emergency response; selection of protective clothing and equipment; statutes and regulations. An expanded discussion follows on each of these uses, along with information on commercial CD-ROM products in these areas.

Industrial hygiene, medical, and safety surveillance records are especially well-suited for CD-ROM. Typically these involve large amounts of data that may legally need to be archived for many years. Archiving these records in paper form creates problems of space, accessibility, durability, and vulnerability. Often the paper on which these records exist deteriorates rapidly, especially heat- or pressure-sensitive recording paper used for audiograms, pulmonary function testing, electrocardiograms, and clinical laboratory data. Even magnetic media is prone to deterioration when undisturbed for long periods of time. Single-paper copies of critical health and safety records are vulnerable to loss by fire or natural disaster.

There are several advantages of transferring these surveillance records to CD-ROM. Archived data can be accessed just as easily as current data. This facilitates and encourages examination of surveillance data for a given individual or department over a long period of time, and also makes group data easier to obtain for epidemiological studies. Multiple copies of the archived data can easily be made; facilitating exchange of data between locations and minimizing the chances of losing data stored at one site. Much less storage space is required than on a mainframe, and the data on CD-ROM discs does not deteriorate as rapidly as magnetic media.

It is important to remember, however, that CD-ROM (or any other form of EDP) cannot totally eliminate the need to archive original medical records in the United States.

Analytical laboratory methods and reference spectra are especially well suited for CD-ROM. Large collections of detailed methods can be compiled, indexed, and integrated with reference spectra chromatography elutin profiles. To the best of our knowledge, detailed laboratory methods are not available on CD-ROM at the present time, but some reference spectra are.

CD-ROM and its variant, laser discs, can also be used for interactive training programs in health and safety. Training, testing, and documentation of training for individual employees can theoretically be accomplished with a single product.

Large collections of MSDSs can be put on CD-ROM. It should be possible to control access to particular MSDSs at the level of a company, location, department, or employee. MSDSs on CD-ROM can be on a network for accessibility in the workplace.

Bibliographic databases also work well on CD-ROM, and heavy use of the CD-ROM versions can result in considerable savings over the same databases on-line. One minor drawback is that bibliographic databases on CD-ROM tend to be somewhat less current than their on-line counterparts.

Compiled databases containing information on toxicology, medical treatment recommendations, HazMat emergency response, and selection of protective clothing and equipment, are generally suitable for CD-ROM. It may be desirable, or even necessary, for this kind of information to be portable or accessible from remote locations, or at the very least as part of a network with access directly from the workplace.

Statutes and regulations on CD-ROM are useful for many purposes, especially where rapid access to legal information and savings relative to on-line access are required. One disadvantage of regulations on CD-ROM, or any other electronic media with periodic rather than continuous updating, is that some of the information may be out of date.

5.6 Advantages and Disadvantages of CD-ROM

One of the main advantages of CD-ROM is the vast amount of information that this technology permits to be stored in a PC environment and the speed at which the information can be obtained. Typically a 120-mm disc can store over 660 megabytes of data, and rapid retrieval and indexing systems allow the user to find the desired information in a matter of seconds. CD-ROMs allow the PC user to have access to mainframe-size files, with the power of a mainframe, at a fraction of the cost.

CD-ROM has several major advantages over other forms of storage for archival purposes. In contrast to magnetic media, files on CD-ROM cannot be intentionally or accidentally erased or changed by computers or exposure to magnetic fields. The overlaying of the pits and lands by a clear plastic layer protects the data from deterioration during use or by exposure to normal environments.

CD-ROM also has a much higher rate of accuracy than electronic media. While estimates of its accuracy have varied and may be a function of the specific hardware and software used to generate a CD-ROM, one figure places the lowest attainable error rate of CD-ROM at 1 in 10^{15} bits (3, Appendix A). This accuracy is higher by several orders of magnitude than floppy or hard disks. Even DNA replications in living cells has an intrinsic error frequency, or spontaneous mutation rate, on any given cell division of the order of 4 in 10^8 to 3 in 10^5 (5, Appendix A).

Storage of information on CD-ROM discs requires much less space than paper, microfiche, or portable magnetic media. With a data density of over 660 megabytes per 0.7-ounce disc, CD-ROM can store approximately 15 gigabytes per pound.

Another way of stating the storage capacity is in terms of bytes per cubic foot: a 4.72-in. disc is approximately 1/16th of an inch thick, representing a total volume of 0.46 in.^{3} or 0.001 ft.^{3} This is equivalent to approximately 600 gigabytes per cubic foot. A book of 100,000 words, or approximately 1 megabyte, formatted on 6 × 9-in. pages and 1-in. thick, would store information at a density of 33 megabytes per cubic foot, more than 20,000 times less than CD-ROM. Finally, $8\frac{1}{2}$ × 11-in. double-spaced typewritten sheets, at 250 words per sheet on 20-lb. paper, would contain roughly 11 megabytes per cubic ft. This is 60,000 times more space than the same information would occupy on CD-ROM.

Another advantage of CD-ROM is its efficient use of space in computers as well as on the bookshelf. Information on CD-ROM is on the same order of magnitude in size as mainframe files, yet CD-ROMs do not tie up storage. Thus, converting very large mainframe files to CD-ROMs, and linking the CD-ROMs to the mainframe with an application server, may be cost-effective in the long run because it may postpone the need to upgrade the mainframe computer.

There are some drawbacks and unknowns in CD-ROM, however, that may make this technology unsuitable for some needs. The cost of producing only one or a few discs is relatively expensive in comparison to magnetic media; therefore CD-ROM is best for applications where many copies of the disc need to be made for security or distribution, and where the information does not change frequently.

Secondly, CD-ROM is much less efficient in storing graphic images than text, but this is due to the nature of graphic images rather than to some intrinsic inefficiency of CD-ROM. Storing graphic images on CD-ROM would not be as efficient in terms of saving space as the examples given above for storing text.

A major limitation with respect to satisfying legal requirements for archiving is that the technology is so new, that not enough time has elapsed to define the archival lifetime of data stored on CD-ROM discs. It is clear, however, that CD-ROM discs have a longer lifetime without degradation or loss of data than magnetic media.

5.7 Examples of CD-ROM Information Products

CD-ROM can be a very cost- and time-effective way of accessing large amounts of published information. At the time of this writing, the potential for delivering information on CD-ROM in health and safety is still in its early stages but changing rapidly.

Appendix A lists some CD-ROM products and suppliers that industrial hygienists, safety engineers, occupational physicians and nurses, and others in health and safety may find useful. The numbers in the righthand column refer to companies listed in the ''CD-ROM Publishers'' section of Appendix A.

CD-ROM products can change rapidly in terms of the available products and their contents. The reader is advised to contact the various publishers of CD-ROM products for current information.

5.8 Summary and Conclusions

CD-ROM is the method of choice for information storage and retrieval when large amounts of data need to be accessed rapidly, and when the data do not change often. CD-ROM can make very large amounts of information available through a personal computer, thus saving the expense of acquiring and maintaining this information in a mainframe environment. It is relatively expensive to produce one or a few copies of a CD-ROM disc, but its excellent archiving properties and accuracy in comparison with magnetic media may justify this cost for companies to convert information to CD-ROM for internal use. While there are already many published sources of information on CD-ROM in health and safety, the field of CD-ROM publishing is still in its infancy.

6 SUMMARY

The intent of this chapter is to provide a useful guide for practicing industrial hygienists in the development of a system for managing the data records with which they work. Considerable gains in productivity are expected by more efficient gathering, searching, comparing, and reporting of this data. It is expected that the automation of the data management process will continue for some time to come, with ever more sophisticated and cost-effective solutions being developed. The leading edge industrial hygienists will want to use these new solutions to enhance their job enjoyment as well as enable them to contribute in even greater ways to their employers or clients.

6.1 Acknowledgments

Many people and companies have contributed to the thinking that this is contained in this chapter. Principal among those is the IBM Corporation, which has provided material and input for the technical aspects of the chapter. Also, Betty Dabney of Micromedex in Denver, Colorado, provided the material on the use of CD-ROMs, Edna Shattuck of General Research Corporation in Vienna, Virginia, has graciously provided the material found in Appendices B–E to be used in selecting software. The section on an instrument-connected system for monitoring of chemicals was contributed by Dana Scott of the Perkin-Elmer Corporation of Palo Alto, California.

REFERENCES

1. C. P. Wrench, *Data Management for Occupational Health and Safety-A User's Guide to Integrating Software*, Van Nostrand Reinhold, New York, 1990.
2. J. D. Woodward, "Environmental Information System—Buy or Build," The Environmental Technology Expo, Apr. 8–11, 1991, Chicago, IL.

APPENDIX A. CD-ROM APPLICATIONS
Examples of Commercially Available CD-ROM Products[a]

Type of Information	Publishers
Bibliographic databases	
Acid Rain Abstracts	4
Administrative (Health)	8
Agricola	31, 35, 36
Aquatic Sciences and Fisheries Abstracts	6
Bibliomed Cardiology Series	12
Bibliomed Citation Series	12
Bibliomed Gastroenterology Series	12
Biological Abstracts on Compact Disc	4
CANCERLIT	2, 8
CDMARK Bibliographic (Library of Congress Catalog)	20
Compact International Agricultural Research Laboratory	9
Ecological Abstracts	6
Energy Information Abstracts	4
Environment Abstracts	4
Environment Periodicals Bibliog	28
Excerpta Medica Abstract Journals	36
Federal Prime Contracts	25
Food Science & Technology Abstracts	6
Life Sciences Collection	6
GPO Monthly Catalog	31, 36
Health and Safety Science Abstracts	6
MEDLINE	2, 6, 8, 11, 12, 13, 36
NIOSHTIC	6, 7
NTIS Database	11, 32, 36
Nursing & Allied Health (CINAHL)	8
Pollution Abstracts	6
Toxicology Abstracts	6
TOXLINE	4, 6
Water Resources Abstracts	28, 31
Wildlife and Fish Worldwide	28
Books	
Agrochemicals Handbook	34
Concise Encyclopedia of Science and Technology	21
Dictionary of Science and Technical Terms	21
Electronic MRI Manual	2
Kirk-Othmer Encyclopedia of Chem. Tech.	10
Physician's Desk Reference	14, 22
Shepard's Catalog of Teratogenic Agents	23
Yearbook of Cardiology	12
Directories of CD-ROM Products	38, 43

[a] Information was believed to be accurate at the time of writing (September, 1991). Publishers listed in the References section may be contacted for current information.

APPENDIX A. (*Continued*)

Type of Information	Publishers
Drugs and Pharmacology	
DRUGDEX System	23
IDENTIDEX System	23
POISINDEX System	23
Emergency Medical Response and Treatment	
EMERGINDEX System	23
POISINDEX System	23
TOMES Plus^{TM} System	23
Environmental	
Acid Rain (Canadian Government Documents)	42
Airborne Antarctic Ozone Experiment	26
Airborne Arctic Stratospheric Expedition	26
Alaska Marine Contaminants Database	29
Arctic & Antarctic Regions Metabase	28
Earth Science Data Directory	31
Ecological Fact Sheets	40
GEOINDEX (U.S. Geologic Maps)	31
Geophysics of North America	27
Global Hypocenter Data Base CD-ROM (Earthquake Database System)	39
HSDB	23
Natural Resources Metabase	28
OHM/TADS	23, 36
TOMES Plus System	23
Toxic Release Chemical Inventory	40
First Aid	
CHRIS	23, 36
TOMES Plus System	23
HazMat Emergency Response	
CHRIS	23, 36
Dangerous Goods	37
DOT Emergency Response Guides	23
HSDB	23
OHM/TADS	23, 26
TOMES Plus^{TM} System	23
Material Safety Data Sheets	
Collections	1, 7, 30, 40
Software for Scanning MSDS's	16
Patents and Trademarks	
ASIST	41
Automated Patent Searching	24
CASSIS/BIB	41
CASSIS/CD-ROM	41
CASSIS/CLSF	41
PraCTis (International)	41
TRADEMARKS	41

APPENDIX A. (*Continued*)

Type of Information	Publishers
Pesticides	
HSDB	23
PESTBANK	36
The Pesticides Disc	34
POISINDEX® System	23
REPRORISK™ System	23
RTECS	23, 36
Shepard's Catalog of Teratogenic Agents	23
TERIS (Teratogen Information System)	23
TOMES Plus System	23
Reference Standards: Biological, Chemical, Physical	
Benchtop PBM Reference Mass Spectra Library Search System	33
Enzymes Database	18
GenBank, EMBO, UGENBANK, UEMBO Genetic reference sequences, with search software	18
LaserGene Software (with DNA and Protein Reference Sequences)	13
NIST Crystal Database Crystal Coordinates Data File	20
NIST/Sandia/ICDD Electron Diffraction Database	20
PDF-2 Database X-Ray Powder Diffraction Reference	20
Swiss-Pro, PIR, KeyBank, KeyTool Protein Reference Sequences, with search software	18
VectorBank Database of genetic vectors	18
Regulatory Information	
Federal Register	10, 11
Statutes and Regulations	7, 15, 17
TOMES Plus System	23
Toxic Release Chemical Inventory	40
Right-to-Know	
New Jersey Fact Sheet	23, 40
Risk Assessment	
IRIS	23
Storage, Shipping, Waste Disposal	
CHRIS	23, 36
Dangerous Goods	37
HSDB	23
TOMES Plus System	23
Toxicology, Health Effects	
CISDOC	36
HSDB	23
HSELINE	36
OHM/TADS	23, 36
POISINDEX® System	23
REPRORISK™ System	23

APPENDIX A. (*Continued*)

Type of Information	Publishers
Toxicology, Health Effects (*Continued*)	
RTECS	23, 36
Shepard's Catalog of Tetragenic Agents	23
TERIS (Teratogen Information System)	23
TOMES Plus™ System	23
Training	
Multi-Media	7, 32

CD-ROM PUBLISHING CORPORATION

1. Aldrich Chemical Company, Inc.; 1001 W. St. Paul Ave.; Milwaukee, WI 53233; (800) 558-9160, (414) 273-3850.

2. Aries Systems Corp.; One Dundee Park; Andover, MA 01810; (508) 475-7200.

3. Biophysica Technologies, Inc.; 902 W. 36th St.; Baltimore, MD 21211; (301) 366-3636.

4. Biosis; 2100 Arch St.; Philadelphia, PA 19103-1399; (800) 523-4806, (215) 587-4800.

5. R. R. Bowker; 205 East 42nd St.; New York, NY 10017; (800) 323-3288.

6. Cambridge Scientific Abstracts; 7200 Wisconsin Ave.; Bethesda, MD 30814; (800) 843-7751.

7. Canadian Centre for Occupational Health & Safety; 250 Main Street East, Hamilton, Ontario L8N 1H6, CANADA; (800) 263-8340, (416) 572-4444.

8. CD Plus; 333 7th Ave., 6th Floor; New York, NY 10001; (212) 563-3006.

9. Consultative Group on International Agriculture Research; 1818 H. St., N.W., Room N5063; Washington, DC 20431; (202) 473-8942.

10. Counterpoint Publishing; 20 William St., Suite G-70; P.O. Box 9135; Wellesley Hills, MA 02181-9135; (617) 235-4667.

11. Dialog Information Services, Inc.; 3460 Hillview Ave.; Palo Alto, CA 94304; (800) 334-2564, (415) 858-3785.

12. Healthcare Information Services, Inc.; 2335 American River Dr., Suite 307; Sacramento, CA 95825; (916) 648-8075.

13. DYNASTAR, Inc.; 1228 S. Park St.; Madison, WI 53715; (608) 258-7420.

14. EBSCO Electronic Information; P.O. Box 13787; Torrance, CA 90503; (800) 888-3272.

15. ERM Computer Services, Inc.; 855 Springdale Dr.; Exton PA 19341; (800) 544-3118, (215) 524-3600.

16. HazMat Control Systems, Inc.; 3409 Lakewood Ave., Suite 2C; Long Beach, CA 90808; (213) 429-9055.

17. IHS Regulatory Products; 15 Inverness Way East; Englewood, CO 80150; (303) 790-0600.

18. IntelliGenetics Inc.; 700 East El Camino Real; Mountain View, CA 94040; (800) 876-9994, (415) 962-7300.

19. International Centre for Diffraction Data; 1601 Park Lane; Swarthmore, PA 19081; (215) 328-9400.

20. Library of Congress; Customer Services Section; Washington, DC 20541; (202) 707-1312.

21. McGraw-Hill Book Company; 11 West 19th St.; New York, NY 10011; (212) 512-2000.

22. Medical Economics Co.; 680 Kinderkamack Rd.; Oradell, NJ 07649; (201) 262-3030.

23. Micromedex, Inc.; 600 Grant St.; Denver, CO 80203-3527; (800) 525-9083, (303) 831-1400.

24. MicroPatent; 25 Science Park; New Haven, CT 06511; (800) 648-6787, (203) 786-5500.

25. MSRS, Inc.; P.O. Box 1794; 10 West Washington St.; Middleburg, VA 22117-1794; (703) 687-6777.

26. NASA/Ames Research Center, c/o Steven Hipskind; MS 245-5; Moffett Field, CA 94035-1000; (415) 604-5076, FTS 464-5076.

27. National Geophysical Data Center; NOAA; 325 Broadway; Boulder, CO 80303-3328; (303) 497-6120.

28. National Information Services Corp.; Suite 6 Wyman Towers; 3100 St. Paul St.; Baltimore, MD 21218; (310) 243-0797.

29. National Oceanic and Atmospheric Administration; c/o Jawed Hameedi; Office of Oceanography and Marine Assessment; Alaska Office; 801 C St.; Anchorage, AK 99513; (907) 271-3580.

30. Occupational Health Services, Inc.; 11 W. 42nd St., 12th Floor; New York, NY 10036; (800) 445-MSDS; (212) 789-3535—NY).

31. Online Computer Library Center, Inc.; 6565 Frantz Rd.; Dublin, OH 43017-0702; (614) 764-6000.

32. Online Products Corp.; 20251 Century Blvd.; Germantown, MD 20874; (800) 922-9204.

33. Palisade Corp.; 31 Decker Rd.; Newfield, NY 14867; (800) 432-7475, (607) 277-8000.

34. Pergamon Press, Inc.; Maxwell House, Fairview Park, Elmsford, NY 10523; in U.K.: Irwin House, 118 Southwark St.; London SE1 OSW U.K. (071) 928-1404.

35. Quanta Press, Inc.; 2550 University Ave. West; Suite 245 N; St. Paul, MN 55114; (612) 641-0714.

36. SilverPlatter Information, Inc.; One Newton Executive Park; Newton Lower Falls, MA 02162-1449; (800) 343-0064, (617) 969-2332.

37. Springer-Verlag; Electronic Media Dept.; 175 Fifth Ave.; New York, NY 10010; (800) SPRINGER, (212) 460-1622.

38. United States Department of Health and Human Services, Public Health Service, Centers for Disease Control, Center for Environmental Health and Injury Control, Information Resources Management Group, Atlanta, GA 30333.

39. United States Geological Survey; National Earthquake Information Center; Denver Federal Center; P.O. Box 25046, MS 967; Denver, CO 80225-0046; (303) 236-1500, FTS 776-1500.

40. United States Government Printing Office; Superintendent of Documents; Washington, DC 20402-9325; (202) 783-3238. (Department of Defense Hazardous Materials Infor-

mation Services (HMIS), 008-000-00567-2; EPA Regulatory and Community Right to Know Toxic Release Chemical Inventory 055-000-00356-4.)

41. United States Patent and Trademark Office; Office of Electronic Data Conversion and Dissemination; Crystal Park 2—1100C; Washington, DC 20231; (703) 557-6154.

42. University of Vermont; c/o Albert Joy; Bailey/How Library; Burlington, VT 05405-0036; (802) 656-8350.

43. The H. W. Wilson Co.; 950 University Ave.; Bronx, NY 10452; (800) 622-4002.

APPENDIX B. DATA REQUIREMENTS FOR OCCUPATIONAL HEALTH INFORMATION SYSTEMS

	Requirement Weight*	Evaluations		
		System A	System B	System C
Medical				
Scheduling				
Demographics				
Problems				
Office Visits				
Immunizations				
Exposures				
Medical History				
Physiological				
Physical Exams				
Vision				
Chemistry				
Urinalysis				
Hematology				
Serology				
Audiometry				
Spirometry				
Cytology				
ECG				
X-ray				
Drug Screen				
Diagnosis				
Recommendations				
Medications				
Mortality				
Wellness Program				
Demographics				
Visits				

	Requirement Weight*	Evaluations		
		System A	System B	System C
Claims				
Case Descriptions				
Medical Costs				
Legal Costs				
Salary Costs				
Award Costs				
Other Costs				
Recovered Costs				
Status Costs				
Environmental Agents				
Descriptions				
Synonyms				
Information Requests				
MSDS Revisions				
Ingredients				
Physical Properties				
Health Hazards				
Control Measures				
Spill Procedures				
Handling, Storage				
Precautions				
Labels				
Transportation				
Toxicology				
Hygiene Limits				
Monitoring Interval				
Workplace				
Description				

Item	Score
Physiology	
Strees Tests	
Objectives	
Workshops	
Follow-up	
Employee Assistance	
Demographics	
Visits	
Health Risk Appraisal	
Demographics	
HRA	
Medical Inventory	
Description	
Suppliers	
Purchases	
Provider	
Demographics	
Certification	
Continuing Education	
Employee	
Job History	
Protective Devices	
Training	
Absences	
Workplace	
Description	
Inventories	
Medical Requirements	

Item	Score
Inventory	
IH Estimates	
Engineering Controls	
Inspections	
Incidents	
Incidents Costs	
IH Sampling	
Description	
Scheduling	
Method	
Details	
Data	
Results	
Recommendations	
Equipment	
Description	
Scheduling	
Calibration	
Inspections	
Customers	
Description	
Orders	
ICD-9 Codes	

Total Scores _____

*The relative weight may be altered if the program does not give the full function sought. Use full weight if all of function is present.

APPENDIX C. REPORTING REQUIREMENTS FOR OCCUPATIONAL HEALTH INFORMATION SYSTEM

	Requirement Weight*	Evaluations		
		System A	System B	System C
Medical				
Employee Medical Report				
Summary Medical Report				
Medical Follow-up				
Work Restrictions				
Health Risk Appraisal				
Hearing Threshold Shift				
Analysis of Abnormal by Employee Groups				
Schedules				
Other:_____				
Industrial Hygiene				
Sample Schedules				
Sample Report				
Sample Summaries				
Sample Follow-up				
Incomplete Samples				
Lab Audit				
Workplace Inspection and Follow-up				

	Requirement Weight*	Evaluations		
		System A	System B	System C
MSDS Updates				
Customer Letter				
Customer Report				
Product Report				
Materials Report				
Supplier Report				
Right-to-Know				
SARA Tier I				
SARA Tier II				
Other:_____				

*The weighting may be altered if the full function does not appear in the program. Use full weight for full function.

Problem Summary _____

Agents Inventory _____

Workplace History of
Environmental Agents _____

Carcinogen List _____

Material Safety Data
Sheet _____

MSDS Revision
Review Sheet _____

Labels _____

Others:_____ _____

**Hazard
Communication**

Training Status _____

Protective Equipment
Assignments _____

First Aid Information _____

Spill Information _____

Carcinogen List _____

Agents Inventory _____

Workplace History
of Environmental
Agents _____

Material Safety Data
Sheet _____

MSDS Revisions _____

Labels _____

441

APPENDIX D. DATA REQUIREMENTS FOR ENVIRONMENTAL INFORMATION SYSTEM

	Requirement Weight*	Evaluations		
		System A	System B	System C
Permits				
Description	___	___	___	___
Scheduling	___	___	___	___
Sources	___	___	___	___
Renewal	___	___	___	___
Limits Parameters	___	___	___	___
Monitoring				
Description	___	___	___	___
Scheduling	___	___	___	___
Sample Details	___	___	___	___
Personnel	___	___	___	___
Collection and Preparation	___	___	___	___
Site Water Measurements	___	___	___	___
Chain of Custody	___	___	___	___
Water Indices	___	___	___	___
Other Results	___	___	___	___
Compliance	___	___	___	___
Waste Profile				
Description	___	___	___	___
Lab Analysis	___	___	___	___
Waste Properties	___	___	___	___
Constituents	___	___	___	___
Manifests				
Description	___	___	___	___
Line Items	___	___	___	___

	Requirement Weight*	Evaluations		
		System A	System B	System C
Environmental Agents				
Description	___	___	___	___
Synonyms	___	___	___	___
Manuf. & Suppliers	___	___	___	___
Ingredients	___	___	___	___
Physical Properties	___	___	___	___
Health Hazards	___	___	___	___
Control Measures	___	___	___	___
Spill Procedures	___	___	___	___
Handling & Storage	___	___	___	___
Labels	___	___	___	___
Transportation	___	___	___	___
Toxicology	___	___	___	___
Information Request	___	___	___	___
MSDS Revision	___	___	___	___
Limits/Regulations				
Agent Description	___	___	___	___
Hygiene Limits	___	___	___	___
Monitor Interval	___	___	___	___
Tank Reconciliation				
Tank ID	___	___	___	___
Tank Contents	___	___	___	___
Inspection	___	___	___	___
Status	___	___	___	___
Scheduling	___	___	___	___
Soil Baseline				
Soil Description	___	___	___	___

Transporters ___
Transporter Tracking ___
TSD Tracking ___
TSD Receipts ___
Discrepancies ___
Subsequent Manifests ___
Cert. of Destruction ___
Manifest Destruction ___
Associated Costs ___

Special Substance Data
Description ___
Status ___
Inspection ___
Follow-up ___

Problem & Event Data
Event ___
Description ___

Organization Data
Description ___
TSD Profile ___
State Permit ID ___

Process/Waste Minimization
Process ID ___
Control Devices ___
Waste Min. Efforts ___
Cost Savings ___

Horizons ___
Soil Limits ___

Sources Processes
Description ___
Inservice Dates ___

Production
Process ID ___
Process Schedule ___
Production ___
Product Transfer ___

Source/Process Inventory
Process ID ___
Scheduling ___
Inventory Date ___
Inventory Results ___
Hazard Estimates ___

Source/Process Inspection
Process ID ___
Scheduling ___
Inspection Results ___

Regulatory Lists ___

*The relative weight may be altered if the program does not give the full function sought. Use full weight if all of function is present.

443

APPENDIX E. REPORTING REQUIREMENTS FOR ENVIRONMENTAL INFORMATION SYSTEM

	Requirement Weight*	Evaluations		
		System A	System B	System C
Chemical Tracking				
Mass-balance				
Quarterly Discharge Monitoring				
Inventory Checklist				
Special Occurrences				
Special Substances Detail				
Agency/Facility Catalog				
(Also includes all following reports except those in permits)				
SARA				
Form R				
Tier I				
Tier II				
MSDS				
Training				
Right-to-Know Compliance				
First Aid				
Spill Procedures				
Agency/Facility Catalog				

(All reports cover ambient air, point source air, surface water, ground water, outfall, biological, tank and soil monitoring)

*The number assigned to a requirement gives it a weight relative to all other requirements. A system alternative is assigned the requirement, or a fraction of the requirement weight as judged appropriate by the evaluator. The system with the highest total score is the closest match to the defined requirements and priorities.

layout = 1.

Waste Management
Uniform Hazardous
  Waste Manifest
Manifest Status
Manifest Summary
Cumulative Waste
  Disposal
Waste Profile
Profile Update
Agency/Facility
  Catalog

Permits
Permit Summary
Permit Detail
Permits Requiring
  Action
Agency/Facility
  Catalog

Monitoring
Quarterly Discharge
  Monitoring
Source Detail
Source Summary
Source Monitoring
Monitoring Reminder
Problem Summary
Agency/Facility
  Catalog

APPENDIX F. FUNCTIONAL AND SYSTEM REQUIREMENTS FOR OCCUPATIONAL HEALTH AND ENVIRONMENTAL INFORMATION SYSTEMS

	Requirement Weight*	Evaluations		
		System A	System B	System C
User-Modifiable data structures	___	___	___	___
User-defined record keys	___	___	___	___
User-defined data editing	___	___	___	___
User-defined normal ranges exposure limits	___	___	___	___
Referential Editing	___	___	___	___
User-defined data entry screens	___	___	___	___
User-defined calculations	___	___	___	___
Other: _____	___	___	___	___
_____	___	___	___	___
_____	___	___	___	___
Interfaces to				
Other hardware software systems	___	___	___	___
Laboratory equipment	___	___	___	___
Statistical packages	___	___	___	___
Personal computers	___	___	___	___

	Requirement Weight*	Evaluations		
		System A	System B	System C
Hardware compatibility	___	___	___	___
Operating system compatibility	___	___	___	___
Other: _____	___	___	___	___
_____	___	___	___	___
_____	___	___	___	___
Report generator	___	___	___	___
Ad hoc reporting	___	___	___	___
User-modifiable standard reports	___	___	___	___

*The number assigned to a requirement gives it a weight relative to all other requirements. A system alternative is assigned the requirement weight if it completely meets the requirement, or a fraction of the requirement weight as judged appropriate by the evaluator. The system with the highest total score is the closest match to the defined requirements and priorities.

Other:_____

Data archiving
  and purging
Foreign language
  prompt translation
On-line help
Custom helps
Other:_____

User-defined data
  access security
User-controlled
  functional
  security
Audit trails
Other:_____

APPENDIX G. JOINT APPLICATION DESIGN (JAD) PROCESS

The purpose of a JAD is to define the objectives, scope, requirements, and external design of a system. By external design, we are speaking of part of the application that the end user sees, but not the coding behind it.

A JAD plan session is used to define system objectives and scope. The purpose of a JAD plan is to define the scope for large applications or groups of applications. A JAD plan provides an organized method for gathering and documenting the data required to estimate, plan, and schedule JAD sessions.

For a JAD plan, individuals with a solid understanding of the functional areas affected by the application are brought together for a brief, but intensive-cooperative effort. These individuals, led by a JAD leader, define the scope of the project by developing a list and description of business functions to be included in the system. They also establish a preliminary list of required screens, reports, and connection points to other systems (system I/Os). This list serves as a basis for estimating the length and complexity of the design process. Any significant issues regarding the application are documented as well.

The product of a JAD plan is a documented and agreed-upon statement of what is to be done. It contains the system objectives, functions, and scope. Such a statement, and the commitment it represents, is an excellent basis for estimating, planning, and obtaining management approval for the projected system.

After what is to be done has been defined, the "how" can be addressed. A JAD design is intended for this purpose.

Once again, individuals directly involved with the proposed system are brought together for brief, but intensive, sessions under the guidance of a JAD leader. The result is a document containing a definition of the system requirements and the external design. This document is the foundation for the implementation and installation of the system.

It is felt that a JAD is an efficient and economical technique for obtaining user and management commitment for the remainder of the application development effort.

The same basic types of activities take place in both a JAD plan and a JAD design. These are:

1. Preparation and kick-off
   a. Establishment of the executive objectives
   b. Orientation to the application
   c. Preparation of materials to be used in the JAD
   d. Kick-off meeting of the executive sponsor and JAD participants to assure a common understanding of the objectives of the project.
2. AD sessions, whether for planning or design purposes, are intensive meetings with a carefully selected user, management, and data processing representatives. These sessions are chaired by the JAD leader and documented by JAD analysts.

Activities during a JAD design session are different from those in a JAD plan. Because the JAD plan session activities have already been described, we will concentrate here on the activities in a JAD design session. These activities deal with the system first on a general level and then go on to specify the system in detail.

Called *system definition*, the general definition activities obtain participant commitment to the following:

Statement of purpose of the system

System functions

Assumptions on which the system is based

Constraints on the system

These topics form the basis of the JAD documentation. Special forms are used to document the system functions during the sessions. These include:

Work flow diagrams of business process and their relationships

Work flow descriptions of inputs, reports, and users of the system

Function flow diagrams of data groups and inputs/outputs of each business function.

When the general overview of the system is complete, the more detailed specification of the system can begin. This is called *external design*. During external design, JAD participants commit to, and document the following:

Definition of data elements (date, time, CAS, etc.)

Screen and report layouts (if a query language is to be used, the query search requirement are identified but screen layouts will be omitted)

Edit and validation requirement for screens and reports

Function descriptions including security and architecture requirements, volumes of transactions, and frequencies

3. Wrap-up: During wrap-up, the conclusions documented during the JAD sessions are edited and checked for consistency and completeness.
4. Executive presentation: At the executive presentation, the JAD participants summarize the results of their JAD sessions and present the completed documentation to the executive sponsor.

GLOSSARY

Only the most usual and frequently encountered definitions are given here. The reader is advised to consult a more comprehensive list for additional material

Baud. (1) A unit of signaling speed equal to the number of discrete conditions or signal events per second; for example, one baud equals one-half dot cycle per second in Morse code, on bit per second in a train of binary signals, and one 3-bit value per second in a train of signals each of which can assume one of eight different states.

Cache Memory. A special buffer storage, smaller and faster than main storage, that is used to hold a copy of instructions and data in main storage that are likely to be needed next by the processor, and that have been obtained automatically from main storage. (T)

Case. Computer-assisted software engineering. A set of tools or programs to help develop complex applications.

Data Dictionary. (1) A database that, for data of a certain set of applications, contains metadata that deal with individual data objects and their various occurrences in data structures. (T) (2) A centralized repository of information about data such as meaning, relationships to other data, origin, usage, and format. It assists management, database administrators, system analyst, and application programmers in planning, controlling, and evaluating the collection, storage, and the use of data. (3) In the System/36 interactive data definition utility, a folder that contains field, format, and file definitions. (4) In IDDU, an object for storing filed, record format, and file definitions.

Database. (1) A collection of data with a given structure for accepting, storing, and providing, on demand, data for multiple users. (2) A collection of interrelated data organized according to a database schema to serve one or more applications. (3) A collection of data fundamental to a system.

EDI. Electronic data interchange. A standard format for the transmission of data and the use of that data.

Fourth-Generation Language (4GL). A high-level language used in programming that has additional tools for screen generation, interfaces, and system design.

Information Engineering. The application of rigorous discipline to the management of data. Usually involves goals, data design, and data relationships among other things.

LAN. Local area network. A LAN is a cabling configuration that is used to link several computing systems at one location.

Object Code. Output from a compiler or assembler that is itself executable machine code or is suitable for processing to produce executable machine code.

Operating System. Software that controls the execution of programs and that may provide services such as resource allocation, scheduling, input/output control, and data management. Although operating systems are predominantly software, partial hardware implementations are possible.

RAM. Random-access memory. The memory found on a computer chip that is in a computing system.

Record. (1) In programming languages, an aggregate that consists of data objects, possibly with different attributes, that usually have identifiers attached to them. In some programming languages, records are called structures. (2) A set of data treated as a unit. (3) A set of one or more related data items grouped for processing.

Relational Database. A database in which the data are organized and accessed according to relations.

RISC. Reduced instruction set computer. A type of operating system that is frequently used in scientific computing. Many manufacturers have their own versions.

ROM. Read-only memory. The memory of a computer that allows reading only. It cannot be written over without being destroyed.

Shared Data. Data that used for more than one applications or purpose of giving efficiency to the total system.

Source Code. The input to a compiler or assembler, written in a source language. Contrast with object code.

SQL. Structured query language. This allows the querying or seeking out of information in a database in a standard form.

Third-Generation Language (3GL). A high-level computer language that uses ordinary English words.

Time Sharing. (1) A method of using a computing system that allows a number of users to execute programs concurrently and to interact with the programs during execution.

UNIX. UNIX (trademark of UNIX Systems Laboratories, Inc.) operating system. An operating system developed by Bell Laboratories that features multiprogramming in a multiuser environment. The UNIX operating system was orginally developed for use on minicomputers but has been adapted for mainframes and microcomputers.

Upwardly Compatible. The capability of a computer to execute programs written for another computer without major alteration, not vice versa.

Utility Program. (1) A computer program in general support of computer processes; for example, a diagnostic program, a trace program, a sort program; synonymous with service program. (2) A program designed to perform an everyday task such as copying data from one storage device to another.

WAN. Wide area network, a network that extends over many locations.

CHAPTER TEN

Statistical Design and Data Analysis Requirements

Nelson A. Leidel, Sc.D. and Kenneth A. Busch, M.S.

1 INTRODUCTION

The work of professionals in industrial hygiene and allied disciplines such as environmental health can substantially benefit from use of airborne exposure study design and data analysis methodologies that are based in mathematical statistics and probability theory. It has been said that "the science of statistics deals with making decisions based on observed data in the face of uncertainty" (Bowker and Lieberman, 1972, p. 1).

Section 2 of this chapter discusses some major areas of industrial hygiene practice where statistical methods perform an important role. The need for statistically sound study designs for both experimental and observational studies is discussed in Section 2.1. Brief discussions appear in Section 2.2 concerning statistical methods used for occupational epidemiological studies and in Section 2.3 concerning estimating possible threshold levels and low-risk levels for occupational exposures. Finally, Section 2.4 introduces the area of application for statistics that is of primary interest and receives most attention in this chapter. This is the estimation of occupational exposures to airborne contaminants and calculation of error limits for such estimates. Nine possible objectives of occupational exposure estimation are discussed, which have their own special requirements for study design strategies.

Occupational exposure study designs and related data analysis methods have come to be broadly called sampling strategies (Leidel et al., 1977). These sampling strat-

Patty's Industrial Hygiene and Toxicology, Third Edition, Volume 3, Part A, Edited by Robert L. Harris, Lewis J. Cralley, and Lester V. Cralley.
ISBN 0-471-53066-2 © 1994 John Wiley & Sons, Inc.

egies are plans of action, based on statistical theory used to determine a logical, efficient framework for application of general scientific methodology and professional judgment. These are broadly outlined in Section 2.5.

Section 3 presents some basic statistical theory relevant to occupational exposure data. Distributional models are given that identify the contributions of various sources of variation to the overall (net) random error in occupational exposure estimates. The National Institute for Occupational Safety and Health (NIOSH) nomenclature for exposure data is first given in Section 3.1. Then in Section 3.2 a model is given for the contributions of the various components of variation to the net random error in occupational exposure measurements (due to the measurement procedure used). Section 3.3 extends the model for total error to include random and systematic variations in true exposure levels (over times, locations, or workers doing similar work).

Section 3 also includes information on the mathematical characteristics of basic distributional models. This is the starting point for deriving sampling distributions of industrial hygiene exposure data taken by various sampling strategies. General properties of the normal distribution model are given in Section 3.4, and of the lognormal distribution model (both two-parameter and three-parameter) in Section 3.5. Section 3.6 discusses the adequacy of normal and lognormal distribution models for certain general types of continuous variable data (specifically occupational exposure measurements). These data models are then used to apply special-interest applications of statistical theory to occupational exposure study designs (Section 5) and specialized data analyses (Section 6).

Section 4 discusses basic principles of statistically sound study design and data analysis that apply to all industrial hygiene surveys, evaluations, or studies. Section 4.1 presents general study design principles for experimental or observational studies intended to estimate means, variances, quantiles, tolerance limits, or proportions below a designated value for a single population and for those studies seeking to compare any of these parameters between two study groups [e.g., between an exposed ("treatment") group and an unexposed ("control") group]. Section 4.2 then discusses general principles of data analysis. Particular emphasis is given to the necessity for appropriate selection, estimation, and verification of a distributional model (or models) for the study data.

Section 5 gives particular study designs to be used for collecting data to estimate individual occupational exposures and exposure distributions. The section first discusses the cornerstone concept of worker target populations. Section 5.1 discusses in detail another important concept, the determinant variables affecting occupational exposure levels experienced by a target population. Section 5.2 discusses exposure measurement strategies selected to measure a short- or long-period, time-weighted average (TWA) exposure of an individual worker on a given day. Both practical and statistical considerations are discussed for long-term and short-term exposure estimates. Lastly, exposure monitoring strategies are presented in Section 5.3 for measuring multiple exposures (e.g., multiple workers on a single day, a single worker on multiple days, or multiple workers on multiple days). Eight possible elements of monitoring programs are discussed in detail, with examples of both exposure screening and exposure distribution monitoring programs.

Section 6 presents specialized applied methods for formal statistical analysis of occupational exposure data generated by the study designs discussed in Sections 4 and 5. The first portion of Section 6 (Sections 6.1 through 6.5) covers methods for computing confidence intervals for true exposures of individual workers that have been estimated using exposure measurement strategies. In addition, statistical significance tests of hypotheses are presented for classifying individual exposure estimates relative to an exposure control limit. The middle portion of Section 6 (Sections 6.6 through 6.10) covers inferential methods for computing tolerance limits, tolerance intervals, and point estimates of exposure distribution fractiles. The last portion of Section 6 (Sections 6.11 through 6.13) covers graphical techniques for presenting lognormally distributed data and their associated tolerance limits.

2 GENERAL AREAS OF APPLICATION FOR STATISTICS IN INDUSTRIAL HYGIENE

In the practice of industrial hygiene and allied professions, statistical theory is applicable to both:

1. *Experimental* studies (i.e., studies with planned intervention on determinant factors suspected of altering the phenomenon under study) and
2. *Observational* surveys and studies (i.e., studies with no deliberate human intervention).

With statistically valid study design and data analysis, both experimental and observational studies can validly and reliably identify causes of occupational health problems, screen workplaces for excessive exposure conditions, estimate worker exposure levels, evaluate the effectiveness of engineering controls, and determine the protection levels afforded by personal protective equipment. Of course, depending on factors such as relative cost, conditions amenable to experimentation, availability of relevant laboratory models, and interpretability of available field data, one or the other of these two general types of studies is usually a clear choice for any given research objective. However, experimentation generally has several fundamental advantages over observational surveys and studies. These are discussed in Section 2.1 in the context of health effects studies related to workplace contaminant exposures. An advantage of observational studies is that subjects are humans; epidemiologic studies are discussed in Section 2.2.

2.1 Need for Statistically Valid Study Design

Typical industrial hygiene *experimental* studies include estimating distributions of worker exposures for selected conditions, determining the efficacy of exposure control systems, and estimating the accuracy and variability of measurement methods. Experiments may be done to study the effects of determinant variables on worker exposure levels, on the efficacy of control measures, or on exposure measurement procedures. In these experiments, the "treatments" are usually a set of controlled conditions under which an exposure level (or other physical or biological response) is measured as the response variable. Thus, "treatments" are usually exposure groups in this type of experiment. If a factor is "controlled" through deliberate selection of particular existing conditions, one has a *quasi-experimental* study in a field setting. In contrast to experimental studies, one conducts an *observational* study when conditions must be taken as found in a field setting (i.e., without opportunity to experiment with preselected, controlled levels for the determinant variables).

An inherent problem for industrial hygienists is that "safe" exposure levels for many substances are unknown and must be determined using the best available evidence (Lowrance, 1976). To do this, pertinent exposure level and health effects data are collected and statistically analyzed. One approach is to expose suitable animal species and observe biological effects that may occur. A second important approach to estimation of acceptable exposure levels, the epidemiological study on exposed workers, is discussed in a limited manner in this section, and references to more comprehensive presentations are supplied.

For either type of industrial hygiene study, experimental or observational, the resulting data will have stochastic components (i.e., random or chance variations) that cannot be ignored when evaluating the data. Different types of biological data may have fundamentally different statistical properties. Population health effects may be measurable in:

1. The average amount of change in a quantitative biological parameter measured on a continuous scale. Examples of continuous variable measurements are lung volume, heart rate, and body weight.
2. The presence or absence of a qualitative biological abnormality. An example would be a pathological condition such as a tumor.
3. Values that exist only at a limited number of discontinuous (i.e., discrete) points on an ordered scale. For example, severity of lung histopathology has been graded on a 6-point rating scale.

The first and second types of data are the most frequently encountered, and they will be discussed in Section 4 in relation to principles of study design and data analysis.

The results from any industrial hygiene study should be evaluated both in terms of internal validity and external validity. With regard to *internal validity*, suppose a researcher were to conclude that the performance of control method B is "significantly better" than the performance of method A. If the conclusion was erroneous because of random errors that had affected the exposure measurements used to evaluate performances of both methods, then the study conclusion would not have internal validity. That is, chance would have led to an incorrect conclusion that there was a true difference in performance. As will be discussed later, variability exists in exposure results for numerous reasons. This variability can then cast doubt on the internal validity of any conclusions drawn from the exposure results.

It is also essential to examine the *external validity* of exposure results from any given industrial hygiene study. That is, how valid are the study results *outside of the study sample*? As before, suppose a researcher were to conclude that the performance of control method B is "significantly better" than the performance of method

A. If the conclusion were based on worker activities or control method conditions that are irrelevant to tasks and circumstances in the real world, then the conclusion and research findings would have no external validity. External validity also includes topics such as possible *nonrandom sampling errors* (i.e., "mistakes") and *biases* in exposure results. Biases are systematic errors that deprive a statistical result of representativeness by distorting its expected value. Unlike internal validity, for which there are objective statistical computations to justify conclusions, evaluating external validity must sometimes be a subjective matter that relies largely on professional judgment.

For those researchers who wish to generalize their exposure results to larger populations or other settings (workplaces), a two-stage process is involved during which external validity problems can arise (Bracht and Glass, 1968; Cook and Campbell, 1979). First, researchers must define a *target population* of persons, settings, and times (see Section 5). Second, researchers must draw a sample of exposed workers to represent the target population. However, these samples usually cannot be drawn in a formal randomized manner. Instead, sampled workers are usually selected merely in a convenient manner that gives an intuitive impression of representativeness. However, the sample of workers selected, as well as the settings, and conditions of any given workplace study, particularly observational studies, may severely hamper the *generalizability* of subsequent exposure estimates.

Cook and Campbell (1979, p. 71) have suggested that it is useful to distinguish between (1) target populations, (2) formally representative samples that correspond to known populations, (3) samples actually achieved in field research, and (4) achieved populations. They have noted:

> To criticize the study because the achieved sample of settings was not formally representative of the target population may appear unduly harsh in light of the fact that financial and logistical resources were limited, and so sampling was conducted for convenience rather than formal representativeness. . . . *it is worth noting that accidental samples of convenience do not make it easy to infer the target population*, nor is it clear what population is actually achieved.

The objectives of an occupational health study should be the primary determinant of its statistical design, *not* expedient considerations of "available" specialized experimental facilities or presence of "experts" on a given professional staff. Availability of specialized research staff and personal interests of these researchers can be powerful incentives for inappropriately designing a research study.

Finney (1991) discusses a professional ethics code for professional statisticians that includes maintenance of an attitude of scientific objectivity that is not biased by the interests of an employer or client, preservation of confidentiality of data furnished for analysis and of the results of that analysis, an obligation to screen raw data for possible mistakes or outliers, an obligation to perform data analyses using best available methods (e.g., most powerful significance tests) unless simpler methods are believed to produce equivalent results, and an obligation not to use a "canned" data analysis protocol (e.g., personal computer or mainframe software) unless its underlying assumptions and correct interpretation of results are understood. Finney also discusses the responsibility of a client to furnish the consulting statistician with complete and accurate raw data, along with adequate background information on the sampling strategy and environmental circumstances surrounding collection of the data, so that appropriate data analyses and valid interpretation of results can be performed.

It is the responsibility of the researcher, not the statistician, to assure that determinant factors such as the species used as animal models, exposure techniques, biological parameters that can be accurately measured, and exposure measurement procedures are given consideration and are appropriately incorporated into the study protocol. The statistician then addresses such tactical problems as sample size, allocation of experimental units (subjects) to treatments (exposure groups), schedules for sampling, and data analysis techniques. Proper consideration and selection of study design and data analysis parameters and methodologies will assure the researcher that the study will have adequate (but not excessive) statistical power and technical capability to detect the anticipated effects that are the focus of the research.

Sometimes technically appropriate experimental facilities or situations are available, but the feasible sample sizes are only marginal insofar as the production of definitive results is concerned. Nevertheless, an expedient decision may be made to proceed with the study under the misguided rationale that "some information is better than none." This incorrect and wasteful practice often results in studies that are predestined to be inconclusive and possibly misleading to subsequent investigators. The potential waste of time and resources can be minimized by first securing a statistician's evaluation of (or better yet, assistance with) the study design and data analysis plan. The statistician can usually warn the researcher if the planned study has low statistical power or for other reasons (e.g., bias due to confounded factors) lacks the statistical capability to detect the desired size of effect of a determinant factor, if indeed the effect exists.

If some species of experimental animal could serve as a perfect biological model for health effects on humans caused by workplace exposure, then there would be no question that controlled, animal exposure studies would yield better information than observational studies of humans. The deficiencies of human observational studies are analogous to those that would occur in an animal study for which subjects were constrained within a measured but uncontrolled laboratory environment for 8 hr each day, but allowed to roam freely through the streets and alleys outside their laboratory, eating and breathing whatever they encountered, during the other 16 hr.

Unfortunately, the perfect biological model for human health effects does not exist, but nevertheless the advantages of the carefully controlled, animal exposure experiment are several. First, in an experimental animal study one can *control* the primary study factor(s) (e.g., exposure levels, exposure duration, exposure schedule) and determine its direct effect, at differing levels of interest, on one or more response variables. In observational studies of humans, the primary study factor(s) can be observed, but not controlled, and toxicological and mathematical extrapolation of observed results to other exposure levels of interest must be performed.

Second, with animal studies there is more freedom to design experimental studies for complete elimination of bias (we shall put aside for now the interspecies extrapolation bias that may exist) and for high sensitivity to small effects. Given large enough sample sizes, an experimental study can be designed to have suitably high statistical power (probability) of detecting a small effect (if detecting this small effect is worth the required time and resources). Even a much larger observational study may not be able to detect a statistically significant, moderate-sized exposure effect, since there are unknown, or uncontrolled and unmeasured (even unmeasurable), secondary factors that operate within the workplaces and home environments to modify or distort the workers' biological responses to workplace exposures. Here, making a simple comparison of response variable values between exposed and unexposed groups will yield an imprecise estimate of the exposure effect (i.e., a wide confidence interval for the true magnitude of the effect) and probably will also yield a biased (substantially inaccurate) estimate of the effect. Bias can occur because of unequal levels of secondary factors in the control and exposed groups, so that we observe the joint (net) effect due to both the primary factor and to unintended differences in levels of the secondary factors. This joint effect is said to consist of *confounded effects* (i.e., joint effect of two or more factors confused with each other).

In observational studies for which levels of some secondary factors are known, methods of formal statistical analysis can be useful to adjust for the part of the bias due to confounded effects of these factors. For secondary factors that have effects that are additive to each other and to the effect of the primary factor, several methods for bias adjustment are available. For continuous response variables, these include *covariance analysis* (ANCOVA), and *analysis of variance* (ANOVA) for randomized blocks. For discrete response variables, there are procedures such as the *chi-square test for matched data* (Cochran's Q test) and *two-way analysis of variance of ranks*.

In observational studies no statistical adjustment can usually be made for any bias due to unknown secondary factors or due to covariates for which the levels were not measured. However, in experimental studies this dilemma of unknown or uncorrectable bias need never occur if the experiments are statistically well designed. Bias can be prevented in experiments by using statistical design features, such as selection of subjects at random from the same pool for assignment to control and test groups, or restricted random selection with matching on a secondary factor or within blocks (e.g., age groups, genders, weight ranges). Randomizing within blocks not only prevents bias but also improves precision. Similar blocking can be employed in observational studies during the data analysis, but only regarding the known secondary factors. It is *random assignment* of subjects to the exposure groups that is the best protection against bias due to effects of *unknown* secondary factors. Random assignment can be used in experimental studies, but not in observational studies for which there will always be some bias (one hopes it will be small) due to unequal representation of secondary factors (i.e., those not taken account of in the data analysis) in the exposed and unexposed groups.

In spite of the inherent limitations and deficiencies of observational studies on workers, they are an invaluable source of information on health effects due to work-related diseases. To some research scientists, the fact that the subjects of observational studies are workplace-exposed humans is an overriding consideration that transcends the statistical advantages and experimental flexibility afforded by animal exposure studies. Effects observed in workplace observational studies can be strongly suggestive of effects due to the primary factor(s) studied, but the results cannot be considered definitive by themselves. Replication of the result found for a given toxic agent, preferably in other studies of different worker target populations, for several occupational settings and several cultural, ethnic, or socioeconomic groups, affords much higher credibility and substantiation to the initial findings. This is true because the universal presence of a similar observed effect in different occupational settings (that involve different secondary factors) would tend to implicate the primary factor as the cause, not the variety of secondary factors.

On the other hand, a single controlled animal experiment that has been toxicologically and statistically well-designed (with balance and randomization used in all phases of the work) and competently conducted, theoretically can yield a definitive result (within the limits of random error, which can be governed by its experimental design). Even so, an experimental scientist feels better after the initial findings have been replicated in an independently conducted study. Also, the criticism can be made, "What good is a 'definitive' result for animals if it does not apply to humans?" An extrapolation from animals to humans must only be defended on the basis of the similarities of their appropriate biological mechanisms and structures, supported presumptively by empirical species similarities previously observed. A good statistical experimental design does not assure validity of the interspecies extrapolation; however, it does assure validity of the findings for the animal species used.

Finally, observational studies in workplaces can be an economical source of estimates of chronic effects due to long-term, low-level workplace exposures (although the estimates are often somewhat imprecise and ill-defined). Observational studies use existing data (although these data sometimes are not readily accessible), whereas a chronic exposure animal study requires at least several months and sometimes several years to conduct.

2.2 Epidemiological Studies

In the occupational epidemiological study, an attempt is made to associate an observed incidence or severity of adverse health effects in groups of human workers with factors such as industry, job type, or some measure of exposure to potentially toxic materials. The last type of study may attempt to estimate a *dose–response* relationship between level of exposure to the material and prevalence or incidence of health effect being studied. However, the type of dose–response relationships utilized for occupational settings generally are more correctly referred to as exposure–response functions since the effective dose at the critical site(s) within each individual worker's body is never really known (Hatch, 1968). Ulfvarson (1983, p. 285) has noted:

> The concept of exposure of an employee to a substance in the work environment may denote at least two things. It may indicate the dose of the substance absorbed in the body. It may also merely indicate the presence of the employee in an environment in which there is a more or less-well determined concentration of the substance, from which an uptake of the substance is deduced.

For practical reasons, the latter concept of exposure is currently used by industrial hygienists.

Additionally, the *exposure variable* utilized may be either an average exposure level over some long period (such as a worker's total working lifetime) or a time-integrated (cumulative) exposure [e.g., ppm-years, (fibers/cm^3)-years, (mg/m^3)-years]. Ford et al. (1991) define a cumulative exposure (CE) estimate and a lifetime, weighted-average, exposure (LWAE) estimate. The CE estimate is computed by quantitative integration of the terms: (1) exposure duration multiplied by (2) matching exposure level (concentration) measured in a subject's various working zones. The CE estimate assumes that a given working zone has an exposure level that is *uniform over time*, and the exposure level estimate for that working zone is taken to be the geometric mean (GM) of (a supposed random sample of) exposure levels for that zone. The LWAE estimate is then defined to be LWAE = CE/(total duration of exposure). It is interesting that the Ford et al. study included neurobehavioral test data that had a better interworker correlation with average exposure level (LWAE) than with cumulative exposure (CE) or with total duration of exposure.

Estimation of these or similar exposure–response relationships is desirable since they can be used in the process of selecting exposure-control limits for workers in occupational environments (Hatch, 1972, 1973). Recognizing the statistical nature of the problem, Roach (1953) first suggested that an occupational hazard (silica dust exposure leading to silicosis) should be analyzed as an exposure–response function similar to the dose–response function utilized in toxicological research.

A major obstacle in estimation of exposure–response relationships is accurate and sufficiently precise estimation of worker exposure levels and their relation to doses delivered to the body. Gamble and Spirtas (1976) have presented a systematic approach for using occupational titles to classify the "effect" and "exposure" of workers in retrospective, exposure–response estimation. Esmen (1979) has proposed a process for reconstruction of the integrated exposure of one or more agents over a reasonably long period of time. He presents the basis for a model and a simplified procedure for what he calls a *retrospective industrial hygiene survey*. Roach (1966, 1977) has proposed sampling strategies that consider the biological half-times of the measured substances.

Combined with the substantial advantage of having human workers as subjects, occupational observational studies generally carry the frustrating disadvantages of lack of control of the exposure conditions, uncontrolled effects of the exposures outside the occupational environment, and interactions between exposure effects and demographic and socioeconomic factors. These design problems for observational studies were discussed at length in Section 2.1. Statistically it should be possible to estimate a set of exposure–response relationships for any work-related disease given that a reasonably homogeneous and large enough group of workers is available. As a practical matter, it is difficult to obtain homogeneous and unbiased worker populations of adequate size for which one has adequate, accurate exposure information (Buchwald, 1972). As a result, there have been a minimal number of exposure–response functions estimated in the last 30 years for occupational populations. These include functions estimated by Hatch (1955) for silicosis in miners and other dusty trades, Lundin et al. (1977) for respiratory cancer in uranium miners, Berry et al.

(1979) for asbestosis in asbestos textile workers, and Dement et al. (1982) for lung cancer and other nonmalignant respiratory diseases in chrysotile asbestos textile workers. Berry et al. (1979) has an appendix with a particularly good discussion of measures of exposure and occupational dose–response relationships.

Hornung (1991) discusses use of exposure matrices to determine exposure levels for the various combinations of job factors and personal factors that are relevant in an epidemiological study. In comparison to the larger uncertainties due to job and personal factors, and measurement method biases, sampling and analytical error contribute relatively little to the uncertainty of exposure level estimates for missing job matrix cells (i.e., cells for which there are health effects data but for which no direct exposure–measurements data exist). Approaches used to ''fill in'' the missing cells include:

1. Interpolation between surrounding cells.
2. Use of mathematical prediction models based on principles of physics, data on production levels, and environmental factors.
3. Use of statistical models fitted to measurement data in surrounding cells.

In experimental clinical studies of volunteer subjects, better control over, and accurate measurement of, short-term human exposures can be obtained. Of course, an overriding necessity is safety of the human subjects, and this usually precludes use of exposure levels high enough, or exposures long enough, to produce the chronic toxicity that can occur in the workplace.

Ulfvarson (1983) has reviewed and extensively discussed the limitations to the use of typical worker exposure data in epidemiologic studies. He concluded that there are a considerable number of possible biases in the estimation of uptake of a substance into the bodies of a group of workers, when body uptake is uncritically derived from typical sources of airborne exposure levels. He defined a positive bias as where the uptake is overestimated in comparison to the true uptake. He also rated the validity of his conclusions regarding the probable sign of the bias as follows: {3} = self evident, {2} = a conclusion with some reservation, and {1} = an educated guess. Some of the possible biases listed by Ulfvarson (with his own ratings {in braces} of the validity of their sign) follow:

1. Positive bias {2} due to the use of a measurement strategy intended to collect a sufficient mass of contaminant for the purpose of exceeding the minimum detectable level for the analytical method.
2. Positive bias {3} due to the use of a ''worst-case,'' biased measurement strategy.
3. Positive or negative bias due to the use of daily TWA exposure results that do not represent the unsampled workers to whom they are applied.
4. Positive bias {2} due to the use of data from area (static) sampling devices that were deliberately located to yield high results such as would be obtained with a source sampling strategy.

5. Negative bias {2} due to failure to obtain repeat measurements when the first result demonstrates compliance, but may have been unusually low.

6. Positive bias {2} due to use of exposure data from establishments that do not represent the unsampled establishments to which they are applied.

7. Positive and negative biases due to seasonal variations in exposure levels.

8. Positive bias {3} due to rotation of workers to unexposed work areas, when the exposure measurements are only taken during the work operation.

9. Positive bias {3} due to the use of effective respirators by workers.

10. Negative bias {2} due to existence of unfavorable exposure patterns creating a ''resonance'' between intermittent airborne levels and levels in the body.

11. Negative bias {2} due to increased lung ventilation resulting from hard physical labor.

Ulfvarson (1983) recommends that an epidemiologist use the list of possible biases as a checklist and try to find out the premises of the available data and thus the most probable sign of bias.

In connection with human risk assessment, Kimbrough (1991) makes similar severe criticisms of the exposure assessment methods used in both animal exposure studies and human epidemiological studies:

1. Indirect exposure assessments are said to have doubtful accuracy; if possible, direct measurements of body burdens are preferable. In assessments of inhalation or ingestion exposures, physical and chemical properties that affect ''bioavailability'' should not be ignored (e.g., particle size and presence of trace elements to which the toxin may be bound).

2. Kimbrough also criticizes the validity and accuracy of some types of extrapolations of toxicologic data from animal exposure studies to humans. An expedient basis for extrapolation has sometimes been to make an assumption that humans are at least as sensitive as the most sensitive animal species tested; this is not necessarily true.

3. Multiple causes exist for many chronic diseases, which can lead to confounding (of the primary deterministic variable or environmental factor that is under study) with correlated causative agents that are not part of the study.

4. Biological half-lives of toxins are sometimes shorter at high dosages; therefore, making the usual assumption of a constant excretion rate can lead to underestimation of exposures that (a) were sustained at an earlier time, (b) resulted in presently observed body burdens, and (c) caused chronic biological effects.

5. Animal dose–response data, calculated from dosages on the basis of body weight or body surface area, are not necessarily predictive of biological effects in all humans receiving similar dosages. In an epidemiological study, the average exposure level and correlated response rate assigned to a cohort group may differ from rates for selectively exposed or specially sensitive cohort

subgroups (e.g., smokers and nonsmokers working under the same conditions).

General statistical methodology for epidemiological studies is given in texts such as those by MacMahon and Pugh (1970), Mausner and Bahn (1974), Friedman (1974), Lilienfeld and Lilienfeld (1980), Schlesselman (1982), and Rothman and Boice (1982). Only since about 1980 have specialized epidemiologic texts appeared that deal specifically with studies of occupational groups. One is Monson (1990), which is now in its second edition, a second is Chiazze et al. (1983), and a third is Karvonen and Mikheev (1986). In addition, state-of-the-art general methodology for epidemiological studies in occupational populations is reported in journals such as *American Journal of Epidemiology, American Journal of Industrial Medicine, Archives of Environmental Health, British Journal of Industrial Medicine, International Archives of Occupational and Environmental Health,* and *Journal of Occupational Health.* Both theoretical and practical problems of performing occupational health field studies are being solved, thanks to the extensive field experience that investigators have accrued in occupational observational studies. The March 1976 issue of *Journal of Occupational Medicine* deals entirely with occupational epidemiology, including an article by Enterline (1976) warning of the pitfalls in epidemiological research. Potential problems in occupational studies include inaccurate or uninterpretable cause-of-death statements on death certificates, improper control groups, lack of reliable and accurate quantitative exposure estimates, overlapping exposure and follow-up periods, and competing (but sometimes unknown) causes of death.

The research needs in epidemiology have been summarized by an authoritative Second Task Force appointed by Department of Health, Education, and Welfare. In Chapter 15 of their 1977 report (U.S. Department of Health, Education, and Welfare, 1977), a subtask force chaired by Dr. Brian MacMahon made an assessment of the state-of-the-art of epidemiology and other statistical methods used in environmental health studies. Among the specific recommendations were the following:

1. For clinical environmental research, guidelines are needed for the protection of human subjects and special attention should be given to statistical design and analysis of such studies.
2. For epidemiological studies, better exposure data are needed. Health professionals should help make decisions about what environmental data are collected by governmental agencies. Also, routine surveillance is needed of disease incidence in occupational groups along with surveillance of exposure levels.
3. Animal studies should be used to identify biochemical or physiological early-indicator effects of serious chronic disease in worker groups believed to be exposed to potentially hazardous agents.
4. In all areas of environmental health research, more powerful statistical techniques are needed in areas of multivariate analysis, time-series and sequential analysis, and nonparametric methods. Better dose–response models for mix-

tures of toxic agents are needed, as are better models for animal-to-human extrapolation (particularly for carcinogenesis).

Another useful reference to more specific methodological techniques is the "Steelworker Series" of 10 epidemiological reports by J. W. Lloyd and C. K. Redmond, published between 1969 (Lloyd and Ciocca, 1969) and 1978 (Collins and Redmond, 1978) in *Journal of Occupational Medicine*. These reports are considered by some to constitute the evolution to the present state-of-the-art of modern methodology for epidemiological studies of an occupational population. A paper by Kupper et al. (1975) discusses methods for selecting suitable samples of industrial worker groups and valid control groups.

2.3 Threshold and Low-Risk Exposure Levels

Ideally, the industrial hygienist would like to be able to assess the net risk of an adverse health effect as a function of a worker's past, present, and future exposures and in relation to age, race, gender, and other individual susceptibility factors. If such comprehensive toxicological knowledge were available in relation to past exposures, appropriate limits on future exposures of an individual could be recommended to control the chances of the worker experiencing adverse chronic health effects.

However, achieving such a high level of toxicological understanding is not realistic—one would have to know all the "exposure–time–response" relationships for the toxic material of interest (i.e., the relationships between level of exposure, length of exposure, pattern of intermittent exposures, and the incidence and severity of health effects that occur to some or all of an exposed population). In most cases the industrial hygienist has only an approximate estimate of some "low-risk" exposure level at which the incidence of an adverse health effect *appears* to be low or absent in one or more working populations.

Better yet, if possible, one would like to know a "safe" level of exposure below which all adverse health effects other than the most minor and transient ones are absent in *almost all* workers. This type of level is often called an *exposure threshold level*, or simply *threshold level*. The existence of such thresholds is a controversial question, which we will not attempt to answer here. This concept has been used as the basis of *threshold limit values* (TLV) recommended by the American Conference of Governmental Industrial Hygienists in their annual publication exemplified by ACGIH (1991). Some references to discussions of the question of existence of thresholds are Elkins (1948), *American Industrial Hygiene Association Quarterly* (1955), Stokinger (1955, 1962, 1972), Smyth (1956, 1962), Hatch (1971), Bingham (1971), Hermann (1971), and Thomas (1979).

Statistical models have been developed that can aid us in extrapolating to "acceptably low-risk" exposure levels from higher-risk, dose–response data. Since a low- or minimal-risk exposure level always exists (regardless of whether a true threshold exists), a low-risk level of exposure can usually be selected that is "sufficiently safe." Note that the determination of "sufficiently safe" involves cost–benefit considerations of a socioeconomic and political nature and is not solely a scientific determination. Hartley and Sielken (1977) have reviewed the technical aspects of statistically estimating ''safe doses'' in carcinogenesis experiments. They leave the definition of an ''acceptable'' increase in the risk of carcinogenesis to the regulatory agencies. Acceptable risk levels that have been suggested are 10^{-8} (1 in 100 million) and 10^{-6} (1 in 1 million). Either of these low-risk exposure levels is effectively impossible to determine by direct experimentation with animals because of the enormous sample sizes that would be required to distinguish such minuscule tumor incidence differences. The problem becomes particularly acute when there is a ''normal'' or ''background'' (control) tumor incidence.

Therefore, various procedures have been proposed for extrapolation to low-risk exposure levels using a mathematical model for an exposure–response curve that has been fitted to higher-level, exposure–response data. Among these mathematical modeling procedures are the Mantel–Bryan (1961) procedure (Mantel and Bryan, 1961; Mantel et al., 1975) that is based on extrapolation using a conservative slope on logarithmic-normal probability graph paper. Other models are reviewed by Armitage (1982). Some of the models are more flexible than others. For example, Hartley and Sielken (1975a,b) use a polynomial instead of a straight line to extrapolate. Other models are derived from assumed biological mechanisms. Crump et al. (1977) assume a multistage biological mechanism for development of cancer. The Crump model is fitted to simple incidence versus dose data, whereas the Hartley–Sielken model can be fitted to time-to-tumor data. Krewski and Brown (1981) have provided a comprehensive list of references for carcinogenic risk assessment, which are grouped by carcinogen bioassay, carcinogenicity screening, quantitative risk assessment, and regulatory considerations.

It is difficult to determine exposure–time–response relationships, from limited and unstructured epidemiological data such as are usually available for humans, that reflect a disease (adverse health effect) state that can be attributed to working environments. Nonetheless, exposure control limits can be estimated from carefully conducted, extensive epidemiological studies of workers. For example, the lower limit of a range of average personal exposure levels for a *large* group of workers in a given plant could be taken as a conservative control limit (i.e., more worker protective), if the ''health profile'' of these workers is found to be similar to that of a cohort group of unexposed but otherwise similar individuals. The condition that the unexposed cohort group be ''otherwise similar'' is often difficult to attain; relevance of this necessary condition to validity of the estimated health effect has been previously discussed in Section 2.1. But in any case, this approach is considerably less desirable than use of the sought after, but generally unattainable, exposure–response curves. With the latter approach the exposure experience of the entire exposed work population related to the occurrence of some adverse health effect is used to estimate an appropriate low-risk, exposure control limit. With the former approach that usually is based on only a small fraction of the total exposed group, an exposure control limit is usually set on the basis of the estimated lower range of exposures (see above) in the highest exposure group that did not show the health effect. However, such failure to find the health effect in a small sample of workers may have limited statistical significance. That is, it is possible to fail to observe a health effect in a small random sample of workers even though they had been selected from a larger group containing an unacceptable proportion who would show the adverse health effect. Thus this procedure may yield an estimated control limit that is set too high for the workforce. Also, the susceptibility characteristics of a small sample of workers, who work and live in close proximity, may be different from the rest of the workforce and different from other similarly exposed groups.

2.4 Objectives of Occupational Exposure Estimation

Statistical methodologies can make substantial contributions toward achieving the three goals vital to the objective of effective worker protection: *recognition*, *evaluation*, and *control* of chemical and physical stresses to workers. Industrial hygienists are called on to examine the work environment and recognize workplace stresses and factors that have the potential for adversely affecting worker health. Then they must evaluate the magnitude of the stresses and interpret the results to be able to give an expert opinion regarding the general healthfulness of the workplace, either for short periods or for a lifetime of worker exposure. Finally, there must be a determination of the need for, or effectiveness of, control measures that minimize adverse health effects of workplace exposures.

The practical aspects of achieving these three goals require answers to two initial questions:

1. Are exposure measurements necessary?
2. If so, what type of measurements are needed in relation to our reasons for taking the exposure measurements?

Attempting to answer these two broad questions will require answering additional questions such as the following that will influence the choice of study design strategy:

1. Do we need only rough estimates or precise estimates of exposure levels?
2. Do we need only worst-case estimates of the higher exposure levels or do we need estimates of exposure distributions?
3. What agents will we be measuring?
4. Which workers will be sampled?
5. How many samples will we be taking?
6. When will we take the samples?
7. At what locations will we take the samples?
8. What actions can be taken based on results of the data analysis?

Regarding our data analysis strategies we must ask: How will we analyze our data and reach decisions regarding our research hypotheses in order to decide if the exposure levels are acceptable?

Answers to the preceding questions must be arrived at by first clearly defining the *objective(s)* of estimating worker exposures. Clear definition of the objective(s) will facilitate the formulation of appropriate study designs and data analyses. Typically the objective(s) of estimating worker exposures will be one or more of the following nine.

2.4.1 Hazard Recognition

Hazard recognition is the identification of hazardous agents present in the workplace that are used in such a fashion that exposures create a possible health hazard. It involves the (1) identification of the hazardous agents present in the workplace, (2) identification of the job activities and work operations that involve their use, and (3) determination of whether the hazardous agents are intermittently or routinely released into the workplace air. The existing control measures that might reduce the hazardous exposures must be identified. It must also be determined if hazardous conditions could occur during an irregular episode or during an accident. These include hazardous agents, explosive concentrations, or insufficient oxygen.

2.4.2 Hazard Evaluation

Hazard evaluation is similar to, but wider in scope, than hazard recognition. Where hazardous agents arise from industrial operations, it must be determined if it is possible for a hazardous exposure to occur. This can involve an evaluation of the severity of health hazard(s) from hazardous agents, explosive concentrations, or insufficient oxygen due to industrial operations. It may involve determining if worker complaints or health problems could be due to hazardous exposure levels in the workplace.

Hazard evaluation can include estimating the distribution(s) of worker exposure levels. Specifically, the shape, location, and dispersion of the exposure distribution could be estimated. Percentiles of exposure distributions (along with associated tolerance limits) could be estimated as a function of average-exposure level. The important determinant factors affecting worker exposure levels might be identified. These are the qualitative or quantitative variables or factors that influence or affect worker exposure levels. Periods of abnormally high exposure during the work shift should be looked for.

2.4.3 Control Method Evaluation

The adequacy of existing *control measures* to substantially reduce the probability of unacceptable worker exposures would be determined. The adequacy of newly installed controls and their control effectiveness could be compared to the precontrol exposure situation. The important design and operating parameters affecting the efficacy of control methods could be identified and evaluated.

2.4.4 Exposure Screening Program

This is a limited exposure monitoring program that is designed to identify target populations of workers (see Section 5) with other-than-acceptable, exposure level distributions (i.e., those worker groups with substantial proportions of exposures exceeding an exposure control limit or regulatory standard). Such groups will then receive additional exposure monitoring. The program uses an Action Level as a screening cut-off to select appropriate target populations for inclusion in a limited exposure surveillance program that provides for either: (1) minimal periodic exposure measurements to be made on only a few workers or (2) a more extensive exposure distribution monitoring program. An *exposure screening program* uses the least resources but provides for reasonable protection of the worker target population. This type of a limited program, as given in Leidel et al. (1977), was recommended by NIOSH for use in regulatory monitoring programs (see Section 2.4.6).

2.4.5 Exposure Distribution Monitoring Program

This is a more extensive exposure monitoring program designed to quantify target population exposure distributions over an initial temporal base period (e.g., one day, several weeks, months, or years). Initial exposure distribution estimates would be periodically updated with routine exposure monitoring of the target population(s). Accomplishing this objective requires the appropriate definition of target populations and the identification of those determinant variables that affect the exposure distributions (see Section 5).

2.4.6 Regulatory Monitoring Program

Federal or state regulations may require the establishment of exposure monitoring programs [e.g., Occupational Safety and Health Administration (OSHA), Mine Safety and Health Administration (MSHA)]. The primary goal of this objective is "acceptable quality maintenance" regarding worker exposure levels. That is, the goal is to ensure that all worker exposure levels meet applicable permissible exposure limits (PELs) set by the regulatory agency. This program may involve elements from both an exposure screening program and an exposure distribution monitoring program. However, most OSHA-type monitoring programs are basically exposure screening programs.

2.4.7 Epidemiological Studies

When appropriate data are available, epidemiologists generally attempt to associate an observed incidence or severity of adverse health effects in target populations with exposure levels of hazardous agents. Demonstrating that an increasing response is associated with increasing exposure gives strong support to a hypothesis of disease causation due to exposure to the agent investigated. Using occupational exposure estimation for this purpose generally requires valid, reasonably precise, retrospective estimation of exposure levels for individual workers during all periods of their working lives. In exploratory studies, sometimes population average or population minimum exposures must be used as rough approximations to exposure levels of individual workers. However, this can lead to serious misclassification errors.

Monson (1990) has suggested seven types of criteria to follow when interpreting occupational epidemiologic data:

1. Consistency
2. Specificity
3. Strength of association

4. Dose–response relationships
5. Coherence
6. Temporal relationship
7. Statistical significance

However, when using the criterion of dose–response relationships, Monson (1990, p. 100) has cautioned:

> The lack of a dose–response relationship is fairly weak evidence against causality. The measure of exposure may be misclassified, there may be a threshold necessary for the exposure to cause the disease, there may be bias in the measure of exposure. The presence of a dose–response relationship is relatively strong evidence for causality.

2.4.8 Measurement Method Comparison

Often in industrial hygiene one desires to demonstrate the equivalency of a new exposure measurement method with currently used "standard" method. This approach is less desirable than defining a performance standard for monitoring a workplace contaminant, and then performing a measurement method validation for a proposed new method (see Section 2.4.9). However, a methods comparison may be necessary if the "control" method has been used in past epidemiological studies or is the basis of a regulatory exposure level [e.g., Mine Research Establishment (MRE) horizontal elutriator for coal dust, vertical elutriator for cotton dust, USPHS/NIOSH method for asbestos fibers]. This objective involves determining if the new method can be used as an adequately precise predictor of results that would be obtained if the standard method were used.

2.4.9 Measurement Method Validation

To properly evaluate workplace exposure results, it is highly advisable to estimate the accuracy of the selected measurement method and also identify the causes and magnitude of the random error components in the method. Then one can estimate the following for the method: (1) the *repeatability* (total precision error measured as variability in replicate measurements on the same sample by the same analyst) and (2) the *reproducibility* (variability between laboratories).

It is also highly advisable to estimate two important "detection limits" for the measurement method. The first of these is *limit of detection* (LOD), which is the minimum detectable level that can be reliably differentiated from "background noise." The LOD value should be reported for all observed values at or below the LOD (i.e., "zero" values *should not* be reported).

The second important "detection limit" is the *limit of quantitation* (LOQ), which is the minimum value for which the method yields quantitative estimates that have acceptably low uncertainty. The LOQ is greater than the LOD. The LOQ is the minimum measurement level for which an interval estimate should be reported (e.g., confidence interval, confidence limits). Sometimes the accuracy and precision estimates for the method will be compared to a performance standard to determine the acceptability of the measurement procedure (e.g., the "$\pm 25\%$ accuracy at a 95% confidence level" criteria often used by NIOSH and OSHA as discussed in Leidel et al., 1977).

2.5 Sampling Strategies for Occupational Exposure Estimation

It is important to recognize that *sampling strategies* describe a major portion of a series of *exposure assessment* steps used to achieve the goal of worker protection from adverse health effects due to hazardous workplace exposures. Figure 10.1 presents an overview of exposure sampling strategies as they relate to personal and group exposure assessments. Most exposure estimation sampling strategies have the following four common elements indicated in Figure 10.1:

1. Establishing the *purposes and objectives* for estimating occupational exposures (e.g., see Section 2.4).
2. Determining the *study design* for both the preliminary and full-scale studies (the principles of study design and data analysis are discussed in Section 4 and specific study designs are given in Section 5).

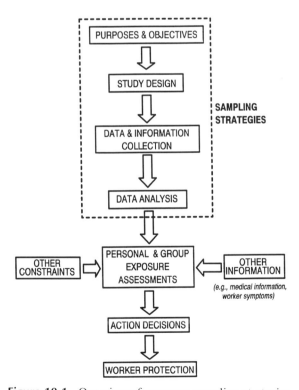

Figure 10.1 Overview of exposure sampling strategies.

Table 10.1 Sampling Strategies for Exposure Estimation Objectives

● Primary strategy
○ Secondary strategy

Column objectives:
1. Hazard recognition
2. Hazard evaluation
3. Control-method evaluation
4. Exposure-screening monitoring program
5. Exposure-distribution monitoring program
6. Regulatory monitoring program
7. Epidemiological study
8. Measurement-methods comparison
9. Measurement-method validation

Strategy	1	2	3	4	5	6	7	8	9
STUDY-DESIGN STRATEGIES									
1. Workplace observations	●	○							
2. Estimate exposure levels from material-usage rates	○								
3. Measuring worst-case exposures	●	○							
4. Exposure-screening monitoring program to identify higher-risk workers or target populations for additional monitoring	○			●		●			
5. Exposure-distribution monitoring program to quantify target-population exposure distributions	○				●	○	●	○	
6. Identifying and quantifying determinant variables for exposure-level distributions	○	●				○	●	○	
DATA-ANALYSIS STRATEGIES									
1. Using an Action Level as a screening cut-off value to identify medium- to high-risk workers or target populations for additional monitoring				●		●			
2. Performing hypothesis testing for classifying individual TWA-exposure point estimates		○			○	○	○		
3. Calculating interval estimates for individual TWA-exposure estimates (e.g., confidence limits)	●	●	○		●	●	●	●	●
4. Calculating point and interval estimates for proportions and fractiles of exposure distributions (e.g., tolerance limits)				●	○	●	●	●	●
5. Using lognormal probability paper for analyzing and summarizing worker or target-population exposure distributions	○	●			●	○	●	○	○
6. Using control charts or semilogarithmic paper for analyzing and summarizing daily TWA-exposure levels and distribution parameters				○	●	●			

3. Collecting the experimental or observational data and other necessary information.

4. Performing the *data analysis* (see Section 6).

Figure 10.1 also indicates that sampling strategy results are then combined with other information (e.g., medical information, worker symptoms) and other constraints, and these are the three elements of personal and group *exposure assessments*. These assessments then lead to *action decisions* necessary for *worker protection* from hazardous exposures.

General types of occupational exposure sampling strategies suitable for the nine objectives of exposure estimation discussed in Section 2.4 are summarized in Table 10.1. Solid bullets (•) indicate primary strategies and hollow bullets (○) indicate secondary strategies in Table 10.1. *Study design strategies* can include:

1. Making workplace observations.
2. Estimating possible airborne levels from material usage rates.
3. Measuring worst-case exposures.
4. Conducting an exposure screening monitoring program to identify higher-risk workers or worker target populations for additional monitoring (see Section 5.3).
5. Conducting an exposure distribution monitoring program to quantify target population exposure distributions (see Section 5.3).
6. Identifying and quantifying determinant variables for exposure level distributions (see Section 5.1).

Data analysis strategies can include:

1. Using an action level as a screening cut-off value to identify medium- to high-risk workers or target populations for additional monitoring.
2. Performing hypothesis testing for classifying individual TWA exposure point estimates (see Sections 6.1.2, 6.2.2, 6.3.2, 6.4.2, and 6.5.2).
3. Calculating interval estimates for individual TWA exposures (see Sections 6.1.1, 6.2.1, 6.3.1, 6.4.1, and 6.5.1).
4. Calculating point and interval estimates for proportions and fractiles of exposure distributions (see Sections 6.7, 6.8, and 6.10).
5. Using lognormal probability paper for analyzing worker or target population exposure distributions (see Sections 6.12 and 6.13).
6. Using control charts or semilogarithmic paper for analyzing daily TWA exposure levels and distribution parameters (see Section 6.11).

3 STATISTICAL THEORY: COMPONENTS OF VARIATION, DISTRIBUTIONAL MODELS FOR OCCUPATIONAL EXPOSURE DATA, AND TESTS OF SIGNIFICANCE USED FOR COMPLIANCE/ NONCOMPLIANCE DECISION MAKING

To plan an assessment study for measuring occupational exposures and properly analyzing the resulting worker exposure data, one must first understand how various types of errors can affect individual exposure measurements. Evaluating the errors that affect an individual exposure measurement is analogous to evaluating how the sizes of individual trees of the same type and age vary randomly within sections of a forest. Adequate distributional models for errors in occupational exposure results, and the sources of variation affecting a single-exposure measurement, will be discussed in Sections 3.2 and 3.3. Next, one must understand the patterns in which groups of data will occur. For example, values of the location parameter (mean values) of the within-group distributions may differ between groups due to assignable causes. This is analogous to interpreting differences between average tree sizes obtained from different forests as being due to such factors as soil fertility, rainfall, pollution, and so on. Models for describing the behavior of such data families will be discussed in Sections 3.4 through 3.6.

3.1 Nomenclature for Exposure Concentration Data

Exposure concentration data from occupational settings can be reported as single measurements or as other estimates (e.g., time-weighted average concentrations). An *exposure measurement* is defined as the measured *airborne concentration* of a material in a single air sample taken near a worker such that it is a valid measurement of that worker's exposure. For the sake of brevity, phrases such as *exposure, exposure data*, and *exposure measurement(s)* will be used for the longer and more accurate phrases *exposure concentration, exposure intensity, exposure concentration data, exposure intensity data, exposure concentration measurement(s)*, and *exposure intensity measurement(s)*. Use of the single word *exposure* will not refer to the dose of contaminant or agent received by a worker's body; rather, the term *biological dose* will have that meaning.

Technical Note: The use of the word *sample* in both the statistical sense and the industrial hygiene physical sense in this chapter can present a source of confusion. Conceptually, a *statistical sample* consists of one or more items selected from a parent population, each of which has some characteristic measured. However, in the physical sense an *industrial hygiene sample* or exposure sample is an *exposure measurement* determined, for example, from a measured amount of an airborne material collected on a physical device (e.g., filter, charcoal tube, passive dosimeter). Such industrial hygiene sampling is usually performed by drawing a measured volume of air through a filter, sorbent tube, impingement device, or other instrument to trap and collect the airborne contaminant. Passive dosimeters rely on diffusion to move the contaminant to the collecting media. In the sense of this chapter, an occupational exposure sample and accompanying act of sampling combine both the concept of a *statistical sample* (i.e., one result among many that could occur under the same conditions) and the *physical sample*, which is a physical agent that is measured or material sample that is chemically analyzed or interpreted.

An *exposure estimate* is an estimate of a workplace exposure concentration or intensity over a specified time period that is calculated from one or more exposure measurements. An exposure estimate may be obtained for a period as short as a few seconds or represent a period from minutes to years. In the latter case it is known as a TWA exposure, which is the time-integrated instantaneous exposure (e.g., the cumulative concentration) divided by the length of time for the exposure period. If cumulative exposure could be estimated from a single air sample, for example, it would be the quotient of the weight of the material in the air sample divided by volume of air sampled during the measurement period.

Typically, the duration of TWA exposure estimates for chemical agents will be about 15 min or less to estimate short-term, acute-exposure risk; 8-hr to estimate a workday exposure risk; and 40-hr or longer to estimate prolonged, chronic exposure risk. If it is appropriate from a toxicological standpoint, longer time-averaging exposure periods (a month or longer) can be used to estimate chronic exposure risk. Certain nomenclature has been developed by Leidel et al. (1977) to describe several different types of TWA exposure estimates. These are illustrated in Figure 10.2 and include:

1. *Full-period single-sample estimate:* A single exposure measurement taken for the full duration of the desired time-averaging period (e.g., 40 hr for a workweek TWA, 8 hr for an 8-hr workday TWA, or 15 min for a 15-min short-term TWA).

2. *Full-period consecutive samples estimate:* The TWA of a continuous series of exposure measurements (equal or unequal, nonoverlapping time intervals) covering the full duration of the desired time-averaging period.

3. *Partial-period consecutive samples estimate:* The TWA of a series (continuous or noncontinuous) of exposure measurements (equal or unequal, nonoverlapping time intervals) obtained for a total duration less than the desired time-averaging

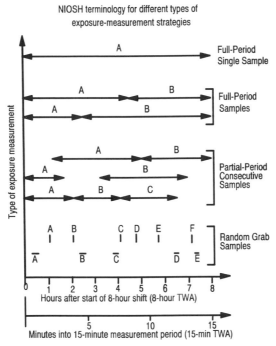

Figure 10.2 Types of exposure measurements that could be taken for an 8-hr TWA or 15-min short-term average exposure standard (from Leidel et al., 1977).

period. For an 8-hr TWA exposure estimate, this would mean that exposure samples assumed to represent the entire 8-hr exposure would be selected to cover about 4 to less than 8-hr. Several samples totaling less than 4 hr (e.g., eight 15-min samples) can usually be better treated as grab samples for the purposes of statistical analysis.

4. *Grab-samples estimate:* The average of several short-period samples taken during random intervals of the time-averaging period. Sometimes it is not feasible, due to technical limitations in measurement methods (e.g., direct reading instruments, some colorimetric detector tubes), to obtain a type 1 or type 2 exposure estimate. In such situations, grab samples may be taken during several short intervals (e.g., seconds or several minutes up to less than about 30 min) within the desired longer time-averaging period such as 15 min or 8 hr.

Adequate distributional models for describing variability of the preceding four types of exposure estimates will be presented in Section 3.6. Selection of particular intervals to be sampled is discussed in Section 5.2.2.

3.2 Net Error Model for Exposure Measurements

Suppose a worker's exposure to a workplace contaminant is to be measured. Assume that an appropriate measurement method is available that *on the average* can give valid (i.e., representative and accurate) determinations of airborne concentrations. The sampling equipment and laboratory instruments must be properly calibrated to reduce systematic errors or biases. This does not imply that every sample will give an exactly correct answer (i.e., the true value of the airborne concentration at the time and place where the sample is taken). To the contrary, random analytical error causes every sample result to differ somewhat from the respective true average exposure that existed during the time period of the sample. "Exposure" is used here synonymously with "exposure concentration" because it is assumed that no difference exists between the *true concentration* measured and the *true exposure concentration* intended to be measured (e.g., as the concentration in a worker's "breathing zone"). The discrepancies between the reported results and the unknown true exposures are termed *random errors* because they are assumed to vary in magnitude and direction in a random manner from sample to sample. Random errors, within limits, are inherent to any measurement method and equipment. The presence of random error does not imply that the method has been improperly used (i.e., that mistakes have been made). Of course, a discrepancy outside the usual range of variability (an excessively imprecise result) for the method could indicate that a mistake has been made and such data might be discarded, especially if the suspect result can be associated with an identifiable irregularity that occurred during the sampling procedure or in the sample analysis.

To systematically approach the statistical treatment of random errors, we use a mathematical statistical model. The true average concentration at the spatial location and temporal period of the exposure sample is denoted by the symbol μ and the particular reported result from the sample is denoted by the symbol X. Thus the total error of a single sample ϵ_T is given by

$$\epsilon_T = X - \mu \tag{1}$$

The total (net) error ϵ_T is the algebraic sum of independent measurement errors, which typically result from the component sampling and analytical steps in the measurement procedure. For example,

$$\epsilon_T = \epsilon_S + \epsilon_A \tag{2}$$

where ϵ_S is a positive or negative random sampling error and ϵ_A is a positive or negative random analytical error. For *independent analytical errors*, the size and sign of the error does not depend on the size or sign of the sampling error. All ϵ's have the same concentration units as the reported result X (e.g., ppm, mg/m^3).

Thus any exposure estimate X that is both an industrial hygiene and statistical *sample* can be represented as the following algebraic sum of the true concentration μ and the net sampling and analytical error for the particular sample:

$$X = \mu + \epsilon_S + \epsilon_A \tag{3}$$

If multiple samples could be taken at *exactly the same point in space and over the same time period* and *if the true value μ were identical for all samples*, they would be true *replicate samples*. Note that in actual industrial hygiene sampling it may be difficult to obtain duplicate samples that are true replicates. Nevertheless, a given sample must be thought of as a random sample from the hypothetical population of all replicate samples that might have been obtained under exactly the same exposure conditions. Seim and Dickeson (1983) have suggested a device for collecting actual replicate samples in the workplace.

3.3 Sources of Variation in Exposure Results

Routinely a population of exposure sample measurements from a given sampling strategy (e.g., grab samples for a worker during a work shift, 8-hr TWA exposures for several workers on a work shift, a series of 8-hr TWA exposures for a worker on several work shifts) will exhibit variability (i.e., scatter or dispersion). An important part of the interpretation of such results is an analysis of the pattern of variation, or distribution, of the data. From the sample results one usually attempts to draw inferences about the population distribution of exposure levels (i.e., about the pattern of all levels occurring for the same conditions under which the sample results were obtained). When analyzing sample data, it is important to understand the sources of variation that combine to create observed total (net) variability (due to total errors measured as differences between measurement results and true exposure levels). The sizes of these variations are a function of both the exposure levels and the measurement method. Both *random* and *systematic* errors can be affected by both the *exposure levels* being measured (Sections 3.3.1 and 3.3.3) and the *measurement procedure* used to obtain the sample results (Sections 3.3.2 and 3.3.4).

3.3.1 Random Variation in Workplace Exposure Levels

An elementary mathematical model for component random errors and related net random variation in replicate sample results was discussed in Section 3.2. Our recognition of random influences on the measurement process can be extended to a recognition of other random influences that affect the true value of what is measured. This extension of the model will enable us to better understand the sources of variation in exposure results at different sites and times. Random changes in the determinant variables affecting the workplace exposure levels can lead to random variation in the exposure results. Exposure *determinant variables* are qualitative or quantitative variables, factors, or parameters that influence or affect true airborne exposure levels. The types of determinant variables are discussed in detail in Section 5. For now, note that random variation in determinant variables can result in any or all of the following types of variability in real exposure levels in the workplace:

1. *Intra*day (within a day) exposure level fluctuations within the same workplace,
2. *Inter*day (between days) exposure level fluctuations within the same workplace,
3. *Inter*worker (between workers) exposure level differences between different workers within a job group or occupational category.

It is important to realize that random variation in exposure levels and in subsequent exposure measurements can be accounted for (but not prevented) by appropriate statistical procedures. Generally the magnitude of random exposure variations over space, time, or workers cannot be quantified or predicted before making numerous exposure measurements, since often it is not possible to predict how the many determinant variables will affect workplace exposure levels.

3.3.2 Random Variation in the Measurement Procedure

In the previous section, we discussed sources of random variation in true workplace exposure levels. Random errors also occur in the exposure level measurement procedure, and these physical variations lead to random variations in corresponding exposure measurements data. Examples of possible sources of random physical errors in the *process* of exposure measurement are:

1. Random changes in pump flow rate (or mass flow rate with passive dosimeters) during sample collection.
2. Random changes in collection efficiency of the sampling device.
3. Random changes in desorption efficiency of the samples during analysis.

It is important to realize that variations in multiple-measurement results obtained during an hour, a day, or over many days that are due to fluctuations in the actual workplace exposure levels will usually exceed measurement procedure variation by a substantial amount (often by factors of 10 or 20). Thus the predominant component of variation in these results will be due to the considerable variation in what is being measured and *not* due to the measurement method itself. This is illustrated in Figure 10.3.

However, in contrast to random variation in the true exposure levels, the rela-

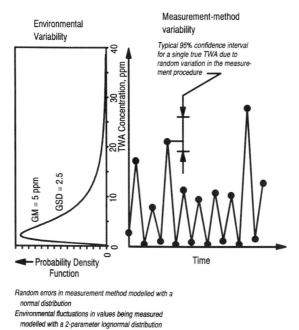

Actual Workplace TWA Exposures: Comparison of variability from:
1. Measurement method (sampling and analytical steps)
2. Environmental fluctuations in actual workplace-exposure levels

Figure 10.3 Variation components in multiple measurement results.

tively smaller random measurement errors can generally be quantified before making the measurements (i.e., ranges of error can be estimated probabilistically from methods evaluation experiments performed before the exposure measurements). Then, the effects of the known distribution of random measurement errors on forthcoming exposure results can be minimized by the application of sampling strategies and programs based on statistical principles.

3.3.3 Systematic Variation in Worker Exposure Levels

In contrast to random variations, *systematic variations* in workplace exposure levels or systematic errors in the measurement procedure cannot be predicted with statistical methodologies based in probability theory. Instead, the study design must anticipate and make provision for systematic errors. During a data analysis performed to compare exposure results between two groups, the comparison should be made within blocks (i.e., within appropriate subgroups that are homogeneous in all other respects), or the measurements should be corrected for possible systematic errors due to extraneous factors *before* any statistical analyses are performed.

Systematic biases or shifts in the determinant variables affecting the workplace exposure levels will lead to systematic shifts in the exposure results. For example, some systematic shifts in determinant variables, and their consequences, are:

1. Changes in a worker's exposure situation (such as several different jobs or operations during a work shift or over several days) can result in intraday or interday shifts in worker exposure.

2. Production or process changes can cause shifts in worker exposure levels (intraday or interday).

3. Control procedure or control system changes can cause shifts in worker exposure levels (intraday or interday).

3.3.4 Systematic Errors in the Measurement Procedure

Besides the random errors in a measurement procedure, there can also be *systematic errors* that occur during the measurement procedure and lead to systematic errors (biases) in exposure results. Examples are:

1. Mistakes in pump calibration and drops in pump battery voltage leading to systematic errors in air flow rate.

2. Use of the sampling device at temperature or altitude conditions substantially different than the calibration conditions (see Technical Appendix G of Leidel et al., 1977).

3. Physical or chemical interferences during sample collection.

4. Sample degradation during storage before analysis.

5. *Intra*laboratory errors (similar errors in groups of analyses performed within a laboratory) due to chemical or optical interferences, improper procedures, mistakes in analytical instrument calibration, or failure to properly follow the steps of an analytical procedure.

6. *Inter*laboratory differences due to use of different methods, different equipment, or different training of personnel.

Systematic measurement errors may be identified and their effects minimized with the use of quality assurance programs.

In the statistical sense, a substantial systematic shift or error in either the measurement process or the exposure levels being measured creates a different population with another location (central tendency) on the exposure level scale. If the systematic shift(s) goes undetected, the resulting two (or several) "side-by-side" sample populations can be mistakenly analyzed as a single distribution. *The inferential statistical procedures presented in this chapter will not detect and do not allow for the analysis of highly inaccurate results caused by systematic errors or shifts.* Unfortunately, systematic errors in a measurement procedure or systematic shifts in exposure levels sometimes go undetected and introduce considerably larger errors into the exposure results than would be caused by the usual random variations. This can lead to reporting inferences from the sample results that have erroneously higher uncertainty (less precision) than is stated, if precision is calculated from the known amount of random variability that has previously existed when no biases are present. Since vague or uncertain inferences generally have little value, it becomes critical to identify potential systematic errors and take steps in the study design to eliminate them, or correct for them, if possible, when the results are statistically analyzed.

For example, for organic solvent exposure measurements, Olsen et al. (1991) found larger random variability in analyses of replicate field samples than in analyses of replicate laboratory samples of the same pure compound. A large (5- to 10-fold) systematic error (bias) was also detected that was associated with the use of two different sampling strategies.

3.3.5 Location of the Measurement Device in Relation to the Worker

A most important goal of personal exposure measuring for chemical agents is to obtain *valid* estimates of the concentrations breathed by workers. A valid exposure estimate is one that measures what it is purported to measure. More specifically, criteria for evaluating the validity of worker exposure estimates include:

1. **Relevance.** Is the air concentration in the sample equivalent to the concentration breathed by the worker(s) of interest?
2. **Calibration.** Are the exposure measurements unbiased estimates of the true concentration sampled (i.e., are measurements accurate on the average)?
3. **Precision.** Was adequate exposure information obtained to derive a sufficiently precise exposure estimate for the worker(s) of interest (either on a given day or over some longer period such as several years of employment)?

Concerning relevance of an exposure estimate to a worker's actual exposure, exposure measurements should be taken in the worker's *breathing zone* (i.e., air that would most nearly represent that inhaled by the worker). There are three basic classes of exposure measurement techniques:

1. **Personal.** The measurement device is directly attached to a worker and worn continuously during all work and rest operations. Thus the device collects air from the breathing zone of the worker.
2. **Breathing Zone.** The measurement device is held by a second person who attempts to sample the air in the breathing zone of a worker.
3. **General Air.** The measurement device is placed in a fixed location in a work area. This technique is also called *area sampling*.

If measurements taken by the *general air* technique are to be used, then it is necessary to demonstrate that they are valid estimates of personal exposures. Normally this is difficult to do. Refer to Technical Appendix C of Leidel et al. (1977) for a discussion of this subject.

3.4 The Normal Distribution Model

3.4.1 Descriptive Parameters

A utilitarian mathematical model for the frequency distribution of some types of continuous variable occupational health data is the *normal distribution*. This model has a mathematical formula that can be used to describe and compute the normal distribution curve. In routine practice the formula is rarely directly applied since tables for the distribution are readily available in technical handbooks. This distribution is also available in hand-held calculators or personal computer software (e.g., Lotus 1-2-3 or Microsoft Excel 4.0 spreadsheets or Mathcad 3.1 for Windows). The distributional formula relates ordinates $f(X)$ of the curve to values of the variable X. It is called a *distribution function* or simply *distribution*.

Technical Note: This terminology is not universally used. Our definition of the term *distribution function* is according to Hald (1952), but others refer to $f(X)$ as the *frequency function* or *probability density function*.

The distribution function $f(X)$ for a normal distribution is represented by the special notation $N(X; \mu, \sigma^2)$. That is, for the normal distribution model (a.k.a. normal curve), the ordinate or height of the probability density curve is given by the formula:

$$f(X) = N(X; \mu, \sigma^2) = \frac{1}{\sigma\sqrt{2\pi}} \exp\left(\frac{-\frac{1}{2}(X - \mu)^2}{\sigma^2}\right) \tag{4}$$

Note that two constants (parameters), μ, the mean, and σ, the standard deviation, completely characterize the normal distribution. Thus the notation $N(X; \mu, \sigma^2)$ is statistical shorthand for "the normal distribution of a variable X that has the true mean μ and variance σ^2." All normal curves have the same general appearance—a bell-shaped curve that is symmetrical about its mean. The true mean μ, also called the *expected value* of the random variable X, denoted $E(X)$, is the weighted average value of all values of the random variable X. The weighting function is the distribution function [i.e., each X is weighted by its probability density $f(X)$].

Mathematically, the mean $E(X)$ of any distribution function, say $f(X)$, is the corresponding distribution curve's center of gravity, which is defined by

$$E(X) = \int_{-\infty}^{+\infty} X f(X) \, dX \tag{5}$$

For the mean of the normal distribution the general $f(X)$ in Eq. (5) is replaced by $N(X; \mu, \sigma^2)$ so that

$$E(X) = \int_{-\infty}^{+\infty} \frac{X}{\sigma\sqrt{2\pi}} \exp\left[\frac{-\frac{1}{2}(X - \mu)^2}{\sigma^2}\right] dX = \mu \tag{6}$$

The integration in Eq. (6) is not obvious and the details of its evaluation are not presented here. The point to note is that for the normal distribution, the parameter μ in its formula is the mean of the distribution.

Similarly, it can be shown that for the normal distribution, the weighted average value of squared deviations $(X - \mu)^2$ is σ^2, that is,

$$E(X - \mu)^2 = \int_{-\infty}^{+\infty} \frac{(X - \mu)^2}{\sigma\sqrt{2\pi}} \exp\left[\frac{-\frac{1}{2}(X - \mu)^2}{\sigma^2}\right] dX = \sigma^2 \qquad (7)$$

For any distribution the mean square of deviations from the mean, denoted by $E[X - E(X)]^2$, is known as the *variance* of X. The variance is the square of the standard deviation. For the normal distribution, the variance is equal to its second parameter σ^2, so that the standard deviation is σ. The *mode* of any distribution is the point on the X scale at which the maximum of the distribution function occurs. The *median* is the middle X value, that is, the value exceeded by 50 percent of the area under the distribution curve. Hereafter we may also refer merely to "proportion of the distribution between two values of X," which should be understood to mean "proportion of the area under the distribution curve between two values of X." Since the total area under any distribution curve is unity (1.0), the "proportion of the area between two X values" can also be called the "area between two X values." For the normal distribution the mode, median, and mean are all equal to μ.

The two parameters μ and σ of a normal distribution completely determine its location (central tendency) and shape (dispersion or variability). The location parameter is the mean μ, which is the center point of the curve. The variability parameter (or measure of dispersion) is the standard deviation σ, which indicates how much dispersion there is of the X values about their mean. Table 10.2 gives some examples of relationships between the mean, standard deviation, and proportions of the total distribution that lie within various intervals containing the mean. See Section 6.9 for a procedure to calculate intervals of a normally distributed variable X (with known parameters), which contain designated proportions of the distribution. Such values can be expressed by *fractile* terminology [this is Hald's (1952) terminology, some others use *quantile*]. The fractile X_P is the value of X that has proportion P of the area of the distribution $f(X)$ at or below it.

Values of X (e.g., concentration results for replicate samples) occur within intervals above and below μ with predictable *relative frequencies* (or probabilities) that are equal to corresponding areas under the curve (the *sample distribution* curve). Ordinates of the sample distribution curve *do not* give probabilities of corresponding

Table 10.2 Areas under the Normal Curve

X Interval	z Interval	Percent of Data within Interval
$\mu - \sigma$ to $\mu + \sigma$	-1.0 to 1.0	68.4
$\mu - 1.645\sigma$ to $\mu + 1.645\sigma$	-1.645 to 1.645	90.0
$\mu - 1.960\sigma$ to $\mu + 1.960\sigma$	-1.960 to 1.960	95.0
$-\infty$ to $\mu + 1.645\sigma$	$-\infty$ to 1.645	95.0
$\mu - 2\sigma$ to $\mu + 2\sigma$	-2 to 2	95.4
$\mu - 1.960\sigma$ to $+\infty$	-1.960 to ∞	97.5
$\mu - 2.576\sigma$ to $\mu + 2.576\sigma$	-2.576 to 2.576	99.0
$\mu - 3\sigma$ to $\mu + 3\sigma$	-3 to 3	99.7

Where $z = (X - \mu)/\sigma$ = standard normal deviate.

sample results. Rather, it is the *area* under the sample distribution curve between two values of X that is equal to the relative frequency (proportion) of replicate samples that would occur in that interval. For this area to represent a probability that is a proportion between 0 (impossibility) and 1 (certainty), a distribution curve is *standardized* such that the total area under the curve is exactly unity (1.0).

3.4.2   Coefficient of Variation

On the average, random errors in exposure concentration measurements, due to errors in exposure measurement procedures during sampling (e.g., elapsed time, air flow rate) and during subsequent chemical analyses, are generally proportional to the level of airborne concentration measured. Therefore, it is appropriate to express the magnitudes of these errors as fractions of the concentration levels. In this way, the measurement variability of an exposure measurement procedure can be expressed as a constant value that is independent of the concentration measured. A measure of this proportional variability called the *coefficient of variation* (CV) is defined by: $CV = \{E[X - E(X)]^2\}^{1/2}/[E(X)]$ [i.e., CV = (standard deviation)/(expected measurement)]. Some chemists know it as the *relative standard deviation* (RSD) or (s_r). For the normal distribution, $CV = \sigma/\mu$.

If a measurement procedure is composed of two or more independent steps (e.g., obtaining the sample, subsequent laboratory analysis), it can be shown [see Eq. (2)] that the net error $(\epsilon_T = \epsilon_S + \epsilon_A)$ for the combined steps of the procedure has the following *total coefficient of variation*:

$$CV_T = [CV_S^2 + CV_A^2]^{1/2} = \sigma_T/\mu \tag{8}$$

where the subscript S denotes the sampling step and the subscript A denotes the analytical step. It is important to realize that the CVs are not directly additive; instead, the CV_T increases as the square root of the sum of the squares of the component CVs.

The total relative standard deviation CV_T generally can be treated as constant within the range of concentrations at which the measurement method is routinely applied. At a given concentration μ within the application range, the standard deviation of the measurement error is given by

$$\sigma_T = (\mu)(CV_T) \tag{9}$$

where

$$\sigma_T = [\sigma_S^2 + \sigma_A^2]^{1/2} \tag{10}$$

Note that a random variable that is assumed to be normally distributed can theoretically attain negative values, whereas airborne concentration measurements cannot attain negative values. Nevertheless, the normal distribution model is usually an adequate approximation to the sampling distribution of replicate sample chemical analyses. Most measurement methods used in industrial hygiene have net random errors whose standard deviation is small compared to the true mean airborne concentration. Thus the portion of the normal distribution model lying left of zero has negligible area and the model adequately predicts the distribution of (positive) replicate measurements. This is illustrated in Figure 10.4, which shows the relative frequency of the many possible results one might obtain with one measurement of an 80-ppm true exposure using a $CV_T = 0.10$ measurement method. As indicated in Table 10.2, about 68 percent of the possible results are within the region centered about μ, from 72 ppm ($\mu - \sigma$) to 88 ppm ($\mu + \sigma$). Additionally, there is only a 2.5 percent likelihood that a measurement of the actual 80 ppm will lie less than 64.3 ppm ($\mu - 1.960\sigma$) or greater then 95.7 ppm ($\mu + 1.960\sigma$). [Note: $\sigma = (0.10)(80$ ppm$) = 8.0$ ppm.]

3.5 The Lognormal Distribution Model

A second mathematical model of great utility for several types of industrial hygiene data is the logarithmic-normal, or *lognormal distribution*. The general properties of the several types of lognormal distributions have been extensively discussed by Aitchison and Brown (1957). Section 3.5.1 will discuss a variate whose logarithm is distributed according to normal law (i.e., the case of a *two-parameter lognormal*). This is the simplest case because it involves primarily an interplay of the mathematical properties of the logarithmic function (used as a data transformation) and the well-known statistical properties of the normal distribution. These were discussed in the previous section. In Section 3.5.2, the definition and scope of the lognormal distribution will be extended with the use of a third parameter to shift the origin of the distribution's measurement scale.

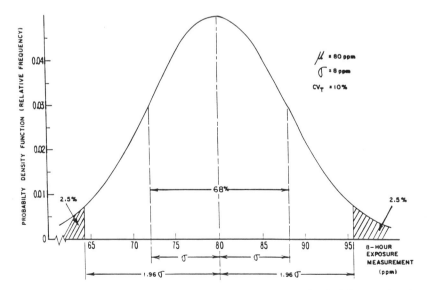

Figure 10.4 Predicted sampling distribution of simultaneous, single 8-hr samples from an employee with a true exposure average μ of 80 ppm.

3.5.1 Two-Parameter Lognormal

A *two-parameter* lognormal curve has the general formula

$$f(X) = \frac{1}{X(\ln \sigma_g)\sqrt{2\pi}} \exp\left(\frac{-\frac{1}{2}(\ln X - \ln \mu_g)^2}{\ln^2 \sigma_g}\right) \tag{11}$$

where $0 < X < \infty$. Equation (11) is called the lognormal probability density function, which is generally shortened to *lognormal distribution function*. Its general structure is similar to Eq. (4) for the normal distribution function. The relationship between these two distributions is that for a random variable X, which is lognormally distributed, the values $\ln X$ or $\log X$ will be normally distributed. As with the normal distribution, the basic two-parameter lognormal distribution is fully characterized by two parameters. However, the lognormal parameters are known as the *geometric mean* (GM) and *geometric standard deviation* (GSD). The true geometric mean μ_g is defined by

$$E(\ln X) = \int_{-\infty}^{+\infty} (\ln X)\, N(\ln X;\, \ln \mu_g,\, \ln^2 \sigma_g)\, d \ln X = \ln \mu_g \tag{12}$$

and the true geometric standard deviation σ_g is defined by

$$E(\ln X - \ln \mu_g)^2 = \int_{-\infty}^{+\infty} (\ln X - \ln \mu_g)^2 N(\ln X;\, \ln \mu_g,\, \ln^2 \sigma_g)\, d \ln X = \ln^2 \sigma_g$$

$$\tag{13}$$

In these equations the notation used is analogous to the $N(X;\, \mu,\, \sigma^2)$ notation defined in Eq. (4) for a normally distributed variable X with mean μ and variance σ^2. Thus μ_g and σ_g are antilogs to the base e of the mean and standard deviation, respectively, of the natural logarithmic transform of X. The interpretation of μ_g and σ_g parameters for a two-parameter lognormal distribution differs somewhat from the interpretation of μ and σ for a normal distribution. The similarity is that the geometric mean is the *location* parameter and the geometric standard deviation is the *variability*, or dispersion, parameter. Note that the distribution function for the basic two-parameter lognormal model originates at zero and does not exist in the region below zero.

In Table 10.2 for a normal distribution, multiples of σ were added to and subtracted from μ to obtain intervals of X that contain specified proportions of the distribution. The factors that multiply σ are standard normal deviates (i.e., values of a normally distributed variable with mean zero and variance one), which are known as Z *values*. Values of Z are listed in tables of the standard normal distribution available in statistical texts and other scientific reference books. A given Z value, denoted Z_P, corresponds to a probability P that a randomly selected value of X will be within the interval $(\mu - Z_P\sigma)$ to $(\mu + Z_P\sigma)$. For example, Table 10.2 shows that $Z_{0.684} = 1.000$, $Z_{0.90} = 1.645$, and $Z_{0.95} = 1.960$. Corresponding intervals for a lognormal distribution are of the form $\mu_g/\sigma_g^{Z_P}$ to $\mu_g\sigma_g^{Z_P}$. Table 10.3 gives examples

Table 10.3 Areas under the Lognormal Curve

X Interval	Percent of Data within Interval
μ_g/σ_g to $\mu_g\sigma_g$	68.4
$\mu_g/\sigma_g^{1.645}$ to $\mu_g\sigma_g^{1.645}$	90.0
$\mu/\sigma_g^{1.960}$ to $\mu_g\sigma_g^{1.960}$	95.0
0 to $\mu_g\sigma_g^{1.645}$	95.0
μ_g/σ_g^2 to $\mu_g\sigma_g^2$	95.4
$\mu_g/\sigma_g^{1.960}$ to $(+\infty)$	97.5
$\mu_g/\sigma_g^{2.576}$ to $\mu_g\sigma_g^{2.576}$	99.0
μ_g/σ_g^3 to $\mu_g\sigma_g^3$	99.7

of two-parameter lognormal intervals corresponding to the intervals for a normal distribution in Table 10.2. See Section 6.10 for a procedure to calculate intervals of a lognormally distributed variable X (with known parameters) that contain designated proportions of the distribution. Such intervals are bounded by *fractiles* X_P (see the definition given in Section 3.4.1).

3.5.2 Three-Parameter Lognormal

The utility of the basic two-parameter lognormal model can be considerably expanded by the introduction of a third parameter that is a change-of-origin parameter. A simple displacement of a random variate X that is *not* lognormally distributed can sometimes be made to define a transformed variate $X_T = (X - k)$ that *is* lognormally distributed. If the range of X is $k < X < \infty$, the range of X_T will be $0 < X_T < \infty$. The two-parameter model can be thought of as a special case of a three-parameter lognormal model, for which $k = 0$. The third parameter k for the lognormal model is sometimes selected to be the lower bound of the known range of values of the original variate X and can be thought of as the threshold of the three-parameter lognormal distribution, just as zero is the threshold for the basic two-parameter model.

The three-parameter lognormal model is useful when the data, such as exposure results, exhibit more skewness to the right than would be expected for a lognormal distribution that is not located close to the zero origin. Such a distribution could result if there were lognormal random additive variations in exposure levels combined with a fixed-background exposure level. An appropriate constant k is subtracted from each data value to create transformed data values that are then analyzed using techniques appropriate for two-parameter lognormally distributed data. Estimated parameters of the transformed distribution, along with appropriate confidence limits, are calculated. Examples are the GM of X_T and its confidence limits, or tolerance limits for the random variable X_T. The constant k is then added to all calculated values of X_T to estimate corresponding values for X relevant to the original data distribution. Details for appropriate estimation of k are presented in Section 6.12, Step 5 of the Solution.

3.6 Adequate Distributional Models for Exposure Results

To design efficient and statistically powerful studies, make rational decisions in hypothesis tests, and make valid inferences regarding expected limits on true occupational exposures, it is necessary to use *adequate* distributional models of exposures for the target populations the samples represent. Adequate distributional models are the keystones for valid use of the parametric statistical methods that will be presented in Sections 5 and 6. The rationale and mathematical statistical properties of models given in this section are also discussed in depth in Busch and Leidel (1988).

The adequacy of a model is dependent on its ability to serve as a workable forecaster of parameters (e.g., mean) or functions of parameters (e.g., CV or fractiles) of the parent population. Moroney (1951) has noted:

> Probably there never is a mathematical function that fits a practical case absolutely perfectly. Nor is it at all necessary that there should be. What we seek is not a *perfect* description of a distribution but an *adequate* one; that is to say, one that is good enough for the purpose we have in view.

Wilkins (1976) has remarked that statistical tests for goodness-of-fit have the characteristic that the test will reject any hypothetical model for a practical data set if there is a sufficient number of samples. The critical value for the test statistic (based on the permissible departure of the observed data distribution from the chosen distributional model) can even be smaller than the precision provided by the measurement methodology used to obtain the data. Even though we might observe a ''statistically significant lack of fit,'' the proposed distributional model may still be *adequate* for our needs. One must also be cautious regarding hypothesis test outcomes of ''no statistically significant lack of fit.'' These outcomes may be due to low statistical power for the tests resulting from small sample sizes.

3.6.1 Applications for the Normal Distribution

The *normal distribution* is usually an adequate model for the following populations of industrial hygiene results.

3.6.1.1 Populations of Replicate Analyses Performed on an Industrial Hygiene Sample (e.g., aerosol filters or charcoal tubes). *Replicate analyses* of a given sample are defined to be repeated analyses with variability equal to that which would exist in analyses of physically different samples, if these could have been obtained *without* sampling errors, of exactly the same concentration in exactly the same setting. In other words, replicate analyses are those with variability that reflects the total random error of the analytical procedure, not just components of error due to some (but not all) steps in the analysis. For an unbiased analytical method (i.e., without systematic error), the expected value of truly replicate analyses of a given physical sample is not the true time-integrated concentration the sample was obtained from. Rather, the expected value of replicate analyses is an air concentration equivalent to the amount actually present in the sample (i.e., the expected value includes the random sampling error for that sample). The sample coefficient of vari-

ation ($\widehat{CV}_A$) computed from replicate analyses is a measure of dispersion for the analytical procedure and is usually taken to be an approximation to the true CV_A measuring the proportional error due to the analytical step of the exposure measurement procedure. Note that the total (net) error of the measurement procedure is $CV_T = (CV_A^2 + CV_S^2)^{1/2}$, where CV_S is the coefficient of variation for *sampling errors* (i.e., those introduced by the physical sampling portion of the measurement procedure during which the contaminated air is moved onto or through the sampling media).

Section 6.6 presents the computation of *tolerance limits* for a variable that is normally distributed with unknown parameters. Based on results for *n* replicate samples, this procedure can be used to compute a *tolerance interval* that we can be 95 percent confident will contain at least 95 percent of analytical results for the same concentration by the same method under the same conditions. Procedures are given in Sections 6.12 and 6.13 that detail the use of logarithmic probability paper to compute and display tolerance limits for lognormally distributed data. These procedures can be modified, where a normal distribution is expected, by substituting normal probability paper (where a linear scale is used for the original data instead of a logarithmic scale).

3.6.1.2 Populations of Replicate Measurements of Calibrated Test Concentrations.

The arithmetic mean of the replicate measurements is the best estimate of a calibrated test concentration. The sample coefficient of variation calculated from a set of *replicate measurements* is an estimate of the measurement procedure's total CV_T (combined CV for the sampling and analytical portions of the method). Refer to Section 3.4.2 and Step 1 of this section. In this context the term *sampling error* refers primarily to physical random error in the volume of air sampled. Here, sampling error does not include intersubject variability, nor does it include the real variations in air concentrations of the contaminant over space and time. Such spatial and temporal variations are indeed appropriate parts of the total error of some special types of exposure estimates (and will be so included, e.g., for an 8-hr TWA estimated from grab samples, see Sections 3.6.2 and 6.4.1).

Industrial hygiene researchers will usually obtain multiple measurements taken simultaneously at sampling locations in a small spatial volume (e.g., a sphere less than 30 cm in diameter), when attempting to estimate the variability (CV) of a measurement method. Unfortunately, unless the sampled workplace atmosphere is truly homogeneous, the sample results may lead to a variability estimate for the measurement method that is erroneously high. It is usually assumed that each of the measurements is a sample of the same true concentration, but this may not be a valid assumption, and the researcher must demonstrate that the sample environment is truly homogeneous.

Figure 10.5 illustrates the use of a normal distribution to model a sampling distribution of replicate measurements of a 5-ppm calibrated test concentration. The figure assumes a measurement method with a $CV_T = 0.10$ method was used. The top portion of the figure also indicates what a typical histogram might look like for relatively few replicate measurements. Note that the histogram is not exactly symmetrical about $\mu = 5.0$ ppm due to random sampling variation. The bottom portion

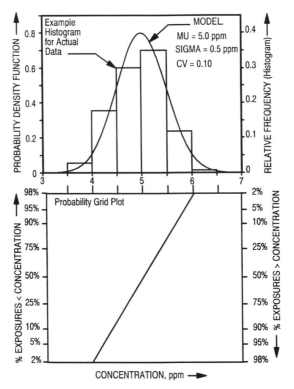

Figure 10.5 Normal distribution model for sampling distribution of replicate measurements of a 5-ppm calibrated test concentration with a $CV_T = 0.10$ method.

of Figure 10.5 illustrates how a normal distribution for this type of sample data can be estimated by fitting a straight line to a plot of the sample cumulative distribution on normal-probability graph paper. On this type of graph paper, the sample distribution should yield a linear plot, aside from the expected deviations accountable to random sampling variations.

It is important to note that normality of replicate exposure measurements (or at least approximate normality that is sufficient to meet the requirements of any inferential statistical methodology one desires to use) is to be expected from theoretical considerations. The total net error of any particular exposure measurement is the net error resulting from many random incremental positive and negative physical influences during the various sampling and analytical stages of a measurement. Determinant variables that can lead to positive and negative errors in measurements include unavoidable technician variations, small environmental variations (e.g., humidity, temperature), and functional variations in component parts of the sampling and analytical equipment (e.g., voltage, pump flow rate, operating temperature). Insofar as these various sources of random additive errors operate independently, their net influences tend to make the net error follow a normal probability density curve. The proof that this is true would be similar to the proof of the central limit theorem from mathematical statistics.

3.6.1.3 Populations of Replicate Full-Period Single-Sample Estimates of Exposure.

The justification for using the normal distribution to model the random errors of replicate exposure measurements was discussed in Section 3.6.1.2. A single full-period sample estimate can be considered a sample of one from a hypothetical parent population of all replicate exposure measurements that could have been obtained at exactly the same point in space and over the same time period. Even though the sample is continuously collecting airborne material from constantly varying levels (i.e., from a distribution of true levels that generally vary substantially), it is assumed that on the average such samples faithfully and accurately integrate all the instantaneous concentration levels. The result of this integration is the mass of formerly airborne material collected in the sample. Thus the single full-period measurement is a time-integrated exposure estimate for the duration of the sample. A previously well-determined total coefficient of variation CV_T for the measurement procedure is used as the measure of dispersion for the population of possible replicate samples from which the one at hand is considered to be a random sample. This CV_T is treated here as known and will be used for inferential decision making and confidence interval calculations for the full-period type of exposure estimate.

Section 6.1 presents applied statistical procedures for computing confidence limits (Section 6.1.1) and classifying exposures relative to an exposure control limit (Section 6.1.2) for the case of exposure estimates based on full-period single samples.

3.6.1.4 Populations of Full-Period Consecutive-Sample Estimates of Exposure.

The justification for using the normal distribution to model the random errors of averages of sets of consecutive samples taken during the time-averaging period of the exposure estimate is an extension of the preceding discussion for the full-period single-sample estimate. It is assumed that the random proportional errors of the set of consecutive measurements are independent and all have the same known total coefficient of variation for the measurement method.

Sections 6.2 and 6.3 present applied statistical procedures for computing confidence limits on the true exposure (Sections 6.2.1 and 6.3.1) and for classifying exposures relative to an exposure control limit (Sections 6.2.2 and 6.3.2) when exposure estimates are based on the average of full-period consecutive samples. The confidence interval and decision-making computations are presented for two different types of situations. *Uniform exposure* methods are given in Section 6.2 to be used when one believes that all consecutively sampled periods had equal true average exposure concentrations. Conservative *nonuniform exposure* methods are given in Section 6.3, to be used if one believes that the sampled periods had substantially different exposure concentrations.

3.6.2 Applications for a Lognormal Distribution

A *lognormal distribution* (either two- or three-parameter) is usually an adequate model for four general types of populations discussed in this section. However, be alert that a population of industrial hygiene data may be a composite of several different distributions. For example, a substantial portion of the data may occur at zero concentration with the remainder occurring in a two-parameter lognormal distribution and one or more three-parameter lognormal distributions.

3.6.2.1 Populations of True Exposure Levels at Different Times during Periods of Hours to Years.

The justification for using the lognormal distribution to model workplace exposure levels at different times for a given worker, or averages for a *target population* of workers at different times (see Section 5), has been presented in Technical Appendix M of Leidel et al. (1977). According to Hahn and Shapiro (1968), the choice of a distributional form for many types of data can be based on theoretical considerations. However, for other types of data where not enough is known about the underlying physical mechanisms to be able to do this, they suggest basing choice of the distributional form on goodness-of-fit tests for observed data. They suggest use of the versatile Johnson and Pearson family of distributions for empirical fitting of approximate distributional models. They say this empirical modeling approach is usually adequate within the range of the sample data used to fit the model.

Stoline (1991) compared goodness-of-fit of lognormal distributional models to the fit of more general Box-Cox distributional models. He found that for most of the data sets reasonable goodness of fit could be achieved with *either* the lognormal or Box-Cox distributional forms. The Box-Cox model fit slightly better, but not statistically significantly better, for a few samples. These empirical results for environmental monitoring data relate strictly only to the types of water contamination data that Stoline examined, but the statistical properties of air monitoring data are likely to be similar. Therefore, given that there is likely to be a choice between nearly equivalent Box-Cox and lognormal distributional alternatives, the lognormal model would be preferable because existing normal distribution theory could be easily applied (to a logarithmic transform).

Conditions conducive to (but not all necessary for) the occurrence of lognormal distributions are found in populations of workplace exposure levels. These conditions include:

1. Physical causes of variability tend to cause the same proportional changes in concentration, irrespective of whatever concentration is present.
2. The true exposure levels cover a wide range of values, often several orders of magnitude. The variation of the true exposure levels is of the order of the size of the exposure levels.
3. The true exposure levels lie close to a physical limit (zero concentration).
4. A finite probability exists of unusually large values (or data ''spikes'') occurring.

Section 6.7 presents an applied statistical procedure for the computation of tolerance limits for a lognormally distributed population with unknown parameters. This procedure is useful for computing, from a sample of exposure levels, a tolerance interval that we can be 95 percent confident will contain at least 95 percent of the population of true exposure levels. Section 6.8 details the computation of a point estimate and confidence limits for the proportion of a lognormally distributed population that exceeds a specified value (such as an exposure control limit). Section 6.11 suggests the use of semilogarithmic graph paper for plotting variables that are lognormally distributed in time, which frequently is the case for true exposure levels at different times during a period. Sections 6.12 and 6.13 present procedures for using logarithmic probability paper to estimate the parameters of logarithmic normal distributions and for displaying tolerance limits for estimated lognormal distributions.

3.6.2.2 Grab-Sample Populations of Intraday Exposure Measurement Populations Consisting of Short-Term Samples.

The duration of each grab-sample is short compared to the total interval from which the samples were obtained (e.g., less than about 5 percent of the interval). Grab-sample exposure estimates reflect the lognormal intraday distribution of true exposure levels that is sampled (i.e., environmental variability). Grab-sample populations also have a component of normally distributed measurement process (sampling and analysis) error (i.e., measurement method variability), but this component generally is negligible compared to the larger amount of lognormal variation of the true concentrations. These two variability contributors are graphically illustrated in Figure 10.6.

For evaluation of a worker's individual health risk, an inference must be made

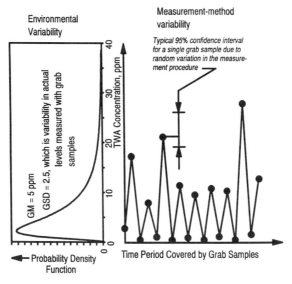

Figure 10.6 Comparison of variability contributions in a grab-sample estimate.

concerning the relation between the TWA exposure control limit and the true arithmetic mean of the entire population of grab samples from which the few samples at hand were selected at random. If the true exposures occurring during the interval could be considered uniform (effectively equal), then the sample arithmetic mean of the grab samples would be an adequate estimate of the TWA exposure. But, if there are subintervals with respectively stable but different exposures, then a sample arithmetic mean for each period of equal exposure should be computed from the grab samples of each period and a TWA estimate then computed from the series of arithmetic means. Confidence limits for such estimates of TWA exposures can be based on the normal distribution, since the only errors are those of the measurement procedure.

However, if there were general lognormal variability among the total set of grab-sample intervals making up the period of the standard, then an estimate of the TWA exposure based on grab samples would require computing a sample geometric mean and converting it to an estimate of the TWA exposure. A procedure for doing this is given in Section 6.4, but this is not a recommended procedure because the correction factor for converting a GM estimate to a *TWA exposure* estimate is a function of the GSD. When the sample geometric standard deviation for each interval of uniform exposure has to be computed (to estimate the variability of the worker's exposure during each interval), this complex estimation technique introduces considerable sampling error into the TWA exposure estimate.

Section 6.7 presents an applied statistical procedure for the computation of tolerance limits for a lognormally distributed population with unknown parameters. This procedure is useful for computing tolerance intervals that we can be 95 percent confident will contain 95 percent of the true short-term exposure levels during a given day for which grab samples were taken. Also, Section 6.11 suggests the use of semilogarithmic graph paper for plotting variables that are lognormally distributed in time, which typically is the case for grab-sample measurements on a given day. Sections 6.12 and 6.13 present procedures for using logarithmic probability paper to estimate the parameters of lognormal distributions and for displaying corresponding tolerance limits.

3.6.2.3 *Populations of Daily 8-hr TWA Exposure Estimates for a Worker.* For evaluation of a worker's individual health risk due to a chronic exposure, the long-term average of daily exposures could be estimated by the arithmetic mean of a random sample of daily TWA estimates, *if* the daily exposures could be considered uniform. If not, then a sample arithmetic mean for each group of daily TWAs from each multiday period of equal exposure would have to be computed and a long-term TWA estimate computed from the series of arithmetic means. In either of these cases (where uniform exposures on each day can be assumed), confidence limits for the long-term TWA exposure would be based on the normal distribution, since the only errors are in the measurement procedure.

The CE (cumulative exposure) and LWAE (lifetime, weighted-average exposure) methods of Ford et al. (1991), described earlier in Section 2.2, have the desirable property of being practical exposure evaluation tools in epidemiological studies because they permit cumulative or weighted-average personal exposure estimates to be made even when individual, personal exposure concentration measurements had not been made. However, for this purpose it is questionable to use a geometric mean as the exposure estimate for a given working zone. Exposure concentrations in a given working zone usually vary over time, at least to some extent; therefore, an arithmetic mean concentration would be theoretically preferable to a geometric mean concentration because an arithmetic mean is directly proportional to the cumulative mass of contaminant to which a worker was exposed while in that zone.

Admittedly, if the true exposure level within a working zone were *really* constant over the exposure period of interest (as assumed by Ford et al., 1991), then *either* the arithmetic mean or the geometric mean would be a consistent estimator of the true exposure concentration for a working zone. (A consistent estimator is one for which, as the sample size becomes very large, the probability approaches 1.0 that an estimate of that type will be infinitesimally close to the true value of the parameter.) But for smaller sample sizes, the arithmetic mean would usually be preferable to the geometric mean as an estimate of the true exposure level. This is so because the arithmetic mean would usually be closer to being an unbiased estimate due to the fact that *average* net errors of sampling and analysis are usually more nearly normally distributed than they are lognormally distributed. A more detailed discussion of the rationale for choice of the normal distribution model for estimates of average exposure is given in Busch and Leidel (1988).

Thus, for the case of day-to-day lognormal random variability (without multiday periods of equal exposures), the sample geometric mean of a random sample of daily exposures would need to be computed and converted to an estimate of the long-term arithmetic mean exposure using a function of the sample geometric standard deviation as a correction factor. This complex estimation procedure for the arithmetic mean of a lognormal distribution has a relative large variance and should be avoided if possible. An alternative procedure, for sufficiently large samples (e.g., at least 30 days), is to compute the sample arithmetic mean and compute its confidence limits under normal distribution assumptions. The justification for this is that the sample means of many random samples are approximately normally distributed even though the single samples are not normally distributed. A detailed discussion of this point is given as a technical note after the next paragraph.

With populations of daily 8-hr TWA exposure estimates, measures of central tendency can be misleading regarding occupational health risk. Analogously, one may report that a river has an average depth of 2 ft, thus inferring that it is safe to wade in, but people can drown in those parts of the river that are more than 5 ft deep. It is important to consider the upper tail (higher values) of any exposure distribution. Tolerance limits provide an indication of the potential upper levels of exposure distributions or an indication of the potential widths of exposure distributions. Often, tolerance limits are the relevant statistical values instead of estimates and confidence limits for central tendency parameters (arithmetic means and medians). Frequently, investigators compute the latter values merely because elementary statistical texts contain equations for estimates of, and confidence limits on, central tendency parameters, while these texts fail to discuss the concept and use of tolerance limits.

Technical Note: The reader should be cautioned that it is possible to erroneously apply the central limit theorem of mathematical statistics and conclude that daily TWA exposure estimates that are calculated from lognormally distributed, intraday exposure levels (e.g., from grab samples) should have an (interday) normal distribution. To the contrary, the theorem merely implies that the means of *n identically distributed* (e.g., lognormally) independent random variables will be approximately normally distributed regardless of the distribution of the individual variables. But note that the sample means (daily TWAs) of the individual variables (all possible instantaneous exposure values over some *multiday* period) *are not* the means of independent random samples obtained from the same lognormal distribution. Different lognormal distributions of intraday exposure levels exist for the various days, and *another* lognormal distribution exists for interday variability of the daily TWAs. Each "mean" (TWA) is then merely a *single* sample from the interday lognormal distribution of daily TWAs.

An appropriate application of the central limit theorem would be to a multiday exposure average computed from a lognormal population of daily TWA exposure estimates. Each daily TWA would constitute a single sample. If *n* randomly selected TWAs were drawn from the lognormal interday population of TWAs and a sample mean calculated, then the distribution of such multiday exposure averages (each estimated from *n* samples) would be approximately normally distributed. The approximation improves as the sample size *n* increases.

3.6.2.4 Populations of Daily 8-hr TWA Exposures for a Group of Workers Having Similar Expected Exposures (e.g., in the same exposure environment, from the same job type or occupational group).

For evaluation of individual worker exposures and attendant health risk, neither the arithmetic mean nor the geometric mean of any such population of multiple exposures of a group of workers is an appropriate parameter, unless the distribution has negligible variation. The difficulty is that a particular worker's individual distribution of exposures may consistently lie in the high (or low) tail of a multiday, multiworker distribution because the true multiday exposure average of that worker may be substantially different from the central tendency of the multiworker exposure distribution. Geometric standard deviations would be the appropriate measure of dispersion for the separate interday and interworker components of the total variation of daily exposures.

Both types of exposure distributions (multiple work shifts for a given worker and multiple workers for a given work shift) are usually approximately lognormal. However, to be able to use a *single* lognormal distribution for exposures of different workers on different work shifts (or days), the following conditions must apply. The *between days for a given worker* and *between workers on a given day* random variations must be independent. Such independence can be assured by randomly selecting the worker–day combinations for which exposures are to be measured. A suitable procedure to do this would be to randomly select the days to be sampled and then randomly select a different worker for measurement on each selected day. This selection procedure is required in order that the resulting data will follow a single lognormal distribution. If the same group of workers were measured on each of the same several days, exposures would be intercorrelated and would not constitute a simple random sample from the same lognormal distribution. The method of analysis of variance (of a logarithmic transform of exposure results) would then have to be used to separate the total variation into components due to worker-to-worker (on the same day) and day-to-day (for the same worker) lognormal variations. This technique is so complex that its complete exposition is inappropriate in this chapter.

Busch and Leidel (1988) present normal and lognormal distributional models for the "components of variance" that are associated with the sources of variability that affect various types of industrial hygiene sampling data. They also give formulas for mean and variance parameters of related distributional models for total (net) error. These formulas are expressed in terms of the distributional parameters for the component errors.

If cross-classified exposure data must be analyzed to determine tolerance limits, a professional statistician's assistance most likely will be needed. Additional discussion is given in Section 6.8 concerning an example computation of lognormal tolerance limits determined from exposures for randomly selected worker–day combinations.

3.7 Decision Values for the Unknown Mean of a Normal Distribution with Known Coefficient of Variation

Frequently, n independently collected consecutive samples X_1, X_2, $\cdots$, X_n are obtained that collectively span the period of a worker's time-weighted average exposure (e.g., 40-hr, 8-hr, 15-min TWA). Such samples are termed *full-period consecutive samples* and the average of the n measurements is used to estimate the TWA exposure. Assume that these measurements have net random errors that are normally and independently distributed with the same total coefficient of variation CV_T. Then the mean $\overline{X}$ of the n measurements will also be normally distributed. This normal distribution of $\overline{X}$ values has the following mean and variance:

$$\mu_{\overline{X}} = E(\overline{X})$$

$$= \left(\frac{1}{n}\right)[E(X_1) + E(X_2) + \cdots + E(X_n)]$$

$$= \left(\frac{1}{n}\right)(\mu_1 + \mu_2 + \cdots + \mu_n) \tag{14}$$

$$\sigma_{\overline{X}}^2 = \left(\frac{1}{n^2}\right)(CV_T^2)(\mu_1^2 + \mu_2^2 + \cdots + \mu_n^2) \tag{15}$$

If a worker's work shift exposure were *uniform* (i.e., all consecutively sampled periods having effectively equal true average concentrations), there would be $\mu_i = \mu$ for $i = 1, 2, \cdots , n$, and the TWA measurement mean exposure $\overline{X}$ could then be treated as a random sample from a normal distribution with mean μ and variance $\sigma^2/n = (1/n)(CV_T^2)(\mu^2)$. (*Note!* Substantially different exposure situations during a workshift generally result in a nonuniform 8-hr TWA exposure.) In 95 percent of such uniform mean measurements, the average $\overline{X}$ would be within an interval $\mu \pm [(1.96)(CV_T)(\mu)/n^{1/2}]$. Equivalently, *two-sided* intervals $\overline{X} \pm [(1.96)(CV_T)(\mu)/n^{1/2}]$ would contain μ for 95 percent of the $\overline{X}$ values. Note that the latter probability intervals are centered about the randomly varying TWA measurement means $\overline{X}$.

To create a decision-making test (i.e., statistical significance test of a null hypothesis of compliance of a reported TWA exposure mean for a worker with an *exposure control limit* or *exposure standard*, denoted ECL), assume that the worker's true TWA exposure level μ is equal to the value ECL. Under this null hypothesis, *decision intervals* surrounding the ECL can be computed that would contain the TWA exposure mean ($\overline{X}$) in *at least* 95 percent of similar cases. Such a decision interval, for the case of *uniform exposures*, would be of the form

$$ECL \pm \frac{(1.96)(CV_T)(ECL)}{\sqrt{n}} \tag{16}$$

The lower bound will be termed the *lower decision value* (LDV) and the upper bound will be termed the *upper decision value* (UDV). Similar *open decision intervals* that are upper-bounded only (by UDV, i.e., *one-sided decision intervals*) can be computed that would contain the TWA measurement mean $\overline{X}$ for *at least* 95 percent of similar cases, given that $\mu \leq ECL$. Such a one-sided decision interval, $\overline{X} \leq [ECL + (1.645)(CV_T)(ECL)/n^{1/2}] = UDV$, is open on the lower side and its upper bound will be denoted as the UDV for $\overline{X}$. This is because the decision interval will be exceeded only by 5 percent or less of $\overline{X}$ values, when the null hypothesis, H_0: $\mu \leq ECL$, is true. Hence the interval's upper bound will be denoted as $UDV_{5\%}$. Therefore, in case $\overline{X} > UDV_{5\%}$, the hypothesis H_0 is considered unlikely to be true, since this occurrence would be infrequent under H_0, and an alternative hypothesis, H_1: $\mu > ECL$, is accepted because H_1 gives the observed $\overline{X}$ a more reasonable probability of having occurred by chance. Specific applied methods with examples will be presented in Sections 6.1 through 6.3 that apply the one-sided $LDV_{5\%}$ and similar $UDV_{5\%}$ concepts to decision making regarding compliance or noncompliance of a particular TWA exposure estimate with an exposure control level or standard.

Instead of decision values, if we desire to calculate a one-sided, 95 percent *lower confidence limit* (LCL $_{1,.95}$) for μ, we could rearrange the following probability statement:

$$P\left\{\overline{X} \leq \left[\mu + \frac{(1.645)(\mu)(CV_T)}{\sqrt{n}}\right]\right\} = 0.95$$

and obtain the $LCL_{1,.95}$ as

$$\mu \geq \frac{\overline{X}}{1 + (1.645)(CV_T)/n^{1/2}}$$

The analogous one-sided 95 percent *upper confidence limit* ($UCL_{1, .95}$) for μ is

$$\mu \leq \frac{\overline{X}}{1 - (1.645)(CV_T)/n^{1/2}}$$

To compute two-sided, 95 percent confidence limits ($LCL_{2, .95}$ and $UCL_{2, .95}$) for an *interval estimate*, use the above formulas with 1.96 substituted in place of 1.645. In summary, for normally distributed and independent measurement errors, these formulas give exact 95 percent confidence limits in cases where CV_T is known and there are uniform exposures in n equal-duration, consecutive, sampling periods. For the general case of unequal sampling durations and nonuniform exposure, approximate confidence limits can be obtained using methods given in Sections 6.2.1 and 6.3.1. Specific cases of confidence limit applications are also discussed with examples in Sections 6.1 through 6.3.

4 PRINCIPLES OF STUDY DESIGN AND DATA ANALYSIS FOR EVALUATING OCCUPATIONAL EXPOSURES

One important goal of research is to make inferences about some population or draw other general conclusions, based on sample survey results or experimental study data. Often an investigator will seek the assistance of a statistician in analyzing nondefinitive results from a research study. Unfortunately, sometimes the results presented for analysis are not only fragmentary, but incoherent, so that next to nothing can be done with them except perhaps compute some trivial descriptive statistics. This does not have to happen. Research dollars do not have to be wasted on unproductive studies. Adherence to statistical principles of study design and related data analysis can produce substantially better results.

A study should be initiated, conducted, and the results evaluated only if the investigator has a clear purpose in mind and a clear idea about the precise way the results will be analyzed to yield the desired information. Far too often studies are conducted in the blithe and uncritical belief that a subsequent "statistical analysis" will yield something useful, especially when a statistician is engaged to "juggle the data."

The methodological tools of statistics cannot extract information or inferences that are not inherent to the data. The use of statistical technique to analyze study results requires asking certain questions concerning the hypotheses to be tested and the parameters to be estimated. Our ability to test an hypothesis depends on the study design used and the circumstances in which the data were collected (Finney, 1982). Statistical distributional assumptions are required for application of most methods of statistical analysis and these assumptions depend partly on the circumstances and pattern of the experimentation or data collection. Besides using the right research tools, we must also use valid study designs and adequate sample sizes in order to have a good chance of detecting changes or effects of practical significance (i.e., those effects large enough to be of interest). The effects of interest are often small enough so that they might be obscured by random errors or hidden by other confounding effects, unless special attention were given to designing a study with sufficient statistical power.

To solve these problems, the investigator must know enough of statistical principles and techniques to be able to recognize when advice is needed from a statistician *before* the study is initiated. It is hoped that review and adherence to the principles of study design and data analysis discussed in this section will provide this basic level of awareness of statistical principles. Many of these principles are common sense, but unfortunately, common sense is frequently uncommon. Altman (1982) believes that the general standard of statistics in medical journals is poor, and the situation is not any better for industrial hygiene journals. Of course, uniform guidelines for statistical design cannot be precisely applied in every study. Section 2 identified *observational* and *experimental* studies as the two major classes of research studies. Special considerations may be involved in subclasses of these study types such as exploratory, methodological, and pilot or preliminary studies, where the primary purpose is to test feasibility or to evaluate alternative approaches or techniques.

4.1   Study Design Principles and Implementation Guidelines

The following guidelines are based on those suggested by Crow et al. (1960), Soule (1973), and Green (1979).

4.1.1   Establish the Study Objectives

Clearly establish the purpose and scope of the study. By whom, for what purpose, and how will the results be used? There should be a complete, clear, and concise statement of the objectives for the study. State how the anticipated results will specifically be used to meet the objectives. Are the results intended to be definitive or will this be a pilot test or feasibility study? Classify the study as *descriptive* (i.e., principal objective is to estimate only basic statistical parameters such as means) or *analytic* or *inferential* (i.e., designed to test a specific research hypothesis involving some process of inference in probability). Any statistical review should, in part, concern itself with how well the study design can meet the stated objectives and study hypotheses, if any. The study results can only be as relevant and productive as is permitted by the initial conception of the research problem.

State the study conditions and parameters used to represent the conditions to which the results will relate. Provide a clear and complete definition of the study target population to which inferences will be made or hypotheses investigated, based on results obtained from a sample. The *target population* is a subset of the general population that is both subject to the study exposure(s) and at risk of the development of the occupational disease or adverse health effect(s). Any given *sample* is a member of the *sample population* that consists of all samples that could have been selected from the target population. Also, include ranges of the determinant variables and define the reporting units (e.g., individuals, job types, establishments, industries).

Chapter 2 of the American Industrial Hygiene Association (AIHA) manual edited by Hawkins et al. (1991) discusses the concept of a *homogeneous exposure group* (HEG) and outlines several methods for defining HEGs: (1) task-based approach, (2) job-description-based approach, (3) chemical-based approach, (4) process/job/agent/task-based approach, and (5) data-analysis-based approach. HEGs are recommended for use in determining exposure assessment priorities and as an element in the selection of related exposure monitoring sampling strategies.

4.1.2 Formulate the Design for a Preliminary Study

Examine the precision afforded by different study sizes, with consideration for the benefits of a statistically powerful study versus the disadvantages of a less powerful study. Weak or equivocal results carry appreciable risk of wrong decisions and resultant limited conclusions. Plan to take replicate samples within each "treatment" (i.e., within each combination of time, location, and any other determinate variable). Statistically significant differences between "treatments" can only be demonstrated by comparison of the measured differences to variability within treatments. Attempt to obtain an equal number of randomly selected replicate samples for each combination of determinant variables. Taking measurements of "representative" or "typical" experimental units (i.e., reporting units) *does not* constitute random sampling, although random sampling would usually tend to be both of these. The element of random sampling is essential to methods of statistical inference that are based in the mathematical theory of probability. To test whether a treatment (exposure) has an effect, collect samples both where the condition is present and where the treatment condition is absent, but all other determinant variables are the same. Usually, an effect can *only* be demonstrated by comparison with such a proper control. However, in some cases the only "controls" also differ from the exposed group with respect to another variable. If this extra-experimental variable has been measured for individual subjects, it may be possible to correct for the extra-exposure part of the total exposure group difference by means of such statistical techniques as covariance analysis or comparison within blocks.

4.1.3 Review the Design with All Collaborators

Discuss the design with collaborators, reach an understanding, and keep notes about what decisions hinge on each outcome. Collaborators should anticipate and discuss all determinant variables that might affect the results. Review the study design in sufficient detail to discover any procedures that might lead to bias in the results. Obtain a pertinent peer review of the study design and use the comments as if the reviewers were collaborators in the study. Review and discuss the *robustness* of the chosen statistical methods (i.e., the tendency for inferences based on the methods to remain valid despite a violation of one or more assumptions underlying the theoretical development of the method). Provide for examining the study results to detect errors caused by serious violations of the methodology assumption. Discuss the objectives, goals, and study design with representatives of management and labor, as appropriate to their respective interests. Review how the results should be reported and to whom.

4.1.4 Conduct the Preliminary Study

Those who skip a preliminary study due to "not enough time" usually end up wasting time by attempting to analyze a study with trivial or equivocal results. This is the opportunity to test the feasibility of the study design before substantial resources are committed to the research. Obtain sufficient data to provide adequate estimates of the variance components that will be encountered; these can be used as a basis to develop an efficient design for the more definitive main study to follow. Obtain adequate information to evaluate the adequacy of the measurement equipment, personnel training and readiness, facilities that will be needed for statistically valid data analysis, and to determine appropriate ranges for the study variables. Verify that the chosen exposure measurement method is adequate and appropriate for the entire range of study conditions and determinant variables anticipated. Experience from past similar studies can sometimes be used as a substitute for a preliminary study or pretest. However, a careful critique of the relevancy of past studies should be made before using the data to design the new study.

4.1.5 Complete the Study Design

Use the variability estimates from the preliminary study to estimate the power of the study to detect the size of effect to which the study must be sensitive. If possible, specify the basis for the sample size calculation in a statement such as:

> To detect a true difference of _____ (units or percent) between group A and group B, with _____ percent statistical power and a _____ percent probability of making a type-I error, _____ trials are needed.

Present the design in clear terms to assure that its provisions can be followed without confusion. Include the intended data analysis methods as part of the design, including validity checks on the governing statistical assumptions. Review the principles of data analysis given in Section 4.2. Consider the necessity for a data transformation (see Section 4.2.1).

4.1.6 Conduct the Study

During the study, maintain communication among all collaborators, so that problems and intermediate results may be evaluated and dealt with, in keeping with the study objectives and the related study design previously agreed on. If measurements are to be taken in the plant, advise representatives of management and labor in advance and during your sampling.

4.1.7 Conduct the Data Analysis

Using the principles presented in Section 4.2, follow the data analysis methods that were previously selected for the study. Stay with the results from these methods because they were previously chosen as the most appropriate statistical methods to test the desired hypotheses. In a well-designed study, with proper attention to randomization and balance, unexpected or undesirable results *are not* valid reasons for subsequent rejection of the statistical methodology and then hunting for a "better" one. Remember that overinterpretation is an attempt to compensate for underplanning.

4.1.8 Prepare a Report

Report results of the study in relation to its original objectives and goals. Discuss both statistical and practical significance of the results. Present related conclusions based on inferences indicated and supported by the results for test data. State necessary limitations on the inferences in your discussion and conclusions. Present summary data, statistics, and results in clear graphs and tables. In general, graphs are superior to tables for portraying trends, correlations, scatter, outliers, and so forth. If the results suggest a need for further studies, outline the course that such studies should take.

Just as exposure measurements should be accurate and precise estimates of worker exposure, so should reports be accurate and precise communications of your study objectives, results, and conclusions. Study and use of guides such as those by Bates (1978), Crews (1974), and the CBE Style Manual Committee (1983) can contribute greatly to the quality of your written reports.

4.1.9 Implement Appropriate Follow-up

Discuss study results with appropriate representatives of management and labor so that corrective action to reduce health hazards can be implemented, if necessary, along with additional exposure monitoring, biological monitoring, and medical surveillance programs as required. Are such evaluations needed for air pollution, water pollution, hazardous waste disposal, or safety?

4.2 Principles of Data Analysis

4.2.1 Choose a Distributional Model

Choose a distributional model (see Section 3.6) for the target population that the sample data represent. Note that a data *transformation* may be necessary (i.e., converting the data into such form that they follow a common distribution with known properties and readily available analytical methodology). Data transformations may be needed for other reasons than just giving the transformed results a convenient distributional form. Murphy (1982) has discussed six objectives of transformation:

1. Normalization
2. Stabilization of the variance
3. To make the effects linear and additive
4. To make the mean a good measurement of "the typical value"
5. To linearize a relationship between two or more variables
6. To remodel the distribution into a more familiar one

4.2.2 Review Statistical Assumptions

Review assumptions that underlie the statistical methods to be used. If a specific distributional model can be assumed, proceed to Section 4.2.6. The parametric methods outlined therein have the benefits of (1) lower sample sizes (hence lower costs) for the same statistical power or (2) moderate gains in power for the same sample size, when compared to the more robust nonparametric methods. (The "distribution free" methods do not depend on a specific probability model with one or more parameters for the distributional form of the parent population.) However, if the sample data are not *adequately* fitted by the assumed parametric distribution (see Section 3.6) or if other assumptions (e.g., independence) of the inferential methods are suspect, then one has unknown risks of incorrect hypothesis testing decisions, inaccurate confidence intervals, or inaccurate tolerance limits. The sizes of such inaccuracies depend on the robustness of the methods used. If sample size is insufficient (less than about 10) to examine the pattern of the data (see next section) and test the distributional model, proceed to Section 4.2.6. Then one may be in the position of having to use parametric methods without even weak verification of assumptions. A larger sample size (e.g., 30 or greater) is needed to empirically estimate distributional fractiles from ranked sample data without assuming any mathematical model for the distributional form (see Section 4.2.5).

4.2.3 Qualitatively Examine the Data

Plot the grouped sample data as a histogram or individually on appropriate probability paper (e.g., normal, lognormal, Weibull) to qualitatively examine data regarding the distributional assumptions and to investigate unusual patterns in the data (see Section 6.12). Data that are lognormally distributed in time can be plotted on semilogarithmic graph paper so that the data plot is symmetrical about the distribution's geometric mean. This is a qualitative way of looking for trends in the data over time (see Section 6.11).

Gross errors due to mistakes in sampling or chemical analytical procedures can contaminate a data sample. Barnett and Lewis (1978) have surveyed most of the extensive statistical theory that has been developed concerning detection of data "outliers" and concerning methods for performing analyses of data sets that may be contaminated with outliers. In particular, they give "robust" methods for estimating the mean of an underlying distribution when a sample supposedly taken entirely from that distribution contains one or more outliers. Methods are given for several types of data models (e.g., normal and lognormal distributions), for various numbers and types of outliers (e.g., outliers in left tail, in right tail, or in both tails), for known and unknown variances, and for different objectives of the data analysis.

4.2.4 Estimate Sample Distributional Parameters

Use the sample data to calculate estimates of the parameters of the assumed distributional model [e.g., arithmetic mean and standard deviation for a normal distribution (see Section 3.4.1), geometric mean and geometric standard deviation for a two-parameter lognormal distribution (see Section 3.5.1), and if necessary the third parameter (for origin translocation) for a three-parameter lognormal distribution (see Section 3.5.2)].

When the data sample contains data values that are too low (or too high) to be quantitatively measured, tables of factors given in Cohen (1961) can be used to compute an optimal maximum-likelihood estimate (MLE) of the mean of a normal (or lognormal) distribution. The Cohen tables include factors for both censored samples and truncated samples. Formulas for the MLE are given in Hald (1952), based on theory for sampling from a censored normal distribution.

Hornung and Reed (1990) advocate use of an easier-to-compute estimator of the geometric mean when there are nondetectable concentration values present in a sample taken from an assumed lognormal distribution. They merely replace values below the limit of detection L by $L/\sqrt{2}$ and then calculate the estimated geometric mean. In a practical sense, they say that this estimate of central tendency is sufficiently accurate for the usual sample sizes used in industrial hygiene sampling. The maximum-likelihood estimate is admittedly more precise, but Hornung and Reed judge the Hald (1952) technique to be not usually worth the effort of its complex calculation.

4.2.5 Verify Distributional Model

To qualitatively test a distributional model, use the sample estimates of parameters from the previous step to plot the estimated target population distribution on the appropriate probability paper and compare to the actual sample data distribution. Mage (1982) has suggested an objective method for testing normal distributional assumptions using probability paper.

With larger sample sizes (at least 30), a histogram of the sample data can be compared to the shape of the fitted distributional function. To quantitatively verify the distributional model, the classic chi-square test can be applied to the histogram's interval frequencies. This goodness-of-fit test usually is the first to come to mind, but it generally requires substantially larger sample sizes than occur in industrial hygiene. Lilliefors (1967) has adapted the Kolmogorov–Smirnov test when the mean and variance of the distributional model are unknown (as is typically the case with occupational exposure data). This modified goodness-of-fit test can be used with small sample sizes (e.g., 10 to 20), for which the accuracy of the chi-square test would be questionable. Also, it is claimed to be a more powerful test than the chi-square test for any sample size. Iman (1982) has provided graphs for use with the Lilliefors test.

Waters et al. (1991) stress that although a probability plot may be quite useful to interpret the *nature* of a confirmed nonlognormal distributional condition, a preliminary formal test of significance is needed to confirm the fact that any seeming nonlinearity viewed in a probability plot is not in fact merely a random sampling variation. They propose use of an approximate test based on use of their ''ratio metric'' goodness-of-fit test statistic, which is the ratio of the arithmetic sample mean divided by the maximum-likelihood estimate of the arithmetic mean. The ratio metric test is presented as a simpler to compute, but nearly as accurate, alternative to the established and powerful Shapiro–Wilk W test (1965). Our view is that the savings in computational effort for the Waters et al. (1991) test for nonlognormality seems not important enough to warrant use of an approximate test instead of the powerful and more rigorous Shapiro–Wilk W test.

If the fit is judged inadequate, return to Section 4.2.1 and choose another data transform or distributional model.

4.2.6 *Apply Chosen Statistical Methodology to the Sample Data Analysis*

If the fit of the distributional model is judged adequate (or assumed to be adequate in which case one has been able to delete the steps following Section 4.2.2) or if the statistical methodology is sufficiently robust for its intended application, proceed with the data analysis. Calculate appropriate confidence intervals and tolerance limits, perform hypothesis tests, and so on based on the assumed distributional model. If the data or data transforms are not adequately fitted by available continuous-variable, parametric distribution models, certain nonparametric (distribution-free) methods are available. However, nonparametric methods generally require sample sizes larger than occur with sample sets used for exposure estimation; therefore, for these types of data, the distributional forms have been investigated using prior data collected for that purpose, and these forms can usually be assumed to apply to similar samples taken thereafter.

4.2.7 *Report Results and Interpretation*

If a report or journal article is written, it should include statements covering (1) the statistical rationale used to develop the study design and (2) the related protocol used for the statistically valid data analysis. These statements should contain sufficient detail so that a reader with statistical expertise would be able to duplicate the results if supplied with the investigator's raw data. If a personal computer or mainframe statistical program was used, a precise reference to it would usually preclude the need for presenting additional computational details. If the statistical analysis had been performed by a consultant or collaborator, their assistance would usually be needed in the writing, or at least in the editing, of the final report.

5 STUDY DESIGNS FOR ESTIMATING OCCUPATIONAL EXPOSURES AND EXPOSURE DISTRIBUTIONS

Study design considerations, including sample size estimation techniques, for obtaining occupational exposure estimates and their distributions will be discussed from two perspectives in this section. The first perspective, presented in Section 5.2, discusses *exposure measurement strategies* for making individual *exposure estimates* (see Section 3.1). The discussion examines different approaches to measuring exposures for individual workers on a single day.

The second perspective covers *monitoring strategies* for obtaining exposure estimates for worker target populations. Monitoring strategies are built on appropriate measurement strategies. *Exposure monitoring* will be defined as a series of steps necessary for estimating multiple exposures of a target population. This material is presented in Section 5.3. Fundamental to exposure assessments and exposure monitoring programs is the concept of a worker *target population*. A target population is usually defined as a function of:

1. Numbers of workers exposed under the conditions that are being studied (e.g., process factors and determinant variables)
2. Exposure measurement averaging times
3. Time period to be represented by the target population exposure distribution
4. Ranges of determinant variables affecting the exposure levels to which the target population is exposed.

The number of workers can be as few as one worker, measurement averaging times typically are either 15 min or 8 hr, and the time period can be as short as a few hours on a given day. Determinant variables will be discussed in Section 5.1. Typical examples of target populations for several combinations of temporal periods and numbers of workers include the following:

1. Daily 8-hr exposure estimates for a single worker over several days to years
2. Eight-hour exposure estimates for several workers on a single day
3. Daily 8-hr exposure estimates for many workers over several days to years

The elementary case of a single worker on a single day is governed by exposure measurement considerations, which are discussed in Section 5.2. For this situation there is but one exposure, but the parent population is considered to be all measurements (with their errors) that could have been made of that one exposure on the same day.

The other two types of target populations are considered under Section 5.3, Exposure Monitoring Strategies. In these cases the parent populations consist of multiple daily exposures and/or multiple workers exposed, as well as the exposure measurement populations for these workers at these times.

The relationship between target populations and determinant variables is graphically shown in Figure 10.7. This figure is a flowchart for exposure assessments, which are less comprehensive than exposure monitoring programs. Exposure assessments generally include the following elements:

1. Defining an appropriate target population of workers
2. Selecting a sample population from the target population for exposure measuring
3. Selecting one or more exposure measurement strategies
4. Obtaining exposure measurement samples for the sample population
5. Calculating exposure point estimates and interval estimates
6. Possibly estimating an exposure distribution for the target population
7. Comparing the exposure estimates (and possibly the exposure distribution) to decision criteria for acceptability of exposure levels

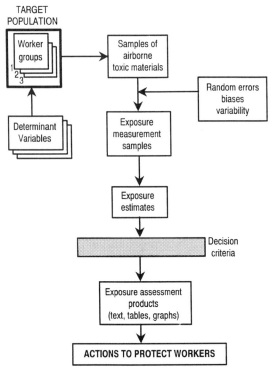

Figure 10.7   Exposure assessment flow chart.

8. Reporting the exposure assessment products for the target population (e.g., text, tables, graphs)
9. Taking any actions necessary to protect workers in the target population

5.1 Determinant Variables Affecting Occupational Exposure Levels

Both exposure measurement strategies and exposure monitoring strategies must consider the *determinant variables* that affect the true exposure levels that are to be estimated with worker exposure measurements. Determinant variables are the qualitative or quantitative factors that determine, or at least are associated with and affect, the actual worker exposure levels. They generally can be classified as process, environmental, temporally associated, behavioral, and incidental. Note that a failure to identify and consider significant determinant variables generally has more deleterious consequences for a study than identifying and investigating too many determinant variables, including some that have little effect on exposure levels. *One cannot make reliable inferences from exposure results beyond the number and range of determinant variables represented by the sampling distribution.*

The following groups of typical determinant variables are presented as examples only and are not inclusive. The significant determinant variables for any given group of workers and time period must be identified, from experience with similar situations if possible, or using a research study if necessary.

5.1.1 Process Factors for Chemical Agents

These are factors related to primary contaminant levels or to the control of emissions and/or exposure levels.

1. Process type and operation (see page 27 of Leidel et al., 1977; Burgess, 1981; and Cralley and Cralley, 1982)
2. Chemical composition of material used in operation
3. Physical state and properties of material used (e.g., vapor pressure, size distributions of particulates and aerosols)
4. Rate of operation (e.g., mass or volumetric rate, revolutions per minute, linear rate, items in a given time period)
5. Energy conditions of operation (e.g., temperature, pressure)
6. Degree of process automation
7. Emissions from adjacent operations
8. Airflow patterns around workers (e.g., from exhaust ventilation, from adjacent operations)
9. Heating and ventilation airflows
10. Exposure control methods (e.g., local exhaust ventilation, respirators)

The scope of this chapter does not include mathematical modeling of exposure levels as a function of process factors (e.g., using multiple regression analysis). The assumptions underlying the usual regression analysis methods include an assumption that the independent variables are measured without error, but this assumption would not be valid for most quantitative process factors. The textbook by Fuller (1987) brings together previously published rigorous statistical theory that is needed for the correct application of regression analysis to industrial hygiene problems in which *both* the dependent and independent variables are measured with error.

5.1.2 Environmental Determinant Variables

These are environmental factors, variations of which can modify exposure levels.

1. Meteorological conditions
2. Age, size, and physical layout of plant
3. Job category (e.g., responsibilities, work operations, work areas, time spent at each) (see Corn and Esmen, 1979)

5.1.3 Temporally Associated Determinant Variables

Time is an independent variable correlated with worker exposure levels. Time cannot be a direct cause of changes in workplace exposure levels but may be useful to predict exposure levels that follow time cycles, have systematic time trends, or have autocorrelation between present and past levels. Time series models are useful when the causative determinant variables are either unknown or unmeasured.

1. Contaminant buildup in the workplace air from morning to afternoon
2. Exponential clearance due to air flushing and dilution during nonworking hours
3. Cyclical or trending process operations with respect to work shift, season of year, and year or decade

5.1.4 Behavioral Determinant Variables

Behavioral factors affect work habits that in turn affect exposure levels. Effects of these factors cannot be modeled mathematically, but the factors can be helpful in explaining (or at least rationalizing) observed exposure differences and effects.

1. Worker job practices, movements, and habits
2. Worker training
3. Worker attitudes
4. Management and supervisory attitudes
5. Presence of exposure measurement equipment, industrial hygiene personnel, or supervisory personnel

5.1.5 Incidental Determinant Variables

Irregular changes in exposure levels can occur due to episodes, incidents, accidents, and otherwise unintended happenings such as described by Crocker (1970):

1. Spills due to falls, punctures, tears, corrosion, and so forth
2. Equipment maintenance or lack thereof
3. Failure of process equipment prone to corrosion and leakage (e.g., pump packings, tank vents)
4. Interruption of utilities to process equipment or exposure control systems
5. Interaction due to accidental mixing or simultaneous release of two vessels' contents
6. Interruption or increase in flow of one or more process streams
7. Vessel failure
8. Accidental overpressurization, overheating, or overcooling of process equipment
9. Sudden plant flooding, violent storms, or earthquakes
10. Operator errors or instrument failure

5.2 Exposure Measurement Strategies

Exposure measurement strategies deal with the considerations necessary to measure individual worker exposures on a given day to obtain short- or long-period TWA exposure estimates. Monitoring strategies for measuring multiple exposures (e.g., multiple workers on a single day or a single worker on multiple days) are presented in Section 5.3.

5.2.1 Practical Considerations

The adequate and preferable strategies for a given measurement situation are governed by both practical and statistical considerations. An adequate measurement strategy is one that is good enough for the purpose we have in mind. Thus one should clearly identify the objective of the intended exposure estimation (see Section 2.4) and the required precision of the estimates. A measurement strategy can then be selected that meets this precision requirement.

Most of the following discussion will concern measurements obtained for estimating individual worker exposures to chemical agents; most of the concepts presented are applicable to estimating individual workers' exposures to physical agents as well. Generally these exposure estimates are then compared to exposure control limits to determine the relative hazard for individual workers or the acceptability of the workplace exposure levels (e.g., in relation to OSHA PELs or ACGIH TLVs). It should be noted that there are other specialized purposes for airborne contaminant measurements. These include:

1. *Source sampling* at potentially hazardous operations. Hubiak et al. (1981) have discussed the utility of short-term source sampling for identifying work operations that need additional engineering effort to reduce contaminant emissions (and related worker exposures). Sometimes these can be considered *worst-case exposure* levels, which are used in screening strategies for hazard evaluations.

2. *Evaluating work practices* to determine exposure variability and detect hazardous levels due to inappropriate work practices. This technique can also assist in differentiating between exposure levels due to work practices and those due to inadequate engineering controls. With direct-reading measurements it is possible to make on-the-spot recommendations for improving work practices and to recognize the appropriate direction for engineering control research.

3. *Worker training* to improve the effectiveness of work practices and engineering controls. Selected workers can perform the work operation while others observe the work practices and note the resulting exposure levels. The goal is to have all workers approach the results of the best or ''cleanest'' worker. Showing workers their exposure profile recorded on a strip chart will help them to better understand what a change in work practices or use of engineering controls can achieve.

4. *Evaluating individual job tasks* for their relative contribution to a worker's 8-hr TWA exposure. This may lead to developing appropriate administrative controls, improved work practices, engineering controls, or at least to identifying the need for personal respiratory controls.

5. *Continuous monitoring* to detect extraordinary exposure levels, so that a warning can be provided before a serious hazard develops. Generally, fixed sampling systems are used with a central analyzer. These systems may sample from multiple points in the workplace or in the return air of recirculation systems. Holcomb and Scholz (1981) have reported an evaluation of typical continuous monitoring equipment.

6. *Screening measurements* for qualitative detection of airborne hazards during emergencies and uncontrolled releases. A screening strategy may also be used for

quantitative determination of airborne hazards for hazard evaluation and spot check-
ing of exposure levels. Typically, detector tubes and direct-reading meters are used
for molecular size contaminants and aerosol monitors are used for particulates.
Schneider (1980) and Leichnitz (1980) have discussed the use of colorimetric detector tubes for screening. King et al. (1983) have described a simultaneous direct-
reading indicator tube system for rapid qualitative measurements.

Some nonstatistical considerations affecting the selection of measurement strategies follow:

1. Amount of information available regarding the nature and concentration of
   airborne contaminants to be measured and possible interfering chemicals
2. Availability and cost of sampling equipment (e.g., pumps, filters, detector
   tubes, direct-reading meters, passive dosimeters)
3. Availability and cost of sample analytical facilities (e.g., for filters, charcoal
   tubes, dosimeters)
4. Availability and cost of personnel to take the measurements
5. Location of work operations and workers to be sampled
6. The need for obtaining results immediately, within a day or two, or after several weeks

5.2.2 Statistical Considerations for Long-Term Exposure Estimates

Generally the statistical design of exposure measurement strategies is concerned with
reducing the limit of random error in an exposure estimate calculated from several
exposure measurements, which also has the effect of increasing the power of a related hypothesis test for compliance or noncompliance with an exposure control
limit. The imprecision (uncertainty) of a long-term (e.g., 8 hr) exposure estimate is
governed by three classes of factors:

1. Random variation in the measurement procedure (i.e., the precision of the air
   sampling/chemical analysis method, see Sections 3.3.2, 3.4.2, and 3.6.1) or
   the random variation in the true exposure levels during the estimation period
   (see Sections 3.3.1 and 3.6.2)
2. Sample size (i.e., number of measurements obtained during the time-averaging period for the TWA estimate) and the duration of each measurement
3. Sampling period selection (i.e., whether random or systematic sampling is to
   be used to select subintervals for sampling during the TWA period of interest,
   or a cumulative sample is to be taken over the entire period)

It should be realized that there is no "best" measurement strategy for all situations. However, some strategies are more desirable than others. The following discussion points out the statistical considerations for the four different types of TWA
exposure estimates (see Section 3.1). Remember that the word *period* refers to the
duration of the desired time-averaging period (e.g., 15 min for a 15-min TWA estimate, 8 hr for an 8-hr TWA estimate).

The *full-period consecutive-samples estimate* is the "best" strategy in that it yields an estimate with the least uncertainty (narrowest confidence interval). This is because the uncertainty of this type of exposure estimate is a function only of the measurement method error (see Section 3.2) and is independent of the substantial variability in the actual exposure levels measured (see Section 3.6.1). There are only moderate statistical benefits to be gained (Technical Appendix E of Leidel et al., 1977) from increased sample sizes (e.g., eight 1-hr samples versus four 2-hr samples), but with the substantially increased analytical costs (per 8-hr TWA exposure estimate) the practical benefits are negligible. Figure 10.8 illustrates the effects of increased sample size. Generally two consecutive samples for the full time-averaging period provide sufficient precision for most exposure estimation purposes. For a method with very large CV_T, more samples may be warranted.

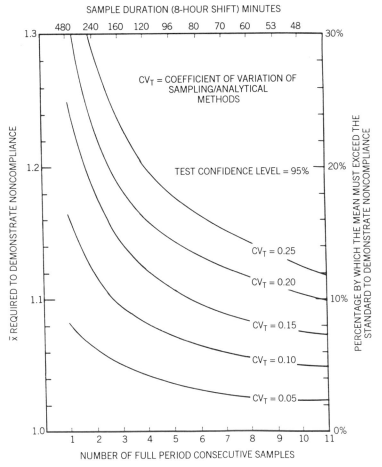

Figure 10.8 Effect of sample size for a full-period, consecutive-samples estimate on noncompliance demonstration when test power is 50 percent and test confidence level is 95 percent; CV_T is coefficient of variation of exposure measurement method.

The *full-period single-sample estimate* (e.g., one 8-hr sample) is the "second-best" strategy, if an appropriate measurement method is available. An exposure estimate calculated from one 8-hr measurement is almost as precise as an estimate computed from two 4-hr measurements, since both strategies employ full-period sampling. The disadvantage of a single measurement is that a bias or mistake in the measurement is difficult to detect. Also, substantial differences in exposure levels during the measurement period are not revealed, but this feature is also an advantage in that temporal variability is "integrated out" by the cumulative, physical sampling procedure itself.

The *partial-period consecutive-samples estimate* is substantially less desirable than the preceding two estimates. The major problem created by this strategy is that exposure levels are unknown during the unsampled portion of the TWA period. Strictly speaking, the measurement samples are representative only of the periods that are actually sampled. For example, if one desires to estimate an 8-hr TWA from a sample or samples spanning only 5 hr, then a problem is created of assuring that the 5-hr period results represent the entire 8-hr period. Reliable knowledge or professional judgment may sometimes be used to extrapolate the 5-hr TWA to an 8-hr estimate. This should be done only after considering the effect on the real 8-hr TWA of any substantially higher or lower actual exposure levels during the unsampled period. Figure 10.9 illustrates the effect that a liberal assumption of zero exposure (i.e., less worker protective), for the unsampled portion of the TWA period has on the value required to demonstrate noncompliance at a statistical test power of 50 percent.

The *grab-samples estimate* is the least preferable strategy for estimating an 8-hr TWA exposure. This exposure estimate has substantially larger uncertainty than the first two types of estimates. This is because the uncertainty of a grab-samples estimate is dominated by the considerable variability of the exposure levels measured, which is substantially larger than the measurement method variability. This is illustrated in Figure 10.6 which compares the relative contributions to variability in a grab-samples estimate. Regarding sample size, Figure 10.10 shows that a reasonable number of grab samples for an exposure estimate is between 8 and 11, taken at random intervals during the TWA period. However, this applies only if the worker's exposure levels are adequately uniform during the TWA period. If the worker is at several work locations or operations during the TWA period, then at least 8 to 11 grab-sample measurements should be obtained during *each* period of anticipated uniform exposure that substantially contributes to the TWA exposure. If one has to take fewer than 8 to 11 measurements during each uniform exposure period, then allocate the total number of measurements in proportion to the duration of each period. That is, take more measurements during the longer periods of anticipated uniform exposure.

If grab samples are taken, the duration of each measurement need be only long enough to collect sufficient mass of contaminant to reach the minimum level of detection for the analytical method. That is, any increase in sample duration beyond the minimum time to collect a sufficient mass of contaminant is unnecessary and unproductive.

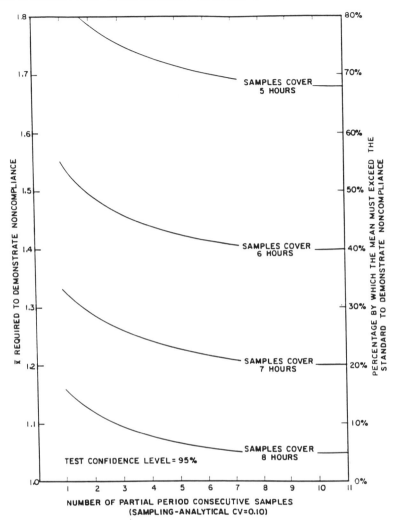

Figure 10.9 Effect of (1) total time covered by all samples and (2) sample size for a partial-period, consecutive-samples estimate on noncompliance demonstration when test power is 50 percent and test confidence level is 95 percent.

For grab samples it is desirable to choose the sampling periods in a statistically random fashion. The accuracy of the probability levels for the statistical methodologies presented in Section 6 for testing hypotheses of compliance or noncompliance depends on implied assumptions regarding the lognormality and independence of the grab-sample results that are averaged. These assumptions are not unduly restrictive if precautions are taken to avoid bias when selecting the sampling times during the period of the exposure estimate.

For grab sampling, a TWA estimate represents a period longer than the total

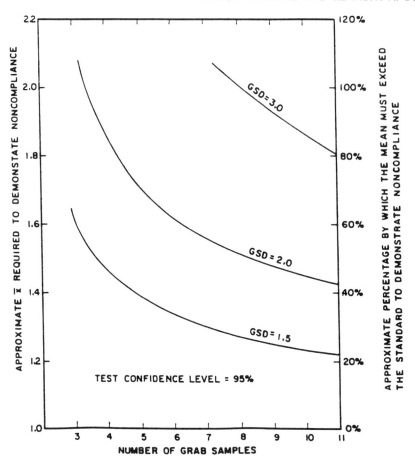

Figure 10.10 Effect of sample size for a grab-samples estimate on noncompliance demonstration when test power is 50 percent and test confidence level is 95 percent; GSD is the geometric standard deviation for intraday variation in the airborne environmental concentrations.

interval measured, but an unbiased estimate of the true average is obtained when samples are taken at random intervals. It is valid to sample instead at equal intervals if the series is known to be stationary with contaminant levels varying randomly about a constant mean, and if exposure fluctuations are of short duration compared to the length of the sampled interval. However, if the average exposure level and its confidence interval were calculated from samples taken at equally spaced intervals, biased results could occur if there were exposure cycles in the operation that were in phase with the intervals sampled. The important benefits of random sampling are that subsequent results are unbiased even if cycles and trends occurred during the period of the exposure estimate.

The word *random* refers to the method used for selecting the sample. A *random sample* is one chosen in a manner such that each possible sample has a fixed and determinate probability of selection. A practical way of defining a random sample for an exposure measurement is one obtained such that any portion of the TWA exposure estimate period has the same chance of being sampled as any other. Ordinary haphazard or seemingly purposeless choices generally do not guarantee true randomness. Devices such as random number tables, or random numbers generated by hand-held calculators or personal computer software (e.g., Lotus 1-2-3 or Microsoft Excel 4.0 spreadsheets or Mathcad 3.1 for Windows) can be used to prevent subjective biases that tend to be inherent in arbitrary personal choices. Technical Appendix F of Leidel et al. (1977) details a formal statistical method for choosing random sampling periods. Rather than using the random number table discussed in the method, it would be more efficient to generate the random numbers for the procedure with a hand-held calculator or personal computer software.

5.2.3 Statistical Considerations for Short-Term Exposure Estimates

Short-term exposure estimates (e.g., 15 min or less) generally are obtained only for determination of peak exposures during short periods. These estimates may be used for comparison with *ceiling* exposure control limits designed to prevent acute health effects. Short-term samples taken to measure short-term exposures are statistically analyzed in a manner similar to short-term samples taken to measure long-term exposures. However, two important differences should be noted.

The first difference is that the measurements taken for estimation of peak exposures are best taken in a *nonrandom* fashion. That is, all available knowledge relating to the exposure level determinant variables such as work area, worker, work practices, and type of operation should be utilized to obtain samples during periods of maximum expected exposure.

The second difference is that measurements obtained for short-term estimates are generally taken to represent a much shorter period than short-term samples taken for estimating 8-hr TWA exposures. Each short-term measurement usually consists of a single instrument reading (if a direct-reading device is available) or a 5- to 15-min sample if a minimum mass of material needs to be collected (e.g., on a charcoal tube or filter). A series of samples spanning 15 min could also be taken and the measurements averaged.

Leidel et al. (1977) recommend that a minimum of three short-term exposure estimates be obtained on any given work shift for a worker and the highest of all estimates be used as an estimate of the worker's peak exposure for that work shift. This recommended minimum number of estimates is not based on statistical considerations, but on practicality. Taking at least three samples increases the probability of detecting an exposure close to the highest exposure. Three samples also facilitate the detection of gross mistakes or biased measurements. However, usually only the highest value (not the average of the three or more) would be compared to a ceiling exposure limit. If measurements are obtained for evaluation with a short-term TWA limit, such as a *threshold limit values–short-term exposure limit* (TLV–STEL) of the American Conference of Governmental Industrial Hygienists (e.g., ACGIH, 1991), the total sampling time should equal the time-averaging period for the limit, which is typically 15 min. Thus, for some colorimetric detector tubes, it might be necessary to take several consecutive samples and average the results.

Although short-term measurements taken for estimation of peak exposures are usually best taken in a nonrandom (biased) fashion, random sampling may be useful in some work situations where the exposure levels appear uniform during a work shift. Professional judgment may be unable to identify particular periods with a risk of higher-than-usual exposure. For this case, a statistical procedure is given below that can be used as a peak exposure detection strategy. The sample size recommendations given are based on combinatorial probability formulas detailed in Technical Appendix A of Leidel et al. (1977).

PURPOSE. Provide a sample size to assure (i.e., have a probability of at least 90 or at least 95 percent) that at least one randomly sampled period will be from the higher exposures present during the work shift or other total duration examined (i.e., from the highest 10 percent or highest 20 percent of the work shift exposure distribution).

ASSUMPTIONS. No limiting assumptions are required. The derivation of this method is based on the hypergeometric sampling distribution (i.e., sampling from a finite population without replacement). No mathematical model is assumed for the distribution of exposure levels present in the finite-size population of possible measured periods; therefore, these sample sizes may be larger than would be needed if a parametric distributional form of the exposures were confidently known. However, this nonparametric procedure is useful when the form of the exposure distribution is irregular or unknown. Tables 10.4, 10.5, and 10.6 give the required sample sizes for 32, 48, and 96 possible sampling periods in a total time period whose highest short-term exposure levels are to be estimated.

EXAMPLE. For a target population of 32 consecutive 15-min periods in an 8-hr work shift for a worker, determine the necessary sample size such that there is at least 90 percent confidence that at least one sampled period will be from those periods with the highest 20 percent of exposures occurring during the work shift.

SOLUTION. There are 32 discrete, nonoverlapping, 15-min periods in an 8-hr work shift. Table 10.4 indicates that a random sample of 10 of the 32 15-min periods will have at least 90 percent probability of containing one or more of the 6 periods during

Table 10.4 Required Sample Size for Detecting at Least One of the Higher Exposures Among 32 Periods in the Total Duration (15-min Sample Periods in an 8-hr Work Shift)

To Obtain at Least One Sample Period from the	At a Confidence Level of	Random Sample from At Least
Top 20%	0.90	10 periods
Top 20%	0.95	12 periods
Top 10%	0.90	17 periods
Top 10%	0.95	20 periods

Table 10.5 Required Sample Size for Detecting at Least One of the Higher Exposures Among 48 Periods in the Total Duration (10-min Sample Periods in an 8-hr Work Shift)

To Obtain at Least One Sample Period from the	At a Confidence Level of	Random Sample from at Least
Top 20%	0.90	10 periods
Top 20%	0.95	13 periods
Top 10%	0.90	21 periods
Top 10%	0.95	25 periods

which the 20 percent highest exposures will occur. The number 6 is the largest integer representing 20 percent or less of 32.

Where the short-term time-averaging period is 10 min, there would be 48 such periods in an 8-hr work shift and the sample sizes in Table 10.5 would be appropriate. For example, to have at least 90 percent probability that the sampled periods include at least one of the four periods that have 10 percent or less of the highest exposures, 21 of the 48 periods should be selected at random and sampled.

Less than 10-min time-averaged measurements may sometimes be obtained, as with a 3-minute colorimetric tube or spot readings with a direct-reading meter. Then the sample sizes in Table 10.6 are appropriate.

5.3 Exposure Monitoring Strategies

The exposure monitoring strategies presented in this section are guidelines for measuring multiple exposures (e.g., multiple workers on a single day, a single worker on multiple days, multiple workers on multiple days). These strategies should be used with the exposure measurement strategies (for measuring an individual's exposure on a single occasion) treated in the previous section. Monitoring strategies for monitoring programs should always consider eight elements, which will be presented in further detail in the following sections:

Table 10.6 Required Sample Size for Detecting at Least One of the Higher Exposures Among 96 or More Sample Periods Less Than 5 min Each from an 8-hr Work Shift

To Obtain at Least One Sample Period from the	At a Confidence Level of	Random Sample From at Least
Top 20%	0.90	10 periods
Top 20%	0.95	13 periods
Top 10%	0.90	21 periods
Top 10%	0.95	27 periods

A review report by Rappaport (1991) supports a school of thought for certain industrial hygienists who believe that OSHA PELs should not be formulated as upper bounds on 8-hr TWA exposures of single employees. Rather, the report says that the entire *exposure distribution* should be controlled for the purpose of controlling both the long-term average exposure and the proportion of high 8-hr TWA exposures in the right tail of the temporal exposure distribution. The report says that in practice the tail area of the exposure distribution can also be controlled by merely controlling the long-term mean of the distribution.

Admittedly, it is primarily the long-term cumulative exposure that must be controlled because it is believed to be correlated with risk of chronic toxicity. The usual OSHA 8-hr TWA permissible exposure limits have in fact been used as a means to control corresponding long-term arithmetic mean exposures. This type of OSHA PEL usually has its basis either:

1. In dose–response relationships seen in chronic, animal exposure experimental studies (i.e., different groups of animals exposed for a long time to a range of carefully controlled, constant, repeated daily doses) or

2. In effects in observational studies of human workers whose corresponding ''working lifetime'' cumulative exposures have been estimated in connection with epidemiological studies.

Therefore, the argument is sometimes made that it would be ''logical'' to protect human workers against anticipated biological effects of chronic exposure by formulating OSHA PELs that are upper limits on long-term averages of daily TWA exposure levels.

But an argument against use of long-term average PELs is that it would not be feasible to enforce such exposure limits. This is true because it would be necessary to sample a worker's full range of exposures using multiday random sampling strategies and a sufficiently large sample size to determine a reasonably precise estimate of the long-term mean. This could be done validly (but perhaps not feasibly) only for those types of exposure measurements data that do not have patterns of temporal autocorrelation or systematic (average) interworker differences. For those cases the following comments concerning lack of validity of limited period multiday exposure averages as legal exposure limits would not apply. In general, unless random variability (with no time trends or cycles) exists, a group of consecutive sampling days cannot be treated as a random sample (or even as a representative sample) of a given employee's long-term distribution of daily exposures. Any number of working days, whether sampled randomly or consecutively during a limited period of time (e.g., one year), would not be representative of the entire distribution of past and future working days during a working lifetime. Production levels and operations change in practice, and this produces related changes in work tasks and exposure levels. Resulting exposure levels may vary systematically (e.g., cyclically), rather than randomly, during extended periods of time. Thus, arguments against use of long-term average exposure limits for individual workers relate mainly to infeasibility of the sampling strategies that would be needed for meaningful enforcement.

There are even stronger arguments against taking an average of *multiple worker exposure levels* to compare to an exposure control limit. An OSHA PEL is intended to apply to every individual worker, and interworker exposure variability typically exceeds variability of physical sampling and chemical analysis. Simple indiscriminate (unbalanced) mixing of interworker and interday variability is invalid because the pattern of interworker differences may be similar at different times. That is, interworker differences at different points in time may be autocorrelated, rather than fully random, and a multiworker average would be biased as an estimate of the average exposure of any individual worker. It would be biased high for some and low for others. Of course, if it can be validly assumed that there are no consistent interworker exposure differences, then any worker's exposure measurement would apply to any other worker in the target population.

On the other hand, current OSHA PELs based on single-day, 8-hr TWAs do permit valid enforcement decisions (compliance, no decision, or noncompliance) to be made. With 8-hr TWA PELs, random sampling plans are possible that properly support related statistical decision theory that is given in this report. Of course, it would be pointless to make a statistically valid enforcement decision about compliance with a PEL that is biologically meaningless. However, this is not the case for current 8-hr TWA PELs. When it is determined that compliance with an 8-hr TWA PEL exists on several occasions (or on a deliberately selected "worst occasion"), it is usually possible to conclude that the related long-term mean exposure is also at a compliance level (unless operations change).

The preference expressed here for 8-hr TWA PELs does not imply that there is no need to monitor exposures of multiple employees at different times. The remainder of this section deals with strategies for such exposure monitoring.

Section 2.4 listed two major types of monitoring programs as possible objectives of exposure estimation. The first type is an *exposure screening program*, which is a limited exposure monitoring program designed to identify target populations of workers with other-than-acceptable exposure distributions for follow-up periodic monitoring. The program uses minimal resources consistent with reasonable protection for workers. This type of program uses an action level as a screening cut-off to identify appropriate target populations for inclusion in a limited exposure surveillance program or a more extensive *exposure distribution monitoring program*. The latter program is a more extensive one intended to quantify exposure distributions of target populations. Generally this is first done for an initial base period; then the initial estimates of the exposure distributions are periodically updated with more current estimates from routine exposure monitoring.

There are several commonalities between exposure screening and exposure distribution monitoring programs. Guidelines noting the similarities and differences in the two types of programs will be presented in the following sections, which detail eight possible elements for these two types of monitoring programs.

5.3.1 Need for Exposure Measurements

Both exposure screening and exposure distribution monitoring programs need to begin by determining the need for exposure estimates. Desirable predecessors to these programs, which may negate the need for exposure measurements, include:

1. Conducting a workplace materials survey to determine if potentially harmful materials are being used in the workplace (see page 21 of Leidel et al., 1977).

2. Conducting a walkthrough survey to identify process operations that may be potentially hazardous and to determine if workers may be exposed to hazardous airborne concentrations of materials released into the workplace (see page 24 of Leidel et al., 1977).

3. Estimating airborne concentrations based on the amount of material released into the workplace air. This may be useful for contaminants that are low to moderate hazards (see pages 28–30 of Leidel et al., 1977). However, this technique usually requires a substantial safety factor to account for uncertainty, which may limit its usefulness.

4. Source sampling at potentially hazardous operations. This may be useful for screening strategies, such as may be used for hazard evaluations. The usual assumption is that the resulting exposure estimates are worst-case exposure levels and all worker exposures will be less than the values found at the source(s) of the contaminant (see Section 5.2.1 above and pages 24 and 27 of Leidel et al., 1977). This assumption may be invalid if there are more than one or two contaminant sources.

5. Preparing a written determination of the need (or lack of need) for exposure measurements. The written determination would consider:

    a. Any information, observation, or calculation that might indicate worker exposures

    b. Any measurement taken

    c. Any worker remarks of symptoms that may be due to exposure to workplace materials or operations

    d. Any possible changes in production, process, or controls that could result in hazardous increases in airborne levels of contaminants or render control procedures inadequate

5.3.2 Airborne Chemical(s) to Be Measured

The next element, in both exposure screening and exposure distribution monitoring programs, is determination of the airborne materials to be measured. This is best done by considering the information acquired from the various steps in the previous program element. It may be necessary to first do screening measurements (see Section 5.2.1.6), if prior information is unavailable and qualitative detection is promptly required (e.g., emergencies and uncontrolled releases).

5.3.3 Strategy for Initial Monitoring

Initial monitoring is the first monitoring program element where an exposure screening program differs considerably from an exposure distribution monitoring program. For an exposure screening program, the objective of initial monitoring is to selectively obtain exposure estimates only for "maximum-risk" workers. These can be defined as those workers "believed to have the greatest exposure."

For exposure screening monitoring, the most efficient approach to sampling is a nonrandom selection of the highest-risk workers. The selection process must use competent professional judgment that relies on experience and knowledge of the exposure level determinant factors pertinent to the target population. Some factors to consider in selecting the maximum-risk worker are given on pages 33–34 of Leidel et al. (1977). Related determinant variables affecting the exposure levels of the target populations are discussed above in Section 5.1. The important point with this approach is to sample only those workers whose exposures are believed to represent the higher exposures of the target population. However, this approach is subject to frailties of professional judgment that could lead to erroneous conclusions regarding the exposure levels for the highest-risk workers.

For an exposure screening program, if maximum-risk workers cannot be identified for each operation or target population with reasonable confidence, a second approach is random sampling. The same sample size theory applies here as was discussed in paragraph 5.2.3 for random sampling of a homogeneous-risk target population, based on the combinatorial probability formulas detailed in Technical Appendix A of Leidel et al. (1977). This approach is less efficient than the first, which relies on professional judgment to select a nonrandom sample. However, the results are independent of any mistakes that might occur in professional judgment.

PURPOSE. Provide a *large enough sample size* to assure (i.e., have a probability of at least 90 or at least 95 percent) that at least one randomly sampled worker will have a high exposure relative to most other workers in the target population (i.e., be from the highest 10 percent or highest 20 percent of the specified target population exposure distribution).

ASSUMPTIONS. The sample of workers to be measured is assumed to be randomly chosen from the specified target population. The derivation of this method is based on the theory of random sampling without replacement (see Section 5.2.3). No mathematical form is assumed for the exposure distribution of the target population. Thus these sample sizes may be larger than necessary for some situations where the exposure distribution is confidently known, such as when a two- or three-parameter lognormal distribution can be reliably used as a model. However, these sample size recommendations are robust.

EXAMPLE. Estimate an appropriate sample size for a 26-worker target population such that there is at least 90 percent probability that the sample will include at least one higher-risk worker from the top 10 or lesser percent of the 26-worker exposure distribution.

SOLUTION. Table 10.7 indicates that for the $N = 26$ workers, a random sample of size $n = (N - 8) = 18$ workers will have at least 90 percent probability of containing one or more of the $N_0 = 2$ workers in 26 who represent no more than 10 percent of the highest exposures (7.7 percent in this example). The value 2 for N_0, the number of high values in the population, is the largest integer representing no more than 10 percent of $N = 26$ workers.

COMMENTS. At least 18 of the 26 workers in the target population need to be randomly sampled to be at least 90 percent sure to "catch" one worker from the two that constitute no more than the upper 10 percent of the target population. With this small target population (and extremely small number of highest-exposure workers) it is necessary to measure almost 70 percent of the total. This method is inefficient for small target populations but works considerably better with larger ones. For a target population of 100 workers, the necessary sample would be about 20 (or only 20 percent of the total) to have at least 90 percent probability of sampling one worker from the 10 workers that constitute the highest 10 percent of the exposure distribution. Necessary sample size tables for other upper tail percentages of exposure distributions or other confidence levels can be computed from the material in Technical Appendix A of Leidel et al. (1977).

Table 10.7 Necessary Sample Size n for $P \geq 90\%$ that One or More Items Will Be from the Highest 10% or Less (i.e., Will Be from the N_0 Highest Items in N)

N—Size of Target Population	n—Necessary Sample Size	N_0—Number of "Highest 10% or Less" Values
10–19	$(N - 1) = 9$ to 18	1
20	$(N - 6) = 14$	2
21–24	$(N - 7) = 14$ to 17	2
25–27	$(N - 8) = 17$ to 19	2
28–29	$(N - 9) = 19$ to 20	2
30–31	$(N - 14) = 16$ to 17	3
32–33	$(N - 15) = 17$ to 18	3
34–35	$(N - 16) = 18$ to 19	3
36–37	$(N - 17) = 19$ to 20	3
38–39	$(N - 18) = 20$ to 21	3
40–41	$(N - 23) = 17$ to 18	4
42–43	$(N - 24) = 18$ to 19	4
44–45	$(N - 25) = 19$ to 20	4
46	$(N - 26) = 20$	4
47–48	$(N - 27) = 20$ to 21	4
49	$(N - 28) = 21$	4
50	$(N - 32) = 18$	5

Since the target population is finite (26 in this example), only discrete probability levels P are attainable with the sampling plan. The sample size n is chosen to have at least 90 percent or greater probability ($P \geq 0.90$) that the n items in the sample will include one or more of the N_0 highest items. Since integer numbers N_0 of highest values cannot be chosen to *exactly* represent 10 percent of the population size N (unless $N = 10$, 20, 30, etc.), N_0 is chosen to be the largest integer for which $N_0 \leq [(0.10)(N)]$.

Compared to an exposure screening program, the sampling procedure for an exposure distribution monitoring program is considerably different. An objective of the latter program is to estimate the exposure distribution of the target population over an initial base period, which may range from one day to several years. The necessary sample size for initial monitoring is influenced by the desired precision of the exposure distribution estimate for the target population.

One approach to exposure distribution estimation is to plot a sample of exposure estimates on lognormal probability graph paper. This technique is presented in Section 6.12. An expanded discussion is given in Technical Appendix I of Leidel et al. (1977). An example of this approach is given in Paik et al. (1983) and the *American Industrial Hygiene Association Journal* (1983). For this approach, a *minimum* sample of about 6–10 TWA exposure estimates generally is most cost effective. That is, one should randomly sample at least 6–10 8-hr TWA exposure estimates for the defined target population over the desired temporal base period. The rationale underlying the recommendation for 6–10 TWA estimates is illustrated in Figure 10.11. If one is considering the computation of *one-sided tolerance limits* for a lognormally distributed variable, such as exposure estimates from a target population of workers

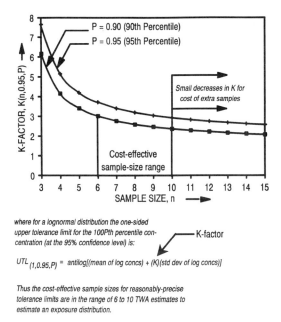

Figure 10.11 Cost-effective sample sizes for estimating an exposure distribution.

(see Section 3.6.2.4) as will be discussed in Section 6.7, this figure indicates that the K coefficient in the tolerance limit algorithm decreases relatively little after n values in the range of 6–10. That is, there will be small decreases in K after $n = 10$ compared to the costs incurred for the additional samples.

Note that 6 to 10 exposure samples will only provide the roughest estimate of the *shape* of the actual exposure distribution. It will tell one almost nothing about the goodness-of-fit of a particular distributional model, such as lognormal. Data presented by Daniel and Wood (1971), and discussed in Appendix I of Leidel et al. (1977), indicates that considerably larger samples (such as 30–60) are necessary to obtain an estimate with low uncertainty of the central 80 percent of the target population exposure distribution. Even with these sample sizes, unusual behavior in the 10 percent upper and lower tails still cannot be confidently determined. However, as mentioned in Section 3.6, an exposure distribution estimate with moderate uncertainty may suffice because it is an *adequate* estimate. The adequacy of any distributional estimate is dependent on its ability to serve as a workable forecaster of the unsampled portions of the exposure distribution for the target population. That is, is the estimate good enough for our purpose? This reinforces the point that the purposes of exposure measurements and monitoring need to be clearly defined *before* the measurements are obtained.

Besides providing information on the range and frequency of exposures, the exposure distribution monitoring strategy has another important advantage. The initial monitoring can also yield an estimate of the variability parameter of the distribution, to be used as an indicator to judge if the target population has been adequately defined. Remember that a target population of exposures generally is defined in terms of (1) the numbers and types of workers, (2) the temporal period to be covered, and (3) ranges of determinant variables that affect the exposure levels to which the population is exposed. If the variability of the sample exposure distribution exceeds a geometric standard deviation (GSD) of about 3.0, then the suitability of the factors used to define the target population should be reexamined (this approximate criterion is based on the professional judgment of one of the authors, NAL). Other factors may be needed that define more limited population groupings in order to achieve less variability for the exposure distribution. Such a process may need to be repeated several times until one achieves an exposure distribution variability parameter that is small enough to permit meaningful exposure monitoring.

If too many workers from populations with several different exposure distributions having substantially different median-exposure levels are pooled into one target population, then a pooled exposure distribution of excessive variability can result. A highly disperse exposure distribution generally is unsatisfactory because exposure estimates for individual workers will have a considerable amount of uncertainty. For example, for a two-parameter lognormal distribution with an arithmetic mean of 100 ppm and a GSD of 2.5 (GM = 65.7 ppm), the 5th-percentile worker exposure is about 15 ppm and the 95th-percentile worker exposure would be about 300 ppm! Note that 1 in 10 of the worker exposures would lie either below 15 ppm or above 300 ppm. For this exposure distribution, the best one could say for any individual worker is that there is 90 percent confidence that during the sampled temporal period the worker's exposure was between 15 and 300 ppm. However, this highly imprecise estimate might be adequate if the exposure control level was substantially higher than the 95th-percentile exposure level, such as 1000 ppm.

5.3.4 Criteria for Decision Making

Decision making is a monitoring program element where the objectives of an exposure screening program and an exposure distribution monitoring program are similar, but the techniques used differ considerably because the available information is different. The two types are discussed under separate headings below.

In this section, criteria for decision making about exposure distributions cannot be discussed in a definitive framework of ''legal'' and ''illegal'' (or other black and white definitions of acceptable or unacceptable) parameters of exposure distributions. Generally, well-accepted toxicological interpretations are unavailable regarding parameters such as percentages of days workers are (or possibly are) exposed above an 8-hr, TWA exposure control limit, when the long-term (multiple day) mean is below the control limit. Therefore, we will not attempt to provide explicit definitions of *acceptable* and *unacceptable* exposure distributions. These definitions must be derived for individual work exposure situations in the context of the particular toxicological, regulatory, and administrative considerations of a given occupational exposure environment. The technical discussions of this section are intended to assist the competent professional in estimating and interpreting worker exposure distributions. The statistical tools for making controlled-risk decisions and judgments about the acceptability of an exposure distribution, based on an estimate of that distribution, are to be found in Section 6. In particular, procedures are presented in Sections 6.7 and 6.13 for computing and displaying tolerance limits for a lognormal distribution of exposures, and in Section 6.8 for computing a point estimate and confidence limits for the percentage of lognormally distributed values exceeding an exposure control limit.

DECISION MAKING IN AN EXPOSURE SCREENING PROGRAM. The main objective of exposure screening is to identify target populations with unacceptable exposure distributions for follow-up exposure monitoring or other actions (e.g., worker medical surveillance, worker training programs, engineering exposure controls, personal exposure controls, or other specialized measurement actions such as noted in Section 5.2.1). The decision-making technique used is to compare the exposure estimate with an appropriate action level that serves as a first screening cut-off value. A secondary objective would be to subclassify the unacceptable exposure distributions into subgroups that require either (1) follow-up monitoring at a normal frequency (e.g., at least every 2 months) or (2) follow-up monitoring at an increased frequency (e.g., at least monthly). The second screening cut-off value for these decisions is the exposure control limit chosen for adequate worker protection. For target populations judged to have acceptable exposure distributions, generally the indicated action for an exposure screening program would be to redetermine the need for exposure measurements (see Section 5.3.1) each time there is a change in production, process, or control measures that could result in a substantial increase in exposure levels.

The original action level was developed by NIOSH for regulations promulgated by the Occupational Safety and Health Administration (OSHA) (Leidel et al., 1975; also see pages 10–11 and Technical Appendix L of Leidel et al., 1977). For that action level, only single-day exposure estimates for the maximum-risk workers from each examined target population are compared to an action level set at 50 percent of the regulatory exposure standard for the particular substance. The NIOSH action level concept uses single-day exposure estimates to reach decisions regarding the acceptability of possible exposure levels on unmeasured days. Single-day estimates are the sole basis for deciding whether further exposure monitoring or other actions should be performed for the target population's workers.

Note that a required assumption for the application of an action level is that the companion exposure control limit (ECL) is sufficiently protective of workers. That is, the use of an action level to reduce employer costs of exposure monitoring, medical surveillance, worker training, and so forth presumes that exposures below the ECL create minimal or acceptable risks to workers. The rationale of using an action level is to screen worker exposures so that there is a low probability that a minimal proportion of daily exposures will exceed the ECL linked to the action level. Thus it is implicit that a small proportion of exposures below the ECL lead to minimal or acceptable risks to workers. If this is not the case, then it is not appropriate to use an action level as a justification or screening value to reduce or eliminate exposure monitoring, medical surveillance, and so forth.

It is also important to note that the *50 percent action level* should be thought of as a *regulatory action level*, since:

1. It is an approximate value developed for simplicity of application.
2. It was a value intended to reduce employer exposure monitoring burdens.
3. It incorporated feasibility considerations for regulatory monitoring requirements.

During the development of the 50 percent action level for regulatory purposes, the premise used was that "the employer should try to limit to 5 percent probability, that no more than 5 percent (or greater) of an employee's actual (true) daily-exposure averages exceed the standard" (Leidel et al., 1975, p. 29). This *is not meant to imply* that a work situation with as many as 5 percent of overexposure days is acceptable to OSHA. The stated premise is part of the "risk tradeoff" that is characteristic of statistical decision theory. Measurements subject to random errors are used to make decisions about overexposure. The limit of 5 percent overexposure days cited above is merely the minimum of hypothetical overexposure incidences that, if they occurred, would be discovered at least 95 percent of the time they were measured when the GSD of the exposure distribution is 1.22 or less.

The Leidel et al. (1975) report demonstrated that an action level appropriately computed to achieve the stated goal of a low 5 percent probability level is a variable value that is a function of the interday variability of the true daily TWA exposure levels. Regarding the interday variability measured by a geometric standard deviation (GSD) of an assumed two-parameter lognormal distribution, Leidel et al. (1975,

p. 29) noted: "Higher GSDs require lower fractional action levels. A GSD of 2.0 requires an action level as low as 0.115 of the standard!" However, NIOSH decided to recommend to OSHA a *single-value* action level computed as 50 percent of an exposure standard. This pragmatic decision was intended (1) to increase the probability of employer compliance with regulatory monitoring requirements as a result of the simplicity of applying a single action level, (2) to lower implementation costs, and (3) to substantially increase feasibility. The alternative would have been recommending a procedure that required an employer to obtain a GSD estimate of the interday exposure level variability for each exposure situation before computing an appropriate action level.

The decision to recommend a single-value action level (50 percent) keyed to a single GSD of 1.22 did lead to a regulatory screening strategy with some limitations. The potential problems and limited performance characteristics of the "bare bones" regulatory screening strategy (Leidel et al., 1977) have been comprehensively analyzed by Tuggle (1981, p. 493) who noted: "The intent of the NIOSH decision scheme is to give indications about periodic monitoring: whether to terminate (TERM), initiate/increase (INCR), or continue (CONT) exposure measurements." Tuggle (1981, p. 495) concluded the following:

> For environments with a very low fraction of exposures above the PEL, the NIOSH scheme renders the correct, TERM decision with a high probability—for all variabilities. For low variability environments, the relative probability of a TERM decision decreases sharply (the relative probability of an INCR decision increases sharply) as the fraction of exposures above the PEL increases. In other words, at low variability, the NIOSH scheme is very "sensitive" to an increasing fraction of overexposures. However, as exposure variability increases, this sensitivity falls off, and the NIOSH decision scheme assumes an increasing probability of incorrect, TERM decisions for high-exposure risk environments for which monitoring should definitely be continued or even increased. For example, consider environments with moderate-to-high variability and with 25 percent of all exposures above the PEL—one exposure in four an overexposure: Figure 3 shows that the NIOSH decision scheme produces incorrect, TERM decisions for such environments, from about 20 percent to well over 50 percent of the time.

Tuggle (1981) noted the following limitations of a simplistic regulatory screening strategy and suggested some desirable characteristics for an augmented exposure screening program:

1. The regulatory screening strategy bases a decision on, at most, only two measurements: the last one and possibly the one preceding. An augmented strategy should use *all* the data collected.
2. The regulatory screening strategy considers only a qualitative aspect of the exposure estimate, that is, whether the exposure is below the 50 percent action level or above the exposure standard. An augmented strategy should also consider the quantitative aspect of *how much* an exposure estimate is below the action level or above the exposure standard.

3. The regulatory screening strategy does not determine or account for the range of exposure level variabilities actually encountered. An augmented strategy should, since exposure variability (GSD) is a factor equally as important as median exposure (GM) in evaluating the occurrence of overexposures.

4. Lastly, the regulatory screening strategy produces incorrect decisions to terminate exposure sampling, particularly with substantial proportions of over-exposures and moderate- to high-variability exposure level environments. An augmented strategy should limit such incorrect decisions to an acceptably low probability that is independent of exposure variability.

Tuggle (1981) suggested a simple modification to the regulatory screening strategy given by Leidel et al. (1977) to reduce incorrect decisions to terminate exposure monitoring that result from analyzing initial monitoring results. He recommended that the first decision criterion require, in all instances, two (instead of one) consecutive exposure estimates below the 50 percent action level to terminate routine exposure monitoring. He demonstrated that this would substantially reduce the probability of incorrect decisions to terminate monitoring. Tuggle (1982) subsequently recommended an augmented monitoring strategy based on one-sided tolerance limits for a distribution of daily exposure estimates assumed to follow a two-parameter lognormal distribution.

It should be noted that the NIOSH ''employee exposure determination and decision strategy'' recommended by NIOSH to OSHA for incorporation in regulations (Figure 1.1 of Leidel et al., 1977) does maintain the objective of a 5 percent limit on the risk of making incorrect decisions of noncompliance during the actual periods and for the actual workers that were measured for TWA exposures. However, the NIOSH regulatory screening strategy does not achieve 95 percent probability of terminating a process that has 5 percent or more overexposure days, unless the GSD of the process is at or below the value 1.22 to which the 50 percent action level is keyed. The limitations of a constant-value action level that is keyed to a single GSD were also discussed by NIOSH in Leidel et al. (1975). However, a constant value was selected for practical reasons stated earlier. Tuggle's mathematical statistical analysis of the NIOSH-recommended action level of 50 percent is correct, but practical remedies are still not at hand.

DECISION MAKING IN AN EXPOSURE DISTRIBUTION MONITORING PROGRAM. Compared to an exposure screening program, the decision-making process for evaluating initial monitoring results from an exposure distribution monitoring program is considerably different. However, the objectives of the two monitoring programs are similar.

For an exposure distribution monitoring program, the objective of the decision-making element is to classify the exposure distributions or target populations into one of the following three classes: acceptable, marginal, or unacceptable. An *acceptable exposure distribution* is one for which we are confident that essentially all exposures are less than the desired exposure control limit (ECL). A *marginal exposure distribution* is one for which we are confident that at least a moderate proportion of exposures exceed the desired exposure control limit. An *unacceptable exposure distribution* is one for which we are confident that at least a substantial proportion of worker exposures exceed the desired exposure control limit.

If the exposure distribution for a target population is judged acceptable, generally the follow-up action would be limited to routine exposure monitoring at an appropriate future time. The objective of follow-up action would be to monitor and revise the initial or previous estimate of the exposure distribution for the target population. If the judgment for the exposure distribution is marginal, then the indicated action generally would be more definitive exposure monitoring, which could include the specialized measurement actions noted in Section 5.2.1. This additional monitoring should identify the marginally and unacceptably exposed workers and could assist in implementing overexposure prevention measures. Lastly, if the exposure distribution is judged unacceptable, the indicated action would follow the same course specified for a target population identified with an action level screening approach as unacceptably exposed.

Two statistical techniques that are useful in an evaluation of exposure distribution estimates for target populations of workers are (1) probability plotting, including computation and graphical display of tolerance limits, and (2) distribution fitting to estimate one or more quantiles (a formal technique based in statistical estimation theory). The first technique utilizes the type of lognormal probability plot presented in Sections 6.12 and 6.13 and Technical Appendix I of Leidel et al. (1977). Exposure estimates are plotted on lognormal probability paper along with lines representing the desired ECL exposure limit and tolerance limits for the true exposure distribution. Rough estimates of overexposure "risks" (stated in percentages of overexposed workers or workdays with their attendant uncertainties) can then be obtained from such exposure probability plots. The estimates are given by the indicated probabilities at the intersection of the exposure estimate distribution plot and the ECL exposure limit line and the exposure values above and below the ECL.

This analytical approach is illustrated on lognormal probability paper in Figure 10.12. The two dashed lines indicate the estimated exposure distributions (see Section 6.12) and the two thick solid lines bounding the shaded areas represent upper

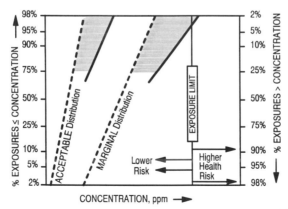

Figure 10.12 Graphical criteria for acceptable, marginal, or unacceptable exposure distributions.

tolerance limits (one-sided upper confidence limits for differing percentiles P of the parent distribution of exposures). The shaded regions represent the uncertainty in the estimated exposure distribution, since these are areas in which the true exposure distribution could lie.

Note that the qualitative definitions acceptable, marginal, or unacceptable exposure distribution involve both objective quantitative and subjective qualitative aspects of decision making using competent professional judgment based on relevant training and experience. The professional judgment exercised should consider at least the following information: (1) individual aspects of the target population, (2) chemical(s) creating the exposures and their health effects at different exposure levels, (3) possible exposure levels (both routine and episodic), (4) time patterns of exposure, (5) knowledge of exposure level determinant factors pertinent to the target population, (6) the overexposure risk estimates, (7) slope (variability) of the exposure distribution, and (8) experience with similar exposure situations. Note that this technique should be used with caution and decisions made with some degree of conservatism that favors protection of worker health in the absence of definitive information (i.e., err in the direction of marginal or unacceptable exposure distribution decisions by placing the "burden of proof" on acceptable decisions).

The second evaluation technique, formal point estimation, complements the other approach, which is graphical. The two techniques can be used independently, but the authors believe that both techniques are useful for evaluation of any exposure distribution data set. The quantitative inferential approach is to calculate the best estimate (point estimate) for the proportion of the exposure distribution exceeding some desired value (e.g., exposure control limit, exposure standard). In addition, confidence limits for the true value of the parameter whose point estimate was obtained should also be calculated to indicate the uncertainty of the estimate due to limited sample sizes. Approximate confidence limits could be either one or two sided and typically will be calculated at the 95 percent confidence level. The details for this technique are presented in Section 6.8.

5.3.5 Strategy for Periodic Monitoring for Continuing Hazard Evaluation

Periodic follow-up exposure monitoring applies both to an exposure screening program and to an exposure distribution monitoring program and has a similar objective for both programs. The objective is that individual workers or target populations of workers with potentially other-than-acceptable (i.e., marginal or unacceptable, see Section 5.3.4) exposure distributions be periodically monitored, often enough to promptly detect unacceptable exposure situations. In addition, an exposure distribution monitoring program should periodically monitor at a frequency adequate to identify substantial changes in the target populations' exposure distributions.

Control charts are commonly used in quality control operations, but Hawkins and Landenberger (1991) discuss some limitations of conventional control charting techniques when they are applied to industrial hygiene data. They point out that the usual industrial-hygiene sample sizes are adequate only for the most sophisticated types of control chart analysis, such as moving average and moving range methods. However, if one does desire to apply quality assurance procedures to lognormally distributed data, review the techniques given by Morrison (1958).

Remember that for an exposure screening program the set of possible decisions suggested (Section 5.3.4) regarding the target population's exposure distribution are acceptable, unacceptable with "normal" frequency follow-up monitoring, and unacceptable with "increased frequency" follow-up monitoring.

For regulatory purposes, NIOSH recommended to OSHA that minimum requirements for "normal" frequency of monitoring for an exposure screening program be set at 2–3 months (Leidel et al., 1977). It should be noted that, in addition to basic health and safety protection considerations, feasibility and simplicity of application considerations played an important role in this recommendation. Lynch et al. (1978, p. 13) have commented for NIOSH:

> Four such exposure measurements during the year are considered minimal to detect any significant fluctuations in the average level of environmental contaminants. Employee-exposure measurements four times per year per exposed employee are a reasonable minimal burden on the employer. The employer should not interpret this to be an absolute minimum that would always be appropriate to determine each employee's exposure. There is information that a waiting time between sampling events (for typical data) of over 1 month results in a relatively low level of confidence in the exposure estimates. However, a maximum waiting time of 3 months between measurements should protect each employee and give some idea of variation without putting an inordinate burden on the employer.

For an "increased frequency" of regulatory follow-up monitoring, NIOSH recommended a minimum frequency of once a month (Leidel et al., 1977). When unexpectedly or unusually high exposure results are obtained by either initial monitoring or periodic monitoring, more intensive monitoring is needed to quickly identify increasing exposure trends that may lead to hazardous exposures.

For other than regulatory monitoring programs, the selection of an adequate frequency for periodic exposure monitoring should be based on experience with similar exposure situations, knowledge of the exposure level determinant factors pertinent to the target population, and use of competent professional judgment. Guidelines for the factors to be considered include:

1. Nature and degree of health hazard from the exposure situation, evaluated at all possible exposure levels, even those that might be unexpected.

2. Ratio of previous exposure levels to the desired exposure-control limit, and possible trends in this ratio.

3. Degree of variability seen in previous exposure distributions, variability between the distributions, and trends over time. These can be studied by plotting the sample data sets from several longitudinal points in time on semilogarithmic paper (see Section 6.11). Another complementary approach would be the plotting of several sample data distributions on the same lognormal probability graph.

4. Determinant variables affecting the target population's exposure levels (see Section 5.1).

5. Reliability of decision-making techniques used and quantity and quality of exposure estimates available. Exposure screening program decision tech-

niques yield decisions of less reliability than the quantitative techniques available for exposure distribution estimates.

5.3.6 Occasions Requiring Extraordinary Monitoring

With either an exposure screening program or an exposure distribution monitoring program, additional exposure monitoring is indicated whenever a change occurs in production, process, personnel, exposure controls, or any other determinant variable that could lead to substantial increases in worker exposure levels.

5.3.7 Criteria for Termination of Monitoring

This program element concerns a specialized type of decision making. As with monitoring of exposures for the purpose of decision making about exposure distributions, the objectives of termination of routine periodic monitoring are similar in an exposure screening program and in an exposure distribution monitoring program, but the techniques used and available information differ considerably. In some situations it may be desirable to conserve exposure measurement resources by terminating routine monitoring of target populations for which one is confident that the exposure distributions have been consistently acceptable. As with selecting an adequate frequency of periodic monitoring (Section 5.3.5), decisions to terminate monitoring should be based on experience with similar exposure situations, knowledge of the exposure level determinant factors pertinent to the target population, availability of monitoring resources, and availability of competent professional judgment. Use the factors given in Section 5.3.5 when considering termination of routine monitoring.

For *regulatory* purposes in an exposure-screening program, NIOSH recommended to OSHA that regulations allow cessation of monitoring for a worker if two consecutive exposure estimates taken at least one week apart were both less than the 50-percent action level. See Section 5.3.4 for a discussion of an analysis by Tuggle (1981) of this regulatory termination rule and his suggested alternative approach for situations where a two-parameter lognormal distribution of exposures can be assumed.

For exposure distribution monitoring programs, decisions to terminate monitoring should also be based on experience with similar exposure situations, knowledge of the exposure level determinant factors pertinent to the target population, availability of monitoring resources, and use of competent professional judgment. Also use the factors given in Section 5.3.5 when considering termination of routine monitoring.

There is an additional quantitative tool available when considering termination of an exposure distribution monitoring program. Tuggle (1982) has recommended the use of one-sided tolerance limits if one can assume that all available exposure estimates are from a single two-parameter lognormal distribution. The computed one-sided upper tolerance limit for the target population would be compared to the selected exposure control limit (ECL) and monitoring would be continued if the tolerance limit exceeded the ECL. All available exposure estimates would be considered sample data and utilized for computation of a tolerance limit.

However, if the exposure data are not stable over time (i.e., if differing distributions of exposures exist over time for the same target population of workers), competent professional judgment may be used to censor some of the earlier exposure estimates. If this is not done, the computed tolerance limit will probably be larger than it should be, which would cause errors in the decision making that lead toward unjustified continuation of monitoring. Thus if all the data are used, instead of a smaller sample based on the more recent exposure data, the decision making would tend to be conservative (i.e., favor continuation of possibly unnecessary monitoring, but reduce possible risks to workers).

5.3.8 Procedures for Follow-up

This program element should always be used for all types of monitoring programs. One should discuss the planning, conduct, and results of each program element with appropriate representatives of management and labor. If necessary, implement corrective action to reduce potential or identified health hazards. Write periodic evaluation reports as appropriate. Determine if additional exposure monitoring, biological monitoring, or medical surveillance programs are necessary. Determine if areas outside the workplace should be evaluated regarding possible environmental air or water pollution, hazardous waste disposal practices, or safety practices.

6 APPLIED METHODS FOR ANALYZING OCCUPATIONAL EXPOSURE DATA

This sixth and last section of this chapter is devoted to the applied statistical methods that we have suggested be used for analysis of occupational exposure data. Data to be analyzed by these methods can be generated by the study designs discussed in Sections 4 and 5. The first portion of this section (Sections 6.1 through 6.5) covers methods for computing one- and two-sided confidence limits and confidence intervals for the true exposures of individual workers during the same periods as their exposure measurements. In addition, classification tests for individual exposure estimates relative to an exposure control limit are presented. The second portion of Section 6 (Sections 6.6 through 6.10) covers inferential methods for exposure distributions such as computing tolerance limits, tolerance intervals, and point estimates of the proportions of values in distributions that exceed a chosen limit, along with associated confidence limits. The third portion of the section (Sections 6.11 through 6.13) presents graphical techniques for plotting lognormally distributed data and associated tolerance limits. Appropriate sample size recommendations for specialized objectives were given earlier in Sections 5.2.3 and 5.3.3.

It is strongly recommended that the calculations presented in these methods be performed on a hand-held programmable calculator or MS-DOS personal computer (PC) software (e.g., Lotus 1-2-3 or Microsoft Excel 4.0 spreadsheets or Mathcad 3.1 for Windows). By programming the algorithms given in this chapter and then running the numerical examples, one can ensure that the correct equations are being used. Additionally, it is very useful and instructive if one then graphs the results to obtain a better understanding of the results. Piele (1990) presents series of examples demonstrating how to set up and solve statistical problems with either Lotus 1-2-3,

Release 2.01 or higher, or Microsoft Excel, Version 2.0 or higher. His emphasis is on spreadsheet concepts and functions that can be applied to a wide range of quantitative problems.

The methodologies for the exposure classification tests (statistical hypothesis tests) given in Sections 6.1 through 6.5 were originally developed by NIOSH for OSHA in support of their regulatory enforcement procedures. The purpose of the methodologies is to limit the risk of making unjustified noncompliance decisions (i.e., decisions not supported by sufficient accuracy in the exposure measurement data) regarding regulatory exposure limits. The necessity for development of such procedures is a reflection of the nature of the legal system used in the United States.

The presentation of statistical hypothesis tests in this chapter may unfortunately give the erroneous impression that industrial hygienists should view any ECL, AC-GIH TLV, or legal standard as a definitive boundary between ''safe'' and ''dangerous'' exposure levels for all workers, *which is definitely not the case*. The hypothesis tests for individual worker exposure estimates presented in this section should receive routine application only in legal proceedings concerning regulatory issues. That is, an employer should not attempt to judge the acceptability of a workplace environment by statistically comparing only a few imprecise exposure measurements with some exposure limit.

It is strongly recommended that the computation of confidence limits be the preferred procedure for calculating the uncertainty of individual worker exposure measurements. The primary purpose of computing confidence limits should be the estimation of the uncertainty of individual worker exposure estimates due to the random errors of the exposure measurement procedure used. It is important to realize that the confidence limits do not reflect the substantial interday and interworker exposure level variability. When attempting to judge the acceptability of exposures in a workplace, an industrial hygienist should if possible evaluate the multiday and multiworker exposure distributions (see Section 5.3.4). Of course, an employer continually has the legal obligation to maintain exposure levels below applicable regulatory exposure limits on *all* workdays for *all* workers.

6.1 Full-Period Single-Sample Estimate

The methods in this section are applicable to exposure estimates calculated from a single exposure measurement taken for the full duration of the desired time-averaging period (e.g., 40 hr for a workweek TWA, 8 hr for an 8-hr workday TWA, 15 min for a 15-min TWA).

6.1.1 Confidence Limits for a True TWA Exposure

PURPOSE. Compute one- or two-sided *confidence limits* for a true worker exposure that is estimated from a single exposure measurement when the total coefficient of variation for the exposure measurement procedure is known.

ASSUMPTIONS. The sample result is assumed to be a valid measurement of the worker's exposure at the time and place of the sampling (refer to discussions in Sections

3.1 and 3.3.5). The random errors in full-period single-sample estimates X are assumed to be independent and normally distributed with zero mean (see Section 3.6.1.3). This assumption implies that the arithmetic mean of many replicate exposure measurements (i.e., taken at exactly the same point in space and over the same time period) would be equal to the true exposure average. This true mean (or "expected value") is unknown, but the total coefficient of variation for the exposure measurement procedure (denoted by CV_T, see Section 3.4.2) is assumed known. Adequate confidence limits can be computed for the true concentration based on an assumption that the standard deviation is equal to the product of CV_T and the true full-period exposure μ, where μ is estimated by measurement X. The accuracy of the computed confidence limit(s) will be a function of the accuracy of the coefficient of variation used.

EXAMPLE. An exposure measurement procedure with a CV_T of 0.09 was used to obtain an 8-hr TWA exposure estimate for a worker. A single 8-hr measurement yielded an estimate $X = 0.20$ mg/m^3. Compute two-sided 95 percent confidence limits for the worker's true exposure based on the exposure estimate X.

SOLUTION

1. Compute the two-sided 95 percent upper and lower confidence limits ($UCL_{2,0.95}$ and $LCL_{2,0.95}$) for the worker's true exposure μ from:

$$UCL_{2,0.95} = \frac{X}{1 - (1.96)(CV_T)}$$

$$LCL_{2,0.95} = \frac{X}{1 + (1.96)(CV_T)}$$

In this example

$$UCL_{2,0.95} = \frac{0.20}{1 - (1.96)(0.09)}$$

$$= \frac{0.20}{0.824}$$

$$= 0.24 \text{ mg/m}^3$$

$$LCL_{2,0.95} = \frac{0.20}{1 + (1.96)(0.09)}$$

$$= \frac{0.20}{1.176}$$

$$= 0.17 \text{ mg/m}^3$$

Thus, in this example, the two-sided confidence interval at the 95 percent confidence level for the true worker exposure that was estimated by the single-sample exposure measurement of 0.20 mg/m^3 is from $LCL_{2,0.95} = 0.17$ mg/m^3 to $UCL_{2,0.95} = 0.24$ mg/m^3. That is, one can be 95 percent confident that the *true exposure* μ is between 0.17 and 0.24 mg/m^3.

2. To compute either one-sided 95 percent confidence limit ($UCL_{1,0.95}$ or $LCL_{1,0.95}$) for the true exposure, use the formula for the corresponding two-sided limit with a Z value of 1.645 substituted for the multiplier 1.96. (The multiplier 1.96 corresponds to 2.5 percent of the area in a standard normal distribution allocated to each tail, whereas 1.645 corresponds to the total 5 percent allocated to one tail.)

COMMENTS. Note that, for any given estimate and associated confidence interval, the interval either does or does not contain the true value. However, we can say, in a special sense, that there is a 95 percent *probability* of the true exposure being bounded by the computed confidence interval. That is, if repeated samples were taken, and confidence limits calculated for each sample, 95 percent of the confidence limits would enclose the true worker exposure.

Also, the indicated uncertainty in the exposure estimate reflected in the width of the asymmetrical confidence interval (-15 to $+21$ percent about the point estimate in this example) *does not* incorporate uncertainty due to *interday* exposure level variability (see Sections 3.3.1 and 3.6.2.3). The confidence interval pertains *only* to the worker's true exposure on the particular day and during the period of that day that was actually measured.

6.1.2 Classification of Exposure (Hypothesis Testing)

PURPOSE. Classify an *exposure estimate* based on a single-day exposure measurement as *noncompliance exposure*, *possible overexposure*, or *compliance exposure* relative to an exposure control limit (ECL).

ASSUMPTIONS. The same as used for Section 6.1.1.

EXAMPLE. A charcoal tube and personal pump were used to sample a worker's TWA exposure to α-chloroacetophenone. The analytical laboratory reported a TWA exposure estimate X of 0.040 ppm and stated that the measurement procedure had a CV_T of 9 percent. The applicable ECL was 0.05 ppm. With 95 percent confidence, classify the TWA exposure estimate X relative to this ECL.

SOLUTION

1. For an employer, in this example $X \leq$ ECL. However, if it had been that $X >$ ECL, no statistical significance test (of the null hypothesis of noncompliance against an alternative of compliance) would need to be made because the estimate itself would exceed both the ECL and LDV (lower decision value). In that case it would be a noncompliance exposure for the employer.

However, since $X \leq$ ECL, it is necessary to compute the one-sided LDV to do a hypothesis test for compliance. The decision is to be made with at least 95 percent confidence; that is, with 5 percent *or less* probability of being wrong (in statistical terminology, a 0.05 or less probability of a type 1 error). Thus, a subscript of 5% for LDV will be used to identify the decision test's *level of significance* (also called the *size of the significance test*).

$$LDV_{5\%} = ECL - (1.645)(CV_T)(ECL)$$

$$= (0.05 \text{ ppm}) - (1.645)(0.09)(0.05 \text{ ppm})$$

$$= (0.05 \text{ ppm})(1 - 0.148) = (0.05)(0.852)$$

$$= 0.043 \text{ ppm}$$

Then the exposure estimate X is classified according to the following criteria:

a. If $X \leq LDV_{5\%}$, classify as compliance exposure, (i.e., reject the null hypothesis and accept the alternative hypothesis) or

b. If $X > LDV_{5\%}$, classify as possible overexposure, (i.e., do not reject the null hypothesis).

Since the exposure measurement $X = 0.040$ ppm is less than the $LDV_{5\%}$ of 0.043 ppm, the employer can decide that the measured worker had a compliance exposure on the day and during the period of the measurement.

2. For a compliance officer, because $X \leq ECL$, no statistical test of the null hypothesis of compliance against an alternative of noncompliance need be made because the estimate is already less than the ECL and therefore would be less than the UDV (upper decision value). Hence it would be a compliance exposure for the compliance officer.

However, if the result had been $X > ECL$, it would have been necessary to compute the one-sided UDV test statistic for noncompliance:

$$UDV_{5\%} = ECL + (1.645)(CV_T)(ECL)$$

Then the exposure estimate is classified according to the following criteria:

a. If $X > UDV_{5\%}$, classify as noncompliance exposure, (i.e., reject the null hypothesis and accept the alternative hypothesis), or

b. If $X \leq UDV_{5\%}$, classify as possible overexposure, (i.e., do not reject the null hypothesis).

COMMENTS. Figure 10.13 is a graphical interpretation of decision regions using LDVs for employer decisions and UDVs for compliance officer decisions. The observed value X for the TWA exposure estimate is denoted by TWA inside a diamond in Figure 10.13.

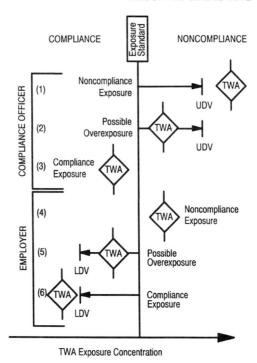

Figure 10.13 Classification of exposure with hypothesis testing (one-sided decision values).

Note the *limited applicability* of these tests. The outcomes of these hypothesis tests apply only to the particular worker's exposure on the day and during the period of the exposure measurement.

The use of CV_T in the classification formulas is equivalent to calculating the standard deviation of X (the single exposure measurement) as $[(CV_T)(ECL)]$ instead of $[(CV_T)(\mu)]$, where μ is the true exposure for which X is the estimate. The use of a one-sided, 5 percent significance level, lower decision value or a one-sided, 5 percent significance level, upper decision value ($LDV_{5\%}$ or $UDV_{5\%}$, see Section 3.7) computed in this manner is correct, since the classification rule selected for use as a compliance officer's test for noncompliance is algebraically equivalent to a significance test of the null hypothesis of compliance (i.e., of H_0: $\mu \leq ECL$). Similarly, the classification rule selected for use as the employer's test for compliance is equivalent to a significance test of the null hypothesis of noncompliance (i.e., of H_0: $\mu > ECL$). In both cases, setting $\mu = ECL$ to compute the decision value serves to keep the size of the test at or below 0.05 ($\leq$ 5 percent probability of incorrectly rejecting the null hypothesis).

The rationale for the hypothesis tests is:

1. For a test at the 5 percent level of significance, calculate a one-sided upper decision value ($UDV_{5\%}$) or lower decision value ($LDV_{5\%}$). These decision values

are in fact *critical values* for the measurement X, under the null hypothesis that the true exposure is equal to the ECL. If the null hypothesis were true, the exposure estimate would not exceed the UDV (or be less than the LDV for the complementary hypothesis test) more than 5 percent of the time for the same true exposure and measurement conditions.

2. For a test of noncompliance, if the exposure estimate exceeds the one-sided upper decision value ($UDV_{5\%}$), reject the null hypothesis of compliance and decide for noncompliance. Similarly, for a test for compliance, if the exposure estimate is less than the one-sided lower decision value ($UDV_{5\%}$), reject the null hypothesis of noncompliance and decide for compliance.

6.2 Full-Period Consecutive-Samples Estimate for Uniform Exposure

The methods in this section are for exposure estimates calculated from a series of consecutive exposure measurements (equal or unequal time duration) that collectively span the full duration of the desired time-averaging period.

6.2.1 Confidence Limits for a True TWA Exposure

PURPOSE. Compute one- or two-sided confidence limits for a true TWA exposure, based on multiple exposure measurements in a uniform exposure situation.

ASSUMPTIONS. The random errors for measurements X_i are assumed to be normally and independently distributed (see Section 3.6.1.3). The true worker exposure average for a sampled period during the total period of the exposure estimate TWA (i.e., true arithmetic mean of the hypothetical parent population of replicate exposure measurements at exactly the same point in space during the same time period of length T_i) is unknown, but is assumed to be about the same for all periods sampled (uniform exposure environment). The coefficient of variation of replicate measurements made by the exposure measurement procedure is assumed known and constant (see Section 3.4.2). The accuracy of the computed confidence limit(s) will be a function of the accuracy of the coefficient of variation used.

It is assumed here that all sampled periods have equal true average concentrations; if it is expected that the samples have significantly different values because of different exposure situations during the workshift, then the conservative procedure in Section 6.3.1 can be used. Where exposures are highly variable over the different sampling periods during the duration of the estimated TWA, the use of the formulas in this section will underestimate the random sampling error in TWA, thus underestimating the width of the confidence interval.

EXAMPLE. An exposure measurement procedure with a CV_T of 0.08 was used to obtain an 8-hr TWA exposure estimate denoted TWA. A personal pump and three charcoal tubes were used to consecutively measure a worker's approximately uniform exposure to isoamyl alcohol. The analytical lab reported the following exposure estimates for the three tubes: $X_1 = 90$ ppm over $T_1 = 150$ min, $X_2 = 140$ ppm over $T_2 = 100$ min, and $X_3 = 110$ ppm over $T_3 = 230$ min. The ECL chosen by the industrial hygienist was 100 ppm. The CV_T for the measurement procedure is 0.08 (8 percent). Compute the one-sided 95 percent upper confidence limit for the true 8-hr TWA exposure. The TWA estimate is 110 ppm.

SOLUTION

1. If the sample durations $(T_1, T_2, \cdots, T_n)$ are approximately equal, the one-sided 95 percent upper confidence limit for the true TWA exposure can be computed from the short equation:

$$UCL_{1,0.95} = TWA/\{1 - [(1.645)(CV_T)]/(n)^{1/2}\}$$

In this example the sample durations are not approximately equal, but for illustrative purposes and as a first approximation:

$$UCL_{1,0.95} = \frac{110 \text{ ppm}}{1 - [(1.645)(0.08)]/(3)^{1/2}}$$

$$= \frac{110 \text{ ppm}}{1 - 0.0760}$$

$$= \frac{110 \text{ ppm}}{0.924}$$

$$= 119 \text{ ppm}$$

Thus an approximate one-sided 95 percent $UCL_{1,0.95}$ for the true TWA exposure is 119 ppm based on the measured 110 ppm TWA estimate.

2. Since the sample durations are not equal, compute the more exact one-sided 95 percent upper confidence limit for the true TWA from the longer equation:

$$UCL_{1,0.95} = \frac{(TWA)(T_1 + \cdots + T_n)}{(T_1 + \cdots + T_n) - [(1.645)(CV_T)(T_1^2 + \cdots + T_n^2)^{1/2}]}$$

In the example:

$$UCL_{1,0.95} = \frac{(110 \text{ ppm})(150 + 100 + 230)}{(480) - [(1.645)(0.08)(150^2 + 100^2 + 230^2)^{1/2}]}$$

$$= \frac{(110 \text{ ppm})(480)}{(480) - (38.5)}$$

$$= 120 \text{ ppm}$$

In this case the short equation gave a confidence limit that was only slightly lower than the more accurate estimate. The $UCL_{1,0.95}$ is 120 ppm based on the 110 ppm TWA estimate. That is, we can be 95 percent confident that the true TWA exposure is less than 120 ppm.

3. To compute a one-sided 95 percent lower confidence limit ($LCL_{1,0.95}$) substitute a plus sign for the minus sign in the denominator of the step 1 or 2 formulas. To compute a two-sided 95 percent confidence interval ($LCL_{2,0.95}$ to $UCL_{2,0.95}$) for the true TWA exposure, substitute a Z value of 1.96 for 1.645.

COMMENTS. This method is an extension of the procedure presented in Section 6.1.1 for a TWA estimate based on a single exposure measurement. Review the comments of that section.

The indicated uncertainty in the TWA estimate reflected in the 10-ppm difference between the TWA of 110 ppm and the $UCL_{1,0.95}$ of 120 ppm (9 percent higher than the TWA estimate in this example) *does not* incorporate uncertainty due to *inter*day exposure level variability (see Sections 3.3.1 and 3.6.2.3). The one-sided confidence limit pertains *only* to the worker's true exposure on the day actually sampled and during the total period of the three measurements.

6.2.2 Classification of Exposure (Hypothesis Testing)

PURPOSE. Classify a TWA exposure estimate that is based on multiple exposure measurements from a uniform exposure situation, as noncompliance exposure, possible overexposure, or compliance exposure, regarding an exposure control limit (ECL).

ASSUMPTIONS. The same as used for Section 6.2.1. The formulas of this section strictly apply only to the case of uniform exposure. Where true exposures are highly variable between the sampling periods over the duration of the TWA estimate (nonuniform exposure situation), the use of the formulas in this section will underestimate the random sampling error in the TWA estimate. This misapplication would therefore increase the chance of deciding a noncompliance exposure (with the test for the compliance officer) or deciding a compliance exposure (with the test for the employer). For a nonuniform exposure situation, use the conservative methods given below in Section 6.3.2.

EXAMPLE. The same as used for Section 6.2.1, except classify with 95 percent confidence the 110-ppm TWA exposure estimate relative to the ECL of 100 ppm.

SOLUTION

1. If the sample periods (T_1, T_2, $\cdots$, T_n) are equal (or, for a somewhat inexact test, approximately equal), classify the TWA estimate using the short equations in step 2 (for an employer) or step 3 (for a compliance officer). For unequal sample periods, classify using the longer equations in step 4 (for an employer) or step 5 (for a compliance officer).

2. For an employer, if TWA $\geq$ ECL, a possible overexposure is indicated by the TWA itself and no statistical test for compliance need be made. This is the case for this example. However, when there are approximately equal sample periods and TWA $<$ ECL, compute the one-sided LDV (lower decision value) to do a test for compliance:

$$LDV_{5\%} = ECL - \frac{(1.645)(CV_T)(ECL)}{(n)^{1/2}}$$

Then classify the TWA exposure estimate based on the multiple measurements according to:

a. If TWA $\leq$ LDV$_{5\%}$, classify as compliance exposure.
b. If TWA $>$ LDV$_{5\%}$, classify as possible overexposure.

3. For a compliance officer, if TWA $\leq$ ECL, it is superfluous to make a statistical test for noncompliance, since the TWA estimate obviously would also be less than the UDV because by definition the ECL is always less than the UDV. However, for TWA $>$ ECL, and where there are approximately equal sampling periods, the compliance officer would compute the one-sided UDV$_{5\%}$ critical value of the TWA:

$$UDV_{5\%} = ECL + \frac{(1.645)(CV_T)(ECL)}{(n)^{1/2}}$$

Then the exposure estimate would be classified according to:

a. If TWA $>$ UDV$_{5\%}$, classify as noncompliance exposure.
b. If TWA $\leq$ UDV$_{5\%}$, classify as possible overexposure.

4. For an employer, when the periods of the measurements are not approximately equal and TWA $<$ ECL, compute the one-sided LDV critical value of the TWA exposure estimate:

$$LDV_{5\%} = ECL - \frac{(1.645)(CV_T)(ECL)(T_1^2 + \cdots + T_n^2)^{1/2}}{T_1 + \cdots + T_n}$$

Then use the classification criteria of step 2 to perform a hypothesis test for compliance.

5. For a compliance officer, when the periods of the measurements are not approximately equal and TWA $>$ ECL, compute the one-sided UDV$_{5\%}$ to be used as the critical value of the TWA exposure estimate:

$$UDV_{5\%} = ECL + \frac{(1.645)(CV_T)(ECL)(T_1^2 + \cdots + T_n^2)^{1/2}}{T_1 + \cdots + T_n}$$

Then use the classification criteria of step 3 to perform a hypothesis test for noncompliance. For this example:

$$UDV_{5\%} = 100 + \frac{(1.645)(0.08)(100)(150^2 + 100^2 + 230^2)^{1/2}}{150 + 100 + 230}$$

$$= 100 + 8.0 = 108 \text{ ppm}$$

Since the TWA $= 110$ ppm exceeds the $UDV_{5\%}$ of 108 ppm, the TWA exposure estimate of 110 ppm is classified as a statistically significant noncompliance exposure. The compliance officer can state that, for the measured worker on the day and during the period of the three measurements, the true 8-hr TWA exposure was a noncompliance exposure. The statistical decision of noncompliance is made with 5 percent or less probability of being wrong.

COMMENTS. The same as for Section 6.1.2, which should be reviewed before the use of this section.

The outcomes of these hypothesis tests are valid *only* for the worker's exposure on the day and during the period of the exposure measurements.

In the example, the measurement results indicate a sufficiently uniform exposure over the duration of the TWA estimation period, so that use of the formulas in this section is appropriate.

6.3 Full-Period Consecutive-Samples Estimate for Nonuniform Exposure

The methods in this section are a generalization of those in Section 6.2. They are for exposure estimates calculated from a series of consecutive exposure measurements (equal or unequal time duration) that collectively span the full duration of the desired TWA exposure period. The following methods are longer than those of Section 6.2 but are not limited by the assumption of nearly equal arithmetic mean exposures during the TWA period (uniform exposure environment). For highly nonuniform exposure situations, use of the less complex methods of Section 6.2 (intended for uniform exposure situations) may slightly underestimate the sampling error in the TWA estimate. However, the conservative methods of this section will usually slightly overestimate the sampling error in the TWA estimate. The $LDV_{5\%}$ of Section 6.3.2 will be slightly lower than that from 6.2.2 and the $UDV_{5\%}$ from Section 6.3.2 will be slightly higher than that of 6.2.2.

6.3.1 Confidence Limits for a True TWA Exposure

PURPOSE. Compute one- or two-sided confidence limits for a true TWA exposure, based on an exposure estimate calculated from multiple exposure measurements from a nonuniform exposure situation.

ASSUMPTIONS. The same as used for Section 6.2.1, except that there is no assumption that arithmetic mean exposures are all essentially equal (for the n sample periods that make up the TWA period).

EXAMPLE. Two charcoal tubes and a personal pump were used to obtain an 8-hr TWA exposure estimate, denoted TWA, for isoamyl alcohol. The results for the two tubes were reported as $X_1 = 30$ ppm over $T_1 = 300$ min and $X_2 = 140$ ppm over $T_2 = 180$ min, with a CV_T of 0.08 for the measurement procedure. The ECL chosen by the industrial hygienist was 100 ppm. For this nonuniform exposure situation, compute the two-sided 95 percent confidence interval for the true 8-hr TWA exposure, based on the following estimated TWA $= 71$ ppm:

$$\text{TWA} = \frac{(300 \text{ min})(30 \text{ ppm}) + (180 \text{ min})(140 \text{ ppm})}{480 \text{ min}} = 71.2 \text{ ppm}$$

SOLUTION

1. If the sample durations $(T_1, T_2, \cdots , T_n)$ were approximately equal, the two-sided 95 percent confidence interval for the true TWA exposure could be computed from the short equation:

$$\text{LCL}_{2,0.95} \text{ and UCL}_{2,0.95} = \text{TWA} \pm \frac{(1.96)(CV_T)(X_1^2 + \cdots + X_n^2)^{1/2}}{(n)(1 + CV_T^2)^{1/2}}$$

In this example the durations are not approximately equal, but for illustrative purposes and as a first approximation:

$$\text{LCL}_{2,0.95} \text{ and UCL}_{2,0.95} = 71.2 \pm \frac{(1.96)(0.08)(30^2 + 140^2)^{1/2}}{(2)(1 + 0.08^2)^{1/2}}$$

$$= 71.2 \pm 11.2 = 60.0 \text{ and } 82.4 \text{ ppm}$$

which is reported as 60 and 82 ppm. Thus an approximate two-sided 95 percent confidence interval for the true TWA exposure is 60 to 82 ppm. The best point estimate for the true exposure is the TWA estimate of 71 ppm.

2. Since the sample durations are not equal, compute the more accurate two-sided 95 percent confidence interval for the true TWA exposure from the longer equation:

$$\text{LCL}_{2,0.95} \text{ and UCL}_{2,0.95} = \text{TWA} \pm \frac{(1.96)(CV_T)(T_1^2X_1^2 + \cdots , + T_n^2X_n^2)^{1/2}}{(T_1 + \cdots + T_n)(1 + CV_T^2)^{1/2}}$$

In the example:

$$\text{LCL}_{2,0.95} \text{ and UCL}_{2,0.95} = 71.2 \pm \frac{(1.96)(0.08)(300^2 \times 30^2 + 180^2 \times 140^2)^{1/2}}{(300 + 180)(1 + 0.08^2)^{1/2}}$$

$$= 71.2 \pm 8.7 = 62.5 \text{ and } 79.9 \text{ ppm}$$

which is reported as 62 and 80 ppm. Thus an improved, more accurate value for the two-sided 95 percent confidence interval for the true average exposure is 62 to 80 ppm with a point estimate TWA of 71 ppm. That is, we can be 95 percent confident that the true TWA exposure is between 62 and 80 ppm.

3. To compute either a lower or upper one-sided 95 percent confidence limit for the true average exposure, substitute a Z value of 1.645 for 1.960 in step 1 or 2.

COMMENTS. Similar to those given in Sections 6.1.1 and 6.2.1.

6.3.2 Classification of Exposure (Hypothesis Testing)

PURPOSE. Classify a TWA exposure estimate that is based on multiple exposure measurements from a nonuniform exposure situation, as noncompliance exposure, possible overexposure, or compliance exposure, relative to an exposure control limit (ECL).

ASSUMPTIONS. The same as used for Section 6.2.1, except that there is no assumption that arithmetic mean exposures are essentially equal during n sample periods within the TWA period.

EXAMPLE. The same as used for Section 6.3.1, except classify the TWA estimate of 71 ppm relative to the 100-ppm ECL.

SOLUTION

1. If the sample periods $(T_1, T_2, \cdots, T_n)$ are approximately equal, classify the TWA estimate using the short equation in step 2 (for an employer) or the equation in step 3 (for a compliance officer). For unequal sample periods, use the longer equations in step 4 (for an employer) or step 5 (for a compliance officer).

2. For an employer, if TWA $\geq$ ECL, no statistical test for compliance need be made. Since by definition the ECL always exceeds the LDV, therefore the TWA must also exceed the LDV. However, when TWA $<$ ECL and if there are approximately equal sample periods, compute the following LDV for use in a one-sided hypothesis test for compliance:

$$\text{LDV}_{5\%} = \text{ECL} - \frac{[(1.645)(\text{CV}_T)(X_1^2 + \cdots + X_n^2)^{1/2}](\text{ECL}/\text{TWA})}{(n)(1 + \text{CV}_T^2)^{1/2}}$$

For this example the two measurement periods are not approximately equal, but as an illustration (and as a first approximation of $\text{LDV}_{5\%}$):

$$\text{LDV}_{5\%} \approx 100 - \frac{[(1.645)(0.08)(30^2 + 140^2)^{1/2}](100/71)}{(2)(1 + 0.08^2)^{1/2}}$$

$$= 100 - 13.2 = 86.8 \text{ ppm}$$

which is reported as 87 ppm after appropriate rounding. Then classify the TWA exposure estimate based on the multiple measurements according to:

a. If TWA $\leq$ $\text{LDV}_{5\%}$, classify as compliance exposure.
b. If TWA $>$ $\text{LDV}_{5\%}$, classify as possible overexposure.

Since the TWA point estimate of 71 ppm is substantially less than the approximate $\text{LDV}_{5\%} = 87$ ppm, the employer can state that, for the measured worker on the day and during the period of the measurement, the exposure estimate was a compliance exposure. The probability of this decision being incorrect is considerably less than 5 percent, since TWA is considerably less than its critical value ($LDV_{5\%}$) for the test of compliance. However, since the example has unequal sampling periods, the equation in step 4 below was used to obtain a more accurate value of $LDV_{5\%}$. The formula in step 4 yields a higher critical value ($LDV_{5\%} = 90$ ppm, see below) for TWA, but it does not change the classification decision. The step 4 procedure yields a more highly statistically significant decision of compliance exposure than is provided by the more conservative step 2 test.

3. For a compliance officer, if TWA $\leq$ ECL (as in this example), it is superfluous to compute a statistical test for noncompliance (see discussion in Section 6.2.2, part 3 of the solution). However, for TWA $>$ ECL, and where there are approximately equal sample periods, the compliance officer would compute the following noncompliance one-sided test statistic:

$$\text{UDV}_{5\%} = \text{ECL} + \frac{[(1.645)(\text{CV}_T)(X_1^2 + \cdots + X_n^2)^{1/2}](\text{ECL}/\text{TWA})}{(n)(1 + \text{CV}_T^2)^{1/2}}$$

Then the exposure estimate TWA, which is based on the multiple measurements, would be classified according to:

a. If TWA $>$ $\text{UDV}_{5\%}$, classify as noncompliance exposure.
b. If TWA $\leq$ $\text{UDV}_{5\%}$, classify as possible overexposure.

4. For an employer, when the periods of the measurements are not approximately equal and TWA $<$ ECL, compute the one-sided LDV critical value of the TWA:

$$\text{LDV}_{5\%} = \text{ECL} - \frac{[(1.645)(\text{CV}_T)(T_1^2 X_1^2 + \cdots + T_n^2 X_n^2)^{1/2}](\text{ECL}/\text{TWA})}{(T_1 + \cdots + T_n)(1 + \text{CV}_T^2)^{1/2}}$$

Then use the classification criteria of step 2. In this example:

$$\text{LDV}_{5\%} = 100 - \frac{[(1.645)(0.08)(300^2 \times 30^2 + 180^2 \times 40^2)^{1/2}](100/71)}{(300 + 180)(1 + 0.08^2)^{1/2}}$$

$$= 100 - 10.3 = 89.7 \text{ ppm}$$

which is reported as 90 ppm.

5. For a compliance officer, when the periods of the measurements are not approximately equal and TWA $>$ ECL, compute the noncompliance one-sided test statistic:

$$\text{UDV}_{5\%} = \text{ECL} + \frac{[(1.645)(\text{CV}_T)(T_1^2 X_1^2 + \cdots + T_n^2 X_n^2)^{1/2}](\text{ECL}/\text{TWA})}{(T_1 + \cdots + T_n)(1 + \text{CV}_T^2)^{1/2}}$$

Then use the classification criteria of step 3.

COMMENTS. The same as for Section 6.1.2, plus the following statistical note applicable to hypothesis testing for nonuniform exposures.

Statistical Note: The following notation is used. Let

μ_i = true exposure during ith exposure interval of the time-averaging period, $i = 1, 2, \cdots, n$

$\mu = (1/T)(T_1\mu_1 + T_2\mu_2 + \cdots + T_n\mu_n)$ = true TWA exposure

T_i = length of ith sampling interval

$T = T_1 + T_2 + \cdots + T_n$ = time-averaging period of TWA and associated ECL

X_i = exposure estimate for the ith exposure interval

To test H_0: $\mu \leq$ ECL (i.e., compliance exposure) against the alternative hypothesis H_1: $\mu >$ ECL (i.e., noncompliance exposure), a critical value must be selected for the measured exposure TWA. The critical value for a test of noncompliance is intended to be the TWA that would be exceeded at most only 5 percent of the time when H_0 is true. The correct critical value (denoted by $UDV_{5\%}$ for the upper decision value) is

$$UDV_{5\%} = ECL + [(1.645)(CV_T)] \left[\frac{\left(\sum_{i=1}^{n} T_i^2 \mu_i^2 \right)}{(T^2)(1 + CV_T^2)} \right]^{1/2}$$

where

$$\left(\frac{1}{T} \right) \sum_{i=1}^{n} T_i\mu_i = ECL$$

is a restriction on the μ_i's. A problem of this nonuniform case is that many combinations of μ_i's exist that all have the desired time-weighted average (ECL) under the null hypothesis H_0. The range of choices for μ_i's is discussed below, and our preferred selection is given for use in the classification decision formulas given elsewhere in Section 6.3.2.

1. Maximum statistical power (probability of rejecting H_0 when H_1 is true) would be obtained by choosing

$$\mu_i = \left(\frac{T}{T_i} \right) \left(\frac{ECL}{n} \right) \qquad i = 1, \cdots, n$$

However, if this choice is made, the probability of falsely rejecting the null hypothesis [i.e., of falsely deciding for noncompliance (H_1) when compliance (H_0) actually exists] would be higher than the intended 5 percent. This procedure might be acceptable for the type of "screening monitoring" that is followed by confirmatory sampling. But usually this choice of μ_i's under H_0 would be unsatisfactory because it gives a liberal test of noncompliance. It could also be shown to give a liberal test of compliance. It would usually not be a good choice for use by either the compliance officer or the employer.

2. Minimum statistical power to reject H_0 would be obtained by choosing

$$\mu_1 = \left(\frac{T}{T_1}\right) (\text{ECL}) \quad \text{and} \quad \mu_i = 0 \quad i = 2, \cdots, n$$

The index value $i = 1$ could be assigned to any of the n intervals. All the exposure is put into this single interval and none into the other intervals. This choice provides a conservative test with *less* than a 5 percent chance of making a type 1 decision error (falsely rejecting H_0). The advantage to making a noncompliance decision with this procedure is that the choice of μ_i's would be incontrovertible. Also, if an employer could show compliance using this procedure, there would be high assurance that the ECL was not exceeded on the day of the exposure measurement. However, the conservatism of this procedure would make it undesirable for routine use by a compliance officer because it provides less statistical power to detect true overexposure situations, and this could compromise worker protection.

3. On balance, it seems best to try to control the size of the hypothesis test as close to the intended 5 percent level as we can. To attempt this, we have substituted the following μ_i estimates into the UDV (or LDV) formulas given earlier in this statistical note:

$$\hat{\mu}_i = \left(\frac{\text{ECL}}{\text{TWA}}\right) (X_i) \quad i = 1, 2, \cdots, n$$

This procedure provides μ_i values that have ECL as their time-weighted average (under H_0), and the $\hat{\mu}_i$ values are in the same ratios to each other as are their sample estimates. The resulting formulas for UDV and LDV have been presented in Section 6.3.2.

6.4 Grab-Sample Estimate, Small-Sample Size (Less than 30 Measurements During the Time-Averaging Period)

Grab samples are samples taken during intervals that are short compared to the duration of the time-averaging period they are drawn from (e.g., intervals of seconds to about 2 min for a 15-min TWA period and up to about 30 min for TWA periods of 8-hr). Grab-sample measurements should be taken at random intervals during the desired time-averaging period (see Technical Appendix F of Leidel et al., 1977).

6.4.1 Confidence Limits for a True TWA Exposure, Small Number of Grab Samples

PURPOSE. Compute one- or two-sided confidence limits for an exposure estimate based on less than 30 grab samples.

ASSUMPTIONS. Unfortunately, there are no simple statistical methods available to determine exact one- or two-sided confidence limits for an arithmetic mean that is estimated from a small sample size (less than 30) of grab samples as an exposure estimate. This is because grab-sample data generally reflect a lognormal distribution of true exposure levels among the short intervals that are sampled (see Section 3.6.2.2). Thus the statistical problem is to estimate confidence limits for the arithmetic mean of a lognormal distribution. It is possible to easily compute one- or two-sided exact confidence limits for the geometric mean of a lognormal distribution of grab-sample results. But it is important to note that a geometric mean exposure is not the most suitable parameter for estimation of worker exposure risk. Therefore, this section will present a well-known approximate procedure for confidence limits on the true arithmetic mean exposure. The geometric mean is also an inappropriate parameter for comparison to an exposure control limit (ECL). Fortunately, an exact exposure classification procedure for grab-sample estimates was made available by NIOSH (Leidel et al., 1977; Bar-Shalom et al., 1975), which is presented in Section 6.4.2. Also, more recently, Armstrong (1992) has provided tables of factors that facilitate the calculation of exact confidence limits for the arithmetic mean of a lognormal distribution. Refer to the notes at the end of Section 6.4.2.

For small sample sizes, Aitchison and Brown (1957) state that statistical theory fails to provide a means of obtaining exact confidence limits or intervals for the arithmetic mean (μ_X) of a lognormally distributed variable X. For large sample sizes, their suggestion is to compute the limits:

$$\overline{X} \pm \frac{Z_P s_X}{n^{1/2}}$$

where $Z_P = 1.96$ for two-sided confidence limits at the $P = 0.95$ confidence level and s_X is the sample standard deviation computed from a random sample of n lognormally distributed X's (e.g., concentrations in n grab samples taken at random intervals). See Section 6.5.1 for our recommended slight modification to the Aitchison and Brown (1957) large-sample ($n \geq 30$) confidence limits. We have modified the formula only slightly by using Student's t statistic, $t_{P, n-1}$, in place of the standard normal deviate, Z_P.

For the small-sample case ($n < 30$), which is the subject of this section, no simple theory for exact confidence limits on μ_X is available. However, such confidence limits are not needed to perform statistical tests of significance under null hypotheses of compliance ($\mu_X \leq$ ECL) and noncompliance ($\mu_X >$ ECL). The hypothesis tests concerning μ_X, the arithmetic mean concentration, can be carried out with the sample mean ($\overline{y}$) and sample standard deviation (s_y) of the logarithmic transformation ($y = \ln X$). Detailed explanations of two special methods for testing hypotheses about μ_X are given in Section 6.4.2. In case confidence limits on μ_X are desired in the small-sample case, an often-used approximate method is outlined here.

SOLUTION

1. First, for the ln-transformed data, compute exact confidence limits on μ_y, the true mean of $y = \ln X$. These exact limits are given by

$$\bar{y} - \frac{t_{P,n-1}s_y}{\sqrt{n}} < \mu_y < \bar{y} + \frac{t_{P,n-1}s_y}{\sqrt{n}}$$

where $t_{P,n-1}$ is the Student's t statistic for $(n-1)$ degrees of freedom for the P level of significance. For example, for two-sided 95 percent confidence limits and a sample size of 10, $t_{0.95,9} = 2.262$.

2. Then, detransform each logarithmic-value confidence limit on μ_y to a corresponding value of μ_X using the relationship:

$$\mu_X = \exp{(\mu_y)} \exp{\left(\frac{\sigma_y^2}{2}\right)} \approx \exp{(\mu_y)} \exp{\left(\frac{s_y^2}{2}\right)}$$

The approximation is due to the necessary use of the sample variance s_y^2 instead of the true variance σ_y^2, to perform the detransformation (i.e., to convert the geometric mean of X [$GM_X = \exp{(\mu_y)}$] to the corresponding arithmetic mean of $X(\mu_X)$). This approximate method is adequate for most practical industrial hygiene applications.

COMMENTS. This approach is graphically illustrated in Figure 10.14.

6.4.2 Classification of Exposure (Hypothesis Testing), Small Number of Grab Samples

PURPOSE. Based on multiple grab samples, classify an exposure estimate as non-compliance exposure, possible overexposure, or compliance exposure relative to an exposure control limit (ECL).

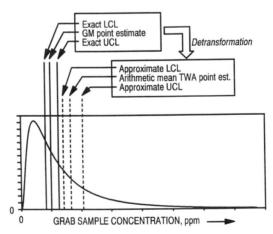

Figure 10.14 Two-sided confidence limits for a true TWA exposure estimated with a grab-samples exposure measurement.

ASSUMPTIONS. Note: *This method cannot process zero data values.* It is assumed that none of the sample results are zero. Refer to Technical Appendix I of Leidel et al. (1977) for a discussion of how to treat a sample population that includes zero values.

The grab-sample measurements are drawn from a single two-parameter lognormal distribution of true exposure levels. If it is suspected that the parent population of exposure levels is more accurately described by a three-parameter lognormal distribution (see Section 3.5.2), then the location constant k should be estimated (see Section 6.12, Step 5 of the Solution) and subtracted from all measurements and the *ECL* before the method in this section is used.

Grab-sample populations also have a component of approximately normally distributed, measurement-procedure error, but this component is assumed negligible compared to the lognormally distributed environmental variations in the true exposure levels sampled. The net effect of measurement errors is to increase slightly the variability of the lognormal distribution without changing its shape appreciably (Busch and Leidel, 1988).

The true arithmetic mean TWA exposure (i.e., true arithmetic mean of a 2-parameter lognormal distribution of all exposure levels that occurred at the same sampling position over the duration of the TWA period) is assumed unknown, and the geometric standard deviation parameter of the parernt lognormal expsoure distribution is also assumed unknown. Both are estimated from the sample population of grab-sample results.

The grab-sample measurements are assumed to be a random sample of all exposure levels that occurred at the sampling position during the TWA period of interest. One should not attempt to estimate an 8-hr TWA-exposure based on short samples selected at random from only a small portion of the TWA period (e.g., the last 2 hr). The sample periods from the TWA period should be chosen as a random sample from the entire TWA period.

The statistical theory for the method in this section is contained in Bar-Shalom et al. (1975).

EXAMPLE. A personal pump and 8 charcoal tubes were used to estimate a worker's 8-hour TWA-exposure to ethyl alcohol. Each tube was placed on the worker for 20 min. The 20-min periods were randomly selected from the 24 possible nonoverlapping periods during the workshift. The 8-hr *ECL* used by the employer is 1000 ppm. The following results were reported by the company laboratory: $X_1 = 1225$ ppm, $X_2 = 800$ ppm, $X_3 = 1120$ ppm, $X_4 = 1460$ ppm, $X_5 = 975$ ppm, $X_6 = 980$ ppm, $X_7 = 525$ ppm, and $X_8 = 1290$ ppm. With 95 percent confidence, classify the exposure estimate, TWA = 1058 ppm, relative to the ECL of 1000 ppm.

SOLUTION

1. Compute a standardized value (x_i) for each X_i sample result by dividing by the ECL, and then compute the base 10 logarithm (y_i) of each x_i. That is, $y_i = \log_{10}(X_i/\text{ECL})$.

2. In this example:

Original Sample Results X_i (ppm)	Standardized Results X_i/ECL	$y_i = \log_{10}(X_i/ECL)$
1225	1.225	0.0881
800	0.800	−0.0969
1120	1.120	0.0492
1460	1.460	0.1644
975	0.975	−0.0110
980	0.980	−0.0088
525	0.525	−0.2798
1290	1.290	0.1106

3. Compute the classification variables, $\bar{y}$, s, and n. First compute the arithmetic mean $\bar{y}$ of the logarithmic values y_i; then compute the standard deviation s of the y_i values. The number of grab samples used in the computations is n.

In this example $\bar{y} = 0.0020$, $s = 0.1400$, and $n = 8$. These three variables (sample estimates of two parent population parameters and the sample size) will be used in the classification procedure.

4. Using the classification chart in Figure 10.15 for the average of several grab-sample measurements, plot the classification point that has coordinates $\bar{y}$ and s. A family of curves form the boundaries of three classification regions. Each pair of boundaries is for a given sample size n. Pairs of classification boundaries for odd values of n ranging from $n = 3$ to 25 are provided (all even sample sizes except for 4 will have to be interpolated).

5. Classify the exposure estimate:

a. If the classification point lies on or above the upper curve of the pair corresponding to the number of measurements n (i.e., the n-curve pair), then classify as noncompliance exposure.

b. If the classification point lies below the lower curve of the n-curve pair, then classify as compliance exposure.

c. If the classification point is between the two curves for sample size n, then classify as possible overexposure.

d. If s exceeds 0.5 (greater than the range of the classification chart), a possible cause is that one or more of the measurements is relatively distant from the main portion of the sample distribution (i.e., an outlier that could be a mistake). Another explanation is that two or more substantially different exposure level distributions are being mistakenly analyzed as a single distribution (see Section 3.3.4).

6. In this example the classification point lies between the upper interpolated $n = 8$ curve and the lower interpolated $n = 8$ curve. Thus the TWA exposure estimate of 1058 ppm, estimated from the 8 grab-sample measurements, is classified as a

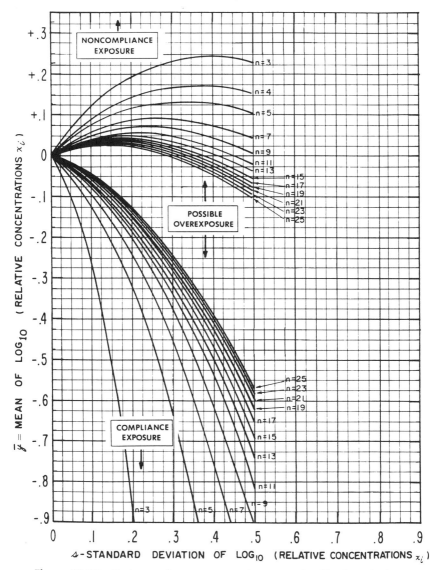

Figure 10.15 Grab-sample, measurement-average classification chart.

possible overexposure relative to the ECL of 1000 ppm. That is, the estimate is not precise enough to be able to confidently say that the true exposure was not in compliance.

COMMENTS. The x_i ratios (X_i/ECL) computed in step 1 of the solution are standardized exposure values that make the concentrations of contaminant independent of the ECL (in concentration units). This enables us to use the same decision chart for any concentration level. All values $x_i = (X_i/\text{ECL})$ are comparable to a single scale of compliance with an ECL of unity for x_i. That is, the ECL for the transformed variable x will always be unity, which corresponds to zero on the $y = \ln(x)$ ordinate scale of the classification chart in Figure 10.15.

Rappaport and Selvin (1987) emphasize that most 8-hr TWA PELs were formulated for the purpose of indirectly limiting corresponding *long-term* average exposures. They present a formal statistical significance test for noncompliance that compares an estimate of the long-term arithmetic average exposure to a permissible exposure level. The test implicitly assumes lognormal distributions of both interday and interworker 8-hr TWA exposures, and the test statistic (denoted by T) is calculated from 8-hr TWA exposure measurements made on a sample of workdays selected at random during a period of one year or so. They state that different employees should be measured on the different sampling days, selected at random from within a "uniform exposure group."

While we do not agree that a long-term average PEL is a feasible regulatory concept, we do note that the T statistic can also be adapted to our present purpose of testing for a significant difference between an 8-hr permissible exposure level (PEL) and the arithmetic mean (μ_c) of a lognormal distribution of grab-sample exposure levels. The proposed test statistic T is given by

$$ T = \frac{\bar{x}_c - \text{PEL}}{s_{\bar{x}_c}} $$

where

$$ \bar{x}_c = e^{[\bar{x}_{\ln(c)} + 1/2(s_{\ln(c)})^2]} \qquad s_{\bar{x}_c} = \sqrt{\frac{(\text{PEL})^2[(s_{\ln(c)})^2 + \frac{1}{2}(s_{\ln(c)})^4]}{n-2}} $$

$\bar{x}_c$ is the maximum-likelihood estimator of true mean μ_c, $s_{\bar{x}_c}$ is the estimated standard deviation of $\bar{x}_c$, and $\bar{x}_{\ln(c)}$ and $s_{\ln(c)}$ are the sample mean and sample standard deviation of a natural logarithmic transform [$\ln(c)$], calculated from n concentration measurements (denoted by c).

With respect to the size of the test (i.e., the probability α of a type I error), for $n > 5$ and GSD ≤ 3.0, an adequate approximation to the distribution of the T statistic is said to be Student's t distribution with $(n - 1)$ degrees of freedom.

The power of this T test [i.e., the probability $(1 - \beta)$ of obtaining a statistically significant difference when a hypothetical true difference exists] was evaluated in Rappaport and Selvin (1987), both for an employer's test and for a compliance officer's test. For a test with 5 percent size ($\alpha = 0.05$), Table 10.8 provides approximate minimum sample sizes (n) that are needed to attain at least 90 percent power [i.e., $(1 - \beta) \geq 0.90$]. The formula below for n is the basis for computed values of n that correspond to central tendency and variability parameters of various assumed lognormal distributions of concentrations.

Table 10.8 Minimum Values of n Needed for $\alpha = 0.05$ and $(1 - \beta) \geq 0.90$

	μ_0/PEL	GSD = 1.5	GSD = 2.0	GSD = 2.5	GSD = 3.0	GSD = 3.5
Compliance	0.10	2	6	13	21	30
	0.25	3	10	19	30	43
	0.50	7	21	41	67	96
	0.75	25	82	164	266	384
Noncompliance	1.25	25	82	164	266	384
	1.50	7	21	41	67	96
	2.00	2	6	11	17	24
	3.00	1	2	3	5	6

$$n \cong \frac{[Z_{(1-\alpha)} + Z_{(1-\beta)}]^2 \{[\ln(\text{GSD})]^2 + \frac{1}{2}[\ln(\text{GSD})]^4\}}{\left(1 - \dfrac{\mu_c}{\text{PEL}}\right)^2}$$

where $\ln(\text{GSD}) = \sigma_{\ln(c)} =$ standard deviation of $\ln(c)$
$\mu_c/\text{PEL} =$ arithmetic mean of c/permissible exposure limit

For a small sample ($n < 20$) of lognormally distributed air concentrations, Selvin and Rappaport (1989) state that the simple arithmetic mean is a less-variable estimator of the population mean than the maximum-likelihood estimator for purposes of estimation of average occupational exposure. For larger sample sizes, even without use of the bias correction factor given in Aitchison and Brown (1957), Selvin and Rappaport state that the maximum-likelihood estimate (MLE) provides a more precise estimator of the arithmetic mean. However, for *all* sample sizes, it is the *MLE that must be used* in order for available statistical decision theory to be applicable to making controlled-risk decisions of compliance and noncompliance with exposure control limits. Additional theory is given in Aitchison and Brown (1957), which shows that for any sample size the MLE, combined with a correction for bias, provides a minimum variance unbiased estimate of the arithmetic mean of a lognormal distribution.

Therefore, although it is less convenient to apply than the T test, we recommend uniform use of the bias-corrected MLE estimate of μ_c for making statistical decisions of compliance or noncompliance in a legal context. The graphical decision chart procedure given in Leidel et al. (1977) (and presented earlier in this section) eases the clerical task of using the bias-corrected MLE estimate of the mean of lognormally distributed grab samples to make tests of compliance or noncompliance.

Armstrong (1992) has published a useful table of factors that facilitate computation of exact confidence limits for the arithmetic mean of a lognormal distribution. Confidence limits for the true arithmetic mean are obtained by multiplying the appropriate factors, which are listed as function of geometric standard deviation (GSD) and sample size (n), times the estimated geometric mean (GM). The factors were determined by Armstrong from theory and tables published by Land (1973, 1988). Armstrong then used the resulting confidence limits to perform a significance test that is equivalent to the graphical exposure classification method given for small numbers of grab samples in Leidel et al. (1977). Individual readers may or may not prefer using an exposure classification method that is based on confidence limits in preference to our graphical decision charts.

Armstrong (1992) compares examples of exact confidence limits with corresponding approximate confidence limits computed by four different methods, including methods presented earlier in this chapter. He concludes that adequate accuracy for practical applications requires calculation of the exact limits, rather than any of the approximate limits, whenever GSD > 1.5 or $n < 25$.

6.5   Grab-Sample Estimates of the TWA Exposure, Large-Sample Size (30 or More Measurements During the Time-Averaging Period)

Usually one collects far fewer than 30 measurements during an 8-hr TWA period, or 15-min TWA period, because of the cost of each measurement (even for inexpensive ones such as with colorimetric tubes) and limited availability of personnel to take the measurements. However, if one has a direct-reading instrument available for the contaminant of interest, then it is feasible to obtain more than 30 samples during the desired TWA period. If the larger number of samples can be taken at random intervals, this strategy is preferable to the small-sample size approach (less than 30) discussed in the previous section, since for larger sample sizes the uncertainty regarding the true TWA exposure is considerably less. Additionally, for sample sizes of 30 or more, the statistical analysis is less complex because the distribution of the average of the exposure measurements is adequately described by a normal distribution. This section will present methods for computing both confidence limits and hypothesis tests for TWA exposure estimates that are calculated from large samples. One does not have to calculate the logarithms of the standardized measurements (as was necessary in Section 6.4 for TWA estimates based on small sample sizes). The hypothesis tests are less complex, and this method can process zero data values, unlike the procedure in Section 6.4.

6.5.1   Confidence Limits for a True TWA Exposure, $\geq$ 30 Grab Samples

PURPOSE. Compute one- or two-sided confidence limits for an exposure estimate based on 30 or more grab samples.

ASSUMPTIONS. This procedure is robust regarding the actual distribution of exposure levels occurring during the TWA period of the grab-sample measurements. Both the true exposure level (i.e., arithmetic mean of the distribution of all exposure levels that occurred at the sampling position over the duration of the TWA period) and the standard deviation of the parent exposure distribution are assumed unknown. Both are estimated from the results obtained from grab samples.

The grab-sample measurements are assumed to be a random sample of all exposure levels that occurred at the sampling position during the TWA period of interest.

One should not attempt to estimate an 8-hr TWA exposure average based on short samples selected at random from only a small portion of the TWA period (e.g., the last 2 hr). The sample periods from the TWA period should be chosen as a random sample from the entire TWA period.

EXAMPLE. A direct-reading ozone meter with strip chart recorder was used to continually measure a worker's exposure to ozone. The following 35 measurement values were read off the strip chart record at 35 times randomly selected within the 8-hr period. All values are given in ppm.

0.084	0.062	0.127	0.057	0.101	0.072	0.077
0.0145	0.084	0.101	0.105	0.125	0.076	0.043
0.079	0.078	0.067	0.073	0.069	0.084	0.061
0.066	0.085	0.080	0.071	0.103	0.075	0.070
0.048	0.092	0.066	0.109	0.110	0.057	0.107

Compute the two-sided 95 percent confidence limits for the true TWA exposure based on the sample mean estimate of 0.0794 ppm ozone.

SOLUTION

1. Compute the arithmetic mean $\overline{X}$ and standard deviation s of the $n = 35$ measurements X_i. Here, $\overline{X} = 0.0794$ ppm and $s = 0.0233$ ppm.

2. Compute the two-sided 95 percent confidence interval (limits $LCL_{2,0.95}$ and $UCL_{2,0.95}$) for the true TWA exposure, based on the TWA exposure estimate $\overline{X}$, from the equations:

$$LCL_{2,0.95} = \overline{X} - \frac{(t_{0.95,n-1})(s)}{\sqrt{n}}$$

$$UCL_{2,0.95} = \overline{X} + \frac{(t_{0.95,n-1})(s)}{\sqrt{n}}$$

Note that the equation factor $t_{0.95,34} = 2.032$ will differ for sample sizes other than 35. See the comments section that follows the solution.

3. In this example:

$$LCL_{2,0.95} = 0.0794 - \frac{(2.032)(0.0233)}{(35)^{1/2}}$$

$$= 0.0794 - 0.0080 = 0.071 \text{ ppm}$$

$$UCL_{2,0.95} = 0.0794 + \frac{(2.032)(0.0233)}{(35)^{1/2}}$$

$$= 0.0794 + 0.0080 = 0.087 \text{ ppm}$$

Thus the two-sided 95 percent confidence interval for the true TWA exposure, based on an exposure estimate of 0.079 ppm obtained from 35 samples, is 0.071 to 0.087 ppm. That is, we can be 95 percent confident that the true TWA exposure is between 0.071 and 0.087 ppm.

COMMENTS. The factor 2.032 in the equations for $LCL_{2,0.95}$ and $UCL_{2,0.95}$ is taken from the Student's t table for $(n - 1) = 34$ degrees of freedom. For the large-sample (≥ 30) n values appropriate for this procedure, the normal distribution Z value of 1.960 can be used as an approximation to Student's t.

The indicated uncertainty in the TWA exposure estimate, as reflected in the width of the confidence interval (± 10 percent of the TWA estimate in this example), does not incorporate uncertainty due to interday exposure level variability (see Sections 3.3.1 and 3.6.2.3). The confidence interval width reflects only the intraday variability of the exposure levels occurring over the 8-hr work shift period of the grab-sample measurements. That is, the computed confidence interval is valid only for the worker's TWA exposure on the day and during the period of the grab samples, at the same sampling position.

6.5.2 Classification of Exposure (Hypothesis Testing), ≥ 30 Grab Samples

PURPOSE. Classify a TWA exposure estimate based on 30 or more grab-sample measurements as noncompliance exposure, possible overexposure, or compliance exposure relative to an exposure control limit (ECL).

ASSUMPTIONS. The same as used for Section 6.5.1.

EXAMPLE. The same data as used for Section 6.5.1, except classify with 95 percent confidence the TWA exposure estimate regarding an ECL of 0.1 ppm.

SOLUTION

1. As in Section 6.5.1, compute the arithmetic mean $\overline{X}$ (the TWA estimate) and standard deviation s of the $n = 35$ measurements X_i. The results are $\overline{X} = 0.0794$ ppm, $s = 0.0233$ ppm.

2. For an employer, if $\overline{X} > ECL$, no statistical test for compliance need be made, since the exposure estimate exceeds the ECL. If $\overline{X} \leq ECL$, compute the one-sided lower decision value for comparison with the sample mean:

$$LDV_{1-P} = ECL - \frac{(t_{P,n-1})(s)}{(n)^{1/2}}$$

For this example, $LDV_{5\%} = 0.1000 - \frac{(1.691)(0.0233)}{(35)^{1/2}}$

$$= 0.093 \text{ ppm}$$

Note that the factor 1.645 from the normal distribution's Z values could have been used as an approximation to the correct Student's t value of 1.691 [for $(n - 1) = 34$ degrees of freedom and $1 - P = 0.05$ significance level for a one-sided test].

Then classify the exposure estimate based on the TWA estimate X of a single worker's TWA exposure on a single day according to:

a. If $\overline{X} \leq LDV_{5\%}$, classify as compliance exposure.
b. If $\overline{X} > LDV_{5\%}$, classify as possible overexposure.

Since the $\overline{X} = 0.079$ ppm is less than the $LDV_{5\%}$ of 0.093 ppm, the employer can state that for the measured worker, on the day and during the period of the TWA exposure estimate, the exposure estimate was a statistically significant compliance exposure at the 5 percent significance level. To check for even higher degrees of statistical significance, $LDV_{1\%}$ and $LDV_{0.1\%}$ could also be computed by using $t_{0.99, n-1}$ and $t_{0.999, n-1}$, respectively, instead of $t_{0.95, n-1}$ in the equation for LDV.

3. For a compliance officer, in this example TWA $<$ ECL, so no statistical test for noncompliance need be made since the TWA estimate $\overline{X}$ is less than the ECL and hence certainly less than the UDV. For $\overline{X} >$ ECL, the compliance officer would compute the noncompliance one-sided upper decision value for comparison with the sample mean:

$$ UDV_{5\%} = ECL + \frac{(t_{0.95, n-1})(s)}{(n)^{1/2}} $$

where $t_{0.95, n-1}$ is the 95th percentile (one-sided) of Student's t distribution with ($n - 1$) degrees of freedom. Note that 1.645 can be used as an approximation for $t_{0.95, n-1}$, since $n \geq 30$ when this procedure is used. Then classify the TWA exposure estimate $\overline{X}$ according to:

a. If $\overline{X} > UDV_{5\%}$, classify as noncompliance exposure.
b. If $\overline{X} \leq UDV_{5\%}$, classify as possible overexposure.

COMMENTS. Figure 10.13 is a graphical interpretation of decision regions using LDVs for employer decisions and UDVs for compliance officer decisions. The observed value $\overline{X}$ for the TWA exposure estimate is denoted by TWA inside a diamond in Figure 10.13.

Note the limited validity of these tests. The outcomes of these hypotheses tests are valid only for the worker's exposure on the day and during the period of the grab-sample exposure measurements, at the same sampling position.

6.6 Tolerance Limits for a Normally Distributed Variable with Unknown Parameters

PURPOSE. Based on a random sample of n observations, compute one- or two-sided tolerance limits that have a desired γ level of confidence (probability) that the tolerance interval they bound will contain a specified proportion P of the normal distribution represented by the random sample.

ASSUMPTIONS. The parent population for which the tolerance limit(s) statement will be made is assumed to be adequately described by a normal distribution.

Both the true mean and true standard deviation parameters of the parent normal distribution are assumed to be unknown. These two distributional parameters are estimated by sample values, $\overline{X}$ and s, computed from a random sample of n observations.

A tolerance limit can be thought of as a confidence limit for a designated percentile of the parent distribution of single observations. However, note that this procedure for estimating tolerance limits is less "robust" (against lack of normality of the parent population) than procedures for estimating confidence limits on the arithmetic mean distributional parameter. This is true because sample means tend to have a "bell-shaped" (nearly normal) distribution even if the single observations were not from a normal distribution (at least for moderate-to-large sample sizes). Thus, one must be cautious and avoid overinterpretation of tolerance limits. That is, tolerance limits should not be used for "fine-line" decision making. A useful rule-of-thumb is that they probably should not be reported to more than two significant figures, which generally is the usual level of precision for industrial hygiene measurements.

The primary use of these methods should be to obtain an indication of the potential upper values of a parent distribution (e.g., higher-valued exposure measurements by computation of a one-sided upper tolerance limit for the upper-95th-percentile exposure measurement) or an indication of the potential width of the parent distribution (e.g., potential range of exposure measurements by computation of a two-sided tolerance interval for the central 90 percent of exposure measurements).

EXAMPLE. An exposure measurement procedure was used to repeatedly measure a calibrated reference concentration of 100 ppm. The six replicate measurements of the 100 ppm were reported as 95.7, 90.9, 109.4, 107.6, 101.1, and 84.7 ppm. Compute a two-sided, $100\gamma = 95$ percent confidence level, tolerance interval for the central $P = 95$ percent of the parent distribution of possible measurements from a "true" concentration of 100 ppm.

SOLUTION

1. The mean will not be assumed known since any measurement method systematic error could cause the true mean of measurements to differ from the true concentration of 100 ppm. Compute the arithmetic mean $\overline{X}$ and standard deviation s of the $n = 6$ measurements X_i. Here, $\overline{X} = 98.2$ ppm and $s = 9.63$ ppm.

2. Compute the two-sided, 95 percent confidence level, upper and lower tolerance limits (UTL, LTL) for the central 95 percent of the parent population from the equations:

$$\text{UTL}_{2,\gamma,P} = \overline{X} + (K)(s) \quad \text{and} \quad \text{LTL}_{2,\gamma,P} = \overline{X} - (K)(s)$$

where K is a factor obtained from appropriate statistical tables for two-sided tolerance limits for a normally distributed variable, which is given as a function of the following three independent variables:

a. Sample size ($n = 6$ in this example)

b. Confidence level (e.g., $\gamma = 0.95$)

c. Proportion of the population (e.g., $P = 0.95$)

3. For this example $K = 4.414$ and the $UTL_{2,0.95,0.95}$ calculations are

$$UTL_{2,0.95,0.95} = 98.2 + (4.414)(9.63)$$

$$= 98.2 + 42.5$$

$$= 140.7 \text{ ppm}$$

which is reported as 141 ppm after appropriate rounding. Then compute

$$LTL_{2,0.95,0.95} = 98.2 - (4.414)(9.63)$$

$$= 98.2 - 42.5$$

$$= 55.7 \text{ ppm}$$

which is reported as 56 ppm after rounding.

4. Thus the two-sided tolerance interval (95 percent confidence level, central 95 percent of the parent population) is 56 to 141 ppm. That is, based on the frugal sample of six replicate measurements of a 100-ppm reference concentration, the best we can say about the measurement procedure is that we can be 95 percent confident that 95 percent of future results with this procedure at 100 ppm will lie between 56 and 141 ppm (under conditions similar to those during which the replicate samples were obtained).

COMMENTS. Tolerance limits (and tolerance interval) can be thought of as γ-level specialized confidence limits (and confidence interval) for a fractile interval containing the designated proportion P of the parent distribution.

Note that we cannot say that there is a 95 percent probability of 95 percent of the parent distribution lying in the tolerance interval 56 to 141 ppm. For any given tolerance interval such as this one, the computed interval either does or does not contain the stated percentage of the parent population. However, since 95 percent (100γ) of such tolerance intervals would each contain at least 95 percent ($100P$) of the parent population, the chances are 19 to 1 that this particular tolerance interval does in fact have 95 percent or more of the distribution.

Frequently researchers or writers will erroneously state that 95 percent of their population of sample results will lie in the interval $(\bar{X} - 2s)$ to $(\bar{X} + 2s)$. (Generally 2.000 is used as an approximation to the more exact value of 1.960 for the standard normal deviate.) This type of inferential statement is true only for large (i.e., in the hundreds) sample sizes. Typically, researchers only have small sample sizes (e.g., less than 30) and the K factors for the appropriate two-sided tolerance limit computations are substantially greater than 2. Thus it is possible to substantially underestimate the width of a parent distribution from small sample sizes if the proper tolerance interval computations are not performed. If one had naively estimated the central 95 percent of the results as bounded by $(\overline{X} \pm 2s)$, then one would have substantially underpredicted the possible range of future measurement method results as about 79 to 117 ppm.

The range of the six results was about 91 to 109 ppm, yet the two-sided tolerance interval for 95 percent of the results was 56 to 141 ppm. This wide tolerance interval is partially due to the frugal sample size used for its computation. However, it also demonstrates how poor an indicator the range of a small sample can be for the span of a normal distribution.

6.7 Tolerance Limits for a Lognormally Distributed Variable with Unknown Parameters

PURPOSE. Based on n samples taken at random, compute one- or two-sided tolerance limits that have a desired γ level of confidence that the tolerance interval they bound will contain a desired proportion P of the lognormal distribution from which the sample was taken.

ASSUMPTIONS. There is a value of the response variable for each member of a large parent population of "sampling units" to which the tolerance limits statement will apply. Examples of sampling units could be replicate workers (same sampling period and same type of work) or repeated exposure periods for a given worker. The corresponding population of exposure measurements is assumed to be adequately described by a single 2-parameter lognormal distribution. If it is suspected that the parent population's exposure measurements are more accurately described by a three-parameter lognormal distribution (see Section 3.5.2), then the location constant k for the parent distribution should be estimated (see Section 6.12, step 5 of the solution) and subtracted from all sample values before the method in this section is used.

Both the geometric mean and geometric standard deviation of the parent population distribution are unspecified and assumed unknown. Both distributional parameters are to be estimated from response variable measurements made on a random sample of n sampling units selected from the parent population.

This method cannot process zero data values. It is assumed that none of the sample results are zero (or less than zero). Refer to Technical Appendix I of Leidel et al. (1977) for a discussion of how to treat a sample population that includes zero values.

Note that this procedure for estimating tolerance limits is less robust than procedures for estimating confidence limits for the geometric mean distributional parameter. The reason for this is explained in Section 6.6 relevant to tolerance limits for a normally distributed variable. Similar comments would apply here, since tolerance limits for a lognormally distributed variable X are merely a detransformation of tolerance limits for the normally distributed variable $y = \ln X$. The robustness that exists for the sample mean $(\overline{y})$ of the log transformation also exists for its detransformation $[\text{GM}_X = \text{antilog}\,(\overline{y})]$.

EXAMPLE. Five workers were randomly selected from a target population of workers. Each selected worker's 8-hr TWA exposure was measured on a different workday, randomly selected within a 6-month period. The five results were reported as 11.1, 10.6, 21.4, 3.9, and 4.9 ppm. Compute the one-sided, 95 percent confidence level, upper tolerance limit for the lower 95 percent of a parent distribution of TWA exposures that is defined by randomly selected workers measured on randomly selected days.

SOLUTION. Assume that general (i.e., group average) exposure levels (X) for the work environment are lognormally distributed among workdays and that percentage differences among individual workers' exposures are similar on each workday. Workers' mean exposure levels are also lognormally distributed. Then, the net variability due to both sources of variability (i.e., due to days and workers) is also lognormally distributed. Under this model, the following analysis is appropriate.

1. Compute the base 10 logarithm, $y_i = \log X_i$, of each sample value. Here the five logarithmic values y_i are 1.045, 1.025, 1.330, 0.591, and 0.690.

2. Compute the sample mean $\bar{y}$ and standard deviation s of the $n = 5$ logarithmic values. Here, $\bar{y} = 0.936$ and $s = 0.298$. As supplementary information, one can compute the sample geometric mean GM of X by taking the antilog$_{10}$ of $\bar{y}$ (i.e., $10^{0.936} = 8.63$ ppm) and the sample geometric standard deviation GSD by taking the antilog$_{10}$ of s (i.e., $10^{0.298} = 1.986$).

3. Compute the one-sided, 95 percent confidence level, upper tolerance limit (UTL$_{1,0.95,0.95}$) for the lower 95 percent of the defined population of worker-day combination exposure levels (or compute LTL$_{1,0.95,0.95}$, the lower 95 percent tolerance limit for the upper 95 percent of the distribution). Use the following equations:

$$\text{UTL}_{1,0.95,0.95} = \text{antilog}_{10} \left[\bar{y} + (K)(s) \right]$$

$$\text{LTL}_{1,0.95,0.95} = \text{antilog}_{10} \left[\bar{y} - (K)(s) \right]$$

where K is a factor obtained from appropriate statistical tables for one-sided tolerance limits for a normally distributed variable.

4. For the values of this example $K = 4.202$ and the calculations are

$$\text{UTL}_{1,0.95,0.95} = \text{antilog}_{10} \left[0.936 + (4.202)(0.298) \right]$$

$$= \text{antilog}_{10} [2.188] = 10^{2.188}$$

$$= 154 \text{ ppm}$$

5. Thus the one-sided upper tolerance limit (95 percent confidence level, lower 95 percent of the defined population) is 154 ppm. That is, based on the frugal sample of only 5 exposure estimates over a 6-month period from the target population of workers, the best we can say is that we are 95 percent confident that 95 percent of the daily exposure averages over the 6-month period were below 154 ppm.

COMMENTS. This type of one-sided upper tolerance limit can be thought of as a one-sided upper confidence limit for the true value of a lognormally distributed random variable that exceeds a specified proportion (percentile) of the population of values. This concept is graphically illustrated in Figure 10.16 for a lognormally distributed exposure population. A companion technique is presented in the following section for computing a point estimate (and associated confidence limits) for the true proportion of a lognormal distribution exceeding a specified value of the random variable (e.g., proportion of all exposures exceeding a specified exposure level such as a PEL or TLV).

Note that we cannot say that there is a 95 percent probability that 95 percent of the parent distribution was below 154 ppm. For any given computed one-sided upper tolerance limit such as this one, the limit either does or does not exceed the stated percentage of the parent population. However, since 95 percent of such upper tolerance limits would each exceed at least 95 percent of the parent population, chances are at least 19 to 1 that this particular upper tolerance limit does in fact exceed 95 percent or more of the distribution.

The range of the five results was 4 to 21 ppm, yet the one-sided upper tolerance limit for the lower 95 percent of the results was 154 ppm. The high $UTL_{1,0.95,0.95}$ is partially due to the frugal sample size used for its computation. Note that the tolerance limit K factor, which is 4.202 in this example, would be only about half as large if n were 50 instead of the 5 in this example. This demonstrates how poor an estimator the range of a small sample is for the span of a lognormal distribution, particularly since the distribution is skewed toward higher values.

To be able to partition the total exposure variability GSD into components due to exposure variability between workers and exposure variability between days, it would be necessary to measure exposures of several workers on each of several days and examine the data by an appropriate analysis of variance method (ANOVA), using a logarithmic transformation of the exposure concentrations as the response variable.

Finally, note that in some real exposure settings, there may be "interaction" (i.e., lack of independence) between interworker exposure variations and interday exposure variations. In such cases, the interworker exposure distributions would

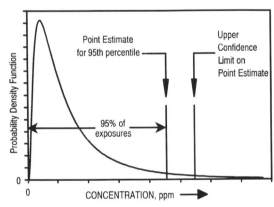

Figure 10.16 One-sided, 95 percent confidence level, upper tolerance limit for the lower 95 percent (95th percentile) of a lognormally distributed variable with unknown parameters.

have to be determined separately for each workday. For the solution to the example, it was assumed that equal percentage differences among individual workers' true exposures exist on each workday, which is equivalent to assuming "no interaction" on the scale of the logarithmic transformation.

6.8 Point Estimate and Confidence Limits for the Proportion of a Lognormally Distributed Population (Unknown Parameters) Exceeding a Specific Value

PURPOSE. Compute a point estimate P and confidence limits for the true proportion π of a two-parameter lognormal distribution exceeding some specified value (e.g., exposure control limit), when given a random sample from the distribution.

ASSUMPTIONS. Values of the response variable (e.g., single or multiple worker 8-hr TWA exposure estimates) for the parent population are assumed to be adequately described by a single two-parameter lognormal distribution. If it is suspected that the parent population is more accurately described by a three-parameter lognormal (see Section 3.5.2), then the location constant k for the parent distribution should be estimated (see Section 6.12, step 5 of the solution) and subtracted from all sample values before the method in this section is used.

Both the geometric mean (GM) and geometric standard deviation (GSD) of the parent population distribution are unspecified and assumed unknown. Both distributional parameters are to be estimated from response variable measurements made on a random sample of n sampling units selected from the parent population.

Note: This procedure cannot process zero data values. It is assumed that none of the sample results are zero (or less than zero). Refer to Technical Appendix I of Leidel et al. (1977) for a discussion of how to treat a sample population that includes zero values. Note that this procedure for computing a point estimate P of the actual proportion π of a lognormal distribution exceeding a specified value is less robust than procedures for estimating the geometric mean distributional parameter. One must be cautious and avoid overinterpretation of these point estimates and their confidence limits. A point estimate of distributional tail area is sensitive to departures from the assumption of lognormality, and these estimates should not be used for "fine-line" decision making. A useful rule-of-thumb is that estimated tail areas (probabilities) should not be reported to more than two decimal places. The primary use of these methods should be to obtain an indication of the potential frequency of high-exposure levels from a parent lognormal distribution.

The procedure assumes that the exposure estimates are from a "stable" unvarying parent distribution. Experience, professional judgment, and knowledge of the exposure level determinant factors must be relied on here for assurance of the validity of this assumption. Only current sample data that represent a stable, unvarying exposure situation should be used in the following computations. One way of assuring that this condition is met is to plot the sample data on semilogarithmic paper (see Section 6.11). If the data are judged as trending upward (or downward) with time, then this procedure *should not be used* because an erroneous point estimate could result. Only if the long-term exposure plot appears "level" (i.e., constant mean after measurement errors have been smoothed out) should one use this procedure.

EXAMPLE. Initial monitoring of a target population of 35 workers exposed to chromic acid mist and chromates was done by sampling 1 worker on each of 6 days. A different worker was selected at random each day for exposure measurement. The monitoring yielded the following six 8-hr TWA exposure estimates for 6 different workers: 0.105, 0.052, 0.082, 0.051, 0.180, and 0.062 mg/m^3. Compute a point estimate P of the true proportion π of 8-hr TWA exposures experienced by the 35 workers that have exceeded an exposure control limit ECL of 0.10 mg/m^3 and the associated one-sided 95 percent upper confidence limit $UCL_{1,0.95}$ for the true proportion π.

SOLUTION. As in the tolerance limit example of Section 6.7, we will assume that true exposures for randomly selected, worker-day combinations are affected by two types of independently distributed, lognormally distributed, random variations:

a. Lognormally distributed, daily geometric means (over all workers in the target population)
b. Lognormally distributed, multiplicative (proportional) factors for differences between workers; ratios between exposures of individual workers are constant from day to day.

To be able to use a single lognormal distribution as an appropriate model for the type of exposure monitoring data collected in these two examples, it is essential to do the exposure monitoring according to a particular scheme whereby each randomly selected worker is sampled on only a single randomly selected day, and only one worker is measured each day. If a given worker were sampled repeatedly, the exposure measurements would be intercorrelated (as opposed to independent) samples and a more complex approach (i.e., ANOVA) would be needed in order to properly identify the day-to-day and worker-to-worker components of total variability.

1. Compute the sample geometric mean (GM) and geometric standard deviation (GSD) for the sample of n results. For the n = six 8-hr TWA exposure estimates, the sample GM = 0.080 mg/m^3 and the sample GSD = 1.627.

2. Compute $g(\text{ECL}) = [\log(\text{ECL}) - \log(\text{GM})]/\log(\text{GSD}) = 0.458$.

3. From a table of areas P under the standard normal curve from Z to $+\infty$, obtain the P value for $Z = g(\text{ECL})$. For $g(\text{ECL}) = 0.458$, $P = 0.32$, which is the integral of the standard normal curve from $g(\text{ECL})$ to plus infinity. Note that this computed P estimate should be about the same as a point estimate for π obtained with an approximate graphical plotting technique such as that described in Section 6.12 (i.e., a P point estimate for π that is the indicated probability at the intersection of an exposure distribution line and an exposure control limit line).

4. To compute confidence limit(s) on the actual proportion π, first decide whether one- or two-sided limits are desired and select the confidence level for the compu-

tation. Then from a table of Z values, obtain the related Z value for the desired confidence level. For example, to compute a one-sided 95 percent $UCL_{1,0.95}$ (or one-sided 95 percent lower confidence limit, $LCL_{1,0.95}$) for π, use a Z value of 1.645 (or -1.645). For two-sided 95 percent limits on π, use ± 1.960, and for two-sided 99 percent limits use ± 2.576.

5. Computing confidence limit(s) for the actual proportion π involves the solution to a quadratic equation. The two necessary quadratic roots are given by

$$U = \frac{-b \pm (b^2 - 4ac)^{1/2}}{2a}$$

where

$$a = \frac{1}{2n-3} - \frac{1}{Z^2} \qquad b = \frac{2g}{Z^2}\left(\frac{2n-3}{2n-2}\right)^{1/2}$$

and

$$c = \frac{1}{n} - \frac{g^2}{Z^2}\frac{2n-3}{2n-2}$$

The two U values are standard normal deviates (i.e., Z values) corresponding to the lower and upper confidence limits for the true proportion π of exposures exceeding ECL. Use of the larger value from the U equation leads to a one- or two-sided (depending on the Z value selected) LCL on the area to the right of ECL (i.e., an LCL on π). Use of the smaller value from the U equation leads to a UCL for π.

In this example, we will select $Z = +1.645$ to obtain a one-sided 95 percent upper confidence limit. Since $n = 6$ and $g = 0.458$ in this example, the three intermediate functional variables for U are $a = -0.258$, $b = 0.321$, and $c = 0.0969$. The resulting two values for U are -0.251 and $+1.494$. The smaller value (-0.251) corresponds to a probability of 0.599 or about 0.60, which is the $UCL_{1,0.95}$ for π.

6. We have computed the point estimate P of 0.32 for π (the actual proportion of overexposures) and the one-sided 95 percent upper confidence limit of 0.60 for π. Given the assumptions of this inferential method, our best estimate is that 32 percent of the 8-hr TWA exposures experienced by the 35 workers (on the 45 days of exposure that were sampled at random) exceeded the exposure control limit of 0.10 mg/m^3. However, this point estimate P for the actual proportion π of overexposure worker-days was based on a frugal sample of only 6 exposure estimates (one worker on each of six days). Note that the one-sided, 95 percent upper confidence limit for π is 60 percent, which indicates the large amount of uncertainty in the estimated proportion P. Thus we should state that the true (actual) percentage of worker-days exceeding the exposure limit could have been as high as 60 percent.

COMMENTS. This approach is graphically illustrated in Figure 10.17. The thick, angled, dashed line represents the estimated exposure distribution (a.k.a. sample

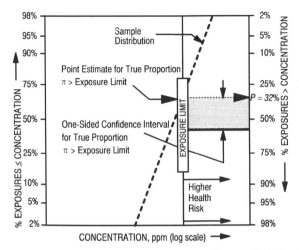

Figure 10.17 Point estimate P and one-sided, upper confidence limit for the actual proportion π of an exposure distribution exceeding an exposure control limit (ECL).

distribution) plotted on lognormal probability paper (see Section 6.12). The thin, horizontal, dotted line at the top of the shaded region points to the point estimate P of 0.32 from the intersection of the estimated exposure distribution line and the exposure limit ECL line. The thick, horizontal solid line at the bottom of the shaded region below the P value line represents the one-sided, upper confidence limit for the true proportion π of exposures exceeding ECL. The shaded region represents the uncertainty in P, since this area indicates the region in which the true exposure distribution and resulting π could lie.

If one had good reason to assume that a particular worker's exposure distribution over some long period (e.g., 2 months) is the same as the target population's distribution, then one could infer that any given worker was overexposed on 32 percent of the exposure days, but could have been overexposed as many as 60 percent of the exposure days. However, one must recognize that *individual worker overexposure risks* (resulting from individual exposure distributions) can be "masked" by the exposure distribution for the target population. For example, the lower portion of the target population exposure distribution might be created by workers with consistently low exposures, perhaps due to better work practices. Then these "lower-exposure tail" workers would have individual overexposure risks *substantially lower* than the 32 percent estimated for the group as a whole. Correspondingly, the upper portion of the group's exposure distribution might be due to workers with consistently high exposures, perhaps due to "dirtier" job locations or poorer work practices such as might result from inadequate training. These "upper-exposure tail" workers would have individual overexposure risks *substantially higher* than the 32 percent estimated for the group as a whole.

A companion technique for computing a one-sided tolerance limit (e.g., the exposure value that has 95 percent confidence of exceeding at least the lower 95 percent of the exposure distribution) was presented in the previous Section 6.7.

6.9 Fractile Intervals of a Normally Distributed Population with Known Parameters

PURPOSE. Compute fractile intervals for a population adequately described by a specified normal distribution.

ASSUMPTIONS. The term *fractile* refers to the value of the random variate below which a stated proportion of the distribution lies. The population for which the fractile interval statement will be made is assumed to be adequately described by a normal distribution. Both the arithmetic mean and standard deviation of the population distribution are assumed known so that neither distributional parameter will be estimated from sample data. Note that these assumptions are rarely met in practice, but this procedure is presented for illustrative purposes to assist one to better understand the normal distribution model. The normal tolerance limits procedure of Section 6.6 should be used where the distributional parameters are estimated from typical industrial hygiene sample sizes (e.g., less than 50). If there is any question whether one has a sufficient sample size to compute fractiles with this procedure, use the normal tolerance limit procedure.

EXAMPLE. Suppose that on a particular day, a worker's true 8-hr TWA exposure is "known" to be 25 ppm. On that day, if the worker's exposure were measured using a procedure with a total coefficient of variation of 10 percent ($CV_T = 0.10$), we would expect the value of the measurement to follow a normal distribution with mean $\mu = 25$ ppm and standard deviation $\sigma = (\mu)(CV_T) = 2.5$ ppm (see Sections 3.4 and 3.6.1). Compute the lower and upper values that bound the central 95 percent of the normal distribution of possible measurement values. These boundaries are known as the 2.5 percent and 97.5 percent *fractiles* of the normal distribution. There is a 95 percent probability that any single random measurement will lie within such a fractile interval.

SOLUTION

1. Table 10.2 indicates that the X interval for the central 95 percent of a normal distribution is from ($\mu - 1.96\sigma$) to ($\mu + 1.96\sigma$).
2. In this example the central 95 percent interval is bounded by 25 ppm $\pm$ (1.96)(2.5 ppm) = 25 ppm $\pm$ 4.9 ppm, which is 20.1 to 29.9 ppm.
3. Thus the central fractile interval enclosing 95 percent of the possible measurements, assuming a normal distribution with mean value equal to the true TWA exposure of 25 ppm and CV_T of 10 percent is 20.1 to 29.9 ppm. We can also say that there is a 95 percent probability that any single random measurement will lie within the interval 20.1 to 29.9 ppm.
4. If we are interested in the 95th percentile, Table 10.2 indicates that 95 percent of the measurements will be below ($\mu + 1.645\sigma$). In this example this 95 percent fractile is bounded by [25 ppm + (1.645)(2.5 ppm)] = [25 + 4.1] ppm = 29.1 ppm.

5. Thus the lower 95 percent of a normal distribution of possible measurements of a true TWA exposure of 25 ppm consists of all measurements at or below 29.1 ppm. We can also say that there is a 95 percent probability that a single random measurement of a true TWA of 25 ppm will be less than 29.1 ppm.

6.10   Fractile Intervals of a Lognormally Distributed Population with Known Parameters

PURPOSE. Compute fractile intervals for a population adequately described by a specified lognormal distribution.

ASSUMPTIONS. The population for which the fractile interval statement will be made is assumed to be adequately described by a known two-parameter lognormal distribution.

Both the population geometric mean and population geometric standard deviation of the population distribution are assumed known so that neither distributional parameter will need to be estimated from sample data. Note that these assumptions are rarely met in practice, but this procedure is presented for illustrative purposes to assist one to better understand the lognormal distribution model. The two-parameter lognormal tolerance limits procedure of Section 6.7 for samples should be used where the distributional parameters are estimated from typical industrial hygiene sample sizes (e.g., less than 50).

EXAMPLE. Suppose that the interday variability of a worker's 8-hr TWA exposures is adequately described by a known two-parameter lognormal distribution with a population geometric mean μ_g of 20 ppm and population geometric standard deviation σ_g of 1.9 (see Section 3.5 and Section 3.6.2). Note that the σ_g is dimensionless (does not have concentration units attached to it) unlike an arithmetic standard deviation σ, which has the same units as the corresponding arithmetic mean μ. Compute the lower and upper daily exposure values bounding the central 90 percent of the daily exposure distribution. These boundaries are the 5 and 95 percent fractiles of the two-parameter lognormal distribution.

SOLUTION

1. Table 10.3 indicates that the X interval for the central 90 percent of a two-parameter lognormal distribution lies in the fractile interval $(\mu_g)/(\sigma_g^{1.645})$ to $(\mu_g)(\sigma_g^{1.645})$.

2. In this example the central 90 percent fractile interval lies between (20 ppm)$/(1.9^{1.645})$ and (20 ppm)$(1.9^{1.645})$, which is 7.0 to 57.5 ppm.

3. Thus the central 90 percent fractile interval, for the two-parameter lognormal distribution of 8-hr TWA exposures with a true geometric mean μ_g of 20 ppm and true geometric standard deviation σ_g of 1.9, is about 7 to 57 ppm. That is, 90 percent of the daily TWAs lie within the interval 7 to 57 ppm. We can also say that there is a 90 percent probability that a single random daily exposure will lie within the

fractile interval 7 to 57 ppm. Note that this central fractile interval of the two-parameter lognormal distribution was computed to be *balanced* in the sense that it has equal percentages (5 percent) in each of the left and right tails of the distribution that lie outside the central fractile interval.

5. If we are interested in the 95th percentile, Table 10.3 indicates that 95 percent of daily TWA exposures will be below $(\mu_g)(\sigma_g^{1.645})$. In this example, this 95th percentile is equal to $(20 \text{ ppm})(1.9^{1.645})$, which is 57.5 ppm.

6. Thus the lower 95 percent portion of the two-parameter lognormal distribution consists of daily 8-hr TWA exposures at or below 57.5 ppm. That is, 95 percent of the daily TWAs will be at or below 57.5 ppm. We can also say that there is a 95 percent probability that a single random daily exposure will be less than 57.5 ppm.

6.11 Use of Semilogarithmic Graph Paper to Make Time Plots of Variables Believed to be Lognormally Distributed in Time

PURPOSE. To *plot data* that is believed to be lognormally distributed over time on semilog graph paper to *check for possible cycles or trends with time*.

ASSUMPTIONS. This procedure is useful to qualitatively look for exposure trends or cycles in time, when the distribution of available data is skewed to the right (toward higher values). No quantitative inferences are made with this procedure. To produce a nearly symmetrical data plot for lognormally distributed data, the mid-point of the logarithmic scale should be given a value close to the geometric mean of the sample data. Data that are approximately described by a two-parameter lognormal model (see Section 3.6.2) should show symmetrical random variability around the geometric mean (i.e., with no apparent trend or other systematic pattern in time).

Note that zero data values cannot be displayed with this procedure.

EXAMPLE. A worker's exposure to chromic acid mist and chromates was measured on 7 February. The 8-hr TWA exposure estimate of 0.105 mg/m³ was judged a possible overexposure relative to a company's exposure control limit (ECL) of 0.1 mg/m³.

Before making major capital improvements to the local exhaust ventilation, the company industrial hygienist decided to explore the interday variability of 8-hr TWA exposures for the worker after the initial measurement, on 5 other days in February and March. The measurement results in mg/m³ in chronological order are: 7 Feb., 0.105; 15 Feb., 0.052; 20 Feb., 0.082; 28 Feb., 0.051; 14 Mar., 0.180; and 25 Mar., 0.062. Plot the data to yield a symmetrical distribution over time.

SOLUTION

1. Use semilogarithmic graph paper with enough cycles and with scaling that is appropriate for the sample data. Semilog paper utilizes a logarithmic scale for one variable and a linear scale for the second variable. In this procedure, the exposure data is plotted on the logarithmic scale, and time is plotted on the linear scale.

Almost 60 types of semilog graph papers are presented by Craver (1980) including a 2-cycle by 36 divisions (3 years divided into months), 2-cycle by 52 divisions (1 year divided into weeks), 1-cycle by 60 divisions (5 years divided into months), and 1-cycle by 366 divisions (1 year divided into days).

2. In this example the range of the exposure measurements covers two decades (0.051 to 0.180 mg/m^3 falls within the interval 0.01 to 1.0 mg/m^3), so that 2-cycle semilog paper would be appropriate. Also, the time variable covers 46 days (almost 8 weeks), but the time values are not evenly spaced so that about 50 or more divisions would be appropriate on the linear scale (for time in days). If the time values were spaced exactly one or more weeks apart (i.e., at multiples of seven days), then a linear scale with about 8 or more major divisions would be appropriate.

COMMENTS. If one desires to apply quality assurance procedures to lognormally distributed data, review the article by Morrison (1958).

6.12 Use of Logarithmic Probability Graph Paper to Fit a Lognormal Distribution to Data

PURPOSE. Graphically estimate a two- or three-parameter lognormal distribution by fitting a straight line to a plot of the sample cumulative distribution on lognormal probability graph paper.

ASSUMPTIONS. It is assumed that the sample data are drawn from a single two-parameter lognormal distribution. The distribution will then yield a linear plot, aside from the expected deviations accountable to random sampling variations. The procedure will also explain how to qualitatively detect three-parameter lognormal distributions as a characteristic type of nonlinearity in the plot and transform the sample results so they plot linearly (see part 5 of the solution). The plotting procedure will explain how to qualitatively detect multimodal distributions, such as mixtures of two or more lognormal distributions (see part 6 of the solution). Neither the true geometric mean μ_g nor true geometric standard deviation σ_g need be known for this plotting procedure.

Note: This method cannot process zero or negative data values. It is assumed that all sample results are nonzero. Refer to Technical Appendix I of Leidel et al. (1977) for a discussion of how to treat a sample population that includes zero values.

The sample results are assumed to adequately represent the parent distribution that is to be estimated.

EXAMPLE. The same as used in the preceding Section 6.11. Plot the data to estimate the long-term lognormal distribution of daily exposures that would exist if the conditions of the 2-month representative period of the measurements persisted and the exposure distribution was stable over the long run (i.e., unchanging μ_g and σ_g).

SOLUTION

1. Only the basic aspects of probability plotting will be presented in this solution. Extensive details for this procedure are presented in Technical Appendix I of Leidel

et al. (1977). Use logarithmic probability paper with enough cycles for the range of sample data. Logarithmic probability paper utilizes one logarithmic scale for the measured variable and the other scale is a (nonlinear) cumulative probability scale. The same configuration of plotted points would exist if the probability scale were replaced by one that is linear in either the Z value or the probit of cumulative probability. The *probit* is a transformed variable equal to 5 plus the Z value (standard normal deviate) corresponding to the cumulative area under a normal distribution that is equal to the probability to be plotted. If the cumulative percentage of a normal distribution were plotted on a linear scale instead of on the special cumulative probability scale, it would form an S-shaped curve called an *ogive*. However, when the ogive curve is plotted against a cumulative probability scale, a linear function (straight line) results. Therefore, in this procedure the values of the exposure are plotted against the logarithmic scale and the expected cumulative percentages determined from positions (ranks) in the ordered (ranked) data are plotted against the probability scale. The latter expected values are given as plotting positions in Table I-1 of Leidel et al. (1977). Craver (1980) presents four types of normal probability paper (with different systems for numbering the linear scale for the random variable such as exposure concentration) and three types of lognormal probability paper (1-, 2-, and 3-cycle logarithmic scales for a random variable such as exposure concentration).

2. In the example data given in Section 6.11, the range of the exposure measurements covers two decades (0.01 to 1.0 mg/m^3), so that 2-cycle logarithmic probability paper would be appropriate for fitting a lognormal distribution as a model for the exposure data distribution.

3. Rank the sample data from lowest-exposure result to the highest-exposure value and obtain the expected cumulative percentage (plotting position) for each of the n = 6 values from Table I-1 of Leidel et al. (1977). For this example the plotting coordinates for each of the 6 coordinate pairs (i.e., measurement in mg/m^3 as ordinate vs. cumulative percentage as the abscissa plotting location on the cumulative probability scale) are: (0.051, 10.3%), (0.052, 26.0%), (0.062, 42.0%), (0.082, 58.0%), (0.105, 74.0%), and (0.180, 89.7%).

4. Using these plotting coordinates, plot a point for each sample value. One can also plot the individual uncertainties for each measurement by first calculating individual confidence limits for each measurement with a procedure selected from Sections 6.1.1, 6.2.1, 6.3.1, or 6.5.1. This will qualitatively aid one in comparing the amount of measurement procedure uncertainty to the uncertainty due to environmental variability of the true exposure levels (see Section 3.3).

5. If the plotted distribution has a substantial "hockey stick" appearance, with a flattening of the curve (approaching zero slope) at substantial portions of lower cumulative probability (the lowest 20 or more percent of the sample), then it might be indicative of a three-parameter lognormal distribution (see Section 3.5.2). Such a distribution can result if there are lognormal random variations that are added to a constant background level of the same contaminant. Such a data plot can be linearized by estimating the third parameter k and subtracting this constant ("background level") from each measurement value before plotting. An adequate k can be estimated from the initial "hockey stick" plot by noting the value that is asymptot-

ically approached by the "blade" of the hockey stick. That is, estimate k from the concentration value that the measurements appear to converge to at the lowest cumulative probabilities. A detailed example is given on pages 103 to 104 of Leidel et al. (1977).

6. If the data plot appears to have one or more "dog legs" or kinks in the central region of the plot, then it may be that the plotted distribution is a mixture of two or more individual lognormal distributions (multimodal data). One should then attempt to classify the sample data into two or more appropriate lognormal distributions by individually examining the determinant variables for each of the sample data. Additional qualitative interpretations of lognormal probability plots are given in Table I-3 on page 102 of Leidel et al. (1977).

6.13 Use of Logarithmic Probability Graph Paper to Display Tolerance Limits for an Estimated Lognormal Distribution

PURPOSE. Graphically display one- or two-sided tolerance limits, at the 95 percent confidence level, for the 75th, 90th, and 95th percentiles of an estimated lognormal distribution. Model the parent distribution of daily TWA exposures by plotting the estimated distribution and associated tolerance limits on lognormal probability paper.

ASSUMPTIONS. The same as used for Section 6.7.

EXAMPLE. The same as used for Section 6.12. Compute the one-sided, 95 percent confidence level, upper tolerance limits for the 75th, 90th, and 95th percentiles of the parent distribution of daily TWA exposures for the worker exposed to chromic acid mist and chromates over the 2-month period. Then plot the three UTL points on the same lognormal probability plot used to display the lognormal distribution estimate obtained in the previous Section 6.12.

SOLUTION

1. Compute the base 10 logarithm of each sample value. Here, the 6 logarithmic values in chronological order from 7 February to 25 March are: -0.979, -1.284, -1.086, -1.292, -0.745, and -1.208.

2. Compute the arithmetic mean $\bar{y}$ and standard deviation s of the $n = 6$ logarithmic values. Here, $\bar{y} = -1.099$ and $s = 0.211$. As supplementary information, one can compute the sample geometric mean (GM) by taking the antilog$_{10}$ of $\bar{y}$ ($10^{-1.099} = 0.080$ mg/m^3) and the sample geometric standard deviation (GSD) by taking the antilog$_{10}$ of s ($10^{0.211} = 1.63$).

3. Compute the three one-sided upper tolerance limits (UTL$_{1, \text{confidence level, percentile}}$) for the parent population from the three equations given below, for the particular case of $n = 6$:

$$\text{UTL}_{1, 0.95, 0.75} = \text{antilog}_{10} [\bar{y} + (1.895)(s)]$$

$$\text{UTL}_{1,0.95,0.90} = \text{antilog}_{10} \, [\bar{y} + (3.006)(s)]$$

$$\text{UTL}_{1,0.95,0.95} = \text{antilog}_{10} \, [\bar{y} + (3.707)(s)]$$

where each tolerance limit factor K was obtained from appropriate tables for one-sided tolerance limits, such as Table A-7 on page T-15 of Natrella (1963).

4. In this example the calculations are (all with 95 percent confidence):

$$\text{UTL}_{1,0.95,0.75} = \text{antilog}_{10} \, [-1.099 + (1.895)(0.211)]$$

$$= 0.20 \text{ mg}/\text{m}^3 \text{ for the 75th percentile}$$

$$\text{UTL}_{1,0.95,0.90} = \text{antilog}_{10} \, [-1.099 + (3.006)(0.211)]$$

$$= 0.34 \text{ mg}/\text{m}^3 \text{ for the 90th percentile}$$

$$\text{UTL}_{1,0.95,0.95} = \text{antilog}_{10} \, [-1.099 + (3.707)(0.211)]$$

$$= 0.48 \text{ mg}/\text{m}^3 \text{ for the 95th percentile}$$

5. Thus the (concentration, percentile) coordinates on the (logarithmic, cumulative probability) axes for the three one-sided upper tolerance limits (95 percent confidence level) are $(0.20 \text{ mg}/\text{m}^3, 75\%)$, $(0.34 \text{ mg}/\text{m}^3, 90\%)$, and $(0.48 \text{ mg}/\text{m}^3, 95\%)$. That is, based on the frugal sample of six daily exposures over a 2-month representative period for the worker, the best we can conclude is that we are 95 percent confident that under these conditions 75 percent of daily exposures would be below $0.20 \text{ mg}/\text{m}^3$, 90 percent would be below $0.34 \text{ mg}/\text{m}^3$, and 95 percent would be below $0.48 \text{ mg}/\text{m}^3$. The three tolerance limit values could then be plotted on the same lognormal probability plot used to graphically display the estimated lognormal distribution obtained in Section 6.12.

COMMENTS. A tolerance limit can be thought of as a confidence limit for a specified fractile of a population. Thus a ''tolerance line'' created by connecting the three tolerance limit points can be thought of as an approximate confidence line for the estimated lognormal population distribution. This approach for graphical analysis and presentation is illustrated in Figure 10.18. The thick, angled, dashed line across the full vertical span of the figure represents the estimated exposure distribution plotted on lognormal probability paper (see Section 6.12). The shorter, thick, angled, solid line at the right of the shaded region represents the many possible one-sided, upper tolerance limits for the actual exposure concentrations exceeding selected proportions P of the actual exposure distribution. The shaded region represents the uncertainty in estimating actual exposure concentrations exceeding selected proportions P, since this area indicates the region in which the true exposure distribution could lie. Percentile point estimates (i.e., those exposure concentrations exceeding specified proportions P of the exposure distribution) are estimated by: (1) selecting a P value on the left ordinate; (2) moving horizontally to the thick, dashed exposure distribution line; and (3) moving vertically to the logarithmic exposure concentration scale. One-sided, upper tolerance limits (95 percent confidence limits)

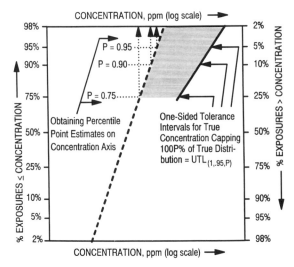

Figure 10.18 One-sided upper tolerance limits for specified concentration percentiles.

for selected percentile point estimates are similarly obtained by moving horizontally to the thick, solid line to the right of the shaded region and then moving vertically to the logarithmic exposure concentration scale.

REFERENCES

ACGIH (1991). *1991–1992 Threshold Limit Values for Chemical Substances and Physical Agents and Biological Exposure Indices*, American Conference of Governmental Industrial Hygienists, Cincinnati, OH.

Aitchison, J. and J. A. C. Brown (1957). *The Lognormal Distribution*, Cambridge University Press, Cambridge, England.

Altman, D. G. (1982). *Statist. Med.*, **1**, 59–71.

Am. Ind. Hyg. Assoc. Quar. (1955). **16**, 27–39.

Am. Ind. Hyg. Assoc. J. (1983). **44**, 697.

Armitage, P. (1982). *Biometrics Supplement: Current Topics in Biostatistics and Epidemiology*, 119–129.

Armstrong, B. G. (1992). *Am. Ind. Hyg. Assoc. J.*, **53**, 481–485.

Barnett, V. and T. Lewis (1978). *Outliers in Statistical Data*, Wiley, Chichester, England.

Bar-Shalom, Y., D. Budenaers, R. Schainker, and A. Segall (1975). *Handbook of Statistical Tests for Evaluating Employee Exposure to Air Contaminants*, National Institute for Occupational Safety and Health, U.S. Department of Health, Education, and Welfare (NIOSH) Publication 75-147, Cincinnati, OH.

Bates, J. D. (1978). *Writing with Precision—How to Write so That You Cannot Possibly Be Misunderstood*, Acropolis Books, Washington, D. C.

Berry, G., J. C. Gilson, S. Holmes, H. C. Lewinsohn, and S. A. Roach (1979). *Br. J. Ind. Med.*, **36**, 98–112.

Bingham, E. (1971). *Arch. Environ. Health*, **22**, 692–695.

Bowker, A. H. and G. J. Lieberman (1972). *Engineering Statistics*, Prentice-Hall, Englewood Cliffs, NJ, p. 1.

Bracht, G. H. and G. V. Glass (1968). *Am. Educ. Res. J.*, **5**, 437–474.

Buchwald, H. (1972). *Ann. Occup. Hyg.*, **15**, 379–391.

Burgess, W. A. (1981). *Recognition of Health Hazards in Industry*, Wiley, New York.

Busch, K. A. and N. A. Leidel (1988). ''Statistical Models for Occupational Exposure Measurements and Decision Making,'' in *Advances in Air Sampling*, ACGIH Industrial Hygiene Science Series, Lewis Publishers, Chelsea, MI, pp. 319–336.

CBE Style Manual Committee (1983). *CBE Style Manual*, 5th ed., Council of Biological Editors, Bethesda, MD.

Chiazze, L., Jr., F. E. Lundin, and D. Watkins, Eds. (1983). *Methods and Issues in Occupational and Environmental Epidemiology*, Ann Arbor Science Publishers, Ann Arbor, MI.

Cohen, A. C. (1961). *Technometrics*, **3**, 535.

Collins, J. F. and C. K. Redmond (1978). *J. Occup. Med.*, **20**, 260–266.

Cook, T. D. and D. T. Campbell (1979). *Quasi-Experimentation—Design and Analysis Issues for Field Settings*, Houghton Mifflin, Boston, pp. 70–80.

Corn, M. and N. A. Esmen (1979). *Am. Ind. Hyg. Assoc. J.*, **40**, 47–57.

Cralley, L. V. and L. J. Cralley, Eds. (1982). *Industrial Hygiene Aspects of Plant Operations, Vol. 1, Process Flows*, Macmillan, New York.

Craver, J. S. (1980). *Graph Paper from Your Copier*, H. P. Books, Tucson, AZ.

Crews, F. (1974). *The Random House Handbook*, Random House, New York.

Crocker, B. B. (1970). *Chem. Engr.*, May 4, 97.

Crow, E. L., F. A. Davis, and M. W. Maxfield (1960). *Statistics Manual*, Dover Publications, New York.

Crump, K. S., H. A. Guess, and K. L. Deal (1977). *Biometrics*, **33**, 437–451.

Daniel, C. and F. S. Wood (1971). *Fitting Equations to Data*, Wiley-Interscience, New York, Appendix 3A.

Dement, J. M., R. L. Harris, M. J. Symons, and C. Shy (1982). *Ann. Occup. Hyg.*, **26**, 869–887.

Elkins, H. B. (1948). *Ind. Hyg. Quar.*, **9**, 22–25.

Enterline, P. E. (1976). *J. Occup. Med.*, **18**, 150–156.

Esmen, N. (1979). *Am. Ind. Hyg. Assoc. J.*, **40**, 58–65.

Finney, D. J. (1982). *Statist. Med.*, **1**, 5–13.

Finney, D. J. (1991). *Biometrics*, **47**, 331–339.

Ford, D. P., B. S. Schwartz, S. Powell, T. Nelson, L. Keller, S. Sides, J. Agnew, K. Bolla, and M. Bleecker (1991). *Am. Ind. Hyg. Assoc. J.*, **52**, 226–234.

Friedman, G. D. (1974). *Primer of Epidemiology*, McGraw-Hill, New York.

Fuller, W. A. (1987). *Measurement Error Models*, Wiley, New York.

Gamble, J. and R. Spirtas (1976). *J. Occup. Med.*, **18**, 399–404.

Green, R. H. (1979). *Sampling Design and Statistical Methods for Environmental Biologists*, Wiley, New York.

Hahn, G. J., and S. S. Shapiro (1968). *Statistical Models in Engineering*, Wiley, New York.

Hald, A. (1952). *Statistical Theory with Engineering Applications*, Wiley, New York, p. 91.

Hartley, H. O. and R. L. Sielken, Jr. (1975a). *A Non-Parametric for "Safety" Testing of Carcinogenic Agent*, Food and Drug Administration Technical Report 1, Institute of Statistics, Texas A & M University, College Station, TX.

Hartley, H. O. and R. L. Sielken, Jr. (1975b). *A Non-Parametric for "Safety" Testing of Carcinogenic Agent*, Food and Drug Administration Technical Report 2, Institute of Statistics, Texas A & M University, College Station, TX.

Hartley, H. O. and R. L. Sielken, Jr. (1977). *Biometrics*, **33**, 1–30.

Hatch, T. H. (1955). *Am. Ind. Hyg. Assoc. Quar.*, **16**, 30–35.

Hatch, T. F. (1968). *Arch. Environ. Health*, **16**, 571–578.

Hatch, T. F. (1971). *Arch. Environ. Health*, **22**, 687–689.

Hatch, T. F. (1972). *J. Occup. Med.*, **14**, 134–137.

Hatch, T. F. (1973). *Arch. Environ. Health*, **27**, 231–235.

Hawkins, N. C. and B. D. Landenberger (1991). *Appl. Occup. Environ. Hyg.*, **6**, 689–695.

Hawkins, N. C., S. K. Norwood, and J. C. Rock, Eds. (1991). *A Strategy for Occupational Exposure Assessment*, American Industrial Hygiene Association, Akron, OH.

Hermann, E. R. (1971). *Arch. Environ. Health*, **22**, 699–706.

Holcomb, M. L. and R. C. Scholz (1981). *Evaluation of Air Cleaning and Monitoring Equipment Used in Recirculation Systems*, National Institute for Occupational Safety and Health, Publication DHHS (NIOSH) 81-113, Cincinnati, OH.

Hornung, R. W. (1991). *Appl. Occup. Environ. Hyg.*, **6**, 516–520.

Hornung, R. W. and L. D. Reed (1990). *Appl. Occup. Environ. Hyg.*, **5**, 46–51.

Hubiak, R. J., F. H. Fuller, G. N. VanderWerff, and M. Ott (1981). *Occ. Health Safety*, **50**, 10–18.

Iman, R. I. (1982). *Am. Statistician*, **36**, 109–112.

Karvonen, M. and M. I. Mikheev (1986). *Epidemiology of Occupational Health*, World Health Organization, Regional Office for Europe, WHO Regional Publications, European Series No. 20, Copenhagen.

Kimbrough, R. D. (1991). *Appl. Occup. Environ. Hyg.*, **6**, 759–763.

King, M. V., P. M. Eller, and R. J. Costello (1983). *Am. Ind. Hyg. Assoc. J.*, **44**, 615–618.

Krewski, D. and C. Brown (1981). *Biometrics*, **37**, 353–366.

Kupper, L. L., A. J. McMichael, and R. Spirtas (1975). *J. Am. Stat. Assoc.*, **70**, 524–528.

Land, C. (1973). *J. Am. Stat. Assoc.*, **68**, 960–963.

Land, C. (1988). "Hypothesis Tests and Interval Estimates," in *Lognormal Distributions. Theory and Applications*, E. Crow and K. Shimizu, Eds., Marcel Dekker, New York, pp. 87–112.

Leichnitz, K. (1980). *Draeger Review*, **46**, 13–21.

Leidel, N. A., K. A. Busch, and W. E. Crouse (1975). *Exposure Measurement Action Level and Occupational Environmental Variability*, National Institute for Occupational Safety and Health, U.S. Department of Health, Education, and Welfare (NIOSH) Publication 76-131, Cincinnati, OH.

Leidel, N. A., K. A. Busch, and J. R. Lynch (1977). *NIOSH Occupational Exposure Sampling Strategy Manual*, U.S. Department of Health, Education, and Welfare (NIOSH) Publication 77-173, Cincinnati, OH.

Lilienfeld, A. M. and D. E. Lilienfeld (1980). *Foundations of Epidemiology*, 2nd ed., Oxford University Press, New York.

Lilliefors, H. W. (1967). *J. Am. Stat. Assoc.*, **62**, 399–402.

Lloyd, J. W. and A. Ciocca (1969). *J. Occup. Med.*, **11**, 299–310.

Lowrance, W. W. (1976). *Of Acceptable Risk—Science and the Determination of Safety*, William Kaufmann, Los Altos, CA.

Lundin, F. E., J. K. Wagoner, and V. E. Archer (1977). *Radon Daughter Exposure and Respiratory Cancer Quantitative and Temporal Aspects*, NIOSH-NIEHS Joint Monograph No. 1, U.S. Department of Health, Education, and Welfare.

Lynch, J. R., N. A. Leidel, R. A. Nelson, and R. F. Boggs (1978). *The Standards Completion Program Draft Technical Standards Analysis and Decision Logics*, National Institute for Occupational Safety and Health, National Technical Information Service Publication PB 282 989, Springfield, VA.

MacMahon, B. and T. F. Pugh (1970). *Epidemiology Principles and Methods*, Little Brown, Boston.

Mage, D. T. (1982). *Am. Statn.*, **36**, 116–120.

Mantel, N. and W. R. Bryan (1961). *J. Nat. Cancer Inst.*, **27**, 455–470.

Mantel, N., N. R. Bohidar, C. C. Brown, J. L. Ciminera, and J. W. Tukey (1975). *Cancer Res.*, **35**, 865–872.

Mausner, J. S. and A. K. Bahn (1974). *Epidemiology, An Introductory Text*, Saunders, Philadelphia.

Monson, R. R. (1990). *Occupational Epidemiology*, 2nd ed., CRC Press, Boca Raton, FL.

Moroney, M. J. (1951). *Facts from Figures*, Penguin Books, Baltimore, p. 261.

Morrison, J. (1958). *Appl. Stat.*, **7**, 160–172.

Murphy, E. A. (1982). In *Biostatistics in Medicine*, Johns Hopkins University Press, Baltimore, Chapter 4.

Natrella, M. G. (1963). *Experimental Statistics*, National Bureau of Standards Handbook 91, Superintendent of Documents, U.S. Government Printing Office, Washington, D.C.

Olsen, E., B. Laursen, and P. S. Vinzents (1991). *Am. Ind. Hyg. Assoc. J.*, **52**, 204–211.

Paik, N. W., R. J. Walcott, and P. A. Brogan (1983). *Am. Ind. Hyg. Assoc. J.*, **44**, 428–432.

Piele, D. T. (1990). *Introductory Statistics with Spreadsheets*, Addison-Wesley, Reading, MA.

Rappaport, S. M. (1991). *Ann. Occup. Hyg.*, **35**, 61–121.

Rappaport, S. M. and S. Selvin (1987). *Am. Ind. Hyg. Assoc. J.*, **48**, 374–379.

Roach, S. A. (1953). *Br. J. Ind. Med.*, **10**, 220.

Roach, S. A. (1966). *Am. Ind. Hyg. Assoc. J.*, **27**, 1–12.

Roach, S. A. (1977). *Ann. Occup. Hyg.*, **20**, 65–84.

Rothman, K. J. and J. D. Boice, Jr. (1982). *Epidemiologic Analysis with a Programmable Calculator*, Epidemiology Resources, Boston.

Schlesselman, J. J. (1982). *Case-Control Studies: Design, Conduct, Analysis*, Oxford University Press, New York.

Schneider, D. (1980). *Draeger Rev.*, **46**, 5–12.

Seim, H. J. and J. A. Dickeson (1983). *Am. Ind. Hyg. Assoc. J.*, **44**, 562–566.

Selvin, S., and S. M. Rappaport (1989). *Am. Ind. Hyg. Assoc. J.*, **50**, 627–630.

Shapiro, S. S. and M. B. Wilk (1965). *Biometrika*, **52**, 591–611.

Smyth, H. F., Jr. (1956). *Am. Ind. Hyg. Assoc. Quar.*, **17**, 129–185.

Smyth, H. F., Jr. (1962). *Am. Ind. Hyg. Assoc. J.*, **23**, 37–44.

Soule, R. D. (1973). "An Industrial Hygiene Survey Checklist," in National Institute for Occupational Safety and Health, *The Industrial Environment—Its Evaluation and Control*, Department of Health, Education, and Welfare, Cincinnati, OH.

Stokinger, H. E. (1955). *Pub. Health Rep.*, **70**, 1–11.

Stokinger, H. E. (1962). *Am. Ind. Hyg. Assoc. J.*, **23**, 45–47.

Stokinger, H. E. (1972). *Arch. Environ. Health*, **25**, 153–157.

Stoline, M. R. (1991). *Environmetrics*, **2**, 85–106.

Thomas, H. F. (1979). *Ann. Occup. Hyg.*, **22**, 389–397.

Tuggle, R. M. (1981). *Am. Ind. Hyg. Assoc. J.*, **42**, 493–498.

Tuggle, R. M. (1982). *Am. Ind. Hyg. Assoc. J.*, **43**, 338–346.

Ulfvarson, U. (1983). *Int. Arch. Occup. Environ. Health*, **52**, 285–300.

U.S. Department of Health Education and Welfare (1977). *Advances in Health Survey Research Methods: Proceedings of a National Invitational Conference*, sponsored by the National Center for Health Services Research, U.S. Department of Health, Education and Welfare (HRA) Publication No. 77-3154.

Waters, M. A., S. Selvin, and S. M. Rappaport (1991). *Am. Ind. Hyg. Assoc. J.*, **52**, 493–502.

Wilkins, P. E. (1976). *J. Air Pollution Control Assoc.*, **26**, 935.

CHAPTER ELEVEN

Detecting Disease Produced by Occupational Exposure

Philip E. Enterline, Ph.D.

1 INTRODUCTION

It is the nature of modern human beings to speculate on the causes of disease. Often this speculation centers on factors associated with work, and the recognition of disease-producing agents in occupational settings has resulted in striking health improvements in some groups of workers. Such conditions as lead intoxication, mercury poisoning, and benzol poisoning have largely disappeared because they could be easily identified with a particular substance, and industrial hygiene measures were relatively easy to apply (Hunter, 1978). For some occupational disease, such as bladder cancer in the dye industry and phossy jaw in the match industry, the use of certain substances was simply stopped in order to protect the health of the worker.

As the more obvious industrial hazards have been identified and brought under control, the task of identifying disease produced by occupational exposure has become increasingly difficult. The skills of many disciplines are now being brought to bear on the problem of identifying hazards that yet exist. Epidemiologists, biostatisticians, biomathematicians, toxicologists, and industrial hygienists have joined occupational physicians in studies designed to detect or predict occupational hazards.

One way of detecting and evaluating disease in working populations is by a well-planned and well-conducted epidemiologic study. As early as 1928 the framework for epidemiologic investigations of the effects of work exposures was laid down by

Patty's Industrial Hygiene and Toxicology, Third Edition, Volume 3, Part A, Edited by Robert L. Harris, Lewis J. Cralley, and Lester V. Cralley.
ISBN 0-471-53066-2 © 1994 John Wiley & Sons, Inc.

Bridge and Henry (1928). In order to detect a causal relationship between exposure and occupational cancer they proposed "that the incidence rate in the occupations under review should exceed that in the general population and that, in the occupation concerned, there should be sufficient association of a worker with a substance proved experimentally to have carcinogenic properties." For scientific workers of that period this proved to be an almost impossible set of conditions. For example, in 1935 a pathologist named Gloyne reported two cases where both asbestosis and lung cancer occurred in asbestos workers coming to autopsy (Gloyne, 1935). In this report he quoted the conditions laid down by Bridge and Henry and recognized that these had not been met. He had no idea as to the incidence of lung cancer in asbestosis cases or in asbestos workers, and no work had been done that proved experimentally that asbestos was a carcinogen. His conclusion was: "It seems worthwhile to record these two cases, not in any attempt to make out an etiological association of these two diseases, but in order to emphasize certain histological points in which one disease appears to bear on the other." Actually it took over 30 years before Bridge and Henry's conditions for establishing a relationship between asbestos and lung cancer were met. Recognition by the scientific community of the cancer hazards of asbestos were probably greatly delayed because of this (Enterline, 1978).

We now have many resources for conducting both experimental and epidemiologic investigations that were not available to Gloyne. We have also expanded on the rules originally proposed by Bridge and Henry for detecting a causal relationship between exposure and occupational disease. The following criteria are now generally recognized as important in concluding that exposure and disease are causally related (U.S. DHEW, 1963):

1. The consistency of an association. The same association found in a variety of settings is strong evidence of a causal relationship. Of particular importance are studies that differ in their potential confounding factors but that agree on a particular association.

2. The specificity of an association. If a substance is associated with not one but many diseases, evidence for a causal relationship is weakened. In occupational studies specific agents are often associated only with a single disease.

3. The strength of an association. A very large excess in a particular disease in the presence of a particular exposure is not likely to be due to the fact that the epidemiologic method does not usually permit control or adjustment for all confounders, that is, disease-related differences between exposed and nonexposed populations. Thus, large excesses are likely to be due to a true association.

4. A dose–response relationship. This is an important criterion for concluding that disease and exposure are causally related.

5. The demonstration of a temporal relationship. Exposure must precede disease. The time period should be a reasonable one for a particular disease. This is the only required condition for causality.

6. Statistical significance. This relates to the size of the study and the magnitude of the excess. Tests of statistical significance are valuable in judging the meaning of a particular association.

7. The existence of experimental data. There are very few agents that produce disease in humans that have not been shown experimentally to produce disease in animals. The existence of experimental data greatly reinforces epidemiologic observations.

Not only are the rules for carrying out epidemiologic studies now better understood than at the time of Bridge and Henry, but we have reasons for carrying out occupational epidemiologic studies that are perhaps more pressing than those that prevailed during that period. Progress in discovering cures for many diseases has been disappointingly slow, and there is a well-recognized need to identify hazardous substances in our environment as a preventive measure. Moreover, there recently has been a special interest in epidemiologic studies on the part of industry since industry is now being held by the U.S. courts to "have the knowledge of an expert" about the health effects of their products (*Borel v. Fibreboard*, 1973). This means that industrial managers need to be among the first to be aware of any health effects associated with their products. One way to do this is to carry out epidemiologic investigations on workers exposed to their products or exposed to ingredients in their products.

This chapter will describe the kinds of studies that relate to detecting diseases produced by occupational exposure along with some guidelines for carrying out and reporting on such studies. It will not deal extensively with statistical methods for evaluating these studies. In an Appendix at the end of this chapter the reader is referred to texts that deal most clearly with this subject.

2 WHAT IS OCCUPATIONAL EPIDEMIOLOGY

Epidemiology has been defined as the study of the distribution and determinants of disease in human populations (MacMahon and Pugh, 1980). Occupational epidemiology simply limits the populations studied to persons employed and focuses on the occupational environment as a possible determinant of disease. Epidemiology focuses on risks of disease in groups of individuals rather than in any single individual and is observational as opposed to experimental.

In order to evaluate epidemiologic data it is important to understand the nature of the experimental method of study. In an experiment involving the role of a particular substance in disease causation, subjects would be randomly assigned to exposed and nonexposed groups, the exposures would be carried out, then both exposed and unexposed groups observed for disease. In making this random assignment all factors related to disease can be assumed, within statistically defined limits, to be identical in both groups except for exposure to the substance of interest. In an epidemiologic investigation, on the other hand, allocation between exposed and nonexposed groups is not random and these groups are likely to differ on many factors related to disease. Some of these factors, such as age, sex, race, and smoking habits, may be known, and matching or adjustment for these factors improves comparability between exposed and nonexposed groups. Other factors are unknown and adjustments for them cannot be made. As Hill (1953) puts it, in constructing an epidemiologic study "one must have the experimental approach firmly in mind."

That is to say, insofar as possible one must consider whether the matching or adjustment adequately deals with all of the important variables. If an important variable has not been properly dealt with, then the study is confounded by that variable and results may be impossible to interpret.

Ideally an epidemiologic study should start with some exposure of interest so that this exposure then becomes a hypothesis that is tested with regard to some disease or diseases. For example, because asbestos is a known cause of lung cancer, and because it is the physical property of asbestos that is the reason for its carcinogenic effects, workers exposed to man-made mineral fibers similar to asbestos have been examined to see if these workers also had a lung cancer excess (Enterline, 1991).

3 TYPES OF EPIDEMIOLOGICAL STUDIES

Epidemiological studies designed to detect disease produced by occupational exposure can be classified into five categories, each of which contributes a different level of inferential knowledge concerning disease etiology. These levels are directly related to the extent to which confounding factors can be taken into consideration:

1. **Ecological Studies.**   These are studies in which an evaluation is made of the spatial or temporal patterns of morbidity or mortality in human populations and where all classifications are made on the basis of aggregates of individuals as distinct from single individuals. In this type of study individuals are not classifiable according to the study parameters, and this type of study does not permit a direct measure of association. Thus, within time–space groupings it is not possible to examine the joint distribution of exposure and disease. An example would be a comparison of cancer mortality in counties classified according to the presence or absence of selected industries and the conclusion that cancer excesses are related to the presence of a particular industry (Blot et al., 1979).

2. **Demographic Studies.**   These are studies in which an evaluation is made of the risk of morbidity or mortality in human populations composed of individuals classifiable by demographic characteristics such as age, sex, occupation, income, or education. In this type of study the source of data often differs for the numerators and denominators used in calculating rates. Here death rates are usually calculated for selected causes and occupations by using counts of persons in a particular occupation as enumerated in a population census (for the denominators) and as recorded on death certificates (for the numerators). An example is the occupational mortality publications for England and Wales (Her Majesty's Stationery Office, 1978).

3. **Cross-Sectional Studies.**   These are studies in which an evaluation is made of the prevalence of disease at a specific time among two or more groups, the individuals of which are classified by exposure, or some index of exposure at that specified time. An example would be comparing the prevalence of X-ray abnormalities in a group of workers exposed to coal dust with a group of workers not so exposed (Enterline, 1967).

4. **Cohort Studies.** These are studies in which an evaluation is made of the incidence of disease among groups of individuals classified by exposure to a specific agent or environment, either inferred or actually measured, and with each group followed over some period of time. An example would be identifying workers in the steel industry and tracing these to see how their mortality experience relates to their jobs (Lloyd, 1971).

5. **Case-Control Studies.** These are studies in which a comparison is made of the past exposure of groups of individuals classified according to whether a specific disease is present (cases) or absent (controls). Controls are usually intended to be representative of the population from which the cases arose. An example would be a study that compares the smoking habits of a group of lung cancer cases entering a hospital with the smoking habits of all or a sample of other hospital admissions (Doll and Hill, 1952).

3.1 Ecological Studies

Geographic variations in mortality are the basis for most ecologic studies and have formed the basis for determining the etiology of disease for a very long time. In John Snow's classic study of the 1849 cholera epidemic in London, it was the variation in mortality from cholera in different districts of London that formed the basis for his observation that cholera was related to contamination of the water supply (Snow, 1965). Then as now, however, there was disagreement as to the cause of variations in mortality. William Farr, regarded by Raymond Pearl as "the greatest medical statistician who ever lived," believed that the geographic variation had something to do with elevation, since death rates were highest in low-lying areas (Pearl, 1940). Fortunately Snow prevailed, pointing out that most elevated towns in England had also suffered excessively from cholera. In London itself, he argued, one area was a full 56 feet above the "Trinity high water mark" and had a death rate for cholera of 55 per 1000, while many other districts "of less than half the elevation did not suffer one third as much." Had Farr looked further into the situation, he would have noted that a third variable was at work—the source of the water supply and that if adjustment were made for this, the geographic variation could be largely accounted for.

As an example of some uses of ecologic data for the United States, in 1957 a special tabulation of deaths and population was prepared for the years 1949–1951 that made possible the comparison of death rates by cause and by county of residence, with appropriate adjustment for age, race, and sex. This formed the basis for data shown in Figure 11.1. This shows age-adjusted death rates for coronary heart disease for each of 166 economic subregions in the United States (Enterline et al., 1960). An economic region is defined as a group of U.S. counties in which people make a living in about the same way, and this seems to be an interesting grid against which to display health-related data. In Figure 11.1 metropolitan counties (urban areas) are shown in black so that all of the variation displayed is in nonurban areas. An ecological study would attempt to explain this kind of geographic variation by associating it with some enumerated factor or factors. The National Cancer Institute published cancer death rates by county for the years 1950–1969, accompanied by

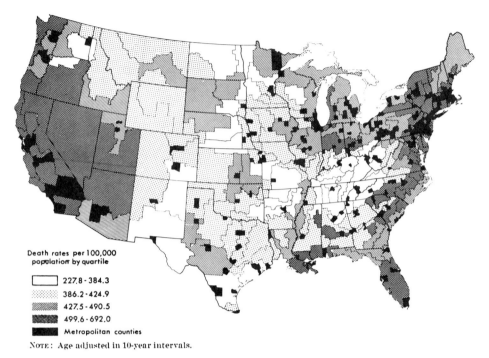

Death rates per 100,000
population by quartile

☐ 227.8 - 384.3
▨ 386.2 - 424.9
▨ 427.5 - 490.5
▨ 499.6 - 692.0
■ Metropolitan counties

NOTE: Age adjusted in 10-year intervals.

Figure 11.1   Coronary heart disease rates for 116 economic subregions, nonmetropolitan areas only, white males aged 45–64 (1949–1951).

an atlas of cancer mortality (U.S. DHEW, 1975). A number of hypotheses have been generated from these data. Currently U.S. race, sex, age, time period, geographic area specific mortality and population data are available from data tapes assembled by the National Center for Health Statistics (Marsh, 1987).

A problem with ecological studies is that they deal with the characteristics of groups of people rather than with individuals. In a classic paper published in 1950 Robinson pointed out that when using states as the ecologic unit, there is a strong positive correlation between the percentage of the population of each state that is foreign born and the income level in each state (Robinson, 1950). From this, one might conclude that the reason is because foreign-born persons have higher incomes than native-born persons. At the time Robinson published his paper just the opposite was true, and some other variables were obviously responsible for the positive association. The New England states had high incomes while the South had a relatively low income. The New England states had a high percentage foreign born relative to the South. Clearly there were large differences between these two groups of states that needed to be accounted for to make an ecologic study meaningful. The many problems with ecological studies have been extensively reviewed by Morgenstern (1982).

Thus far, ecologic studies have not provided much useful information with regard to causal factors in disease. The problem is similar to the one faced by John Snow in 1849, and often we tend to work with knowledge like that possessed by William Farr. Apparently we have been unable to adjust for the right variables. Until these can be identified and measured, progress in applying our knowledge of geographic variations in disease to the problem of disease etiology will be slow.

3.2 Demographic Studies

Although William Farr missed the reason for the geographic variation in cholera observed by John Snow, he was responsible for developing the basis for demographic studies of occupational mortality. In 1839 William Farr was appointed first head of the Office of the Registrar General for England and Wales. Using data available to him in that position he was able to combine counts of workers in various occupations derived from the federal census with counts from an item on occupation that appeared on death certificates. In his annual report for 1851 he made reference to mortality rates by occupation (His Majesty's Stationery Office, 1855). Starting in 1861 mortality rates by occupation for England and Wales have been published every 10 years (except 1941). These publications are excellent references for providing clues to health hazards in various occupational groupings (Her Majesty's Stationery Office, 1978).

In the United States counts of deaths by occupation were published by the United States Bureau of the Census as early as 1890 and 1900, but no rates were computed. Complete tabulations were also made for the years 1910 and 1920 but were not released, probably because of the poor quality of information appearing on death certificates regarding the decedent's occupation. Just prior to 1930, state vital statistics offices took part in an intensive campaign to improve occupation information on death certificates in preparation for a study of mortality rates by occupation for the year 1930. Despite this emphasis, a large proportion of death certificates filed in 1930 contained unknown or nonspecific occupations. For this reason, mortality rates by occupation were published for only 10 states where occupational data on death certificates appeared to be reasonably good (Whitney, 1934). Death rates for selected occupations were derived by relating counts for items regarding occupation and industry appearing on death certificates filed during the year 1930 to population counts by occupation and industry as enumerated in the 1930 decennial census. Subsequent evaluation of the comparability of occupations reported on death certificates and census returns for the same persons showed a lack of correspondence, chiefly because of problems in classifying retired persons. For this reason plans for tabulations were dropped for 1940.

For 1950, state vital statistics offices were again encouraged to improve returns on the occupation and industry items appearing on the death certificate and, for deaths occurring in 1950, occupation and industry items were coded by the National Office of Vital Statistics of the United States Public Health Service for men 20–64 years of age. Earlier studies of the comparability of occupation and industry items on the death certificates, and on the census returns for the same individuals indicated that for males in this age group lack of comparability was not so serious. The only year for which mortality data are available by occupation and industry for the entire United States is 1950 (Guralnick, 1962).

For 1960, an attempt was made to match a sample of death certificates filed during the months of May, June, July, and August with the April 1960 census returns, supplemented by additional information where matches could not be made (U.S. DHEW, 1969). Classification of deaths by occupation and industry as enumerated in the 1960 census was thus possible since the problem of noncomparability of occupation and industry items was largely eliminated. Unfortunately, the number of matches was disappointingly small, and numbers of deaths too small to provide very many useful details with regard to death rates by occupation.

Current interest in occupational mortality centers on improving the quality of information appearing on death certificates and on developing and applying a standardized method for coding industry and occupation (Dubrow et al., 1987). Perhaps the most useful occupational mortality data are calculations of the proportion of all deaths due to a particular cause since this does not require combining data from the U.S. census with those on death certificates (Milham, 1983).

Probably a better source of mortality data, from a technical standpoint at least, are statistics collected by life insurance companies. These records have long been recognized as a valuable source of information on industrial hazards, particularly accidents. Two types of data have been published: data relating to experience with individual life insurance policies, and data relating to group life insurance policies. The earliest intercompany studies related to experience with individual policies. They were made possible by the pooling of records from several companies, coupled with the adoption by insurance companies of a standardized method for classifying occupations. Studies of ordinary life policies provide mortality data by occupation, whereas studies of group policies provide mortality data by industry. Both types of data have been published by the Society of Actuaries (Society of Actuaries, 1967). Insurance data have been used to examine mortality among American coal miners (Enterline, 1972).

Data published by life insurance companies relate deaths to the occupation or industry recorded at the time the policy was issued, and thus do not suffer from the defect of combining occupation and industry items from census returns and death certificates. On the other hand, they have the disadvantage that follow-up for death often relates only to the period an individual works and a life insurance policy is in effect. If persons are selected out of the work force for health reasons, this will introduce bias.

Demographic studies have offered some guidance in detecting disease produced by occupational agents. For example, Kennaway and Kennaway (1936) identified several groups of coal carbonization and coal by-products workers with a high risk of lung cancer. An epidemiological investigation subsequently confirmed these findings (Doll et al., 1965). On the other hand, the 1936 study found not one single case of lung cancer in asbestos workers and may have delayed the recognition of a relationship between asbestos and cancer.

3.3 Cross-Sectional Studies

These are usually studies that relate to employee physical examinations and deal with disease prevalence at a particular point in time as opposed to incidence over some period of time. They are particularly appropriate for occupational diseases of long duration and usually involve relating physical examination findings to the duration and extent of exposure to various substances. Signs of exposure such as blood lead levels, urinary arsenic levels, and skin disorders are discoverable through cross-sectional studies. Signs of chronic respiratory disease including X-ray changes and pulmonary function abnormalities have been studied extensively by cross-sectional methods. The identification of coal workers pneumoconiosis in American coal miners came about as the result of this kind of study (Enterline, 1967).

Employee physical examinations deal with the prevalence of disease, although by linking successive examinations time-related disease incidence can often be estimated. The relationship between prevalence P and incidence I is

$$P = ID$$

where D is duration of disease. Whatever unit is selected for duration of disease becomes the unit of time over which incidence is expressed. For example, if a disease has an incidence rate of 1 per 100 per year and a 5-year duration, then the prevalence of the disease would be 5 per 100. Duration begins with the first detectable sign of disease and ends with death of the individual or the disappearance of the sign. Ecologic and demographic studies can deal with either incidence or prevalence data.

Most diseases have an incidence, a prevalence, and a duration. Some statistics of interest, however, have no time dimension. Deaths and industrial accidents can be studied only as incidence data. A consequence of accidents, disability does have a time dimension and can be presented as the prevalence of persons disabled due to accidents.

Cross-sectional studies have many advantages. One of these is the ability to plan for and collect a considerable amount of data in a fairly brief interval of time using uniform definitions for the characterization of disease and of exposure. In these studies it is also usually possible to enumerate and adjust for differences in the exposed and nonexposed segments of the population so as to prevent confounding and facilitate statements about occupational factors causally related to disease.

One disadvantage in cross-sectional studies of employed populations is that workers available for study at a particular point in time may be the product of a process in which workers have been selected in or out of the work force for health reasons. As a result, disease excesses related to work exposure can be muted or exaggerated.

3.4 Cohort Studies

Far better than ecological, demographic, or cross-sectional studies for diseases of relatively short duration, such as cancer, are cohort studies. These are investigations in which workers are individually identified and followed through time to determine their health experience. The term *cohort* refers to that part of a population born during a particular period and observed for its characteristics as it enters successive time and age intervals. In current usage, this notion has been broadened to describe any designated group of persons who are followed or traced over a period of time. Cohort analysis can be historical or concurrent. Historical cohort analysis involves identification of a cohort at some time in the past and observing its health experience up to the present. Concurrent cohort analysis involves identification of a cohort either historically or currently and observing its health experience on into the future.

Cohort studies usually deal with the concept of life expectancy and can be directed to the question of whether there is any evidence that a particular environment has a life-shortening effect. Secondarily, they deal with the specific causes of death responsible for life-shortening effects and the environmental factors related to these causes. A life table method of analysis is ordinarily used here (Cutler and Ederer, 1958). Where an internal comparison group is available, cohort analysis can also be viewed in the context of a case control study. This kind of analysis involves developing fourfold tables showing deaths and nondeaths for exposed and not exposed workers and summating these across stratifying categories such as age, time, and sex (Mantel and Haenszel, 1959). Cohort studies may deal with morbidity as well as mortality (Cooper et al., 1962). Some areas that need careful consideration in cohort studies are discussed in the following sections.

3.4.1 Selection of a Cohort for Study

Usually only workers for whom a disease excess seems biologically possible should be selected. If, for example, a possible effect is cancer, only populations that can be followed many years after some occupational experience should be studied since cancer is usually manifest only many years after exposure. Cancer occurring less than 5 years after initial employment is almost certainly not due to that employment. It would be pointless, therefore, to include recent hires in a situation where cancer is the health effect of interest unless future follow-up of a cohort is anticipated and recent hires will be followed for a considerable time in the future. Populations selected for study should also:

1. Be a complete enumeration of all persons employed under a particular set of circumstances. Missing members of such a cohort may be those of most interest and result in bias of unknown magnitude and direction.

2. Have had a significant work period. Workers employed only a few days or weeks are usually not of interest except, perhaps, for studying the effects of high acute exposures. Short-term workers may have unusual personal or social characteristics that influence their health experience and make inferences about a particular exposure difficult.

3. Have available reliable demographic work history data and identification data. Job histories at the plant or location studied are important. Job histories prior to that employment are also important, but often not available. Where prior job histories are of possible importance but uncertain, it may be appropriate to emphasize observations on workers hired at younger ages where extensive prior employment is unlikely. Demographic data should include sex, birth date, and if possible race and country of birth. Identification data should include full name, social security number, company badge number, and armed forces identification number.

4. Be traceable. Some working populations such as migratory farm laborers, World War II female industrial workers, foreign nationals, and workers not covered

by the U.S. Social Security System are often difficult to trace. Bias is likely in studies that include large numbers of such workers.

5. Have environmental exposures of sufficient magnitude and specificity to anticipate a biological response. These exposures should be carefully evaluated and hypotheses developed in advance of the study. Such hypotheses might be based on animal investigations, case reports, or other epidemiologic studies. (Measurement of exposure is dealt with elsewhere in this volume.)

6. Be large enough to yield meaningful results. This applies particularly where hypothesis testing is involved. For a particular study size it involves calculation of the probability of detecting (as statistically significant) excesses in mortality or disease of some specified magnitude. These calculations are called power calculations. If the anticipated excess is small, then the study size must be large so that if the anticipated excess is actually found it will be statistically significant. Power calculations are most meaningful before a study is done since after the study results are known, there is no need for postulating an "anticipated" excess (Detsky and Sackett, 1985). A confidence interval around an observed excess (or deficit) in disease combines information about the study size and the study findings in such a way as to indicate the range within which the true excess lies.

3.4.2 Abstracting Records

It is preferable that records of cohort members be selected and abstracted or microfilmed by the person or organization assuming responsibility for the validity and scientific accuracy of a particular investigation. This places the burden of proof as to the validity of observations on a single individual or organization. It also allows careful study of records in a research environment and reexamination of the database as frequently as is necessary. It permits an external audit of data and facilitates further studies where these seem desirable. Abstracts should include job histories in sufficient detail to permit reasonable estimates of the type, duration, and intensity of exposure to the agent or agents of interest.

3.4.3 Verifying Cohort Completeness

As noted above, incomplete cohorts may result in bias of unknown magnitude or direction—either missing excesses or deficits for certain causes of death or identifying excesses or deficits that do not in fact exist. Where the cohort is developed from company-held records several methods for verifying cohort completeness exist:

1. Comparing names in company-held records with names that appeared on Forms 941 (or SS-1) submitted to the federal government by industry for tax withholding purposes (Marsh and Enterline, 1979).
2. Comparing names in company-held records with names on union seniority lists.
3. Searching for missing badge numbers where badge numbers are assigned consecutively, or in some other known order.
4. Comparing hire and termination dates for the cohort with known hiring and

termination patterns, and using hire and termination dates to generate a worker census for comparison with known employment data.

5. Making internal comparisons such as comparing names in personnel files with payroll lists or medical department files.
6. Comparing deaths known to the company with deaths identified by following the cohort.

3.4.4   *Entering Records for Automated Data Processing*

All records need to be put into machine-readable form. This involves coding and data entry. Both coding and data entry should be verified. In addition, all data should be subject to a machine edit looking for inconsistent items in demographic and other data. Where work histories are coded, job or departmental codes need to be developed that are meaningful in the context of the study.

3.4.5   *Follow-up*

Cohort studies involve establishing death or disease status for each cohort member as of some recent date. For cohort members who are not known from company-held records to be living or dead, the most productive resource for follow-up for deaths are records maintained by the U.S. Social Security Administration. An additional source, the National Death Index, is a record system maintained by the U.S. Department of Health and Human Services starting in 1979 (U.S. DHHS, 1990). This is more up-to-date than social security records and may be used in lieu of or in addition to the social security files. Some other sources include clearing with state drivers license bureaus, the Veterans Administration, or personal contact with the individual, his or her friends or relatives by mail or by telephone. The goal is to determine the vital status (living or dead) for every member of the cohort as of the study end date. For known deaths, death certificates must be located and coded by a qualified nosologist. Death certificates not in company-held records may be obtained from state health departments. When deaths are identified through the social security system, place of death is not always certain and a search in several states is sometimes needed to locate death certificates.

If comparisons are to be made with official mortality statistics published or available from state or national governmental agencies, death certificates must be coded as to the underlying cause of death using rules published by the World Health Organization (WHO, 1978). Since these are the rules used by governmental agencies that publish mortality statistics, it is important that they be followed if comparisons are to be made between deaths observed in the cohort and deaths expected. Only information on the death certificate can be used to establish cause of death since this is all that is available to the governmental agencies (Enterline, 1975). Thus, it would be improper to change the cause of death on the death certificate using clinical records since this kind of correction will not have been used when official mortality statistics were compiled, and thus a bias could be introduced. On the other hand, if only within-cohort comparisons are made, correcting all death certificates in the cohort would be proper.

3.4.6 Analysis of Data

Cohort or cross-sectional studies often start with observation of a cluster of cases or deaths from a particular cause and proceed to find out whether an excess truly exists. Usually investigation of disease clusters in this way does not lead to useful information about causation since clusters of cases or deaths that are not related causally can occur by chance, and it is difficult to take this into account in data analysis (Enterline, 1985). Ordinary tests of statistical significance do not apply here. As a general rule cases of disease that give rise to epidemiological investigations cannot be used in the investigative study (Doll, 1985). This is because any epidemiological study that starts with or includes an excess (which a disease cluster almost always represents) cannot be used to find out whether an excess exists.

Some guidelines have been proposed for reporting the results of epidemiologic studies (Anonymous, 1981). It is important to clearly state the purpose of the study including hypotheses, if any, that are being tested. A description of the history of the plant or workplace being studied, the processes used, and of the environment and exposures of interest is needed. The cohort must be defined and some rationale for its selection given. Assurance that the cohort is complete and follow-up adequate is needed. Demographic characteristics of the cohort should be given to the extent possible.

The mortality experience should be presented in relation to an internal control or to some external population adjusted for potentially confounding variables. External comparison groups (usually reflected in official mortality data) are often preferable since they are large enough to provide fairly stable estimates of expected deaths. A life table method is the most commonly used method to calculate deviations from expectations (Cutler and Ederer, 1958). Ordinarily the ratio of observed to expected deaths is calculated with expected deaths the sum of external specific death rates (usually specific for year, age, sex, and race) multiplied by person-years lived by the cohort in each strata. When calculated in this way, it can be said that expected deaths have been adjusted for year, age, sex, and race. The ratio is called a standardized mortality ratio (SMR). Comparisons can also be made between the distribution of deaths by cause in the cohort studied and deaths in a comparison population with adjustments as for SMRs. This is called a standardized proportionate mortality ratio (SPMR).

The external comparison group should be selected so as to represent the population from which the workers being studied were drawn. Most studies define this as the population of the county or state where the workers were located or the entire United States. Official mortality rates for this comparison group are available from the U.S. National Office of Vital Statistics and published for various age, sex, race, geographic, and cause-of-death groupings. Detailed data can be compiled from data tapes provided by the U.S. National Office of Vital Statistics or outputs from this tape can be purchased. An example of the latter is a service provided by the University of Pittsburgh (Marsh, 1987).

There is a problem in making comparisons with official mortality rates because the rules for classifying causes of death are revised at about 10-year intervals. Moreover, some of the code numbers are also changed. This makes calculation of expected deaths for specific causes difficult when a cohort is followed over a time period that spans two or more revisions of the international rules for classification of causes of deaths. As an example of a rule change, under the seventh revision of the rules (introduced in 1958) cancer of the maxillary was coded to bone cancer while in the eighth revision (effective in 1968) it was coded to sinus cancer. As an example of a coding change, prostate cancer was assigned to the code 177 under the seventh revision but to code 185 under the eighth revision.

To overcome these problems it is first necessary, for causes of death of interest, to identify the codes under each revision of the rules that best represents each cause of death. Next, either the death certificates for the study population must be coded to the revisions in effect at time of death or the official mortality data for each cause must be adjusted for differences across revisions caused by changes in the coding rules. The latter is possible using comparability ratios. These ratios are published following each revision to show the effects of changes in coding rules on deaths for selected causes. Currently, federal, state, or county death rates for use in cohort studies can be purchased either in adjusted or unadjusted form (Marsh, 1987).

Coding to the revision of the rules in effect at the time of death is desirable if coders can be found who can code to several revisions. If all death certificates for the study population are coded to a single revision, then expected deaths must be adjusted using comparability ratios. A disadvantage of this method is that comparability ratios are not available for every cause of death or for residual categories. Moreover, comparability ratios are inexact as they are based only on random samples of death certificates.

Official mortality rates can be adjusted to revisions of the rules starting in 1949. If all death certificates for a cohort followed for deaths for 30 years—say from 1955 to 1985—were coded to the revision in effect in 1985 (the ninth revision), then the official mortality rates would need to be adjusted for changes in coding rules over the years 1955–1985. Thus, death rates during the period of the eighth revision (1968–1978) would need to be adjusted to the ninth revision: during the period of the seventh revision (1958–1967) adjusted to the eighth and then the ninth, and during the sixth revision (1949–1957) to the seventh, then the eighth, and then the ninth.

When external comparisons are made, some of the variables that could confound epidemiologic analysis in occupational studies include year of death, sex, age, race, residence, and smoking habits. While it is relatively simple to adjust for year of death, age, sex, race, and residence, information on smoking habits is often difficult to obtain and failure to adjust for this can bias study results. Table 11.1 shows how smoking habits relate to selected cancers. Clearly, if the smoking habits of the study population differ from the comparison population, differences in certain cancers will be observed that could be mistakenly attributed to occupational exposure.

In presenting data, number of workers, number of deaths, and person-years should be shown for each table, and statistically significant excesses or deficits in deaths should be noted and/or confidence intervals shown. A convenient method for testing the statistical significance of an SMR is by use of tables provided by Bailar and Ederer (1964). Both internal and external comparison groups should be used, if possible. For an exposure of interest, exposed and nonexposed members of the co-

Table 11.1 Smoker/Nonsmoker Ratios of Age-Adjusted Cancer Mortality Rates

Site	Ratio
Buccal cavity	8.9
Lung	6.8
Larynx	5.1
Esophagus	3.2
Pancreas	1.7
Bladder	1.0

Source: Hammond (1975).

hort should be compared. Where external comparisons are made, local populations should be used wherever possible due to the kinds of geographic variations in mortality shown in Figure 11.1. There are also urban–rural differences for certain diseases. An example is shown in Table 11.2. Internal comparisons are often more useful than comparisons with external populations; however, numbers may be too small to permit reliable internal comparisons or the entire cohort may be considered exposed so that no internal comparison group is available.

For causes of death that test some hypotheses about exposure, data should be displayed by duration of exposure and, if possible, type and intensity of exposure. Often job or department can be used as a surrogate for type and intensity of exposure, while duration of work in that job or department can be used for duration of exposure. Multiplying duration of exposure by intensity of exposure gives a measure of cumulative exposure, and such measures are widely used in occupational epidemiology. For cancer, time since first exposure or hire should be shown. A useful table is one that displays data by time since first exposure and duration of exposure. Trend tests can be applied to variables that should be related to disease if a causal factor is present. For example, cumulative exposure and time since first exposure (for cancer) are usually related to disease if causation is present. Discussion of dis-

Table 11.2 Urban–Rural Ratios of Age-Adjusted Cancer Mortality Rates, Selected Sites, 1950–1969

Site	Males	Females
All Cancer	1.6	1.4
Esophagus	3.1	2.1
Rectum	2.7	2.1
Bladder	2.1	1.6
Lung	1.9	1.6
Breast	1.8	1.6
Stomach	1.4	1.4
Leukemia	1.1	1.2

Source: Hoover et al. (1975).

ease excesses not related to hypotheses being tested should be examined in the same detail but clearly separated from those that are.

Special consideration in the analysis should be given to the possible consequence of worker selection (Goldblatt et al., 1991). Some companies tend to hire unhealthy workers or workers with habits that lead to poor health, while others tend to hire healthy workers with desirable personal health habits. In addition, workers often select themselves out of employment on health or health-related factors. Thus, working populations may tend to become healthy or unhealthy due to this second kind of selection, and this needs to be considered in the analysis. Wherever possible, information on health-related personal or social factors should be presented and discussed. Because of worker selection in and out of the work environment, causation may be present despite poor associations with measures of exposure (Doll, 1985).

When life table methods are used for the analysis, at least three computer programs are available, and they should be strongly considered in preference to writing an entirely new and untested program (Marsh and Preininger, 1980; Monson, 1974; Waxweiler et al., 1983). Care should be taken in setting appropriate starting points for person-year counts. Person-year accumulations should only take place while populations are observed for death or other events of interest. If mathematical models are to be used for the analysis, care should be taken that the models are appropriate for the data set. Consideration should be given as to whether response should be measured on a relative or an absolute scale.

When external comparisons are made, the overall mortality rates in worker populations will usually be 70–90 percent of the mortality rates in the general population (Monson, 1986). The reason for this is that populations healthy enough to be employed will on average have a lower death rate than the population from which they were drawn since the latter includes institutionalized and other unhealthy subpopulations. If all-cause death rates are less than 60 percent of the mortality in the general population and the study is large, the possibility that some records or deaths were missed should be considered.

3.5 Case Control Studies

Perhaps case control studies are best described by examining the following table where N is the total number of individuals studied and the lower case letters a, b, c, d refer to the numbers of individuals in a particular cell:

	Exposed	Not Exposed	
Diseased (cases)	a	b	$a + b$
Not diseased (controls)	c	d	$c + d$
	$a + c$	$b + d$	N

Usually case control studies start with workers known to have a particular disease and workers not known to have the disease and compare their exposure histories. These studies contrast with cohort or cross-sectional studies that usually start with workers believed to have a particular exposure and workers not believed to have had that exposure and compare the incidence or prevalence of disease in the two groups.

In a cohort or cross-sectional study the incidence or prevalence rate is expressed as:

$$\text{Exposed} \quad a/a + c \qquad \text{Not Exposed} \quad b/b + d$$

and most often, the ratio R of these two rates is calculated:

$$R = \frac{a/a + c}{b/b + d}$$

This ratio is commonly called the relative risk or rate ratio.

In a case control study, the exposure of cases of disease is usually compared with the exposure of controls. The purpose is to see if cases have some common exposure—an exposure that distinguishes them from controls. Controls should be selected so as to be representative of the population that produced the cases of disease. Ordinarily, what is calculated is the ratio of the odds of exposures in the case group (a/b) to the odds of exposure in the control group (c/d). This produces a statistic like a relative risk or rate ratio since it is equal to the ratio of the odds of disease in the exposed group (a/c) to the odds of disease in the not exposed group (b/d). This is called an odds ratio, OR:

$$\text{OR} = \frac{a/b}{c/d} = \frac{ad}{cb} = \frac{a/c}{b/d}$$

The odds ratio is a good estimate of the relative risk or risk ratio since $a/c \div b/d$ approaches $a/a + c \div b/b + d$ when controls are properly selected and the disease or condition is not highly prevalent. When controls are actually a random sample of the population that produced the cases, they are called population based controls. If the sampling ratio (F) is known, case control studies produce an unbiased estimate of the rate ratio or relative risk R:

$$R = \frac{a/a + Fc}{b/b + Fd}$$

where F is the inverse of the fraction of the population that is in the sample.

Where exposure is strongly associated with a particular disease, something less than a population-based set of controls may be adequate. Many case control studies, for example, compare the exposure of persons dying from a particular disease with the exposure of persons dying of other diseases, and in many situations this appears to be adequate.

Case control studies have an advantage in that data collection efforts can be limited for the denominators that make up incidence or prevalence rates. For example, controls might be stratified random samples drawn from these denominators, with the total number of controls only a small multiple (2–5) of the cases and with strat-ification (or matching) on variables that relate to disease. To illustrate, suppose age is related to disease in such a way that cases are older than the population from which they come. Selecting controls distributed by age like the cases would elimi-nate the age difference when cases and controls are compared.

Combining a case control study with a cohort analysis can be very efficient. Such a study would involve collecting a limited amount of information for an entire cohort and much more detailed information on the diseases of interest and on a random sample of the entire cohort. The sample would then be used as a control group for the cases. Such a study is called a nested case control study (Marsh, 1983). Most of the detailed comments on cohort analysis apply to case control studies and, in fact, when it is possible to select population-based controls case control studies can be thought of as efficient cohort studies.

Case control studies have another advantage: Where a cluster of cases was the reason for the study, these cases can usually be included. This is because case con-trol studies focus on measuring exposure excesses rather than the disease excesses measured in cohort or cross-sectional studies. However, case control studies can include the cluster only if the cluster was not identified because of exposure. A case control study is an excellent way of fully evaluating unanticipated disease excesses in a cohort or cross-sectional study. If cases do not differ from controls in terms of suspected causative agents, then the agents are not likely to be causal and some other mechanisms are at work. Cases do not occur by chance, however, each case may be caused by a different mechanism so that they may appear to occur by chance (Rothman, 1990).

4 CREDIBILITY

It is important where there is an industrial sponsor whose products are being ex-amined for health effects, or where a study is conducted in connection with litiga-tion, that study results not be subject to control by the sponsor. This includes control over publication of the results and stop/go rules, where the sponsor is given the option of terminating the project depending on study outcomes. It is important that the study protocol be set in advance and that this be adequate to answer an agreed-upon set of questions. The question to be addressed in the analysis should be con-sidered carefully. For example, the question might be asked: Did the study popu-lation have a death rate higher than some referent population living at the same ages and in the same time period? This is different from the question: Is there any evi-dence that workers are harmed by a particular set of work exposures and conditions?

5 USES OF OCCUPATIONAL EPIDEMIOLOGY

The knowledge that certain diseases are produced by particular kinds of occupational exposures has had an important impact on workers health and has not only served to set exposure limits in the workplace but also in the general environment. For example, inhaled arsenic was discovered as a cause of lung cancer through studies of workers exposed to arsenic. This led not only to strict regulation of arsenic in the workplace but to regulation in the general environment as well. Government regulatory and public health activities in general depend on epidemiological data to predict what health effects will occur under various exposure scenarios. This requires measures of response at various dose levels and some assumptions about the form of the dose–response relationship.

While epidemiologic data are used to predict what will happen for regulatory and public health activities, in recent years they have also been used extensively in workmen compensation cases and in litigation against industry for damage due to use of or exposure to their products (Henderson, 1990). In this context they are used not to determine what will happen but to determine what has happened (Enterline, 1980). Here interest lies in causation where disease has already occurred. Use of epidemiologic data for this purpose is quite different than where the future occurrence of disease is being predicted.

To illustrate how epidemiologic studies are used in litigation, suppose epidemiologic studies show that persons exposed to some substance have a risk of lung cancer of 2 in 100 whereas persons not exposed have a risk of 1 in 100, so that the rate ratio is 2. From this it can be judged that an exposed individual with disease is just as likely to have gotten the disease from exposure as from something else. This is sometimes often called "the rule of 2" and usually the individual in such a case would be awarded some compensation for the disease or injury since the chances are 50–50, or from the standpoint of the court close to "more probable than not" that disease was caused by the exposure. Often, however, other facts can be presented that could effect the case—facts such as whether an individual with lung cancer was a smoker. Epidemiologic studies have shown that smokers are much more likely to get lung cancer than nonsmokers, and this may be taken into consideration in ruling on a particular case. If the exposure of interest were asbestos, however, this would probably be wrong since asbestos seems to multiply the existing background risk. Thus, if asbestos doubles the risk of cancer for smokers and doubles the risk for nonsmokers, then the probability that a smoker has his cancer due to asbestos is 50 percent and the probability that a nonsmoker has his cancer due to asbestos is also 50 percent. This does not mean that smoking is not an important covariate in epidemiologic studies. It just means that when the effects of disease-causing agents are multiplicative, one has no bearing on the other in judging causation in a particular case.

For cases in litigation the courts are often required to make a decision as to whether a particular occupational exposure carries any risk at all. Often these decisions diverge from those of the scientific community and the U.S. regulatory agencies. There appears to be at least two reasons for this. Because the cases before the courts relate to human disease, they tend to draw heavily on epidemiological data whereas regulatory agencies also consider animal or mechanistic data. Further, the courts usually must depend on experts presented by plaintiffs and defendants. Although judges are well aware there are experts willing to testify to the truth of virtually anything, they must somehow put together disparate views from a variety of experts to arrive at an opinion (Byrd and Gawlak, 1991).

6 SUMMARY

Occupational epidemiology presents an opportunity to identify substances that are primary in the causation of human disease. For some diseases, such as certain cancers, identification and elimination of such substances seems to offer the only hope for reducing the burden of these diseases in the next few decades. Some factors that need to be considered when carrying out and analyzing occupational epidemiologic investigations have been presented.

REFERENCES

Anonymous (1981). "Guidelines for Documentation of Epidemiologic Studies," *Am. J. Epid.*, **114**, 609–613.

Bailar, J. C. and F. Ederer (1964). "Significance Factors for the Ratio of a Poisson Variable to Its Expectation," *Biometrics*, **20**, 640.

Blot, W., B. Stone, J. Fraumeni, and L. Morris (1979). "Cancer Mortality in U.S. Counties with Shipyard Industries During World War II," *Environ. Res.*, **18**, 281–290.

Borel v. Fibreboard (1973). 493 Federal Report, 2d series.

Bridge, J. C. and S. A. Henry (1928). "Industrial Cancers," in *Report of the International Conference on Cancer*, London, July 1928, William Wood, Baltimore.

Byrd, D. M. and W. Gawlak (1991). "The Rules of the Game: What Recent Rulings Say about Courts' and Regulators' Differing Approaches to Establishing Causation for Chronic Health Risks," in *The Analysis, Communication, and Perception of Risk*, Plenum Press, New York.

Cooper, W. C., P. E. Enterline, and E. T. Worden (1962). "Estimating Occupational Disease Hazards through Medical Care Plans," *Publ. Hlth. Rep.*, **77**(12), 1065–1070.

Cutler, S. J. and F. Ederer (1958). "Maximum Utilization of the Life Table Method in Analyzing Survival," *J. Chronic Dis.*, **8**, 699–712.

Detsky, A. S. and D. L. Sackett (1985). "When Was a 'Negative' Clinical Trial Big Enough? How Many Patients You Needed Depends on What You Found," *Arch. Intern. Med.*, **145**, 709–712.

Doll, R. (1985). "Occupational Cancer: A Hazard for Epidemiologists," *Int. J. Epid.*, **14**, 22–31.

Doll, R. and H. B. Hill (1952). "A Study of the Etiology of Carcinoma of the Lung," *Br. Med. J.*, **2**, 1271–1286.

Doll, R. (1952). "Causes of Death Among Gasworkers with Special Reference to Cancer of the Lung" *Br. J. Ind. Med.*, **9**, 180–185.

Dubrow, R., L. P. Sestito, H. R. Labich, C. A. Burnett, and L. A. Salg (1987). "Death Certificate-based Occupational Mortality Surveillance in the United States," *Am. J. Ind. Med.*, **11**, 329–342.

Enterline, P. E. (1967). "The Effects of Occupation on Chronic Respiratory Disease," *Arch. Environ. Health*, **14**, 189–200.

Enterline, P. E. (1972). "A Review of Mortality Data for American Coal Miners," *Ann. NY Acad. Sci.*, **200**, 260–272.

Enterline, P. E. (1975). "Pitfalls in Epidemiologic Research," *JOM*, **18**, 150–156.

Enterline, P. E. (1978). "Asbestos and Cancer: The International Lag" (editorial), *Am. Rev. Resp. Dis.*, **118**, 975–978.

Enterline, P. E. (1980). "Attributability in the Face of Uncertainty," *CHEST*, **78**, 377S–379S.

Enterline, P. E. (1985). "Evaluating Cancer Clusters," *J. Am. Ind. Hyg. Assoc.*, **46**, 10–13.

Enterline, P. E. (1991). "Carcinogenic Effects of Man-made Vitreous Fibers," *Ann. Rev. Publ. Health.*, **12**, 459–480.

Enterline, P. E., A. E. Rikli, H. I. Sauer, and M. Hyman (1960). "Death Rates for Coronary Heart Diseases in Metropolitan and Other Areas," *Pub. Health Rep.*, **75**, 759–766.

Gloyne, S. R. (1935). "Two Cases of Squamous Carcinoma of the Lung Occurring in Asbestosis," *Tubercle*, **17**, 5.

Goldblatt, P., J. Fox, and D. Lean (1991). "Mortality of Employed Men and Women," *Am. J. Ind. Med.*, **20**, 285–306.

Guralnick, L. (1962). "Mortality by Occupation and Industry Among Men 20 to 64 Years of Age: United States, 1950," U.S. Department of Health, Education and Welfare, Vital Statistics—Special Reports, Vol. 53, No. 2, September.

Hammond, E. C. (1975). "Tobacco," in *Persons at High Risk of Cancer, Etiology and Control*, J. F. Fraumeni, Ed. Academic Press, New York.

Henderson, T. (1990). "Toxic Torts, Litigation, Medical and Scientific Principles in Causation," *Am. J. Epid.*, **132**, (Suppl. 1), S69–S78.

Her Majesty's Stationery Office (1978). Occupational Mortality. The Registrar General's Decennial Supplement for England and Wales, 1970–72, Series DS No. 1, London.

Hill, A. B. (1953). "Observation and Experiment," *N. Engl. J. Med.*, **248**, 995–1001.

His Majesty's Stationery Office (1855). Fourteenth Annual Report of the Registrar General of Births, Deaths, and Marriages in England, London.

Hoover, R., T. J. Mason, F. W. McKay, and J. F. Fraumeni (1975). "Geographic Patterns of Cancer Mortality in the United States," in *Persons at High Risk*, J. F. Fraumeni, Ed. Academic Press, New York.

Hunter, D. (1978). *The Diseases of Occupations*, 6th ed., Hodder and Stoughton, London.

Kennaway, N. M. and E. L. Kennaway (1936). "The Incidence of Cancer of the Lung and Larynx," *J. Hyg. Camb.*, **36**, 236.

Lloyd, J. W. (1971). "Long-term Mortality Study of Steelworkers. V. Respiratory Cancer in Coke Plant Workers," *JOM*, **13**(2), 53–68.

MacMahon, B. and T. F. Pugh (1980). *Epidemiology: Principles and Methods*, Little, Brown, Boston.

Mantel, N. and W. Haenszel (1959). "Statistical Aspects of the Analysis of Data from Retrospective Studies of Disease," *J. Nat. Cancer Inst.*, **22**, 719.

Marsh, G. M. (1983). "Mortality Among Workers from a Plastics Producing Plant: A Matched Case Control Study Nested Within a Cohort Study," *JOM*, **25**, 219.

Marsh, G. M. (1987). Mortality and Population Data System, Technical Report, University of Pittsburgh, Department of Biostatistics, Pittsburgh.

Marsh, G. M. and P. E. Enterline (1979). "A Method for Verifying the Completeness of Cohorts Used in Occupational Mortality Studies," *JOM*, **21**(10), 665–670.

Marsh, G. M. and M. E. Preininger (1980). "OCMAP: A User Oriented Occupational Cohort Mortality Analysis Program," *Am. Stat.*, **34**, 245.

Milham, S. (1983). Occupational Mortality in Washington State, DHHS Publication No. 83-116, Washington, D.C.

Monson, R. R. (1974). "Analysis of Relative Survival and Proportional Mortality," *Comput. Biomed. Res.*, **7**, 325–332.

Monson, R. (1986). "Observations on the Healthy Worker Effect," *JOM*, **28**, 425–433.

Morgenstern, H. (1982). "Uses of Ecologic Analysis in Epidemiologic Research," *AJPH*, **72**(12), 1336–1344.

Pearl, R. (1940). *Introduction to Medical Biometry and Statistics*, W. B. Saunders, Philadelphia.

Robinson, W. S. (1950). "Ecological Correlations and the Behavior of Individuals," *Am. Sociol. Rev.*, **15**, 351–357.

Rothman, K. (1990). "A Sobering Start for the Cluster Busters Conference," *Am. J. Epid.*, **132** (Suppl. 1), S6–S13.

Snow, J. (1965). *Snow on Cholera*, Hafner Publishing, New York.

Society of Actuaries (1967). *1967 Occupational Study*, Society of Actuaries, Chicago.

U.S. DHEW (1963). Smoking and Health, PHS Publication No. 1103, Department of Health, Education and Welfare, U.S. Government Printing Office, Washington, D.C.

U.S. DHEW (1969). The 1970 Census and Vital Health Statistics—A Study Group Report of the Public Health Conference on Records and Statistics, Public Health Service Publication No. 1000, Ser. 4, No. 10, Department of Health, Education and Welfare, U.S. Government Printing Office, Washington, D.C.

U.S. DHEW (1975). Atlas of Cancer Mortality for U.S. Counties, DHEW Publication No. (NIH), 75-780, Department of Health, Education and Welfare, U.S. Government Printing Office, Washington, D.C.

U.S. DHHS (1990). User's Manual, The National Death Index, DHHS Publication No. (PHS) 90-1148, Department of Health and Human Services, U.S. Government Printing Office, Washington, D.C.

Waxweiler, R. J., J. J. Beaumont, J. A. Henry, D. Brown, C. Robinson, G. Ness, J. Wagoner, and R. Lemen (1983). "A Modified Life-table Analysis Program System for Cohort Studies," *JOM*, **25**, 115–124.

Whitney, L. (1934). Death Rates by Occupation Based on Data of the U.S. Census Bureau, National Tuberculosis Association.

WHO (1978). *Manual of the International Classification of Disease, Injuries and Causes of Death*, World Health Organization, Geneva.

APPENDIX: STATISTICAL METHODS BIBLIOGRAPHY

N. Breslow and N. E. Day, *Statistical Methods in Cancer Research, Vol. 1, The Analysis of Case-Control Studies*, IARC Scientific Publication No. 32. IARC, Lyon, 1980.

N. Breslow and N. E. Day, *Statistical Methods in Cancer Research, Vol. II, The Design and Analysis of Cohort Studies*, IARC Scientific Publication No. 82. IARC, Lyon, 1987.

H. Checkoway, N. E. Pearce, and D. J. Crawford-Brown, *Research Methods in Occupational Epidemiology*, Oxford University Press, New York, 1989.

D. Kleinbaum, L. Kupper, and H. Morgenstern, *Epidemiologic Research*, Lifetime Learning Publication, Belmont, CA, 1982.

K. Rothman, *Modern Epidemiology*, Little, Brown, Boston, 1986.

Health Surveillance Programs in Industry

W. Clark Cooper, M.D., and
Mitchell R. Zavon, M.D.

1 INTRODUCTION

There is no universal agreement on what is included or what should be included in the term *health surveillance in industry*. Some apply a narrow interpretation, for example, making it synonymous with "health effects monitoring" or "the periodic medical-physiological examinations of exposed workers with the objective of protecting health and preventing occupationally related disease. The detection of established diseases is outside the scope of the definition" (Zielhuis, 1985). It was defined as distinct from *biological monitoring*, "the measurement and assessment of workplace agents or their metabolites either in tissues, secreta, excreta, expired air or any combinations of these to evaluate exposure and health risk, compared to an appropriate measure" (Zielhuis, 1985). In our discussion, however, we will not limit the discussion to such a narrow definition of health surveillance; the maintenance and protection of worker health must include evaluations of exposure as well as evidence of biologic effects, if any.

Any discussion of health surveillance in industry must take into account the reasons for such surveillance. First and foremost in the rationale for such a program should be concern with health maintenance and health protection of those in the work force. Any discussion of health surveillance in industry must, however, take into account the important influence of government regulations. These include man-

Patty's Industrial Hygiene and Toxicology, Third Edition, Volume 3, Part A, Edited by Robert L. Harris, Lewis J. Cralley, and Lester V. Cralley.
ISBN 0-471-53066-2 © 1994 John Wiley & Sons, Inc.

dates by the Occupational Safety and Health Administration (OSHA), the Mine Safety and Health Administration (MSHA), the Environmental Protection Agency (EPA), the Equal Employment Opportunities Commission (EEOC), and others. There are also influential guidelines by nonregulatory agencies, such as the National Institute for Occupational Safety and Health (NIOSH). The specific requirements embedded in regulations promulgated by these agencies, and some of the fallacies therein, will be addressed after a more general review of current medical and scientific concepts regarding the role of such surveillance in worker health protection.

There have been several excellent and comprehensive reviews of the topic that should be read by all who are interested (World Health Organization, 1975; Rothstein, 1984). A major conference on medical screening and biological monitoring, sponsored by NIOSH, EPA, and the National Cancer Institute (NCI) held in 1984 had over 70 presentations. The conference proceedings were published in 1986 (Halperin et al., 1986).

While our emphasis will be on hazard-oriented health surveillance aimed at protection of workers, it is still important to regard this as part of a broader program of health maintenance. Health surveillance must not be fragmented! The effective incorporation of special elements into a more general health program is the preferred approach. For this reason it is not feasible to discuss hazard-oriented examinations out of the context of examinations to detect preexisting conditions or the development of abnormalities unrelated to the work environment. Schulte (1991) pointed out similar overlapping when considering biomarkers, that is, indicators of exposure, effect, and susceptibility.

2 OBJECTIVES OF HEALTH SURVEILLANCE

An occupational health and safety program has many elements. Physical examinations and laboratory testing are only parts of a comprehensive program. Perhaps the most valuable and important part of the examination is the history. A self-history questionnaire, of which there are many readily available examples, can be followed up by the examiner pursuing leads opened up by the self-history. But the physical examination and the laboratory testing no more prevent illness or injury than a film badge prevents radiation exposure. All parts of the examination should be evaluated individually. Each part of the examination should only be included if it can be justified by the information it provides. The entire examination should be designed and performed to secure maximum preventive benefits at minimal risk to the worker and minimal cost and inconvenience to the employer. The following sections outline the types of examinations that are commonly performed.

2.1 Preplacement Examinations

Preplacement examinations are those performed on otherwise qualified applicants, persons who have already been offered employment. There should be no implication that an examination is a "preemployment" examination or that an employment offer is contingent on "passing an examination." In the past this examination was done before an offer of employment was made, and it was often stated categorically that an offer of employment would be made only if the applicant "passed" the examination. The objective was to obtain the most physically fit work force that was available and to exclude individuals with any suggestion of medical or psychological problems. Such practices were defended by pointing out the cost of hiring and training individuals who could not perform their jobs, who would work but a brief time, and who could create expensive medical care problems.

For many years this has been recognized as ethically and socially unacceptable (Goerth, 1983). The Americans with Disabilities Act (ADA, 1990), which became operative in 1992, now places severe restrictions on an employer's response. It stipulates that if the examination detects a condition that would jeopardize the health or well-being of the employee or of fellow employees, it is now the responsibility of the employer to modify the work situation so that it is safe for the employee and for fellow employees, if technologically and economically feasible. To decide what is technologically and economically feasible may be very difficult.

During the past two decades, there has also been increasing ability to detect genetic factors that are associated with higher risks for many diseases (genetic predisposition) as well as some that are indicative of hypersusceptibility to specific hazardous agents. In 1983, the Office of Technology Assessment of the Congress of the United States (OTA, 1983) published a 105-page review of the state of the art as of that period as it related to genetic testing. They differentiated *genetic screening*, "a one-time test to determine the presence of particular genetic traits in individuals," from *genetic monitoring*, "the periodic testing of workers to assess damage to their DNA or chromosomes from exposure to hazardous substances or agents."

The ethical and legal implications of genetic screening and indeed of all screening procedures have been the subject of many very spirited discussions in the past decade (Lappé, 1983; Bayer, 1986; Ashford, 1986; Samuels, 1986; Bernard and Lauwerys, 1986; Lappé, 1986; Ashford et al., 1991; Draper, 1991; Schulte, 1991; AMA, 1991). Certainly medical or genetic screening can only be done with full awareness of the requirements of the Americans with Disabilities Act. These requirements can, in turn, have a negative impact on the performance in some industries of any medical examinations at all. They may present problems, not prevent them. The foregoing issues may modify but do not justify the elimination of preplacement examinations. As will be pointed out, they serve an essential function by providing: (1) a record of previous work experiences, (2) information on the state of health prior to joining the work force, (3) in a few situations useful tests for hypersusceptibility, and (4) a baseline for comparison with later observations and measurements. To a core preplacement examination, that is, history, physical examination, and limited clinical laboratory appraisal, there should be added elements tailored to the special potential hazards of the location or job under consideration.

2.2 Preassignment Examinations

When a worker is being transferred from one operation to another and the new job has known potential hazards different from the former operation, it may be necessary to carry out a special examination. This may include inquiries about the present state of health and activities off the work site that might, in some manner unsuspected by the worker, interact with a substance to which there might be exposure at the new job. This examination might also include a special physical examination, clinical laboratory tests, or, in some cases, biological monitoring to provide a baseline. Similar considerations and special examinations may also be justified in case of significant changes in processes or working conditions. And, in instances where some persons are known to be peculiarly susceptible to a chemical agent, if available, test for susceptibility may be justified. Special change of job examinations of this type also offer opportunity for additional worker education about possible health hazards associated with the new job or process, with emphasis on personal protection.

2.3 Periodic Examinations

Periodic examinations, an essential part of health monitoring, are performed to detect changes in physical condition or any signs or symptoms of incipient disease. When appropriate, they may include biological monitoring for absorption of harmful agents and the use of other biomarkers. Although most regulations for specific chemical exposures require that examinations be made available at least annually, when exposures are controlled, such frequent examinations are rarely productive.

Periodic examination also provides a splendid opportunity for reinforcing education regarding good health habits and noting the potentially disastrous effects of bad health habits such as smoking cigarettes, eating a high fat diet, or failure to get a reasonable amount of exercise. The periodic surveillance of a group of workers having similar exposures may identify changes that are not of sufficient magnitude to be notable in an individual, but that suggest an abnormal situation and perhaps an increased risk when analyzed statistically for the group (Zavon, 1963). The absence of positive findings is equally useful as part of the monitoring needed in a preventive program.

2.4 Termination Examinations

Documentation of the health status of the individual worker at the termination of employment is desirable for both the worker and the employer. Evidence of any changes that have occurred during the employment period can be evaluated at the time when exposure data are more readily acquired than at a later date. Changes in the audiogram of significant degree can be evaluated in an attempt to determine if such changes are job related. If corrective action is indicated, it can be recommended immediately. If a compensation claim should be filed, that too can be done immediately.

There is, of course, no way to rule out the possibility that effects may show up at a later date. The long latency before cancer caused by asbestos, cancer caused by smoking cigarettes, radiation-induced leukemia, or numerous other long-delayed disease can make etiologic diagnoses very difficult in some situations.

2.5 Special-Purpose Examinations

Other examinations, which may not fit into the preceding four categories, may be needed for special purposes such as the following:

1. Requirements of regulatory agencies for evaluating the health status of specific types of employment, such as vehicle drivers in interstate commerce or airline pilots.
2. Evaluation of the effects, if any, of accidental exposure to a specific chemical or physical agent or a mixture of chemical and/or physical agents.
3. Evaluation of health status before return to work after absence for disease, injury, or a prolonged layoff.
4. Evaluation of the health status of an employee who has difficulty in performing work satisfactorily, in the absence of any known hazard.
5. Determination of impairment of function or disability after specific worker complaints.
6. Undue complaint of stress associated with employment.
7. Certification of fitness to wear a respirator (discussed later).

3 SURVEILLANCE FOR GENERAL HEALTH MAINTENANCE

3.1 Content and Scope

The presence or absence of long-term benefits and the cost effectiveness of periodic examinations have been subjects for investigation and discussion for many years (Robert, 1959; Siegel, 1966; Morgan, 1969; Jacobs and Chovil, 1983). The American Medical Association in a council report published in 1983 (AMA, 1983) provides general guidelines that includes recommendations for medical evaluations at intervals of 5 years until age 40, with shorter intervals until age 65, when annual examinations are suggested. Most physicians who have given thought to the subject tend to agree that multiphasic screening, if used judiciously, can provide the occupational physician with a valuable tool. Inasmuch as the workplace is the complement of the school where children gather in one place, adults generally gather at their place of work. The workplace serves as a focus where surveillance of blood pressure, detection of diabetes, and numerous other life promoting and life extending health procedures can be implemented with minimum inconvenience to the individual and minimum need for self-initiation. It is important that all screening be subject to rigorous quality control. Each test or testing procedure has its own range of normal values. An understanding is needed of the range of normal values. A relaxed attitude toward minor deviations from the average range and vigorous follow-up of significant findings should be standard operating procedure. Screening should never be undertaken without a written procedure in place specifying precisely how the results will be conveyed to the individual, what will be done about significant findings as far as medical follow-up, and how the cost for follow-up will be borne. This latter issue can cause significant employee relations problems if the results, on recheck with the personal physician, are deemed to be insignificant, yet the employee is presented with a bill for services not covered by insurance.

3.2 Educational Value

As pointed out earlier, as in all medical examinations, the opportunity for education of the patient or the worker is one of the major potential benefits of periodic examinations. The opportunity for general health education is inseparable from education oriented toward occupational hazards.

4 HAZARD-ORIENTED MEDICAL EXAMINATIONS

It is necessary to consider hazard-oriented medical examinations in terms of (1) what is legally required, (2) what has been recommended by official agencies such as NIOSH, and (3) what has persuasive medical and scientific justification. The resulting programs are not necessarily the same. Above all else, the surveillance program should be dictated by ''good practice'' and then reviewed to make sure that it meets all legal and regulatory requirements. Regulations seldom clearly define the objectives of medical surveillance nor consider risks as well as benefits of such programs. Most people would question the wisdom of mandating complex and expensive examinations of thousands of workers whose exposures are at levels so low that hazard-derived abnormalities would be extremely rare (Halperin et al., 1986). Nevertheless, a mandatory examination, if incorporated in a regulatory standard, must be met faithfully and with careful attention to the quality of information and its relationship to work exposures.

Before leaving this subject, a word about ''newly discovered illnesses.'' Cumulative trauma disorder can be used as an example of such an illness. If it exists, does it exist to the extent that is implied by the number of reports of its presence and the number of people compensated for disability? Health surveillance for such newly discovered, newly popularized entities must be very carefully constructed and implemented (Hadler, 1990).

4.1 OSHA Requirements

4.1.1 General Requirements

All permanent standards of the Occupational Safety and Health Administration (OSHA) contain requirements for medical surveillance. As of mid-1992 there were 27 such permanent standards, 5 of which apply to work situations rather than specific chemical agents. In addition, there are several proposed standards. Table 12.1 shows the particular substance or work situation and the standard, with the specific pertinent paragraph in the standard.

4.1.2 Examinations for Ability to Wear a Respirator

Current OSHA regulations state: ''Persons should not be assigned to tasks requiring use of respirators unless it has been determined that they are physically able to

Table 12.1 OSHA Standards Requiring Medical Surveillance

Substance	Standard
Hazardous waste operations	1910.120 (f)
Asbestos	1910.1001 (l) Appendix D (1910.1011 (j)
4-Nitrobiphenyl	1910.1003 (g)
α-Naphthylamine	1910.1004 (g)
Methyl chloromethyl ether	1910.1006 (g)
3,3'-Dichlorobenzidine	1910.1007 (g)
bis-Chloromethyl ether	1910.1008 (g)
β-Napthylamine	1910.1009 (g)
Benzidine	1910.1010 (g)
4-Aminodiphenyl	1910.1011 (d), (g)
Ethyleneimine	1910.1012 (g)
β-Propiolactone	1910.1013 (g)
2-Acetylaminofluorene	1910.1014 (g)
4-Dimethylaminoazobenzene	1910.1015 (g)
N-Nitrosodimethylamine	1910.1016 (g)
Vinyl chloride	1910.1017 (k)
Inorganic arsenic	1910.1018 (h) (3) (iv), (n), (q), Appendix C
Lead	1910.1025 (j), Appendix C
Cadmium	1910.1027 (l)
Benzene	1910.1028 (i), Appendix C
Coke oven emissions	1910.1029 (j)
Blood-borne pathogens	1910.1030 (f),
Cotton dust	1910.1043 (h), Appendix B
1,2-Dibromo-3-chloropropane	1910.1044 (m), (p)
Acrylonitrile	1910.1045 (n), (q)
Ethylene oxide	1910.1047 (i), (k)
Formaldehyde[a]	1910.1048 (i), (o), Appendix C
Laboratory chemicals	1910.1450 (g)
4,4'-Methylenedianiline	1910.1050 (proposed)
1,3-Butadiene	1910.1051 (i) (proposed)
Methylene chloride	1910.1052 (i) (proposed)
Ethylene dibromide	—, — (proposed)

[a] Proposed amendment *Federal Register* 6/15/91 pp. 32302–32318.

perform the work and use the equipment. The local physician should determine what health and physical conditions are pertinent. The respirator user's medical status should be reviewed periodically (for instance annually)'' [29-CFR 1910-134 (b) (10)]. Inasmuch as no medical examination procedure has yet been validated for determining a person's ability to wear a respirator, there are numerous approaches to the problem of complying with this part of the regulation (Harber, 1984; Houdous, 1986). This topic will be discussed in more detail in Section 9.3.

4.2   Mine Safety and Health Administration

The Federal Coal Mine Health and Safety Act of 1969 (FCMHSA, 1969) was land-mark legislation in its requirement for medical examination of coal miners. It set schedules for chest roentgenograms, required the Secretary of Health, Education and Welfare to prescribe classification schemes for radiographic interpretation and tied compensability to film interpretations. Regulations developed under the act added provisions relating to the training and proficiency of physicians who interpreted films and specified methods of measuring pulmonary function, among other detailed requirements. Thus was born the ''B'' Reader.

The legislative mandate, PL 95-164 (FMSHAA, 1977), for the Mine Safety and Health Administration included the provision that its standards shall ''where appropriate, prescribe the type and frequency of medical examinations or other tests which shall be made available by the operator.'' Current MSHA regulations pertaining to medical examinations and the specific requirements are listed in Table 12.2.

4.3   Medical Surveillance Recommendations by the National Institute for Occupational Safety and Health (NIOSH)

The NIOSH criteria documents, which recommend standards for occupational exposures, always include recommendations to OSHA for medical surveillance. In nearly all cases these prescribe mandatory surveillance. The objectives (e.g., for epidemiologic studies, for early detection and treatment, for detection of group risk factors, or merely for good occupational medical practice) are rarely stated. In comparatively few instances have these recommendations been incorporated in the OSHA regulations. Although they do not have the force of regulations, they constitute a powerful coercive influence, suggesting as they do a level of good practice. They are a useful guide but must be viewed critically as to justification and whether implementation is justified by the benefit to be derived.

In 1974 NIOSH commissioned the preparation of so-called mini-criteria documents for nearly 400 chemicals. The NIOSH/OSHA Standards Completion Program, Draft Technical Standards, included recommendations for biologic monitoring and medical examinations. These were included in the Proctor and Hughes handbook *Chemical Hazards in the Workplace* (Proctor and Hughes, 1978), which provided a valuable summary to aid the occupational physician in examinations of workers. This was updated in 1988 and again in 1991 (Proctor et al., 1988; Hathaway et al., 1991). For 198 of 398 substances in their 1988 summary, the authors did not see a need for periodic physical examinations solely on the basis of exposures

Table 12.2 MSHA Regulations and Their Medical Requirements

MSHA Regulation	Requirement
30 CFR Part 70.510	Audiometric examinations for underground
30 CFR Part 71.805	and surface coal miners
30 CFR Part 49.7	Physical requirements for mine rescue teams
42 CFR Part 37	X-ray program for underground coal miners

to a specific agent. They stressed that a brief interim history is usually sufficient for such surveillance.

4.4 Other Sources for Recommendations

It is no longer sufficient to assume that the principal objective of health surveillance is to observe changes in cardiovascular, pulmonary, hepatic, or hematopoietic functions. Reproductive function (Whorton et al., 1977), nervous system function, indeed, any organ or organ system may prove to be critical and therefore must be observed periodically to ensure that adverse effects are not occurring. At least one of the medical specialty societies, the American Thoracic Society, has developed detailed recommendations for surveillance of the respiratory tract when there is a potential for exposure to respiratory hazards (American Thoracic Society, 1982).

4.5 Justification for Hazard-Oriented Medical Surveillance

Monitoring of the environment provides an ongoing indication of exposure to chemical and physical agents of use for engineering control and evaluating medical findings. Medical surveillance of the individuals working or living in that environment provides another level of monitoring while biologic indicators of absorption of chemical agents based on analysis of the agent or a metabolite in expired air, urine, or blood is a monitoring method increasingly used by the industrial hygienist and the medical community (Lauwerys, 1983). The physician can use such measurements as part of the hazard-oriented surveillance program in addition to monitoring minor physiologic changes, reductions of function or pathologic changes that may be early manifestations of toxicity in individual workers or groups of workers. In further justification of periodic medical surveillance it should be noted that every contact with the health establishment provides an opportunity for education of the worker, education that can create a healthy awareness of risk without creating unnecessary fears.

4.5.1 Indicators of Absorption

Biologic monitoring has become increasingly important in medical surveillance (Lowry, 1986; Bernard and Lauwerys, 1986). A substance present in the workplace may also be present in consumer products and in the general environment. Only by measuring the concentration of the substance or a metabolite in body fluid, expired air, or other bodily substance can an evaluation of total absorption be made. Sampling should be done at a specific time for some substances but may be done at any time for long biologic half-lived materials, such as lead. But the conditions of collection of the sample, how it is stored and transported, and how it is to be analyzed should all be determined before collection. Abnormally high or low values, depending on the biological marker being used, if confirmed by additional biologic sampling, may show the need for careful environmental monitoring and increased controls or better hygienic education. A more detailed discussion of monitoring for heavy metals is provided in Volume IIIB, Chapter 3, of this series.

The American Conference of Governmental Industrial Hygienists (ACGIH) Biological Exposure Indices Committee publishes an annual update of information on selected biomarkers (ACGIH, 1991). These biological exposure indices (BEIs) are defined as reference values or guidelines, representing the level of determinants that are most likely to be observed in specimens collected from a healthy worker who has been exposed to chemicals to the same extent as a worker with inhalation exposure to the threshold limit value (TLV). Documentation is published periodically, with the most recent having appeared in 1991 (ACGIH, 1991). Space does not permit a detailed discussion of this subject but a list of the 26 substances included in the 1991–1992 annual report is shown in Table 12.3. The complete report includes sampling times and all of the BEIs that have been adopted as well as new or revised BEIs that have been proposed but not yet adopted.

Table 12.3 Adopted Biological Exposure Determinants 1991–1992

Airborne Chemical	Material Being Measured
Aniline	p-Aminophenol in urine
	Methemoglobin in blood
Benzene	Total phenol in urine
	Benzene in exhaled air
Cadmium	Cadmium in urine
	Cadmium in blood
Carbon disulfide	2-Thiothiazolidine-4-carboxylic acid in urine
Chromium (VI) water-soluble fume	Total chromium in urine
Carbon monoxide	Carboxyhemoglobin in blood
	CO in end-exhaled air
N,N-Dimethylformamide (DMF)	N-Methylformamide in urine
Ethyl benzene	Mandelic acid in urine
	Ethyl benzene in end-exhaled air
Fluorides	Fluorides in urine
Furfural	Total furoic acid in urine
n-Hexane	2,5-Hexanedione in urine
	n-Hexane in end-exhaled air
Lead	Lead in blood, urine
	Zinc protoporphyrin in blood
Methanol	Methanol in urine
	Formic acid in urine
Methemoglobin inducers	Methemoglobin in blood
Methyl chloroform	Methyl chloroform in end-exhaled air
	Trichloroacetic acid in urine
	Total trichloroethanol in urine
	Total trichloroethanol in blood
Methyl ethyl ketone (MEK)	MEK in urine
Nitrobenzene	Total p-nitrophenol in urine
	Methemoglobin in blood
Organophosphorus cholinesterase inhibitors	Cholinesterase activity in red cells
Parathion	Total p-nitrophenol in urine
	Cholinesterase activity in red cells

Table 12.3 (*Continued*)

Airborne Chemical	Material Being Measured
Pentachlorophenol (PCP)	Total PCP in urine
	Free PCP in plasma
Perchloroethylene	Perchloroethylene in end-exhaled air
	Perchloroethylene in blood
	Trichloroacetic acid in urine
Phenol	Total phenol in urine
Styrene	Mandelic acid in urine
	Phenylglyoxylic acid in urine
	Styrene in venous blood
Toluene	Hippuric acid in urine
	Toluene in venous blood
	Toluene in end-exhaled air
Trichloroethylene	Trichloroacetic acid in urine
	Trichloroacetic acid and trichloroethanol in urine
	Free trichloroethanol in blood
	Trichloroethylene in end exhaled air
Xylenes	Methylhippuric acids in urine

Reproduced with permission of the American Conference of Governmental Industrial Hygienists from the 1991–1992 *Threshold Limit Values and Biological Exposure Indices*.

4.5.2 Indicators of Early Effects

Biological monitoring is an indicator of absorption of the substance being monitored but not an indicator of an early effect. The use of the term *subclinical poisoning* or *subclinical effect* has served to confuse the question of what is an early adverse effect versus what is simply evidence of absorption (Lowry, 1986). There are many recommendations regarding when a biological marker for health effects should be a trigger for action, but few of these recommendations are based on specific criteria, and there is little consistency among the recommendations for various substances. Absorption of organophosphorus cholinesterase inhibitors will result in inhibition of red blood cell and plasma cholinesterase. At what point should the worker be removed from further exposure when there is reduction in cholinesterase? (Coye et al., 1986; Gallo and Lawryk, 1991; McConnell et al., 1992). Similar questions can be asked about other biologic markers for lead, cadmium, cotton dust, benzene, and a host of other materials.

A major problem in interpreting early changes is that normal variations occur in virtually all tests. A preexposure baseline and serial follow-up testing is required for many of these biological markers. If there is indication of a change from baseline-determined values, there is then the problem of deciding not only whether removal from exposure is justified but also the decision as to whether a suspected abnormality is sufficient to be reported as an occupational illness on the OSHA log 200.

In a later section (4.5.5) problems associated with the early detection of cancer will be discussed. Additional information on this subject will be found in Volume IIIB, Chapter 3, of this series.

4.5.3   Indicators of Hypersusceptibility

All individuals exposed to a chemical do not respond alike. The differences in response are not necessarily the result of different levels of exposure. Although variation in response is the rule rather than the exception, there is usually a normal range
of response that the clinician can come to expect. Occasionally, a person will prove
to be responsive to such a degree that the individual can be properly labeled "hypersusceptible." This classification is usually one of exclusion. Before it can be
made a careful investigation should be made to determine a possible explanation: Is
there a synergistic or additive effect of occupational exposure plus nonoccupational
exposure? Examples commonly encountered are cigarette smoking and asbestos exposure or ethyl alcohol and chlorinated hydrocarbon solvents. Preexisting disease
may be the cause of hypersusceptibility. There may be inborn errors of metabolism
that interfere with the detoxification of a chemical or augment its effects. Inherited
traits that may predispose to hypersusceptibility were discussed in Section 2.1.1.
Although the reality of such inherited factors is indisputable, it has been difficult to
establish firmly their importance in occupational medicine. Cooper reviewed the
status of the most promising indicators of hypersusceptibility and could not recommend their application in routine screening. The tests he considered were those for
sickle cell trait, glucose-6-phosphate dehydrogenase (G6PD), α-1-antitrypsin, and
cholinesterase deviants. All appeared to be appropriate subjects for controlled research but not for mandatory inclusion in regulations or as positive indicators for
exclusion from particular jobs (Cooper, 1973). In the intervening years, there has
been little practical application of genetic screening, although it has been the subject
of a great deal of debate (see Section 2.1.2). It certainly cannot be applied to discriminate against individuals or ethnic groups.

4.5.4   Tests for Effects on Fertility

The discovery in the late 1970s that dibromochloropropane (DBCP), a nematocide
that had been widely used for more than 20 years, could cause reduction in sperm
or sterility in human males at very low doses caused great discussion and conster-
nation (Whorton et al., 1977). The impact of exposures to lead on the ability of both
males and females to reproduce has been a cause for inquiry and research for many
years. This has not necessarily been only because of its impact on fertility but also
because of the possibility that in utero lead exposure might impact adversely on the
fetus (Rom, 1976). Investigations related to reproduction are fraught with considerable difficulty because of the sensitivity of the subject in our culture and the lack
of generally acceptable monitoring methods (Levine et al., 1980). Questionnaires
of past reproductive experience can be considered only an investigational tool after
the fact. Hormone studies of various types have thus far not become routine monitoring methods. It is likely that for the immediate future we will continue to be
limited to retrospective studies to determine whether a chemical or physical agent
does have an impact on fertility.

4.5.5   Early Detection of Cancer

Cancer has become a major preoccupation of the public as life expectancy has increased. Where workers are exposed to suspected or proven carcinogens, medical

surveillance to detect premalignant changes or early cancer has become a fundamental part of health surveillance (Hulka, 1986). Unfortunately it has so far resulted in very limited benefits in reducing occupational cancer mortality. For example, lung cancer and mesothelioma are major concerns in asbestos-exposed workers. However, periodic examinations of current and former asbestos workers rarely detect evidence of tumors that result in curative intervention (Cooper, 1982). This is consistent with other major studies where large populations of cigarette smokers were observed by three medical centers over 5-year periods, with tests including periodic sputum cytologic tests and chest roentgenograms. No statistically significant effects on lung cancer mortality were observed, even in one of the studies where tests were done every 4 months (Berlin et al., 1984).

Bladder cancers have been recognized for many years as a type of occupational or environmental tumor that might respond well to early detection by screening tests, such as urine cytology. However, there are still many uncertainties as to the best procedures and the actual benefits, as pointed out in an excellent international conference on "Bladder Cancer Screening in High Risk Groups" sponsored by NIOSH in 1989 (Schulte et al., 1990).

The current permanent asbestos standard (U.S. Dept. Labor, 1983 29-CFR-1910.1000) does not address itself to lung cancer or mesothelioma in its medical surveillance requirements, which are limited to annual chest films, measurements of pulmonary ventilatory function, and history.

The standard for vinyl chloride (U.S. Dept. Labor, 1983 29-CFR-1910.1000) includes provisions for a battery of liver function tests, presumably because vinyl chloride is a known hepatotoxin, and it has been suggested that such hepatoxic effects would precede angiosarcomas caused by exposure to vinyl chloride. There is no evidence that this is the case. In any event, if exposures are kept below 1 ppm, it is probable that thousands of workers would be repeatedly examined before any vinyl-chloride-related disorder were discovered. In the meantime, many individuals with liver disorders resulting from other toxins, most commonly ethyl alcohol, would be detected and subject to extensive medical investigation, some involving a real risk. Others are likely to be removed from exposure or advised to terminate employment.

The standards for coke oven emissions and for arsenic include provisions for sputum cytologic examinations. The former also requires urine cytologic studies (U.S. Dept. Labor, 1983 29-CFR-1910.1029). Frequent sputum examinations in those who have worked in high-risk areas for many years may detect an occasional operable lung cancer. This is probably an unproductive exercise in relation to individuals working under controlled conditions, even though it is required.

Urine cytology is also unlikely to be a useful screening procedure in coke oven workers, in view of the relatively low incidence of urinary tract cancers in the group. It can be a useful test in those heavily exposed in the past to proved bladder carcinogens, such as β-naphthylamine and benzidine. But even with these compounds, urine cytology has generally proved of limited benefit (Farrow, 1990).

Specifying medical surveillance for cancer, particularly for chemicals whose effects in humans are speculative and based solely on findings in experimental animals, has produced difficulties for regulatory agencies. The published standards include requirements for preplacement and annual physical examinations. Many stipulate a "personal history of the employee, family and occupational background, including genetic and environmental factors." Some also stipulate that "in all physical examinations the examining physician should consider whether there exist conditions of increased risk including reduced immunological competence, those undergoing treatment with steroids or cytotoxic agents, pregnancy and cigarette smoking."

It is unclear how the examining physician would evaluate all the information obtained or how it might affect employability. As can be seen, to follow some of these rules blindly, and without careful thought as to the consequences, could lead one into a maze of problems.

4.5.6 Biochemical Markers of Cancer

Such biochemical markers of cancer or incipient cancer as carcinoembryonic antigen (CEA), α-fetoprotein (AFP), and prostate specific antigen (PSA), deserve mention. The value of these markers remains controversial. PSA, the most recently used of these antigens, is well established and is used routinely in the investigation and confirmation of prostatic cancer. Whether it should be used as a screening tool is controversial only because of the large number of false positives and false negatives. The cost–benefit ratio has to be evaluated both as regards the economics of the screening and the psychologic cost of both false negatives and particularly of false positives (Drago, 1989). As pointed out by Brandt-Rauf (1988) studies of possible biochemical tools for effective monitoring of occupational cancer are making rapid strides, but their utility still has to be proven.

5 RELATIONSHIP OF MEDICAL SURVEILLANCE AND EXPOSURE DATA

The ready availability of computer power undreamed of only a decade ago has made possible the relatively easy coordination of exposure and medical data. Numerous examples of such coordination in both large and small plant populations are now available. Commercially available software for use with desktop personal computers has reduced the technical and financial constraints on such programs. It is too soon to say that such work will revolutionize health surveillance and prevent adverse health effects on large groups of workers. This is a distinct possibility as medical records and exposure data are increasingly subject to ongoing trend analyses as recommended many years ago (Ott, 1977).

6 U.S. EQUAL EMPLOYMENT OPPORTUNITIES COMMISSION (EEOC)

Another agency that has acquired legal sanctions that impact on health surveillance programs is the EEOC, which has responsibility for enforcing the Americans with Disabilities Act (ADA) of 1990 (ADA, 1990). The employment provisions of this act went into effect July 26, 1992, for employers with 25 or more employees and on July 26, 1994, will be extended to cover those with 15 or more employees.

Included in its many provisions involving public transportation, building accommodations, and other provision for those with disabilities are many stipulations regarding the employment of those with disabilities. What may and may not be included in the medical examination and appraisal of potential employees and many of the requirements for accommodating these employees and potential employees are stipulated in the regulations to implement this act. The requirements involve consideration of hardship to employees and employers (ADA, 1991). Whether accommodation of an individual with a disability is economically and technically feasible is one consideration. How much risk is entailed to the worker and co-workers as a result of placing the worker with a disability in a particular job? As can be easily appreciated, decisions on many of these questions will be highly subjective and often arbitrary. How much expenditure is economically feasible to place a worker with a particular disability? What is technically feasible? But for the health program, the question needs to be posed once again, when should the preplacement examination be given; what can be justified for inclusion in the examination; what are the markers for an unsafe situation in which either the worker or co-workers would be at greater than acceptable risk? At this writing, implementation of the act is too new to know what will be ruled acceptable and what will be ruled unacceptable. Good practice and good will must prevail if the examining physician, the employee, and the employer are all to avoid an attack of acute anxiety as well as legal penalties. The reader is referred for additional reading (Anfield, 1992; St. Clair and Shults, 1992).

7 RECORDING

7.1 Maintenance of Records

Because of the long latent period between exposure and the appearance of chronic effects such as asbestosis or occupationally related cancers, there is need for preservation of medical records for what is essentially the lifetime of the employee. OSHA requires that occupational medical records be retained for the duration of employment plus 20, 30, or 40 years, depending on the particular regulation. Many of the early OSHA regulations specify that, if the employee terminates employment, retires, or dies or if the employer ceases business without a successor, the records must be forwarded to the director of OSHA. It is recommended that all medical records be kept for a minimum of 40 years and that duplicate records be kept in storage on computer-accessible film or microfiche. On occasion we have recommended that employees be invited to return and destroy their own medical records on their 100th birthday.

The use of these many records for epidemiological studies of morbidity and mortality is envisioned as a by-product of the OSHA requirement. How useful they will be in this regard remains to be seen (*Fed. Register*, 1989; CMA, 1991).

7.2 Confidentiality of Records

Records containing personal information on workers must be protected from transmission to, or perusal by, those not responsible for medical services or care. Without assurance of confidentiality, it is difficult to elicit vital medical information and impossible to maintain the professional relationship so important in performing the responsibilities of the physician. The ADA regulations require that medical records must be kept confidential and that the guardian of these confidential records need not be a physician but simply a person so designated and properly instructed (ADA, 1991). Management can be informed of a person's limitations without providing confidential medical information. For example, the manager can be informed that the employee should not work around moving machinery, at heights, or driving a vehicle. The manager need not be told that the person has a seizure disorder that is not under complete control. The activities of a diabetic on insulin might be limited by simply stating that the person should not work alone for any extended period.

Confidentiality of records has been difficult to achieve in small as well as large companies. Curiosity has killed the cat, but it hasn't necessarily harmed the fellow employee who tries to get a look at the medical record or the human resources manager who ''just has to know.'' These are difficult situations. The physician or a designated health person or persons should be in charge of the health records and must be given full authority by top management to protect the confidentiality of the records. Release of records to anyone, except on court order, should only be with the express written consent of the employee. Access to medical records by OSHA, NIOSH, or EEOC personnel should be looked at very carefully and provided only with proper legal authorization.

Provisions in OSHA, ADA, or other regulations regarding accessibility and dissemination of medical records vary. The record guardian should consult the specific regulation before agreeing to release any record without a court order. Whereas the standard for asbestos requires that medical reports be sent to employers, more recent standards have been less explicit (Ashford et al., 1991).

8 PROBLEM AREAS

8.1 Compulsory versus Voluntary Examinations

All regulations so far promulgated provide that medical surveillance be made available by employers, but nowhere is it stated that employees shall be required to take the examinations. Individual employers may, however, make examinations a condition of employment. The issues here are obviously complex and the introduction of the Americans with Disabilities Act must only serve to increase the complexity. If a regulatory agency believes that a given examination or test is so important to workers' health that every employer must be prepared to provide it, how can it not require employees to take the examination if they are to work with a specified chemical agent? Present regulations leave the question of such medical surveillance at the level of negotiation between the workers or their representatives and the industry and necessitate assurances relating to protection of job rights. For that part of the work force that is not organized, the vast majority in the United States, such protection would have to be guaranteed by law.

8.2 The Problem of Small Companies

The majority of new jobs, and in a few years, the majority of jobs, will be in small companies. Medical surveillance in small industry may be aided by the recent development of free-standing occupational medical clinics and similar hospital-based clinics. Major corporations with medical departments gradually adjust to regulations and recommendations requiring medical surveillance. The small employer must meet these requirements without the staff support found in some large organizations. These new clinics and the many consultants now available may help fill the needs of the smaller companies. The small company requirements may actually be merging with those of larger companies that have jettisoned their own in-house medical departments and are relying on contract and consulting services.

8.3 Handling Abnormal Findings

A major problem for the occupational physician in the current medical/legal climate is that of borderline abnormal findings. Regulations stipulate that a physician must certify that a worker has no condition that might be adversely affected by exposures on the job. It can be very difficult for a cautious physician to make such a certification if there are any abnormal findings. The presence of metaplastic cells in the sputum of an asbestos worker or a coke oven worker would make certification difficult, even though current exposures were low and controlled. Similarly, some physicians may find certification of ability to wear a respirator difficult for an individual who years earlier had had a coronary occlusion. The risk, however slight, might give pause. Situations that can be handled easily in a normal doctor–patient relationship become matters with serious medicolegal implications because they have been the subject of certification under a federal regulation. Reviews of these many medicolegal problems as well as the accompanying ethical problems are available (Rothstein, 1984; Ashford et al., 1991).

8.4 Consequences of Overregulation

Inherent in much of the foregoing discussion has been concern over the inclusion of detailed requirements for medical surveillance in governmental regulations. Many practices that are highly desirable for physicians to follow in selected occupational groups, or for plants to carry out with full understanding of their implications, are not necessarily right for mass application mandated by law. When applied to individuals with very low exposures as required by present workplace standards, the usual physical examination is likely to yield few abnormal findings. This becomes a stultifying overuse of physician time, needless expense, and a waste of national resources. Often it is former employees, no longer covered by regulations, who could most benefit by periodic medical examinations. But it is the new employees, with relatively little exposure, who receive the attention. Industrial hygienists and occupational physicians need to be listened to and heard when these regulations are eventually revised.

9 LEGAL CONSIDERATIONS

As will be evident from reading earlier sections in this chapter, there are many legal burdens placed on the physician responsible for the medical surveillance of the worker (Felton, 1978; Ashford et al., 1991). A comprehensive review of such legal strictures is not presented here, but a few areas are discussed briefly.

9.1 Informing the Worker

There is general agreement and a regulatory requirement (U.S. Dept. Labor 1987, 1910.1200) that the worker must be given information about the hazards of the workplace. This is management's responsibility, but the physician must and will have a role. For many hazards the explanation is easily and readily understood by the worker. For carcinogens, particularly for those substances designated as "suspect" carcinogens, the message is more difficult to impart. Does the physician even have available all the data necessary to evaluate whether the evidence for carcinogenicity is solid or weak? The best that can be said is that the physician should be honest in giving an appraisal of the evidence, should create sufficient concern and anxiety to encourage observance of rules for containment and protection against known carcinogens, and should indicate the relative probability of effects from low exposures and for weak or suspect carcinogens. To do this properly the physician should seek the assistance of the industrial hygienist, the toxicologist and any other resources that may be available, to come to reasonable conclusions about the risk under the specific circumstances in which the worker is laboring.

9.2 Certification for Continued Employment

Regulations that require the examining physician to certify that workers will not be adversely affected by continued employment are in many instances asking to certify as to the unknown or the unknowable. There will be a tendency by some physicians to take no chances. There is relatively little information to say whether an individual with an elevated serum glutamic oxaloacetic transaminase (SGOT) or other elevated liver enzyme would be harmed by exposure to 1 ppm of vinyl chloride. The probability is extremely high that it would make no difference. If, however, a worker should develop liver disease after certification, regardless of whether it was due to vinyl chloride, the certifying physician could conceivably be sued. A person certified to work with asbestos, even at low levels, after the finding of moderate metaplastic changes in the sputum, could present a similar problem. The examining physician is best advised to use the best clinical judgment of which he or she is capable, learn what is known about the hazard, attempt to learn something of the actual exposures their patients are experiencing, and realize that depriving a person of a job unnecessarily is a very serious matter.

9.3 Wearing of Respirators

In all current regulations [e.g., 29 CFR, Part 1910, 134 (b) (10)] and in NIOSH criteria documents, physicians are given the responsibility for certifying whether workers are physically able to wear nonpowered respirators. Objective criteria of ability to wear a respirator are limited, as noted earlier in this chapter (Harber, 1984; Hodous, 1986). The translation of pulmonary function test results to respirator use has thus far not been validated. Actual trial of individuals using the respirator that they will be called on to use at work is the best test available. Differentiation must be made between situations when a respirator is worn briefly for emergency situations for a worker's own protection, when it is required for the worker to perform duties essential to the safety and health of others, and where the worker may be required to wear it over long periods of time.

10 SUMMARY

The health surveillance of workers requires a baseline examination, usually the preplacement examination, and periodic examinations aimed both at general health maintenance and the prevention or early detection of effects from specific job hazards. Good practice calls for a comprehensive preplacement evaluation, with an occupational and medical history and review of all systems, baseline laboratory studies (including blood chemistry and urinalysis as well as a hemogram), study of visual and hearing acuity, simple tests of pulmonary ventilatory function, a chest roentgenogram, and a general physical examination. The content of periodic examinations should be based on regulatory requirements with overall consideration of what is good practice. There should be special hazard-oriented studies when indicated, such as biologic monitoring for chemicals to which the worker is known to be exposed, early indicators of toxic or other biological effects, and audiometry when indicated by noise exposures of 80 dB or greater.

Scheduling of examination frequency must consider age, nature of hazards, and regulatory requirements, but it is essential that hazard-oriented surveillance, environmental, and general medical data be closely interlocked.

REFERENCES

ACGIH (1986). *Documentation of the Threshold Limit Values and Biological Exposure Indices*, 5th ed., 1986, American Conference of Governmental Industrial Hygienists, Cincinnati, OH (Sixth Edition, 1991, incomplete).

ACGIH (1991). *Biological Exposure Indices* (1991–1992), American Conference of Governmental Industrial Hygienists, Cincinnati, OH, pp. 57–72.

AMA (1983). Council on Scientific Affairs, "Medical Evaluations of Healthy Persons," *JAMA*, **249**, 1626–1633.

AMA (1991). Council on Ethical and Judicial Affairs, "Council Report: Use of Genetic Testing by Employers," *JAMA*, **266**, 1827–1830, Oct. 2.

American Thoracic Society (1982). "Surveillance for Respiratory Hazards in the Occupational Setting," *Am. Rev. Respir. Dis.*, **126**, 952–956.

ADA (1990). Americans with Disabilities Act, PL 101-336, 42 USC 12101 et seq.

ADA (1991). Americans with Disabilities Act (ADA) Regulations, CFR 29:1630, Fri. July 26.

Anfield, R. N. (1992). "Americans with Disabilities Act of 1990: A Primer of Title 1 Provisions for Occupational Health Care Professionals," *J. Occup. Med.*, **34**(5), 503–509.

Ashford, N. A. (1986). "Policy Considerations for Human Monitoring in the Workplace," *J. Occup. Med.*, **28**(8), 563–568.

Ashford, N. A., D. B. Hattis, C. J. Spadafor, and C. C. Caldert (1991). *Monitoring the Worker for Exposure and Disease. Scientific, Legal and Ethical Considerations in the Use of Biomarkers*, Johns Hopkins University Press, Baltimore, MD.

Bayer, R. (1986). "Biological Monitoring in the Workplace: Ethical Issues," *J. Occup. Med.*, **28**(10), 935–939.

Berlin, N. I., C. R. Buncher, R. F. Fontana, J. K. Frost, and M. R. Melaned (1984). "Screening for Lung Cancer," *Am. Rev. Resp. Dis.*, **130**, 565–570.

Bernard, A. and R. Lauwerys (1986). "Present Status and Trends in Biological Monitoring of Exposure to Industrial Chemicals," *J. Occup. Med.*, **28**(8), 558–562.

Brandt-Rauf, P. W. (1988). "New Markers for Monitoring Occupational Cancer: The Example of Oncogene Proteins," *J. Occup. Med.*, **30**(5), 399–404.

CMA (1991). Epidemiology Task Group. "Guidelines for Good Epidemiology Practices for Occupational and Environmental Epidemiological Research," *J. Occup. Med.*, **33**, 1221–1229.

Cooper, W. C. (1973). "Indicators of Susceptibility to Industrial Chemicals," *J. Occup. Med.*, **15**(4), 355–359.

Cooper, W. C. (1982). Asbestos Protection in the Workplace: Problems of Medical Surveillance, presented at World Symposium on Asbestos, Montreal, Quebec, Canada, May 25.

Coye, M. J., J. A. Lowe, and K. Maddy (1986). "Biological Monitoring of Agricultural Workers Exposed to Pesticides: 1. Cholinesterase Activity Determinations," *J. Occup. Med.*, **28**(8), 619–627.

Drago, J. R. (1989). "The Role of New Modalities in the Early Detection and Diagnosis of Prostate Cancer," *Ca-A Can. J. Clinicians*, **39**(6), 326–336.

Draper, E. (1991). *Risky Business. Genetic Testing and Exclusionary Practices in the Hazardous Workplace. Cambridge Studies in Philosophy and Public Policy*, Cambridge University Press, Cambridge, England.

Farrow, G. M. (1990). "Cytology in the Detection of Bladder Cancer: A Critical Approach," *J. Occup. Med.*, **32**(9), 817–821.

FCMHSA (1969). Federal Coal Mine Health and Safety Act of 1969, PL 91-173, December 30.

FMSHAA (1977). Federal Mine Safety and Health Amendments Act of 1977, PL 95-164, November 9.

Fed. Reg. (1989). Aug. 17; 54:34034. Toxic Substances Control Act. Good Laboratory Practice Standards.

Fed. Register (1989). Aug. 17; 54:34052. Federal Insecticide, Fungicide, and Rodenticide Act. Good Laboratory Practice Standards.

Felton, J. S. (1978). "Legal Implications of Physical Examinations," *West. J. Med.*, **128**, 266–273.

Gallo, M. A. and N. J. Lawryk (1991). "Organic Phosphorus Pesticides," in *Handbook of Pesticide Toxicology*, Vol. 2, W. J. Hayes, Jr., and E. R. Laws, Jr., Eds., Academic Press, New York, pp. 917–1123.

Goerth, C. R. (1983). "Physical Standards: Discrimination Risk," *Occ. Health Saf.*, **52**(6), 33–34.

Halperin, W. E., P. A. Schulte, and D. G. Greathouse, Guest Editors (1986). "Conference on Medical Screening and Biological Monitoring for the Effects of Exposure in the Workplace. Part I," *J. Occup. Med.*, **28**(8), 543–788.

Hadler, N. M. (1990). "Cumulative Trauma Disorders. An Iatrogenic Concept," *J. Occup. Med.*, **32**(1), 38–41.

Harber, P. (1984). "Medical Evaluation for Respirator Use," *J. Occup. Med.*, **26**(7), 496–502.

Hathaway, G. J., N. H. Proctor, J. P. Hughes, and M. L. Fischman (1991). *Proctor and Hughes' Chemical Hazards of the Workplace*, 3rd ed., Van Nostrand Reinhold, New York.

Hodous, T. K. (1986). "Screening Prospective Workers for the Ability to Use Respirators," *J. Occup. Med.*, **28**(10), 1074–1080.

Hulka, B. S. (1986). "Screening for Cancer: Lessons Learned," *J. Occup. Med.*, **28**(8), 687–691.

Jacobs, P. and A. Chovil (1983). "Economic Evaluation of Corporate Medical Programs," *J. Occup. Med.*, **25**(4), 273–281.

Lappé, M. A. (1983). "Ethical Issues in Testing for Differential Sensitivity to Occupational Hazards," *J. Occup. Med.*, **25**(6), 797–808.

Lappé, M. A. (1986). "Ethical Concerns in Occupational Screening Programs," *J. Occup. Med.*, **28**(8), 930–934.

Lauwerys, R. R. (1983). *Industrial Chemical Exposure: Guidelines for Biological Monitoring*, Biomedical Publications, Davis, CA.

Levine, R. J., M. J. Symons, S. A. Balogh, D. M. Arndt, N. T. Kaswandik, and J. W. Gentile (1980). "A Method for Monitoring the Fertility of Workers 1. Method and Pilot Studies," *J. Occup. Med.*, **22**(12), 781–791.

Lowry, L. E. (1986). "Biological Exposure Index as a Complement to the TLV," *J. Occup. Med.*, **26**(8), 578–589.

McConnell, R., L. Cedillo, M. Keifer, and M. Palomo (1992). "Monitoring Organophosphate Insecticide-Exposed Workers for Cholinesterase Depression," *J. Occup. Med.*, **34**(1), 34–37.

Morgan, W. K. C. (1969). "The Annual Fiasco (American Style)," *Med. J. Australia*, **2**(11), 923.

OTA (1983). The Role of Genetic Testing in the Prevention of Occupational Disease, prepared by Advisory Panel on Occupational Genetic Testing, Office of Technology Assessment, Congress of the United States, Washington, D.C.

Ott, M. G. (1977). "Linking Industrial Hygiene and Health Records," *J. Occup. Med.*, **19**(6), 388–390.

Proctor, N. H. and J. P. Hughes (1978). *Chemical Hazards of the Workplace*, J. B. Lippincott, Philadelphia, PA.

Proctor, N. H., J. P. Hughes, and M. L. Fischman (1988). *Chemical Hazards of the Workplace*, 2nd ed., J. B. Lippincott, Philadelphia, PA.

Robert, N. J. (1959). "The Values and Limitations of Periodic Health Examinations," *J. Chronic Dis.*, **9**(2), 95–116.

Rom, W. N. (1976). "The Effect of Lead on the Female and Reproduction: A Review," *Mount Sinai J. Med.*, **43**, 542–552.

Rothstein, M. A. (1984). *Medical Screening of Workers*, The Bureau of National Affairs, Inc., Washington, D.C.

Samuels, S. W. (1986). "Medical Surveillance: Biological, Social and Ethical Concerns," *J. Occup. Med.*, **28**(8), 572–577.

Schulte, P. A., W. E. Halperin, E. M. Ward, and A. M. Ruder, Guest Editors (1990). "Bladder Cancer Screening in High-Risk Groups," *J. Occup. Med.*, **32**(9), 787–945.

Schulte, P. A. (1991). "Contribution of Biological Markers to Occupational Health," *Am. J. Ind. Med.*, **20**, 435–446.

St. Clair, S. and T. Shults (1992). "Americans with Disabilities Act: Considerations for the Practice of Occupational Medicine," *J. Occup. Med.*, **34**(5), 510–517.

Siegel, G. S. (1966). "An American Dilemma—The Periodic Health Examination," *Arch. Env. Health*, **13**(9), 292–295.

U.S. Department of Labor, Occupational Safety and Health Administration (1983). OSHA Safety and Health Standards (29 CFR 1910) OSHA 2206, revised March 11, 1983. Superintendent of Documents, U.S. Government Printing Office, Washington, D.C.

U.S. Department of Labor, Occupational Safety and Health Administration (1987). OSHA Safety and Health Standards (29 CFR 1910:1200) Hazard Communication Standard, August 24.

Whorton, D., R. A. Krauss, S. Marshall, and T. H. Milby (1977). "Infertility in Male Pesticide Workers," *Lancet* Dec., 17, **17**, 1259–1261.

World Health Organization (1975). Early Detection of Health Impairment in Occupational Exposure to Health Hazards. Technical Report Series 571, WHO, Geneva, Switzerland.

Zavon, M. R. (1963). "Methyl Cellosolve Intoxication," *Am. Ind. Hyg. Assoc. J.*, **24**, 36–41.

Zielhuis, R. L. (1985). "Biological Monitoring: Confusion in Terminology," *Am. J. Ind. Med.*, **8**, 515–516.

CHAPTER THIRTEEN

Health Promotion in the Workplace

Joseph L. Holtshouser, CIH, CSP

1 INTRODUCTION AND REVIEW

The previous edition of this work provided an introduction to health promotion and a firm foundation in the subjects of wellness, health risk assessment, and health education programs. It also chronicled the rise of industry's commitment to health promotion, which has viewed it as "the combination of activities designed to support behavior which enhances the health of individuals" (Parkinson and Fielding, 1985).

Industry involvement to keep workers healthy traces back to Kaiser Aluminum & Chemical's work with health maintenance organizations in the 1950s. By the mid-1970s, a growing number of companies began to gently urge their employees to adopt healthier life-styles.

According to a survey by the U.S. Department of Health and Human Services, by 1985, almost two-thirds of all companies with more than 50 employees engaged in at least one health promotion activity (Cordtz, 1991a).

Workplace health promotion programs have not changed significantly over the last 10 years. What has changed is the new perspective with which organizations now view health promotion. Although industry led the initial charge, the support for more available activities and employee involvement has spread to other types of organizations, including the public sector.

Another change has been the tendency for organizations to require individuals to assume more responsibility for their own health and well-being. This has led to some

Patty's Industrial Hygiene and Toxicology, Third Edition, Volume 3, Part A, Edited by Robert L. Harris, Lewis J. Cralley, and Lester V. Cralley.
ISBN 0-471-53066-2 © 1994 John Wiley & Sons, Inc.

controversy in terms of how involved an organization can become in the personal habits of its employees. However, past reported trends in rising health care costs suggest a continued pressure on organizations to reduce these costs through any practical means.

The purpose of this chapter is to update the previous work on health promotion by describing (1) some current program basics, (2) the rise of health care costs over the last 10 years, (3) efforts to control health care costs, (4) controversy over the cost control attempts, (5) some successful examples, and (6) what future trends might be for health promotion programs.

2 HEALTH PROMOTION IN THE NINETIES

The relationship between life-style and disease received closer scrutiny in the 1980s. The idea that humans have a greater impact on their own health by their life-style choices was supported by numerous studies. The popular and fashionable notion that modern origins of disease were predominantly the result of occupational or environmental factors gave way to new explanations based on scientific observation and examination.

A philosophy emerged that suggested individuals, armed with the facts about their own personal health status and the significant effects a positive change in life-style could have, would be inclined to take steps toward improving their health. Through health risk appraisals, smoking cessation programs, diet education, weight and cholesterol control, and exercise regimens the risk of disease could be reduced (Holtshouser et al., 1990).

Admittedly, there are risk factors such as age, sex, and heredity that are not controllable. But the American Heart Association considers smoking, weight, and exercise to be totally controllable factors while blood pressure, cholesterol, and stress are considered partially controllable factors (Overman and Thornburg, 1992).

A study by Milliman & Robertson, Inc. found that people who have high blood pressure are 68 percent more likely to have claims of more than $5000 per year than those who do not have high blood pressure; people not wearing seat belts during an automobile accident incur 54 percent more hospital days; and smokers' overall health costs are 18 percent higher than costs for nonsmokers (Woolsey, 1992a). And the Framingham study, conducted by the National Institutes of Health over the past three decades, links heart disease to smoking, obesity, lack of exercise, high cholesterol, and high blood pressure (Overman and Thornburg, 1992).

Offering basic health promotion programs for industry became commonplace in the 1980s and company managers, in many cases, were able to show improvements in the health status of employees and reductions in health care costs. But the continued escalation of health care costs drove managers to more aggressive programs.

2.1 Leading Programs

Companies looked to cutting health care costs by educating employees about high-risk factors, offering health risk assessments for these factors, and helping employees modify their behavior to reduce the effects of these factors.

During the past 5 years, the number of firms actively promoting healthful life-styles among their workers has grown rapidly, and many now operate comprehensive "wellness programs" that include elaborate fitness centers, professional staffs, and huge annual budgets.

2.1.1 Apple Computer

Apple Computer, based in Cupertino, California, offers a companywide comprehensive program of fitness, health education, preventive medicine, smoking cessation programs, and seminars on nutrition and weight management. Employees also can receive a health assessment that measures blood cholesterol, blood pressure, and resting pulse. Fitness evaluations are also available to determine a cardiopulmonary fitness level, strength, flexibility, body composition, nutritional status, and a medical evaluation that includes a physical exam and an exercise stress test (Overman and Thornburg, 1992).

2.1.2 Glendale, Arizona

The City of Glendale, Arizona, has developed a low-cost wellness program to encourage employees to stay healthy that includes incentives for healthy life-styles and targeted health promotion activities (Caudron, 1992). The city contracted with a local health services firm to conduct health risk assessments that would reveal specific employee health needs.

The initial voluntary screening program consists of a life-style questionnaire, complete blood analysis, flexibility test, hearing and vision exams, skin inspection, body fat measurement, lung function test, blood pressure screening, and step-pulse recovery rate test.

The city distributes *Vitality* magazine free to employees monthly, provides an on-site mobile mammography exam service, and contracts for the use of a fitness center at the Glendale Community College where employees can participate in supervised exercise programs.

2.1.3 Coors

Recreation and employee health care has a long history at the Adolph Coors Company, founded in 1873 along the scenic Clear Creek in Golden, Colorado. Major efforts at improving employee health began in the early 1970s. In 1981 the Coors Wellness Center opened in a converted grocery store covering 25,000 square feet. Its mission is to increase employee's and their families' awareness and knowledge of healthy life-styles and to facilitate and support maintenance of healthy changes with programs and incentives (Bunch, 1992).

The strategies used in the programs offered by the Wellness Center follow a planned process of (1) problem awareness, (2) healthy life-styles education in lay-people's terms, (3) incentives to improve the perception of benefits from life-style change, (4) programs to provide skill training for making changes, (5) self-action strategy to adopt new behaviors, and (6) maintenance programs to provide social support and approval for changed behaviors.

The main areas addressed by the Wellness Center include health hazard appraisal and screening, exercise, smoking cessation, nutrition and weight loss, stress and anger management, and physical and cardiovascular rehabilitation.

A cornerstone of the Coors' Wellness Program is its Health Hazard Appraisal, a voluntary program of health assessment with incentives to encourage employee participation. Employees and spouses who complete the health appraisal receive a 5 percent reduction in their insurance co-payment. And nonsmokers pay lower insurance premiums than do smokers. Key elements offered through the appraisal include:

1. Mammography screen
2. Cholesterol screen
3. Blood pressure screen
4. Skin cancer screen
5. Blood lipid panels
6. Body fat evaluation
7. Vascular disease tests
8. Electrocardiograms
9. Treadmill tests
10. Prostrate cancer screen

The physical fitness center offers an impressive array of exercise equipment including:

1. Elevated indoor jogging track
2. Bicycle ergometers
3. Versa-Climbers
4. Strength equipment
5. Treadmills
6. Stairmaster
7. Rowing machines
8. Free weights

There are also more than 26 aerobic classes per week, depending on the season. In 1990, the center had more than 143,000 uses.

The comprehensive types of programs described here have been implemented by such large companies as IBM, AT&T, Johnson and Johnson, Control Data Corporation, Conoco, Eastman Kodak, Goodyear, Ford Motor, General Electric, Travelers, and McDonnell Douglas, to name a few (Cordtz, 1991a).

2.1.4 Modest Programs

Not all companies can afford to institute the kind of multifaceted programs as Coors, but even the simplest designs are being implemented all over corporate America. For example, the following companies have implemented these programs (Cordtz, 1991b):

Company	*Programs*
Society Bank (Cleveland)	Health literature, videos
	Monthly newsletters
St. Petersburg Times	Workshops on health topics
	Health literature, videos
	Nutrition, weight loss
	Smoking cessation
GE Capital Corporation	Health assessment questionnaire
	Health literature, videos
	Smoking cessation
	800 phone number for
	health care decision counselors
Florida Power & Light	Voluntary health screening
	Weight Watchers
	Smoke Enders
	Health literature, videos
	Physical fitness center
	Healthy cafeteria menus

The development of these programs has produced new organizations whose primary purpose is to support the wellness plans defined by the parent organizations (Cordtz, 1991a). Among them are:

Wellness Resource Center

Association for Fitness and Business

Wellness Councils of America

Health Activation Network

National Wellness Institute

Washington Business Group on Health

Table 13.1 Health Promotion Program Models

Element	Comprehensive	Physical Fitness	Single Component
Health literature	Yes	Yes	Yes
Health screenings	Yes	Yes	Yes
Smoking cessation	Yes	Yes	Yes
Weight control	Yes	Yes	Yes
Cancer pervention	Yes	Yes	Yes
Stress reduction	Yes	Yes	Yes
Exercise facilities	Yes	Yes	No
Social support/encouragment	Yes	Yes	No
Employee assistance programs	Yes	No	No
Safety/first aid classes	Yes	No	No
Parenting	Yes	No	No

The previous outlines of health promotion programs gave illustrations of the three basic models that were described in the previous edition of this work. These models were defined as (1) the comprehensive model, (2) the physical fitness model and (3) the single-component model. The elements of the models are outlined in Table 13.1.

The type and complexity of an organization's health promotion program depends on the goals specified by the organization and the available resources.

3 THE RISE OF HEALTH CARE COSTS

In the 1980s benefits managers generally felt that what an employee did away from the workplace was his or her own business. Not today. Employers are becoming more involved in the life-styles of their employees in an attempt to control climbing health care costs. In a recent study, Hewitt Associates found that 76 percent of 618 employers surveyed offer at least one health promotion activity. And the Hay Group of Philadelphia found that 91 percent of 832 respondents have a program to promote employee health (Woolsey, 1992a).

Figure 13.1 illustrates the rise in health care costs per employee since 1970. The U.S. medical bill for 1992 is expected to reach $738 billion, and about 30 percent of that will be paid by business. According to the Conference Board, during the period of 1965 to 1987, health care expenditures have gone from less than 15 percent of corporate profits to 94 percent. It is no wonder that surveys of top executives consistently find the issue of health care costs at the top of the list of priority concerns (Cordtz, 1991a).

3.1 Why Health Care Costs So Much

There are several reasons for the continued rise of health care costs in the United States. A booklet, published by the Hope Heart Institute of Kalamazoo, Michigan, defines the issue clearly. (Hope Heart Institute Publ., 1992).

First, the U.S. health care system is biased toward technology. Exotic treatments and procedures (such as CAT scanners, MRIs, implants, transplants) play an important role in driving costs higher. We all want the very best treatment in the short-

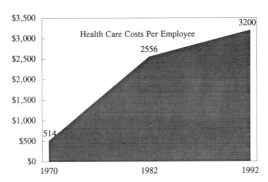

Figure 13.1 Health care costs.

est time, and as long as insurance pays the bill for most of us, we feel somewhat detached from the issue. It seems that there is no way to measure the benefits of medical technology against the overall costs. It is a matter of values and ethics.

Second, U.S. health care is labor intensive, and hospitals and clinics must offer attractive wages to compete with non-health care fields. While expensive, state-of-the-art equipment in non-health care fields tends to reduce labor needs, the opposite is true in the medical care industry. Previously, two nurses would care for a 40-bed ward. Today, those same two nurses may care for only one burn patient. The results are better, but at premium costs.

Third, the United States emphasizes treatment and the individual. Many people will not spend $40 on a bike helmet, but they will spend $400,000 on a head injury case. Other countries rarely try to save premature babies that weigh less than 12 ounces. In the United States, however, doctors routinely try to save much smaller babies at a cost of $120,000 or more per infant for intensive care services.

Britain's national health care system hinders transplant surgery, blood dialysis, and other expensive treatments if a patient is over a certain age or in very poor health. In America, heroic medical efforts are valued irrespective of the costs.

Fourth, Americans like convenience. We do not like to wait for appointments or to travel long distances for medical care. Therefore, there are doctors, hospitals, and high-tech clinics in virtually every urban area. The result is costly duplication of services and equipment.

Fifth, Americans are not very patient. We want our problem solved now. And, if we have health insurance coverage, we tend to frown on the wait-and-see attitude. A patient may demand that a test or procedure be performed immediately, even though the doctor prefers to wait. On the other hand, there are doctors who are in a hurry to use technology and medication when to wait might be more appropriate.

Sixth, some Americans are too patient. We wait too long to see a doctor. Lack of knowledge, self-borne medical expense, and denial of symptoms often cause delay in getting care. We ignore that health problems, both physical and mental, are less costly to treat if caught early.

Seventh, we like to sue people, especially our doctors. Nearly every decision a doctor makes is tempered by the thought of a possible malpractice suit. The result is "defensive medicine" where doctors perform excessive tests to protect against legal liabilities and not solely the need of the patient. Americans have an unrealistic demand for perfection in health care. If any adverse event happens, we believe someone must pay.

3.2 How Health Care Costs Affect Employers

A frequently cited statistic is that about 70 percent of health care claims can be attributed to only 10 percent of a company's employees (Overman and Thornburg, 1992). The expensive health care problems affecting this percentile can be categorized into three broad groups:

1. Cancer, heart disease, and respiratory illness account for 55 percent of hospital claims, according to the National Center for Health Statistics (Overman and Thornburg, 1992).

2. Childbirth-related expense is the largest component of health care cost. The Washington Business Group on Health estimates that as of January, 1989, normal delivery in a hospital averaged $4334, and the uncomplicated Caesarean averaged $7186. Infants born with handicaps that require hospitalization or intensive care cost an average $54,800 per child. The lifetime cost of caring for a low-birth-weight infant can reach $400,000. Yet, for each preventable low-birth-weight delivery between $14,000 and $30,000 could be saved (Overman and Thornburg, 1992).

3. Mental health costs complete the package. Stress costs business about $300 billion a year in the United States, about three times as much as in 1980. About 46 percent of American workers think their job is very stressful and 72 percent experience three or more stress related conditions often. According to Northwestern National Life, the average cost of rehabilitating an employee who is disabled because of stress is $1925. Northwestern estimates that the average employer holds $73,270 in reserve for workers filing disability claims related to stress (Overman and Thornburg, 1992).

Because stress often leads to other illnesses such as ulcers, colitis, hypertension, headaches, low back pain, and cardiac conditions, the cost of stress-related illness is ultimately higher than direct costs indicate. Further manifestations of stress-related illness include absenteeism, rapid turnover, and low productivity. These stress by-products increase employers' costs for training, benefits, quality control, and new product development and the costs associated with response to customer needs.

An estimated 20,000 U.S. corporations offer seminars and activities aimed at relieving tension. Although the incentive to companies is financial, stress management benefits are often seen by employees as goodwill perks that improve morale, loyalty, and productivity.

Another significant cost occurs from a source that many employers see as unfair. Many companies, in an effort to control health costs, fail to provide adequate health insurance or offer employees incentives to obtain coverage elsewhere. This ''cost shifting'' prompts employees to enroll in their spouses' plans. The spouses' employer then takes on the additional burden of providing health care coverage for employees' dependents who might otherwise have coverage through their own employers. While this practice may solve a cost problem for one employer, it merely shifts the cost to another without addressing the underlying problem.

Consultants state that simply identifying where health problems lie within a company is not enough. Successful management of health care costs requires employers to change employees' behavior.

4 THE EFFORTS TO CONTROL COSTS

Employers use several strategies to control health care costs. Benefit managers rely on three basic tactics: (1) cost shifting, (2) utilization review, and (3) managed care. Combinations of these tactics have been successful in saving enormous amounts of money.

4.1 New Program Basics

Cost shifting is a tactic where an employer transfers some of its health care cost to the employee or to another employer.

Utilization review is a system whereby the employer contracts with a medical services company that reviews medical care provided to employees. The utilization review provider serves as a check on costs for medical services. By evaluating medical procedures and physician charges, the provider ensures that employer dollars are well spent and employees receive the appropriate health care.

Managed care is a system in which the employer contracts directly with the health care provider at negotiated medical service rates. Cost savings are realized when the managed care provider controls expenses to below the usual rates paid by the employer while providing quality health care that competes with other area health care providers.

While control efforts have some effect on costs, employers are beginning to focus more attention on prevention rather than cost containment after health problems have developed. The drive to develop wellness programs is an example of this effort. But many employers are uncertain about whether these programs are meeting their objectives.

In a Hewitt Associates survey, 41 percent of responding employers said they were not sure their health promotion activities were effective in improving employee health or containing costs. Further, 32 percent of respondents indicated that these activities were slightly effective or not effective in containing costs (Woolsey, 1992a).

The new wellness programs may be the answer. Unlike their predecessors, the new programs offer health risk appraisals that are tied to financial incentives that encourage employees to get healthy and stay that way. Employers are actually making it financially worthwhile for employees to meet the company's wellness criteria. According to a recent survey, 12 percent of 135 large U.S. firms either offer a discount or impose a surcharge on employee contributions to life or health insurance plans based on certain behaviors (Woolsey, 1992a).

4.2 Variations on a Familiar Theme

An electrical utility holding company in Richmond, Virginia, Dominion Resources Inc., pays its workers an extra $10 per month in pretax flexible benefits credits if they qualify as low-risk in five life-style-related categories (Woolsey, 1992a).

The Moore Company, an electronics firm in Portland, Oregon, has about 65 employees. The company cut its contribution to the employee health plan in half. Now workers must earn back the employer contribution in 25 percent increments by not smoking or using drugs, by keeping their weight within recommended ranges, and by exercising at least twice a week. According to the company, only three or four employees fail to reach 100 percent of the employer's contribution. The plan has paid off. Company health care costs dropped from 8.4 percent of salaries in 1989 to 7.2 percent in 1991 (Overman and Thornburg, 1992).

Based on data showing that smokers' overall health costs were 18 percent higher than costs for nonsmokers, Control Data Corporation began charging smokers 10 percent more in health insurance premiums than nonsmokers, an example of shifting costs to higher risk employees (Woolsey, 1992a). Turner Broadcasting has taken another approach with smokers—they simply refuse to hire them, a policy that has been upheld in court (Cordtz, 1991a).

4.2.1   U-Haul International

In 1989, U-Haul International Inc., of Phoenix, paid $18.6 million in health care expenses for its 18,600 employees and dependents; $7.7 million of that total cost could be attributed to only 392 people. In 1990, U-Haul began requiring tobacco users and overweight individuals to pay $5 each pay period to help offset rising health care premiums. Nonsmokers and employees that meet U-Hauls health criteria are exempt from paying any health insurance premiums. Weight Watchers programs and Jazzercise classes that were previously ignored have gained new life and U-Haul's health care costs decreased 3 percent as of March, 1991 (Woolsey, 1992a).

4.2.2   Birmingham, Alabama

Health care costs for employees of the city of Birmingham, Alabama, have dropped significantly below the national average since workers began participating in a health screening and intervention program. The city began its pilot program in 1985 when health care costs per employee were $2050, above the national average by $300. By 1990 the cost per employee was $2074, almost $1200 below the national average (Whitmer, 1992).

The National Institutes of Health seeded the program with a $1.5 million grant in 1984 and a matching sum from the city. The 5-year partnership was one of the most extensive wellness program studies attempted under closely controlled conditions. The program was divided into two parts: (1) health screening and (2) good health maintenance programs.

The success of the program is based on a 98 percent participation rate in the health screening program, which includes tests for cholesterol levels, triglycerides, cardiac risk, glucose levels, blood pressure, flexibility, resting heart rate, and spirometry. In addition, the screening checks medical and life-style histories. The participation rate is driven by the changes in the city's medical benefits plan that required participation in the health screening program as a prerequisite for eligibility in the city's medical plan (Whitmer, 1992).

4.2.3   Hershey Foods

Hershey Foods Corporation estimates 1992 health care costs at $3100 per employee, well above the national average. The company is trying to encourage workers to adopt healthy life-styles. Management tells employees ''. . . what you do has consequences and now you are going to be more accountable for your behavior'' (Woolsey, 1992c).

An outside consultant identified about 20 modifiable risk factors for Hershey that could help keep medical costs in check. Hershey narrowed the list down to five:

smoking, weight, blood pressure, cholesterol levels, and seat belt use. Using these five criteria, the company introduced a pilot program with some incentives and disincentives that would encourage workers to become healthier.

All Hershey employees paid a flat rate of $5.50 per month toward health insurance. Under the new program the 7 percent of employees who admitted smoking were charged $32 extra per month for their health insurance. Workers with controllable high blood pressure were assessed $35 per month extra, and overweight workers were charged an additional $32 per month. If a worker was in all three categories health insurance became downright expensive. The extra funds collected by Hershey were reallocated back to healthy employees in the form of a flexible benefits credit, making the policy revenue neutral.

Reactions from Hershey employees were divided. Some felt the policy was inappropriate and that it was an intrusion into their personal lives. Others thought it was about time that unhealthy individuals were made accountable for their behavior. The results of the pilot program were positive. Health care costs have been reduced and 2 percent of the smokers gave up smoking (Woolsey, 1992c).

Smoking has been the most often attempted behavior modification because its health benefits are beyond dispute. But the payoff of smoking cessation is long term and hard to identify. There is one health promotion activity whose value is easily and quickly seen. That is prenatal care for pregnant employees.

4.2.4 Haggar Apparel

Haggar Apparel Company of Dallas, Texas, has almost 6000 employees, 93 percent of which are female. At any given time, some 950 of them are pregnant. In 1988, 26 employees had premature babies and each one cost the company almost $29,000 in hospital and doctor bills compared with $2000 for a normal birth. Haggar's medical plan at the time reimbursed only 80% of prenatal expenses. When women were required to pay the costs up front and then get reimbursed, many did not visit a doctor until the seventh month of their pregnancy. By then it was too late and complications had already set in (Cordtz, 1992a).

Haggar began by paying 100 percent of the medical bills directly if a pregnant worker started going to a doctor in the first trimester, but cut the reimbursement to 60 percent if she waited until later. The company also sponsored a series of four one-week courses on the basics of childbearing, healthy families, stress management, and the hazards of drug and alcohol use during pregnancy. The effects were dramatic. Premature births dropped from 26 in 1988 to 11 in 1990 to 3 in 1991. Haggar's medical bills fell below $19,000 per premature birth. Total out of pocket savings were enormous. Absenteeism was lower and productivity increased by reducing the number of worried and distracted mothers of sick infants (Cordtz, 1991a).

All of the health promotion programs discussed so far have a common theme—behavior modification. Whether in the form of financial rewards for healthy behavior or penalties for unhealthy behavior, the net effect is the same—costs are shifted to those with the most health problems. However, the use of incentives and disincentives is a two-edged sword when it approaches the regulation of employee's conduct, particularly if that conduct is legal and private.

5   ARE COST CONTROL PROGRAMS FAIR?

When employers try to regulate employee conduct they run the risk of being sued by employees who believe that what they do on their own time is none of their employer's business. Laws in 21 states prohibit employers from basing employment-related decisions on a worker's off-duty behavior or life-style:

Arizona	Maine	Oklahoma
Colorado	Mississippi	Oregon
Connecticut	Nevada	Rhode Island
Illinois	New Hampshire	South Carolina
Indiana	New Jersey	South Dakota
Kentucky	New Mexico	Tennessee
Louisiana	North Dakota	Virginia

Illinois has a law entitled the "Right to Privacy in the Workplace Act" that prohibits employers in that state from "refusing to hire or from discharging any individual or otherwise disadvantage any individual with respect to compensation, terms, conditions or privileges of employment because the individual engages in lawful activities off the premises of the employer during non-working hours" (Woolsey, 1992b).

The practice of offering "carrots" to healthy employees for their own good health habits while meting out "sticks" to penalize unhealthy employees in the name of health care cost containment, raises a number of concerns from several corners of our society.

Unions generally oppose programs that try to regulate worker life-styles. They are particularly against negative incentive programs regarding health care, such as penalties based on individual characteristics like blood pressure or weight. They argue that workers should be able to live the way they want to and that in some cases individuals may not be able to lose weight or lower their cholesterol level because of genetic factors. In their view, the fact that certain life-styles increase or decrease health care costs has nothing to do with how productive they are in the workplace.

The American Civil Liberties Union stresses that off-duty conduct should not be a concern of the employer if it does not affect job performance (Woolsey, 1992b). In general, employees do have a right to privacy, under the common law. However, there are two principal justifications for employers' interest in off-duty conduct. One is cost containment and the second is safety. The issue of safety is easy to defend. Off-duty conduct may affect employees' on-duty conduct, or their behavior may pose a risk to their co-workers. The rationale for cost containment is problematic.

Among the issues that may jeopardize the fairness of controlling costs through financial incentives are (1) concerns about the voluntariness of health risky behaviors, (2) ambiguity and confusion over the causes of illness and disease, (3) possible discrimination against certain individuals or groups, and (4) the potential arbitrary selection of behaviors targeted for financial incentives.

Voluntariness is a key element. Some have argued that if the behavior associated with an illness is involuntary, then so is the illness. It would be unfair to demand that sufferers of unchosen and involuntary diseases such as cystic fibrosis or other genetic conditions be penalized with higher premium costs.

Proponents of risk-rated health insurance call for penalties for risk takers because they are seen as choosing to risk their health. While most would agree with the reasonableness of this position, opponents would argue that employees who smoke, drink, and engage in other health risky behaviors may not be making voluntary choices. Such unhealthy habits may be supported by strong economic, political, and cultural elements in the general society that precludes or impedes authentic, reasoned choices. Some say this is true for such habits as smoking and drinking, which involve a degree of physical addiction.

An additional argument states that if smokers, for example, are penalized because of their potential for illness, then all others who are equally at fault in causing their own medical needs should be similarly penalized. A whole host of behaviors could be targeted including skiing, high school football, basketball, or even the decision to postpone having children until a woman reaches her late thirties when the risk of contracting certain diseases increases. Under this reasoning few off-duty behaviors would be exempt (Priester, 1992).

To aid in reducing the risk of lawsuits for such employee conduct policies, there are several steps employers should take:

1. After identifying which factors they want to improve, employers need to determine if the reasons for implementing a health risk reduction program are legitimate and defensible. If the goal is related to improving health or safety, then their goals usually are. Consultants say that employers interested in changing employee behavior should try to isolate factors that the employees can control.

2. Next, employers should consider whether there are less restrictive ways to reach their goals. Instead of refusing to hire smokers, they may decide to require smokers to pay higher health insurance premiums.

3. Allow employees to get involved in the program while it is being designed. Focus groups can offer significant and helpful suggestions for implementing such programs. Giving plenty of notice before actual implementation allows employees time to try to change their behavior.

4. Employers should find out what their competitors are doing, gather information on different programs, and find out what works and why. Implement those programs that show the greatest promise for success at the least risk from legal challenges.

Aside from employee lawsuits and individual state laws, the remaining unknown variable is the recent federal law, the Americans with Disabilities Act. This comprehensive act that was effective in July, 1992, prohibits discrimination on the basis of an individual's physical or mental disability in employment, public transportation, public accommodations, and several other areas.

How the act will affect employer incentive programs to change health behavior depends on the interpretation of the term *disability*. It is defined in the act as ''(a) a physical or mental impairment that substantially limits one or more of the major life activities of such individual; (b) a record of such impairment; or (c) being regarded as having such an impairment'' (Priester, 1992).

Courts have previously ruled that possession of a genetic trait for a particular disease, such as Huntington's chorea, would constitute an impairment. Programs targeting behavior that may have some genetic basis may be considered discriminatory under the act.

6 SOME IMPRESSIVE RESULTS

Regardless of the obstacles, a number of employers have instituted health promotion programs, with incentives or disincentives, and have received significant returns on their investments. While health care costs savings are difficult to quantify, many employers have made the effort just to convince themselves that the investment made sense to the bottom line of the business and to the health goals that created the programs.

Over a 5-year period, Blue Cross/Blue Shield of Indiana found, after completing a study of its health promotion program, that the average yearly reduction in health care costs per participant was $519. When the savings were applied to all employees, not just program participants, the average savings per employee became $143. Yearly employee costs for the program were reported at $99 (Sofian, 1991).

Not all companies have tracked the cost savings as detailed as Blue Cross/Blue Shield. But a number of companies have been able to track the costs invested in wellness programs versus the cost savings in their employees' health care plans enough to estimate the return on the dollars they invested in health promotion.

The degree to which the investment is recovered depends on the size of the resources that are allocated to individual programs and how well the programs are managed. Some of the companies discussed previously and the returns they have achieved are:

Organization	Estimated Return for Each $ Invested
Adolph Coors Company	8 to 1
Control Data Corporation	4 to 1
McDonnell Douglas Company	4 to 1
Travelers Insurance Company	3 to 1
Johnson & Johnson Company	2 to 1
UNUM Corporation	2 to 1

Glendale, Arizona, and Birmingham, Alabama, reported more extensive data on the success of their wellness programs and were able to track their returns compared to actual program use rates. Table 13.2 presents their program data.

The successful programs developed by these progressive organizations share an important common trait. They survey their employees to determine the need for the

Table 13.2 Health Promotion Program Returns

Organization	Year Established	Employees Covered	1991 Cost	1991 Use	Estimated Return
City of Glendale, AZ[a]	1982	1400	$148,000	98%	10 to 1
City of Birmingham, AL[b]	1985	4000	$235,000	98%	10 to 1

[a] Caudron, (1992).
[b] Whitmer, (1992).

programs and, once the programs are established, continue to survey them and encourage feedback to determine if the programs are fulfilling employee needs.

7 FUTURE TRENDS

Health care costs in America appear to most observers to be out of control. Facts cited in the Bernstein Research report seem to support that conclusion (Bernstein, 1990). Health care costs have increased about 9 percent annually in the last 5 years while the rate of general inflation was about 5 percent. Projected health care costs are depicted in Figure 13.2 as a percent of the Gross National Product (GNP) (Olson, 1992).

The government, employers, health care providers, and consumers agree that the increase of individual responsibility is an important strategy for health care cost containment in this decade. The extent of individual involvement varies from plan to plan, but, increasingly, consumers are making choices and are more involved in payment for health care services.

Health and Human Services Secretary, Louis M. Sullivan, announced a plan in the fall of 1990 entitled "Healthy People 2000" (Public Health Service, 1990). The goal of this initiative is to improve American's health over the next decade. The

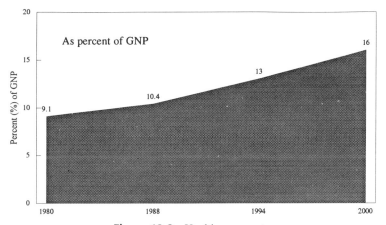

Figure 13.2 Health care costs.

plan focuses primarily on encouraging people to adopt healthful habits to prevent disease. Some of the federal goals set for the year 2000 are:

Exercise	Increase to 30 percent the number of Americans that exercise moderately for at least 30 minutes a day (8 percent increase)
Obesity	Reduce from 26 to 20 percent the proportion of overweight adults
Smoking	Reduce from 29 to 15 percent the proportion of adults who smoke
Pregnancy	Reduce pregnancies of girls under 18 to 50 per 1000; rate in 1985 was 71 per 1000
Auto Safety	Increase to 85 percent from 42 percent the proportion of vehicle occupants who use automobile safety restraints
Infant Mortality	Reduce infant mortality to 7 deaths per 1000; rate in 1991 was about 10 per 1000
Heart Disease	Reduce from 135 per 100,000 to 100 per 100,000 the incidence of coronary heart disease deaths
AIDS	Limit the annual incidence of diagnosed AIDS cases to 98,000. Diagnosed cases in 1991 were between 44,000 and 50,000

Getting employees to be more conscious of health care costs is another important cost containment strategy for the future. A poll of human resources managers listed the following items as having the greatest promise in controlling future health care costs:

1. Getting employees to be more cost conscious.
2. Emphasizing preventative care.
3. Standardizing the costs of medical procedures and treatments.
4. Changing the legal malpractice system.
5. Raising employee deductibles and co-payments.
6. Limiting doctor fees.
7. Involving state and federal government.
8. Limiting procedures, tests, and treatments.
9. Lowering insurance company profits.
10. Reducing current benefits.

In attempts to measure the cost effects of certain health habits, tools are being developed such as the National Center for Health Promotion's Lifestyle Cost Index (Olson, 1992). This will enable employers to estimate costs related to life-style risks and will serve as a basis for passing costs along to consumers who choose to live unhealthy life-styles.

Employers will move to reduce benefits and available employee options to further reduce costs while employees would rather pay more out-of-pocket costs than see benefits reduced. The Bernstein Research report (Bernstein, 1990) predicts that during the next 5 years, employers and employees will gradually accept the practice that complete benefits are only available through limited networks of managed and monitored providers. Bernstein estimates that by 1994, 60 percent of the privately insured population will be using such networks for over 75 percent of their care (Olson, 1992).

The continuous challenge is the search for equitable solutions to provide all Americans affordable, quality health care. In the absence of government intervention, it will be the employers who will find the solutions to reduce their costs. Companies that more fully address the problems of mental health, childbirth expenses, and the risks of cancer, heart disease, and respiratory disease will be rewarded with healthier employees and families and a healthier bottom line.

REFERENCES

Bunch, D. K. (1992). "Coors Wellness Center," *Employee Benefits Journal*, **17**, 14–18.

Bernstein (1990). "The Future of Health Care Delivery in America: Strategic Analysis/Market Forecast," *Bernstein Research Report*, K. S. Abramowitz, Ed., Bernstein & Co., New York.

Caudron, S. (1992). "A Low-Cost Wellness Program," *Personnel Journal*, **71**, 34–38.

Cordtz, D. (1991a). "For Our Own Good," *Financial World*, **160**, 48–54.

Cordtz, D. (1991b). "How FPL Got Well," *Financial World*, **160**, 54–55.

Holtshouser, J. L., M. E. Porter, and R. C. Ruthe (1990). In *Health & Safety Beyond the Workplace*, L. V. Cralley, L. J. Cralley, and W. C. Cooper, Eds., Wiley, New York, pp. 231–246.

Hope Heart Institute (1992). "Why Health Care Costs So Much," *Hope Health Letter*, Hope Publications, Kalamazoo, MI.

Olson, M. I. (1992). "Cost-Containment Scenarios of the Future," *Directors & Boards*, **16**, 25–27.

Overman, S. and L. Thornburg (1992). "Beating the Odds," *HR Magazine*, **37**, 42–47.

Parkinson, R. S. and J. E. Fielding (1985). In *Patty's Industrial Hygiene and Toxicology*, L. J. Cralley and L. V. Cralley, Eds., Vol. 3, 2nd. ed., Wiley, New York, pp. 27–64.

Priester, R. (1992). "Are Financial Incentives for Wellness Fair?" *Employee Benefits Journal*, **17**, 38–40.

Sofian, N. S. (1991). "Health Promotion Can Be a Valuable Strategy to Assist in Cost Containment," *Occupational Health & Safety*, **60**, 26–27.

Public Health Service (1990). U.S. Department of Health and Human Services, *Healthy People 2000*, Washington, D.C., U.S. Government Printing Office.

Whitmer, R. W. (1992). "The City of Birmingham's Wellness Partnership Contains Medical Costs," *Business & Health*, **10**, 60–66.

Woolsey, C. (1992a). "Employers Monitor Lifestyles," *Business Insurance*, **26**, 3–6.

Woolsey, C. (1992b). "Off-Duty Conduct: None of Employer's Business," *Business Insurance*, **26**, 10–11.

Woolsey, C. (1992c). "Linking Wellness to Health Care Costs," *Business Insurance*, **26**, 12.

BIBLIOGRAPHY

Baker, A. (1991). "An Ounce of Prevention," *Business & Health*, **9**, 30–38.

Burcke, J. M. (1992). "Belz Promotes Healthy Habits," *Business Insurance*, **26**, 140.

Caldwell, B. (1992). "Employers Save with Wellness Programs," *Employee Benefit Plan Review*, **46**, 46–51.

Chenoweth, D. (1992). "Soaring Employee Healthcare Costs Make Health Management Critical," *Occupational Health & Safety*, **61**, 36–58.

Claflin, T. (1991). "Woodwill Stretching Program Works for Fitness, Morale, Lower Costs," *Occupational Health & Safety*, **60**, 34–35.

Connors, N. (1992). "Wellness Promotes Healthier Employees," *Business & Health*, **10**, 66–71.

Erfurt, J. (1992). "The Cost-Effectiveness of Worksite Wellness Programs for Hypertension Control, Weight Loss, Smoking Cessation, and Exercise," *Personnel Psychology*, **45**, 5–27.

Eubanks, P. (1991). "Hospitals Offer Wellness Programs in Effort to Trim Health Costs," *Hospitals*, **65**, 42–43.

Eubanks, P. (1992). "Wellness Programs Pay Off for Hospitals and Their Employees," *Trustee*, **45**, 15.

Fisher, S. L. (1992). "Wellness in the Workplace," *Sales & Marketing Management*, **144**, 88–89.

Fuller, W. D. (1992). "Current Efforts at Cost Control," *Directors & Boards*, **16**, 20–23.

Gelb, B. D. and J. M. Bryant (1992). "Designing Health Promotion Programs by Watching the Market," *Journal of Health Care Marketing*, **12**, 65–70.

Grobman, M. (1991). "Helping Employees Stay Healthy," *Business & Health*, Oct., 10–11.

Hyland, S. L. (1992). "Health Care Benefits Show Cost-Containment Strategies," *Monthly Labor Review*, **115**, 42–43.

Kelly, T. (1992). "Modern Problems," *Quality Progress*, **25**, 17–23.

Kenkel, P. J. (1992). "Financial Incentives in Wellness Plans Aimed at Reducing Insurance Costs by Helping Workers Shed Unhealthy Habits," *Modern Healthcare*, **22**, 38.

Lombino, P. (1992). "An Ounce of Prevention," *CFO: The Magazine for Senior Financial Executives*, **8**, 15–22.

Mudrack, P. E. (1992). "Work or Leisure? The Protestant Work Ethic and Participation in an Employee Fitness Program," *Journal of Organizational Behavior*, **13**, 81–88.

Perreca, J. S. (1992). "What Makes a Healthy Wellness Program?" *Human Resources Professional*, **4**, 25–28.

Polakoff, P. L. (1991). "Proper Nutrition Gaining Importance in Workplace Health-Promotion Effort," *Occupational Health & Safety*, **60**, 36.

Polakoff, P. L. (1991). "Report on Worksite Fitness Programs Suggests Criteria to be Considered," *Occupational Health & Safety*, **60**, 54.

Siegelman, S. (1991). ''Employers Fighting the Battle of the Bulge,'' *Business & Health*, **9,** 62–73.

Sigman, A. (1992). ''The State of Corporate Healthcare,'' *Personnel Management*, **24,** 24–31.

Tarkan, L. (1991). ''Stress Relief: The '90s Perk,'' *Working Woman*, **16,** 76–78.

Woolsey, C. (1992d). ''Varied Paths Lead to Common Goal of Wellness,'' *Business Insurance*, **26,** 16–18.

Occupational Health Nursing

Bonnie Rogers, DrPH, COHN, FAAN, and Maria W. Lyzen, MS, RN, COHN

1 INTRODUCTION

Many factors have influenced the evolution of occupational health nursing practice. Among them are the changing population and workforce, the introduction of new chemicals and work processes into the work environment with a concomitant increase in hazards in the workplace, technological advances and regulatory mandates, an increased interest in health promotion and illness/injury prevention, and an increase in health care costs and workers' compensation claims. Because of the dynamic nature of occupational health nursing and the complexities of work-related health problems, it is important that the nurse utilize a multidisciplinary approach to address the health needs of the workforce (Rogers, 1993). Occupational health practice is guided by public health principles, that is, by examining trends in health, illness, and injury in populations in order to protect the health of individuals (Cox, 1989). While individual care is provided to all injured workers, the emphasis in occupational health nursing is on providing programs and services to maintain, monitor, enhance, and protect the health of the aggregate workforce.

Occupational health nursing practice has expanded considerably in recent years. Occupational health nurses practice within the context of a prevention framework in order to maintain, protect, and promote worker health and improve the health and safety of the work environment. The definition of occupational health nursing has recently been revised by the American Association of Occupational Health Nurses (1993) as:

Patty's Industrial Hygiene and Toxicology, Third Edition, Volume 3, Part A, Edited by Robert L. Harris, Lewis J. Cralley, and Lester V. Cralley.
ISBN 0-471-53066-2 © 1994 John Wiley & Sons, Inc.

the specialty practice that provides for and delivers health care services to workers and worker populations. The practice focuses on promotion, protection and restoration of workers' health within the context of a safe and healthy work environment. Occupational health nursing practice is autonomous, and occupational health nurses make independent nursing judgements in providing occupational health services. The foundation for occupational health nursing practice is research-based with an emphasis on optimizing health, preventing illness and injury, and reducing health hazards.

2 PREVENTION FOCUS

The occupational health nurse practices at all levels of prevention with an emphasis on cost containment while preserving and improving quality health services.

Leavell and Clark (1965, p. 20) classified levels of prevention as primary, secondary, and tertiary as follows:

> Primary prevention may be accomplished in the prepathogenesis period by measures designed to promote general optimum health or by the specific protection of human beings against disease agents or the establishment of barriers against agents in the environment.

> As soon as the disease is detectable, early in pathogenesis, secondary prevention may be accomplished by early diagnosis and prompt and adequate treatment. When the process of pathogenesis has progressed and the disease has advanced beyond its early stages, secondary prevention may also be accomplished by means of adequate treatment to prevent sequelae and limit disability. Later, when defect and disability have been fixed, tertiary prevention may be accomplished by rehabilitation.

Using the Leavell and Clark framework, Wachs and Parker-Conrad (1990) describe occupational health programmatic activities that emphasize preventive strategies (Fig. 14.1). Primary prevention strategies are directed toward health promotion and risk reduction. Health promotion is aimed at maintaining and improving the physical and mental well-being of workers or groups of workers, and improving the concept of health in the workplace. Risk reduction measures are designed to eliminate or avert health hazards in order to reduce the risk of disease and prevent injuries. For example, occupational health team members can conduct walk-through surveys to identify workplace hazards and to observe all work practices so that appropriate control strategies can be instituted. Specific primary prevention activities include, for example, immunization programs, utilization of personal protective equipment, and stress management.

Secondary prevention is aimed at early disease detection, prompt treatment, and prevention of further limitations. In the occupational health setting, early detection includes preplacement and periodic screening examinations for health monitoring and surveillance activities so that illnesses and injuries will be identified as soon as possible, appropriate referrals for management of the problem can be made, and the hazard situation can be eliminated. Screening programs such as mammography and analysis of injury trend data are additional examples of early detection strategies.

Tertiary prevention is intended to restore the injured individual's health as fully

Comprehensive Occupational Health Program

Primary Prevention		Secondary Prevention	Tertiary Prevention
Health Promotion	Disease Prevention	Preplacement, Yearly and Termination exams	Early Back to Work
Nutrition	Injury Prevention		Modified Duty
	Accident Investigation	Health Surveillance	
Fitness/Exercise	Disease Prevention		Work Hardening
		Triage System	
Coping Enhancement	Health Risk Appraisal		Onsite Therapy
	Health Education	Employee Record System	
Recreation	Smoking Cessation		Chronic Illness Monitoring
		Accident Reporting	
Parenting Skills	Weight Control		
	Stress Management	Injury Diagnosis and Treatment	

Figure 14.1 A variety of programs at each level of prevention constitutes a comprehensive occupational health program. (From Wachs and Parker-Conrad, 1990, based on a framework developed by Leavell and Clark, 1965. Used with permission of Appl. Occup. Environ. Hyg.

as possible. Rehabilitation strategies for substance-abusing employees, return to work/work hardening programs, and chronic disease monitoring are examples of tertiary preventive measures.

Working with other health care professionals such as industrial hygienists and physicians, the occupational health nurse will want to design programs to identify vulnerable workers who are symptomatic, remove them from exposure to prevent further insult, observe and sample the work environment to determine the exposure source(s), and reduce or eliminate the exposure agent. Utilizing a multidisciplinary approach can increase alternatives for problem solving, thus adding to both the effectiveness and efficiency of programmatic interventions.

3 SCOPE OF PRACTICE

As part of the occupational health team, the occupational health nurse utilizes many skills to improve and foster worker health and safety. The scope of occupational health nursing practice, as shown in Table 14.1, is broad and comprehensive (American Association of Occupational Health Nurses, 1993; Babbitz, 1986; McKechenie, 1985; Rogers, 1993; Wachs and Parker-Conrad, 1990) and includes the following:

3.1 Health Promotion/Protection

Health promotion and health protection activities are designed to improve employees' general health and well-being and to increase employees' awareness and knowledge about toxic exposures in the workplace, life-style risk factors related to health and illness, and strategies to altering behaviors that contribute to health hazard risk. In addition, organizational strategies to enhance workplace health must be emphasized.

Table 14.1 Scope of Practice in Occupational
Health Nursing

Health promotion/protection
Worker health/hazard assessment and surveillance
Workplace surveillance and hazard detection
Primary care
Counseling
Management and administration
Research
Legal/ethical monitoring
Community orientation

3.2 Worker Health/Hazard Assessment and Surveillance

Worker health and hazard assessment and surveillance activities are designed to identify workers' health problems and the states of workers' health in order to match employees with jobs and to protect the workers from work-related health hazards. Various assessments/examinations can be performed such as preplacement, periodic, and return-to-work assessments and examinations. Preplacement examinations also help to establish baseline data for comparison with future health monitoring results.

3.3 Workplace Surveillance and Hazard Detection

Workplace surveillance and hazard detection activities are designed to identify potential and actual hazards that are harmful to workers' health. Familiarity with the nature of the work and with work processes, and the conduct of comprehensive walk-through assessment surveys, including observation of work practices and use of personal protective equipment, are essential to effective hazard detection.

If a hazard is identified, services of an industrial hygienist or a safety specialist will probably be needed to measure levels of exposure of specific substances, or to help with job task analyses. In collaboration with a physician and other health care professionals, the occupational health nurse will need to review data obtained in the hazard assessment in order to recommend health surveillance activities and to participate in the implementation of control strategies. Multidisciplinary collaboration is key to the development and implementation of a successful workplace surveillance and hazard control program (Rogers, 1993).

3.4 Primary Care

Primary care activities incorporate direct health care services to ill and injured workers and include nursing diagnosis, treatment, referral for medical care and follow-up, and emergency care. Nonoccupational health care may also be provided for minor health problems and chronic disease monitoring for employees with stable conditions. The occupational health nurse must work collaboratively with the physician to develop appropriate health care guidelines where appropriate. It is imperative that the nurse be thoroughly familiar with her or his state nurse practice legislation that defines the legal scope and limitations of practice.

3.5 Counseling

Counseling activities are intended to help employees clarify health problems and to provide for strategic interventions to deal with crisis situations and appropriate referrals. Counseling activities can relate to such areas as behavioral, marital, social, and work-related situations and events. The occupational health nurse needs to recognize when to refer an employee for additional assistance so he or she can get appropriate help and can remain productive on the job.

3.6 Management and Administration

Management and administration activities involve overseeing the functioning and service delivery of the occupational health unit. The occupational health nurse is often the health care manager at the work site with responsibilities for program planning and goal development; budget planning and management; organizing, staffing and coordinating activities of the unit, including development of policy, procedures, and protocol manuals; and evaluating unit performance based on achievement of goals and objectives. Cost effectiveness and cost containment of health care services are an integral part of the scope of management responsibilities.

3.7 Research

Research activities are directed toward identifying practice-related health problems and participating in research activities to identify contributing factors and ultimately recommend corrective actions. For example, understanding human behavior and motivation relative to use of personal protective equipment can help determine barriers or facilitators that affect health protecting strategies. As part of a research team, the occupational health nurse can participate in the design, data collection, analysis, and reporting phases of research studies.

3.8 Legal–Ethical Monitoring

Legal and ethical monitoring activities involve knowledge and integration of laws and regulations governing nursing practice and occupational health (e.g., Occupational Safety and Health Act; Hazard Communication). The occupational health nurse must be fully aware of the legal scope of practice embodied within the nurse practice legislation of the state in which she or he practices, must meet the accepted standards for professional practice, and must be certain to practice good recordkeeping for documentation purposes. Both legal and ethical issues may arise in areas such as confidentiality of employee health records, which may cause conflict with other agencies or other levels of management. The occupational health nurse can be guided by the American Association of Occupational Health Nurses (AAOHN) or American Nurses Association (ANA) code of ethics to help with decision making.

3.9 Community Organization

Community organization activities involve developing a network of resources to efficiently and effectively provide services to workers and employers. For example, agencies such as the American Heart Association, American Lung Association, or the American Cancer Society can offer valuable materials, information, and expertise to help with health programs and/or referrals. In addition, working with state and local health departments and with hospitals in relation to employee return to work can be mutually beneficial to the worker, the worker's family, and the employer.

4 PROFESSIONALISM

The AAOHN is the national professional association for registered nurses who provide on-the-job health care for the nation's workers. The AAOHN has published a set of standards for the practice of occupational health nursing (American Association of Occupational Health Nurses, 1988), which is currently under revision.

The mission of the AAOHN is to advance the profession of occupational health nursing:

- Promoting professional excellence to achieve workers' health and safety through education and research
- Establishing professional standards of practice and code of ethics
- Influencing legislative and regulatory issues that impact health and safety
- Fostering internal and external communications in order to facilitate AAOHN's goals and objectives

The association has made a commitment for excellence in the occupational health nursing profession by providing a forum for networking opportunities and a broad range of services for its members in the areas of:

Continuing and academic education
Career development and career opportunities
Professional research
Governmental affairs (federal and state)
Communications (newsletter, journal, publications)
Leadership development

5 INTERDISCIPLINARY FUNCTIONING

Occupational health nurses are important team members in the occupational health arena with their thorough knowledge of, and skills in, occupational health nursing and related knowledge in areas of occupational safety, industrial hygiene, toxicology, and epidemiology. For example, occupational health nurses may be asked to collaborate with safety and engineering professionals in identifying jobs where there is an ergonomics problem and in developing a solution by applying knowledge of human anatomy and physiology. Additionally, nurses may be involved in gathering injury statistics and applying principles of epidemiology to determine causal or associated risk relationship.

In order to prevent and reduce injuries, the occupational health nurse may consult with the safety officer and physician in evaluation of the preplacement physical examination to ensure safe job placement. This type of multidisciplinary collaboration is essential for a dynamic health-oriented occupational health service. Opportunities for occupational health nurses to participate as team members are in the areas of personnel and finance where employers are interested in cost containment of health care for active employees, their spouses and other dependents, and retirees. Increasingly, occupational health nurses will assume a greater role in the overall management of the occupational health unit within a collaborative framework.

6 SUMMARY

The practice of occupational health nursing is guided by public health principles in places of work among populations of workers. Occupational health nurses are professional health care providers who are specifically trained in recognition of the associations between conditions of work and human health. They serve with physicians, industrial hygienists, safety specialists, toxicologists, and others as vital members of the team of professionals who deal with problems in occupational health and safety.

Occupational health nursing is a dynamic and evolving professional field. Continuing study by its practitioners is necessary to stay abreast of research findings and new developments pertinent to its practice.

Occupational health nurses are very often in the forefront of any effort to protect and enhance the health of workers. Detection of the first hint of a work-related health problem may depend on the astute observations of the occupational health nurse. For many workers, the occupational health nurse is the primary source of guidance for health maintenance, behavior modification, and life-style for promotion of health.

Occupational health nurses are health professionals with some clearly definable professional obligations. The occupational health nurse must be familiar with federal laws and regulations governing occupational health and safety, and the nurse practice act for the state in which she or he practices. The American Association of Occupational Health Nurses has adopted a code of ethics that provides ethical guidance for occupational health nurses in their professional practice.

REFERENCES

American Association of Occupational Health Nurses (1988). *Standards of Occupational Health Nursing Practice*, Atlanta, GA.

American Association of Occupational Health Nurses (1993). Atlanta, GA.

Babbitz, M. (1986). "The Practice of Occupational Health Nursing in the United States," *Occupational Health Nursing*, **31,** 23–25.

Cox, A. (1989). "Planning for the Future of Occupational Health Nursing," *AAOHN Journal*, **37,** 356–360.

Leavell, H. R. and E. G. Clark (1965). *Preventive Medicine for the Doctor and His Community*, 3rd ed., McGraw-Hill, New York.

McKechenie, M. (1985). "A Descriptive Study of the Scope of Practice of Occupational Health Nurses in One-Nurse Units," *Occupational Health Nursing*, **31,** 18–22.

Rogers, B. (1993). *Occupational Health Nursing: Concepts and Practice*, W. B. Saunders, Philadelphia (in press).

Wachs, J. and J. Parker-Conrad (1990). "Occupational Health Nursing in 1990 and the Coming Decade," *Appl. Occup. Environ. Hyg.*, **5,** 200–203.

CHAPTER FIFTEEN

Occupational Hazards of Archaeology

Lynne E. Christenson, Ph.D.

1 INTRODUCTION

In her 1990 survey entitled *Surviving Fieldwork*, Nancy Howell (1990) reported that anthropologic fieldwork is hazardous to the health and safety of field-workers. This comes as no surprise to any anthropologist who has conducted fieldwork. The focus of this chapter is on archaeology, one of the four anthropologic subdisciplines. Archaeology and the other three subdisciplines—cultural, biological, and linguistic anthropology—were determined to be equally hazardous. Twenty-nine percent of the respondents in the survey reported by Howell (1990) were archaeologists. The specifics of fieldwork hazards differ by subdiscipline and the area of the world in which work is conducted. Archaeologic fieldwork occurs more frequently within the United States than does fieldwork for other subdisciplines (Howell 1990), and that will be the focus of this chapter.

The object of this chapter is twofold. The first is to define the breadth of archaeological fieldwork, since most people still think of archaeology as being conducted only in Egyptian pyramids. The second is to explain the numerous and highly varied health and safety concerns encountered by archaeologists. Within the profession these concerns are largely suppressed, and few people understand what fieldwork entails, much less the hazards associated with it (Waldron, 1985).

Archaeological safety is a relatively unexplored issue. Very little has been written about it (Howell, 1990; Waldron, 1985). Consequently, the tone of this chapter is

Patty's Industrial Hygiene and Toxicology, Third Edition, Volume 3, Part A, Edited by Robert L. Harris, Lewis J. Cralley, and Lester V. Cralley.
ISBN 0-471-53066-2 © 1994 John Wiley & Sons, Inc.

not as scientific in reporting facts and data as others in this volume, and the information is primarily anecdotal. It is with this beginning that the author hopes to alert the industrial hygiene profession to the problems and to offer some suggestions for problem resolution.

2 ARCHAEOLOGY BACKGROUND

As practiced within the United States, archaeology has two divisions. Academic archaeology is conducted by college and university faculty on a seasonal basis with funding from grants. These projects are usually at large, well-defined sites that are not in danger of destruction. Projects at these sites last for several seasons, with student and volunteer laborers on site. Cultural resource management (CRM) archaeology (also known as salvage, contract, and public archaeology) is conducted on sites in danger of damage or destruction. Funding is from the project owner or developer, and there is limited time and money. Archaeologists are employed by a government agency or by an environmental firm hired by the developer. Although hazards are abundant in each category, there are fewer uncertainties with academic archaeology because, after the first year, the potential problems can be identified. Fieldwork and research is an inherent part of both categories of work.

With the advent of cultural resource protection laws in the 1960s and 1970s, archaeology changed from a profession of pith-helmeted men searching for gold and mummies to a profession of committed and dedicated people at all levels of government and the private sector intent on preserving the cultural heritage of the United States (Thomas, 1979). Inherent in the scope of cultural heritage is the prehistoric Native American culture as well as the historic Euro-American, African-American, and Asian-American heritage of over 50 years ago. Archaeologists are currently employed by the federal government (in the Bureau of Land Management, U.S. Forest Service, National Park Service, and the Army, Navy, and Air Force), by state governments (in departments of transportation, state parks departments, departments of forestry, and state historic preservation offices), as well as county and city governments. Environmental analysis firms intent on meeting government regulations for environmental impact reports also employ archaeologists. Many colleges and universities have archaeologists on their faculties, as well as other individuals who manage the cultural resource interests of the institution. The work conducted by these people is varied and the safety concerns are numerous.

Environmental laws require that the presence or absence of cultural resources be determined early in the development of a project. Before brush-clearing or identification of hazards, the archaeologist must search for eligible resources. Archaeologists and biologists are often the first people to actually walk a project area, and they are thus the first to encounter hazards. This adds to the uncertain and unsafe nature of the discipline. Archaeology fieldwork is hard, physical work. Conducted in the natural environment, in all forms of weather, it consists of walking long distances, climbing, fording streams and creeks, as well as digging, screening, and carrying heavy objects.

2.1 Fieldwork

After establishment of a research design and review of pertinent resources, the archaeologist conducts an on-foot survey of the project area. While this might be slightly modified depending on terrain (Meighan and Dillon, 1989), the most common method is to have a string of people at 10- to 20-m intervals walk over the area in question. In this manner all sites, either prehistoric or historic, can be identified, photographed, and mapped. Sites are usually registered with a state-designated clearinghouse where each is assigned an individual number. For example, CA-SDI-8125 is the 8,125 site registered in San Diego County, California.

Where the slope is flat, the vegetation sparse, and the landowners or tenants friendly, this may be a relatively easy and hazard-free procedure. However, this ideal condition is seldom the case. Surveying for a new trail for the Forest Service requires trampling over, under, and through brush along steep canyons and across streams, usually to the distress of indigenous insects, reptiles, and other animals, not to mention indignant pot (marijuana) farmers. Surveys for new housing projects at the edge of inhabited areas can entail avoiding protective dogs, edgy homeowners with guns, or feral cattle, along with encountering illegally dumped hazardous wastes, old mines, or wells.

The next phase after the on-foot survey involves subsurface testing of the sites. Again, depending on the scope of the project, the research design, and the nature of the site, some sites may be visited only once, and others two or more times. Archaeologists often forge new roads bringing equipment to sites for testing. Testing entails digging, either in the form of trenches, 1- × 1-m units, or shovel test pits. Often, the excavated dirt is screened in order to find the smallest cultural or ecological remains. These remains are then cataloged, bagged, boxed, and removed from the site. Excavated units are most often backfilled.

A major excitement, as well as problem, for archaeologists is the uncertainty of what is beneath the ground. While the pieces of the cultural puzzle are what they are searching for, intrusive factors may provide dangers. Buried hazardous wastes, gas tanks, dens or nests of snakes, scorpions, or sick or injured animals cannot always be anticipated. Deep excavations can also be major safety hazards, with the potential for side wall collapse.

After the fieldwork, the analysis of the excavated artifacts and biological remains is conducted. This analysis may range from simple weighing and measuring to intricate procedures such as reconstitution of fecal remains. The final stage is to write the report and disseminate pertinent information.

3 HAZARDS OF ARCHAEOLOGY

No single archaeologist is constantly being confronted with safety problems or hazardous situations. The incidents discussed in this section are culled from experiences of many different professionals.

The itinerant nature of the profession must be considered. Crew members frequently travel from one project to the next, working in one city or county for 3 to 6 weeks, then moving to the next city or county. This unfamiliarity with regional terrain and history increases their exposure to potential hazards.

Another important factor is the uncertain nature of the type or extent of the hazards to the profession as a whole. Not only are hazards often unknown prior to encountering them, incidents where hazards were known but not conveyed to the field-worker are not uncommon. In a recent discussion, an archaeologist from New Mexico was in southern California to conduct a survey along the international border. He happened to be with a group of local archaeologists the day before beginning the survey. In the ensuing discussion it became clear that he was unaware of locally known hazards of his assignment. The government agency that had hired him had neglected to tell him that he would encounter bandits with knives and guns, that any women with him were in danger of rape by these bandits, and that several of the areas he would go into were popular open target shooting areas.

Before discussing the safety problems and possible solutions, two attributes of archaeology and archaeologists should be mentioned. First, Howell (1990) reported that the lack of discussion of safety was a major problem. Of three introductory textbooks on archaeology checked by the author, none contains any section on dangers, hazards, or safety factors (Fagan, 1978; Hole and Heizer, 1973; Thomas, 1979). Respondents in Howell's (1990) survey indicated that only some of them discussed safety with students. Most archaeologists don't spend time on the subject. Possible reasons for this are, first, that the hazards are numerous and varied, and it is difficult to anticipate which will be encountered. And second, there is the fear that students and potential workers might not want to pursue this project or profession (Howell, 1990). This lack of documentation of the fieldwork hazards is in part what prompted Howell (1990) to write her book, *Surviving Fieldwork*.

With professional archaeologists, however, the situation is different. Archaeologists often take on a cavalier and danger-loving attitude. This cavalier attitude is crucial to understanding not only the hazards of archaeology but risk assessment as well. That this is a hazardous profession is not only well known but is worn by some like a badge. One colleague called archaeologists ''cowboys,'' referring to the wild west attitude of daring and taunting danger. Discovery and preservation of important cultural resources is the motivating factor for archaeologists. If these cowboys don't find the sites, or extract the information, the site will be destroyed by the freeway, housing project, skyscraper, or sewer project. Archaeologists have already lost too much information and too many sites, and their attitudes are to let nothing stand in their way of conducting projects. While the story of Indiana Jones is fictional, the attitude of the protagonist is familiar to all professional archaeologists. No situation is too dangerous or too foolish for some. As a profession, archaeologists have yet to understand how to negotiate with the owner/government agency, so that they can meet their objectives with a minimum of hazards. The attitude of the archaeologists is the first level in risk assessment or safety considerations.

3.1 Hazards of Exposure

Since fieldwork is conducted outdoors, archaeologists are familiar with the sudden changes in weather. Whether they are prepared for these changes depends on the circumstances. When surveying, the team may be miles away from the vehicle, in areas far from roads or structures. Lightening, tornadoes, and downpours are commonly encountered hazards that cannot always be anticipated (Howell, 1990). When dangerous storms strike, most professionals find shelter where they can, and wait it out.

Problems with heat or cold can be regular dangers (Horvath, 1985). For archaeologists in the western United States, it is well known that desert surveys and excavations take place only in the summer, usually by people who are not acclimated to the heat. Recent surveys of the Yuma Proving Grounds outside Yuma, Arizona, were conducted in May and June. Temperatures in the open desert were over 120 degrees. Although the survey team members were instructed to carry plenty of water, were met at regular intervals by a water replacement truck, and surveyed only from 6 a.m. to 2 p.m., this was less than ideal survey conditions. One member of the team who felt he was acclimated to the desert from previous surveys (months before), and therefore did not need to wear protective clothing or drink as much water as others, suffered acute heat illness. Recently, some archaeologists have begun recognizing the dangers and have set rules forbidding work in the desert in July or August (Shaefer, 1991).

Sunburn and heat exhaustion are problems on surveys and excavations. Academic archaeology is usually conducted in the summer, between school terms. Students who have spent the preceding fall and winter in classrooms often misjudge what 5 days, 8 hr a day sun exposure can do to the skin. Howell (1990) states that 54 percent of the anthropologists surveyed know of team members who had sunburn, and 29 percent had heat stroke or heat exhaustion. Everyone on a new project is excited to find something, and none want to appear incapable of sharing equal work load. The fact that acclimatization to heat has not been achieved does not deter them.

Associated with heat is the problem of dehydration. Archaeologists are constantly encouraged to take plenty of water, but their ability to judge what is adequate is faulty. Water is heavy, and when surveying in the heat, one wants to reduce the pack load. Surveyors are also not aware of amounts of fluids required in relation to hard work, heat, and humidity. Surveying in extreme summer heat and humidity can require a person to drink between 1 and 2 gal of water per shift. This constitutes 8 to 16 lb of additional weight, if the surveyors are away from their supply vehicles for an entire day. The extra weight can be hazardous itself when the work involves activities such as climbing up mountains, over tree trunks, under bushes, or over fences. It is essential that supervisors assure that sufficient water and food are provided for each day's work either on an individual or team basis.

Cold is also a factor for archaeologists. While projects should be planned for favorable weather, they often run over or have deadlines that preclude waiting for good weather. Just as people ignore the heat, they tend to ignore wind chill factors. Archaeologists spent 6 years working year round in the field on the Delores Dam project in southwestern Colorado. This occasionally entailed digging in frozen ground, with the potential for frozen or frostbitten fingers, toes, and faces a real threat. Protective clothing helped, but it can be difficult to dig in a 1- × 1-m unit with thick sweaters, overcoats, hats, and gloves. Hot water and heaters were used to thaw out not only the ground but the workers as well.

3.2   Hazards of Insects, Reptiles, and Other Animals

Wild animals, including bobcats, coyotes, foxes, wild dogs, and occasionally mountain lions and bears can be encountered during survey operations. Rabid animals are also occasionally seen. Surveying for the Bureau of Land Management, Forest Service, or Department of Defense puts archaeologists in uninhabited areas where wild animals can be found. Howell (1990) notes that 17 percent of respondents surveyed had bites or injuries from animals. Caves or rock shelters are great places for animals to live. They were also great places for prehistoric people to live, conduct ceremonies, or leave paintings on the walls. Surveying to identify the sites can require crawling into such caves to assess their cultural value. Precautions can be taken to alert any animals inside, but the problems are obvious. Menstruating women are often attractive to carnivores like bears. They have been known to follow such women.

Encounters with domesticated, or partially domesticated, animals are also potentially hazardous. For example, attack dogs may be used to protect property. In rural areas, archaeologists can be on a survey property when the dogs from the neighboring areas intervene. In southern California where marijuana farms and methamphetamine labs are hidden in back country canyons and forests, vicious attack dogs are used for protection. Once while a survey team that included this author was walking along a county road a pack of pit bull dogs ran from a driveway in pursuit. Team members quickly climbed onto rocks and were obligated to wait for the dogs to leave. It was later learned that the local people suspected the property owners of running drugs. Bulls are also a hazard. These large and aggressive animals roam on many U.S. Forest Service or Bureau of Land Management lands. Most of the time they will stay away from humans, but on occasion they have been known to charge.

Insects are a continuous nuisance during fieldwork and can be a hazard. On one project in the Laguna Mountains of San Diego County, the work area was infested with fire ant nests. Constant vigilance on the part of the crew kept the ant attacks to a minimum; however, after suffering bites over a period of about 6 months, one member developed an allergic sensitivity to the bites. The project was one hour away over dirt roads from the nearest telephones, and the two-way truck radios were useless in valleys or drainages. In addition, directing rescue operations to the area would have been extremely difficult.

Bees are a continuous hazard. The summer of 1991 had a particularly aggressive strain of wasps and bees. On one project on a single day two people were stung, one twice, and six others were chased by bees. This particular group of bees tended to burrow into people's hair. When digging with a back hoe, the possibility of finding wasp nests should always be considered. There are numerous stories of archaeologists being attacked by swarms that have been dislodged from ground hives.

Experienced field archaeologists know to wear protective clothing. Long sleeves and pants are the uniform of choice, along with sun glasses and a broad-brimmed hat. However, flies have been known to bite through long-sleeved shirts. No scented oils, perfumes, makeup, or scented lotions should be worn in the field. However, it is difficult to find shampoos and deodorants that do not have some scent. Many insects are attracted to these. Sun-blocking agents also attract insects. In addition, there often is a crew member who wears lotion to promote skin tanning or who wears a scent in anticipation of social activity. By so doing these members jeopardize not only themselves but others working in the same unit.

As mentioned earlier, snakes are a constant hazard. Howell (1990) noted that among archaeologists, North American workers had the highest rate of contact with snakes (7 percent of respondents). While snake bite is rare, the encounters are not. Surveying on cool mornings or cool wet days provides the perfect weather for sluggish snakes. People commonly believe that all snakes hibernate in the winter, regardless of climate. In many southern states, field-workers are shocked to learn that they may encounter snakes during any season. Excavators often find snakes, scorpions, spiders, and other hazardous creatures in their units in the morning. Units are often covered with a piece of wood at night so that unsuspecting animals or people don't fall into them. However, insects and snakes may also choose them as resting places.

3.3 Hazards from Humans

Howell (1990) documents cases of robbery, rape, assault, hostage taking, military attack, assassination, and police arrests. Incidents such as these happened more frequently to archaeologists in other countries than to those in North America. Howell's questionnaire did not include gang violence or being used for target practice. These two factors are dangers of recent times.

The increase in urban crime also affects archaeologists. When a city decides to rejuvenate its center city area, the archaeologist is there to assess which buildings are culturally significant and to excavate in areas where historic buildings used to stand. This is long before the new buildings are built or protections are taken for the new tenants. Robbery or confrontation by people with knives and guns is not uncommon in such projects.

Rural areas are not much different. The popularity of guns and target practice make rural surveys or excavations potentially hazardous. There have been repeated incidents in which archaeologic fieldwork has been interrupted or even temporarily abandoned because of indiscriminate, or even directed, weapons fire.

On a recent excavation project conducted by a local company, golf balls were the hazard. The project was to test a site located between two fairways on a golf course. This project took over a week to conduct, and the golfers continued to play. The archaeologists were required to wear hard hats as protection against errant golf balls.

The advent of migrant workers who live as squatters in make-shift shelters has added additional hazards. Some of these people are aggressive and violent. They can be encountered in many rural or semirural drainages, where they live when not employed in agriculture or as day laborers. As one example, archaeologists from a national environmental firm located a large Archaic Period site in an extensive migrant workers camp. The mitigation project called for many units. While the migrant workers were away during the day, the archaeologists put in their units. This was amid rotting food, human excrement, old refrigerators, general junk as well as personal possessions. Because the project investigator was sensitive to the hazards of human waste and rotting food, the crew was encouraged to wear gloves and masks. The gloves were accepted; the masks were not (Kyle, 1991).

Burial sites traditionally were the focus of archaeology fieldwork. Now, with changes in federal and state laws, archaeologists do not target burials or cemeteries. However, buried human remains can be encountered during any dig. Pathogens are the major hazard when excavating human remains (Waldron, 1985; Lee, 1986). It is unknown how long some spores can last and whether individuals buried in crypts or coffins would still carry these. It has even been suggested that the mysterious illness suffered by people who excavated Egyptian pyramids might be from infections contracted during excavation (Waldron, 1985).

Some human-generated hazards of fieldwork are related to military activity. On the previously mentioned Yuma Proving Grounds survey, the heat was only one of the concerns. Live ordnance was another. Sectors to survey were designated for each given day. While these sectors were being surveyed, guns and artillery were firing nearby. On subsequent days surveying would begin where the firing had occurred, often before the live ordnance people had had time to remove the dangerous material. This was also a testing area, so the ordnance was not always familiar in shape or texture. Laser goggles were worn by ground personnel so they would not be blinded by the laser tracking devices of helicopters flying overhead. In another instance archaeologists surveying in an area where cluster bombs were being tested found pieces of what appeared to be plastic scattered over the surface. These turned out to be unexploded bombs (May, 1991).

Hazardous wastes are a constant threat to archaeologists. One archaeologist conducted a survey at a military installation at a time when she was pregnant. During the survey areas of denuded vegetation were noted. When asked about the lack of vegetation, the military liaison person told the survey personnel not to worry. After delivery of her child, and while still nursing, the archaeologist was informed that she had spent numerous days surveying in an agent orange tested area. She immediately notified her pediatrician, who had never encountered the possibility of agent orange contaminated milk. Fortunately, neither the mother nor her child have experienced any adverse side effects (Walter, 1991).

During the 1950s much aboveground testing of nuclear bombs was conducted at the Nevada test site. One result is that large areas of desert are still radioactive. Archaeologists working in this area have become accustomed to remarks about glowing in the dark. It has been determined that the test-induced radioactivity is located in the dirt and has not penetrated the artifacts. When excavating, field-workers remove the dust from the artifacts in the field before returning them to the laboratory.

3.4 Botanical Hazards

Poison oak, ivy, and sumac are concerns for most archaeologists. It is not always possible to avoid these when surveying. Canyons and drainages have been seen filled with poison oak. When surveying, one cannot travel 2 miles out of the way, up or down the canyon, to avoid the plant. Long pants and shirt sleeves provide some protection but are often inadequate. Excavation at one site was begun in the middle of winter. It was only after digging had commenced that it was discovered that the area was infested with poison oak. The site had not been visited during the time of year when the distinctive leaves of poison oak were out. Contact with twigs and roots resulted in a rash for the exposed field-worker in inconvenient areas of the body.

Cacti and certain other succulents are common hazards. Often hidden around bushes or under leaf litter, these easily become attached to pants and boots. Even though archaeologists on a survey are looking at the ground, such prickly plants are often missed.

3.5 Pathogenic Hazards

The incident that began the author's interest in the hazards of archaeology is a typical story. As part of a cooperative multiyear project with a local government agency, research was conducted at a historic adobe ranch house. Each year another building was renovated, and the archaeology was conducted prior to the renovation. Excavations had been conducted inside the main structure after removal of the floors in several wings, outside in areas where the sewer was to be placed, and under an old sidewalk. The focus for the following season was a building whose original function was unknown. Most recently it had been used as a poultry house. The plan was to excavate down to the floor inside the building to help determine its previous functions. The archaeologists were aware that diseases could be transmitted from the chicken droppings to people. When contacted about the expected exposures, the Department of Health said to not worry about them. Not content with this, the archaeologists contacted industrial hygiene experts, and were instructed at length about the hazards of chickens. Spores were the culprits, resulting in the possibilities of coccidioidomycosis and other fungi. The excavation is now progressing, with every effort made to eliminate the dust. Water is poured on the soil prior to digging; the soil is carried in buckets to a water screening device, where it is completely saturated and pushed through the screen. All excavators and screeners wear masks and gloves.

In many western states, coccidioidomycosis is an occupational hazard for archaeologists. Werner et al. (1972) noted an outbreak in 103 archaeology students at one project. Symptoms were defined and suggestions made as to how to avoid future outbreaks. Since the infective agent seems to be found in rodent burrows, it has been suggested, naively (Waldron, 1985), that archaeologists avoid rodent burrows.

3.6 Physical Hazards

When discussing safety problems in archaeology with members of the industrial hygiene profession, the question most frequently asked is about the hazards of wall collapse. This is a concern, and where the site is deep, units are made into step trenches or supports are added. Loose, wet midden (black, greasy soil high in organic content) is always in danger of collapse. Other physical hazards come from using digging equipment like shovels, picks, and hatchets. Many field-workers suffer back pain from improper body mechanics when picking up screens loaded with soil. Other injuries such as cuts, bruises, broken bones, and eye injuries are common on any project (Waldron, 1985). It is not always possible to place units on flat ground in an open field. Prehistoric sites are found on steep hillsides, close to the edge of canyons, under rock overhangs, and even under water.

4 DISCUSSION

As can be seen from the previous list, the hazards confronting archaeologists are numerous and varied. Issues to be addressed by archaeologists consist of the following:

First, a risk assessment of the known or anticipated hazards is necessary. Howell (1990) presents a good list of assessment factors to consider, and NIOSH et al. (1985) provide a good format for mitigating hazards. Until the hazards actually encountered are documented, the methods of addressing them will still be speculative. Any given geographic area is likely to have a limited number of hazards. Live ordnance is most likely found on military installations; border bandits only in the southwestern states; bears in areas where they currently reside; or freezing temperatures in higher elevations or northern states. In addition, common hazards, like snakes, poison oak, sumac, and ivy, guard dogs, and excavation wall falls can be included in any assessment.

Second, the archaeologist must become educated regarding hazards and safety concerns. Many do not recognize safety problems or the potential for problems. For example, first aid kits are absent from many projects, and when present, are often incomplete. Because they are generally nature lovers and strong environmentalists, archaeologists do not see the natural environment as hazardous. Consequently, risk assessments by archaeologists may neglect natural hazards and are unlikely to include the potential for bee stings, snake bites, or poison oak.

Third, occupational health and safety professionals need to become aware of the level and variety of hazards in archaeology. This chapter is one attempt to alert these professions. It is important to recognize that just because archaeologists work for a company that follows the Occupational Safety and Health Association (OSHA) rules, it does not necessarily follow that the archaeologists themselves are aware of, or implement, the rules.

5 SUMMARY

Anthropology and archaeology are hazardous professions. The specific hazards are dependent on the type of research project, the environment, and the archaeologist's attitude. On any project an archaeologist may encounter safety and health problems with weather, animals, bugs, people, disease, and their own lack of understanding. It is time for archaeologists and industrial hygiene professionals to work together to reduce the hazards and increase awareness.

REFERENCES

Fagan, B. M. (1978). *In the Beginning*, 3rd ed., Little, Brown, Boston.

Hole, F. and R. Heizer (1973). *An Introduction to Prehistoric Archaeology*, 3rd ed., Holt, Rinehart & Winston, New York.

Horvath, S. M. (1985). In *Patty's Industrial Hygiene and Toxicology*, L. J. Cralley and L. V. Cralley, Eds., 2nd ed., Vol III, Wiley, New York, pp. 481–500.

Howell, N. (1990). *Surviving Fieldwork A Report of the Advisory Panel on Health and Safety in Fieldwork*, American Anthropological Association, Special Publication 26, Washington, D.C.

Kyle, C. (1991). Personal communication, Gallegos and Associates, San Diego, CA.

Lee, W. R. (1986). In *Hunter's Disease of Occupations*, Section V, P. A. B. Raffle, W. R. Lee, R. I. McCallum, and R. Murray, Eds., Hodder and Stoughton, London, pp. 745.

May, R. (1991). Personal communication, County of San Diego, Department of Planning and Land Use.

Meighan, C. W. and B. D. Dillon (1989). In *Practical Archaeology Field and Laboratory Techniques and Archaeological Logistics*, B. D. Dillon, Ed., Institute of Archaeology, UCLA, pp. 113–136.

NIOSH, OSHA, USCG, and EPA (1985). *Occupational Safety and Health Guidance Manual for Hazardous Waste Site Activities*, 1st ed., Superintendent of Documents, Washington, D.C.

Shaefer, J. (1991). Personal communication, Brian F. Mooney and Associates, San Diego, CA.

Thomas, D. H. (1979). *Archaeology*, 1st ed., Holt, Rinehart, Winston, New York.

Waldron, H. A. (1985). *Br. J. Ind. Med.*, **42**. 793–794.

Walter. S. (1991). Personal communication, Walter Enterprises, San Diego, CA.

Werner, S. B., D. Papagianis, I. Heindl, and A. Mickel (1972). *N. Engl. J. Med.*, **286,** 507–512.

CHAPTER SIXTEEN

Risk Analysis in Industrial Hygiene

Robert G. Tardiff, Ph.D., A.T.S.

1 INTRODUCTION

Occupational exposures to chemicals, often inevitable, are generally viewed as without consequence to the health of the workers, provided that the doses are no greater than limits set by some recognized authority.

Risk analysis[1] is often vital to decisions on controlling risks and to communicating about the significance of risks before and after abatement of emissions. By its nature, the risk analysis process enables the systematic evaluation of data and the quantitative presentation of complex information. It facilitates both comparisons among alternatives and incorporation of seemingly disparate information (e.g., technological choices, economics, and policy goals) into complex decision-making processes.

Applied to emissions of chemicals in the work environment, risk analysis (1) characterizes injurious effects that might be associated with an activity, (2) enables comparison of health risks of existing technologies with health risks or benefits of proposed replacement or supplemental technologies, (3) helps to distinguish between major and minor sources of risks and their relative impact on health for resource allocations, and (4) permits establishment of priorities in cases presenting multiple potential health problems with limited resources to address them. With an

[1]Also referred to as *risk estimation*.

Patty's Industrial Hygiene and Toxicology, Third Edition, Volume 3, Part A, Edited by Robert L. Harris, Lewis J. Cralley, and Lester V. Cralley.
ISBN 0-471-53066-2 © 1994 John Wiley & Sons, Inc.

understanding of the extent and seriousness of potential health effects, reasoned choices can be made about allocation of resources and about engagement in specific activities.

Risk analysis provides an orderly method to communicate risks associated with complex processes. For instance, workers may fear chemical injury from continuous or accidental releases of numerous chemicals into their work environment by industrial practices. Risk analysis can help people understand the nature and magnitude of those risks in a context to which they can relate. When used to foster understanding of the magnitude of risks, such understanding can serve either to dispel unfounded fears or to indicate a genuine basis for concern and hence for risk avoidance behavior.

The estimation of the degree of risk (and conversely, safety) to human health from substances in the air—whether workplace or ambient—has been used to achieve a variety of objectives.

The most prominent use of risk assessment for substances in air breathed by humans is the setting of tolerable levels of exposure (usually as concentrations in air). By taking into account the toxic properties of a substance and applying predetermined norms of safety (such as the magnitude of some margin of safety), one or more concentrations deemed to be tolerable can be derived. To become a standard, such a concentration needs to be tempered by technological and economic feasibility. Using this approach in part, standards have been set by the Occupational Safety and Health Administration (OSHA) not only to limit concentrations of chemicals (e.g., benzene) in workplace atmospheres but also to prescribe rules for the use of protective equipment by workers. Likewise, limits have been set for pollutants (e.g., ozone) in ambient air and in other media with which humans come into contact.

Selections of one technological approach over another to reduce emissions of chemicals to the atmosphere can be greatly influenced by an understanding of the degree of health safety or health risk associated with each option. Furthermore, knowledge of the health benefits from a specific selection can be shared with those exposed to ease the concern for their well-being.

This chapter first provides an overview of risk analysis and describes some detailed procedures applied to the interpretation of physical, chemical, and toxicity data to estimate risks (Section 2). Next, an illustration of the estimation of risks from exposure to a carcinogen in the workplace is presented (Section 3). That is followed by a comparable illustration to demonstrate the derivation of the degree of safety from workplace exposures to reproductive toxicants (Section 4). A variety of resource documents are available to the reader to obtain more detailed information about risk analysis procedures and their applicability; some include Tardiff and Rodricks (1988), NRC (1977, 1980, 1983), USEPA (1986, 1992b), and WHO (1978, 1984, 1986).

2 THE SUM AND SUBSTANCE OF RISK ANALYSIS

Risk analysis is the process of determining quantitatively the likelihood of adverse effects resulting from exposures, and alternatives for managing them. The risk analysis paradigm widely acknowledged by the U.S. scientific community to be the current standard used in industrial risk management and public policy settings is that first articulated by the U.S. National Academy of Sciences in its report, *Risk Assessment in the Federal Government: Managing the Process* (NRC, 1983).

Health risks are estimated for chemicals known to have a potential for systemic effects in the body after repeated and prolonged exposure (referred to as chronic exposure). Further assessments can be conducted for noncarcinogenic effects, such as respiratory irritation, that can be manifested after relatively brief (i.e., acute) exposure to elevated concentrations of chemicals.

The four basic elements of a risk analysis, as presented in Figure 16.1, are hazard identification, exposure assessment, dose–response evaluation, and risk characterization. Each is described below. A more detailed depiction of the complexity and integration within these elements is presented in Figure 16.2.

2.1 Hazard Identification

In hazard identification a determination is made of whether a causal relationship exists between a chemical and an injurious effect on health. It involves gathering and evaluating toxicity data on the types of health injury or disease that may be produced by a chemical and on the conditions of exposure under which injury or disease is produced.[2] It may also involve characterization of the behavior of a chemical within the body and the interactions it undergoes with organs, cells, or even parts of cells. Data of the latter types may be of value in answering the ultimate question of whether the forms of toxicity known to be produced by a chemical agent in one population group or in laboratory animals are also likely to be produced in the human population group of interest. Risk is not assessed at this stage. Hazard identification is conducted to determine whether and to what degree it is scientifically correct to infer that toxic effects observed in one setting will occur in other settings (e.g., are chemicals found to be carcinogenic or teratogenic at high doses in experimental animals also likely to be so in significantly exposed humans?).

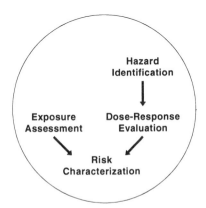

Figure 16.1 Components of a risk analysis.

[2]Hazards other than toxicity (e.g., flammability, explosivity) are not considered in this type of review.

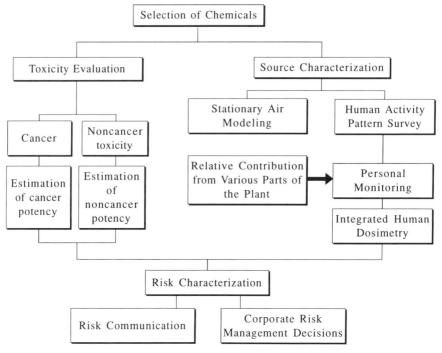

Figure 16.2 Process in the analysis of risks for chemicals of interest.

Within the framework of risk analysis, hazard identification is conducted to fulfill two objectives:

- To ascertain from empirical scientific literature the critical health effects associated with specified chemical(s), including the particular location(s) in, and functions of, the human body that are targets of potential injury. This step includes a determination of health effects associated with both long-term (years to decades) and short-term (24 hr or less) exposures.
- To determine, from all that is known about chemical(s) of interest, the weight of evidence for cause-and-effect relationship(s) between specified chemicals and human injury.

The hazardous properties of compounds are assessed by review of the human epidemiological and toxicological data derived from scientific studies. Epidemiology studies are reports of human disease associated with environmental chemical exposure. Toxicology studies are controlled experiments examining the effects of compounds on experimental subjects (generally laboratory animals). The results of these studies are judged in total to assess the hazardous properties of the compounds and the weight of evidence of the results.

Epidemiology studies have an advantage in that they reflect the effects of chemicals directly on human health. These studies are particularly valuable to determine causation in acute or rare chronic diseases. However, the power of epidemiology to assign causation in more common chronic disease is less, due to its relative insensitivity. In addition, individuals are seldom exposed to only one chemical but rather are exposed to a number of potentially hazardous substances virtually at the same time, both in the workplace and in the home. An additional difficulty in interpreting the findings of epidemiology studies is that precise determinations of worker exposure levels are generally very difficult.

Laboratory animal studies of toxicity have the advantage that an investigation's experimental design can control many of the variables that cannot be controlled in the real setting. In particular, exposure concentration and duration can be determined and maintained with considerable precision. Furthermore, many subtle pathological conditions can be examined in experimental animals. Results from toxicity studies can be used to verify hypotheses derived from epidemiologic associations.

The health effects from exposure to a chemical are typically classified as acute or chronic. Acute health effects can range from subtle and reversible changes in the body (e.g., a temporary rise in an enzyme level) to debilitating, long-term irreversible effects (e.g., stripping of cells lining the lung after a rapid, large exposure to a caustic respiratory irritant). Acute effects occur relatively quickly, vary in severity, and may be reversible. By contrast, chronic injuries generally follow repeated and prolonged exposures (years) to relatively low doses. Injurious effects appear gradually and may be partially or wholly reversible over time once exposure ceases, although less so than effects from most acute exposures. Cancers and chronic obstructive lung disease are illustrations of chronic effects.

Toxicity data, whether from humans or laboratory animals, need to be analyzed comprehensively to determine the strength of associations between exposures and specific pathologies. A number of schemes exist to systematically weight the evidence for causation. Such approaches are specific to a pathology or injury. The toxicity for which weight-of-evidence schemes are most prevalent are cancer and reproductive injury.

The procedures for interpreting these data are those that have been developed over several decades and used to make numerous types of decisions on the safe use of chemicals in food, air, and water and to differentiate between significant and insignificant risks to human health. For the most part, these procedures and techniques have been either developed or sanctioned by the National Academy of Sciences in advising various branches of the federal government about the health risks associated with diverse chemical activities (NRC, 1977, 1980, 1983).

For chemicals suspected of being able to cause cancer, weight-of-evidence criteria have been promulgated by the International Agency for Research on Cancer (IARC, 1987) and by USEPA (1986). In each instance, three sets of data (epidemiology, toxicity, and ancillary but relevant findings) are first evaluated separately. Next, conclusions as to the strength of evidence from human observations and from laboratory animal studies are combined in a numerical rank that reflects the extent to which a substance is likely to cause cancer in humans. The IARC and USEPA schemes are conceptually similar; however, they differ in some of the details of ranking.

For chemicals suspected of being able to cause reproductive injury, Brent (1987) and USEPA (1989b) have each proposed a weight-of-evidence scheme to deal with one specific form of toxicity: birth defects. Brent (1987) proposed a five-step procedure for identifying potential human teratogens:

1. Several well-controlled epidemiologic studies consistently demonstrating an increased incidence of a particular congenital malformation in an exposed human population
2. Secular trends demonstrating a relationship between the incidence of a particular malformation and exposures in the human population
3. An animal model that mimics the human malformation at "clinically comparable exposures" including:
   a. No evidence of maternal toxicity
   b. No reduction in food or water intake
   c. Careful interpretation of malformations that occur in isolation
4. A dose–response relationship for teratogenic effects
5. An understanding of the mechanism of teratogenesis

Although this set of guidelines is useful, several practical difficulties exist in meeting some of the criteria. For instance, if exposure to a teratogen is rare and the defect caused by the teratogen is common, then it would be very difficult to detect a secular trend between exposure and incidence. Furthermore, the general population is exposed to so many chemicals that for an epidemiology study to isolate exposure to only one substance would be rare indeed. Thus, these criteria are generally considered too restrictive to be of practical value.

On the other hand, with USEPA's weight-of-evidence approach for qualitatively assessing teratogenic risk of chemical exposure, findings from human and animal studies are separately evaluated for their scientific quality. Based on USEPA guidelines, the quality of the studies is evaluated, and the subject chemical receives an evaluation of the likelihood that it is a reproductive toxicant in animals or humans. The results of these evaluations are combined, and the chemical receives a rating of the *qualitative* evidence for teratogenicity.

With either approach, great attention must be given to the methods used in reproductive toxicity studies to ensure that the appropriate scientific conclusions can be drawn with confidence based on the data. In particular, greater confidence is given to studies in which (a) an adequate number of doses of the test substance are used, (b) a sufficient number of animals is in each dose group to ensure adequate power to detect the ability to cause injury, and (c) evaluation of the data is complete (to ensure that a potential effect is not missed due to failure to examine a target organ).

The determination of the degree to which exposures in the workplace are likely to be protective of reproductive injury is a function of the extent to which a chemical's toxicity database has been investigated and the degree to which one understands the circumstances under which toxicity is likely to be modified (Kaplan et al., 1987). The confidence in the protective values is dependent on factors such as those listed in Table 16.1.

All relevant studies need to be carefully reviewed to ascertain the presence, or

Table 16.1 Guidelines for Confidence in Toxicity Database

High confidence
Chemical tested in more than one species and sex
Adequate design of studies
Consistent response within species
Consistent response among species
Fetotoxicity present at maternally nontoxic doses
Medium confidence
Adequate design of studies
Chemical tested in only one species or one sex
Fetotoxicity present at maternally nontoxic doses
Low confidence
Inconsistent response within species
Inconsistent response among species
Inadequate study design
Minor or marginal effects that could have been due to other causes
Fetotoxicity present only at maternally toxic doses

absence, of injurious effects at doses in humans lower than those used to treat experimental animals. For instance, if a compound produces no reproductive toxicity at a specified dose in humans or animals, the absence of effect may be due either to its inability to damage the function of the organs or to the administration of too low a dose of the chemical. Finally, if the subcellular mechanism by which a chemical causes injury in an experimental animal is unique to that species and is not present in the human population, then justification exists for not using such data as sentinels of human risk.

Several factors should always be considered when selecting one of the many analytic approaches available for evaluating the total database. The critical ones are *replication* of results, *reproducibility* of results, and *concordance* of results. It is also necessary to consider the *general body of scientific knowledge* concerning the particular toxic endpoint under evaluation.

These concepts are defined here as follows:

- An experimental result is said to be *replicated* if it is found in experiments of identical design.
- An experimental or epidemiological finding is said to be *reproducible* if it is produced in the same species under different conditions, such as different sexes, strains, dose groups, and routes of exposure.
- Experimental and epidemiological findings are said to be *concordant* if they are consistent across species.

For a given set of data pertinent to a given toxic endpoint (assuming each study has been critically evaluated), it is advisable to describe the *degree* of replication, reproducibility, and concordance. In general, as the degree of data replication, reproducibility, and concordance increases, it becomes more certain that a substance possesses the capacity to cause the specific toxic effect under review.

Moreover, it is necessary to consider the degree of correspondence between observations in experimental animals and expected responses in humans for a given form of toxicity, particularly when the only data available derive from animal studies and a judgment must be made about expected effects in humans. If certain types of animal carcinogens are likely to be similarly active in humans, inferences from animal data can be drawn in specific cases. Consequently, risk assessors must maintain their awareness of emerging scientific knowledge so they can make well-informed inferences from data on animals and other types of experimental information.

It is always important to separate true from apparent lack of data on replicability, reproducibility, and concordance. For example, several data sets may, upon superficial examination, appear to lack reproducibility, but more careful examination may reveal that this lack is not substantive. An attempt should be made to learn whether the apparent lack is due to differences in study design and conduct or to true differences in response. It is generally inappropriate to characterize toxicity tests as "positive" or "negative" and then to conclude that the degree of reproducibility of a given effect is low or absent simply because some tests yielded "positive" results and others yielded "negative" results. This oversimplification of test results can be avoided if each study has been critically evaluated so that the differences in various tests (e.g., in extent of examination of test animals, sample size, duration, and magnitude of exposure) are fully known. Once this examination has been completed, it becomes possible to assess the true degree of reproducibility (or lack thereof) for a given effect of a compound. This type of careful examination should also be applied when assessing degree of replication and concordance. Judging the degree of a response's concordance among various species is probably the most difficult aspect of data evaluation, so that the critical evaluation becomes especially important in order to avoid the oversimplification described above.

A final factor to be remembered when judging total data sets is the difference between absence of data and absence of replication, reproducibility, and concordance. A distinction should be made between the presence of positive evidence showing that an effect is not reproducible and the absence of negative evidence because no attempt has been made to reproduce an effect. Clearly, the failure to reproduce an effect (assuming that it is a true failure) may raise serious doubts about the toxic capacity of a substance, whereas the absence of data concerning the reproducibility of an effect should not raise such suspicions. Again, this same principle of evaluation applies not only to reproducibility but also to judgments about replication and concordance. The actual steps in this process are described elsewhere (Tardiff and Rodricks, 1988).

2.2 Dose–Response Assessment

This step describes toxic potency or the quantitative relationship between the amount of exposure to a chemical and the extent of toxic injury or disease.

Data are derived from animal studies or, less frequently, from studies in exposed human populations. Many dose–response relationships may exist for a chemical agent, depending on conditions of exposure (e.g., single versus repeated and prolonged exposures) and the variety of response (e.g., cancer, birth defects) being considered. This process is highly complex, taking into account diverse information about the body's ability to create metabolites more toxic than the parent compound, its ability to detoxify potentially toxic compounds or metabolites, variations in sensitivity to doses of toxic substances, and differences between the mechanisms of toxicity in test organisms (e.g., laboratory rodents) and in human target organs. In many cases, the features of a dose (e.g., duration, frequency, and route) have a great impact on the degree of toxic potency. Specialized procedures must be employed to assure that later characterization of toxic risk is as scientifically defensible as possible.

The presence of a chemical in workplace air does not guarantee that an adverse effect will result. Among the many factors that influence the ability of a chemical to cause harm, the magnitude and duration of the dose[3] determine to the largest extent whether any adverse reaction is possible. Other factors include the nature of the chemical species, route of exposure, and modifying factors such as inborn (i.e., genetic) predispositions and preexisting disease. Yet, during the course of a day, every individual is exposed to a variety of chemicals with no resulting harm, largely because the doses are well within the range of the body's ability to avoid injury to vital functions, through either effective detoxification or efficient and rapid repair of molecular alterations.

Dose–response assessment characterizes the relationship between the dose of an agent administered and the incidence and severity of adverse health effects in an exposed population. The process leads to a determination of *toxic potency*, which is defined as the dose of a substance needed either to cause a specific incidence of injury (e.g., 50 percent of group manifesting an effect, such as eye irritation) or a specific incidence of severity (e.g., 50 percent of a group manifesting mild or severe gastrointestinal upset).

Dose–response relationships can be developed from data obtained in epidemiological studies or experimental toxicology studies. In some cases, epidemiological and human experimental data contain insight into dose–response relationships. These studies are especially useful to determine causation in acute and rare chronic disorders, particularly since toxicity is assessed directly on human health. Their disadvantages to establish dose–response relationships relate to (1) exposures to mixtures of chemicals rather than single substances, thereby complicating the identification of the causative agent(s) in the workplace or elsewhere; (2) determining the dose of a substance that an individual receives; and (3) to the presence of external factors (e.g., smoking cigarettes) that greatly influence the manifestations of toxicity. Even when relevant epidemiological data are available, their findings generally require

---

[3]For convenience, *exposure* is defined as the opportunity for a dose, such as concentrations in food, air, or water, and is generally reported in units such as ppm or ppb or mg/L, etc. *Dose* is defined as the amount received by the body of the target organism (e.g., humans) or a target organ (e.g., the liver or kidneys); it is generally reported in units of weight of the substance (e.g., mg or μg) per body weight of an individual (e.g., kg) per unit of time (e.g., hours or days).

extrapolation from observed exposures to substantially different ones encountered elsewhere to estimate risks to populations of interest (NRC, 1983).

In the absence of, or in addition to, reliable epidemiological data, experimental toxicity data (usually in rodents) are used for dose–response assessments. The inference that results from animal bioassays are generally applicable to humans is fundamental to toxicological research and risk assessment. This premise has been extended from experimental biology and medicine into the experimental observation of carcinogenic effects. Situations exist where observations in animals may not be of relevance in humans, but generally laboratory animal studies have proved to be a reliable indicator of the carcinogenic potential of many chemicals (NRC, 1983).

Many variables, not readily controllable in epidemiological studies, are controlled by design in animal studies. By regulating the route and duration of exposure to a specific chemical, the dose given to experimental animals can be determined quite precisely. Animals can be observed during the course of studies; upon completion of the study, they can be examined for pathological conditions.

Most animal bioassays have been designed primarily to determine which organ is injured by a substance, rather than to determine dose–response relationships. Frequently, cancer bioassays provide little or no data about responses in animals at doses encountered in the workplace or the general environment. Studies with dose–response structure require extrapolation from high-to-low doses using mathematical modeling that incorporates to varying degrees information about physiologic processes in the body (NRC, 1983).

Dose–response functions are often grouped into two classes based on two distinct mechanisms of toxicity: (1) those adverse effects expected to have a biological threshold and (2) those unlikely to have such a threshold. A threshold for a particular toxic response is defined as the dose rate below which the response attributable to the specific agent is virtually impossible (Brown, 1987). Acute toxic responses, for instance, have long been associated with thresholds (Klaassen and Eaton, 1991). In many cases, the biological basis for thresholds for acute and chronic responses can be demonstrated based on mechanistic information, that is, how a substance elicits injury at the molecular level (Aldridge, 1986).

The existence of thresholds for cancer-causing chemicals (at least those called ''initiators'') is, at least hypothetically, unlikely (Klaassen and Eaton, 1991); hence, risk is imputed at all doses, no matter how small. The reason for the presence of risk of cancer even at very low doses of a carcinogen stems from observations that initiators irreversibly damage genetic material (DNA) of a cell causing uncontrolled cell division; that uncontrolled cell replication is viewed as a self-sustaining process, in theory no longer requiring the presence of the toxicant that started the process. Biologically, the absence of a threshold is plausible, but it cannot be confirmed experimentally (NRC, 1983). The choice of an appropriate mathematical model to extrapolate to low doses is based partially on knowledge that a chemical is not likely to have a threshold for this form of pathology.

Each of these issues plays a significant role in determining how useful the data obtained from epidemiological and/or experimental toxicity studies are to estimate the likelihood that the general human population is likely to experience any form of toxicity observed under very different circumstances. Since animal bioassay data are available more often than epidemiological data, the issues of extrapolating animal data from high-to-low doses and from animals to humans are two of the most crucial decisions made concerning the estimation of risk from exposures to a particular chemical.

If a substance causes cancer, the slope of the cancer dose–response curve is used as the unit to describe potency and is called the cancer potency factor (CPF, also technically designated as q_1^*). This approach is used because chemical carcinogens are often viewed as having no demonstrable biological threshold, although some empirical evidence exists to suggest that biological thresholds may well be present (e.g., in humans and laboratory animals, such defense mechanisms as detoxification of carcinogens and efficient repair of molecular lesions).

If a substance causes any form of chronic toxicity other than cancer, the highest dose level that causes no observed adverse effects (NOAEL) serves as an index of toxic potency by delineating doses above which some injury might occur and below which none is likely because of the existence of efficient defense mechanisms in the body. When a NOAEL is obtained from laboratory animal data, the findings are then extrapolated to humans by mathematically taking into account the physiologic and pharmacokinetic differences between the two species. The result of that process produces either a "reference dose" or "reference concentration" (RfD or RfC; terms used by some U.S. regulatory agencies) or an "acceptable daily intake" (ADI; a term used mostly by the World Health Organization and Western European countries).

2.2.1 Quantitative Extrapolation of Dose–Response Curves for Cancer as "Nonthreshold" Toxicity

Possible incidences in humans of cancer can be estimated for doses far below those in the range of observations only by the use of appropriate mathematical models that extend dose–response curves taking into account biological considerations (NRC, 1983). Because such dose–response functions cannot be determined empirically, the actual shapes of such dose–response curves at the lowest ends of the spectrum are unknown.

Since biological thresholds are thought not to exist for most carcinogens, only certain mathematical models are used to predict carcinogenic responses at low doses (nonthreshold models). If a chemical has no threshold for adverse effects, the dose–response curve will be forced to pass through the origin. In the case of nonthreshold chemicals, it is assumed that there is no dose (except zero dose) that corresponds to zero risk of injury. In practical terms, any dose of a carcinogen results in an incremental increase in the risk of cancer (NRC, 1983). Conversely, if a chemical is known to have a threshold for adverse effects, the dose–response curve will not be restricted to passing through the origin. In practical terms, this means that below some exposure level there is no anticipated risk of chronic injury.

Many types of models have been developed to assess the effects of low doses of carcinogens. They may be broadly classified into three classes: statistical, mechanistic, and enhancements. The statistical models assume that each individual in a population has a threshold below which cancer will not occur, and the distribution of thresholds in a population is distributed as a probability function. In contrast, the mechanistic models assume that no threshold exists for carcinogenic effects and that any exposure to a carcinogen results in an incremental risk of cancer. The enhancement models modify the mechanistic by incorporating experimental data on the behavior of the chemical in the body and data on the mechanisms of carcinogenesis.

The most frequently used cancer risk model is a mechanistic type, the linearized multistage model. The key tenet of this model is that it assumes that the production of a malignant cell is a multistep process. Furthermore, at low doses, the risk of cancer is directly proportional to the dose of the carcinogen. At the time that this model was chosen by some regulatory agencies (USEPA, 1986), its theoretical basis most closely approximated current thought on the mechanism of cancer induction (Armitage and Doll, 1961). Since then considerable interest has been placed on a two-stage model that seems to describe well some forms of chemically induced cancers (Moolgavkar and Knudson, 1981). Furthermore, several important refinements have been implemented, such as the application of life-table analyses combined with consideration of the mechanisms of cancer causation (e.g., initiation vs. promotion) in relation to age at onset of first exposure. As an illustration, the LTL model permits a more comprehensive evaluation of less-than-lifetime exposure to a carcinogen by allowing consideration of the mechanism of carcinogenic action of the agent and the effect of exposure during different periods of a person's lifetime (Ginevan, 1989).

Such models generate the slope of a dose–response curve and its upper 95% confidence limit (UCL), which are often referred to as *cancer potency factors*. Such estimates assume exposure to be constant for a working lifetime of 45 years (70 years for environmental exposures). By its nature, UCL overestimates risk, and is, therefore, suited more to standard setting than to defining risks to a specific population. Quantitative calculations of risk are not likely to be higher than those derived by this approach and may be lower or zero.

A major limitation with low-dose extrapolation models is that they all often fit the data from animal bioassays equally well, and it is not possible to determine their validity based on goodness of fit. Each model may fit experimental data equally well, but they are not all equally plausible biologically. The dose–response curves derived from different models diverge substantially in the dose range of interest (NRC, 1983). Therefore, low-dose extrapolation is more than a curve-fitting process, and considerations of biological plausibility of the models must be taken into account before choosing the best model for a particular set of data.

For several decades, regulatory agencies in the United States have applied risk assessment techniques to establish tolerable levels of exposure in all media (air, water, soil) in industrial and commercial settings, and for foods, drugs, and other consumer products. The application of this tool for regulatory standard setting is different from its use in a nonregulatory setting where the objective is to define the risks to a specific target population for a defined chemical exposure. In regulation, conservative "upper-bound" conditions are used for standard setting as a way of indicating that, even in the extreme, risks are not likely to be greater than the selected value. In the nonregulatory setting, "most-likely" conditions are used to estimate the most probable risk to a target population (e.g., the residents of a town). Thus, the regulatory process strives for worst-case exposures and risk; the nonregulatory process strives for likely case exposures of risk.

This difference in application of risk assessment methods can often lead to confusion because of a failure to accurately describe the basis for the results. For instance, the use of upper-bound conditions should lead to a conclusion that the risk "may be as high as" some magnitude of risk (e.g., one in a million) with no indication of how large the risk actually is, whereas the use of most-likely conditions should lead to a conclusion that the risk "is" some magnitude (e.g., one in 10 million), with an indication of the degree of variance.

2.2.2 Extrapolating Animal Data to Humans

The physiological differences between rodents (rats and mice) and humans are at times responsible for differences in toxic potency of the same compound. The dose unit often governs the degree of correspondence in toxic potency between species. Several methods are used to make this adjustment and assume that animal and human risks are equivalent when doses are measured as milligrams per kilogram per day, as milligrams per square meter of body surface area, as parts per million in air, diet or water, or as milligrams per kilograms per lifetime (NRC, 1983).

Scaling factors are used to correct the differences between species when making comparisons among species. On a mechanistic level, scaling factors consider two independent physiological processes: (1) differences in toxicokinetics, which determine the actual dose delivered to target tissues, and (2) differences in tissue sensitivity between species to an identical delivered dose. On a practical level, scaling factors use dose adjustments across species that are based on a normalizing factor such as body weight or total body surface area. The most commonly used scaling factor is body weight, but the most appropriate scaling factor to use in interspecies extrapolations for carcinogenicity has been debated. Surface area, which is approximately equivalent to (body weight)$^{2/3}$, is an alternative to using body weight as a scaling factor between species (Klaassen and Eaton, 1991). Surface area may be more accurate than body weight used directly (Davidson et al., 1986), although some data supports the use of $\frac{3}{4}$ as the exponent for a better fit of the data (Travis and White, 1988; USEPA, 1992a). No single scaling factor is best used in all circumstances. The inclusion of toxicokinetics in the creation of case-specific factors provides the most scientifically defensible cross-species extrapolations (Brown et al., 1988).

2.2.3 Extrapolating Data from One Route to Another Route

The route by which an individual is exposed to a chemical may influence its toxic potency. Such differences in toxic potency may be due to degrees or rates of absorption, to the sequence of organs that are exposed to the substance, or to the nature and rate of the metabolism of the compound. Hence, when risk is to be estimated for a substance by a particular route of administration, experimental data by that same route are preferred to that from another route. However, when experimental data from the actual route of administration are absent, then data from another route must be extrapolated to the one route of interest.

In route-to-route extrapolation, equivalent exposure concentrations from two routes (e.g., ingestion and inhalation) that yield the same absorbed dose are calculated by (1) first determining the amount of a chemical absorbed via the experimental route (e.g., ingestion), and then (2) calculating the applied dose that would yield the same absorbed dose by the route of interest (e.g., inhalation). This procedure uses information about the physical-chemical properties of a substance and about the physiological characteristics of membranes.

To be confident that a specific route-to-route extrapolation is scientifically justified, several criteria should be satisfied: (1) a chemical injures the same organ regardless of route—but to varying degrees (exceptions include corrosives and irritants, which primarily affect local tissue and are not systemically absorbed to any appreciable extent); (2) the toxicity occurs at a site distant from the portal of entry; and (3) after absorption, the behavior (e.g., metabolism and excretion) of a substance in the body is similar by alternative routes of contact.

2.2.4 Risk Extrapolation for "Threshold" (Noncancer) Chronic Toxicity

Many compounds are capable of causing injury to human health provided that sufficiently large daily doses are obtained repeatedly over long periods of time. However, not all chronic toxicants can cause cancer. This section deals solely with substances that are capable of causing chronic toxicity other than cancer.

The ability to cause chronic, noncancer toxicity is generally the result of either the accumulation of damage in susceptible tissues or the accumulation of chemical that ultimately reaches a critical concentration and precipitates injury. Affecting the dose at sensitive tissue sites are processes such as rates of absorption and distribution, redistribution to storage sites, metabolism to more or less toxic substances, and rates of excretion.

Historically, the accepted approach to determine the degree of chronic toxic risk of a substance is to identify that threshold below which no adverse effects are likely to occur. This inflection point in a dose–response curve is known as the no-observed-adverse-effect level (NOAEL).[4] NOAELs may be obtained from studies of humans; however, more commonly, they are derived from studies with laboratory animals. When a NOAEL is obtained from laboratory animals, it usually must be converted to a no-adverse-effect level (NAEL) for humans by taking into account differences between the experimental conditions and actual human exposures and physiological differences between test and target (i.e., human) species.

When a NOAEL is based on data in humans whose circumstances are appreciably different from those in the human setting of interest, some uncertainty also exists about the location of toxicologic thresholds. The NOAEL derived from human observations must also be adjusted to produce a corresponding NAEL.

In either case, uncertainties stem from several factors, each of which can be used to quantitatively modify a NOAEL and estimate a NAEL for humans. The degree of confidence in the NAEL influences directly the degree of confidence in margins of safety (MOS) that are estimated in the risk characterization step described below.

---

[4]This term underscores the realization that the studies from which the observations are made have inherent limitations in their ability to detect adverse effects; that is, the studies are known to have a finite sensitivity to detect toxic responses.

The major adjustments of modifications address the following considerations, some or all of which may apply to a specific chemical exposure:

1. Interspecies Variation in Susceptibility

A factor of between 1 and 10 is used when data from laboratory animals are used to derive a NAEL in humans; if the toxic potency of a compound is similar in humans and experimental animals, then a factor less than 10 may be justified.

2. Intraspecies Variation in Susceptibility

Tests conducted in a homogeneous laboratory animal population (or even a small human volunteer group) often does not account fully for the heterogeneous human population. Individuals may vary considerably in their susceptibility to chemical insult due to genetic factors, lifestyle factors, age, and hormonal status (e.g., pregnancy). To take into account the diversity of human populations, a factor between 1 and 10 is used; when sufficient experience with a compound in humans indicate the existence of a narrow range of susceptibilities exist, a factor less than 10 may be justified.

3. Interspecies Differences in Size and Other Physiological Characteristics

When differences in toxic potency between species represent less the innate toxicity of the chemical and more the differences in the species themselves (e.g., surface area or rate of blood flow), then physiological or allometric scaling factors are justified to provide accurate comparability of doses; one such procedure restates the dose (expressed in mg/kg body weight) in laboratory rodents to one of mg/m^2 surface area in humans.

4. Route-to-Route Variability in Toxic Potency

When the toxicologic information is obtained from one route of administration (e.g., ingestion) yet workers are exposed by another route of contact (e.g., inhalation), resulting differences in toxic potency are taken into account by applying a factor between 1 and 10; if evidence indicates little or no such variability, a factor less than 10 may be justified.

5. Weight of Evidence

The quality of studies on which risk and safety conclusions are to be based for humans may vary considerably, and thus influence confidence in the conclusions. To take into account the reliability of the experimental results, the range of species tested for toxicity, differences in durations of exposure, the inability to identify a NOAEL (in which case, a lowest-observed-adverse-effect level, or LOAEL, is used), and the degree of reproducibility of findings, an uncertainty factor between 1 and 10 may also be applied (Klaassen and Eaton, 1991).

Although methods for quantitative extrapolation of laboratory animal data to estimated human exposures have focused primarily on cancer risks, efforts have been made to develop dose–response models for extrapolation of noncarcinogenic health effects such as developmental toxicity. These models need to be characterized more fully before they can be used in risk assessment (Klaassen and Eaton, 1991).

2.3 Exposure Assessment

This process is used to describe the nature (i.e., distribution of age, sex, and unique conditions such as pregnancy, preexisting illness, and lifestyle) and size of the various populations exposed to a chemical agent, and the magnitude and duration of their exposures. The assessment might include past, current, and exposures anticipated in the future.

As conceptualized, exposure assessment should include two major steps: (1) determining concentrations at the breathing zone and the membrane interface (skin and gastrointestinal tract) and (2) ascertaining the dose to the target tissue, usually at some distance from the point of entry into the body. The first step relies on data obtained from industrial hygiene monitoring activities including data gathered with stationary and personal monitors. At times, these can be supplemented by the use of air dispersion modeling using computer simulations or dose reconstruction modeling using physical models. The second step records work activities that influence rates of degree of absorption into the body and the physical properties of the chemical that influence the rate of absorption into the bloodstream. The outcome of this step provides reliable and defensible quantitative expressions of the relationship between chemical concentrations in the body and associated injury to tissues, organs, and cells. This two-step approach permits the delineation of the contribution of each work area to total dose, and the amalgamation of total dose through a workday in which a worker may move from one location to another. The details of each step are presented below.

Estimates of human doses in the workplace are generally derived for (1) prolonged and continuous exposures over a protracted time period (i.e., years, particularly for exposure to carcinogens) and (2) single peak short-term exposures of 24 hr or less.

The determination of inhaled doses requires the use of appropriate physiological factors,[5] such as body weight, skin surface area, and breathing rates (Table 16.2). Since dose is expressed in terms of body weight (occasionally as surface area), appropriate body weights and surface areas for adults must be selected. When estimating cancer risks applicable to adults (because of long latency, particularly at doses much lower than those used in experimental studies), 70 kg is a standardized body weight for an adult male and 62 kg is that for an adult female. Surface areas of skin for adults along with inhalation rates for men and women are found in Anderson et al. (1985) and USEPA (1989a).

---

[5]Exposure factors are derived from empirical data describing the distributions of human weight, skin surface areas, breathing rates, food and water ingestion rates, activity patterns, and life expectancy as a function of gender and age (USEPA 1989a).

Table 16.2 Summary of Human Inhalation Rates by Activity Level $(m^3/h)^a$

Subject	Resting[b]	Light[c]	Moderate	Heavy[d]
Adult, male	0.7	0.8	2.5	4.8
Adult, female	0.3	0.5	1.6	2.9
Adult, average	0.5	0.6	2.1	3.9

[a] Exposure factors are derived from empirical data describing the distributions of human weight, skin surface areas, breathing rates, food and water ingestion rates, activity patterns, and life expectancy as a function of gender and age (USEPA 1989a).
[b] Includes sitting or standing at an assembly line or sitting at a desk.
[c] Includes activities such as level walking at 2–3 mph, stacking firewood, simple construction, and pushing a wheel barrow with a 15-kg load.
[d] Includes vigorous physical exercise such as cross-country skiing, stair climbing with a load, and chopping wood with an axe.

2.3.1 Atmospheric Concentrations as a Screening Tool

Most often only the first step (measurement of air concentrations in breathing zones) is accomplished, and these values are treated as surrogates for actual dose to the body and to the target organs. This abbreviated approach is of pragmatic value when health-protective standards exist and are expressed as concentrations in air $(mg/m^3$ or ppm). Comparisons between the two can be used to demonstrate degrees of compliance; however, such comparisons cannot be used to ascertain reliably the degree of health risk or safety associated with the actual air concentrations.

2.3.2 Estimation of Human External Dose

Many chemicals exert their chronic toxic effects in organs distant from the portals of entry (i.e., the lungs, skin, and gastrointestinal tract). The body burden, or actual human dose, for systemic toxicants is not solely dependent on the concentration in the air but also on the amount that enters the body. Expressing dose on a mg of substance per mass of body weight per unit of time (e.g., mg/kg day) is often most reflective of the body burden to the individual exposed. Conversion of air concentrations to actual doses entails multiplying the air concentration for a specified time by the breathing rate of the individual, and by dividing the product by the individual's body weight. If groups of individuals are involved, then group averages and ranges can be used. If workers spend parts of the day in work locations whose concentrations of the substance vary considerably, then the contribution of each work location to the body burden would be added together to obtain an expression of the body burden on a daily basis.

External doses are calculated for either of two situations: (1) average daily dose over a year for chronic exposure to noncarcinogens and (2) lifetime average daily dose for repeated and prolonged exposure to carcinogens. In either case, the biological mechanisms of action for either type of toxicant dictate a somewhat different approach to the expression of dose that is most directly related to toxic potency. The steps are as follows.

Yearly Average Daily Dose (ADD$_y$). The yearly average daily dose is computed in three or four steps. Average or maximum daily air concentrations based on duration-specific samples are each aggregated into yearly values to develop weighted annual averages (one for mean concentrations and one for maximum concentrations).

Step 1. Average daily concentration (in mg/m^3) for each individual (ADC$_i$) is calculated by the following equation:

$$ADC_i = \frac{C_1 + C_2 + \cdots + C_n}{D} \tag{1}$$

where C_n = measured daily concentration in mg/m^3 as 8- or 10- or 12-hr average[6] for an individual on the nth day

D = number of days on which measurements were made

Step 1a. ADC$_i$ can be adjusted for the number of months (or weeks) an individual spends at various locations of a workplace by the following equation to obtain the A-ADC$_i$ (in mg/m^3):

$$A\text{-}ADC_i = ADC_{i1}(T_1) + ADC_{i2}(T_2) + \cdots + ADC_{in}(T_n) \tag{2}$$

where T_n = fraction of time a worker spends in location S_n

S_n = location within a facility represented by ADC$_{in}$

Step 2. The average daily dose per person (ADD$_i$; in mg/kg of body weight/day for a year) is calculated by the following equation:

$$ADD_i = \frac{(A\text{-}ADC_i)(m^3 \text{ inhaled air per workday})}{BW} \tag{3}$$

where BW = body weight, kg.

Step 3. The average daily dose for the workforce (ADD$_f$; in mg/kg day for a year is calculated by the following equation:

$$ADD_f = \frac{ADD_{i1} + ADD_{i2} + \cdots + ADD_{in}}{N} \tag{4}$$

where ADD$_{in}$ = average daily dose of nth worker

N = total number of workers

Lifetime Average Daily Dose (ADD$_i$ or LADD). The LADD can be calculated in one or two additional steps, depending on whether the dose is for an individual or

---

[6]Averages can be replaced by the maximum concentration to obtain an indication of the upper bound.

a group of workers. Average or maximum daily air concentrations based on duration-specific samples are each aggregated into lifetime values to develop weighted lifetime averages (one for mean concentrations and one for maximum concentrations).

Step 1. To calculate the LADD for an individual ($LADD_i$), the following equation is used:

$$LADD_i = A\text{-}ADC_i \left(\frac{y}{Y} \right) \tag{5}$$

where y = years worked
      Y = years of total (usually 70) or partial lifetime

Step 2. To calculate the LADD for the workforce ($LADD_f$), the following equation is used:

$$LADD_f = \frac{LADD_{i1} + LADD_{i2} + \cdots + LADD_{in}}{N} \tag{6}$$

where $LADD_{in}$ = lifetime average daily dose for nth worker
      N = total number of workers

2.3.3 Estimation of Human Internal Doses

In some circumstances, dose estimates could be refined by estimating the bodily dose that could reach susceptible organs. Such estimates are obtained by modifying the ADD or LADD estimated in Section 2.3.2 by the degree of absorption via the portal of entry. For instance, if a systemic dose of 10 mg/kg–day had been calculated for a chemical of interest, and the degree of absorption through the lungs were known to be 50 percent and through the skin to be 10 percent, then the absorbed dose would be 6 mg/kg day [= (10 × 50%) + (10 × 10%)].

Determination of *internal* human doses at specific sites in the body where chemicals are likely to exert their effects relies on techniques based on concepts of:

- Distribution of exposure assumptions (e.g., inhalation rates associated with various activities and the time spent on them)
- Dosimetry (e.g., the rate of a chemical's absorption, once inhaled, from the lung into the bloodstream)
- Toxicokinetics (e.g., distribution of a chemical throughout the body, and its rates of metabolism and excretion)

In combination, such approaches are used to establish internal human dose and present an opportunity to understand risks to human health with increased accuracy. This refinement is only applicable when absorption data are available for both humans and the species in which the toxicity data were obtained for use to judge safety or risk from workplace exposure.

In conclusion, although air concentrations may be used as an index of workplace exposure to chemicals, the accuracy of human doses can be enhanced greatly through detailed consideration of factors that influence systemic doses, such as activities that modify breathing rates and those that control the degree and rate of absorption through the lungs and other portals into the bloodstream.

2.4 Risk Characterization

This final step involves integration of the data and analyses from the other three steps of risk assessment to determine the likelihood that specified workers will experience any of the various forms of toxicity associated with a chemical under its known or anticipated conditions of exposure. This step includes estimations of risk for individuals and population groups and a full exposition of the uncertainties associated with the conclusions. Scientific knowledge is usually incomplete, so that inferences about risk are inevitable. A well-constructed risk assessment relies on inferences that are most strongly supported by general scientific understanding, and, to the extent feasible, do not include assumptions derived solely from risk management or public policy directives.

2.4.1 Cancer Risks

The approach used to estimate risks from exposure to carcinogens involves extrapolation of observations of cancer at relatively high doses to much lower doses anticipated or measured in the workplace. The risks for a specified chemical in a defined set of circumstances are estimated by combining the cancer potency factor and the various measures of dose. The risk is expressed as a probability, for instance, as the number of individuals, among all individuals exposed to a cancer-causing agent, that might contract (or die from) the disease attributable to that agent over a specified time, usually a lifetime of 70 years. A specific example would be a lifetime risk of one in a million, meaning one person may get cancer from among a million identically exposed persons. Risk is frequently expressed in a notational format of 10^{-3}, 10^{-4}, 10^{-5}, or 10^{-6}, meaning risks of one in one thousand, one in ten thousand, one in one hundred thousand, or one in a million, respectively. Such cancer estimates reflect the chance that an event may occur; however, because of limitations in knowledge about the processes of cancer causation, it is also possible that the risk may be zero and that no cancer would ensue from a specified exposure. Depending on the quality of the data that underlie an estimated risk, the uncertainties may be small or large; such uncertainties are to be described explicitly and comprehensively at this stage.

2.4.2 Noncancer Risks

For adverse health effects other than cancer, a margin-of-safety (MOS) approach is used to establish whether a potential human dose is lower than a theoretical limit on exposure to provide an acceptable safety margin. The MOS can be determined for a variety of exposure scenarios and for different toxic health effects, such as toxicity to the nervous system or the liver. Using this procedure, first the human NAEL[7] is obtained either from epidemiologic observations or laboratory animal findings. When the NAEL is derived from human data, the measured NOAEL for one or more studies in humans is adjusted for variability in individual susceptibilities, which generally appears to be no greater than one order of magnitude (i.e., between 1 and 10-fold). For instance, a human NOAEL of 10 mg/kg day for a hypothetical compound can be converted to a human NAEL by dividing the NOAEL by 2 or 10 to account for greater variability in sensitivity in an exposed population than in the study group; the result would be a human NAEL of 5, or 1 mg/kg day, respectively.

When a human NAEL is derived from results of animal studies, the human NAEL must account not only for the range of individual susceptibilities (i.e., 1 to 10-fold) but also the possible differences in sensitivity to toxic substances between humans and the laboratory species in which the toxicity had been measured (also ranging between 1 and 10-fold). For example, an animal NOAEL of 100 mg/kg–day for a hypothetical compound could be converted to a human NAEL by dividing the animal NOAEL by 100 to account for differences in sensitivity between test animals and humans and for the degree of variability in sensitivity in an exposed human population; the result would be a human NAEL of 1 mg/kg–day, respectively.

The human NAEL is divided by the daily dose obtained by the population of interest; the result is a MOS. If the margin of safety is greater than 1, a dose unlikely to cause human injury is obtained and may serve as a basis for setting tolerable levels of exposure in the workplace or elsewhere. The larger the MOS, the greater the certainty that no injury will occur.

Human MOS can be estimated for either acute or chronic toxicity by relying on the relevant toxicity and exposure data. For instance, to determine an MOS for acute exposure, peak doses are to be compared to acute NAELs for humans. Depending on the quality of the data that underlie an estimated risk, the uncertainties may be small or large; such uncertainties are to be described explicitly and comprehensively at this stage.

The concept of margins of safety and their application to the determination of acceptable daily intakes (ADIs) was established by the World Health Organization (WHO). Because the terms ''safety'' and ''acceptable'' are value-laden and inherent in the ADI/safety factor approach to determine regulatory values for threshold responses, USEPA has modified this procedure by standardizing ''uncertainty factors'' and a ''modifying factor'' as an additional uncertainty factor that allows for professional judgment on the level of confidence about a NOAEL. The terminology of ''acceptable'' level has been replaced with the terms reference dose (RfD) and reference concentration (RfC). An RfD or RfC is calculated by dividing the NOAEL obtained from an animal bioassay by the product of the uncertainty factors and the modifying factor (Klaassen and Eaton, 1991).

---

[7]In some cases, all examined exposures produce an effect, and only a LOAEL is available. In such cases, the effects occurring at the lowest doses in the most sensitive species/strain/sex are generally used as the basis for estimating a NOAEL.

2.4.3 Mixtures

Since workers may be exposed to more than one compound simultaneously, the possibility of interactions among these chemicals can be assessed. One compound may affect the toxicity of another compound in three possible ways.

1. *Additivity* results when the effect of combined exposure to a combination of compounds equals the sum of the toxicity that would be expected from each of the compounds acting independently.
2. *Synergism* results when the effect of combined exposure to a combination of compounds is greater than the sum of the toxicity that would be expected from each of the compounds acting independently.
3. *Antagonism* results when the effect of combined exposure to a combination of compounds is less than the sum of the toxicity that would be expected from each of the compounds acting independently.

For compounds affecting the same target organ and for which no mechanistic data reject the concept of additive toxicity, their toxicities are assumed to be additive. The effect of additivity may be assessed by calculating a hazard index:

$$\text{HI} = \sum_{n=1}^{i} \frac{\text{dose}_n}{\text{NAEL(H)}_n} \tag{7}$$

where NAEL(H) = the human NAEL. If the hazard index is less than 1, reasonable certainty exists that no adverse effects will occur.

To illustrate the application of risk assessment to the workplace, two case studies are presented, one for a cancer (benzene) and another for reproductive toxicity (benzene, *p*-chlorotoluene, and toluene).

3 ASSESSING THE CANCER RISKS OF BENZENE—A CASE STUDY

In this instance, benzene is present in the atmosphere of workers at a hypothetical petrochemical facility. In this illustration, the data from industrial hygiene monitoring indicate that workers are exposed to benzene at levels that are below the OSHA permissible exposure limits (PEL) of 3.2 mg/m^3 (1 ppm) as a result of application of state-of-the-art practices in industrial hygiene. Nevertheless, both workers and the medical department personnel are asking whether a risk to human health exists and, if so, its magnitude.

3.1 Hazard Identification

Benzene is a volatile, colorless, and flammable liquid aromatic hydrocarbon with a characteristic odor. Benzene is used primarily as a raw material in the synthesis of styrene, phenol, cyclohexanes (nylon), aniline, maleic anhydride (polyester resins), alkyl benzenes (detergents), chlorobenzenes, and other products used in the production of drugs, dyes, insecticides, and plastics. Benzene is also a component of crude oil and gasoline.

3.1.1 Chronic Toxicity

The only two documented forms of chronic toxicity of benzene are cancer (usually acute myelogenous leukemia in adults) and injury to the hematopoietic system (questions have been raised that benzene can cause reproductive injury; that is dealt with in the illustration in Section 4). Because cancer is viewed as a toxic phenomenon unlikely to have a threshold, some risk can be estimated for every chronic dose. For hematopoietic injury, a biological threshold is expected to be present, below which no injury to this organ system is likely to occur. Some investigators have put forth the proposition that damage to the hematopoietic system is a necessary antecedent to the production of leukemia; however, this has not been proven.

Benzene, which can be readily absorbed through skin, lungs, and gastrointestinal tract, is known to be converted in the body to one or more metabolites that are likely to cause the cellular damage in the bone marrow. Thus, although benzene is rapidly excreted and does not accumulate in the body, some forms of subcellular damage may accumulate, as reflected by a sense that, at least hypothetically, cancer risk is cumulative with repeated exposures. As a result, the risk of cancer is often the one that is expected to persist even at relatively low doses. Only the carcinogenic property of benzene is addressed in this illustration.

Benzene seems to produce few forms of other toxicity, except at doses so high as to be encountered in unusual circumstances. This is the case for reproductive toxicity in laboratory animals in which benzene is not known to injure a developing fetus except when mothers are exposed to atmospheric concentrations much higher than those permitted in workplaces (addressed in Section 4).

The information on which these conclusions are based is summarized below.

Observations in Humans. The chronic toxicity of benzene has been investigated extensively (ATSDR, 1989; IARC, 1982). Chronic exposure to concentrations of benzene greater than 200 ppm (640 mg/m^3) in air can seriously damage the bone marrow causing pancytopenia and aplastic anemia. Benzene is a confirmed human carcinogen and is associated, in most cases, with acute myelogenous leukemia at workplace concentrations as low as 35 ppm (112 mg/m^3). Benzene may be capable of causing cancer at lower concentrations; however, epidemiology studies are unable to detect an increased risk of leukemia, due to their relative insensitivity (Rinsky et al., 1981; IARC, 1982, 1987, pp. 120–122).

An association between benzene and leukemia was suggested as early as 1928 (Delore and Borgomano, 1928). Since then, a number of case reports have suggested that benzene can cause leukemia, most notably a series of leukemias reported by Aksoy (1977, 1980) and Aksoy et al. (1974) among Turkish shoemakers. These case reports have been confirmed by epidemiology studies, most notably in a cohort of rubber workers (Infante et al., 1977; Rinsky et al., 1981, 1987) and chemical workers (Wong et al., 1983). Based on such results, IARC has classified benzene as a confirmed human leukemogen (1982, 1987, pp. 120–122).

Aksoy (1980) examined a series of 34 cases of leukemia among shoe workers in Istanbul. Based on a government estimate of the number of shoe workers in Istanbul, the investigator estimated the incidence of acute leukemia, or "preleukemia," among these workers was 13 per 100,000 compared with an estimated incidence of 6 per 100,000 in the general population (relative risk = 2). Since the shoe workers labored in small shops, obtaining reliable estimates of benzene exposure for the population as a whole was not possible. The general exposure was probably high; peak exposures to benzene were reported to be in the range of 210–650 ppm (679–2075 mg/m^3) in some poorly ventilated shops. The duration of exposure was estimated to be 1–15 years with a mean of 9.7 years. Due to the uncertainties in the population at risk and levels and durations of exposure, neither reliable estimates of the risk of cancer in this group of exposed individuals nor an estimate of individual exposure could be obtained (IRIS, 1991). Although these studies provide little quantitative information on the relationship between benzene exposure and leukemia, they do support a positive qualitative relationship between benzene exposure and leukemia.

Infante et al. (1977) and Rinsky et al. (1981) made a retrospective cohort analysis of 748 workers occupationally exposed to benzene between 1940 and 1959 in two factories producing rubber hydrochloride. Vital status was determined for workers up to 1975. Expected rates of death due to leukemia were obtained from U.S. white male mortality statistics. A statistically significant increase in mortality from all leukemias was observed (observed/expected = 7/1.25; SMR = 560). In a follow-up study, Rinsky et al. (1987) extended this cohort to employees hired between 1940 and 1964. In addition, vital status was determined up to 1981. A statistically significant increase in mortality from all leukemias were observed (observed/expected = 9/2.27; SMR = 337). Both studies (Rinsky et al., 1981; 1987) attempted to characterize the exposure of the workers. Workers were classified by job title and, based on industrial hygiene records, the exposures of the individual workers were estimated. A significant correlation between the intensity and duration of benzene exposure and the risk of leukemia was observed, further supporting the assumption that benzene is the etiologic agent.

Wong et al. (1983) examined the mortality of 4062 workers exposed to benzene in the chemical industry in Canada. Workers were classified by job description as to the duration and intensity of benzene exposure. Control subjects were not exposed to benzene in the workplace. A dose-dependent increase in the incidence of leukemia, lymphatic cancer, and hematopoietic cancer was observed in benzene-exposed individuals. The observed increase in incidence of leukemia may have been due in part to a lower than expected incidence of leukemia among control workers, making this study of questionable significance.

Ott et al. (1978) examined the mortality of 594 workers at a Dow Chemical plant. Three leukemias were observed. The increase in cancer was not statistically significant, and no dose–response was observed. On the other hand, the small number of workers exposed makes this study relatively insensitive statistically. In addition, the benzene exposures were relatively low (generally less than 15 ppm), which decreased the probability of detecting a carcinogenic effect. Leukemia cases in this study may have been underreported, since one death certificate listed leukemia as a secondary rather than a primary cause of death. Listing leukemia as a primary cause of death would have added the case to the three cases previously identified, which would have made this study's increased incidence of cancer statistically significant. Because of these factors, this study is considered to be of borderline value.

In addition to causing neoplasms, benzene can alter the function of the bone marrow. Exposure to high concentrations of benzene can result in the development of aplastic anemia. This lesion is characterized by a progressive decrease in the number of circulating formed elements in the blood. The bone marrow in these cases is necrotic with fatty replacement of the functional bone marrow (Snyder and Kocsis, 1975). The lowest benzene concentration associated with the development of aplastic anemia is 60 ppm (192 mg/m^3) (ACGIH, 1991b).

Acute exposure to benzene in humans can result in headache, lassitude, and weariness at concentrations greater than 50 ppm (160 mg/m^3 = 3.8 mg/kg) for 5 hr. At higher concentrations, benzene can depress the central nervous system and even lead to death.

Observations in Animals. In a series of studies, Cronkite and colleagues examined the carcinogenic and hematotoxic effects of benzene. Cronkite (1986) exposed CBA/Ca male mice to benzene at 100 and 300 ppm (320 and 960 mg/m^3) for 6 hr/day, 5 days/week, for 16 weeks, followed by observation for the remainder of their lives. These doses of benzene caused a significant increase in the incidence of leukemia in exposed mice. Cronkite also observed anemia, a decrease in stem cell content of bone marrow, and a decrease in marrow cellularity at doses of 100 ppm (320 mg/m^3) for 6 hr/day, 5 days/week, for 2 weeks. Cronkite et al. (1984) observed increased incidence of leukemia in another strain of mouse (female C57B1/6) following doses of 300 ppm benzene for 6 hr/day, 5 days/week for 16 weeks. In addition, Cronkite et al. (1989) exposed CBA/Ca mice to a range of inhaled concentrations of benzene. A NOAEL was observed at 10 ppm (32 mg/m^3) after benzene exposure 6 hr/day, 5 day/week for 2 weeks. A dose-dependent decrease in blood lymphocytes, bone marrow cellularity, and marrow content of spleen colony-forming units was observed for concentrations above 25 ppm (80 mg/m^3).

Maltoni et al. (1982) exposed Sprague-Dawley rats for 15 weeks to 200 ppm (640 mg/m^3) benzene for 4 hr/day, 5 days/week for 7 weeks and then for 7 hr/day, 5 days/week for 85 weeks. Exposure at this concentration was associated with an increase in total malignant tumors and carcinomas when compared to controls. Tumors and carcinomas were found in the zymbal glands, the oral cavity, mammary glands, and liver. Since humans have no zymbal glands, the relevance of the zymbal gland tumors observed in this study are of questionable significance to human health.

In addition to causing neoplasms, benzene is also hemotoxic in experimental animals. Snyder et al. (1980) exposed male AKR/J mice exposed to 100 ppm (320 mg/m^3) via inhalation after 6 weeks of age for 6 hr/day, 5 days/week for life. A significant increase in the incidence of anemia, lymphocytopenia, and bone marrow hypoplasia was observed. Rozen and Snyder (1985) exposed mice to benzene concentrations of 300 ppm (960 mg/m^3) for 6 hr/day, 5 days/week for 23 weeks. Mice were examined for lymphoid parameters at 1, 6, and 23 weeks. A progressive reduction in the response of T- and B-cells to mitogenic stimuli, a reduction in the numbers of B-lymphocytes in the bone marrow and spleen, and a decrease in the number of T-lymphocytes in the thymus and spleen was observed. At 23 weeks, mitogen-induced proliferation of bone marrow and splenic B-lymphocytes was eliminated. The authors concluded that 300 ppm benzene exposure results in severe depression of B- and T-lymphocyte numbers and response to mitogens. These alterations may be a factor in the carcinogenic response to benzene.

3.1.2 Genotoxicity

Significant increases in chromosomal aberrations of bone marrow cells and peripheral lymphocytes from workers with high exposure to benzene have been cited by numerous investigators (IARC, 1982). High doses of benzene have also induced chromosomal aberrations in bone marrow cell from mice (Meyne and Legator, 1980) and rats (Anderson and Richardson, 1979). Positive results in the mouse micronucleus have been reported (Meyne and Legator, 1980). Benzene was not mutagenic in several bacterial and yeast systems, in the mouse lymphoma cell forward mutation assay, or the sex-linked recessive lethal mutation assay with *Drosophila melanogaster* (IARC, 1982).

3.1.3 Toxicokinetics

The toxicokinetics of benzene are being actively studied (Travis et al., 1990; Snyder et al., 1989), and several physiologically based pharmacokinetic (PB-PK) models have been proposed recently for benzene metabolism (Medinsky et al., 1989; Bois et al., 1991; Spear et al., 1991). Following absorption into the body, benzene is widely distributed to tissues (Travis et al., 1990; Rickert et al., 1979). Because of its lipid solubility, benzene is preferentially distributed to fat (Rickert et al., 1979). In rats, the half-life of benzene in fat is longer than the half-life in other tissues (1.6 hr versus half-lives between 0.4 and 0.8 hr for other tissues) (Rickert et al., 1979). The bone marrow, lung, and kidney are also major sites of benzene deposition in experimental animals (as indicated by higher benzene concentrations in these tissues than in the blood). The half-life of benzene in humans is approximately 0.75 hr at inhaled doses of less than 10 ppm for 8 hr and approximately 1.2 hr at doses of 100 ppm for 1 hr (Srbova et al., 1950; Sherwood, 1988).

It is generally accepted that benzene must be metabolized before it can cause injury (Snyder et al., 1981). Metabolism of benzene occurs principally in the liver; however, extrahepatic metabolism appears to contribute significantly to benzene toxicity, especially in the bone marrow (Travis et al., 1990). In the liver, benzene is metabolized via cytochrome P_{450} to form benzene epoxide. Benzene epoxide undergoes further metabolism to form phenol, catechol, and hydroquinone as primary metabolites (Kalf, 1987). These ring metabolites may be conjugated with sulfate or glucuronic acid and excreted in the urine (Travis et al., 1990). Open-ringed metabolites are also formed and may eventually become the potentially toxic metabolites *trans, trans*-muconaldehyde and *trans, trans*-muconic acid (Parke and Williams, 1953; Witz et al., 1990). Variations in the route of exposure do not significantly alter the production of benzene metabolites (Henderson et al., 1989).

The toxicity of benzene is due to the action of one or more of its metabolites rather than of benzene itself (Parke and Williams, 1953; Snyder et al., 1981). In several studies, the modification of benzene metabolism has led to altered benzene toxicity. For example, toluene, a competitive inhibitor of benzene metabolism, decreased benzene hematotoxicity in mice (Andrews et al., 1977). In another study, Longacre et al. (1981a,b), showed that benzene metabolites were found at higher concentrations in benzene-sensitive DBA/2 mice than in benzene-resistant C57B1/6 mice.

Absorption: Inhalation. For continuous doses of 50–100 ppm benzene for several hours, the absorption of benzene by humans is approximately 50 percent (IARC, 1982; Snyder et al., 1981). Nomiyama and Nomiyama (1974a,b) reported that men and women who were exposed to 52–60 ppm benzene for 4 hr had an estimated respiratory retention (the difference between respiratory uptake and excretion) of 30 percent of the dose with little difference between the sexes.

Inhalation studies in animals confirm that benzene is rapidly absorbed through the lungs. Sabourin et al. (1986) reported that uptake of benzene vapor in rats and mice is inversely proportional to the air concentration. The percentage of inhaled benzene that was absorbed and retained during a 6-hr exposure period decreased from 33 to 15 percent in rats and from 50 to 10 percent in mice as the exposure increased from 10 to 100 ppm (31.9 to 390 mg/m^3).

Absorption: Skin. Benzene can be absorbed through the skin but at a greatly decreased rate when compared to inhalation exposure. Blank and McAuliffe (1985) calculated that an adult working in an ambient air containing 10 ppm (31.9 mg/m^3) benzene would absorb 6.6 mg/hr from inhalation and 1.32 mg/hr from whole body (2 m^2) exposure. The authors propose that because of the water solubility of benzene, diffusion through the stratum corneum is the rate-limiting step for dermal absorption.

3.1.4 Summary

Exposure to benzene is known to cause leukemia after workers have been exposed to high concentrations of the solvent over prolonged periods of time. Benzene or its metabolites is also carcinogenic in experimental animals. Based on results obtained from epidemiology studies, IARC has classified benzene as a known human carcinogen (IARC, 1982). Benzene or its metabolites is genotoxic, causing chromosomal aberrations. Limited information precludes any conclusion about the potential for alterations of the immune and central nervous systems resulting from long-term benzene exposure.

3.2 Dose–Response Characterization

The quantitative descriptor of benzene's cancer potency is the cancer potency factor (CPF). The critical chronic toxic endpoint for benzene exposure is ordinarily acute myelogenous leukemia in adult humans. Since the studies associating cancer with benzene exposure contain data on worker exposure to benzene, USEPA based its CPF on these epidemiologic data. Three studies were used to model the dose–response relationship: Rinsky et al. (1981), Ott et al. (1978), and Wong et al. (1983).

A total of 21 estimates were made using six models and various combinations of the epidemiologic data, with a mathematical correction for the findings of Wong et al. (1983). Evidence exists that benzene is incompletely absorbed by inhalation (Snyder et al., 1981; Nomiyama and Nomiyama, 1974a,b); consequently, USEPA assumed in calculating an inhaled (absorbed) CPF of 2.9×10^{-2} (mg/kg day)$^{-1}$ that 50 percent of inhaled benzene is absorbed (USEPA, 1989c).

For acute toxicity, the critical toxic endpoint is central nervous system depression (headaches, lassitude, and weakness). These symptoms are associated with benzene concentrations between 160 and 480 mg/m^3. The acute NOAEL in humans for these effects was identified as 160 mg/m^3 (Sandmeyer, 1981). The human NAEL is obtained by dividing the human NOAEL of 3.7 mg/kg by an uncertainty factor of 2 to account for the intraspecies differences in sensitivity; the result is a dose of 1.9 mg/kg.

3.3 Exposure Assessment

At this hypothetical facility, 100 individuals are employed to perform assorted tasks in either of two chemical production areas (A or B). From employment records, workers generally perform tasks approximately 40 percent of the workday in area A, and approximately 60 percent of the workday in area B. Length of service at the facility ranges from 2 to 40 years, with 26 years as a median value. Daily work schedules frequently last 10 hr.

Information about levels of benzene in each work area has been collected by two means: (1) 6 area monitors (3 in area A and 3 in area B) collecting data hourly during the workday, and (2) personal monitors affixed to 20 plant workers during four 10-hr shifts. After 2 years of employment, workers were at the plant 48 weeks of the year.

To enable the calculation of doses of benzene for each employee, physiological factors have been obtained. To determine approximate breathing rates of these employees that would in turn influence the magnitude of each employee's body burden of benzene, a questionnaire has been administered to 40 employees. This information provides a basis for determining the absorbed dose of benzene. The body weight of each employee has been obtained from recently obtained records of periodic medical examinations.

Chronic Doses. Chronic doses (in the unit of lifetime average daily dose, or LADD) have been calculated to estimate the chance of developing cancer from benzene at this facility. Personal monitoring data have indicated that the concentration in the breathing zones of the workers averages 0.6 mg/m^3 (±0.08 mg/m^3) integrated across areas A and B.

According to survey responses, while in area A, workers performed moderate to heavy activity, whereas relatively light activity was performed in area B. The average body weight of the employees was 66 kg.

The lifetime average daily dose can be calculated for the workers using Eqs. (5) and (6). The results ranged from 2 μg/kg-day for an individual working at the facility for 2 years to 53 μg/kg-day for the 40-year veteran; for the median 26-year duration of employment, the estimated dose was calculated to be 35 μg/kg–day (Table 16.3). Since these data have been collected solely during one year of operation, the findings are extrapolated to 26 and 40 years, assuming that conditions for the median and maximum of employee work lives have remained constant during the working lifetime.

The area monitoring has yielded annual average concentrations of 0.9 mg/m³ ($\pm$0.01 mg/m³) in area A and 0.2 mg/m³ ($\pm$0.01 mg/m³) in area B. The records of year-long monitoring indicate relatively few incidences of excursions greater than one standard deviation from the mean. Area monitoring data during the time of the personal monitoring are compared to those for the balance of the year-long data collection. The results indicate no statistically significant differences (Table 16.3); consequently, the results of the personal monitoring are deemed to be reasonably representative of repeated and prolonged exposure, at least for the 12 months during which data have been collected. The doses obtained from these results were 3, 43, and 66 μg/kg–day for the 2, 26, and 40 years of employment, respectively.

Acute Doses. To estimate acute doses for which the MOS can be estimated, data from either personal or area monitors can be used. Specifically, the highest measured concentrations of one-hour exposure durations are selected. As such, the exposures calculated by this method are plausible worst case and are consistent with the conditions that are capable of causing acute toxicity. The results indicate that the peak concentrations measured in area A range from 1.6 to 2.4 mg/m³ slightly less than 1 percent of the time. Peak concentrations in area B were considerably less. The highest concentration measured in area A is equivalent to a dose of 0.1 mg/kg.

3.4 Risk Characterization

The cancer risk associated with exposures to benzene in this hypothetical workplace were estimated by combining the cancer potency factor with the calculated doses. The results are presented in Table 16.4. The estimated risks steadily increased from approximately 7 in 10,000 to 2 in a 1000 as a function of years of employment. The risks estimated from either fixed or personal monitoring data were not significantly different from one another; however, exposures calculated from the personal monitoring data are deemed to be somewhat more accurate than those obtained from the fixed or area monitors.

The MOS for the potential for acute toxicity was derived by dividing the human

Table 16.3 Chronic Doses of Benzene

| Years of Work | LADD (μg/kg–day) | |
	Personal	Fixed
2	2.6	3.3
26	3.5	4.3
40	5.3	6.6

Table 16.4 Estimated Cancer Risk from Hypothetical Exposure to Benzene in the Workplace

Years of Work	Estimated Cancer Risk	
	Personal	Fixed
2	7.5×10^{-5}	9.2×10^{-5}
26	9.7×10^{-4}	1.2×10^{-3}
40	1.5×10^{-3}	1.9×10^{-3}

NAEL for acute toxicity by the peak dose in work areas A and B. The results indicate that the MOS is 19. The magnitude of the MOS is sufficient to assure that acute toxicity is unlikely to occur.

3.4.1 Analysis of Degrees of Uncertainty

Exposure assessments employ a series of assumptions, each with its own range of uncertainty that, when combined, can produce an exposure estimate with widely separated numerical bounds. A number of sources of variability and uncertainty impinge upon any given exposure assessment:

- *Measurement error*, which arises from random and systematic error in a given measurement technique
- *Sampling error*, which arises from uncertainties relating to the representativeness of the actual population being measured
- *Variability*, which arises from the natural differences in exposure-related parameters among individuals, such as inhalation rate
- *Limitations in applicability of generic or indirect data* used as surrogate data for actual exposure levels, when these are unknown
- *Professional judgment*, which, if used to assign values to exposure parameters that may not be accurately known, may lead to misleading results, or widely varying results if different values are assigned by different assessors.

Ideally, exposure assessments can be based on measurements of actual concentrations, residues in biological fluids and tissues, data on emissions sources, exposure duration, exposure frequency, and knowledge of environmental transport, transformation, fate, and toxicokinetics. However, it is rare that an exposure can be precisely defined for a given chemical and exposure scenario, because some key data, ranging from detailed knowledge of source terms to characteristics of the exposed population (which would influence the choice of exposure parameters), may be missing. Thus, any value for an input parameter for estimating individual or population exposures via an exposure model or algorithm has an associated level of uncertainty.

Uncertainties Associated with Measured Air Concentration Data. Many of the current air monitoring methods have inherent variability. For example, a high-quality measured data set may have measured concentrations that typically are accurate to within ± 25 percent. A fixed monitoring program with less rigorous control may be accurate to within a factor of 2 or more. Thus, all measured concentrations contain some degree of sampling and analysis errors. These uncertainties should be considered when using measured air quality data for exposure or risk assessment.

Uncertainties Associated with Personal Monitoring and Time–Activity. Attempts have been made to increase certainties in the exposure and risk estimates by applying the result of the personal monitoring studies and time–activity surveys. Personal monitoring increases certainties by accounting for sources of exposure that can effectively be accounted for in no other way; time–activity surveys increase certainties by providing information on activities that can be used indirectly to define more precisely the daily inhalation rates of individual members of the population. Time–activity data also enhance the accuracy of estimates regarding the time spent in different microenvironments; from these findings, it is known that the average individual spends 40 percent of the time in area A and 60 percent in area B.

Uncertainties Associated with Exposure Parameters. The use of accurate and appropriate exposure parameter values provides an important step in increasing certainties in exposure assessment. The results of a recent study conducted for the Chemical Manufacturers Association (RiskFocus, 1990) underscores that the use of accurate and representative exposure factors (such as were obtained through the time–activity survey) increase the certainties in the resulting exposure estimate.

4 ASSESSING THE RISKS OF REPRODUCTIVE TOXICITY—A CASE STUDY

At a hypothetical chemical manufacturing facility, both female and male employees are working under conditions in which three compounds (benzene, toluene, and *p*-chlorotoluene) are present in large quantities in the workplace atmosphere. Despite the demonstration that the facility meets federal standards of industrial hygiene practice and of atmospheric concentrations (i.e., PELs), workers and management have questioned whether these standards of operation are sufficient to protect against impairment of the reproductive performance in women and men. Particular concern surrounds the possibility that developing fetuses of female workers may be at risk of injury from chemical exposures to the mothers. In this instance, the standards do not make clear the extent to which they protect against injury to the reproductive process; hence, risk assessment, relying on industrial hygiene monitoring data, provides some means to address this matter in a systematic manner.

A risk assessment was performed to evaluate the significance to reproductive health of the occupational exposures. The analysis first examined each compound individually and then collectively. The results of the evaluation are summarized below. For each compound, the workers' levels of exposure were compared to the derived tolerable exposure concentration, and to the current standards of the Occupational Safety and Health Administration (OSHA) and the American Conference of Governmental Industrial Hygienists (ACGIH).

As background, the individuals work 8 hr per day, 5 days per week, 50 weeks per year. All of the activity on the job is considered light for purposes of estimating breathing rates. The air concentrations of benzene were derived from fixed monitoring data collected by the firm's Industrial Hygiene Department for the period 1977 to 1989. Nine job codes were identified for 30 workers.

4.1 Benzene

The data for benzene were analyzed according to the four steps in risk assessment. Reproductive toxicity is often divided into development toxicity (injury to the developing fetus, with particular emphasis on birth defects, also called terata[8]) and fertility dysfunction (injury to the reproductive function of either male or female that results in impairment of the ability to reproduce). This distinction is important because laboratory animal studies can be designed to address either one or the other, rarely both; hence the existence of a teratogenicity study may shed little light on the ability of a substance to alter the reproductive process, and vice versa.

4.1.1 Hazard Identification

Evidence exists that benzene can cause injury at levels of exposure below 200 ppm; however, none of these are related to reproductive toxicity. ACGIH (1991a) cites several reports of workplace exposures to benzene at air concentrations around 40 ppm having an adverse effect on blood components. This is supported by results from Deichmann et al. (1963), which suggests that inhalation exposure to 44–47 ppm (131–150 mg/m^3) can affect the bone marrow function in rats.

Nine studies examined the possibility that benzene may adversely alter reproductive outcome in humans (ATSDR, 1989). In a study by Holmberg (1979), a statistically significant association was found between exposure to various organic solvents (benzene, toluene, etc.) and CNS defects in newborns. However, only one case reported exposure to benzene, and the number of controls with benzene exposure was not reported. A serious flaw with this study is that no control was made for other factors that may influence pregnancy outcome. Thus, this study provides insufficient information to draw any conclusions on the potential adverse effects of benzene on pregnancy outcome in humans.

In the eight other studies of pregnancy outcome in benzene-exposed women, none found a significant increase in the frequency of birth defects in the offspring of women occupationally or environmentally exposed to benzene. The benzene doses were not specified, although two of the women were suffering from benzene-induced anemia suggesting that the repeated doses were likely to have been quite high. According to ACGIH (1991a), no blood dyscrasias—believed to be the most sensitive index of noncancer toxicity—occur below an 8-hr time-weighted average (TWA) of 25 ppm (80 mg/m^3).

Several reproductive toxicity studies using laboratory animals have been reported.

---

[8]The term *teratogenesis* is used to refer to the process by which birth defects (frequently anatomical) are produced; and a *teratogen* is one that causes birth anomalies.

Kuna and Kapp (1981) exposed pregnant rats to 0, 10, 50, and 500 ppm (0, 32, 160, and 1600 mg/m^3) benzene in air for 7 hr per day on gestation days 6 through 15. Dams exposed to benzene had a dose-dependent decrease in weight gain between days 5 and 15, which was significant only at 50 and 500 ppm. No overt signs of benzene-related toxicity were observed in exposed rats. Benzene treatment had no effect on pup viability, but there was significantly decreased pup weight at 50 and 500 ppm. At both 50 and 500 ppm, pups had delayed skeletal development as evidenced by lagging ossification; but current scientific thinking holds that this manifestation is not considered teratogenic (Kimmel and Wilson, 1973). However, several developmental abnormalities were noted in the group exposed to 500 ppm, including exencephaly, angulated ribs, dilated lateral and third ventricles of the brain, and forefoot ossification out of sequence. Only a few of the pups had these defects (no more than 4 for each category out of 98 fetuses); however, these anomalies are extremely rare in historical controls suggesting a causal relationship for birth defects. Thus some evidence exists to indicate that benzene is fetotoxic and, at relatively high doses (i.e., 500 ppm and above), may be teratogenic.

Green et al. (1978) exposed rats to 100, 300, and 2200 ppm (320, 960, and 7040 mg/m^3) benzene in air for 6 hr daily on days 6 through 15 of gestation. Rats exposed to 2200 ppm benzene had decreased weight gain and showed signs of narcosis during exposure, suggesting the presence of acute toxicity. No effect on weight gain was noted in rats exposed to 100 or 300 ppm benzene. At 2200 ppm benzene, decreased litter weight and significant increase in skeletal anomalies (primarily missing sternebra and delayed ossification of the sternebra) were observed. At the lower benzene doses, no effect was observed on fetal weight; however, increases in skeletal anomalies were found, but with no clear dose–response in the incidence of the anomalies: The incidence of missing sternebra was highest in the 100- and 2200-ppm benzene exposed group; but the 300-ppm benzene group had no significant increase in missing sternebra. These data indicate that benzene is fetotoxic at doses at which no overt effects on maternal well-being are present.

Murray et al. (1979) exposed pregnant mice and rabbits to 500 ppm (1600 mg/m^3) benzene in air for 7 hr/day on gestation days 6 through 15 and 6 through 18, respectively. No compound-related toxicity was observed in dams. Benzene was slightly embryotoxic as evidenced by decreased fetal weight in the benzene-exposed mice compared to control. In addition, benzene caused delayed development of the skeletal system, suggesting an embryotoxic effect.

Tatrai et al. (1980) continuously exposed rats to 0 and 124 ppm (0 and 300 mg/m^3) benzene in air on days 7 through 14 of gestation. Exposed rats had decreased weight gain, indicative of maternal toxicity. The pups had decreased mean fetal weight with retarded bone development compared to controls. In a similar study, rats were exposed to 0 and 313 ppm (0 and 1000 mg/m^3) benzene in air continuously on days 9 through 14 of gestation (Hudak and Ungvary, 1978). In this study, the dams manifested decreased weight gain, and the pups had decreased birth weight with retarded bone development.

Watanabe and Yoshida (1970) injected pregnant mice with 2600 mg/kg benzene subcutaneously on days 11, 12, 13, 14, and 15 of gestation and sacrificed them on day 19. No controls were used in this study. They observed an increase in the incidence of fetuses with cleft palate and decreased jaw size in mice injected on day 13 of pregnancy compared to other days of injection. This dose of benzene caused a significant decrease in the white blood cell count of mice.

Confidence in Database. Consistently among nine epidemiologic studies, benzene has not been reported to cause birth defects in the offspring of exposed women, even when the women themselves were suffering from signs of benzene-induced chronic damage to the blood-forming organs (e.g., pancytopenia). This observation suggests that benzene has, at worst, only a small ability to cause birth defects in humans. The inability of benzene to cause developmental toxicity in experimental animals has been reported in numerous studies; these findings are consistent with the observations in humans. In these studies, high doses of benzene were embryotoxic (without being teratogenic) in three species of laboratory animals (rats, mice, and rabbits) at doses that would have, or would have been expected to, adversely affect maternal health.

Since the results of these studies in humans and various species of laboratory animals have been consistent, strong assurance is provided to conclude that benzene is not likely to be teratogenic. However, benzene appears to have the potential to be fetotoxic at high doses (>160 mg/m^3). From all of the evidence (see Table 16.5), one can conclude that exposure to benzene at levels that do not affect maternal well-being would have no adverse effects on the offspring.

4.1.2 Dose–Response Assessment

Epidemiological evidence from workers exposed to benzene suggest that the threshold for all noncancer adverse effects is around 40 ppm (128 mg/m^3 = 19.7 mg/kg–8-hr day) and that no noncancer toxicity is likely to occur at or below 25 ppm (80 mg/m^3 = 12.3 mg/kg–8-hr day) (ACGIH, 1991a). To obtain a human NAEL that accounts for the possibility of the existence of a wider range of susceptibility to the potential developmental toxicity of benzene than was observed among those in the published studies, the dose was divided by 3 to obtain a human NAEL of 8.3 ppm (26.6 mg/m^3 = 4.1 mg/kg–8-hr day). In light of the absence of reproductive toxicity in human studies of benzene (despite limitations in those investigations), the NAEL may be conservatively low, implying that somewhat higher doses could well be tolerated without the possibility of reproductive injury.

Table 16.5 Summary of Results[a]

Compound	Developmental Toxicity		Fertility Dysfunction	
	Conducted	Result	Conducted	Result
Benzene	+	T$_\pm$; F$_+$	−	0
p-Chlorotoluene[b]	+	T$_\pm$; F$_+$	−	0
Toluene	+	T$_\pm$; F$_+$	+	−

[a] + = positive; ± = equivocal; − = negative; 0 = no data; T = tetratogenicity, F = fetotoxicity secondary to maternal toxicity.
[b] Based on *o*-chlorotoluene.

4.1.3 Workplace Exposure

Before reproductive risk can be characterized, the magnitude and nature of human exposure and dose must be estimated.

The time-weighted-averages (TWA) were plotted by job code and year against the (TWA) concentration. The scatter plots were examined for temporal trends. Since the data ranged over several orders of magnitude, the method of presentation of choice was in the form of semilog graphs. Estimates of worker exposures were made using cumulative frequency distributions. Due to the small number of workers in certain job categories, it was decided to pool the industrial hygiene data for the cumulative frequency distributions. Since workers were exposed to a wide range of benzene concentrations, the 80th percentile (i.e., the air concentration at or below which 80% of the workers were exposed) was selected as a realistic upper-bound exposure scenario. This value is realistic in as much as reproductive toxicity is generally related to high concentrations for relatively brief periods (i.e., a year) rather than to decade-long averages.

Information provided by the Industrial Hygiene Department suggests that for most processes, the worker activity would be of light exertion. The average lightly working male breathes 0.8 m^3/hr or 6.4 m^3/workday; the average female breathes 0.5 m^3/hr or 4 m^3/workday. The average male weighed 78 kg and the average female weighed 65 kg. Pulmonary absorption was obtained from published values. Since, by definition, the manufacturing process is closed, dermal exposure to the workers is discounted.

The time-weighted-average (TWA) benzene concentrations for the various job codes are depicted in Figure 16.3, along with the current OSHA permissible exposure limit (PEL) of 1 ppm (3.2 mg/m^3) (OSHA, 1987). During the 8-year monitoring period, only three samples exceeded the current OSHA PEL. Inspection of the plot reveals no obvious trends in exposure level with time for any job category.

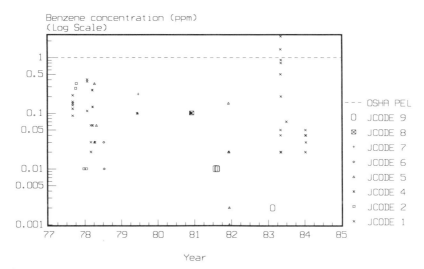

Figure 16.3 Scatter plot of benzene TWA concentration by job code and year.

The 80th percentile exposure level was determined to be 0.5 ppm or 1.47 mg/m^3 (dose = 230 μg/kg–day).

4.1.4 Risk Characterization

The evidence from several observations in humans suggests that even at maternally toxic doses the development of the fetus may not be altered. However, observations in laboratory animals are compelling that excessive exposure to benzene may have adverse effect on the fetus. Based on human observations, such adverse effects are unlikely to occur below 8.3 ppm (26.6 mg/m^3; dose = 4.1 mg/kg–day). The 80th percentile workplace exposure to benzene is about 0.5 ppm (1.5 mg/m^3 = 230 μg/kg–day) at this facility. The MOS for the fetuses of pregnant women exposed to benzene is, therefore, 17.7 (26.6 $\div$ 1.5 mg/m^3). Since OSHA's current PEL is 1 ppm (3.2 mg/m^3) (OSHA, 1987), the MOS for benzene developmental toxicity at the PEL is at least 2 (3.2 $\div$ 1.47 mg/m^3). Therefore, meeting the OSHA standard should be protective against developmental toxicity.

No data were available on whether benzene can alter fertility. Therefore, no conclusion can be reached on whether the PEL of 1 ppm is protective against fertility dysfunction.

4.2 Toluene

The data for toluene were analyzed according to the four steps in risk assessment.

4.2.1 Hazard Identification

The toxicity of toluene has been well studied over the years (WHO, 1985). The primary effects of toluene are on the central nervous system. Short-term exposure to toluene (750 mg/m^3) can cause minor signs of central nervous system toxicity (e.g., headache). Higher exposure levels can result in narcosis (drowsiness). Finally, repeated chronic deliberate inhalation of very high concentrations of toluene (glue sniffing) can result in brain damage.

Chronic toluene abuse associated with glue sniffing has been shown to injure the fetus. Infants born to mothers who sniffed glue during pregnancy have had low birth weights, and some have had a condition resembling fetal alcohol syndrome (Hersh et al., 1985; Goodwin, 1988), but no systematic reports of pregnancy outcomes in glue sniffers have been made. At present, no studies have been reported on pregnancy outcome by toluene vapors at levels encountered in the workplace.

Toluene has been examined for its effects on pregnant rats, mice, and rabbits. When rats were exposed to 6000 mg/m^3 toluene for 24 hr/day on days 9–14 or 9–21 of pregnancy, no birth defects were noted. However, toluene was toxic to the embryos, as evidenced by increased embryo death and decreased birth weight of the pups (Hudak et al., 1977). Additional studies in mice and rabbits have confirmed that toluene can cause low birth weights and delayed development of the bones, although it has not been found to cause birth defects (i.e., not teratogenic). In addition, spontaneous abortions were reported in rabbits exposed to 1000 mg/m^3 toluene, 24 hr/day on days 6–20 of gestation (the rabbit gestation period is 33 days)

(Ungvary and Tatrai, 1985). By contrast, rats exposed to lower concentrations of toluene (375–1500 mg/m^3) for 6 hr per day on days 6–15 of pregnancy showed no signs of developmental toxicity (Litton Bionetics, 1978).

Toluene has also been examined for its effects on reproductive organs. CIIT (1980) reported no adverse pathology in the ovaries and testes of rats exposed to 1100 mg/m^3 for 24 months. In addition, mice exposed to 375 and 1100 mg/m^3 had no altered reproductive behavior, although mice exposed to 7000 mg/m^3 had decreased growth but no effect on reproductive parameters (API, 1985).

The actual dose received by the mice in the API (1985) study can be calculated by a similar set of assumptions as were used for calculation of human dose. In this case, the exposure is assumed to have lasted for 6 hr/day, and a mouse is assumed to breathe 0.0018 m^3/hr (0.011 m^3/6 hr) (Gold et al., 1984) and weigh 30 g. Without data on pulmonary absorption, pulmonary absorption is assumed to be 50 percent. The dose is calculated to be 200 mg/kg–day (1100 mg/m^3)(0.011 m^3)(0.5)/(0.03 kg).

Confidence in Database. The effects of toluene on pregnant animals are generally consistent across species with one exception (i.e., the spontaneous abortion in rabbits exposed to toluene is inconsistent with observations in the majority of the reports). Nevertheless, the results consistently demonstrate that the adverse effects of toluene on pregnancy appeared only at doses that were maternally toxic. Teratogenic effects have been observed in women who chronically abuse toluene; however, in these cases, these women had a history of malnutrition and often alcohol abuse, each of which is known to injure the developing fetus. Thus, these cases are not considered indicative of workplace exposures. Based on these considerations, toluene is expected to have low teratogenic potential. Likewise, the evidence in experimental animals suggests that toluene is fetotoxic, but only at doses sufficient to injure maternal health.

Thus, all of the evidence indicates that exposure to toluene at levels that do not affect maternal well-being would have no adverse effects on the developing offspring. In two species of laboratory animals and in appropriately designed studies, toluene consistently did not alter fertility.

4.2.2 Dose–Response Assessment

The available evidence in experimental animals suggest that toluene is fetotoxic only at doses that are greater than those injurious to maternal health. Therefore, exposure to toluene at levels that do not affect maternal well-being would not have any adverse effects on the offspring. The highest dose of toluene that was found to have no adverse effect on pregnancy outcome was 1500 mg/m^3 (Litton Bionetics, 1978); however, because of the toxicity to rabbits at 1000 mg/m^3, the most confident NOAEL is 375 mg/m^3 (dose = 68 mg/kg–day). Since this value is lower than NOAELs in other studies, this may be considered a worst-case scenario.

The human NAEL of 2.7 mg/kg day is obtained by dividing the NOAEL in laboratory animals (68 mg/kg–day) by 25 (i.e., 5 × 5) to account for inter- and intraspecies variability in response. If the results in rabbits overstate the toxic potency of toluene (which may well be the case), the human NAEL could justifiably be four times higher.

4.2.3 Workplace Exposure

A scatter plot of the TWA toluene concentrations for the various job codes are depicted in Figure 16.4. The ACGIH threshold limit value (TLV) is 100 ppm. Inspection of the plot reveals no obvious trends in exposure level with time for any job category. The frequency distribution of worker exposure (Fig. 16.5), shows that the 80th percentile exposure level was 0.32 ppm, or 1 mg/m^3. The estimated dose is 0.041 mg/kg–day $\{[(1 \text{ mg/m}^3)(6.4 \text{ m}^3/\text{day})(0.5)]/78.1 \text{ kg}\}$ for males and 0.031 mg/kg–day $\{[(1 \text{ mg/m}^3)(4.0 \text{ m}^3/\text{day})(0.5)]/65.4 \text{ kg}\}$ for females.

4.2.4 Risk Characterization

The evidence is compelling that exposure to toluene may have adverse effect on the fetus, but only at very high toluene concentrations that are toxic to the pregnant animal (Table 16.5). Dividing the human NAEL of 2.7 mg/kg–day by the actual dose experienced in the facility (0.31 mg/kg–day) produces a MOS of 8.7, which is sufficiently large to protect against fetotoxicity. The actual MOS may be four times larger if the data from rabbits are not applicable to humans.

In two species of laboratory animals and in appropriately designed studies, toluene consistently did not alter fertility. Therefore, it is inappropriate to calculate a NOAEL, NAEL, or MOS for toluene. Consequently, the TLV of 100 ppm should provide ample protection against dysfunction in fertility.

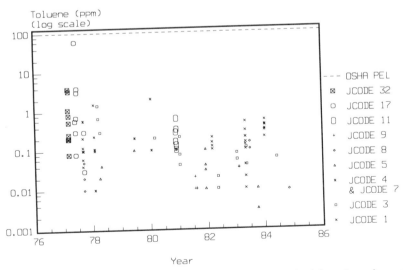

Figure 16.4 Scatter plot of toluene TWA concentration by job code and year.

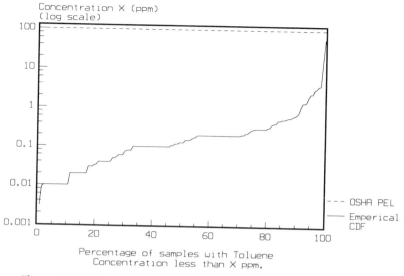

Figure 16.5 Cumulative distribution of toluene TWA concentrations.

4.3 *p*-Chlorotoluene (PCT)

This compound represents a special case in which the safety or risk to the reproductive functions can be estimated solely from the data for a chemical congener because no appropriate data are available for the compound of interest. The data for this compound were analyzed according to the four steps in risk assessment.

4.3.1 Hazard Identification

Because no data were available on the reproductive toxicity of PCT, the reproductive toxicity data for chemical congener of PCT, *o*-chlorotoluene (OCT), were evaluated.

In a study conducted by Huntingdon Research Centre (1983a), rabbits were exposed to 1.5, 4, and 10 g/m³ OCT in air for 6 hr per day from days 6–29. Toxicity was noted in the highest exposure group (ataxia and lachrymation and/or salivation); 5 of the 15 pregnant rabbits died or were sacrificed. In the middle exposure group (4 g/m³), slight ataxia occurred during exposure; there was no mortality among the 13 rabbits during treatment. Two out of twelve rabbits in the control and the 1.5 g/m³ groups died during the study. The authors attributed the mortality to factors other than OCT. No significant effect on mean values for litter size, pre- and post-implantation loss, or litter and mean fetal weight were noted. Thus, OCT produced no reproductive toxicity in pregnant rabbits.

In another study performed at the Huntingdon Research Centre (1983b), rats were exposed to air concentrations of 1, 3, and 9 g OCT/m³ for 6 hr per day from days 6–19 of pregnancy. At the highest level of exposure, rats were ataxic with lachrymation and/or salivation in some animals. At 3 g/m³, some ataxia was noted in exposed rats. No overt signs of toxicity were noted in the rats exposed to 1 g/m^3 OCT. There was a dose-dependent decrease in food consumption and body weight gain in rats exposed to 3 and 9 g/m^3. At 9 g/m^3 the mean litter and mean fetal weights were significantly reduced. In addition, an increase was noted in the incidence of fetal malformations (brachydactyly of a single fore or hind paw). Also an increase was observed in sternebrae variants that was attributed to decreased fetal weight. On the other hand, there was no increase in visceral anomalies in rats exposed to 9 g/m^3. No adverse effects were noted at 1 or 3 g/m^3.

The actual dose at a concentration of 3 g/m^3 received by the pregnant rats can be calculated by a similar set of assumptions as were used for calculation of human dose. In this case, it is assumed that a rat breathes 0.006 m^3/hr (0.036 m^3/6 hr) (Gold et al., 1984) and weighs 278 g [taken from gestation day 14 body weights (Huntingdon Research Centre, 1983b)]. Without data on pulmonary absorption, pulmonary absorption is assumed to be 50 percent. The dose is calculated as 1.9 g/kg day {[(3 g/m^3)(0.36 m^3)(0.5)]/0.278 kg}.

Confidence in Database. The studies followed standard design for developmental toxicity studies. The doses of OCT were high since there was significant maternal toxicity at the two highest doses. The deleterious effects of OCT on the offspring can be attributed to the toxic effect of the chemical on maternal well-being. The absence of signs of teratogenic effects at 1 and 3 g/m^3 supports the supposition that OCT would not be expected to be teratogenic in humans (Table 16.5).

4.3.2 Dose–Response Assessment

The study in rats is best suited for determining the toxic potency of the OCT (Huntingdon Research Centre, 1983b) because of the absence of mortality among treated animals. Since 9 g/m^3 OCT adversely affected fetal development, possibly due to effects on maternal health, this dose is a LOAEL. The NOEL for laboratory animals is 3 g/m^3, which is equivalent to a dose of 1900 mg/kg–day. The human NAEL is calculated by dividing the animal NOAEL of 1900 mg/kg–day by an uncertainty factor of 200 (i.e., 10 × 10 × 2) to account for inter- and intraindividual variability in sensitivity and for the possibility that PCT may have somewhat greater toxic potency than OCT. The result is a human NAEL of 9.5 mg/kg–day.

4.3.3 Workplace Exposure

A scatter plot of the TWA PCT concentrations for the various job codes are depicted in Figure 16.6. The ACGIH TLV for OCT of 50 ppm is also noted. Inspection of the plot reveals no obvious trends in exposure level with time for any job category. The cumulative frequency distribution of PCT TWA concentration is shown in Figure 16.7. It was determined that the 80th percentile exposure level was 0.35 ppm, or 1.8 mg/m^3. Since reproductive data were available only for females, no dose derivation was made for male workers. Since the workers are working at light activity when exposed to PCT, the volume of air the female workers breathed was 4.0 m^3/day. The estimated dose for female workers is 0.055 mg/kg–day {[(1.8 mg/m^3)(4.0 m^3/day)(0.5)]/65.4 kg}.

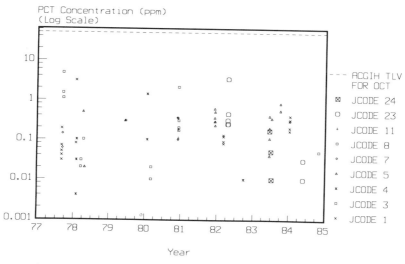

Figure 16.6 Scatter plot of *p*-chlorotoluene by job code and year.

4.3.4 Risk Characterization

The studies suggest that the congener OCT has a low order of toxicity with no apparent effects on fetal development at doses up to 3 g/m³ for 6 hr per day, which is equivalent to a dose of 1900 mg/kg–day. The human NAEL is 9.5 mg/kg–day.

The MOS for OCT exposure by females is approximately 170 (9.5/0.055). This finding indicates that PCT is unlikely to have any effect on fetal development at this

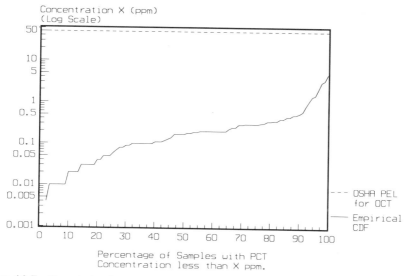

Figure 16.7 Cumulative frequency distribution of *p*-chlorotoluene TWA concentrations.

or lower doses, even if PCT were 2 or 3 times more potent toxicologically than OCT.

Neither PCT and OCT have been tested for effects on fertility in males or females; consequently, no conclusion about their potential for injury to such functions can be reached.

4.4 Assessment of Exposure to Mixtures

Because of the possibility that workers are likely to encounter two or three of these compounds during the workday, it is advisable to determine whether the combined exposures might be of some consequence to the health of the fetus (there are no corresponding data on fertility dysfunction to undertake such an analysis for that toxic endpoint). Such an analysis was performed using Eq. (7). The result was a hazard index of 0.025 when the higher human NAEL for benzene was used and 0.933 when the lower human NAEL was used. In either case, the index is less than 1, and hence indicative that fetal injury is unlikely to occur at the concentrations measured in the workplace.

5 SUMMARY AND CONCLUSIONS

Risk analysis is a systematic means of interpreting complex information to estimate risks to human health from chemical exposures. Applied initially to exposures to substances in the general environment, risk analysis has a role in estimating the impact of chemicals in the workplace, even in the presence of health-based standards.

As currently fashioned, risk analysis examines information in four steps. First, they toxic properties of substances are evaluated to determine what types of injury they are capable of producing in humans and under what circumstances—a process designated hazard identification. Next the amount or dose of substances—that is, their toxic potency—that is needed to cause varying rates of injury or degrees of severity among exposed individuals: the dose–response assessment. Exposure assessment is the third step that characterizes the doses of specified chemicals obtained by target populations (e.g., individuals in a specific facility). Risk characterization amalgamates the information into estimates of possible cancer incidence and of the likely absence of noncancer injury.

The process contains numerous uncertainties that influence confidence in the conclusions. Some of the predominant ones include (1) the degree to which laboratory animal data are predictive of human responses, (2) the variations among humans in their susceptibility to chemical toxicity, and (3) the amount of concordance between the findings in a database and the actual situation for which the risks are being estimated. To the extent that understanding of the influence of such factors for specific substances in the workplace is increasing continually, confidence is enhanced incrementally in the validity of conclusions derived from the process. However, it is important to recognize that, for some chemicals and some forms of exposure, the data are sufficiently incomplete to provide wide error bands surrounding the risk estimates. Such ranges in data quality are apparent in the two illustrations in this chapter. For example, little doubt remains that benzene causes an adult form of leukemia at high doses; less certainty exists about its ability to do so at low doses, although prudent public health policy infers that it may do so. Similarly, confidence is relatively high that benzene is unlikely to cause birth defects; yet, without further study, no comparable claim can be made about the possibility, or lack thereof, that benzene might alter reproductive success.

Despite its limitations, risk analysis is a tool finding increasing use in the workplace to inform workers and management of the extent to which they may be at risk or relatively safe in the implementation of control technologies and in the setting of policies by private industry and government. Used wisely, risk analysis can indeed be a suitable guide to the safe use of chemicals.

REFERENCES

ACGIH (1991a). *Documentation of the Threshold Limit Values and Biological Exposure Indices*, 6th ed., American Conference of Governmental Industrial Hygienists, Cincinnati, OH.

ACGIH (1991b). *1991–1992 Threshold Limit Values for Chemical Substances and Physical Agents and Biological Exposure Indices*, American Conference of Governmental Industrial Hygienists, Cincinnati, OH.

Aksoy, M. (1977). "Leukemia in Workers Due to Occupational Exposure to Benzene," *New Istanbul Contrib. Clin. Sci.*, **12**, 3–14.

Aksoy, M. (1980). "Different Types of Malignancies Due to Occupational Exposure to Benzene: A Review of Recent Observations in Turkey," *Environ. Res.*, **23**, 181–190.

Aksoy, M., S. Erdem, and G. DinCol (1974). "Leukemia in Shoe-workers Exposed Chronically to Benzene," *Blood*, **44**, 837–841.

Aldridge, W. N. (1986). "The Biological Basis and Measurement of Thresholds," *Ann. Rev. Pharmacol. Toxicol.*, **26**, 39–58.

Anderson, D. and C. R. Richardson (1979). "Chromosome Gaps Are Associated with Chemical Mutagenesis," *Environ. Mutat.*, **1**, 179 (Abs. No. Ec-9).

Anderson, E., N. Browne, S. Duletsky, J. Ramig, and T. Warn (1985). *Development of Statistical Distribution or Ranges of Standard Factors Used in Exposure Assessments. Final Report*, U.S. Environmental Protection Agency, Office of Health and Environmental Assessment, Washington, D.C. Available from National Technical Information Service, Springfield, Virginia 22161, as PB85-242667.

Andrews, L. S., E. W. Lee, C. M. Witmer, J. J. Kocisis, and R. Snyder (1977). "Effects of Toluene on the Metabolism, Disposition and Hemopoietic Toxicity of [^{3}H]benzene," *Biochem. Pharmacol.*, **26**, 293–300.

API (1985). *Two-Generation Reproduction/Fertility Study on a Petroleum-Derived Hydrocarbon [i.e., Toluene]*, Vol. 1, American Petroleum Institute, Washington, D.C.

Armitage, P. and R. Doll (1961). "Stochastic Models for Carcinogenesis," in *Proceedings of the Fourth Berkeley Symposium on Mathematical Statistics and Probability*, Vol. 4, University of California Press, Berkeley, pp. 19–38.

ATSDR (1989). *Toxicological Profile for Benzene*, Report No. ATSDR/TP-88/03. Prepared by Clement Associates, Inc., Fairfax, Virginia. Agency for Toxic Substances and Disease Registry, Atlanta. Available from National Technical Information Service, Springfield, Virginia 22161, as PB89-209464.

Blank, I. H. and D. J. McAuliffe (1985). "Penetration of Benzene through Human Skin," *J. Invest. Dermat.*, **85**, 522–526.

Bois, F. Y., M. T. Smith, and R. C. Spear (1991). "Mechanisms of Benzene Carcinogenesis: Application of a Physiological Model of Benzene Pharmacokinetics and Metabolism," *Toxicol. Lett.*, **56**, 283–298.

Brent, R. L. (1987). "Etiology of Human Birth Defects: What Are the Causes of the Large Group of Birth Defects of Unknown Etiology?" in *Developmental Toxicology: Mechanisms and Risk. Banbury Report 26*, J. A. McLachlan, R. M. Pratt, and C. L. Markert, Eds., Cold Spring Harbor Laboratory, Cold Spring Harbor, New York, pp. 287–303.

Brent, R. L. and M. I. Harris (1976). *Prevention of Embryonic, Fetal and Perinatal Disease*, DHEW Publication (NIH)76-853, U.S. Department of Health and Human Services, National Institutes of Health, Bethesda, MD.

Brown, C. C. (1987). "Approaches to Interspecies Dose Extrapolation," in *Toxic Substances and Human Risk. Principles of Data Interpretation*, R. G. Tardiff and J. V. Rodricks, Eds., Plenum Press, New York, pp. 237–268.

Brown, S. L., S. M. Brett, M. Grough, J. V. Rodericks, R. G. Tardiff, and D. Turnbill (1988). "Review of Interspecies in Risk Comparisons," *Reg. Toxicol. Pharmacol.*, **8**, 191–206.

CIIT (1980). A Twenty-Four Month Inhalation Toxicology Study in Fischer-344 Rats Exposed to Atmospheric Toluene, Chemical Industry Institute of Toxicology, Research Triangle Park, NC.

Cronkite, E. P. (1986). "Benzene Hematotoxicity and Leukemogenesis," *Blood Cells*, **12**, 129–137.

Cronkite, E. P., J. Bullis, T. Inoue, and R. T. Drew (1984). "Benzene Inhalation Produces Leukemia in Mice," *Toxicol. Appl. Pharmacol.*, **75**, 358–361.

Cronkite, E. P., R. T. Drew, T. Inoue, Y. Hirabayashi, and J. E. Bullis (1989). "Hematotoxicity and Carcinogenicity of Inhaled Benzene," *Environ. Health Perspect.*, **82**, 97–108.

Davidson, I. W. F., J. C. Parker, and R. P. Beliles (1986). "Biological Basis for Extrapolation Across Mammalian Species," *Reg. Toxicol. Pharmacol.*, **6**, 211–237.

Deichmann, W. B., W. E. MacDonald, and E. Bernal (1963). "The Hemopoietic Tissue Toxicity of Benzene Vapors," *Toxicol. Appl. Pharmacol.*, **5**, 201–224.

Delore, P. and C. Borgomano (1928). "Acute Leukaemia Following Benzene Poisoning. On the Toxic Origin of Certain Acute Leukaemias and Their Relation to Serious Anaemias," *J. Med. Lyon (Fr.)*, **9**, 227–233.

Ginevan, M. (1989). "Appendix A. Explanation and Sensitivity Analysis of the Modified Armitage and Doll Model (LesLife®)," in *Airliner Cabin Environment: Contaminant Measurements, Health Risks, and Mitigation Options*, N. L. Nagda, R. C. Fortmann, M. D. Koontz, S. R. Baker, and M. E. Ginevan, Eds., Report to U.S. Department of Transportation, Washington, D.C., by Geomet Technologies, Germantown, MD, pp. A1–A10.

Gold, L. S., C. B. Sawyer, R. Magaw, G. M. Backman, M. de Veciana, R. Levinson, N. K. Hooper, W. R. Havender, L. Bernstein, R. Peto, M. C. Pike, and B. N. Ames (1984). "A Carcinogenic Potency Database of the Standardized Results of Animal Bioassays," *Environ. Health Perspect.*, **58**, 9–319.

Goodwin, T. M. (1988). "Toluene Abuse and Renal Tubular Acidosis in Pregnancy," *Obstet. Gynecol.*, **71**, 715–718.

Green, J. D., B. K. J. Leong, and S. Laskin (1978). "Inhaled Benzene Fetotoxicity in Rats," *Toxicol. Appl. Pharmacol.*, **46**, 9–18.

Henderson, R. F., P. F. Sabourin, W. E. Bechtold, W. C. Griffith, M. A. Medinisky, L. S. Birnbaum, and G. W. Lucier (1989). *Environ. Health Perspec.*, **82**, 9–17.

Hersh, J. H., P. E. Podruch, G. Rogers, and B. Weisskopf (1985). "Toluene Embryopathy," *J. Pediatr.*, **106**, 922–927.

Holmberg, B. (1979). "Central Nervous System Defects in Children Born to Mothers Exposed to Organic Solvents," *Lancet*, **2**, 177–179.

Hudak, A. and G. Ungvary (1978). "Embryotoxic Effects of Benzene and Its Methyl Derivatives: Toluene, Xylene," *Toxicology*, **11**, 55–63.

Hudak, A., K. Rodics, I. Stuber, and G. Ungvary (1977). "Effects of Toluene Inhalation on Pregnant CFY Rats and Their Offspring," *Munkavedelem*, **23**(1–3), Suppl., 25–30.

Huntingdon Research Centre (1983a). Effect of 2-chlorotoluene Vapour on Pregnancy of the New Zealand White Rabbit, Report to Occidental Chemical Corporation, August 3, 1983, Huntingdon Research Centre, Huntingdon, England.

Huntingdon Research Centre (1983b). Effect of 2-chlorotoluene Vapour on Pregnancy of the Rat, Report submitted to Occidental Chemical Corporation, August 6, 1983, Huntingdon Research Centre, Huntingdon, England.

IARC (1982). "Benzene," in *IARC Monographs on the Evaluation of the Carcinogenic Risk of Chemicals to Humans, Vol. 29, Some Industrial Chemicals and Dyestuffs*, World Health Organization, International Agency for Research on Cancer, Lyon, France, pp. 93–148.

IARC (1987). *IARC Monographs on the Evaluation of Carcinogenic Risks to Humans, Suppl. 7, Overall Evaluations of Carcinogenicity: An Updating of IARC Monographs Volumes 1 to 42*, World Health Organization, International Agency for Research on Cancer, Lyon, France.

Infante, P. F., R. A. Rinsky, J. K. Wagoner, and R. J. Young (1977). "Leukaemia in Benzene Workers," *Lancet*, **2**, 76–78.

IRIS (1989) (retrieved September 1991). "Benzene," Toxicology Data Network (TOXNET), National Library of Medicine, Bethesda, MD.

Kalf, G. F. (1987). "Recent Advances in the Metabolism and Toxicity of Benzene," *Crit. Rev. Toxicol.*, **18**, 141–159.

Kaplan, N., D. Hoel, C. Portier, and M. Hogan (1987). "An Evaluation of the Safety Factor Approach in Risk Assessment," in *Developmental Toxicology: Mechanisms and Risk*, J. A. McLachlan, R. M. Pratt, and C. L. Markert, Eds., Banbury Report, Cold Spring Harbor Laboratory, Cold Spring Harbor, NY, pp. 335–346.

Kimmel, C. A. and J. G. Wilson (1973). "Skeletal Deviations in Rats: Malformations or Variations?" *Teratology*, **8**, 309–316.

Klaassen, C. D. and D. L. Eaton (1991). "Principles of Toxicology," in *Casarett and Doull's Toxicology: The Basic Science of Poisons*, 4th ed., C. D. Klaassen, M. O. Amdur, and J. Doull, Eds., Macmillan, New York, pp. 12–49.

Kuna, R. A. and R. W. Kapp (1981). "Embryotoxic/Teratogenic Potential of Benzene Vapor in Rats," *Toxicol. Appl. Pharmacol.*, **57**, 1–7.

Litton Bionetics (1978). *Teratology Study in Rats. Toluene*, LBI Project No. 20847, Final report to American Petroleum Institute, by Litton Bionetics Inc., Kensington, Maryland, American Petroleum Institute, Washington, D.C.

Longacre, S. L., J. J. Kocsis, C. M. Witmer, E. W. Lee, D. Sammett, and R. Snyder (1981a). "Toxicological and Biochemical Effects of Repeated Administration of Benzene in Mice," *J. Toxicol. Environ. Health*, **7**, 223–237.

Longacre, S. L., J. J. Kocsis, and R. Snyder (1981b). "Influence of Strain Differences in Mice on the Metabolism and Toxicity of Benzene," *Toxicol. Appl. Pharmacol.*, **60**, 398–409.

Maltoni, C., G. Cotti, L. Valgimigli, and A. Mandrioli (1982). "Zymbal Gland Carcinomas in Rats Following Exposure to Benzene by Inhalation," *Am. J. Ind. Med.*, **3**, 11–16.

Medinsky, M. A., P. J. Sabourin, G. Lucier, L. S. Birnbaum, and R. F. Henderson (1989). "A Physiological Model for Simulation of Benzene Metabolism by Rats and Mice," *Toxicol. Appl. Pharmacol.*, **99**, 193–206.

Meyne, J. and M. S. Legator (1980). "Sex-related Differences in Cytogenetic Effects of Benzene in the Bone Marrow of Swiss Mice," *Environ. Mutat.*, **2**, 43–50.

Moolgavkar, S. H. and A. G. Knudson, Jr. (1981). "Mutation and Cancer: A Model for Human Carcinogenesis," *J. Nat. Cancer Inst.*, **66**, 1037–1052.

Murray, F. J., J. A. John, L. W. Rampy, R. A. Kuna, and B. A. Schwetz (1979). "Embryotoxicity of Inhaled Benzene in Mice and Rabbits," *Am. Ind. Hyg. Assoc. J.*, **40**, 993–998.

Nomiyama, K. and H. Nomiyama (1974a). "Respiratory Retention, Uptake, and Excretion of Organic Solvents in Man. Benzene, Toluene, *n*-hexane, Trichloroethylene, Acetone, Ethyl Acetate, and Ethyl Alcohol," *Int. Arch. Arbeitsmed.*, **32**, 75–83.

Nomiyama, K. and H. Nomiyama (1974b). "Respiratory Elimination of Organic Solvents in Man. Benzene, Toluene, *n*-hexane, Trichloroethylene, Acetone, Ethyl Acetate and Ethyl Alcohol," *Int. Arch. Arbeitsmed.*, **32**, 85–91.

NRC (1977). "Chemical Contaminants: Safety and Risk Assessment," in *Drinking Water and Health*, National Research Council, National Academy of Sciences, Washington, D.C., pp. 19–62.

NRC (1980). "Problems of Risk Estimation," in *Drinking Water and Health*, Vol. 3, National Research Council, National Academy Press, Washington, D.C., pp. 25–65.

NRC (1983). *Risk Assessment in the Federal Government: Managing the Process*, National Research Council, National Academy Press, Washington, D.C.

OSHA (1987). "Occupational Exposure to Benzene," *Fed. Reg.*, **52**, 34460–34578 (September 11, 1987).

Ott, M. G., J. C. Townsend, W. A. Fishbeck, and R. A. Langner (1978). "Mortality Among Workers Occupationally Exposed to Benzene," *Arch. Environ. Health*, **33**, 3–10.

Parke, D. V. and R. T. Williams (1953). "Studies in Detoxification. 49. The Metabolism of Benzene Containing [^{14}C]Benzene," *Biochem. J.*, **54**, 231–238.

Rickert, D. E., T. S. Baker, J. S. Bus, C. S. Barrow, and R. D. Irons (1979). "Benzene Disposition in the Rat after Exposure by Inhalation," *Toxicol. Appl. Pharmacol.*, **49**, 417–423.

Rinsky, R. A., R. J. Young, and A. B. Smith (1981). "Leukemia in Benzene Workers," *Am. J. Ind. Med.*, **2**, 217–245.

Rinsky, R. A., A. B. Smith, R. Hornung, T. G. Filloon, R. J. Young, A. H. Okun, and P. J. Landrigan (1987). "Benzene and Leukemia: An Epidemiologic Risk Assessment," *N. Eng. J. Med.*, **316**, 1044–1050.

RiskFocus (1990). Analysis of the Impact of Exposure Assumptions on Risk Assessment of Chemicals in the Environment. Phase I. Evaluation of Existing Exposure Assessment

Assumptions, Report to Chemical Manufacturers Association by RiskFocus® Division, Versar, Inc., Springfield, VA.

Rozen, M. G. and C. A. Snyder (1985). "Protracted Exposure of C57BL/6 Mice to 300 ppm Benzene Depresses B- and T-lymphocyte Numbers and Mitogen Responses; Evidence for Thymic and Bone Marrow Proliferation in Response to the Exposures," *Toxicology*, **37**, 13–26.

Sabourin, P., B. Chen, R. Henderson, G. Lucier, and L. Birnbaum (1986). "Effect of Dose on Absorption and Excretion of ^{14}C Benzene Administered Orally or by Inhalation," *Toxicologist*, **6**, 163.

Sandmeyer, E. E. (1981). "Aromatic Hydrocarbons," in *Patty's Industrial Hygiene and Toxicology*, Vol. 2, 3rd ed., G. D. Clayton, and F. E. Clayton, Eds. Interscience, New York, pp. 3253–3283.

Sherwood, R. J. (1988). "Pharmacokinetics of Benzene in a Human after Exposure at about the Permissible Limit," *Ann. NY Acad. Sci.*, **534**, 635–647.

Snyder, R. and J. J. Kocsis (1975). "Current Concepts of Chronic Benzene Toxicity," *CRC Crit. Rev. Toxicol.*, **3**, 265–288.

Snyder, R., B. D. Goldstein, A. Sellakumar, I. Bromberg, S. Laskin, and R. E. Albert (1980). "The Inhalation Toxicology of Benzene: Incidence of Hematopoietic Neoplasms and Hematotoxicity in ADR/J and C57BL/6J Mice," *Toxicol. Appl. Pharmacol.*, **54**, 323–331.

Snyder, R., S. L. Longacre, C. M. Witmer, J. J. Kocsis, L. S. Andrews, and E. W. Lee (1981). "Biochemical Toxicology of Benzene," *Rev. Biochem. Toxicol.*, **3**, 123–153.

Snyder, R., E. Dimitriadis, R. Guy, P. Hu, K. Cooper, H. Bauer, G. Witz, and B. G. Goldstein (1989). "Studies on the Mechanism of Benzene Toxicity," *Environ. Health Perspect.*, **82**, 31–35.

Spear, R. C., F. Y. Bois, T. Woodruff, D. Auslander, J. Parker, and S. Selvin (1991). "Modeling Benzene Pharmacokinetics across Three Sets of Animal Data: Parametric Sensitivity and Risk Implications," *Risk Anal.*, **11**, 641–654.

Srbova, J., J. Teissinger, and S. Skramovsky (1950). "Absorption and Elimination of Inhaled Benzene in Man," *Arch. Ind. Hyg. Occup. Med.*, **2**, 1–8.

Tardiff, R. G. and J. V. Rodricks, Eds. (1988). *Toxic Substances and Human Risk. Principles of Data Interpretation*, Plenum Press, New York.

Tatrai, E., K. Rodics, and G. Y. Ungvary (1980). "Embryotoxic Effects of Simultaneously Applied Exposure of Benzene and Toluene," *Folia Morphol.*, **28**, 286–289.

Travis, C. C. and R. K. White (1988). "Interspecific Scaling of Toxicity Data," *Risk Anal.*, **8**, 119–125.

Travis, C. C., J. L. Quillen, and A. D. Arms (1990). "Pharmacokinetics of Benzene," *Toxicol. Appl. Pharmacol.*, **102**, 400–420.

Ungvary, G. and E. Tatrai (1985). "On the Embryotoxic Effects of Benzene and Its Alkyl Derivatives in Mice, Rats, and Rabbits," *Arch. Toxicol. (Suppl.)*, **8**, 425–430.

USEPA (1986). "Guidelines for Carcinogen Risk Assessment," *Fed. Reg.*, **51**, 33992–34003 (September 24).

USEPA (1989a). *Exposure Factors Handbook*, Report No. EPA/600/8-89/043, Exposure Assessment Group, Office of Health and Environmental Assessment, U.S. Environmental Protection Agency, Washington, D.C.

USEPA (1989b). "Proposed Amendments to the Guidelines for the Health Assessment of Suspect Developmental Toxicants," *Fed. Reg.*, **54**(42), 9386–9403 (March 6).

USEPA (1989c). *Health Effects Assessment for Benzene*, Report No. EPA/600/8-89/086, Environmental Criteria and Assessment Office, Office and Research and Development, U.S. Environmental Protection Agency, Cincinnati, OH.

USEPA (1992a). "Draft Report: A Cross-species Scaling Factor for Carcinogen Risk Assessment Based on Equivalence of $mg/kg^{3/4}/day$; Notice," *Fed. Reg.*, **57**(109), 24152–24173 (June 5).

USEPA (1992b). "Guidelines for Exposure Assessment," *Fed. Reg.*, **57**(104), 22888–22938 (May 29).

Watanabe, G.-I. and S. Yoshida (1970). "The Teratogenic Effect of Benzene in Pregnant Mice," *Acta Med. Biol.*, **17**, 285–291.

WHO (1978). *Principles and Methods for Evaluating the Toxicity of Chemicals. Part I*, Environmental Health Criteria 6, World Health Organization, Geneva.

WHO (1984). *Principles for Evaluating Health Risks to Progeny Associated with Exposure to Chemicals During Pregnancy*, Environmental Health Criteria 30, International Programme on Chemical Safety, World Health Organization, Geneva.

WHO (1985). *Toluene*, Environmental Health Criteria 52, International Programme on Chemical Safety, World Health Organization, Geneva.

WHO (1986). *Principles and Methods for the Assessment of Neurotoxicity Associated with Exposure to Chemicals*, IPCS Environmental Health Criteria 60, International Programme on Chemical Safety, World Health Organization, Geneva.

Witz, G., T. A. Kirley, W. M. Maniara, V. J. Mylavarapu, and B. D. Goldstein (1990). "The Metabolism of Benzene to Muconic Acid, a Potential Biological Marker of Benzene Exposure," in *Biological Reactive Intermediates*, Vol. IV, C. M. Witmer et al., Eds., Plenum Press, New York, pp. 613–618.

Wong, O., R. W. Morgan, and M. D. Whorton (1983). Comments on the NIOSH Study of Leukemia in Benzene Workers, Technical report submitted to Gulf Canada, Ltd., by Environmental Health Associates.

CHAPTER SEVENTEEN

Job Safety and Health Law

Martha Hartle Munsch, J. D., and Jacqueline A. Koscelnik, J. D.

The principal legislation relating to job safety and health is the federal Occupational Safety and Health Act of 1970 (OSHA). However, certain industries and/or portions of industries are subject to regulation by federal statutes other than OSHA. In addition, OSHA does not entirely preclude regulation of job safety and health by the states or their political subdivisions. Nevertheless, OSHA is clearly the most comprehensive legislative directive relating to workplace safety and health; accordingly, this chapter and the next focus primarily on developments and requirements pursuant to this act.

1 LEGISLATIVE HISTORY AND BACKGROUND OF OSHA

The Occupational Safety and Health Act of 1970 was enacted by Congress on December 17, 1970, and became effective on April 28, 1971. It represents the first job safety and health law of nationwide scope (1). Passage of the act was preceded by a dramatic and bitter labor–management political fight. The legislative history of OSHA is summarized in *The Job Safety and Health Act of 1970* [(1), pp. 13–21].

Congress enacted OSHA for the declared purpose of assuring "so far as possible every working man and woman in the Nation safe and healthful working conditions" [§ 2(b)]. The act is intended to *prevent* work-related injury, illness, and death.

1.1 Agencies Responsible for Implementing and Enforcing OSHA

The Department of Labor is responsible for implementing OSHA. On the date the act became effective, the Department of Labor created the Occupational Safety and

Patty's Industrial Hygiene and Toxicology, Third Edition, Volume 3, Part A, Edited by Robert L. Harris, Lewis J. Cralley, and Lester V. Cralley.
ISBN 0-471-53066-2 © 1994 John Wiley & Sons, Inc.

Health Administration (OSH Administration or OSHA) to carry out such responsibilities. The OSH Administration is headed by an assistant secretary of labor for occupational safety and health and is responsible, among other things, for promulgating rules and regulations, setting health and safety standards, evaluating and approving state plans, and overseeing enforcement of the act (2).

Section 12(a) of the act establishes the Occupational Safety and Health Review Commission (OSAHRC) as an independent agency to adjudicate enforcement actions brought by the secretary of labor. The commission is composed of three members appointed by the president for six-year terms. The chairperson of the commission is authorized to appoint such administrative law judges as he or she deems necessary to assist in the work of the commission [§12(e)].

Sections 20 and 21 of the act give the secretary of health, education and welfare (HEW) [now Health and Human Services (HHS)] broad authority to conduct experimental research relating to occupational safety and health, to develop criteria for and recommend safety and health standards, and to conduct educational and training programs (3). Section 22 establishes the National Institute for Occupational Safety and Health (NIOSH) to perform the functions of the secretary of health and human services under Sections 20 and 21.

The act specifically directs NIOSH to develop criteria documents that describe safe levels of exposure to toxic materials and harmful physical agents and to forward recommended standards for such substances to the secretary of labor (4). The act also directs NIOSH to publish at least annually a list of all known toxic substances and the concentrations at which such toxicity is known to occur [§20(a)(6)].

Section 7(a) establishes a National Advisory Committee on Occupational Safety and Health (NACOSH), whose basic functions are to advise, consult with, and make recommendations to the secretary of labor and the secretary of health and human services on matters relating to the administration of the act. NACOSH consists of 12 members who represent management, labor, occupational safety and occupational health professions, and the public. The members are appointed by the secretary of labor, although four members are to be designated by the secretary of health and human services.

The secretary of labor is authorized by Section 7(b) to appoint other advisory committees to assist him or her in the formulation of standards under Section 6. For example, ad hoc advisory committees have been used to assist in developing standards for exposure to asbestos and coke oven emissions. The secretary of labor has also appointed various standing advisory committees (5).

1.2 Scope of OSHA's Coverage

The Occupational Safety and Health Act of 1970 applies to every private employer engaged in a business affecting commerce, regardless of the number of employees. It applies with respect to employment performed in a workplace in any of the 50 states, the District of Columbia, Puerto Rico, the Virgin Islands, American Samoa, Guam, the Trust Territory of the Pacific Islands, Wake Island, the Outer Continental Shelf Lands, Johnston Island, and the Canal Zone [§§3(5) and 4(a)].

The act's definition of ''employer'' does not include the states, political subdivisions of the states, or the United States. However, Section 19 directs the head of each federal agency to establish and maintain an effective and comprehensive occupational safety and health program that is consistent with the standards required of private employers (6).

2 REGULATION OF JOB SAFETY AND HEALTH BY FEDERAL STATUTES OTHER THAN OSHA

Section 4(b)(1) of OSHA states that nothing in the act shall apply to working conditions of employees with respect to which other federal agencies exercise statutory authority to prescribe or enforce standards or regulations affecting occupational safety or health. Thus federal agencies other than OSHA that are authorized by statute to regulate employee safety and health can continue to do so after the effective date of OSHA; in fact, the *exercise* of such authority preempts OSHA from regulating with respect to such working conditions.

Although Section 4(b)(1) seems to be self-defining, it has generated a tremendous volume of litigation. Three major interpretive questions have been raised:

1. What constitutes a sufficient exercise of regulatory authority to preempt OSHA regulation?
2. Does the exercise of authority by another federal agency in substantial areas of employee safety exempt the entire industry from OSHA standards?
3. Must the other federal agency's motivation in acting have been to protect workers?

It appears to be well settled that the mere existence of statutory authority to regulate safety or health is not sufficient to oust OSHA's regulatory scheme; some exercise of that authority is necessary. Furthermore, at least three federal courts of appeals have taken the position that speculative pronouncements of proposed regulations by a federal agency are not sufficient to warrant preemption of OSHA standards. Rather, it has been held that Section 4(b)(1) requires a concrete exercise of statutory authority (7).

The same courts of appeals have also rejected the notion that the exercise of statutory authority by another federal agency creates an industrywide exemption from OSHA regulations. Rather, the courts have agreed that the term "working conditions" in Section 4(b)(1) refers to something more limited than every aspect of an entire industry. Ambiguity remains, however, with respect to the scope of the displacing effect of another agency's regulation of a working condition.

For example, in *Southern Pacific Transportation Company* (7), the Fifth Circuit explained that the term "working conditions" has a technical meaning in the language of industrial relations; it encompasses both a worker's surroundings and the hazards incident to the work. The court stated that the displacing effect of Section 4(b)(1) would depend primarily on the agency's articulation of its regulations (8):

Section 4(b)(1) means that any FRA [Federal Railroad Administration] exercise directed at a working condition—defined either in terms of a "surrounding" or a "hazard"—displaces OSHA coverage of that working condition. Thus comprehensive FRA treatment of the general problem of railroad fire protection will displace all OSHA regulations on fire protection, even if the FRA activity does not encompass every detail of the OSHA fire protection standards, but FRA regulation of portable fire extinguishers will not displace OSHA standards on fire alarm signaling systems.

The Fourth Circuit defined "working conditions" as "the environmental area in which an employee customarily goes about his daily tasks." The court in *Southern Railway Company* (7) explained that OSHA would be displaced when another federal agency had exercised its statutory authority to prescribe standards affecting occupational safety or health for such an area (9).

The courts of appeals seem to indicate, at least implicitly, that regulation of a working condition by another federal agency need not be as effective or as stringent as an OSHA standard to preempt the OSHA standard (7). But it remains unclear whether a decision by another federal agency that a particular aspect of an industry should not be regulated at all would preempt or preclude OSHA regulation of that same aspect. Resolution of these and other issues involving the scope of the displacement effect under Section 4(b)(1) awaits future litigation or legislation.

Finally, the commission has held that to be cognizable under Section 4(b)(1), "a different statutory scheme and rules thereunder must have a policy or purpose that is consonant with that of the Occupational Safety and Health Act. That is, there must be a policy or purpose to include employees in the class of persons to be protected thereunder" (10). In *Organized Migrants in Community Action, Inc. v. Brennan* (11), the U.S. Court of Appeals for the District of Columbia, although not deciding the issue, implicitly rejected the argument that preemption under Section 4(b)(1) exists only where the allegedly preempting statute was passed *primarily* for the protection of the employees (12).

3 OVERVIEW OF FEDERAL REGULATORY SCHEMES OTHER THAN OSHA

The following material represents an overview of the major federal regulatory schemes other than OSHA that deal with or relate to job safety and health. The listing *is by no means exhaustive*, and employers are urged to consult specific statutory schemes in substantive areas relating to their respective industries.

3.1 Mine Safety and Health Legislation

Occupational safety and health matters with respect to the nation's mining industry are regulated pursuant to the Federal Mine Safety and Health Act of 1977, which became effective in March 1978 (13). The Department of Labor is responsible for enforcing that legislation. The secretary of labor has delegated its enforcement authority to the Mine Safety and Health Administration (MSHA).

Prior to the effective date of the 1977 act, safety and health matters with respect to the mining industry were covered by two separate statutes. The Metal and Non-Metallic Mine Safety Act of 1966 covered mines of all types other than coal mines. Safety and health matters concerning coal mines were regulated by the Coal Mine Health and Safety Act of 1969. The 1977 Mine Safety and Health Act is now the single mine safety and health law for all mining operations. The 1977 act directs the secretary of labor to, *inter alia*, "develop, promulgate, and revise as may be appropriate improved mandatory health or safety standards for the protection of life and prevention of injuries in coal or other mines" (14). The secretary is also empowered to enforce those standards.

3.2   Environmental Pesticide Control Act of 1972

The Federal Environmental Pesticide Control Act of 1972 (FEPCA) regulates the use of pesticides and makes misuse civilly and criminally punishable. The Court of Appeals for the District of Columbia has held that FEPCA authorizes the Environmental Protection Agency (EPA) to promulgate and enforce occupational health and safety standards with respect to farm workers' exposure to pesticides. The EPA has exercised that authority, and thus has preempted OSHA from regulating in that area [*Organized Migrants in Community Action* (11), (15)].

3.3   Federal Railroad Safety Act of 1970

The Federal Railroad Safety Act of 1970 authorizes the Federal Railroad Administration (FRA) within the Department of Transportation (DOT) to promulgate regulations for all areas of railroad safety, including employee safety. To date, however, DOT has not adopted railroad occupational safety standards for all railroad working conditions or workplaces (16). The Department of Labor (OSHA) retains jurisdiction over safety and health of railroad employees with respect to those "working conditions" for which DOT has not adopted standards (17).

3.4   Federal Aviation Act of 1958

The Federal Aviation Act of 1958, as amended, empowers the Federal Aviation Administration (FAA) within the Department of Transportation to promote safety of flight of civil aircraft in air commerce (18), as well as to establish minimum safety standards for the operation of airports that serve any scheduled or unscheduled passenger operation of air carrier aircraft designed for more than 30 passenger seats (19). If the congressional mandate in that statute is deemed to include the safe working conditions of airline and/or airport employees, OSHA would be precluded from exercising its jurisdiction with respect to the working conditions regulated by the FAA. At least one administrative law judge has determined that the FAA's mandate encompasses the safe working conditions of airline ground crews when performing aircraft maintenance (20).

3.5 Hazardous Materials Transportation Act

The Hazardous Materials Transportation Act (HMTA) authorizes the secretary of transportation to issue regulations governing any safety aspect of the transportation of materials designated as hazardous by the secretary (21). The act encompasses shipments by rail, air, water, and highway. If worker safety is deemed to be a purpose of the HMTA, safety standards promulgated by DOT under the statute would trigger a preemption of OSHA jurisdiction with respect to the working conditions covered by such standards.

3.6 Natural Gas Pipeline Safety Act of 1968

The Natural Gas Pipeline Safety Act of 1968 (NGPSA) authorizes the secretary of transportation to establish minimum federal safety standards for pipeline facilities and the transportation of gas in commerce. In *Texas Eastern Transmission Corp.* (22), the Occupational Safety and Health Review Commission determined that the NGPSA was intended to affect occupational safety and health. Thus employers engaged in the transmission, sale, and storage of natural gas would be exempt from OSHA with respect to working conditions covered by DOT standards promulgated under NGPSA (23).

3.7 Federal Noise Control Act of 1972

Although a health and safety standard adopted pursuant to OSHA governs the level of noise to which a worker covered by the act may be exposed in the workplace (see Section 5.3), other federal statutes deal with noise abatement and control as well (24).

The first such enactment, a 1968 amendment to the Federal Aviation Act, required the administrator of the FAA to include aircraft noise control as a factor in granting type certificates to aircraft under the act (25). To the extent that the administrator of the FAA denies certification to an aircraft that produces noise in excess of the standards or prohibits the operation of a certified aircraft in a manner that violates regulations, the general environmental noise level in workplaces covered by OSHA and located adjacent to airports and landing patterns is correspondingly reduced.

The first attempt to deal with noise on a nationwide basis, however, was the federal Noise Control Act of 1972. The control strategy of that act is generally as follows: The administrator of the EPA is required to develop and publish criteria with regard to noise, reflecting present scientific knowledge of the effects on public health and welfare that are to be expected from different quantities or qualities of noise. The administrator then must identify products or kinds of products that in his or her opinion are major sources of environmental noise, and publish noise emission regulations where it is feasible to limit the amount of noise produced by such products (26).

Under the act, the administrator is further charged with publishing regulations identifying ''low noise emission products.'' Such products, once so designated, must thereafter be purchased by federal agencies in preference to substitute products, provided the "low noise emission product" costs no more than 125 percent of the price of the substitute.

As the administrator of the EPA identifies more and more products as major sources of noise and subjects those products to noise emission standards adopted under the Noise Control Act of 1972, the noise levels found in workplaces covered by OSHA and in which such products are used should decrease.

3.8 Federal Toxic Substances Control Act of 1976

The Toxic Substances Control Act of 1976 establishes a broad, nationwide program for the federal regulation of the manufacture and distribution of toxic substances (27). The act divides all "chemical substances" and "mixtures" into two categories: the old and the new. The administrator of the EPA is charged under Section 8(b) with the gargantuan task of compiling and publishing in the *Federal Register* an inventory or list of all chemical substances manufactured or processed in the United States.

Manufacturers of substances that appear on that inventory are at liberty to continue to manufacture and distribute such substances unless the administrator by rule promulgated under Section 4 of the act first requires that a designated substance be tested and data from the tests be submitted to the EPA. If EPA thereafter makes a determination under Section 6 that the continued manufacture, processing, or distribution in commerce of the substance presents an unreasonable risk of injury to health or the environment, the administrator may either prohibit altogether the manufacture of the chemical substance or may impose restrictions (limitations on the quantity manufactured, the use to which the chemical may be put, the concentrations in which it may be used, the labels and warnings that must accompany its sale, etc.).

Section 5 of the act provides a different treatment with respect to a "new chemical substance" or a "significant new use" of a substance that appears on the Section 8(b) inventory. The manufacturer is not at liberty to commence manufacture or distribution of such a "new" substance but must first submit a notice to the administrator of intention to manufacture such a new substance or to engage in a significant new use. Next, testing data relating to the substance's toxicity and the effect on health and on the environment must be submitted. If the administrator does not act within 90 days, the manufacturer is at liberty to proceed with manufacture or distribution. During the initial period, however, the administrator may extend his or her time for action an additional 90 days. If during the original period (or its extension) the administrator believes that the information available is inadequate to make a reasoned finding that the proposed new substance or use does not present an unreasonable risk of injury to health or the environment, he or she may prohibit or limit the manufacture of the substance and obtain an injunction in court for that purpose. It would appear that in the absence of testing data submitted in compliance with Section 4, this injunction against manufacture or distribution of the new substance or use would continue indefinitely. If, however, the administrator finds, based on information provided, that the proposed new substance or use does present an unreasonable risk to health and safety, the administrator must proceed by means of the provisions of Section 6 to prohibit manufacture or to impose restrictive conditions (28).

The administrator of the EPA has promulgated a myriad of regulations implementing the Toxic Substances Control Act. These regulations can be found at 40 CFR Parts 702–766 and 790–799. Besides detailing the procedural and chemical inventory requirements of the act, the regulations set forth guidelines for the manufacture, processing and distribution of polychlorinated biphenyls and fully halogenated chlorofluoroalkanes (29). Moreover, the regulations require local education agencies to identify friable asbestos-containing material in public and private school buildings (Part 763). It is likely that the Toxic Substances Control Act will continue to be a major weapon in the federal health, safety, and environmental arsenal, and significant developments under it should be expected (30).

3.9 Federal Consumer Product Safety Act

The Consumer Product Safety Act of 1972 was drafted to apply only to "consumer products." That term is defined in the act in a manner that serves to exclude most products destined principally for use in workplaces covered by OSHA. Nevertheless, the act promises to provide increased protection of the American worker from hazardous products that by their nature are "consumer products" within the meaning of the act, yet are frequently found in the workplace (31).

The act created a Consumer Product Safety Commission and empowered that agency to promulgate "consumer product safety standards" applicable to consumer products found by the commission to present an unreasonable risk of injury. Such standards may be performance standards or they may require that products not be sold without adequate warnings or instruction. Where no feasible safety standard that could be promulgated would eliminate an unreasonable risk of injury, the commission is empowered to ban the consumer product altogether from interstate sale or distribution.

In addition to publishing safety standards, the commission is empowered to file suit and seek the seizure of a consumer product believed to be "imminently hazardous," regardless of whether the product in question is covered by already promulgated consumer product safety standards. The commission is also authorized to find, after hearing, that a consumer product presents a "substantial product hazard." In the event of such a finding, the commission may order the manufacturer, distributor, or retailer of the product to give public notice of that finding, and to repair, replace, or refund the purchase price of the product affected (32).

3.10 Hazardous Substances Act

The federal Hazardous Substances Act provides a mechanism by means of which the Consumer Product Safety Commission may find that a substance distributed in interstate commerce is "hazardous." After such a finding the commission may either impose packaging and labeling requirements to protect public health and safety or, in the cases of hazardous substances intended for the use of children or likely to be subject to access by children, or substances intended for household use, prohibit distribution altogether ("banned hazardous substance") (33). Insofar as safety in the American workplace is concerned, the effect of the act is that hazardous substances distributed in interstate commerce and utilized by the American worker will arrive safely packaged and accompanied by appropriate warnings.

3.11 The Atomic Energy Act of 1954 and Other Statutory Sources of Radiation Control

The Department of Labor has published occupational health and safety standards regulating exposure to ionizing (i.e., alpha, beta, gamma, X-ray, neutron, etc.) radiation and nonionizing (i.e., radiofrequency, electromagnetic) radiation. However, the primary federal law regulating human exposure to radiation is not OSHA but rather the Atomic Energy Act of 1954, as amended. Exercising power under that statute, the Nuclear Regulatory Commission (NRC) has published "Standards for Protection Against Radiation" (34).

A detailed discussion of those regulations is beyond the scope of this chapter. The operation of the standards can be summarized briefly as follows, however. Any person holding a license issued under the Atomic Energy Act of 1954 and using "licensed material" (i.e., radioactive or radiation-emitting material) may not permit the exposure of individuals within a "restricted area" (i.e., an area in which radioactive materials are being used) to greater doses of radiation than are set forth in the regulations (35).

Although the primary thrust of the NRC regulations is to control ionizing radiation within the "restricted area," the NRC has also published regulations on permissible levels of radiation in unrestricted areas, in effluents discharged into unrestricted areas, and for the disposal of radioactive materials by release into sanitary sewerage systems (36).

The administrator of the EPA, exercising authority under the Atomic Energy Act of 1954 (authority acquired by means of the Reorganization Plan No. 3 of 1970), has also promulgated regulations limiting exposure of the general population to ionizing radiation produced during the operation of nuclear power plants licensed by the NRC (37).

There are additional federal agencies empowered to set standards for ionizing radiation control within areas under their jurisdiction. The Department of Labor, for example, has promulgated regulations regarding radiation exposure in underground mines (38). The Department of Labor has also issued radiation standards for uranium mining conducted under the Walsh–Healey Public Contracts Act (39).

Radiation generated by devices and products that are not governed by the Atomic Energy Act and licensed by the NRC is regulated by the federal Radiation Control for Health and Safety Act of 1968 (40).

3.12 Outer Continental Shelf Lands Act

The Outer Continental Shelf Lands Act authorizes the head of the department in which the Coast Guard is operating to promulgate and enforce "such reasonable regulations with respect to lights and other warning devices, safety equipment, and other matters relating to the promotion of safety of life and property" on the lands and structures referred to in the act or on the adjacent waters (41). If worker safety is deemed to be within the mandate of this statute, OSHA jurisdiction may be preempted with respect to working conditions that are governed by standards issued by the Coast Guard pursuant to this act.

4 REGULATION OF JOB SAFETY AND HEALTH BY THE STATES

One of the primary factors that induced Congress to enact the Occupational Safety and Health Act was the failure of many of the states adequately to regulate workplace safety and health (42). In passing OSHA, Congress hoped to ensure at least a minimum level of regulation of the conditions experienced by workers throughout the country.

The Occupational Safety and Health Act of 1970 preempts state regulation of job safety and health with respect to matters that OSHA regulates, even when a state has a more stringent regulation with respect to a particular hazard (43). However, OSHA does not totally ban the states from developing and enforcing occupational safety and health standards. Pursuant to Section 18(b) of the act, a state may regain jurisdiction over development and enforcement of occupational safety and health standards by submitting to the federal government an effective state occupational safety and health plan. Final approval of a state plan can lead ultimately to exclusive authority by a state over the matters included in its plan.

The process of regaining jurisdiction over the regulation of occupational safety and health begins with the submission of a plan that sets forth specific procedures for ensuring workers' safety and health. According to the regulations of the secretary of labor, the states can submit either of two types of plan: a complete plan or a developmental plan.

A "complete" plan (44) is a plan that, upon submission, satisfies the criteria for plan approval set forth in Section 18(c) of the act, as well as certain additional criteria outlined by the secretary of labor in administrative regulations (45). Complete plans are given "initial" approval by the secretary of labor upon submission. For at least 3 years following the "initial" approval, the secretary of labor will monitor the state plan to determine whether on the basis of the actual operations of the plan, the criteria set forth in Section 18(c) are being applied. If this determination [the "Section 18(e) determination"] is favorable, the state plan will be granted "final approval" and the state will regain exclusive jurisdiction with respect to any occupational safety or health issue covered by the state plan. Federal (i.e., OSHA) standards continue to apply to hazards not covered by the state program; thus, state plans need not address all hazards and yet gaps in protection are avoided.

A "developmental" plan (46) is a plan that, upon submission, does not fully meet the criteria set forth in the statute or in the regulations. A developmental plan may receive initial approval upon submission, however, if the plan contains "satisfactory assurances" that the state will take the necessary steps to bring its program into conformity within 3 years following commencement of the plan's operation.

If the developmental plan satisfies all the statutory and administrative criteria within the 3-year "developmental period," the secretary of labor will so certify and will initiate an evaluation of the actual operations of the state plan for purposes of making a Section 18(e) determination. The evaluation must proceed for at least one year before such a determination can be made.

Plans that have received final approval will continue to be monitored and evaluated by the secretary of labor pursuant to Section 18(f) of the act, which authorizes the secretary to withdraw approval if a state fails to comply substantially with any provision of the state's plan.

Although a state does not regain exclusive jurisdiction over matters contained in its plan until the plan receives final approval, a state with initial plan approval may participate in the administration and enforcement of the act prior to final approval by satisfying the following four criteria (47):

1. The state must have enacted enabling legislation conforming to that specified in OSHA and the regulations.
2. The state plan must contain standards that are found to be at least as effective as the comparable federal standards.
3. The state plan must provide for a sufficient number of qualified personnel who will enforce the standards in accordance with the state's enabling legislation.
4. The plan's provisions for review of state citations and penalties (including the appointment of the reviewing authority and the promulgation of implementing regulations) must be in effect.

If the criteria above are met, the state plan is deemed to be "operational." Thereupon the federal government enters into an operational agreement with the state whereby the state is authorized to enforce safety and health standards under the state plan (47). During this period the act permits, but does not require, the federal government to retain enforcement activity in the state (48). Thus during this period an employer could be subject to enforcement activities by both the state and federal authorities. However, the secretary's regulations provide that once a plan (either complete or developmental) becomes "operational," the state will conduct all enforcement activity, including inspections in response to employee complaints, and accordingly, the federal enforcement activity will be reduced and the emphasis will be placed on monitoring state activity (47).

As of 1991, plans had been *approved* for the Virgin Islands and 15 states: Alaska, Arizona, Connecticut, Hawaii, Indiana, Iowa, Kentucky, Maryland, Minnesota, New York, South Carolina, Tennessee, Utah, Virginia, and Wyoming. In addition, the secretary of labor had *certified* the (developmental) plans of Puerto Rico and 8 states: California, Michigan, Nevada, New Mexico, North Carolina, Oregon, Vermont, and Washington (49).

Also as of 1991, 13 states or territories plus the District of Columbia had submitted plans and were awaiting initial approval by the secretary of labor: Alabama, American Samoa, Arkansas, Delaware, Florida, Guam, Idaho, Massachusetts, Missouri, Oklahoma, Rhode Island, Texas, and West Virginia (49).

The following 11 states submitted plans at one time but have withdrawn them:

Colorado, Georgia, Illinois, Maine, Mississippi, Montana, New Hampshire, New Jersey, North Dakota, Pennsylvania, and Wisconsin. Five states (Kansas, Louisiana, Nebraska, Ohio, and South Dakota) have never submitted plans to the Department of Labor (49).

Most states that are presently operating approved and/or certified plans have adopted standards that are substantially similar, if not identical, to the federal standards. At least five state plans, however, contain certain provisions that vary from the federal standards and have been approved as being "at least as effective" as the federal standards. The states are California, Hawaii, Michigan, Oregon, and Washington (49). In some instances these states have adopted standards that are more stringent than the analogous federal standards. Employers who are operating in more than one state should be aware that they may have to deal with different regulations, different enforcement procedures, and perhaps different interpretations of similar standards for purposes of complying with the applicable occupational safety and health laws (50).

4.1 State Jurisdiction in Areas Regulated by Federal Legislation Other than OSHA

If state legislation regulates a job safety or health issue that OSHA does not cover, the state may continue to enforce its relevant standards unless other applicable federal law has preempted state enforcement.

In some areas a federal regulatory scheme permits concurrent federal and state regulation of job safety and health. For example, the Federal Mine Safety and Health Act of 1977 states that no state law that was in effect on December 30, 1969, or that may become effective thereafter, shall be superseded by any provisions of the federal mine act unless the state law is in conflict with the mine act. State laws and rules that provide for standards more stringent than those of the mine act are deemed not to be in conflict (51).

On the other hand, the Federal Railroad Safety Act of 1970 has essentially preempted the states' regulation of railroad safety and health except in cases of a state having a more stringent law because of the need to eliminate or reduce an essentially local safety hazard (52).

5 EMPLOYERS' DUTIES UNDER THE OCCUPATIONAL SAFETY AND HEALTH ACT OF 1970

A private employer's primary duties under the Occupational Safety and Health Act of 1970 are found in Section 5(a), which provides that each employer:

1. Shall furnish to each employee employment and a place of employment that are free from recognized hazards that are causing or are likely to cause death or serious physical harm to employees.

2. Shall comply with occupational safety and health standards promulgated under the act (53).

5.1 The General Duty Clause [§ 5(a)(1)]

The essential elements of the so-called general duty clause of OSHA are the following (54):

1. The employer must render the workplace ''free'' of hazards that arise out of conditions of the employment.
2. The hazards must be ''recognized.''
3. The hazards must be causing or likely to cause death or serious physical harm (55).

5.1.1 Failure to Render Workplace "Free" of Hazard

It is fairly well settled that Congress did not intend to make employers strictly liable for the presence of unsafe or unhealthful conditions on the job. The employer's general duty must be an achievable one. Thus the term ''free'' has been interpreted by the courts and the commission to mean something less than absolutely free of hazards. Instead, the courts and the commission have held that the employer has a duty to render the workplace free only of hazards that are preventable (56).

The determination of whether a hazard is preventable generally is made in the context of an enforcement proceeding under the act when an employer asserts the inability of preventing the hazard as an affirmative defense to a proposed citation. The employer often contends that: (1) compliance is impossible or infeasible (57) because either the technology does not exist to prevent the hazard or the cost of the technology is prohibitive; (2) compliance with a standard will result in a greater hazard to employees than would noncompliance; or (3) the hazard was created by an employee's misconduct that was so unusual that the employer could not reasonably prevent the existence of the hazard.

When technology does not exist to prevent a hazard, OSHA does not require prevention by shutting down the employer's operation. Rather, the secretary of labor must be able to show that ''demonstrably feasible'' measures would have materially reduced the hazard [*National Realty and Construction Company* (56)]. Similarly, it seems that measures that, even though technologically feasible, would have been so expensive as to bankrupt the employer are not ''demonstrably feasible.'' (For further discussion relating to feasibility, see Section 5.2.3 of this chapter.)

To establish the ''greater hazard'' defense, the employer must show that: (1) the hazards of compliance are greater than the hazards of noncompliance; (2) alternative means of protecting employees are unavailable; and (3) a variance either cannot be obtained or is inappropriate (58).

In addition, the courts and the commission have recognized that certain isolated or idiosyncratic acts by an employee that were not foreseeable by the employer could result in unpreventable hazards for which the employer should not be held liable. For example, in *National Realty and Construction Company*, the court of appeals stated:

> Hazardous conduct is not preventable if it is so idiosyncratic and implausible in motive
> or means that conscientious experts, familiar with the industry, would not take it into

account in prescribing a safety program. Nor is misconduct preventable if its elimination would require methods of hiring, training, monitoring, or sanctioning workers which are either so untested or so expensive that safety experts would substantially concur in thinking the methods infeasible (59).

The court in *National Realty* emphasized, however, that an employer does have a duty to attempt to prevent hazardous conduct by employees. Thus the employer must adopt demonstrably feasible measures concerning the hiring, training, supervising, and sanctioning of employees to reduce materially the likelihood of employee misconduct (60).

There is disagreement among the courts of appeals as to who bears the burden of proving employee misconduct. Most courts have held that the burden is appropriately placed on the employer. Others place the burden on the secretary to disprove unforeseeable employee misconduct (61).

5.1.2   The Hazard Must Be a "Recognized" Hazard

The general duty clause does not apply to all hazards but only to the hazards that are "recognized" as arising out of the employment. The test for determining a "recognized" hazard is whether the hazard is known by the employer or generally by the industry of which the employer is a part (62). This test involves an objective determination and does not depend on whether the employer is aware in fact of the hazard (63).

In *American Smelting and Refining Company v. OSAHRC* (64), the Eighth Circuit held that the general duty clause is not limited to recognized hazards of types detectable only by the human senses but also encompasses hazards that can be detected only by instrumentation.

5.1.3   The Hazard Must Be Causing or Likely to Cause Death or Serious Physical Harm

It is not necessary that there be actual injury or death to trigger a violation of the general duty clause. The purpose of the act is to prevent accidents and injuries. Thus, violation of the general duty clause arises from the existence of a statutory hazard, not from injury in fact (65).

Proof that a hazard is "causing or likely to cause death or serious physical harm" does not require a mathematical showing of probability. Rather, if evidence is presented that a practice could eventuate in serious physical harm upon other than a freakish or utterly implausible concurrence of circumstances, the commission's determination of likelihood will probably be accorded considerable deference by the courts (66).

The term "serious physical harm" is defined neither in the act nor in the secretary's regulations, but OSHA's Field Operations Manual defines it to mean:

(i) Permanent, prolonged, or temporary impairment of the body in which part of the body is made *functionally useless* or is *substantially reduced in efficiency* on or off the job. Injuries involving such impairment would require treatment by a medical doctor, although not all injuries which receive treatment by a medical doctor would necessarily involve such impairment. Examples of such injuries are amputations; fractures (both simple and compound); deep cuts involving significant bleeding and which require extensive suturing; disabling burns and concussions.

(ii) Illnesses that could *shorten life* or *significantly reduce physical or mental efficiency* by inhibiting the normal function of a part of the body, even though the effects may be cured by halting exposure to the cause or by medical treatment. Examples of such illnesses are cancer, silicosis, asbestosis, poisoning, hearing impairment and visual impairment.

5.2 The Specific Duty Clause [§ 5(a)(2)]

Section 5(a)(2) of OSHA imposes on employers a duty to comply with the occupational safety and health standards promulgated by the secretary of labor. These standards constitute the employers' so-called specific duties under the act. Specific promulgated standards preempt the general duty clause, but only with respect to hazards expressly covered by the specific standards (67).

5.2.1 Processes for Promulgating Standards

The act established processes for promulgating three types of occupational safety and health standards: interim, permanent, and emergency.

Interim standards consist of standards derived from (1) established federal standards or (2) national consensus standards that were in existence on the effective date of OSHA (68). Section 6(a) of the act directed the secretary of labor to publish such standards in the *Federal Register* immediately after the act became effective (i.e., April 28, 1971) or for a period of up to 2 years thereafter. These standards became effective as OSHA standards upon publication without regard to the notice, public comment, and hearing requirements of the Administrative Procedure Act (69).

The intent of the interim standards provisions was to give the secretary a mechanism by which to promulgate speedily standards with which industry was already familiar and to provide a nationwide floor of minimum health and safety standards (70). The secretary's 2-year authority to promulgate interim standards expired on April 29, 1973.

Pursuant to Section 6(b) of the act, the secretary of labor is authorized to adopt "permanent" occupational safety and health standards to serve the objectives of OSHA. With respect to standards relating to toxic materials or harmful physical agents, Section 6(b)(5) specifically directs the secretary to set the standard "which most adequately assures, to the extent feasible, on the basis of the best available evidence, that no employee will suffer material impairment of health or functional capacity even if such employee has regular exposure to the hazard dealt with by such standard for the period of his working life."

The promulgation of these "permanent" occupational safety and health standards requires procedures similar to informal rule making under Section 4 of the Administrative Procedure Act. Upon determination that a rule should be issued promulgating such a standard, the secretary must first publish the proposed standard in the

Federal Register. Publication is followed by a 30-day period during which interested persons may submit written data or comments or file written objections and requests for a public hearing on the proposed standard. If a hearing is requested, the secretary must publish in the *Federal Register* a notice specifying the standard objected to and setting a time and place for the hearing. Within 60 days after the period for filing comments, or, if a hearing has been timely requested, within 60 days of the hearing, the secretary must either issue a rule promulgating a standard or determine that no such rule should be issued (71). Once a rule is issued, the secretary may delay the effective date of the rule for a period not in excess of 90 days to enable an affected employer to learn of the rule and to familiarize itself with its requirements.

Section 6(c)(1) of OSHA authorizes the secretary to issue emergency temporary standards if he or she determines: (1) that employees are exposed to grave danger from exposure to substances or agents determined to be toxic or physically harmful or from new hazards and (2) that such emergency standard is necessary to protect employees from such danger.

An emergency temporary standard may be issued without regard to the notice, public comment, and hearing provisions of the Administrative Procedure Act. It takes effect immediately upon publication in the *Federal Register*.

The key to the issuance of an emergency temporary standard is the necessity to protect employees from a grave danger, as defined in *Florida Peach Growers* (72). After issuing an emergency temporary standard, the secretary must commence the procedures for promulgation of a permanent standard, which must issue within 6 months of the emergency standard's publication in accordance with Section 6(c)(3).

5.2.2 Challenging the Validity of Standards

Any person who may be adversely affected by an OSHA standard may file a petition under Section 6(f) challenging its validity in the United States Court of Appeals in the circuit wherein such person resides or has the principal place of business. The petition may be filed at any time prior to the 60th day after the issuance of the standard. Unless otherwise ordered, the filing of a petition does not operate as a stay of the standard.

Section 6(f) of the act directs the courts to uphold the secretary of labor's determinations in promulgating standards if those determinations are "supported by substantial evidence in the record considered as a whole" (73). In practice, the courts have generally declined to apply a strict "substantial evidence" standard of review. Instead, the courts have chosen to apply two different standards depending on whether the agency determination to be reviewed is one of fact or policy. They have essentially taken the position that only the secretary's findings of fact should be reviewed pursuant to a substantial evidence standard, while the secretary's policy determinations should be substantiated by a detailed statement of reasons, which are subject to a test of reasonableness (74). The courts have adopted this approach with respect to emergency temporary standards as well as permanent standards [e.g., in *Florida Peach Growers* (71)].

In addition to a direct petition for review under Section 6(f), a majority of the courts of appeals have held that both procedural and substantive challenges to the validity of OSHA standards may be raised in enforcement proceedings under Section 11 (75). In fact, the Third Circuit in *Atlantic & Gulf Stevedores* (75) stated that the validity of a standard may be challenged not only in a federal court of appeals as part of an appeal from an order of the commission but in the commission proceedings themselves. The court explained, however, that in an enforcement proceeding invalidity is an affirmative defense to a citation and the employer bears the burden of proof on the issue of the reasonableness of the adopted standard. To carry its burden the employer must produce evidence showing why the standard under review, *as applied to it*, is arbitrary, capricious, unreasonable, or contrary to law (76).

5.2.3 Economic and Technological Feasibility of Standards

In enacting OSHA, Congress did not intend to make employers strictly liable for unavoidable occupational hazards. Accordingly, feasibility of compliance is a factor the secretary of labor must consider in developing occupational safety and health standards. As the U.S. Supreme Court explained in *American Textile Mfrs. Inst. Inc. v. Donovan* (77), OSHA's legislative history makes clear that any standard that is not economically or technologically feasible would *a fortiori* not be "reasonably necessary or appropriate" as directed by Section 3(8) of the act (78). Thus, in enacting OSHA, "Congress does not appear to have intended to protect employees by putting their employers out of business" (79).

In analyzing economic feasibility, the secretary has tried to determine whether proposed standards threaten the competitive stability of an affected industry (80). The Supreme Court has not yet decided whether a standard that actually does threaten the long-term profitability and competitiveness of an industry would be "feasible" (81).

The Supreme Court has expressly decided, however, that in promulgating a toxic material and harmful physical agent standard under Section 6(b)(5), the secretary is not required to determine that the costs of the standard bear a reasonable relationship to its benefits [*American Textile Mfrs. Inst. Inc. v. Donovan* (77)]. Rather, Section 6(b)(5) directs the secretary to issue the standard that "most adequately assures . . . that no employee will suffer material impairment of health," limited only by the extent to which this is economically and technologically feasible, or, in other words, capable of being done (82). The Supreme Court left open the possibility, however, that cost–benefit analysis might be required with respect to standards promulgated under provisions other than Section 6(b)(5) of the act (83). The Court also left open the question of whether cost–benefit balancing by the secretary might be appropriate for deciding between issuance of several standards regulating different varieties of health and safety hazards (84).

Finally, in cases of violations of standards caused by employee disobedience or idiosyncratic behavior, the decisions of the courts and the commission have been similar to those rendered under the general duty clause (85) (see also discussion in Section 5.1.1 of this chapter).

For example, in *Brennan v. OSAHRC & Hendrix (d/b/a Alsea Lumber Co.)* (86), an employer was cited for violation of OSHA standards requiring workers to wear certain personal protective equipment. The record established that the violations resulted from individual employee choices, which were contrary to the employer's instructions. The Ninth Circuit affirmed the commission's decision vacating the citations, explaining as follows (87):

> The legislative history of the Act indicates an intent not to relieve the employer of the general responsibility of assuring compliance by his employees. Nothing in the Act, however, makes an employer an insurer or guarantor of employee compliance therewith at all times. The employer's duty, even that under the general duty clause, must be one which is achievable. See *National Realty, supra*. We fail to see wherein charging an employer with a nonserious violation because of an individual, single act of an employee, of *which the employer had no knowledge* and which was contrary to the employer's instructions, contributes to achievement of the cooperation [between employer and employee] sought by the Congress. Fundamental fairness would require that one charged with and penalized for violation be shown to have caused, or at least to have knowingly acquiesced in, that violation [emphasis added, footnote omitted].

Nevertheless, even though Congress did not intend the employer to be held strictly liable for violations of OSHA standards, an employer is responsible if it knew or, with the exercise of reasonable diligence, should have known of the existence of a violation. Thus, in *Brennan v. Butler Lime & Cement Co. and OSAHRC* (88), the Seventh Circuit drew on general duty clause concepts from the *National Realty* case (56) and explained that a particular instance of hazardous employee conduct may be considered preventable even if no employer could have detected the conduct or its hazardous nature at the moment of its occurrence, where such conduct might have been precluded through feasible precautions concerning the hiring, training, or sanctioning of employees (89).

In *Atlantic & Gulf Stevedores, Inc.* the Third Circuit held that such feasible precautions include disciplining or dismissing workers who refuse to wear protective headgear, even where such employer action could subject the company to wildcat strikes by employees adamantly opposed to the regulation (90).

There are some limits to the employer's obligations, however. In *Horne Plumbing and Heating Company v. OSAHRC and Dunlop*, the Fifth Circuit found that an employer had taken virtually every conceivable precaution to ensure compliance with the law, short of remaining at the job site and directing the employees' operations himself. The court held that the final effort of personally directing the employees was not required by the act, and that such an effort would be a "wholly unnecessary, unreasonable and infeasible requirement" (91).

5.2.4 Environmental Impact of Standards

Section 102(2)(C) of the National Environmental Policy Act of 1969 (NEPA) (92) requires all federal agencies, including OSHA, to prepare a detailed environmental impact statement in connection with major federal actions significantly affecting the quality of the human environment. The secretary of labor has identified the promulgation, modification, or revocation of standards that will significantly affect air, water or soil quality, plant or animal life, the use of land or other aspects of the human environment as always constituting such major action requiring the preparation of an environmental impact statement (93). On the other hand, promulgation, modification, or revocation of any safety standard, such as machine guarding requirements, safety lines, or warning signals, would normally qualify for categorical exclusion from NEPA requirements because "[s]afety standards promote injury avoidance by means of mechanical applications of work practices, the effects of which do not impact on air, water or soil quality, plant or animal life, the use of land or other aspects of the human environment" (94). The secretary's regulations regarding the procedure for preparation and circulation of environmental impact statements can be found at 29 CFR §11.1 *et seq.*

5.2.5 Variances from Standards

Section 6(d) of OSHA provides that any affected employer may apply to the secretary of labor for a variance from an OSHA standard. To obtain a variance, the employer must show by a preponderance of the evidence submitted at a hearing that the conditions, practices, means, methods, operations, or processes used or proposed to be used will provide employees with employment and places of employment that are as safe and healthful as those that would prevail if the employer complied with the standard.

If granted, the Section 6(d) variance may nevertheless be modified or revoked on application by an employer, employees, or by the secretary of labor on his or her own motion at any time after 6 months from its issuance. Affected employees are to be given notice of each application for a variance and an opportunity to participate in a hearing.

The act also provides mechanisms to enable employers to obtain variances of a more temporary nature than those sought under Section 6(d). Section 6(b)(6)(A) provides for "temporary" variances upon application when an employer establishes, after notice to employees and a hearing, that (1) he or she is unable to comply with a standard by its effective date because of unavailability of professional or technical personnel or of materials and equipment needed to come into compliance with the standard or because necessary construction or alteration of facilities cannot be completed by the effective date, or (2) he or she is taking all available steps to safeguard employees against the hazards covered by the standard and has an effective program for coming into compliance with it.

Section 6(b)(6)(C) authorizes the secretary to grant a variance from any standard or portion thereof whenever he or she determines, or the secretary of health and human services certifies, that the variance is necessary to permit the employer to participate in an experiment approved by the secretary of labor or the secretary of health and human services designed to demonstrate or validate new and improved techniques to safeguard the health or safety of workers.

Finally, Section 16 permits the secretary of labor, after notice and an opportunity for hearing, to provide such reasonable limitations and rules and regulations allowing "reasonable variations, tolerances, and exemptions to and from any or all provisions of" the act as he or she may find necessary and proper to avoid serious impairment of the national defense.

5.3   Overview of Occupational Safety and Health Standards (95)

The bulk of federal job safety and health standards deal with occupational *safety* rather than with occupational *health* (96).

The occupational safety standards promulgated by the secretary of labor pursuant to OSHA are voluminous, encompassing hundreds of pages in the Code of Federal Regulations. A comprehensive analysis of these safety standards, many of which were adopted in 1971 as interim standards, is accordingly beyond the scope of this chapter (97).

For purposes of simplification, however, OSHA's safety standards can be broken down into the following general categories: (1) requirements relating to hazardous materials (e.g., compressed gas, acetylene, hydrogen, oxygen) and related equipment; (2) requirements for personal protective equipment and first aid; (3) requirements for means of egress, fire protection standards and the national electrical code; (4) general environmental controls (e.g., control of hazardous energy or ''lockout/tagout''); (5) design and maintenance requirements for industrial equipment and walking-working surfaces; and (6) operational procedures and equipment utilization requirements for certain hazardous industrial operations such as welding, cutting and brazing, and materials handling. Certain industries are also subject to specialized safety standards (98).

Like the safety standards, the bulk of OSHA's health standards were promulgated in 1971 as interim standards under Section 6(a). At that time, previously established federal standards were used to set workplace exposure limits for approximately 400 chemical and hazardous substances. These were referred to as threshold limit values (TLVs) and were expressed in terms of milligrams of substance per cubic meter of air and/or parts of vapor or gas per million parts of air. TLVs were defined as representing conditions under which it was believed nearly all workers may be repeatedly exposed day after day without adverse effects (99).

Amidst ongoing criticism that the limits set in the original start-up standard were too lenient, in January 1989, pursuant to Section 6(b), OSHA issued a massive rule revising its air contaminants standard (100). The rule revised permissible exposure limits (PELs) for many of the approximately 400 contaminants that were the subject of the original standard. In addition, PELs were established for substances not previously regulated (101). Compliance with the revised exposure limits through the combined use of engineering controls, work practices, and personal protective equipment was required by September 1, 1989. According to the rule, however, engineering controls (OSHA's preferred method of compliance) need not be in place until at latest December 31, 1992 (102).

Immediately following promulgation, the revised standard on air contaminants was challenged in court by various labor and industry groups. Eleven such cases were consolidated, and on July 7, 1992, in a surprise move, the Eleventh Circuit vacated the standard (*AFL-CIO v. OSHA*, No. 89-7185). The Eleventh Circuit ruled that OSHA had not established a significant risk of material health impairment for each substance regulated nor that a PEL for a given substance was feasible for the affected industry. The Eleventh Circuit rejected the secretary's request for reconsideration, and as of November, 1992, the secretary was seeking a stay of the

Eleventh Circuit's decision to vacate the standard pending a possible appeal to the U.S. Supreme Court (103). Unless it is appealed and reversed, the Eleventh Circuit's ruling would mean a return to the original 1971 limits until OSHA promulgates new standards or until some legislative intervention.

The January, 1989, standard on air contaminants was directed only to general industry. On June 12, 1992, shortly before the Eleventh Circuit decision vacating the general industry standard, OSHA published in the *Federal Register* a proposed rule revising air contaminants exposure limits for construction, maritime, and agriculture industry sectors (104). This rule may face the same fate as the standard for general industry since OSHA took the same approach to rule making with both.

OSHA has promulgated other permanent health standards pursuant to Section 6(b) (e.g., regarding occupational exposure to bloodborne pathogens, discussed below at Section 5.3.13). Each of the permanent standards adopted under Section 6(b) can be viewed as a "complete" standard since each provides not only a specific value for the level of exposure to the toxic substance but also specifications as to monitoring, engineering controls, personal protective equipment, record keeping, medical surveillance, and other matters.

Compliance with promulgated standards is mandatory. In February, 1984, OSHA formally removed 153 provisions that included the advisory word "should" instead of the mandatory "shall" (105). The agency did not replace these provisions with rules promulgated under Section 6(b). Instead, the agency has relied on the general duty clause or mandatory language in general industry standards to enforce the deleted standards (106).

A summary of several of the major health standards under OSHA follows.

5.3.1 Asbestos

OSHA issued a revised asbestos rule on June 20, 1986 (107). The revised rule actually includes two simultaneously issued asbestos standards: one for general industry and another for construction (108). The new standards, which cover not only asbestos but also tremolite, actinolite, and anthophyllite (109), reduce the 8-hr time-weighted average (TWA) from 2 fibers per cubic centimeter of air to 0.2 fibers per cubic centimeter of air (110).

The revised standards were challenged by the Asbestos Information Association, North America (AIA), and by the Building and Construction Trades Department (BCTD) of the AFL-CIO. The unions challenged OSHA's refusal to set a lower 8-hr TWA of 0.1 fiber per cubic centimeter of air. OSHA claimed it promulgated the higher limit because the lower limit was not feasible in the entire industry, and OSHA had discretion to decide what industries should be grouped together for regulatory purposes. The Court of Appeals for the District of Columbia disagreed and held that if OSHA was concerned with the administrative problems involved in desegregating industrial sectors for purposes of the TWA, it must make specific findings on the issue. Therefore, the D.C. Circuit remanded to OSHA on this issue, as well as on whether a short-term exposure limit (STEL) should be established (111).

On September 14, 1988, OSHA established an asbestos STEL of 1 fiber per cubic centimeter of air over a 30-min sampling period (112). OSHA decided that the 8-hr

TWA issue would require additional rule making, and the D.C. Circuit Court approved the OSHA decision on October 30, 1989 (113).

In July, 1990, OSHA proposed new rule making to amend the 1986 asbestos standards. Included is a proposal to lower the permissible exposure limit to 0.1 fiber per cubic centimeter (114). Public hearings were scheduled in January, 1991 (115). The agency intends to issue a final asbestos rule in 1992, which may include a controversial requirement that employers undertake inspections of commercial buildings for the presence of asbestos (116).

5.3.2 Vinyl Chloride

On April 5, 1974, the secretary promulgated an emergency standard for vinyl chloride. One of its requirements was that no worker be exposed to concentrations of vinyl chloride in excess of 50 parts per million parts of air over any 8-hr period. The prior TLV had set an exposure limit of 500 ppm (117).

A permanent standard was promulgated on October 1, 1974. It set a permissible exposure limit for vinyl chloride at no greater than 1 ppm averaged over 8 hr (118). The Court of Appeals for the Second Circuit upheld the standard in *Society of the Plastics Industry*, and it became effective in April 1975 (119).

Generally, the standard requires feasible engineering and work practice controls to reduce exposure below the permissible level wherever possible. Specifically, the employer is required to provide respirator protection and protective garments for employees (120). The employer must provide, for each employee engaged in vinyl chloride operation, a training program "relating to the hazards of vinyl chloride and precautions for its safe use" (121). Entrances to regulated areas must be posted with signs warning of the cancer-suspect nature of vinyl chloride (122). Moreover, the employer must institute a program of medical surveillance for each employee exposed to vinyl chloride in excess of the action level without regard for use of respirators (123). Finally, the employer must maintain accurate medical records to measure employee exposure to vinyl chloride (124).

5.3.3 Carcinogens

The Department of Labor excluded the American Industrial Hygiene Association's (ACGIH's) list of carcinogenic chemicals from its interim standards package, which it promulgated in 1971 (99). After approximately one year of consulting with NIOSH and receiving data and commentary from interested groups, the secretary promulgated emergency temporary standards on May 3, 1973, for a list of 14 chemicals found to be carcinogenic. Permanent standards for the 14 carcinogens were issued on January 29, 1974 (125).

In *Synthetic Organic Chemical Manufacturers Ass'n v. Brennan*, the Third Circuit upheld all the carcinogens standards except one and except for the provisions pertaining to medical examinations and to laboratory usage of said chemicals (126). The standard for 4,4'-methylene bis(2-chloraniline) (MOCA) was subsequently revoked by OSHA (127).

In January 1980, OSHA promulgated a generic carcinogen policy for identifying, classifying, and regulating workplace carcinogens (128). The policy includes a process for screening chemicals and establishing priorities for rule making. It places substances in two categories: potential carcinogenicity and suggestive potential. Model permanent and emergency temporary standards are included (129). Lists of carcinogen candidates were to be published by OSHA periodically pursuant to the policy, but the listing requirements have been administratively stayed since 1983 (130). Since its issuance and amendment in 1981, little has been done to implement and utilize the policy due to legal challenges, pressure from outside interests to amend the policy, and inactivity at OSHA (131).

5.3.4 Coke Oven Emissions

In October 1976, the secretary of labor issued a permanent standard regulating workers' exposure to coke oven emissions. The standard defines coke oven emissions as the benzene-soluble fraction of total particulate matter present during the destructive distillation of coal for the production of coke. It limits exposure to 150 μg of benzene-soluble fraction of total particulate matter per cubic meter of air averaged over an 8-hr period (132).

The standard mandates specific engineering controls and work practices that were to be in use by January 20, 1980. For example, the employer is required to provide protective clothing and equipment (133), hygienic changing rooms (134), and lunchrooms with a filtered air supply (135). The employer must ensure that in the regulated area food and beverages are not consumed, smoking is prohibited, and cosmetics are not applied (136). The employer must provide training to employees regarding the dangers of the regulated area (137). Additionally, the employer must post precautionary signs and labels (138). The employer must institute a medical surveillance program for all those employed in the regulated area at least 30 days per year (139). The employer also must maintain accurate medical records to measure employee exposure to coke oven emission (140).

The steel industry challenged the validity of the coke oven standard, but it was upheld by the Third Circuit in *American Iron and Steel Institute v. OSHA* (141).

5.3.5 Lead

The OSH Administration issued a permanent standard regulating occupational exposure to lead in November 1978 (142). The standard set a permissible exposure limit of 50 μg/m^3 of air over an 8-hr period (143). In addition, the standard required, among other things, the use of respirators and protective work clothing and equipment whenever lead exposure exceeds the PEL, compliance with vigorous rules on housekeeping and hygiene, biological monitoring, and medical surveillance. The standard also required a controversial medical removal protection (MRP) provision pursuant to which certain workers must be removed from the exposed workplace without loss of earnings, benefits, or seniority for at least 18 months. The standard further required employers to create safety and health training programs for their workers exposed to lead; to keep detailed records on environmental (workplace) monitoring, biological monitoring, and medical surveillance; and to make those records available to workers, certain of their representatives, and the government.

Virtually every aspect of the lead standard was challenged by the industry and by organized labor. The U.S. Court of Appeals for the District of Columbia rejected those challenges and upheld the standard as to certain industries (144). However, with respect to 38 other industries, the court held that OSHA failed to present substantial evidence or adequate reasons to support the feasibility of the standard for those industries and thus remanded to the secretary of labor for reconsideration of the technological and economic feasibility of the standard as to those industries (145).

OSHA subsequently amended the standard to require employers that cannot reach the PEL to reduce exposure only to the lowest feasible level (146); it also found that the standard of 50 $\mu g/m^3$ was technologically and economically feasible for all but nine of the remand industries (147). OSHA continued to gather information as to these nine industry sectors, and on July 11, 1989, OSHA issued a statement of reasons why the 50-μg standard was feasible through engineering controls for eight of the remaining nine sectors (148). For the ninth industry sector, nonferrous foundries, OSHA found the 50-μg standard was technologically but not economically feasible for small foundries and therefore promulgated a bifurcated standard for large and small nonferrous foundries (149).

In March 1990, the D.C. Circuit upheld the standard with respect to all but six industries (150), and finally, on July 19, 1991, the court lifted its existing stay of the standard's engineering and work practice control requirements with respect to all of these six industries except the brass and bronze ingot manufacturing industry (151).

5.3.6 Cotton Dust (152)

In June 1978, OSHA promulgated its original cotton dust standard (153). Extensively revised in 1985, the standard sets different PELs for different industries: 200 $\mu g/m^3$ of air averaged over an 8-hr period for exposures in yarn manufacturing and cotton washing operations; 500 $\mu g/m^3$ in textile mill waste house operations and yarn manufacturing from lower-grade washed cotton; and 750 $\mu g/m^3$ in slashing and weaving operations (154). The action levels are set at one-half the PELs (155).

The standard commands compliance with the PELs through a mix of engineering and work practice controls, except to the extent that employers can establish that such controls are infeasible (156). Specifically, the standard requires, among other things the provision of respirators, the monitoring of cotton dust exposure, medical surveillance, medical examinations, employee education and training, and the posting of warning signs (157).

Upon revision of the standard in 1985, OSHA exempted the cottonseed processing and cotton waste processing industries from all but the medical surveillance provisions; the cotton warehousing industry was totally exempted. OSHA also retained, in limited form, a provision that guaranteed transfer to another available position having dust exposure at or below the PEL, without loss of wage rate or benefits, for employees unable to wear respirators (158).

5.3.7 DBCP

In March 1978, OSHA promulgated a permanent standard regarding 1,2-dibromo-3-chloropropane (DBCP) (159). The standard, which became effective April 17,

1978, sets a PEL of one part per billion parts of air over an 8-hr period. Where engineering controls and work practices are not sufficient to reduce exposure to permissible limits, respirators may be used as a supplement in order to achieve the required protection. Protective clothing is required where eye or skin contact may occur. The standard does not apply to the use of DBCP as a pesticide or when it is stored, transported, or distributed in sealed containers (160).

5.3.8 Acrylonitrile

In September 1978, OSHA issued a permanent standard governing workplace exposure to acrylonitrile (161). The standard, which became effective November 2, 1978 (except as to training programs, which were to be set up by January 2, 1979, and engineering controls, which were to be installed by November 2, 1980), sets a permissible exposure limit of 2 parts per million parts of air over an 8-hr period. The standard also sets a ceiling limit of 10 ppm for any 15-min period and an action level of 1 ppm. Exposure above the action level triggers periodic monitoring requirements, medical surveillance, protective clothing and equipment requirements, employee information and training, and housekeeping. Skin or eye contact with the substance is prohibited (162).

5.3.9 Benzene

OSHA's first permanent standard for benzene, which limited occupational exposure to 1 part benzene per million parts of air, was invalidated by the Fifth Circuit in 1978. On appeal, the U.S. Supreme Court affirmed, finding that OSHA had failed to show that the 1 ppm exposure limit was presented a "significant risk of material health impairment" to workers (163).

In 1987, OSHA issued a revised benzene standard (164). The revised standard sets the 8-hr TWA for benzene at 1 ppm, and the STEL at 5 ppm for a 15-min sampling period (165). Under the revised standards, the employer must provide a medical surveillance program to monitor the health of employees exposed to benzene in amounts over the action level of 0.5 ppm for more than 30 days per year. Additionally, OSHA requires a medical removal plan for temporary and permanent removal of employees showing adverse health effects from benzene exposure. The employer is required to provide 6 months of medical removal protection benefits to a removed employee, unless the employee has been transferred to a comparable job with benzene exposure below the action level (166).

Union and industry petitions for review of the new standard were withdrawn from the Third and D.C. Circuits in November 1987, shortly before the effective date of the new standard (167).

5.3.10 Inorganic Arsenic

The OSH Administration issued a permanent standard for inorganic arsenic in May 1978 (168). The standard established a PEL of 10 $\mu g/m^3$ of air over an 8-hr period and also specified various other requirements such as engineering and work practice controls, respiratory protection, and employee monitoring and training (169).

The standard was challenged by industry in *ASARCO, Inc. v. OSHA* (170) and was remanded to OSHA by the Ninth Circuit for reconsideration in light of the Supreme Court's decision invalidating the benzene standard in *Industrial Union Dept. v. American Petroleum Institute* (163). In 1983, OSHA published a final risk assessment indicating that the 10-μg limit reduced the risk of lung cancer by about 98 percent from the previous TLV of 500 μg and that such a reduction satisfies the Supreme Court's "significant risk" test (171). The Ninth Circuit later upheld the standard as supported by substantial evidence of a significant health risk (172).

The standard applies to all occupational exposures to inorganic arsenic except employee exposures in agriculture or industries involving pesticide application, the treatment of wood with preservatives, or the use of wood preserved with arsenic (173).

5.3.11 Ethylene Oxide

OSHA issued a final rule for ethylene oxide in 1984. Pursuant to the rule, OSHA established an 8-hr TWA of 1 ppm but deferred setting an STEL (174). Several months later, OSHA concluded that an STEL was not warranted by the available health evidence. This decision was challenged in court, and the D.C. Circuit, while upholding the validity of the standard, ruled that OSHA's decision regarding the STEL should be remanded for further consideration (175). Finally, in 1988, OSHA set an ethylene oxide STEL of 5 ppm for a 15-min period (176).

The ethylene oxide standard requires the familiar use of engineering and work practice controls, when feasible, as well as medical surveillance, protective clothing and equipment, and employee training and warnings (177). A research study on the effectiveness of the ethylene oxide standard found that the standard was instrumental in significantly reducing worker exposure at a cost much less than what OSHA itself had projected. The study, which was released in 1992, credited the standard with increasing public awareness of the substance's toxicity and in spurring the development of new and inexpensive control technology (178).

5.3.12 Formaldehyde

OSHA's formaldehyde standard, issued in 1987 (179), was significantly revised in 1992 to, among other things, lower the existing PEL from 1 part formaldehyde per million parts air to 0.75 ppm for the 8-hr TWA (180). The revisions were prompted by a decision of the D.C. Circuit to remand the 1987 standard for OSHA's reconsideration of several issues (181).

The standard as amended has medical removal protection (MRP) provisions that supplement its medical surveillance requirements. It also requires specific hazard labeling and annual employee training. The amendments became effective at various dates in 1992 (182).

OSHA has announced that the standard's amendments will provide additional protection primarily to workers in the apparel, furniture, and foundry industries (183).

5.3.13 Bloodborne Pathogens

In one of its most controversial rule makings, OSHA has promulgated a standard aimed at controlling occupational exposure to bloodborne pathogens such as the human immunodeficiency virus (HIV) and the hepatitis B virus (HBV). The final rule was issued in late 1989 following a comment period of several years (184).

Application of the standard is triggered by an employee's reasonably anticipated contact with blood or other potentially infectious materials (defined by a list of body tissues and fluids). Affected employers must develop written exposure control plans that identify exposed employees and tailor the standard's work practices and engineering controls to the particular work environment. Central to the standard is the mandatory use of "universal precautions," a practice by which all blood and other potentially infectious materials are treated as if known to be infected with blood-borne pathogens. The standard has stringent requirements regarding the disposal of wastes and used needles, the use of personal protective equipment, employee training, and labels and warnings. Employers must offer exposed employees the hepatitis B vaccine free of charge and must also provide free post-exposure medical evaluation and testing. Additionally, extensive record keeping, including documented follow-up of exposure incidents, is mandated by the standard.

The standard is aimed at protecting over 5 million workers, many of whom are outside the health care industry (185). Challenges to the standard have been filed (186), and since most of its provisions became effective in July 1992, the standard's effects are to date largely unknown.

5.3.14 Hazard Communication

OSHA issued a hazard communication standard in 1983 (187). The original standard required manufacturers to establish a method to communicate the hazards of chemicals to those who worked with them, mainly through the use of labels on containers, materials safety data sheets (MSDS), and training programs. The Third Circuit upheld the standard in most respects (188).

On August 24, 1987, OSHA expanded the scope of the standard to encompass nonmanufacturing employment (189). The Third Circuit denied subsequent petitions for review of the expanded standard (190).

Identical hazard communication standards exist for specific industries: construction [29 CFR §1926.59], shipyard employment [29 CFR §1915.99], marine terminals [29 CFR §1917.28], and longshoring [29 CFR §1918.90] (191).

5.3.15 Noise

OSHA's standard for occupational exposure to noise specifies a maximum permissible noise exposure level of 90 dbels (dB) for a duration of 8 hr (192). Employers are required to use feasible engineering or administrative controls, or a combination of both, whenever employee exposure to noise in the workplace exceeds the permissible exposure level. Personal protective equipment may be used to supplement the engineering and administrative controls where such controls are not able to reduce the employee exposure to within the permissible limit.

In 1983, OSHA issued a final hearing conservation amendment to the noise standard (193). Under the amendment, the employer must establish a baseline audiometric measurement from which it must compare subsequent annual measurements to determine if a shift in hearing capability has occurred for any employees whose exposures equal or exceed an 8-hr TWA of 85 dB (194). To determine if the exposure equals or exceeds this action level, employers are required to use personal sampling of noise where factors, such as high worker mobility or significant variation in sound levels, make area monitoring generally inappropriate (195). The employer must make hearing protectors available to all employees whose exposure equal or exceed an 8-hr TWA of 85 dB, and must ensure that hearing protectors are worn by any employee who has experienced a shift in hearing capability (196). The Fourth Circuit has upheld the amendment (197).

In 1987, OSHA sought public comments as to whether and to what extent the information collection requirements of the noise standard could be reduced without lessening the standard's effectiveness in preventing hearing loss. OSHA later announced, however, that no such changes would be made (198).

5.3.16   Proposed Standards

OSHA has several standards regarding occupational health on its 1992 agenda. Among those is a rule to protect workers from exposure to cadmium. The notice of proposed rulemaking for cadmium was published in February 1990. OSHA has been ordered by the Third and D.C. Circuits to issue the final rule by August 31, 1992 (199).

OSHA's proposed standard on methylene chloride was published on November 7, 1991 (200). The rule, which would cover some 186,000 workers, contains a proposal to change the current 8-hr TWA for methylene chloride exposure from 500 ppm parts air to 25 ppm. It would also establish a 12.5 ppm action level, a STEL of 125 ppm, and other protective measures. Public hearings are scheduled for late 1992, and OSHA has extended the public comment period (201).

In May 1989, OSHA proposed a standard to limit exposure to 4,4'-methylene-dianiline (MDA) to 10 ppm. A draft final rule is under administrative review in 1992. MDA, which is found in various chemicals and plastics such as polyurethane foam, adhesives and sealants, has been associated with cancer and liver damage (202).

OSHA is also considering rule making in other areas of occupational health. On August 3, 1992, OSHA published an advance notice of proposed rule making on ergonomics (203). With this ANPRM, OSHA is addressing the occupational risk of cumulative trauma disorders (CTDs), which are defined as disorders of the musculoskeletal and nervous systems that are caused by repetitive motions, forceful exertions, vibrations, or sustained awkward positioning while working. Examples of such disorders are carpal tunnel syndrome and back strains (204).

6   EMPLOYEE RIGHTS AND DUTIES UNDER JOB SAFETY AND HEALTH LAWS

6.1   Employee Duties

Section 5(b) of OSHA requires employees to ''comply with occupational safety and health standards and all rules, regulations, and orders issued pursuant'' to the act that are applicable to their own actions and conduct.

The act does not, however, expressly authorize the secretary of labor to sanction employees who disregard safety standards and other applicable orders. The U.S. Court of Appeals for the Third Circuit has held that although Section 5(b) would be devoid of content if not enforceable, Congress did not intend to confer on the secretary or the commission the power to sanction employees (205).

Section 110(g) of the Federal Mine Safety and Health Act of 1977 not only requires mine employees to comply with health and safety standards promulgated under that statute, it also authorizes the imposition of civil penalties on miners who "willfully violate the mandatory safety standards relating to smoking or the carrying of smoking materials, matches, or lighters."

6.2 Employee Rights

The Occupational Safety and Health Act grants numerous rights to employees and/ or their authorized representatives. The most fundamental employee right is the right set forth in Section 5(a) to a safe and healthful employment and place of employment. Other significant rights of employees and/or their authorized representatives include the following:

1. The right to request a physical inspection of a workplace and to notify the secretary of labor of any violations that employees have reason to believe exist in the workplace [§8(f)].
2. The right to accompany the secretary during the physical inspection of a workplace [§8(e)].
3. The right to challenge the period of time fixed in a citation for abatement of a violation of the act and an opportunity to participate as a party in hearings relating to citations [§10(c)].
4. The right to be notified of possible imminent danger situations and the right to file an action to compel the secretary to seek relief in such situations if he or she has "arbitrarily and capriciously" failed to do so [§§13(c) and 13(d)].
5. Various rights, including the right to notice, regarding an employer's application for either a temporary or permanent variance from an OSHA standard [§6(b)(6) and §6(d)].
6. The right to observe monitoring of employee exposures to potentially toxic substances or harmful physical agents, the right to records thereof, and the right to be notified promptly of exposures to such substances in concentrations which exceed those prescribed in a standard [§8(c)].
7. The right to petition a court of appeals to review an OSHA standard within 60 days after its issuance [§6(f)].

Section 11(c) of OSHA makes it unlawful for any person to discharge or in any manner discriminate against an employee because the employee has exercised his or her rights under the act. This provision is designed to encourage employee participation in the enforcement of OSHA standards (206).

Employees who believe they have been discriminated against in violation of Section 11(c) must file a complaint with the secretary of labor within 30 days after such violation has occurred. The secretary will investigate the complaint, and if he or she determines that Section 11(c) has been violated, he or she is authorized to bring an action in federal district court for an order restraining the violation and for recovery of all appropriate relief, including rehiring or reinstatement of the employee to his or her former position with back pay. The statute authorizes only the secretary of labor to bring an action for violation of Section 11(c) (207).

The Federal Mine Safety and Health Act of 1977 also contains a broad antiretaliation provision and grants other rights to mine employees as well (208).

An employee has no explicit right, under OSHA, to refuse a work assignment because of what he or she feels is a dangerous working condition. The secretary of labor, however, has issued an administrative regulation that interprets the act as implying such a right under certain limited circumstances (209). The U.S. Supreme Court, in *Whirlpool Corp. v. Marshall* (210), has held the promulgation of that regulation to be a valid exercise of the secretary's authority under the act. The Court observed that despite the detailed statutory scheme for speedily remedying dangerous working conditions, circumstances may arise when an employee justifiably believes that the statutory scheme will not sufficiently protect him:

> [S]uch a situation may arise when (1) the employee is ordered by his employer to work under conditions that the employee reasonably believes pose an imminent risk of death or serious bodily injury, and (2) the employee has reason to believe that there is not sufficient time or opportunity either to seek effective redress from his employer or to apprise OSHA of the danger (211).

In holding that the regulation conformed to the fundamental objective of the act to prevent occupational deaths and injuries, the Court observed that the regulation also served to effectuate the general duty clause of §5(a)(1) (212).

It is fairly well settled that OSHA does not create a private right of action for damages suffered by an employee as the result of an employer's violation of the act (213). Finally, Section 4(b)(4) states that the act does not supersede or in any manner affect any worker's compensation law.

NOTES AND REFERENCES

1. See Bureau of National Affairs (BNA), *The Job Safety and Health Act of 1970*, Washington, D.C., 1971, p. 13. The Occupational Safety and Health Act is codified in Title 29 of the United States Code (USC), §§651–678. References to the act in this chapter and the next are to the appropriate section of the statute itself and do not include a corresponding citation to the United States Code.

2. N. Ashford, *Crisis in the Workplace*, M.I.T. Press, Cambridge, MA, 1976, pp. 141, 236–237.

3. See American Bar Association, *Report of the Committee on Occupational Safety and Health Law*, ABA Press, Chicago, 1975, p. 107.

4. For a discussion of the weight to be accorded by the secretary of labor to the NIOSH

recommendations, see *Industrial Union Department, AFL-CIO v. Hodgson*, 499 F.2d 467, 476–77; 1 OSHC 1631 (D.C. Cir. 1974). [References to ''F.2d'' designate the volume (e.g., 499) and page (e.g., 467) of the *Federal Reporter, Second Series*, which contains the official reported decisions of the U.S. Courts of Appeals as published by the West Publishing Company. References to ''OSHC'' designate the same decision as reported in the ''Occupational Safety and Health Cases'' published by the Bureau of National Affairs (BNA).

5. See Ashford, *Crisis in the Workplace* (Ref. 2), pp. 249–251. Section 27(b) of the act established a National Commission on State Workmen's Compensation Laws, which was directed to study and evaluate such laws to determine whether they provide an adequate, prompt, and equitable system of compensation for injury or death arising out of, or in the course of, employment. The commission's tasks were completed in July 1972, and the commission was disbanded. For a description of the activities of the commission and an evaluation of its work, see Ashford, *Crisis in the Workplace*, pp. 246, 289–292.

6. Numerous proposals had been introduced in Congress throughout the 1970s to amend the scope of OSHA's coverage, but all of those bills died either in committee or on the House or Senate floor. The scope of OSHA's coverage has not been legislatively amended since OSHA's enactment.

7. *Southern Pacific Transportation Co. v. Usery and OSAHRC*, 539 F.2d 386; 4 OSHC 1693 (5th Cir. 1976), *certiorari denied*, 434 U.S. 874; 5 OSHC 1888 (1977); *Southern Railway Company v. OSAHRC and Brennan*, 539 F.2d 335; 3 OSHC 1940 (4th Cir.), *certiorari denied*, 429 U.S. 999; 4 OSHC 1936 (1976); *Baltimore & Ohio Railroad Co. v. OSAHRC*, 548 F.2d 1052; 4 OSHC 1917 (D.C. Cir. 1976).

8. 539 F.2d at 391; 4 OSHC at 1696.

9. The Third Circuit has also adopted this definition of ''working conditions.'' See *Columbia Gas of Pennsylvania, Inc. v. Marshall*, 636 F.2d 913; 9 OSHC 1135 (3rd Cir. 1980).

10. *Fineberg Packing Co.*, 1 OSHC 1598, 1599 (Rev. Comm. 1974).

11. 520 F.2d 1161; 3 OSHC 1566, 1572 (D.C. Cir. 1975).

12. American Bar Association, *Report of the Committee on Occupational Safety and Health Law*, ABA Press, Chicago, 1976, pp. 247–248.

13. The Federal Mine Safety and Health Act of 1977 is codified at 30 USC §§801 *et seq.*

14. 30 USC §811(a). Comprehensive regulations have been promulgated by the Mine Safety and Health Administration. Those regulations appear in title 30 of the Code of Federal Regulations at Parts 1 to 100. The Code of Federal Regulations is hereinafter cited as ''CFR.'' The mandatory health standards can be found at 30 CFR Parts 70 *et seq.* References to specific provisions of the mine safety and health legislation and standards promulgated pursuant thereto are made throughout this chapter and the next.

15. The FEPCA is codified at 7 USC §§136 *et seq.* It is a comprehensive revision of the Federal Insecticide, Fungicide and Rodenticide Act of 1970, 7 USC §§135 *et seq.* (1970). The regulations promulgated by EPA to protect farm workers from toxic exposure to pesticides are found in 40 CFR §§170.1 *et seq.*

16. See, for example, *PBR, Inc. v. Secretary of Labor*, 643 F.2d 890, 896; 9 OSHC 1357, 1361 (1st Cir. 1981). The secretary of labor has acknowledged that under the Federal Railroad Safety Act the Department of Transportation (DOT) has *authority* to regulate all areas of employee safety for the railway industry. *Southern Railway Company v. OSAHRC* (7), 539 F.2d at 333; 3 OSHC at 1941. However, the scope of DOT's sta-

tutory authority to regulate matters relating to worker health is still unsettled. *Southern Pacific Transportation Company v. Usery* (7), 539 F.2d at 389; 4 OSHC at 1694, n. 3.

The Federal Railroad Safety Act is codified at 45 USC §§421 *et seq.* Other federal statutes dealing with railway safety include the Safety Appliance Acts, 45 USC §§1–16; the Train Brakes Safety Appliance Act, 45 USC §9; the Hours of Service Act, 45 USC §§61 *et seq.*; and the Rail Passenger Service Act, 45 USC §§501 *et seq.* Department of Transportation regulations relating to railway safety can be found in 49 CFR, Chapter II.

The Department of Transportation also has statutory authority to regulate safety in modes of transportation other than rail. For example, the secretary of transportation may prescribe requirements for "qualifications and maximum hours of service of employees of, and safety of operation and equipment of, a motor carrier." 49 USC §3102 (Revised Special Pamphlet 1992). For additional areas of DOT jurisdiction, see Sections 3.4–3.6 of this chapter.

17. See Section 2 of this chapter.

18. The provisions of the statute regarding safety regulation of civil aeronautics are codified at 49 USC §§1421 *et seq.*

19. See 49 USC §1432(a).

20. See decision in *Usery v. Northwest Orient Airlines, Inc.*, 5 OSHC 1617 (E.D. N.Y. 1977). ["E.D. N.Y." refers to the federal district court in the Eastern District of New York. Reported decisions of federal district courts dealing with job safety and health matters are reported in BNA's *Occupational Safety and Health Cases* (OSHC), and many are also reported in West Publishing Company's Federal Supplement (F. Supp.)]. The administrative law judge determined that the FAA had exercised its authority by requiring each air carrier to maintain a maintenance manual that must include all instructions and information necessary for its ground maintenance crews to perform their duties and responsibilities with a high degree of safety.

21. The HMTA is codified at 49 USC App. §§1801 *et seq.* A table of materials that have been designated as hazardous by the secretary can be found at 49 CFR Part 172. The secretary's regulations prescribe the requirements for shipping papers, package marking, labeling, and transport vehicle placarding applicable to the shipment and transportation of those hazardous materials.

22. 3 OSHC 1601 (Rev. Comm. 1975).

23. The secretary of transportation has exercised statutory authority under NGPSA to promulgate safety standards for employees at natural gas facilities. These regulations can be found at 49 CFR Part 192. The NGPSA is codified at 49 USC App. §§1671 *et seq.* See also *Columbia Gas of Pennsylvania, Inc. v. Marshall* (Ref. 9) (regulation by DOT requiring operators to take steps to minimize danger of accidental ignition of gas while employees were performing a "hot tap" on existing gas main while installing auxiliary natural gas pipeline preempted authority of OSHA over the matter).

24. There also exist thousands of state and local laws regulating noise, discussion of which is beyond the scope of this work. See *Compilation of State and Local Ordinances on Noise Control*, 115 *Cong. Rec.* 32178 (1969).

25. See 49 USC App. §1431. In carrying out this task the administrator of the FAA has adopted and published aircraft noise standards and regulations. See 14 CFR Part 36.

Section 7 of the Noise Control Act of 1972, discussed below, amended the noise abatement and control provision of the Federal Aviation Act to provide generally that standards adopted with respect to aircraft noise must have the prior approval of the administrator of the EPA. In addition, the Noise Control Act was amended by the Quiet Communities Act of 1978, P.L. 95-605, 92 Stat. 3079, to provide for, among other things, a unified effort among state, local, and federal authorities to develop an effective noise abatement control program with respect to aircraft noise associated with airports.

26. The Noise Control Act is codified at 42 USC §§4901 *et seq.* The products with which the administrator is statutorily authorized to deal must come from among four categories: construction equipment, transportation equipment (including recreational vehicles and related equipment), any motor or engine (including any equipment of which an engine or motor is an integral part), and electrical or electronic equipment.

The administrator has promulgated comprehensive regulations under the Noise Control Act, which appear at 40 CFR Subchapter G, Parts 201–211. To date these regulations cover interstate rail carriers (Part 201), motor carriers engaged in interstate commerce (Part 202), procedure and criteria for determining low-noise-emission products (Part 203), construction equipment including portable air compressors (Part 204), and transportation equipment, including medium and heavy trucks (Part 205).

Pursuant to Section 8 of the Noise Control Act, the administrator has also published product noise labeling requirements, which can be found at 40 CFR Part 211.

27. The Toxic Substances Control Act of 1976 is codified at 15 USC §§2601 *et seq.* Prior to 1976 there existed no enactment that authorized the federal government to regulate toxic substances generally. Although regulations published under OSHA did govern worker exposure to toxic substances in the workplace (see Section 5.3, this chapter), federal law lacked a general authority to prohibit or restrict the manufacture of such substances.

28. Section 7 of the act empowers the administrator of EPA to commence a civil action in federal court for the purpose of seizing an "imminently hazardous chemical substance," defined in the act as one that "presents an imminent and unreasonable risk of serious or widespread injury to health or the environment."

29. Polychlorinated biphenyls (PCBs) are the only group of chemicals with which the administrator is statutorily obligated to deal. See Section 6(e) of the act. For regulations concerning fully halogenated chlorofluoroalkanes, see 40 CFR Part 762.

30. The regulations issued by the EPA in 1979 dealing with PCBs were judicially reviewed by the U.S. Court of Appeals for the District of Columbia in *Environmental Defense Fund v. EPA*, 636 F.2d 1267 (D.C. Cir. 1980). In that proceeding the Environmental Defense Fund (EDF) challenged: (1) the determination by the EPA that certain uses of PCBs were "totally enclosed" and hence exempt from regulation under the Toxic Substances Control Act; (2) the applicability of the regulations to materials containing concentrations of PCBs greater than 50 parts per million (50 ppm); and (3) the decision of the EPA to authorize the continued availability of 11 nontotally enclosed uses of PCBs. The court upheld the regulations regarding the continued availability of the 11 nontotally enclosed PCB uses, but it set aside the regulations classifying certain PCB uses as "totally enclosed" and the 50-ppm cutoff figure for materials containing PCBs. In response to this decision the EPA revised and amended the Part 761 regulations on PCBs and published new regulations at 47 *Federal Register* 37342-60 (August 25, 1982). (The *Federal Register* is hereinafter cited as *Fed. Reg.*)

31. The Consumer Product Safety Act, as amended, is codified at 15 USC §§2051 *et seq.*
32. Provision is made in the act for suit by "any person who shall sustain injury by reason of any knowing (including willful) violation of a consumer product safety rule or any other rule or order issued by the Commission" against the responsible party in a federal district court. This right to sue is in addition to existing common law, federal, and state remedies.

    Comprehensive regulations promulgated by the Consumer Product Safety Commission can be found at 16 CFR Parts 1000 *et seq.*
33. The Hazardous Substances Act is codified at 15 USC §§1261–1276. Regulations published under this statute can be found at 16 CFR Subchapter C, Parts 1500 *et seq.*
34. The Atomic Energy Act is codified at 42 USC §§2011 *et seq.* The Nuclear Regulatory Commission, an independent executive commission, was created by the Energy Reorganization Act of 1974, 42 USC §5841(a), and all licensing and related regulatory functions of the Atomic Energy Commission were then transferred to the NRC. See 42 USC §5841(f) and (g). The NRC's "Standards for Protection Against Radiation" can be found at 10 CFR Part 20.
35. The permissible dosage per calendar quarter within such a "restricted area" is 1.25 rems to the whole body, head, and trunk, active bloodforming organs, lens of the eyes, or gonads; 18.75 rems to hands and forearms, feet and ankles; 7.5 rems to the skin of the whole body. Dosage standards are also set forth for the inhalation of radioactive substances. See 10 CFR §20.101–103. Detailed personnel monitoring and reporting requirements are also included in the regulations.
36. See 10 CFR §§20.105, 20.106, and 20.303.
37. See 40 CFR Part 190 *et seq.*
38. See 30 CFR Part 57 *et seq.* These regulations are revisions of regulations previously promulgated by the secretary of interior under the Metal and Non-Metallic Mine Safety Act, which was repealed by the Federal Mine Safety and Health Amendments Act of 1977 (see Section 3.1 of this chapter).
39. The Walsh–Healey Act is codified at 41 USC §§35 *et seq.* The relevant regulations can be found in 41 CFR §50-204. These standards were later promulgated by the secretary of labor as established federal standards under Section 6(a) of OSHA. See Section 5.2.1 of this chapter.
40. That statute amended the Public Health Service Act and is codified at (21 USC §§360gg-ss). The regulations promulgated by the Food and Drug Administration under this statute can be found at 21 CFR Parts 1000–1050 (Radiological Health). See also 21 CFR §1020.30, the standard applicable to diagnostic X-ray systems.
41. The Outer Continental Shelf Lands Act is codified at 43 USC §§1331 *et seq.*
42. See Ashford, *Crisis in the Workplace* (Ref. 2), pp. 47–51.
43. See D. Currie, "OSHA," *Am. Bar Found. Res. J.*, 1976, pp. 1107, 1111. Section 18(a) of OSHA explicitly directs, however, that the states may assert jurisdiction under state law with respect to occupational safety or health issues for which no federal standard is in effect. See Section 4.1 of this chapter.
44. A so-called complete plan is described in an administrative regulation issued by the secretary of labor and codified at 29 CFR §1902.3.
45. Pursuant to Section 18(c), the state plan must:
    a. Designate a state agency or agencies as the agency or agencies responsible for administering the plan throughout the state.

b. Provide for the development and enforcement of safety and health standards relating to one or more safety or health issues, which standards (and the enforcement of which standards) are or will be at least as effective in providing safe and healthful employment and places of employment as the standards promulgated under Section 6 of OSHA, which relate to the same issues.

c. Provide for a right of entry and inspection of all workplaces subject to this chapter, which is at least as effective as that provided in Section 8 of OSHA and include a prohibition on advance notice of inspections.

d. Contain satisfactory assurances that such agency or agencies have or will have the legal authority and qualified personnel necessary for the enforcement of such standards.

e. Give satisfactory assurances that such state will devote adequate funds to the administration and enforcement of such standards.

f. Contain satisfactory assurances that such state will, to the extent permitted by its law, establish and maintain an effective and comprehensive occupational safety and health program applicable to all employees of public agencies of the state and its political subdivisions, which program is as effective as the standards contained in an approved plan.

g. Require employers in the state to make reports to the secretary in the same manner and to the same extent as if the plan were not in effect.

h. Provide that the state agency will make such reports to the secretary in such form and containing such information as the secretary shall from time to time require.

The additional criteria outlined in the secretary's regulations can be found at 29 CFR §§1902.3 and 1902.4.

46. 29 CFR §1902.2(b).

47. 29 CFR §1954.3, 1954.10.

48. The commission has held that OSHA is not precluded from exercising its own enforcement authority during this period. *Par Construction Co., Inc.*, 4 OSHC 1779 (Rev. Comm. 1976); *Seaboard Coast Line Railroad Co. and Winston-Salem Southbound Railway Co.*, 3 OSHC 1767 (Rev. Comm. 1975).

49. A chart on the status of state plans can be found in BNA, *OSHR*, Reference File at 81:1003.

50. See P. Hamlar, "Operation and Effect of State Plans," *Proceedings of the American Bar Institute on Occupational Safety and Health Law*, ABA Press, Chicago, 1976, pp. 42–45.

51. See Section 303(e) of the Federal Mine Safety and Health Act of 1977.

52. See Section 205 of the Federal Railroad Safety Act.

53. An employer's general duty to provide a safe workplace may not necessarily be discharged simply by the employer's compliance with specific OSHA standards. In *UAW v. General Dynamics*, 815 F.2d 1570, 1577; 13 OSHC 1201, 1206–07 (D.C. Cir.), *certiorari denied*, 484 U.S. 976 (1987), the court stated:

[I]f . . . an employer knows a particular safety standard is inadequate to protect his workers against the specific hazard it is intended to address, or that the conditions in his place of employment are such that the safety standard will not adequately deal with the hazards to which his employees are exposed, he has a duty under section 5(a)(1) to take whatever measures may be required by the Act, over and above those mandated by the safety standard, to safeguard his workers. In sum, if an employer knows that a specific standard will not protect his workers against a particular hazard, his duty under section 5(a)(1) will not be discharged no matter how faithfully he observes that standard. Scienter is the key.

54. There is language in the legislative history of OSHA to indicate that the general duty clause merely restates the employer's common law duty to exercise reasonable care in providing a safe place for his employees to work. However, the courts have generally characterized such statements as "misleading." For example, in *REA Express v. Brennan*, 495 F.2d 822, 825; 1 OSHC 1651, 1653 (2d Cir. 1974), the Second Circuit could not "accept the proposition that common law defenses such as assumption of the risk or contributory negligence will exculpate the employer who is charged with violating the Act."

55. At a multiemployer construction worksite, this duty may extend to hazardous conditions that the employer neither creates nor fully controls, under what have become known as the *Anning-Johnson/Grossman* rules, unless the employer can show that it took realistic or reasonable measures to protect its employees or that it neither knew or reasonably could have known of the violation. *D. Harris Masonry Contracting, Inc. v. Dole*, 876 F.2d 343; 14 OSHC 1034 (3d Cir. 1989); *Dun-Par Engineered Form Co. v. Marshall*, 676 F.2d 1333, 1335–1336; 10 OSHC 1561, 1562 (10th Cir. 1982); *Electric Smith, Inc. v. Secretary of Labor*, 666 F.2d 1267, 1268–1270; 10 OSHC 1329, 1330–1332 (9th Cir. 1982); *DeTrae Enterprises, Inc. v. Secretary of Labor*, 645 F.2d 103, 104; 9 OSHC 1425, 1426 (2d Cir. 1981); *Bratton Corp. v. OSAHRC*, 590 F.2d 273, 275; 7 OSHC 1004, 1005 (8th Cir. 1979).

56. *National Realty and Construction Company, Inc. v. OSAHRC*, 489 F.2d 1257; 1 OSHC 1422 (D.C. Cir. 1973); *Brennan v. OSAHRC and Canrad Precision Industries*, 502 F.2d 946; 2 OSHC 1137 (3d Cir. 1974).

57. In *Dun-Par Engineered Form Co.*, 12 OSHC 1949, 1956 (Rev. Comm. 1986), the review commission modified the defense of impossibility to one of infeasibility ("We overrule Commission precedent that requires employers to prove that compliance with a standard is 'impossible' rather than 'infeasible.'"). In deciding the subsequent appeal, the Eighth Circuit upheld the commission on this point, although the court did reverse the commission's analysis with respect to allocation of the burden of proof following a showing of infeasibility. *Brock v. Dun-Par Engineered Form Co.*, 843 F.2d 1135; 13 OSHC 1652 (8th Cir. 1988).

58. See *Dole v. Williams Enterprises, Inc.*, 876 F.2d 186, 188; 14 OSHC 1001, 1003 (D.C. Cir. 1989); *Donovan v. Williams Enterprises, Inc.*, 744 F.2d 170, 178, n. 12; 11 OSHC 2241 (D.C. Cir. 1984); *True Drilling Co. v. Donovan*, 703 F.2d 1087, 1090; 11 OSHC 1310, 1311 (9th Cir. 1983); *Carlyle Compressor Co. v. OSAHRC*, 683 F.2d 673, 677; 10 OSHC 1700 (2d Cir. 1982); *PBR, Inc. v. Secretary of Labor*, 643 F.2d 890, 895; 9 OSHC 1357, 1360 (1st Cir. 1981).

59. 489 F.2d at 1266; 1 OSHC at 1427.

60. 489 F.2d at 1266–1267; 1 OSHC at 1427; see also *Capital Electric Line Builders of Kansas, Inc. v. Marshall*, 678 F.2d 128, 130; 10 OSHC 1593, 1594 (10th Cir. 1982); *H. B. Zachry Co. v. OSAHRC*, 638 F.2d 812, 818; 9 OSHC 1417, 1421–1422 (5th Cir. 1981); *General Dynamics Corp. v. OSAHRC*, 599 F.2d 453, 458; 7 OSHC 1373, 1375 (1st Cir. 1979). See also the discussion in Section 5.2.3 of this chapter.

61. See, for examples, *Brock v. L.E. Myers Co.*, 818 F.2d 1270; 13 OSHC 1289 (6th Cir.) (burden on employer to prove employee misconduct), *certiorari denied*, 484 U.S. 989 (1987); and *Pennsylvania Power & Light Co. v. OSHRC*, 737 F.2d 350, 11 OSHC 1985 (3d Cir. 1984) (burden on secretary).

62. *McLaughlin v. Union Oil of California*, 864 F.2d 1039, 1044; 13 OSHC 2033, 2036 (7th Cir. 1989); *Kelly Springfield Tire Co. v. Donovan*, 729 F.2d 317, 321; 11 OSHC 1889, 1891 (5th Cir. 1984); *Pratt & Whitney Aircraft v. Secretary of Labor*, 649 F.2d 96, 100; 9 OSHC 1554, 1557 (2d Cir. 1981); *Continental Oil Co. v. OSHRC*, 630 F.2d 446, 448; 8 OSHC 1980, 1981 (6th Cir. 1980), *certiorari denied*, 450 U.S. 965 (1981); *Brennan v. OSAHRC and Vy Lactos Laboratories*, 494 F.2d 460, 464; 1 OSHC 1623, 1625 (8th Cir. 1974); *National Realty* (Ref. 55), 489 F.2d at 1265 n. 32; 1 OSHC at 1426.

63. *National Realty* (Ref. 56); *Pratt & Whitney Aircraft* (Ref. 62); and *Kelly Springfield Tire* (Ref. 62).

64. 501 F.2d 504; 2 OSHC 1041 (8th Cir. 1974).

65. *Brennan v. OSAHRC and Vy Lactos Laboratories* (Ref. 62), 1 OSHC at 1624; R. Morey, ''The General Duty Clause of the Occupational Safety and Health Act of 1970,'' 86 *Harv. Law Rev.* 988, 991 (1973). The same is true for violations of the specific duty clause.

The fact that a violation is based on the existence of a hazard rather than an injury precludes an employer from arguing lack of ''proximate cause.'' In *Dye Construction Co. v. OSAHRC*, 698 F.2d 423; 11 OSHC 1104 (10th Cir. 1983), the court rejected a defense by an employer that the alleged hazard was not the proximate cause of an employee's injury. The court held that ''[t]he relevant inquiry is not the proximate cause of this particular accident but the risk of accident or injury as a result of the alleged violations and the seriousness of the potential injuries.'' 698 F.2d at 426; 11 OSHC at 1106.

66. *National Realty* (Ref. 56), 489 F.2d at 1265; 1 OSHC at 1426, n. 33; *Babcock & Wilcox Co. v. OSAHRC*, 622 F.2d 1160, 1165; 8 OSHC 1317, 1319 (3d Cir. 1980); *Illinois Power Co. v. OSAHRC*, 632 F.2d 25, 28; 8 OSHC 1512, 1514–1515 (7th Cir. 1980); *Titanium Metals Corp. of America v. Usery*, 579 F.2d 536, 541; 6 OSHC 1873, 1876–1878 (9th Cir. 1978). ''[T]he 'likely to cause' test should be whether reasonably foreseeable circumstances could lead to the perceived hazard's resulting in serious physical harm or death—or more simply, the proper test is plausibility, not probability.'' Morey, ''The General Duty Clause of the Occupational Safety and Health Act of 1970'' (Ref. 65), pp. 997–998.

In *Kelly Springfield Tire Co. v. Donovan* (Ref. 62), 729 F.2d at 324; 11 OSHC at 1894, the Fifth Circuit rejected an argument to change the ''likely to cause'' requirement in general duty cases to the more lenient ''significant risk of harm'' test in specific duty cases.

67. 29 CFR §1910.5(f).

68. An ''established Federal standard'' is defined in Section 3(10) of the act as ''any operative occupational safety and health standard established by any agency of the United States and presently in effect, or contained in any Act of Congress in force on the date of enactment of this Act.'' Section 4(b)(2) of the act listed several federal statutes from which established federal standards were to be derived, including the Walsh–Healey Act, 41 USC §§35–45, the Service Contract Act of 1965, 41 USC §§351–357, and the National Foundation on Arts and Humanities Act, 20 USC §§951–960.

A ''national consensus'' standard is defined in Section 3(9) of the act as any occupational safety and health standard, which "(1) has been adopted and promulgated by a nationally recognized standards producing organization under procedures, whereby it can be determined by the Secretary that persons interested and affected by the scope or provisions of the standard have reached substantial agreement on its adoption, (2) was formulated in a manner which afforded an opportunity for diverse views to be considered, and (3) has been designated as such a standard by the Secretary, after consultation with other appropriate Federal agencies." The principal sources for national consensus standards were the American National Standards Institute (ANSI) and the National Fire Protection Association. *American Federation of Labor v. Brennan*, 530 F.2d 109, 111 at n. 2; 3 OSHC 1820, 1821 at n. 2 (3d Cir. 1975).

69. The Administrative Procedure Act was enacted by Congress in 1946 to impose some coherent system of procedural regularity on the growing regulatory bureaucracy of the federal government. It provides procedures for administrative "rule making" and administrative "adjudication," among other things. See generally H. Linde and G. Bunn, *Legislative and Administrative Processes*, Foundation Press, Mineola, NY, 1976, p. 814. The Administrative Procedure Act is codified in 5 USC §§551 *et seq.*

The courts have held that OSHA does not have the right to change advisory national consensus standards ("should") to mandatory standards ("shall") upon adoption as OSHA standards without following formal rule-making procedures. Absent such rule making, citations issued to employers pursuant to these standards have been vacated. *Usery v. Kennecott Copper Corp.*, 577 F.2d 1113, 1117–1118; 6 OSHC 1197, 1199 (10th Cir. 1977); *Marshall v. Pittsburgh–Des Moines Steel Co.*, 584 F.2d 638, 644; 6 OSHC 1929, 1933 (3d Circ. 1978). See also *Marshall v. Anaconda Co.*, 596 F.2d 370, 376–377; 7 OSHC 1382, 1385–1386 (9th Cir. 1979).

70. *The Job Safety and Health Act of 1970*, BNA, Washington, D.C., 1971, p. 23.

71. See *Florida Peach Growers Association v. U.S. Department of Labor*, 489 F.2d 120, 124; 1 OSHC 1472 (5th Cir. 1974) and Sections 6(b)(1) to 6(b)(4) of OSHA. In *National Congress of Hispanic American Citizens v. Usery*, 554 F.2d 1196; 5 OSHC 1255 (D.C. Cir. 1977), the Court of Appeals for the District of Columbia held that the statutory deadlines in Sections 6(b)(1) to 6(b)(4) for the promulgation of permanent standards were discretionary rather than mandatory as long as the secretary's exercise of discretion was honest and fair.

72. *Florida Peach Growers Association* (Ref. 71), 489 F.2d at 124; 1 OSHC at 1475; see also *Industrial Union Dept. AFL–CIO v. American Petroleum Institute*, 448 U.S. 607, 651 n. 59; 8 OSHC 1586, 1602 (1980).

73. The "substantial evidence" standard of judicial review is traditionally conceived of as suited to adjudication or formal rule making. OSHA, however, calls for informal rule making, which under the Administrative Procedure Act generally entails judicial review pursuant to the less stringent "arbitrary and capricious" test. This apparent anomaly can be explained historically as a legislative compromise. The Senate OSHA bill called for informal rule making, but the House version specified formal rule making and substantial evidence review. The House receded on the procedure for promulgating standards, but the substantial evidence standard of review was adopted. *Industrial Union Department, AFL–CIO v. Hodgson* (Ref. 4), 499 F.2d at 473; 1 OSHC at 1635. For a more detailed discussion of these legislative events, see *Associated Industries of New York State, Inc. v. U.S. Department of Labor*, 487 F.2d 342, 1 OSHC 1340 (2d Cir. 1973).

74. See B. Fellner and D. Savelson, "Review by the Commission and the Courts," *Pro-*

ceedings of the American Bar Association Institute on Occupational Safety and Health Law, ABA Press, Chicago, 1976, pp. 113–114. This approach has been summarized as one requiring the reviewing court to determine whether the agency (1) acted within the scope of its authority; (2) followed the procedures required by statute and by its own regulations; (3) explicated the bases for its decision; and (4) adduced substantial evidence in the record to support its determination. *United Steelworkers of America, AFL–CIO, v. Marshall and Bingham*, 647 F.2d 1189, 1206; 8 OSHC 1810, 1816 (D.C. Cir. 1980), *certiorari denied*, 453 U.S. 913 (1981). See also *Texas Independent Ginners Assoc. v. Marshall*, 630 F.2d 398, 404–405; 8 OSHC 2205, 2209–2210 (5th Cir. 1980); *American Iron and Steel Institute v. OSHA*, 577 F.2d 825, 830–31; 6 OSHC 1451, 1455 (3d Cir. 1978); *Society of the Plastics Industry, Inc. v. OSHA*, 509 F.2d 1301, 1304; 2 OSHC 1496, 1498 (2d Cir. 1975).

75. *Atlantic & Gulf Stevedores, Inc. v. OSAHRC*, 534 F.2d 541; 4 OSHC 1061 (3d Cir. 1976); *Arkansas–Best Freight Systems, Inc. v. OSAHRC and Secretary of Labor*, 529 F.2d 649; 3 OSHC 1910 (8th Cir. 1976); *Deering Milliken, Inc. v. OSAHRC*, 630 F.2d 1094, 1099; 9 OSHC 1001, 1004 (5th Cir. 1980); *Marshall v. Union Oil Co. and OSAHRC*, 616 F.2d 1113, 1117–1118; 8 OSHC 1169, 1173 (9th Cir. 1980); *Daniel International Corp. v. OSAHRC and Secretary of Labor*, 656 F.2d 925, 928–930; 9 OSHC 2102, 2104–2106 (4th Cir. 1981). But see *National Industrial Contractors v. OSAHRC*, 583 F.2d 1048, 1052–1053; 6 OSHC 1914, 1916–1917 (8th Cir. 1978), in which the Eighth Circuit held that procedural challenges to an OSHA standard must be brought in a pre-enforcement proceeding pursuant to Section 6(f) within 60 days from the challenged standard's effective date.

76. *Atlantic & Gulf Stevedores* (Ref. 75), 534 F.2d at 550–552; 4 OSHC at 1067–1068.

77. 452 U.S. 490; 9 OSHC 1913 (1981).

78. 452 U.S. at 513, n. 31; 9 OSHC at 1922, n. 31. Section 3(8) of OSHA contains the general definition of an occupational safety and health standard. It provides as follows:

> The term "occupational safety and health standard" means a standard which requires conditions, or the adoption or use of one or more practices, means, methods, operations, or processes, *reasonably necessary or appropriate* to provide safe or healthful employment and places of employment. (Emphasis added.)

For standards dealing with toxic materials or harmful physical agents, Section 6(b)(5) imposes the following additional requirements:

> The Secretary, in promulgating standards dealing with toxic materials or harmful physical agents under this subsection, shall set the standard which most adequately assures, *to the extent feasible*, on the basis of the best available evidence, that no employee will suffer material impairment of health or functional capacity, even if such employee has regular exposure to the hazard dealt with by such standard for the period of his working life. (Emphasis added.)

79. *Industrial Union Department, AFL–CIO v. Hodgson* (Ref. 4), 499 F.2d at 478; 1 OSHC at 1639, cited approvingly by the Supreme Court in *American Textile Mfrs. v. Donovan* (Ref. 79), 452 U.S. at 513, n. 31; 9 OSHC at 1922, n. 31. In *Industrial Union Department*, the Court of Appeals for the District of Columbia applied Section 6(b)(5) in a case challenging OSHA's standard for exposure to asbestos dust and held that the secretary, in promulgating the standard, could properly consider problems of both economic and technological feasibility.

80. *American Textile Mfrs. v. Donovan* (Ref. 77), 452 U.S. at 530, n. 55; 9 OSHC at 1928–1929, n. 55; *United Steelworkers of America, AFL–CIO v. Marshall and Bingham* (Ref. 74), 647 F.2d at 1265; 8 OSHC at 1864.

81. *American Textile Mfrs. v. Donovan* (Ref. 77), 452 U.S. at 530, n. 55; 9 OSHC at 1928–1929, n. 55.

82. *American Textile Mfrs. v. Donovan* (Ref. 77), 452 U.S. at 509. *American Textile Mfrs.* involved a challenge by the textile industry to OSHA's standard governing occupational exposure to cotton dust. The industry contended, among other things, that the act required OSHA to demonstrate that its standard reflected a reasonable relationship between the costs and benefits associated with the standard. The Supreme Court rejected that argument and upheld the validity of the entire cotton dust standard except for a wage guarantee requirement for employees who are transferred to another position when they are unable to wear a respirator.

83. *American Textile Mfrs. v. Donovan* (Ref. 77), 452 U.S. at 513, n. 32. In *Donovan v. Castle & Cooke Foods and OSAHRC*, 692 F.2d 641; 10 OSHC 2169 (9th Cir. 1982), the Ninth Circuit held that the Supreme Court's holding concerning cost–benefit analysis in the *American Textile Manufacturers* case applied only to standards promulgated under Section 6(b) of OSHA and did not apply to standards promulgated under Section 6(a), such as the noise standard at issue in *Castle & Cooke*, which had been originally promulgated as an established federal standard under the Walsh–Healey Act. But compare *Sun Ship, Inc.*, 11 OSHC 1028 (Rev. Comm. 1982), where the Review Commission applied the reasoning of *American Textile Manufacturers* to the noise standard and rejected the application of cost–benefit analysis in enforcing that standard. See BNA, *OSHR*, Current Report for February 3, 1983, pp. 735–736.

84. *American Textile Mfrs. v. Donovan* (Ref. 77), 452 U.S. at 509, n. 29.

85. Ashford, *Crisis in the Workplace* (Ref. 2), p. 169.

86. 511 F.2d 1139; 2 OSHC 1646 (9th Cir. 1975).

87. 511 F.2d at 1144–1145; 2 OSHC at 1650–1651. See also *Daniel International Corp. v. OSAHRC and Secretary of Labor*, 683 F.2d 361; 10 OSHC 1890 (11th Cir. 1982).

88. 520 F.2d 1011; 3 OSHC 1461 (7th Cir. 1975).

89. Accord, *H. B. Zachry Company v. OSAHRC*, 638 F.2d 812; 9 OSHC 1417 (5th Cir. 1981) (defense of employee negligent misconduct fails because of the employer's inability to establish to the satisfaction of the fact-finder that it effectively communicated and enforced work rules, which were necessary to ensure compliance with OSHA standards).

90. *Atlantic & Gulf Stevedores* (Ref. 75), 534 F.2d at 555; 4 OSHC at 1068–1069. See Note, "Employee Noncompliance with OSHA Safety Standards," 90 *Harv. Law Rev.* 1041 (1977).

91. 528 F.2d 564, 570; 3 OSHC 2060, 2064 (5th Cir. 1976).

92. 42 USC §4332.

93. 29 CFR §11.10(a)(3).

94. 29 CFR §11.10(a)(1).

95. Health and safety standards promulgated by the secretary of labor pursuant to OSHA can be found at 29 CFR Part 1910. A standards digest (OSHA Publication 2201) outlining the basic applicable standards is published in BNA, *OSHR*, Reference File at 31:4001.

Federal health and safety standards for the construction industry were initially promulgated under the Contract Work Hours and Safety Standards Act, 40 USC §§327 *et seq.* These standards were incorporated by reference under OSHA, are enforceable under both laws, and can be found at 29 CFR Part 1926. A standards digest (OSHA Publication 2202) outlining the basic applicable construction standards is published in BNA, *OSHR*, Reference File at 31:3001. Health and safety standards for ship repairing, shipbuilding, shipbreaking, and longshoring were initially promulgated pursuant to the Longshoremen's and Harbor Worker's Compensation Act, 33 USC §§901 *et seq.* These standards were incorporated by reference by OSHA, are enforceable under both laws, and can be found in 29 CFR Parts 1915 and 1918. Health and safety standards originally promulgated under the Walsh–Healey Public Contracts Act, the McNamara–O'Hara Service Contract Act of 1965, and the National Foundation on the Arts and Humanities Act of 1965 can be found in 41 CFR Part 50-204, 29 CFR Part 1516, and 20 CFR Part 505. These were adopted and are enforceable by OSHA. Standards promulgated under the aforementioned statutes will be superseded if corresponding standards that are promulgated under OSHA are determined by the secretary of labor to be more effective. See Section 4(b)(2) of OSHA.

Federal health and safety standards for coal mines were promulgated by the Department of Interior pursuant to the federal Coal Mine Health and Safety Act of 1969. Standards under the 1969 act were adopted without change by the Federal Mine Safety and Health Act of 1977. CCH, *Empl. Safety and Health Guide* ¶¶5924, 5931. Health standards for underground coal mines can be found in 30 CFR Part 70; health standards for surface work areas of underground coal mines and surface coal mines are codified in 30 CFR Part 71. Requirements for approval of coal mine dust personal sampler units designed to determine the concentrations of respirable dust in coal mine atmospheres can be found in 30 CFR Part 74; health standards for coal miners with evidence of pneumoconiosis are codified in 30 CFR Part 90. The safety standards for underground coal mines can be found in 30 CFR Part 75, and the safety standards for surface coal mines and surface work areas of underground coal mines are codified in 30 CFR Part 77.

The secretary of interior promulgated health and safety standards for metal and nonmetallic mines pursuant to the federal Metal and Non-Metallic Mine Safety Act of 1966. Mandatory standards adopted under the 1966 act were adopted without change by the Federal Mine Safety and Health Act of 1977. However, advisory metal and nonmetallic standards did not become mandatory standards under the 1977 act. A committee later reviewed these advisory standards and made recommendations for conversion to mandatory standards, which MSHA accepted in August 1979 by converting scores of noncoal advisory standards to mandatory standards. CCH, *Empl. Safety and Health Guide* ¶¶5924, 5931. The standards for surface metal and nonmetal mines and for underground metal and nonmetal mines are found at 30 CFR Parts 56 and 57, respectively.

96. Safety standards generally focus on the time that an employee is actually working. The harm created by a safety hazard is generally immediate and violent. An occupational health hazard, on the other hand, is slow acting, cumulative, irreversible, and complicated by nonoccupational factors. Ashford, *Crisis in the Workplace* (Ref. 2), pp. 68–83.

97. For a listing of the initial package of national consensus and established federal standards, see 36 *Fed. Reg.* 10466–10714 (1971).

98. Industries covered by specific OSHA regulations include pulp, paper and paperboard mills, textiles, bakery equipment, laundry machinery and operations, sawmills, pulpwood logging, telecommunications, agriculture, grain handling facilities, commercial diving, construction, ship repairing, shipbuilding, shipbreaking, marine terminal employment, and longshoring. The mining industry is subject to comprehensive safety regulations issued pursuant to the Federal Mine Safety and Health Act of 1977 (formerly the Federal Coal Mine Health and Safety Act of 1969 and the Federal Metal and Non-Metallic Mine Safety Act of 1966). See also Section 3 of this chapter.

99. The TLVs had been developed principally in 1968 by the American Conference of Governmental Industrial Hygienists (ACGIH) and were subsequently incorporated into the Walsh–Healey Act. See Ashford, *Crisis in the Workplace* (Ref. 2), p. 154. The secretary of labor did not include the ACGIH's carcinogen standards in his Section 6(a) package, but instead preferred to develop his own standards regarding carcinogens. Ashford, pp. 154, 247–48.

100. 54 *Fed. Reg.* 2332 (January 19, 1989).

101. The PELs established by OSHA can be found at 29 CFR §1910.1000, Tables Z-1-A, Z-2, and Z-3. Pursuant to the January 1989 rule making, OSHA made 212 PELs more protective, established 162 new PELs for previously unregulated substances, and left other PELs unchanged. In so doing, OSHA stated that it relied heavily on the widely accepted 1987–1988 TLVs published by ACGIH and the Recommended Exposure Limits (RELs) developed by the National Institute for Occupational Safety and Health (NIOSH). 54 *Fed. Reg.* 2332, 2333–35 (January 19, 1989).

102. 54 *Fed. Reg.* 2332, 2335 (January 19, 1989); BNA, *OSHR*, Current Report for January 18, 1989, p. 1475.

103. BNA, *OSHR*, Current Reports for July 15, 1992, pp. 219–221 and November 4, 1992, p. 1145.

104. 57 *Fed. Reg.* 26002 (June 12, 1992).

105. 49 *Fed. Reg.* 5318 (1984).

106. BNA, *OSHR*, Current Report for March 17, 1983, p. 894 (stating that OSHA intended to issue general duty clause citations for serious hazards to replace the revoked standards).

107. 51 *Fed. Reg.* 22612 (June 20, 1986). The original asbestos standard was issued on June 7, 1972, in a Section 6(b) rule making that retained the emergency temporary standard and limited the 8-hr time-weighted average (TWA) of airborne concentration of asbestos to 5 fibers greater than 5 μm in length per milliliter of air (the ''five-fiber standard'') for 4 years, then required 2 fibers. 37 *Fed. Reg.* 11322 (1971). In *Industrial Union Department, AFL-CIO v. Hodgson*, 449 F.2d 467; 1 OSHC 1631 (D.C. Cir. 1974), the Court of Appeals for the District of Columbia upheld the standard with two exceptions. OSHA had to reconsider the effective date of the two-fiber standard and the standard's record-keeping provision. In response, OSHA initiated a new rule-making proceeding and issued a proposed revised asbestos standard in 1975, but this proposal was never promulgated. OSHA issued an emergency temporary standard for asbestos in 1983 [48 *Fed. Reg.* 51086 (1983)], and this was vacated by the Fifth Circuit in 1984. *Asbestos Information Ass'n v. OSHA*, 727 F.2d 415; 11 OSHC 1817 (5th Cir. 1984).

108. The general industry standard is codified at 29 CFR §1910.1001; the construction standard is codified at 29 CFR §1926.58.

109. 29 CFR §1910.1001(b). Shortly after issuing the rule, however, the agency partially stayed the revised standard insofar as it applies to "nonasbestiform" tremolite, actinolite, and anthophyllite. 51 *Fed. Reg.* 37002 (October 17, 1986). During the stay, the 1972 standard applied to the minerals. In June 1992, the agency announced it would no longer regulate nonasbestiform minerals under the asbestos standards, stating that substantial evidence was lacking to demonstrate that the minerals pose the same health hazards as asbestos. OSHA intends to regulate the nonasbestiform minerals under the portion of the air contaminants standard (in the Z-1-A Table) that covers particulates not otherwise regulated. The particulates standard is less stringent compared to the asbestos standard. 57 *Fed. Reg.* 24310 (June 8, 1992); BNA, *OSHR*, Current Report for June 10, 1992, p. 55.

110. 29 CFR §1910.1001(c)(1).

111. *Building & Construction Trades Dep't, AFL–CIO v. Secretary of Labor*, 838 F.2d 1258 (D.C. Cir. 1989).

112. The STEL is now codified at 29 CFR §1910.1001(c)(2).

113. BNA, *OSHR*, Current Report for November 8, 1989, p. 1062.

114. 55 *Fed. Reg.* 29712 (July 20, 1990).

115. BNA, *OSHR*, Current Report for January 16, 1991, p. 1246.

116. BNA, *OSHR*, Current Report for January 8, 1992, p. 1142.

117. See *Society of the Plastics Industry v. OSHA* (Ref. 74), 509 F.2d at 1306; 2 OSHC at p. 1500.

118. The standard can be found at 29 CFR §1910.1017.

119. *Society of the Plastics Industry v. OSHA* (Ref. 74), 509 F.2d at 1311; 2 OSHC at 1504.

120. 29 CFR §1910.1017(h)(1)(ii).

121. 29 CFR §1910.1017(j).

122. 29 CFR §1910.1017(l).

123. 29 CFR §1910.1017(k).

124. 29 CFR §1910.1017(m).

125. See *Dry Color Manufacturers' Association, Inc. v. U.S. Department of Labor*, 486 F.2d 98; 1 OSHC 1331, 1332 (3d Cir. 1973). The 14 chemicals included in the carcinogen standards were 4-nitrobiphenyl, alpha-naphthylamine, 4,4'-methylene bis (2-chloroaniline), methyl chloromethyl ether, 3,3'-dichlorobenzidine (and its salts), bis-chloromethyl ether, beta-naphthylamine, benzidine, 4-aminodiphenyl, ethyleneimine, beta-propiolactone, 2-acetylaminofluorene, 4-dimethylaminoazobenzene, and N-nitrosodimethylamine. The permanent standards were codified at 29 CFR §§1910.1003–1910.1016.

126. 500 F.2d 385; 2 OSHC 1402 (3d Cir. 1974), *certiorari denied*, 423 U.S. 830 (1975).

127. 29 CFR §1910.1005 was deleted in 41 *Fed. Reg.* 35184 (August 20, 1976).

128. OSHA's Cancer Policy is found at 29 CFR Part 1990.

129. CCH, *Empl. Safety and Health Guide* ¶1031.

130. 48 *Fed. Reg.* 242 (January 4, 1983); see also CCH, *Empl. Safety and Health Guide* ¶1031.

131. BNA, *OSHR*, Current Report for December 23, 1987.

132. The text of the coke ovens standard can be found at 29 CFR §1910.1029.

133. 29 CFR §1910.1029(h). This includes, but is not limited to, flame-resistant pants, jackets, gloves, footwear, and face shields.

134. 29 CFR §1910.1029(i)(1).

135. 29 CFR §1910.1029(i)(3).

136. 29 CFR §1910.1029(i)(5).

137. 29 CFR §1910.1029(k).

138. 29 CFR §1910.1029(l).

139. 29 CFR §1910.1029(j).

140. 29 CFR §1910.1029(m).

141. 577 F.2d 825; 6 OSHA 1451 (3d Cir. 1978), *certiorari dismissed*, 448 U.S. 917 (1980).

142. CCH, *Empl. Safety and Health Guide* ¶1198. Before issuance of the separate standard, occupational exposure to lead was regulated pursuant to the standard on air contaminants.

143. The lead standard can be found at 29 CFR §1910.1025. For a detailed early history of the standard, see *United Steelworkers of America, AFL–CIO v. Marshall*, 647 F.2d 1189; 8 OSHC 1810 (D.C. Cir. 1980), *certiorari denied*, 453 U.S. 913 (1981).

144. *United Steelworkers of America v. Marshall* (Ref. 143).

145. For a listing of the industries as to which the standard was remanded, as well as for a summary of the court's order, see *United Steelworkers of America v. Marshall* (Ref. 143), 647 F.2d at 1311; 8 OSHC at 1901–1902.

146. See BNA, *OSHR*, Current Report for December 17, 1981, pp. 539–540.

147. 46 *Fed. Reg.* 60758 (December 11, 1981).

148. 54 *Fed. Reg.* 29142 (July 11, 1989).

149. 55 *Fed. Reg.* 3146 (January 30, 1990). OSHA set the standard for small foundries at 75 μg and at 50 μg for large foundries.

150. CCH, *Empl. Safety and Health Guide* ¶1198. Those six industries were: nonferrous foundries, secondary copper smelters, brass and bronze ingot manufacturers, independent collectors and processors of scrap lead (including independent battery breakers), leaded steelmaking operations, and lead chemical manufacturers.

151. BNA, *OSHR*, Current Report for July 24, 1991, pp. 228–229. The court remanded the standard as to the brass and bronze industry for an economic feasibility determination of the 50-μg standard.

152. Cotton that has been washed in accordance with outlined processes is exempt from the standard. 29 CFR §1910.1043(n). Some forms of cotton dust exposure not covered by the separate cotton dust standard are regulated pursuant to the air contaminants standard, 29 CFR §1910.1000.

153. The cotton dust standard is found at 29 CFR §1910.1043. The original standard was generally upheld by the Supreme Court in *American Textile Mfrs. Inst. v. Donovan*, 452 U.S. 490; 9 OSHC 1913 (1981).

154. 29 CFR §1910.1043(c)(1).

155. 29 CFR §1910.1043(c)(2).

156. 29 CFR §1910.1043(e).

157. 29 CFR §1910.1043(f), (h), (i) and (j).

158. CCH, *Empl. Safety and Health Guide* ¶1202.

159. The DBCP standard can be found at 29 CFR §1910.1044.

160. CCH, *Empl. Safety and Health Guide* ¶1203.

161. The acrylonitrile standard can be found at 29 CFR §1910.1045.

162. See CCH, *Empl. Safety and Health Guide* ¶1203.

163. *Industrial Union Dept. v. American Petroleum Institute*, 448 U.S. 607; 8 OSHC 1586 (1980). OSHA was required to review other standards (e.g., the arsenic standard) in light of this Supreme Court decision.

164. 52 *Fed. Reg.* 34460 (September 11, 1987). The benzene standard was codified at 29 CFR §1910.1028.

165. The 1-ppm limit is identical to that set in the 1978 standard. The 1987 standard, however, is followed by a risk assessment analysis from which OSHA determined the benzene level that posed a significant risk. 52 *Fed. Reg.* 34460 (September 11, 1987). The 1978 standard was based on a feasibility determination after OSHA simply assumed that no level of benzene exposure would be safe for human. 43 *Fed. Reg.* 5918, 5946–47 (1978).

166. 29 CFR §1910.1028(i).

167. See CCH, *Empl. Safety and Health Guide* ¶1199.

168. The arsenic standard can be found at 29 CFR §1910.1018.

169. There is no ceiling limit because of technical inability to control occasional exceedances. See CCH, *Empl. Safety and Health Guide* ¶1197.

170. 647 F.2d 1; 9 OSHC 1508 (9th Cir. 1981).

171. 48 *Fed. Reg.* 1864 (January 14, 1983).

172. The arsenic standard was validated in *Asarco, Inc. et al. v. OSHA*, 746 F.2d 483 (9th Cir. 1984).

173. CCH, *Empl. Safety and Health Guide* ¶1197.

174. 49 *Fed. Reg.* 25734 (June 22, 1984). The ethylene oxide standard is codified at 29 CFR §1910.1047.

175. *Public Citizen Health Research Group v. Tyson*, 796 F.2d 1479; 12 OSHC 1905 (D.C. Cir. 1986).

176. 53 *Fed. Reg.* 1724 (January 8, 1988). The STEL appears at 29 CFR §1910.1047(c)(2).

177. CCH, *Empl. Safety and Health Guide* ¶1204. The standard prohibits rotating employees in and out of exposure areas as a means of exposure control, because of the risk to other employees.

178. BNA, *OSHR*, Current Report for July 22, 1992, p. 245. The study analyzed the use of ethylene oxide in hospitals, which is the largest industry sector affected because ethylene oxide is used as a sterilant.

179. 52 *Fed. Reg.* 46168 (December 4, 1987). The standard appears at 29 CFR §1910.1048.

180. 57 *Fed. Reg.* 22290 (May 27, 1992).

181. *United Auto Workers v. Pendergrass*, 878 F.2d 389; 14 OSHC 1025 (D.C. Cir. 1989).

182. CCH, *Empl. Safety and Health Guide* ¶1205.

183. BNA, *OSHR*, Current Report for June 3, 1992, p. 7.

184. 56 *Fed. Reg.* 64004 (December 6, 1991). The standard is codified at 29 CFR §1910.1030.

185. BNA, *OSHR*, Current Report for December 11, 1991, p. 875.

186. BNA, *OSHR*, Current Report for May 13, 1992, p. 1628. Litigation is pending.

187. The hazard communication standard for general industry is codified at 29 CFR §1910.1200.

188. *United Steel Workers of America v. Auchter*, 763 F.2d 728; 12 OSHC 1337 (3d Cir. 1985).

189. 52 *Fed. Reg.* 31852 (August 24, 1987).

190. *Associated Builders and Contractors, Inc. v. OSHA*, 862 F.2d 63; 13 OSHC 1945 (3d Cir. 1988).

191. CCH, *Empl. Safety and Health Guide* ¶1301.

192. The noise standard can be found at 29 CFR §1910.95.

193. 29 CFR §1910.95(c)–(p). The amendment exempts employers engaged in oil and gas well drilling and service operations. 29 CFR §1910.95(o).

194. 29 CFR §1910.95(g). The agency review commission has ruled that the action level of 85 decibels excludes "impulse noise," defined as sharp noise peaks lasting less than one second and spaced more than one second apart. *Collier-Keyworth Co.*, 13 OSHC 1208, *remanded* 13 OSHC 1269 (Rev. Comm. 1987), *on remand* 13 OSHC 1940 (Rev. Comm. 1988).

195. 29 CFR §1910.95(d)(1)(ii).

196. 29 CFR §1910.95(i)(1), (2)(ii)(B). Notwithstanding the fact that an employer provides hearing protectors, employers may receive citations for failing to ensure their use. See, for example, *United States Container Corp.*, 13 OSHC 1415 (Rev. Comm. 1987).

197. *Forging Indus. Ass'n v. Donovan*, 773 F.2d 1436 (4th Cir. 1985).

198. 53 *Fed. Reg.* 26437 (July 13, 1988).

199. BNA, *OSHR*, Current Report for April 29, 1992, p. 1558.

200. 56 *Fed. Reg.* 57036 (November 7, 1991).

201. BNA, *OSHR*, Current Report for June 10, 1992, p. 55.

202. BNA, *OSHR*, Current Report for January 15, 1992, p. 1148.

203. 57 *Fed. Reg.* 34192 (August 3, 1992).

204. Prior to the ANPRM for ergonomics in general industry, in August 1990 OSHA issued ergonomics guidelines for meatpacking plants.

205. *Atlantic & Gulf Stevedores* (Ref. 75), 534 F.2d at 553; 4 OSHC at 1069–1070.

206. *Dunlop v. Trumbull Asphalt Company, Inc.*, 4 OSHC 1847 (E.D. Mo. 1976).

207. See *Powell v. Globe Industries, Inc.*, 431 F. Supp. 1096; 5 OSHC 1250 (N.D. Ohio 1977). The National Labor Relations Board (NLRB) has concurrent jurisdiction over Section 11(c) cases. In 1975 the general counsel of the NLRB and the secretary of labor entered into an understanding for the procedural coordination of litigation arising under Section 11(c) of OSHA and Section 8 of the National Labor Relations Act, to avoid duplicate litigation. See J. Irving, "Effect of OSHA on Industrial Relations and Collective Bargaining," in *Proceedings of the ABA National Institute on Occupational Safety and Health Law*, ABA Press, Chicago, 1976, pp. 125–127. See also, 40 *Fed. Reg.* 26083 (June 20, 1976).

208. The general antiretaliation provision in the 1977 Mine Safety and Health Act is set forth in Section 105(c) of the statute. The 1977 mine act also provides for immediate inspection of a coal mine at the request of a miner [§103(g)], the right of employees to accompany the inspector on his walk-around inspection of the coal mine [§103(f)], and limited payments to miners when a safety violation closes the mine [(§111)]. Black lung (coal worker's pneumoconiosis) benefits are provided to totally disabled coal miners and surviving dependents of coal miners whose deaths were due to black lung disease [§§401 *et seq.*].

209. 29 CFR §1977.12(b)(2).

210. 445 U.S. 1; 8 OSHC 1001 (1980).

211. 445 U.S. at 10-11; 8 OSHC at 1004.

212. With respect to work stoppages over safety disputes in the context of collective bargaining agreements and the National Labor Relations Act, see *Gateway Coal Company v. United Mine Workers*, 414 U.S. 368; 1 OSHC 1461 (1974) and *National Labor Relations Board v. Tamara Foods, Inc.*, 692 F.2d 1171 (8th Cir. 1982).

213. See, for example, *Skidmore v. Travelers Ins. Co.*, 483 F.2d 67; 1 OSHC 1294 (5th Cir. 1973). See also, *Taylor v. Brighton Corp.*, 616 F.2d 256; 8 OSHC 1010 (6th Cir. 1980).

CHAPTER EIGHTEEN

Compliance and Projection

Martha Hartle Munsch, J. D., and Jacqueline A. Koscelnik, J. D.

1 INVESTIGATIONS AND INSPECTIONS

With the enactment of the Occupational Safety and Health Act of 1970 (OSHA), Congress authorized the secretary of labor to enter, inspect, and investigate places of employment to discover possible violations of the employer's general and specific duties under the act.

Section 8(a) authorizes the secretary, upon presenting appropriate credentials to the owner, operator, or agent in charge:

1. To enter without delay and at reasonable times any factory, plant, establishment, construction site, or other area, workplace or environment where work is performed by an employee or employer.

2. To inspect and investigate during regular working hours and at other reasonable times, and within reasonable limits and in a reasonable manner, any such place of employment and all pertinent conditions, structures, machines, apparatus, devices, equipment, and materials therein, and to question privately any such employer, owner, operator, agent, or employee.

The Federal Mine Safety and Health Act of 1977 directs authorized representatives of the secretary to make frequent, unannounced inspections and investigations in coal or other mines each year. The purposes of these visits include determining whether an imminent danger exists and whether there is compliance with the mandatory health and safety standards issued under that statute (1).

Patty's Industrial Hygiene and Toxicology, Third Edition, Volume 3, Part A, Edited by Robert L. Harris, Lewis J. Cralley, and Lester V. Cralley.
ISBN 0-471-53066-2 © 1994 John Wiley & Sons, Inc.

1.1 Inspection Procedures—Warrants

Section 8(a) of OSHA on its face gives the secretary of labor the unqualified right to enter and inspect, in a ''reasonable manner,'' any place of employment upon the presentation of credentials. The United States Supreme Court in 1978 held Section 8(a) unconstitutional insofar as it authorized nonconsensual warrantless inspections at an employer's establishment. In *Marshall v. Barlow's, Inc.* (2), the Court held that an employer may refuse entry to an OSHA compliance officer unless a warrant is obtained, reasoning that employers have a reasonable expectation of privacy in their commercial property and are therefore guaranteed the right, under the Fourth Amendment, to be free from unreasonable official intrusions (3).

The Supreme Court explained, however, that the secretary need not make a showing of probable cause in the criminal sense to obtain a warrant. Instead, probable cause authorizing an administrative inspection may be based either on specific evidence of a violation or on a showing that reasonable legislative or administrative standards for conducting an inspection are satisfied with respect to a particular establishment (4). By way of illustration of the latter basis, the Court stated that a warrant could properly be issued upon a showing that a particular business was chosen for inspection pursuant to a general enforcement plan derived from ''neutral sources'' (5).

An inspection may proceed without a warrant, of course, if the employer or an authorized third party consents to the inspection (6). It is OSHA's policy to seek preinspection warrants only when it is likely that employers will not allow OSHA inspectors access to areas or records (7).

1.1.1 Probable Cause Necessary for Issuance of Warrants

Following the *Barlow's* decision, federal courts have addressed the question of what constitutes a showing of probable cause necessary to obtain an OSHA inspection warrant. The Supreme Court stated in *Barlow's* that probable cause authorizing an OSHA inspection may be based on evidence that a specific violation exists at the establishment to be inspected. Thus, probable cause for warrant issuance may be established by OSHA's receipt of an employee complaint (8), although courts of appeals have held that the nature of the violation complained of must be described in the warrant application so the magistrate issuing the warrant may make an independent determination that probable cause exists (9). In addition, the majority of federal courts addressing the issue have concluded that a warrant based on specific employee complaints is overly broad if it purports to authorize a ''wall-to-wall'' inspection of the entire plant. Instead, the scope of the warrant and resulting inspection must bear a reasonable relationship to the specific violations complained of (10).

There appears to be some question whether a past history of OSHA violations satisfies the requirement of probable cause for issuance of a warrant. The Second Circuit Court of Appeals has upheld the issuance of a warrant based on an employer's record of past violations (11). The Seventh Circuit has upheld a warrant based on a showing that the employer had been cited for a violation at an old plant and that the employer had tried to abate the violation by moving the unsafe operations to a new plant (12). The same court, however, has stated that a history of past violations, standing alone, is insufficient to establish probable cause (13).

Finally, the Supreme Court in *Barlow's* stated that an inspection warrant could be issued pursuant to an inspection plan based on "neutral criteria." Under this standard, the warrant application must describe the administrative plan being used, so the magistrate can make an independent determination that it is based on "neutral criteria" (14). Thus, warrant applications describing OSHA's general inspection program, in which employers are randomly chosen for inspection from a list of firms in industries with above-average lost workday rates, have been held to satisfy probable cause (15).

In addition, the Seventh Circuit has held that an inspection warrant was properly issued on the basis of OSHA's National Emphasis Program targeting high-hazard industries, even without evidence in the warrant application showing why a particular firm within the industry was selected (16). On the other hand, district courts in Pennsylvania and New Jersey have ruled that an employer's involvement in a high-hazard industry is insufficient, by itself, to establish probable cause for issuance of a warrant. Instead, these courts have held that the warrant application must contain information from which the magistrate can conclude that the particular firm within the high-hazard industry was chosen for inspection on the basis of "neutral criteria" (17).

1.1.2 Ex Parte *Warrants*

Additional questions raised by the Supreme Court's decision in *Barlow's* involve the proper procedure for obtaining, enforcing, and contesting OSHA inspection warrants. Although the Fifth Circuit Court of Appeals has held that the U.S. district courts do not have jurisdiction to issue injunctions compelling employers to submit to OSHA inspections (18), it is well settled that both the district courts and the U.S. magistrates have the authority to issue administrative inspection warrants (19). A more difficult question has been presented regarding OSHA's right to obtain an *ex parte* inspection warrant: that is, a warrant obtained without prior notice to the employer. In *Barlow's*, the Supreme Court suggested that the secretary could, by appropriate regulation, provide for *ex parte* warrants (although the regulation then in force called instead for "compulsory process") (20). When the original regulation (21) was held not to include *ex parte* warrants (22), an "interpretive rule" was issued stating that the term "compulsory process" was intended to include *ex parte* warrants (23). In *Cerro Metal Products, Division of Marmon Group, Inc. v. Marshall* (24), the Third Circuit held the "interpretive rule" invalid because it was inconsistent both with OSHA's prior interpretations of the "compulsory process" regulation and with the Supreme Court's dictum in *Barlow's*. Thus the court held that *ex parte* warrants were unavailable to the secretary. The Seventh, Ninth, and Tenth Circuits disagreed, holding that the secretary's "interpretive rule" properly authorized *ex parte* warrants (25). The controversy became moot when OSHA subsequently promulgated a new regulation (26) that clearly states that OSHA may obtain an inspection warrant without prior notice to the employer (27).

1.1.3 Challenging Validity of Warrant

A final question that faced the federal courts as a result of the *Barlow's* decision was the proper procedure for an employer to use in challenging the validity of an OSHA inspection warrant. In general, it appears that an employer who wishes to contest a warrant has two choices: (1) He or she may refuse to obey the warrant and may move to quash it in district court or (2) he or she may allow the inspection to proceed and challenge the warrant's validity before the Review Commission if the employer contests any citations issued pursuant to the inspection.

In *Babcock & Wilcox Company v. Marshall*, the Third Circuit indicated that an employer may obtain a district court hearing on a warrant's validity prior to inspection by refusing to obey the warrant (thereby risking contempt), moving to quash the warrant, and promptly appealing if the motion is denied (28). However, the Third Circuit held that if the employer obeys the warrant, he must "exhaust his administrative remedies" before the Review Commission prior to obtaining judicial review of his objections to the warrant (29).

The overwhelming majority of federal appellate courts have agreed with the Third Circuit that the doctrine of exhaustion of remedies precludes an employer from challenging an executed warrant in federal court (30). In contrast, the Seventh Circuit has held that in certain circumstances an employer may obtain district court review of a warrant even after an inspection has been completed (31). The same court, however, recently modified its stance on post-execution challenges when it joined the other circuits insofar as requiring employers who challenge completed OSHA inspections *on Fourth Amendment grounds* to first go to the Review Commission before turning to the federal courts (32).

1.2 Other Inspection Matters

It is well settled that OSHA inspections must be made at reasonable times, in a reasonable manner, and within reasonable limits pursuant to the act (33). The act also requires the inspector to present his or her credentials to the employer before beginning the inspection (34). The Fifth Circuit Court of Appeals has held, however, that even if the inspector fails to present credentials, such failure cannot operate to exclude evidence obtained in the inspection when there is no showing that the employer was prejudiced thereby in any way (35).

Section 8(e) of OSHA requires that a representative of the employer and an authorized employee representative be allowed to accompany the OSHA inspector during the physical inspection of any workplace. In *Chicago Bridge & Iron Company v. OSAHRC and Dunlop* (36), the Seventh Circuit held the dictates of Section 8(e) to be mandatory rather than merely directory. The court refused, however, to hold that the absence of a formalized offer of an opportunity to accompany the compliance officer on his inspection rendered the citations for violations observed during that inspection void *ab initio*. Rather, the court explained that when there has been substantial compliance with the mandate of the act regarding walk-around rights and the employer is unable to demonstrate that prejudice resulted from his nonparticipation in the inspection, citations issued as a result of the inspection are valid (37).

In *Accu-Namics* (35), the Fifth Circuit did not reach the question of whether the language of Section 8(e) was mandatory or directory, but it did refuse to adopt a rule that would exclude all evidence obtained illegally, no matter how minor or technical the government's violation and no matter how egregious or harmful the employer's safety violation.

The Ninth and Tenth Circuits have also concluded that minor violations of Section 8(e)'s inspection procedures do not justify dismissing a citation (38) or suppressing evidence gained from an inspection (39), at least as long as there has been substantial compliance with the act and the employer's defense on the merits has not been prejudiced.

The Fourth and Eighth Circuits have gone even further, holding as a matter of law that regardless of whether there has been substantial compliance by the inspector, Section 8(e) violations do not affect the validity of citations issued or evidence obtained unless the employer can show that he or she has been prejudiced in preparing or presenting a defense on the merits. In *Pullman Power Products, Inc. v. Marshall and OSAHRC* (40), the Fourth Circuit held that, in the absence of such prejudice, the validity of citations was not affected by an inspector's alleged failure to properly present his credentials, conduct opening and closing conferences, and provide walk-around rights. In so holding, the court declined to reach the issue of substantial compliance, stating broadly that the employer's inability to show prejudice bars its attack on the validity of the citations. The Eighth Circuit reached a similar conclusion in *Marshall v. Western Waterproofing Co., Inc.* (41). Although stating that the requirements of Section 8(e) are not merely directory, the court nevertheless held that, in the absence of prejudice to the employer, an inspector's failure to comply with these requirements will not justify suppression of evidence, regardless of whether there has been substantial compliance (42).

Applicable regulations authorize OSHA consultants to make inspections at work sites in order to help employers comply with OSHA standards (43). The consultants do not report violations to the compliance officers. There is also a voluntary protection program (VPP), under which OSHA recognizes exemplary workplaces by removing participating employers from OSHA's general inspection lists (44). The VPP was introduced in 1982 (45).

Statistics regarding OSHA inspections show a trend toward fewer inspections, although the number of violations cited by compliance officers has increased and there are larger proposed penalties (46).

2 RECORD KEEPING AND REPORTING

Section 8(c)(1) of OSHA requires employers to make, keep, and preserve such records regarding their OSHA-related activities as the secretary of labor, in cooperation with the secretary of health and human services (HHS), may prescribe as necessary or appropriate for the enforcement of the act or for developing information on the causes and prevention of occupational accidents and illnesses. These records must also be made available to the secretaries of labor and/or HHS (47).

Section 8(c)(2) more specifically directs the secretary of labor, in cooperation with the secretary of HHS, to issue regulations requiring employers to maintain accurate records of, and to make periodic reports on, work-related deaths, injuries, and illnesses (other than minor injuries requiring only first aid treatment and not involving medical treatment, loss of consciousness, restriction of work or motion, or transfer to another job).

Regulations promulgated by the secretary of labor to implement Sections 8(c)(1) and (2) require employers to keep the following records (or the equivalents thereof):

OSHA Form 200	A log and annual summary of all recordable occupational injuries and illnesses.
OSHA Form 101	A supplementary record for each occupational injury or illness (48).

The regulations further require that a copy of Form 200, summarizing the year's occupational illnesses and injuries, be posted in each establishment in a conspicuous place or places where notices to employees are customarily posted. The required records must be retained in each establishment for 5 years following the end of the year to which they relate (49).

Employers have reporting as well as record-keeping obligations. The current regulations provide that within 48 hr after an on-the-job accident that is fatal to one or more employees or that results in hospitalization of five or more employees, the employer must report the accident either orally or in writing to the nearest office of the area director of the OSH Administration. These requirements may soon become more stringent. Under a rule proposed by OSHA in May 1992, employers would be required to report occupational accidents resulting in hospitalization of *three* or more employees and would have to report reportable accidents within *8 hr* of their occurrence (or within *8 hr* of such time that the employer learns of the accident) (50).

In addition, Section 24 of the act directs the secretary of labor, in consultation with the secretary of HHS, to develop and maintain a program of collection, compilation, and analysis of occupational safety and health statistics. The secretary of labor has given the commissioner of the Bureau of Labor Statistics (BLS) the authority to develop and maintain such a program. This program requires employers to participate in periodic surveys of occupational injuries and illnesses. The survey form is OSHA Form 200S, and an employer who receives such a form has a duty to complete and return it promptly (51).

Small employers are exempt from many, but not all, of OSHA's record-keeping requirements. For example, employers who had no more than 10 employees at any time during the calendar year preceding the current one are exempt from the requirements of keeping OSHA Forms 200 and 101. Such employers, however, are not exempted from the requirement of reporting accidents resulting in fatalities or multiple hospitalizations. In addition, small employers may be selected to participate in the BLS periodic surveys and, if selected, must maintain a log and summary on Form 200 for the survey year, and must make the required reports on survey Form 200S (52).

Finally, with respect to toxic materials or harmful physical agents, the secretary of labor, in cooperation with the secretary of HHS, is directed by Section 8(c)(3) of the act to issue regulations requiring employers to maintain accurate records of employee exposure to potentially toxic materials or harmful physical agents that are required to be monitored or measured under Section 6. These regulations must guarantee employees or their representatives an opportunity to observe the required monitoring or measuring, and to have access to the records of employee exposure. Further, the regulations must ensure employees and former employees access to such records as will indicate their own exposure to such substances (53).

These statutory directives have been implemented by the secretary of labor in a rule governing employee exposure and medical records (54). This rule requires employers to maintain exposure and medical records pertaining to their employees' exposure to toxic substances and harmful physical agents. The exposure records must be made accessible to exposed and potentially exposed employees, as well as to designated employee representatives and to OSHA. Access to medical records must also be ensured for the employee and for OSHA; because of the privacy interests involved, however, an employee's medical records are open to a collective bargaining representative only with the employee's consent (55).

3 SANCTIONS FOR VIOLATING SAFETY AND HEALTH LAWS

3.1 Citations

The Occupational Safety and Health Act authorizes the secretary of labor to issue citations and proposed penalties to employers who are believed to have violated the act or its implementing regulations. Section 9(a) directs the secretary to issue citations "with reasonable promptness" following an inspection or investigation. According to Section 9(c), no citation may be issued after the expiration of 6 months from the occurrence of any violation (56).

Each citation is to be in writing and must "describe with particularity the nature of the violation," including a reference to the provision of the statute, standard, rule, regulation, or order alleged to have been violated (57). Section 9(a) states that each citation must also fix a reasonable time for the abatement of the violation (58).

Section 9(b) requires employers to post each citation prominently at or near each place where a violation referred to in the citation occurred. The mechanics of how, when, where, and how long to post the citations are set forth in regulations issued by the secretary (59).

3.2 Penalties

Within a reasonable time after a citation has been issued, the secretary is directed by Section 10(a) to notify the employer by certified mail of the penalty, if any, that will be assessed for the violation. Violations fall into the following general categories: serious, nonserious, *de minimis*, willful, repeated, and criminal, and the penalty will be based at least in part on the nature of the violation (60).

With the Omnibus Budget Reconciliation Act of 1990, Congress amended OSHA to provide for dramatically increased penalties for occupational safety and health violations. In general, maximum penalty levels were increased sevenfold, and a $5000 statutory minimum penalty for willful violations was established. The new penalty structure, instituted primarily to raise revenue, affects all violations occurring on or after November 5, 1990 (the statute's effective date) (61).

3.2.1 Serious Violations

Section 17(k) of OSHA defines a serious violation as follows:

[A] serious violation shall be deemed to exist in a place of employment if there is a substantial probability that death or serious physical harm could result from a condition which exists, or from one or more practices, means, methods, operations, or processes which have been adopted or are in use, in such place of employment unless the employer did not, and could not with the exercise of reasonable diligence, know of the presence of the violation.

The *probability* of an *accident* occurring need not be shown to establish that a violation is serious (62). Rather, as the Ninth Circuit court ruled in *California Stevedore and Ballast* (62), a serious violation exists if any accident that should result from a violation would have a substantial probability of resulting in death or serious physical harm (63). No actual death or physical injury is required to establish a serious violation (64).

Employer knowledge is clearly an element of a serious violation. The knowledge requirement in Section 17(k) deals with actual or constructive knowledge (65) of practices or conditions that constitute violations of the act; it is not directed to knowledge of the law (66). The burden of proof is on the secretary to prove knowledge (67) as well as the other elements of a serious violation of the act.

Since 1990, Section 17(b) provides that an employer who has received a citation for a serious violation of the act *must* be assessed a civil penalty of up to $7000.

3.2.2 Nonserious Violations

The original Senate version of the occupational safety and health bill treated all violations as "serious." As finally enacted, however, OSHA incorporated a House proposal for violations "determined not to be of a serious nature" (68).

The statute does not describe the elements of a nonserious violation and provides no guidelines for determining when a violation is not serious. The Fifth Circuit, however, has described nonserious violations as violations that do not create a substantial probability of serious physical harm (69). The commission has explained that serious and nonserious violations are distinguished on the basis of the seriousness of injuries that experience has shown are reasonably likely to result when an accident does arise from a particular set of circumstances (70). At least one federal court of appeals has held that employer knowledge is an element of a nonserious violation (71).

When a violation is determined not to be serious, the assessment of a penalty is discretionary rather than mandatory (72). Section 17(c) states that the employer *may* be assessed a civil penalty of up to $7000 for each nonserious violation.

Section 110(a) of the Federal Mine Safety and Health Act of 1977 does not allow discretionary penalties for so-called nonserious violations, but instead requires the secretary of labor to assess a civil penalty of up to $50,000 for each violation of a mandatory health or safety standard under the act.

3.2.3 De Minimis *Violations*

If noncompliance with an OSHA provision or standard presents no direct or immediate threat to the safety or health of employees, the violation is *de minimis* and the secretary of labor may issue only a notice—not a citation—to the employer (73). The notice contains no proposed penalty (74). The secretary has explained that a violation should be considered *de minimis* if: (a) an employer complies with the intent of a standard but deviates from its particular requirements in a way that has no direct or immediate relationship to safety and health; or (b) an employer complies with a proposed amendment to a standard and the amendment provides equal or greater safety and health protection than the standard itself; or (c) an employer's workplace is "state of the art"—that is, it is technically advanced beyond the requirements of a standard and provides equal or greater safety and health protection (75).

3.2.4 *Willful or Repeated Violations*

The Occupational Safety and Health Act provides more stringent civil penalties for employers who "willfully or repeatedly" violate the act or any regulations promulgated pursuant thereto. Under Section 17(a) of the act, willful or repeated violations are subject to penalties of up to $70,000 for each violation. Additionally, there is now a *minimum* penalty of $5000 for each such violation (76).

The act contains no definition of either "willful" or "repeated" as applied to violations. Thus it is not surprising that there has been some disagreement on the elements of these types of violations.

In *Frank Irey Jr., Inc. v. OSAHRC* (77), the Third Circuit initially held that "[w]illfulness connotes defiance or such reckless disregard of consequences as to be equivalent to a knowing, conscious, and deliberate flaunting of the Act. Willful means more than merely voluntary action or omission—it involves an element of obstinate refusal to comply." The majority of the circuits, as well as the Review Commission, declined to follow the *Frank Irey* definition of "willfulness." The First and Fourth Circuits interpreted a willful action as a "conscious, intentional, deliberate voluntary decision," regardless of venial motive (78). The Review Commission agreed that no showing of malicious intent is necessary to establish "willfulness," (79) defining a willful violation as one "committed with either an intentional disregard of, or plain indifference to, the Act's requirements" (80). Most courts of appeals have either adopted the Review Commission's standard or have embraced similar definitions that do not require a showing of a bad motive (81).

Thus, a conflict developed between the Third Circuit's view, as expressed in

Frank Irey, and the majority approach. The Court of Appeals for the District of Columbia Circuit characterized this conflict as more apparent than real. In *Cedar Construction Co. v. OSAHRC and Marshall* (82) that court indicated that the two approaches were likely to yield the same results in particular cases, since there is little practical difference between "obstinate refusal to comply" and "intentional disregard" of the act. The Third Circuit later agreed when it again addressed the "willfulness" question. In *Babcock & Wilcox Co. v. OSAHRC* (83), the court explained its holding in *Frank Irey* as follows (84):

> [T]he supposed conflict among the circuits on this point has been generated by several courts of appeals reading into our *Irey* definition a requirement that the employer act with "bad purpose." Read in this fashion, *Irey* has not been followed by some circuits. . . . To our way of thinking, an "intentional disregard of OSHA requirements" differs little from an "obstinate refusal to comply;" nor is there in context much to distinguish "defiance" from "intentional disregard." . . . We also believe, as does the District of Columbia Circuit, that the same results would likely be reached in various cases, including the one here, regardless of the verbiage utilized. . . . It is not unusual that different words are used to describe the same basic concept.

This clarification by the Third Circuit thus resulted in general agreement among the circuits that the Review Commission's "intentional disregard" standard is the correct one, and that no malicious intent need be shown to establish a willful violation.

The interpretation of "repeated" violation has also generated disagreement among employers, the courts, and the commission. Once again, the controversy has centered on a Third Circuit ruling. In *Bethlehem Steel Corp. v. OSAHRC and Brennan* (85), the Third Circuit held that the commission can find a repeated violation only when the evidence shows the employer consciously ignored or "flaunted" the requirements of the act and was cited for a similar violation on at least *two prior* occasions. The court further explained that in applying the "repeatedly" portion of the act, the commission must determine that the acts themselves "flaunt" the requirements of the statute, but need not determine whether the acts were performed with an *intent* to "flaunt" the requirements of the statute.

Most other federal courts refused to adopt the Third Circuit's interpretation of a "repeated" violation. The Fourth Circuit, in *George Hyman Construction Company v. OSAHRC* (86), reasoned that the requirement of flagrant misconduct to establish a repeated violation fails to "recognize a meaningful distinction between willful and repeated violations." Other circuits have agreed that an employer need not have a particular state of mind or motive for "flaunting" the act, nor otherwise exhibit an aggravated form of misconduct to be guilty of a repeated violation (87). Most of these courts also rejected the Third Circuit's conclusion that a repeated violation must be based on at least two prior violations, requiring instead only one prior and substantially similar infraction.

The Review Commission also agreed that no "flaunting" of the act need be shown to establish a repeated violation. In *Potlach Corporation* (88), the commission defined a "repeated" violation as follows (89):

A violation is repeated under section 17(a) of the Act if, at the time of the alleged repeated violation, there was a Commission final order against the same employer for a substantially similar violation (90).

With respect to the length of time for which a previous violation may serve as the basis for a subsequent, ''repeated'' violation, it is OSHA policy to issue a citation as a repeated violation if:

(a) The citation is issued within 3 years of the final order of the previous citation, or

(b) The citation is issued within 3 years of the final abatement date of that citation, whichever is later (91).

3.2.5 Criminal Sanctions

Job safety and health laws also provide criminal sanctions for certain specified conduct (92). The most stringent criminal sanctions are those set forth in the Federal Mine Safety and Health Act of 1977. Section 110(d) states that a mine operator shall be subjected to a fine of up to $25,000 or a prison term of up to one year (or both) for willfully violating mandatory mine health or safety standards or for knowingly failing to comply with certain orders issued under that statute. That section further provides for increased maximum penalties of $50,000 and 5 years in prison for a second conviction (93).

Under OSHA, willful violations (94) of the act (or of relevant standards, rules, regulations or orders) that cause death to an employee shall be punishable upon conviction by a fine of up to $10,000 or imprisonment for up to 6 months or both. Second convictions carry maximum penalties of $20,000 and one year in jail (or both) (95).

Criminal penalties may also be imposed for: (a) knowingly making any false statement, representation, and so forth, in any document filed pursuant to or required to be maintained by OSHA or by the Mine Safety and Health Act of 1977 (96); (b) giving advance notice of an OSHA or Mine Safety and Health inspection without the authority of the secretary of labor (97); (c) knowingly distributing, selling, and so on, in commerce any equipment for use in coal or other mines that is represented as complying with the Mine Safety and Health Act and does not do so (98); (d) killing an OSHA inspector or investigator on account of the performance of his or her duties (99).

3.2.6 Failure to Abate a Violation

The Occupational Safety and Health Act does not specify fixed periods within which violations must be remedied, but Section 9(a) does require that each citation ''fix a reasonable time for the abatement of the violation'' (100). The abatement period does not begin to run until the date of the final order of the commission affirming the citation, as long as the review proceeding, if any, initiated by the employer was in good faith and not solely for delay or avoidance of penalties (101). An employer who fails to correct a violation within the period specified in the citation may receive an additional citation pursuant to Section 10(b) for failure to abate. Failure to abate may result in the assessment of civil penalties of not more than $7000 per day for each day the violation continues (102).

Notices of violations under Sections 104(a) and 104(b) of the Federal Mine Safety and Health Act of 1977 must similarly specify time periods for abatement of violations. Failure to abate under that statute can result in an order directing all persons to be withdrawn from the affected area of the mine until a representative of the secretary determines that the violation has been abated.

3.3   Contesting Citations and Penalties

Section 10(a) of OSHA and the regulations promulgated thereunder provide a means for contesting citations and proposed penalties. After the employer has been notified of the penalty proposed by the secretary, the employer has 15 working days to notify the secretary that he or she wishes to contest the citation or the proposed assessment of penalty. A failure to notify the secretary within 15 days of intent to contest the citation or proposed penalty will render the citation or penalty ''a final order of the Commission and not subject to review by any court or agency'' (103).

The secretary's regulations (104) instruct the employer that ''[e]very notice of intention to contest shall specify whether it is directed to the citation or to the proposed penalty, or both.'' Similarly, the courts have construed the OSHA enforcement scheme as mandating a distinction between contesting a citation and contesting a proposed penalty (105). Thus, the Fifth Circuit held (106) that an employer's letter that contested the proposed penalty but failed to contest the citation (in fact, the letter affirmatively admitted the violation) constituted waiver of the employer's right to challenge the citation on appeal. However, the commission has stated that it will construe notices of contest that are limited to the penalty to include a contest of the citation as well if the cited employer indicates later that it was his or her intent to contest the citation (107).

If an employer files a timely notice of contest (or if within 15 working days of the issuance of a citation, a representative of the employees files a notice challenging the period of abatement specified in the citation), the secretary must immediately advise the Occupational Safety and Health Review Commission of the intent to contest. The commission then must afford an opportunity for an administrative hearing (108).

A commission hearing is conducted pursuant to the Administrative Procedure Act and is presided over by a single administrative law judge employed by the commission. After taking testimony, the judge writes an opinion, which is subject to review by the full three-member commission at its discretion. An aggrieved party may petition for discretionary review before the full commission (109), and any commission member may direct review of a case on his or her own motion (110). If no commissioner directs review, or if a timely petition for review is not filed, the administrative law judge's decision becomes a final order of the commission (111).

Section 10(c) authorizes the commission to review either the citation or the proposed penalty or both:

The Commission shall thereafter [i.e., after hearing] issue an order, based on findings of fact, affirming, modifying or vacating the Secretary's citation or proposed penalty, or directing other appropriate relief, and such order shall become final thirty days after its issuance.

Section 17(j) further empowers the commission to assess appropriate civil penalties, giving due consideration to the size of the business of the employer being charged, the gravity of the violation, the good faith of the employer, and the history of previous violations (112). The commission has taken the position that it may exercise its power to *increase* the secretary's proposed penalty after considering the factors outlined above, and courts of appeals have expressed the view that the commission may act in this manner (113).

Courts have also sanctioned the commission's right to increase the degree of a violation from nonserious to serious (114). The commission has taken the view that it can reduce the degree of a violation as well (115).

The final stage of an OSHA enforcement proceeding is review in a court of appeals (and thereafter discretionary review by the Supreme Court). Any person adversely affected or aggrieved by the commission's disposition (116) may obtain review in an appropriate court of appeals pursuant to Section 11(a) of the act (117). Section 11(b) provides that the secretary of labor may also obtain review or enforcement of any final order of the commission by filing a petition for such relief in the appropriate court of appeals. The reviewing court is bound by Section 11(a) to apply the ''substantial evidence test'' to the commission's findings of fact (118). The same section empowers the court to direct the commission to consider additional evidence if the evidence is material and reasonable grounds existed for a party's failure to admit it in the hearing before the commission. Regarding the penalty imposed by the commission, the reviewing court may inquire only whether the commission abused its discretion because the assessment of a penalty is not a finding of fact but rather the exercise of a discretionary grant of power (119).

3.4 Imminent Danger Situations

Section 13(a) of OSHA confers jurisdiction on the U.S. district courts, upon petition of the secretary of labor, to restrain hazardous employment conditions or practices if they create an imminent danger of death or serious physical harm that cannot be eliminated through the act's other enforcement procedures.

As originally reported out of the House committee, the act contained a provision that would have permitted an OSHA inspector to close down an operation for up to 72 hr without a court order if he or she found that an imminent danger existed. The original Senate version of the bill also contained a provision allowing an inspector to close down an operation for 72 hr, but this provision was revised so that no shutdown may occur unless a federal district judge grants an application for a temporary restraining order (120).

The Federal Mine Safety and Health Act of 1977, however, permits coal or other mine operations to be shut down without a restraining order from a court. Section

107(a) of the act provides that when a federal inspector finds that an imminent danger is present in a mine, he or she shall order the withdrawal of all other persons from a part or all of that mine until the imminent danger no longer exists (121).

3.5   Challenges to the OSHA Enforcement Scheme

Not surprisingly, the citation and penalty scheme of OSHA has been subject to challenges on several fronts. For example, in *Atlas Roofing Company, Inc. v. OS-AHRC* (122), a cited employer contended that the act was constitutionally defective because: (a) civil penalties under OSHA are really penal and call for the constitutional protections of the Sixth Amendment and Article III; (b) even if the penalties are civil, OSHA violates the Seventh Amendment because of the absence of a jury trial for fact finding; (c) the act denies the employer his or her right to a Fifth Amendment "prejudgment" due process hearing since commission orders are self-executing unless the employer affirmatively seeks review; and (d) the overall penalty structure of OSHA violates due process because it "chills" the employer's right to seek review of the citation and penalty. The Fifth Circuit rejected all the employer's constitutional contentions. The Supreme Court granted a petition for *certiorari* in that case, limited to the Seventh Amendment issue (123), and subsequently upheld the act's provision for imposition of civil penalties without fact finding by a jury (124).

In late 1992, the Review Commission was considering challenges to OSHA's "egregious case" policy—the agency's method of citing alleged violations on an instance-by-instance basis (with resulting substantially higher proposed penalties) in particularly flagrant cases (125). Employers targeted for these mega-fines have challenged the policy, arguing that the secretary has exceeded her authority (1) by essentially creating a new category of violations and/or penalties not established by Congress in the act or (2) by not following formal administrative rule-making procedures before instituting such a policy. It has even been suggested that the policy could violate the constitutional prohibition on excessive penalties. The secretary has taken the position that such policy is well within her prosecutorial discretion (126). A decision by the Review Commission is pending (127).

4   THE FUTURE OF JOB SAFETY AND HEALTH LAW

It is fairly certain that change is on the horizon for job safety and health law for a number of reasons, not the least of which is the Democratic administration of Bill Clinton. The number of citations contested before the OSHA Review Commission is predicted to continue to rise as a result of the sevenfold increase in penalties enacted by Congress in 1990, coupled with OSHA's use of its egregious case policy (128).

Most significantly, OSHA may well be affected in 1993 by legislative changes, as "OSHA reform" legislation is debated in Congress (129). Highlights of the comprehensive proposed legislation include: extension of coverage to public sector employees, required employer/employee participation in job safety and health programs, deadlines for OSHA's rule making, protection of whistleblowers, and expanded criminal penalties to cover serious injury as well as fatality situations (130).

NOTES AND REFERENCES

1. See Section 103(a) of the Federal Mine Safety and Health Act of 1977. For a discussion of "imminent danger," see Section 3.4 of this chapter. Underground coal or other mines must be inspected at least four times each year. Each surface coal or other mine must be inspected at least twice a year.

2. 436 U.S. 307; 6 OSHC 1571 (1978). Several state courts have likewise held the inspection provisions of their respective state occupational safety and health acts unconstitutional or have held that inspections are permissible under those provisions only pursuant to a warrant. For example: *Woods & Rohde, Inc., d/b/a Alaska Truss & Millwork v. State*, 565 P.2d 138; 5 OSHC 1530 (Alaska Sup. Ct. 1977) (provision authorizing warrantless inspections violates state constitution); *State v. Albuquerque Publishing Co.*, 571 P.2d 117; 5 OSHC 2034 (New Mexico Sup. Ct. 1977), *certiorari denied*, 435 U.S. 956; 6 OSHC 1570 (1978) (nonconsensual inspection pursuant to state occupational safety and health act requires warrant satisfying administrative standards of probable cause); *Yocom v. Burnette Tractor Co., Inc.*, 566 S.E.2d 755; 6 OSHC 1638 (Kentucky Sup. Ct. 1978).

3. By contrast, the Supreme Court has upheld the constitutionality of warrantless inspections under the Federal Mine Safety and Health Act of 1977 on the grounds that mining has long been a pervasively regulated industry, in which a businessman can have no reasonable expectation of privacy against official inspections. In *Donovan v. Dewey*, 452 U.S. 594 (1981), the Court reasoned further that the Federal Mine Safety and Health Act inspection provisions are more narrowly drawn and provide for less administrative discretion concerning inspections than does OSHA, thus making such inspections reasonable under the Fourth Amendment.

4. 436 U.S. at 320-21; 6 OSHC at 1575-76. Several state courts have held that this relaxed administrative standard of probable cause is sufficient to support an inspection warrant pursuant to their respective occupational safety and health acts. See, for example, *Yocom v. Burnette Tractor Co., Inc.*, (2); *State v. Keith Manufacturing Co.*, 6 OSHC 1043 (Oregon Ct. of Appeals 1977) (provision of Oregon Safe Employment Act authorizing inspection warrant based on administrative standard of probable cause is constitutional); *State v. Kokomo Tube Co.*, 426 N.E.2d 1338; 10 OSHC 1158 (Indiana Ct. of Appeals 1981) (administrative probable cause standard articulated in *Barlow's* is applicable to inspection warrants under Indiana OSH Act); *Salwasser Mfg. Co. v. Occupational Safety and Health Appeals Bd.*, 14 OSHC 1278 (California Ct. of Appeal 1989).

5. 436 U.S. at 321; 6 OSHC at 1575-76. Such "neutral sources," according to the Court, could include statistics indicating "dispersion of employees in various types of industries across a given area, and the desired frequency of searches in any of the lesser divisions of the area . . ." 6 OSHC at 1576.

6. See, for example, *Donovan v. A. A. Beiro Constr. Co.*, 746 F.2d 894; 12 OSHC 1017 (D.C. Cir. 1984) (all contractors at the site consented to the inspection of a site owned by the District of Columbia); *J. L. Foti Constr. Co. v. Donovan*, 786 F.2d 714; 12 OSHC 1737 (6th Cir. 1986) (general contractor at work site consented); *National Eng'g*

& *Contracting Co. v. Department of Labor*, 687 F. Supp. 1219; 13 OSHC 1793 (S.D. Ohio 1988) (U.S. Army Corps of Engineers consented to inspection), *affirmed*, 902 F.2d 34; 14 OSHC 1621 (6th Cir.), *certiorari denied*, 111 S. Ct. 344; 14 OSHC 1920 (1990).

7. OSHA Instruction CPL 2.45B.III.B.3.a (June 15, 1980), amended by CPL 2.45 CH-1, (December 3, 1990), in BNA, *OSHR*, Reference File at 77:2301.

8. See *Marshall v. Chromalloy American Corp. (Gilbert & Bennett Manufacturing Co.)*, 589 F.2d 1335; 6 OSHC 2151 (7th Cir.), *certiorari denied*, 444 U.S. 884; 7 OSHC 2238 (1979); *Marshall v. W and W Steel Co., Inc.*, 604 F.2d 1322; 7 OSHC 1670 (10th Cir. 1979); *Burkart Randall Div. of Textron, Inc. v. Marshall*, 625 F.2d 1313; 8 OSHC 1467 (7th Cir. 1980); *Martin v. Intern. Matex Tank Terminals—Bayonne*, 928 F.2d 614, 14 OSHC 2153 (3d Cir. 1991). On the other hand, it has been held that the occurrence of an accident on the employer's premises is insufficient evidence of a specific violation to establish probable cause for an inspection. See, for example, *Donovan v. Federal Clearing Die Casting Co.*, 655 F.2d 793; 9 OSHC 2072 (7th Cir. 1981) (newspaper reports of accident do not satisfy "specific evidence of existing violation" standard); *Marshall v. Pool Offshore Co.*, 467 F. Supp. 978; 7 OSHC 1179 (W.D. La. 1979) (while actual fatalities do not constitute evidence of specific violation, an OSHA policy of investigating all fatalities may satisfy administrative "neutral criteria" standard).

9. *Weyerhaeuser Co. v. Marshall*, 592 F.2d 373; 7 OSHC 1090 (7th Cir. 1979); *Marshall v. Horn Seed Co., Inc.*, 647 F.2d 96; 9 OSHC 1510 (10th Cir. 1981). The Seventh Circuit has concluded that, since probable cause in the strict criminal sense is not required in OSHA inspection cases, the warrant application need not establish the reliability of the complainant or the basis of his complaint. *Marshall v. Chromalloy American Corp. (Gilbert & Bennett Manufacturing Co.)* (Ref. 8); *Burkart Randall Div. of Textron, Inc., v. Marshall* (Ref. 8). The Tenth Circuit, on the other hand, held in *Horn Seed Co.* that a warrant application based on an employee complaint must contain evidence showing the complaint was actually made by a complainant who was sincere in asserting a violation existed and who had some plausible basis for his belief. *Accord, Martin v. Inter. Matex Tank Terminals—Bayonne* (Ref. 8) (citing *Horn Seed* test).

10. *Marshall v. North American Car Co.*, 626 F.2d 320; 8 OSHC 1722 (3d Cir. 1980) (rejecting the secretary's argument that 29 CFR §1903.11, which provides that an inspection based on an employee complaint is not limited to matters complained of, permits a wall-to-wall inspection); *Donovan v. Sarasota Concrete Co.*, 693 F.2d 1061; 11 OSHC 1001 (11th Cir. 1982); *Establishment Inspection of Asarco, Inc.*, 508 F. Supp. 350; 9 OSHC 1317 (N.D. Tex. 1981). However, in *In re Establishment Inspection of Cerro Prods. Co.*, 752 F.2d 280, 283; 12 OSHC 1153, 1155 (7th Cir. 1985), a full-scope warrant based on an employee complaint was upheld where: (1) there was no evidence of harassment in the filing of the complaint; (2) there was evidence that the employer had a high-hazard workplace in a high-hazard industry (based on a high industry lost workday incident or LWDI rate); (3) there had been no inspections in the previous fiscal year; and (4) OSHA's limited resources would be conserved by allowing a full-scope inspection. The Seventh Circuit thus followed the approach of the Eighth Circuit, which, in *Appeal of Carondelet Coke Corp.*, 741 F.2d 172; 11 OSHC 2153 (8th Cir. 1984), held that whether a general warrant should issue must be determined on a case-by-case basis. A wall-to-wall inspection may also be permissible if the complained-of violation affects the entire plant. For example, see *In re Establishment Inspection of Seaward International, Inc.*, 510 F. Supp. 314 (W.D. Va. 1980), *affirmed*

without opinion, 644 F.2d 880 (4th Cir. 1981) (complaint involving exposure to carcinogens); *Hern Iron Works, Inc. v. Donovan*, 670 F.2d 838; 10 OSHC 1433 (9th Cir.), *certiorari denied*, 103 S. Ct. 69 (1982) (complaint pertaining to ventilation system).

11. *Marshall v. Northwest Orient Airlines, Inc.*, 574 F.2d 119; 6 OSHC 1481 (2d Cir. 1978).

12. *Pelton Casteel, Inc. v. Marshall*, 588 F.2d 1182; 6 OSHC 2137 (7th Cir. 1978).

13. *Marshall v. Chromalloy American Corp. (Gilbert & Bennett Manufacturing Co.)* (Ref. 8). *Accord, Marshall v. Weyerhaeuser Co.*, 456 F. Supp. 474; 6 OSHC 1920 (D.N.J. 1978) (past violations do not establish probable cause when a follow-up inspection of those violations revealed they had been corrected).

14. *In re Establishment Inspection of Northwest Airlines, Inc.*, 587 F.2d 12; 6 OSHC 2070 (7th Cir. 1978); *Marshall v. Weyerhaeuser Co.* (Ref. 13).

15. *Peterson Builders, Inc.*, 525 F. Supp. 642; 10 OSHC 1169 (E.D. Wis. 1981); *Donovan V. Athenian Marble Corp.*, 10 OSHC 1450 (W.D. Okla. 1982); *Erie Bottling Corp.*, 539 F. Supp. 600; 10 OSHC 1632 (W.D. Pa. 1982); *Urick Foundry Co. v. Donovan*, 542 F. Supp. 82; 10 OSHC 1765 (W.D. Pa. 1982); *Brock v. Gretna Mach. & Ironworks*, 769 F.2d 1110; 12 OSHC 1457 (5th Cir. 1985); *Donovan v. Trinity Indus.*, 824 F.2d 634, 636; 13 OSHC 1369, 1371 (8th Cir. 1987); *Pennsylvania Steel Foundry & Mach. Co. v. Secretary of Labor*, 831 F.2d 1211; 13 OSHC 1417 (3d Cir. 1987); *Industrial Steel Prods. Co. v. OSHA*, 845 F.2d 1330; 13 OSHC 1713 (5th Cir.), *certiorari denied*, 488 U.S. 993 (1988).

16. *Marshall v. Chromalloy American Corp. (Gilbert & Bennett Manufacturing Co.)* (Ref. 8). *Accord, Marshall v. Multi-Cast Corp.*, 6 OSHC 1486 (N.D. Ohio 1978); *The Fountain Foundry Corp. v. Marshall*, 6 OSHC 1885 (S.D. Ind. 1978).

17. *Urick Foundry*, 472 F. Supp. 1193; 7 OSHC 1497 (W.D. Pa. 1979) (warrant would be sufficient if it showed, e.g., that the particular firm chosen for inspection had a history of prior OSHA violations, or that it was selected at random from the list of firms in the high-hazard industry); *Marshall v. Weyerhaeuser Co.* (Ref. 13) (fact that employer was member of high-hazard industry does not establish probable cause in the absence of additional facts explaining why this particular employer was selected).

18. *Marshall v. Gibson's Products, Inc. of Plano*, 584 F.2d 668; 6 OSHC 2092 (5th Cir. 1978). Based on its decision in *Gibson's Products*, the Fifth Circuit has further held that the district court has no jurisdiction to issue an injunction enforcing a warrant that has already been issued. *Marshall v. Shellcast Corp.*, 592 F.2d 1369; 7 OSHC 1239 (5th Cir. 1979). In *Marshall v. Pool Offshore Co.* (Ref. 8), however, the district court held that, despite *Gibson's Products*, it had jurisdiction to impose sanctions on an employer for its refusal to permit an inspection because the proceeding was for civil contempt for failure to obey a warrant rather than for an injunction to compel an inspection.

19. There are statutory bases, for example, 28 U.S.C. §1337 and §1345, that generally give the district courts jurisdiction. Some federal courts have held that the district courts (and the U.S. magistrates) have the authority to issue warrants simply because it would be inconsistent with the rationale of *Barlow's* to hold otherwise. For example, *The Fountain Foundry Corp. v. Marshall* (Ref. 16); *Marshall v. Weyerhaeuser Co.* (Ref. 13); *Marshall v. Huffhines Steel Co.*, 478 F. Supp. 986; 7 OSHC 1850 (N.D. Tex. 1979). The Seventh Circuit has held that federal magistrates have the authority to issue OSHA inspection warrants pursuant to 28 U.S.C. §636, which permits district judges

to assign to magistrates any duties not inconsistent with the constitution or federal law. See *Marshall v. Chromalloy American Corp. (Gilbert & Bennett Manufacturing Co.)* (Ref. 8); *Pelton Casteel, Inc. v. Marshall* (Ref. 12). *Accord, Marshall v. Multi-Cast Corp.* (Ref. 16).

20. 436 U.S. at 316-317; 6 OSHC at 1575.

21. The regulation permitting the secretary to obtain "compulsory process" to compel an inspection was published at 29 CFR §1903.4.

22. See *Cerro Metal Products v. Marshall*, 467 F. Supp. 869, 872; 7 OSHC 1125, 1126-1127 (E.D. Pa. 1979), *affirmed*, 620 F.2d 964; 8 OSHC 1196 (3d Cir. 1980).

23. 43 *Fed. Reg.* 59838 (1978); 29 CFR §1903.4(d). See discussion in *Cerro Metal Products* (Ref. 22), 7 OSHC at 1127.

24. 620 F.2d 964; 8 OSHC 1196 (3d Cir. 1980). *Accord, Marshall v. Huffhines Steel Co.*, 7 OSHC 1910 (N.D. Tex. 1979), *affirmed without opinion sub nom. Donovan v. Huffhines Steel Co.*, 9 OSHC 1762 (5th Cir. 1981).

25. *Rockford Drop Forge Co. v. Donovan*, 672 F.2d 626; 10 OSHC 1410 (7th Cir. 1982); *Stoddard Lumber Co. v. Marshall*, 627 F.2d 984; 8 OSHC 2055 (9th Cir. 1980); *Marshall v. W & W Steel Co., Inc.* (Ref. 8).

26. 45 *Fed. Reg.* 65916–65924 (1980).

27. The new regulation, specifically held valid by the court in *Donovan v. Blue Ridge Pressure Castings, Inc.*, 543 F. Supp. 53; 10 OSHC 1217 (M.D. Pa. 1981), reads as follows [29 CFR §1903.4(d)]:

    (d) For purposes of this section, the term compulsory process shall mean the institution of any appropriate action, including *ex parte* application for an inspection warrant or its equivalent. *Ex parte* inspection warrants shall be the preferred form of compulsory process in all circumstances where compulsory process is relied upon to seek entry to a workplace under this section.

28. 610 F.2d 1128, 1135–1136; 7 OSHC 1880, 1884 (3d Cir. 1979). An employer's right to a pre-inspection hearing in federal court on a warrant's validity was reaffirmed by the Third Circuit in *Cerro Metal Products v. Marshall* (Ref. 24), holding an employer could bring an action in district court, prior to inspection, seeking to enjoin OSHA from obtaining an *ex parte* warrant. The Seventh Circuit has also held an employer may, in limited circumstances, raise a pre-inspection warrant challenge in district court. See *Blocksom & Co. v. Marshall*, 582 F.2d 1122; 6 OSHC 1865 (7th Cir. 1978) (employer may raise invalidity of warrant as a defense in action seeking to hold it in contempt for failure to obey warrant).

29. 610 F.2d at 1135–1137; 7 OSHC at 1883–1885. *Accord, Marshall v. Whittaker Corp., Berwick Forge & Fabricating Co.*, 610 F.2d 1141; 6 OSHC 1888 (3d Cir. 1979); *Establishment Inspection of the Metal Bank of America, Inc.*, 700 F.2d 910; 11 OSHC 1193 (3d Cir. 1983).

30. *In re Worksite Inspection of Quality Products, Inc.*, 592 F.2d 611; 7 OSHC 1093 (1st Cir. 1979); *Marshall v. Central Mine Equipment Co.*, 608 F.2d 719; 7 OSHC 1907 (8th Cir. 1979); *Baldwin Metals Co., Inc. v. Donovan*, 642 F.2d 768; 9 OSHC 1568 (5th Cir.), *certiorari denied sub nom. Mosher Steel Co. v. Donovan*, 454 U.S. 893 (1981); *In re J. R. Simplot Co.*, 640 F.2d 1134 (9th Cir. 1981), *certiorari denied*, 455 U.S. 939 (1982); *Robert K. Bell Enters. v. Donovan*, 710 F.2d 673 (10th Cir. 1983), *certiorari denied*, 464 U.S. 1041 (1984); *Establishment Inspection of Gould Publishing Co.*, 934 F.2d 457; 15 OSHC 1073 (2d Cir. 1991).

31. *Weyerhaeuser Co. v. Marshall*, 592 F.2d 373; 7 OSHC 1090 (7th Cir. 1979); *Federal*

Casting Div., Chromalloy American Corp. v. Donovan, 684 F.2d 504; 10 OSHC 1801 (7th Cir. 1982) (following *Weyerhaeuser*).

32. *In re Establishment Inspection of Kohler Co.*, 935 F.2d 810 (7th Cir. 1991).

33. *Dunlop v. Able Contractors, Inc.*, 4 OSHC 1110 (D. Mont. 1975), *affirmed sub nom. Marshall v. Able Contractors, Inc.*, 573 F.2d 1055; 6 OSHC 1317 (9th Cir.), *certiorari denied*, 439 U.S. 826 (1978).

34. See Section 8(a). The secretary's regulations provide in relevant part [29 CFR §1903.7(a)]:

> At the beginning of an inspection, Compliance Safety and Health Officers shall present their credentials to the owner, operator, or agent in charge at the establishment; explain the nature and purpose of the inspection; and indicate generally the scope of the inspection and the records . . . which they wish to review.

By regulation, employers are entitled to a copy of the requests for inspection if the inspection follows an employee's complaint [29 CFR §1903.11]. Employers also have the right not to have trade secrets disclosed as a result of the inspection [29 CFR §1903.9]. The secretary's regulations further provide for a conference at the conclusion of the inspection. During this conference the compliance safety and health officer advises the employer of any apparent safety or health violations disclosed by the inspection, and the employer is afforded an opportunity to bring to the attention of the officer any pertinent information regarding conditions in the workplace [29 CFR §1903.7(e)].

35. *Accu-Namics, Inc. v. OSAHRC and Dunlop*, 515 F.2d 828; 3 OSHC 1299 (5th Cir.), *rehearing denied*, 521 F.2d 814 (5th Cir. 1975), *certiorari denied*, 425 U.S. 903 (1976).

Another controversial question involving the exclusion of evidence has arisen as a result of the Supreme Court's decision in the *Barlow's* case: whether the exclusionary rule, borrowed from the field of criminal law, applies in OSHA hearings to prohibit the use of evidence obtained without a valid search warrant. The Ninth Circuit, in *Todd Shipyards Corp. v. Secretary of Labor*, 586 F.2d 683; 6 OSHC 2122 (9th Cir. 1978), suggested that the rule should not apply because OSHA hearings are civil rather than criminal in nature. The Tenth Circuit indicated the contrary in *Savina Home Industries, Inc. v. Secretary of Labor and OSAHRC*, 594 F.2d 1358; 7 OSHC 1154 (10th Cir. 1979), holding that the rule should apply in order to deter improper OSHA inspections. The Eleventh Circuit, while not deciding whether the exclusionary rule must be applied in OSHA enforcement proceedings, has stated that the OSHA Review Commission is free to apply the rule if it sees fit, and need not allow an exception permitting the use of evidence obtained by OSHA "in good faith." *Donovan v. Sarasota Concrete Co.* (Ref. 10). On the other hand, the Seventh Circuit held in *Donovan v. Federal Clearing Die Casting Co. and OSAHRC*, 695 F.2d 1020; 11 OSHC 1014 (7th Cir. 1982), that the commission must apply a "good faith" exception to the exclusionary rule where an OSHA search was made pursuant to a warrant upheld by the district court, even though the warrant was later invalidated on appeal. In 1984, in *INS v. Lopez-Mendoza*, 468 U.S. 1032, the Supreme Court held that the exclusionary rule does not apply in one civil context, deportation proceedings. Subsequently, in *Pennsylvania Steel Foundry and Machine Co. v. Sec'y of Labor*, 831 F.2d 1211; 13 OSHC 1417 (3d Cir. 1987), the Third Circuit cited *Lopez-Mendoza* in finding the exclusionary rule did not apply to an OSHA proceeding when the invalidity of the warrant was not based on constitutional violations but merely on a violation of OSHA regulations. Similarly, in *Smith Steel Casting Co. v. Brock*, 800 F.2d 1329; 12 OSHC 2121 (5th Cir. 1986), the Fifth Circuit stated that the exclusionary rule should not be used to exclude evidence obtained through an invalid warrant for purposes of preventing the secretary from ordering correction of OSHA violations involving unsafe working conditions; however, the illegally obtained evidence could be excluded for purposes of assessing penalties to punish past violations (subject to the good faith exception). As these conflicting decisions indicate, the applicability of the exclusionary rule in OSHA enforcement proceedings remains an unresolved question.

36. 535 F.2d 371; 4 OSHC 1181 (7th Cir. 1976).

37. 535 F.2d at 377; 4 OSHC at 1185. The Seventh Circuit found substantial compliance in the *Chicago Bridge & Iron Co.* case because of the on-site representative of Chicago Bridge & Iron was informed of the pending inspection and was given literature setting forth the directives of the act.

38. See *Marshall v. C. F. & I. Steel Corp. and OSAHRC*, 576 F.2d 809; 6 OSHC 1543 (10th Cir. 1978), in which the inspector failed to give the employer formal notice that its facilities—in addition to those of its subcontractor—were the target of the inspection. Because the employer's personnel manager did in fact accompany the inspector on his rounds, the court held that dismissal of the citation would be a grossly excessive sanction in relation to the inspector's minor violations of Section 8(e).

39. See *Hartwell Excavating Co. v. Dunlop*, 537 F.2d 1071; 4 OSHC 1331 (9th Cir. 1976), in which the court found substantial compliance even though the employer's superintendent was not notified of the inspection until it had been partially completed, because the inspector had made an unsuccessful attempt to locate the supervisor earlier.

40. 655 F.2d 41; 9 OSHC 2075 (4th Cir. 1981).

41. 560 F.2d 947; 5 OSHC 1732 (8th Cir. 1977).

42. In *Leone v. Mobil Oil Corp.*, 523 F.2d 1153 (D.C. Cir. 1975), the District of Columbia Circuit held that neither OSHA nor the Fair Labor Standards Act of 1938, 29 U.S.C. §203(o), requires an employer to pay wages for time spent by employees in accompanying OSHA inspectors on walk-around inspections of the employer's plant. However, in 1977, OSHA announced an amendment of its administrative regulations (29 CFR §1977.21) to reflect a new policy that employees should be paid by their employers for time spent on walk-arounds. The Chamber of Commerce of the United States filed suit challenging the policy, and the Court of Appeals for the District of Columbia Circuit ultimately held the regulation invalid because OSHA had failed to promulgate it properly under the notice and comment requirements of the Administrative Procedure Act. *Chamber of Commerce of United States v. OSHA*, 636 F.2d 464; 8 OSHC 1648 (D.C. Cir. 1980). The regulation was subsequently revoked [45 *Fed. Reg.* 72118 (1980); BNA, *OSHR*, Current Report for October 30, 1980, pp. 593–594], and a new one was proposed and issued [45 *Fed. Reg.* 75232 (1980); 46 *Fed. Reg.* 3852 (1981); BNA, *OSHR*, Current Reports for: November 20, 1980, pp. 653–658; January 22, 1981, pp. 845–853]. After the Reagan administration took office, however, implementation of the new walk-around pay rule was delayed in response to a memorandum from the president [BNA, *OSHR*, Current Report for February 5, 1981, p. 1225], and was ultimately withdrawn [46 *Fed. Reg.* 28842 (1981); BNA, *OSHR*, Current Report for June 4, 1981, pp. 21–24].

43. 29 CFR §1908.6.

44. See OSHA Instruction TED 8.1 (Nov. 10, 1986), in BNA, *OSHR*, Reference File at 77:4001.

45. See 47 *Fed. Reg.* 29025 (July 2, 1982); OSHA Instruction CPL 5.1.II.

46. BNA, *OSHR*, Current Report for December 19, 1990, p. 1163. It is OSHA's policy to inspect companies with higher than average injury rates. OSHA Notice CPL 2.82,

reported in CCH, *Empl. Safety & Health Guide Developments* ¶ 9627 (Mar. 25, 1988). Further, unionized workplaces are more likely to receive safety and health inspections (and pay higher fines) than nonunionized workplaces. BNA, *OSHR*, Current Report for April 16, 1991, p. 1244.

47. Record-keeping and reporting requirements under the Federal Mine Safety and Health Act of 1977 are set forth in subsections 103(c)–103(e) and 103(h) of that statute.

48. These regulations are found in 29 CFR Part 1904. The secretary has defined "recordable occupational injuries or illnesses" as those that result in [29 CFR §1904.12(c)]:

   (1) Fatalities, regardless of the time between the injury and death, or the length of the illness; or
   (2) Lost workday cases, other than fatalities, that result in lost workdays; or
   (3) Nonfatal cases without lost workdays, which result in transfer to another job or termination of employment, or require medical treatment (other than first aid) or involve: loss of consciousness or restriction of work or motion. This category also includes any diagnosed occupational illnesses that are reported to the employer but are not classified as fatalities or lost workday cases.

   Employers should resolve doubts about whether an injury is recordable—for example, because it is unclear whether an injury required first aid or medical treatment or whether the injury is occupationally related—in favor of recording. See, for example, *General Motors Corp., Inland Div.*, 8 OSHC 2036 (Rev. Comm. 1980); *General Motors Corp. Warehousing and Distribution Division*, 10 OSHC 1844 (Rev. Comm. 1982).

49. 29 CFR §1904.6. Effective January 1, 1991, the responsibility for establishing record-keeping requirements for occupational illnesses and injuries was transferred from the Bureau of Labor Statistics (BLS) to OSHA and its newly created Office of Record-keeping and Data Analysis [BNA, *OSHR*, Current Report for January 23, 1991, pp. 1269–1270]. OSHA subsequently has undertaken revision of its record-keeping regulations (BNA, *OSHR*, Current Report for March 6, 1991, p. 1444). Slated for formal proposal in 1992, the new regulations are reported to be geared toward simplifying the required paperwork and improving the quality of information elicited (BNA, *OSHR*, Current Report for January 8, 1992, p. 1136). Pursuant to the anticipated rule making, the OSHA Form 200 Annual Summary Log and Form 101 Supplementary Record Form are being revised (BNA, *OSHR*, Current Reports for February 19, 1992, p. 1254 and March 25, 1992, p. 1421).

50. 57 *Fed. Reg.* 21222 (May 19, 1992).

51. 29 CFR §§1904.20–1904.21.

52. 29 CFR §1904.15.

53. Subsection 103(c) of the Federal Mine Safety and Health Act of 1977 contains a similar requirement guaranteeing employees access to toxic materials exposure records.

54. 29 CFR §1910.20. The medical records rule is augmented by the specific permanent standards issued by the secretary of labor for certain toxic materials and harmful physical agents. See, for example, the OSHA standard regarding employee exposure to coke oven emissions, 29 CFR §1910.1029(m).

55. Despite these privacy interests, OSHA is specifically granted access to employee medical records. 29 CFR §1910.20(c)(10). However, additional regulations require OSHA to observe administrative procedures designed to protect the confidentiality of this information. 29 CFR Part 1913. In a case involving similar privacy concerns, the Supreme Court refused to review the Sixth Circuit's ruling that NIOSH may subpoena

employee medical records in the course of a health hazard inquiry. In *General Motors Corp. v. NIOSH*, 636 F.2d 163; 9 OSHC 1139 (6th Cir. 1980), *certiorari denied*, 454 U.S. 877; 10 OSHC 1032 (1981), the court held that, as long as there is no public disclosure of this medical information, NIOSH may obtain access to it without violating the employees' privacy rights. At least one federal district court has reached a similar conclusion regarding OSHA's medical records access rule. *Louisiana Chemical Assoc. v. Bingham*, 550 F. Supp. 1136; 10 OSHC 2113 (W.D. La., 1982), *affirmed*, 731 F.2d 280, 11 OSHC 1992 (5th Cir. 1984).

56. The Review Commission has held that this 6-month limitation period does not apply if the secretary's inability to discover the violation was caused by the employer's failure to report a fatal accident as the secretary's regulations require [29 CFR §1904.8]. The commission reasoned that allowing an employer to escape a citation because of its own failure to comply with OSHA's reporting requirements would reward it for its own wrongdoing. *Yelvington Welding Service*, 6 OSHC 2013 (Rev. Comm. 1978).

57. See Section 9(a). Interpretations of the "particularity" requirements have generally dealt with: (a) the precision of the reference to the standard allegedly violated, and (b) the adequacy of the description of the alleged violation. Several federal courts have held that, to meet the particularity requirement, a citation must provide the employer with "fair notice" of the violation sufficient to enable it both to prepare its defense and to correct the cited hazard. *Marshall v. B.W. Harrison Lumber Co. and OSAHRC*, 569 F.2d 1303; 6 OSHC 1446 (5th Cir. 1978); *Whirlpool Corp. v. OSAHRC and Marshall*, 645 F.2d 199; 7 OSHC 2059 (9th Cir. 1980), *certiorari denied*, 457 U.S. 1132 (1982); *Noblecraft Industries, Inc. v. Secretary of Labor and OSAHRC*, 614 F.2d 199; 7 OSHC 2059 (9th Cir. 1980); *Whirlpool Corp. v. OSAHRC and Marshall*, 645 F.2d 1096; 9 OSHC 1362 (D.C. Cir. 1981). See also *Otis Elevator Co.*, 13 OSHC 1791 (Rev. Comm. 1988) [holding that the citation was not too vague because (1) the basic elements of the charge describing the hazard were present, and (2) there was extensive pretrial discovery], *affirmed*, 871 F.2d 155; 13 OSHC 2085 (D.C. Cir. 1989).

58. The Fifth Circuit has held that an employer may raise a citation's lack of particularity in a later failure-to-abate action by the secretary. In so holding, the court reasoned that a citation that is too vague to give notice of the action necessary to correct the cited hazard also makes it impossible for the Review Commission to determine whether the hazard has been abated. See *Marshall v. B.W. Harrison Lumber Co. and OSAHRC* (Ref. 57).

59. The secretary's regulations can be found at 29 CFR §1903.16. The employer must post notice of OSHA citations at or near the place of the violation immediately upon receipt of the notice [29 CFR §§1903.14, 1903.16]. If it is not possible to post the notice of citation at the violation site, employers must post it in a "prominent place where it will be readily observable by all affected employees" [29 CFR §1903.16(a)]. If the employer settles with OSHA (to reduce the fines), a notice of settlement must also be posted [29 CFR §2200.100(c)]. Section 17(i) mandates penalties of up to $7000 for each violation of the posting requirements.

60. Penalties are prescribed by Section 17 of OSHA. Applicable regulations on proposed penalties are found at 29 CFR §1903.15. In addition to the gravity of the violation, the size of the business, the employer's good faith, and the employer's history of previous violations are also taken into account in determining the amount of the proposed penalty [29 CFR §1903.15(b)].

61. Following passage of the new law, OSHA amended its Field Operations Manual with respect to its instructions on penalties. See BNA, *OSHR*, Current Report for January

30, 1991, p. 1289. The penalties section of the Field Operations Manual [OSHA Instruction CPL 2.46B, as amended] is found in BNA, *OSHR*, Reference File at 77:2701.

62. *Dorey Electric Co. v. OSAHRC*, 553 F.2d 357; 5 OSHC 1285 (4th Cir. 1977); *California Stevedore and Ballast Co. v. OSAHRC*, 517 F.2d 986; 3 OSHC 1174 (9th Cir. 1975); *Shaw Construction, Inc. v. OSAHRC*, 534 F.2d 1183, 1185 & n. 4; 4 OSHC 1427, 1428 & n. 4 (5th Cir. 1976). The Review Commission has held that the probability of an accident, while not necessary to establish a serious violation, is a factor to be considered in determining the gravity of the violation for penalty assessment purposes. Thus, an employer's accident-free history, while not a defense to a citation for a serious violation, may justify reducing the proposed penalty. *George C. Christopher & Sons, Inc.*, 10 OSHC 1436, 1446 (Rev. Comm. 1982).

63. *Accord, Usery v. Hermitage Concrete Pipe Co. and OSAHRC*, 584 F.2d 127; 6 OSHC 1886 (6th Cir. 1978); *Kent Nowlin Construction Co., Inc. v. OSAHRC*, 648 F.2d 1278; 9 OSHC 1709 (10th Cir. 1981); *Central Brass Mfg. Co.*, 13 OSHC 1609, 1611 (Rev. Comm. 1987).

64. *Brennan v. OSAHRC and Vy Lactos Laboratories Inc.*, 494 F.2d 460; 1 OSHC 1623 (8th Cir. 1974).

65. An employer has an obligation to inspect the workplace to discover and prevent possible hazards, and its failure to do so can support a finding that it should have known of the violations it failed to discover. *Joseph J. Stolar Construction Co., Inc.*, 9 OSHC 2020 (Rev. Comm.), *appeal denied*, 681 F.2d 801; 10 OSHC 1936 (2d Cir. 1981). However, the existence of an effective employer inspection and maintenance program, designed to discover and remedy hazardous conditions, may preclude a finding that the employer should reasonably have known of violations it actually failed to discover. *Cullen Industries, Inc.*, 6 OSHC 2177 (Rev. Comm. 1978). See, however, *East Texas Motor Freight, Inc. v. OSAHRC and Donovan*, 671 F.2d 845; 10 OSHC 1457 (5th Cir. 1982) (because employer's inspection program was not effective in discovering and repairing defective machinery, employer was charged with knowledge of undiscovered defects).

66. *Mid-Plains Construction Co.*, 3 OSHC 1484 (Rev. Comm. 1975); *Southwestern Acoustics & Specialty, Inc.*, 5 OSHC 1091 (Rev. Comm. 1977); *Cleveland Consol.*, 13 OSHC 1114 (Rev. Comm. 1987); *B.B. Riverboats, Inc.*, 13 OSHC 1350 (Rev. Comm. 1987).

67. *Brennan v. OSAHRC and Hendrix (d/b/a Alsea Lumber Co.)*, 511 F.2d 1139, 1144; 2 OSHC 1646, 1648–1649 (9th Cir. 1975). *Accord, Ocean Electric Corp. v. Secretary of Labor and OSAHRC*, 594 F.2d 396; 7 OSHC 1149 (4th Cir. 1979); *Diversified Industries Div., Independent Stave Co. v. OSAHRC and Marshall*, 618 F.2d 30, 31 n. 8; 8 OSHC 1107, 1108 n. 8 (8th Cir. 1980).

68. Conference Report 91-1765, 1970, *U.S. Code Congressional and Administrative News*, p. 5237; OSHA §17(c).

69. *Ryder Truck Lines, Inc. v. Brennan*, 497 F.2d 230, 233; 2 OSHC 1075, 1077 (5th Cir. 1974).

70. *Standard Glass and Supply Co.*, 1 OSHC 1223–1224 (Rev. Comm. 1973). See also *Consolidated Rail Corp.*, 10 OSHC 1564, 1568 (Rev. Comm. 1982) (where the evidence did not show the violation created a substantial probability of serious injury in case of an accident, the citation was affirmed as nonserious rather than serious); *Pace Constr. Co.*, 12 OSHC 1830, 1831 (Rev. Comm. 1986) (because the operators near an unguarded hand grinder machine wore goggles and stood at a distance from the

machine and there had been no prior accidents, the violation was only nonserious); *Cuthers Corp. d/b/a Woodland Constr.*, 13 OSHC 1906 (Rev. Comm. 1988) (there was no probability of death because the machine moved very slowly).

In determining whether a violation is serious or nonserious, the commission has held that a number of nonserious violations may be grouped together to form a serious violation if the cumulative effect of the violations could result in an accident causing death or serious injury. *H.A.S. & Associates*, 4 OSHC 1894, 1897–1898 (Rev. Comm. 1976). When the violation poses a risk of illness through cumulative exposure to a toxic substance, the relevant question is whether the secretary can show a substantial probability of serious harm resulting from the degree and length of actual employee exposure. *Bethlehem Steel Corp.*, 11 OSHC 1247, 1252 (Rev. Comm. 1983); *Texaco, Inc.*, 8 OSHC 1758, 1761 (Rev. Comm. 1980).

71. *Brennan v. OSAHRC and Hendrix (d/b/a Alsea Lumber Co.)* (67); *National Steel and Shipbuilding Co. v. OSAHRC*, 607 F.2d 311, 315–316 n. 6; 7 OSHC 1837, 1840 n. 6 (9th Cir. 1979). See also *Dunlop v. Rockwell International*, 540 F.2d 1283, 1291; 4 OSHC 1606, 1611–1612 (6th Cir. 1976), in which the court upheld a divided Review Commission's affirmance of a nonserious, rather than a serious, violation where employer knowledge was not established. The court, in dictum, approved of the reasoning of the Ninth Circuit in the *Alsea Lumber* case. *Contra, Arkansas–Best Freight Systems, Inc. v. OSAHRC*, 529 F.2d 649, 655 n. 11; 3 OSHC 1910, 1913 n. 11 (8th Cir. 1976); *Brennan v. OSAHRC and Interstate Glass Co.*, 487 F.2d 438, 442 n. 19; 1 OSHC 1372, 1375 n. 19 (8th Cir. 1973).

72. Section 17(i) of the act requires that a penalty be assessed for violation of OSHA's posting requirements (Ref. 59). However, the commission has stated that such a violation is nevertheless considered ''nonserious.'' *Thunderbolt Drilling, Inc.*, 10 OSHC 1981 (Rev. Comm. 1982).

73. *Lee Way Motor Freight, Inc. v. Secretary of Labor*, 511 F.2d 864, 869; 2 OSHC 1609, 1612 (10th Cir. 1975). See §9(a) of the act and the regulations promulgated thereunder at 29 CFR §1903.14.

74. It is OSHA policy to document *de minimis* violations in the same way as other violations except that *de minimis* violations are not included on the citation. OSHA Instruction 2.45 B, as amended, is found at BNA, *OSHR*, Reference File at 77:2501. The passage pertaining to *de minimis* violations is found at 77:2512–13.

75. OSHA Instruction 2.45 B, as amended, BNA, *OSHR*, Reference File at 77:2512–13. For some examples of *de minimis* violations: *Cleveland Consol. Inc.*, 13 OSHC 1114, 1118 (Rev. Comm. 1987) (failure to post hospital telephone numbers was a *de minimis* violation where hospital was located within 300 yards of the work site); *Daniel Constr. Corp.*, 13 OSHC 1128, 1131 (Rev. Comm. 1986) (failure to properly label an air receiver in compliance with the Boiler and Pressure Vehicle Code was a *de minimis* violation because it had no bearing on the health and safety of employees).

76. The Federal Mine Safety and Health Act of 1977 [Section 110(d)] does not provide more stringent civil penalties for willful violations of that act's safety and health standards but does impose criminal liability on mine operators who are found guilty of such willful conduct. The act does impose [Section 110(g)] a civil penalty on miners who willfully violate the safety standards relating to smoking or the carrying of smoking materials.

77. 519 F.2d 1200, 1207; 2 OSHC 1283, 1289 (3d Cir. 1974), *affirmed on other points sub non. Atlas Roofing Co., Inc. v. OSAHRC*, 430 U.S. 442 (1977).

78. *Messina Construction Corp. v. OSAHRC*, 505 F.2d 701; 2 OSHC 1325 (1st Cir. 1974); *Intercounty Construction Co. v. OSAHRC*, 522 F.2d 777, 780; 3 OSHC 1337, 1339 (4th Cir. 1975), *certiorari denied* 423 U.S. 1072 (1976).

79. *Kent Nowlin Construction, Inc.*, 5 OSHC 1051, 1055 (Rev. Comm. 1977), *affirmed in relevant part* 593 F.2d 368, 369; 7 OSHC 1105, 1108 (10th Cir. 1979).

80. *Kus-Tum Builders, Inc.*, 10 OSHC 1128, 1131 (Rev. Comm. 1981); *A. Schonbek & Co., Inc.*, 9 OSHC 1189, 1191 (Rev. Comm. 1980), *affirmed* 646 F.2d 799; 9 OSHC 1562 (2d Cir. 1981). Several state tribunals have also adopted the "intentional disregard and indifference" test for "willfulness." See *Monadnock Fabricators, Inc.* (Docket No. RB491), a ruling of the Vermont Occupational Safety and Health Review Board (summarized in BNA, *OSHR*, Current Report for June 24, 1982, p. 96). The New Mexico Health and Safety Review Commission applied the same standard in *Environmental Improvement Division v. Stearns-Roger, Inc.* (Docket No. 82-3) (summarized in BNA, *OSHR*, Current Report for November 18, 1982, pp. 491–492).

81. *A. Schonbek & Co., Inc. v. Donovan and OSAHRC*, 646 F.2d 799, 9 OSHC 1562 (2d Cir. 1981), *affirming*, 9 OSHC 1189 (Rev. Comm. 1980); *Mineral Industries & Heavy Construction Group (Brown & Root, Inc.) v. OSAHRC and Marshall*, 639 F.2d 1289, 9 OSHC 1387 (5th Cir. 1981); *Georgia Electric Co. v. Marshall and OSAHRC*, 595 F.2d 309; 7 OSHC 1343 (5th Cir. 1979); *Empire–Detroit Steel Div., Detroit Steel Corp. v. OSAHRC and Marshall*, 579 F.2d 378; 6 OSHC 1693 (6th Cir. 1978); *Western Waterproofing Co., Inc. v. Marshall and OSAHRC*, 576 F.2d 139; 6 OSHC 1550 (8th Cir.), *certiorari denied*, 439 U.S. 965 (1978); *National Steel & Shipbuilding Co. v. OSAHRC* (Ref. 63); *Kent Nowlin Construction Co. v. OSAHRC and Marshall*, 593 F.2d 368; 7 OSHC 1105 (10th Cir. 1979), *affirming in relevant part* 5 OSHC 1051 (Rev. Comm. 1977).

82. 587 F.2d 1303; 6 OSHC 2010, 2012 (D.C. Cir. 1978).

83. 622 F.2d 1160; 8 OSHC 1317 (3d Cir. 1980).

84. 622 F.2d at 1167–1168; 8 OSHC at 1322 (footnotes omitted).

85. 540 F.2d 157; 4 OSHC 1451 (3d Cir. 1976).

86. 582 F.2d 834, 840–41; 6 OSHC 1855, 1859 (4th Cir. 1978).

87. See *Todd Shipyards Corp. v. Secretary of Labor*, 586 F.2d 683; 6 OSHC 2122 (9th Cir. 1978); *Bunge Corp. v. Secretary of Labor and OSAHRC*, 638 F.2d 831; 9 OSHC 1312 (5th Cir. 1981); *J.L. Foti Construction Co. v. OSAHRC and Donovan*, 687 F.2d 853; 10 OSHC 1937 (6th Cir. 1982); *Dun-Par Engineered Form Co. v. Marshall and OSAHRC*, 676 F.2d 1333; 10 OSHC 1561 (10th Cir. 1982).

88. 7 OSHC 1061 (Rev. Comm. 1979).

89. 7 OSHC at 1063. Except for those within the Third Circuit, Review Commission judges continue to find repeated violations after only one prior citation. For example: *Carl Thomas Constr. Corp.*, 13 OSHC 1671 (Rev. Comm. 1988); *Bat Masonry Co.*, 13 OSHC 1876, 1877 (Rev. Comm. 1988). Compare *Mellon Stuart Co.*, 13 OSHC 1025, 1026 (Rev. Comm. 1987).

90. The definition of "substantially similar" violations and the proper allocation of the burden of proving such similarity is not settled. For a discussion of these problems, see *Bunge Corp. v. Secretary of Labor and OSAHRC* (Ref. 87). In Chapter IV of its Field Operations Manual (pertaining to citing violations), OSHA sets forth its position on various issues as to what constitutes a repeated violations (e.g., whether the violations must be of the identical standard; whether the violations must have occurred at

the same site for a single employer with multiple workplace locations; and whether the time between violations affects a "repeated" finding). See BNA, *OSHR*, Reference File at 77:2501, 2511–12.

91. Chapter IV of OSHA's Field Operations Manual, printed in BNA, *OSHR*, Reference File at 77:2501, 2512.

92. The U.S. Supreme Court recently decided not to review the issue of whether OSHA preempts state criminal proceedings for health and safety violations. In *Pymm v. New York*, the state convicted two corporate officers on charges of conspiracy, falsifying business records, assault, and reckless endangerment. 151 A.D.2d 133, 546 N.Y.S.2d 871; 14 OSHC 1297 (1989), *affirmed*, 76 N.Y.2d, 561 N.Y.S.2d 687, 563 N.E.2d 1; 14 OSHC 1833 (1990), *certiorari denied*, 111 S. Ct. 958 (1991). The verdict was set aside by the trial court on the grounds that federal OSHA preempted state prosecution. The appellate court and the state high court reinstated the verdict, concluding that the OSH act does not preempt state criminal prosecutions. A Texas court has reached the same result in *Sabine Consol. Inc. v. Texas*, 806 S.W.2d 553, 14 OSHC 2049 (Tex.Ct.Cr.App. 1991), and those convicted are also requesting review by the U.S. Supreme Court on another variation of this preemption question. BNA, *OSHR*, Current Report for July 15, 1992, p. 224.

93. 30 U.S.C. §820(d).

94. For an interpretation of "willful" in the context of Section 17(e) of OSHA, 29 U.S.C. §666(e), see *United States v. Dye Construction Co.*, 510 F.2d 78 (10th Cir. 1975); *United States v. Dye Constr. Co.*, 510 F.2d 78; 2 OSHC 1510 (10th Cir. 1975).

95. 29 U.S.C. §666(e). Currently there are efforts in Congress to increase the criminal sanctions for OSHA violations. For example, the Workplace Protection Act of 1991 is proposed legislation that would, among other things, expand the scope of OSHA criminal sanctions (to include employers whose willful conduct resulted in employee injury, not just death), as well as increase the magnitude of the sanctions (to maximum prison terms of 10 to 20 years). BNA, *OSHR*, Current Report for June 3, 1992, p. 5.

96. See Section 17(g) of OSHA, 29 U.S.C. §666(g) and Section 110(f) of the Federal Mine Safety and Health Act of 1977, 29 U.S.C. §820(f).

97. See Section 17(f) of OSHA, 29 U.S.C. §666(f) and Section 110(e) of the Federal Mine Safety and Health Act of 1977, 29 U.S.C. §820(e).

98. See Section 110(h) of the Federal Mine Safety and Health Act of 1977.

99. See the United States Criminal Code, 18 U.S.C. §1114.

100. 29 U.S.C. §658(a). In determining a reasonable abatement period, OSHA's Field Operations Manual advises consideration of the following factors: (1) the gravity of the alleged violation; (2) the availability of needed equipment, material and/or personnel; (3) the time required for delivery, installation, modification, or construction; and (4) training of personnel. BNA, *OSHR*, Reference File at 77:2325.

101. 29 U.S.C. §659(b).

102. 29 U.S.C. §666(d).

103. Despite the requirements of §10(a), the Review Commission may in limited circumstances entertain a late notice of contest pursuant to Federal Rule 60(b). This rule permits a federal court to grant relief from a final order for a number of reasons, including a party's mistake, surprise or excusable neglect, the presence of newly discovered evidence, the fraud or misconduct of an adverse party, and so on. Rule 60(b), which was held applicable to the Review Commission by the Third Circuit in *J.I. Hass*

Co. v. OSAHRC, 648 F.2d 190; 9 OSHC 1712 (3d Cir. 1981), has been viewed by the commission as affording a possible basis for considering an untimely notice of contest. *Special Coating Systems of New Mexico, Inc.*, 10 OSHC 1671 (Rev. Comm. 1982). But see *J.F. Shea Co.*, 15 OSHC 1092 (Rev. Comm. 1991) (refusing to allow a notice of contest that had been filed eight days late due to an administrative mishap in the company's mailroom); *Medco Plumbing, Inc.*, 15 OSHC 1325 (OSHRC J. 1991) [dismissing notice of contest that was one day late because illness of secretary and being busy with other matters were not considered excusable neglect under Rule 60(b)].

104. 29 CFR §1903.17.

105. *Brennan v. OSAHRC and Bill Echols Trucking Co.*, 487 F.2d 230; 1 OSHC 1398 (5th Cir. 1973); *Dan J. Sheehan Co. v. OSAHRC and Dunlop*, 520 F.2d 1036; 3 OSHC 1573 (5th Cir. 1975), *certiorari denied*, 424 U.S. 956 (1976).

106. Dan J. Sheehan (105), 520 F.2d at 1038–1039; 3 OSHC at 1575.

107. *Turnbull Millwork Co.*, 3 OSHC 1781 (Rev. Comm. 1975); *State Home Improvement Co.*, 6 OSHC 1249 (Rev. Comm. 1977). The Seventh and Eighth Circuits have expressed their approval of the commission's Turnbull rule, at least in situations in which the employer is a layman proceeding *pro se*. *Penn-Dixie Steel Corp. v. OSAHRC and Dunlop*, 553 F.2d 1078; 5 OSHC 1315 (7th Cir. 1977); *Marshall v. Gil Haughan Construction Co. and OSAHRC*, 586 F.2d 1263; 6 OSHC 2067 (8th Cir. 1978). The commission has gone further and has applied the rule even where the employer is represented by counsel. *Nilsen Smith Roofing & Sheet Metal Co.*, 4 OSHC 1765 (Rev. Comm. 1976). However, the commission has made it clear that the same lenient policy does not apply to an employer who understands the difference between contesting a penalty and a citation, and who nevertheless limits his notice of contest to the penalty. *F.H. Sparks of Maryland, Inc.*, 6 OSHC 1356 (Rev. Comm. 1978).

108. See Section 10(c) of OSHA. The rules governing practice before the commission can be found at 29 CFR §§2200.1–2200.212. In addition to the extensive, formal rules of practice, these rules include a subpart (at §§2200.200–2200.212) to permit any party to request simplified proceedings before an ALJ in cases that do *not* involve alleged general duty violations or alleged violations of certain enumerated standards. Procedures are simplified in several ways: the pleadings are limited, discovery is generally not permitted, the Federal Rules of Evidence do not apply, and interlocutory appeals are not permitted.

109. The commission's rules of procedure provide that a petition for discretionary review must state specific grounds for relief. The commission normally limits review to cases in which a party asserts that: (a) a finding of material fact is not supported by a preponderance of the evidence; (b) the ALJ's decision is contrary to law or to the rules or decisions of the commission; (c) an important question of law, policy, or discretion is involved; (d) the administrative law judge committed a prejudicial procedural error or an abuse of discretion; or (e) review by the full commission will resolve a question about which individual ALJs have rendered differing opinions. 29 CFR §2200.91(d).

110. 29 CFR §§2200.91 and 2200.92. Any member of the commission clearly has the authority to direct review of a case on his own motion, even if no party has requested review. *GAF Corp.*, 8 OSHC 2006 (Rev. Comm. 1980). However, the commission's rules limit the grounds on which a member may independently order review. Except in extraordinary circumstances, review is limited to the issues raised by the parties before the administrative law judge, 29 CFR §2200.92(c). Thus, it is normally preferable for a party to obtain review by filing a petition. Indeed, a party's failure to file a petition

may foreclose later judicial review of any objections to the administrative law judge's decision. 29 CFR §2200.91(f). See also *Keystone Roofing Co., Inc. v. OSAHRC and Dunlop*, 539 F.2d 960; 4 OSHC 1481 (3d Cir. 1976); *McGowan v. Marshall*, 604 F.2d 885; 7 OSHC 1842 (5th Cir. 1979).

111. 29 CFR §2200.90(d).

112. The Review Commission's failure to adequately consider these statutory penalty assessment factors can result in the Court of Appeals remanding the case to the commission for more particularized findings regarding the penalty. *Astra Pharmaceutical Products, Inc. v. OSAHRC and Donovan*, 681 F.2d 69; 10 OSHC 1697 (1st Cir. 1982).

113. *REA Express v. Brennan*, 495 F.2d 822; 1 OSHC 1651 (2d Cir. 1974); *Brennan v. OSAHRC and Interstate Glass Co.*, 487 F.2d 438; 1 OSHC 1372 (8th Cir. 1973); *California Stevedore & Ballast Co. v. OSAHRC* (Ref. 62). Reduction of the secretary's proposed penalty is also a matter within the commission's discretion. *Western Waterproofing Co., Inc. v. Marshall and OSAHRC* (Ref. 81). Compare *Dale M. Madden Construction, Inc. v. Hodgson*, 502 F.2d 278; 2 OSHC 1236 (9th Cir. 1974), where the court held that the commission has no authority to modify *settlements* made by the secretary and the cited employer. See also *Marshall v. Sun Petroleum Products Co. and OSAHRC*, 622 F.2d 1176; 8 OSHC 1422 (3d Cir. 1980), *certiorari denied*, 449 U.S. 1061, holding that the commission may review and modify settlements, but only for the limited purpose of determining the reasonableness of the abatement period when it has been challenged by employees.

114. See *California Stevedore* (Ref. 62).

115. See, for example, *Dixie Roofing and Metal Co.*, 2 OSHC 1566 (Rev. Comm. 1975). See also *Consolidated Rail Corp.*, 10 OSHC 1564, 1568 (Rev. Comm. 1982), in which the commission reduced the degree of a violation from serious to nonserious without specifically discussing its authority to do so. The secretary, however, has recently argued that the commission's authority does *not* include reclassifying a citation as *de minimis*. BNA, *OSHR*, Current Report for October 28, 1992, p. 1127.

116. Although §11(a) permits "any person" aggrieved by a commission order to obtain judicial review, the Third Circuit has held that the role of employees or their representatives in a proceeding initiated by the employer is strictly limited to challenging the reasonableness of the abatement period pursuant to §10(c) of the act. In *Marshall v. Sun Petroleum Products* (Ref. 113), the court indicated that this limitation should also apply to the right of employees to initiate judicial review. On the other hand, the District of Columbia Circuit has concluded that employees, although prohibited from *instituting* a commission action on matters other than the reasonableness of the abatement period, may nevertheless participate fully as parties in employer-initiated proceedings. The right of employees to participate in enforcement proceedings, the court reasoned, must include the right to appeal from an unfavorable commission decision. *Oil, Chemical, and Atomic Workers v. OSAHRC*, 671 F.2d 643; 10 OSHC 1345 (D.C. Cir.), *certiorari denied sub nom. American Cyanamid and Atomic Workers*, 459 U.S. 905 (1982).

117. Petitions for review of a commission order must be filed within 60 days of issuance of the order. Review is obtained in either the court of appeals where the violation is alleged to have occurred, where the employer has its principal office, or in the Court of Appeals for the District of Columbia Circuit. 29 U.S.C. §660(a).

118. The Sixth Circuit has defined "substantial evidence" in the context of OSHA proceedings as "such relevant evidence as a reasonable mind might accept as adequate to

support a conclusion." *Martin Painting & Coating Co. v. Marshall*, 629 F.2d 437; 8 OSHC 2173, 2174 (6th Cir. 1980), *certiorari denied*, 449 U.S. 1062, quoting *Dunlop v. Rockwell International*, 540 F.2d 1283, 1287; 4 OSHC 1606, 1608 (6th Cir. 1976). While the commission's factual findings are conclusive if supported by substantial evidence, its interpretations of statutory language are legal conclusions, which are not accorded the same deference on review. *Usery v. Hermitage Concrete Pipe Co. and OSAHRC* (Ref. 63). The commission's resolution of questions of credibility, on the other hand, are insulated from reversal on appeal unless they are contradicted by "uncontrovertible documentary evidence or physical facts." *Super Excavators, Inc. v. OSAHRC and Secretary of Labor*, 674 F.2d 592; 10 OSHC 1369, 1370 (7th Cir. 1981), *certiorari denied*, 457 U.S. 1133; 11 OSHC 1304 (1982), quoting *International Harvester Co. v. OSAHRC*, 628 F.2d 982, 986; 8 OSHC 1780, 1783 (7th Cir. 1980).

119. *Secretary v. OSAHRC and Interstate Glass* (Ref. 71), 487 F.2d at 442; 1 OSHC at 1375.

120. See *Usery v. Whirlpool Corp.*, 416 F. Supp. 30, 34; 4 OSHC 1391, 1392–1393 (N.D. Ohio 1976), *reversed on other points sub nom. Marshall v. Whirlpool Corp.*, 593 F.2d 715; 7 OSHC 1075 (6th Cir. 1979), *affirmed*, 445 U.S. 1 (1980). The U.S. Supreme Court in the *Whirlpool* case was faced with the question whether an employee is protected from retaliation by his employer if he walks off the job in an imminent danger situation. The secretary of labor, in 1973, promulgated a regulation that protected employees from discrimination if they refused in good faith to work under life-threatening conditions. The rule applied only if the employee reasonably believed that a real danger of death or serious injury existed, and that the danger was too immediate to be eliminated through the act's normal enforcement channels. 29 CFR §1977.12(b)(2). In *Whirlpool Corp.*, the Supreme Court upheld §1977.12(b)(2) as a valid exercise of the secretary's authority under the act. Thus, employees are afforded a limited right to refuse to work in imminent danger situations. See also G. Scarzafava and F. Herrera, Jr., *Workplace Safety—The Prophylactic and Compensatory Rights of the Employee*, 13 St. Mary's L.J. 911, 931–933 (1982).

121. Imminent danger is defined by Section 3(j) of the Federal Mine Safety and Health Act of 1977 as "the existence of any condition or practice in a coal or other mine that could reasonably be expected to cause death or serious physical harm before such condition or practice can be abated. . . ." The definition of "imminent danger" in the Coal Mine Health and Safety Act of 1969 was almost identical to that which now appears in the 1977 act. In a case arising under the 1969 provision, the Seventh Circuit affirmed the Board of Mine Operations Appeals' holding that an "imminent danger" situation exists if, in a reasonable man's estimation, it is at least as probable as not that continuation of normal coal extraction operations in the disputed area will result in the occurrence of the feared accident or disaster before the danger is eliminated. *Freeman Coal Mining Co. v. Interior Board of Mine Operations Appeals*, 504 F.2d 741; 2 OSHC 1308 (7th Cir. 1974). *Accord, Old Ben Coal Corp. v. Interior Board of Mine Operations Appeals*, 523 F.2d 25; 3 OSHC 1270 (7th Cir. 1975); *Eastern Assoc. Coal Corp. v. Interior Board of Mine Operations Appeals*, 491 F.2d 277 (4th Cir. 1974).

122. 518 F.2d 990; 3 OSHC 1490 (5th Cir. 1975); *affirmed*, 430 U.S. 442; 5 OSHC 1105 (1977).

123. The Supreme Court also granted *certiorari* in *Frank Irey, Jr., Inc. v. OSAHRC and Brennan*, 519 F.2d 1215; 3 OSHC 1329 (3d Cir. *en banc* 1975) to review the same issue. *The Atlas Roofing and Frank Irey* cases were decided by the Court in consolidated proceedings.

124. 430 U.S. 442; 5 OSHC 1105 (1977).

125. The "egregious case" policy is referenced in the Field Operations Manual's Chapter VI, relating to penalties, at section B.9.d. (BNA, *OSHR*, Reference File at 77:2704). According to that section, OSHA considers as egregious cases those "willful, repeated and high gravity serious citations and failures to abate." The policy and its implementation are explained in detail in OSHA Instruction CPL 2.80 (October 1, 1990), reprinted in BNA, *OSHR*, Reference File at 21:9649.

126. The policy has been challenged in, among other cases, *Sec'y of Labor v. Caterpillar, Inc.*, OSHRC Docket No. 87-0922 and *Sec'y of Labor v. Kaspar Wire Works, Inc.*, OSHRC Docket No. 90-2775. See BNA, *OSHR*, Current Reports for April 10, 1991, p. 1576, and June 19, 1991, pp. 52–54.

127. BNA, *OSHR*, Current Report for August 19, 1992, p. 391.

128. BNA, *OSHR*, Current Report for November 11, 1992, pp. 1193–1194.

129. OSHA reform legislation is embodied in two bills—House No. 3160 and Senate No. 1622. Some consider OSHA reform one of a few top priority laws for the Clinton administration in 1993. See BNA, *OSHR*, Current Report for November 11, 1992, p. 1188.

130. BNA, *OSHR*, Current Reports for January 8, 1992, p. 1133, and March 18, 1992, p. 1391–92.

CHAPTER NINETEEN

Professional Liability and Litigation

Robert B. Weidner, J. D.

1 INTRODUCTION

There are no two words in the English language that strike more fear into the heart of any professional than "professional liability." In this day and age of myriads of liability suits being filed, professionals may well be reminded of the old saying, "An ounce of prevention is worth a pound of cure"—that is, giving no cause for a liability suit is better than defending one's self in a lawsuit. In past years industrial hygienists may have thought that the only professionals who had to worry about being sued for liability were the physicians and nurses. However, "[N]o man . . . is so high that he is above the law" (1).

It would not be reasonable to illustrate all of the possible ways that an industrial hygienist could be held professionally liable. Suffice it to say that they run the gamut from negligence because an employee was not told that a material with which he or she was working is toxic to an intentional tort because an operation was not shut down even though the industrial hygienist knew that the ventilation system was not operating properly. How an industrial hygienist's liability is established depends on how he or she fits in the picture. Generally, employers are liable for the negligence of employees, whereas independent contractors are sued separately.

The nature of the duty determines who has to pay for failure of the parties to meet that duty and, generally, whether liability is based on no-fault workers' compensation or on common law negligence. Recoveries for damages under common

Patty's Industrial Hygiene and Toxicology, Third Edition, Volume 3, Part A, Edited by Robert L. Harris, Lewis J. Cralley, and Lester V. Cralley.
ISBN 0-471-53066-2 © 1994 John Wiley & Sons, Inc.

law typically are much larger than those under workers' compensation acts; however, a greater burden of proof is required of the injured party.

Common law presumes that each person has a right to be free from unreasonable harm or injury. Should a person be harmed, he or she is to be compensated by the party causing the harm for any damages experienced as long as the harmful conduct was in violation of the common law. Negligence is based essentially on three elements:

1. There must be a legal duty owed by the actor (2). Usually this is a duty to act as a reasonable person would act under the same set of circumstances.
2. It must be shown that the actor in some way breached that duty.
3. It must be shown that the breach of the duty caused the damage that resulted to the plaintiff.

Every industrial hygienist has the responsibility of acting with the same degree of care in discharging his or her duties as a reasonable industrial hygienist would. This standard of care or duty does not mean that the industrial hygienist has to display above-average performance. Generally speaking, this standard of care is essentially the same for any part of the country. If an industrial hygienist does not meet this standard of care in a given situation, then it could be considered negligence.

Proof that an industrial hygienist has not met the standard of care requirement is not grounds in itself for a suit. The plaintiff must also show that actual damage resulted because of an act. The committed act or acts must have substantially contributed to the harm the plaintiff experienced.

Negligence per se is a legal doctrine that applies to an industrial hygienist who is acting in violation of a statute or regulation and is therefore presumed not to be satisfying the required standard of care.

Under the Occupational Safety and Health Administration (OSHA) Regulation of August 1980, an employee has the right to any records that relate to his or her exposure. All employers who deal with toxic materials or harmful agents must make records available to an employee within 15 days of request (3). Records that must be given to an employee include his or her medical tests, environmental data about the workplace, medical opinion relative to the employee, and any medical complaints and prescriptions that are given. Please note that these records must be kept by the employer for at least 30 years.

Union or employee representatives may, with the employee's consent, examine an employee's medical and exposure records. Federal and state governments also may compel production of medical records by statute requiring the reporting of occupational injury or illness, by judicial process, or by quasi-judicial administrative process. Before turning over health information under compulsion of process, the appropriateness of the process itself should be determined. The federal courts have held that the National Institute for Occupational Safety and Health (NIOSH) can obtain individual medical records without an employee's consent only when the employee cannot be identified from the material (4). Case law and judicial trends indicate that revealing employee health and exposure records outside the corporation, unless specifically for the care of the employee or with a carefully worded release by the employee, can be the basis of an action against whoever releases the records. The trend of the law at present is to suggest that employee health and exposure records are not to be disclosed to individuals outside the corporation without a properly signed authorization from the employee.

The role of the industrial hygienist has dramatically changed since the early days of the profession. At one time the industrial hygienist had to be concerned only about being able to recognize the environmental factors and stresses associated with work and work operations and to understand their effects on humans and their well-being; to evaluate, on the basis of experience and with the aid of quantitative measurement techniques, the magnitude of these stresses in terms of ability to impair health and well-being; and to prescribe methods to eliminate, control, or reduce such stresses when necessary to alleviate their effects.

No longer is it sufficient to do what we have previously considered to be an "adequate" job. An industrial hygienist must now perform his or her duties with the ever-present thought that whatever is done may become the subject of litigation. More and more often what we have is one of the most interesting areas of professional interaction between law and science. Striking similarities and sharp disparities exist between the scientific method and the legal process. Law is normative while science is objective. There are some similarities, however; they both rely on precedent, assume the burden of proof, and, to some extent, the evolution of evidence.

It is vital that industrial hygienists be aware not only of the regulations with which they are expected to comply in professional practice but of other aspects of law as they relate to professional liability and to other kinds of litigation in which they may be called upon to play a part.

2 COMMUNICATIONS

2.1 Potential Problems

One area of concern is the communication problems that can exist between the legal community and industrial hygienists. A good industrial hygienist can do wonders to aid an attorney in synthesizing the knowledge of specialists into a usable and understandable body of general knowledge to solve problems. The industrial hygienists who wish their work to have an impact must be capable of communicating their knowledge in an understandable and usable form.

This becomes complicated because certain dissimilarities between the disciplines of law and science do exist. Many writers have emphasized the contrast between the inductive process utilized by science and the deductive process utilized by law. But this is an oversimplification; by no means does it explain the conflicts we hear about. In any event, both the industrial hygienist and the attorney should recognize and respect the differences in the way each approaches and thinks about a problem. Each needs to learn more of, and to understand better, the objectives and techniques of the other.

2.2   Role of the Industrial Hygienist in Litigation

The dissatisfaction that industrial hygienists feel when they are called upon to participate as witnesses in legal proceedings also stems from another contrast—the contrast between the role that they visualize for themselves and the role they often actually play. Too often industrial hygienists consider themselves as giving highly specific answers to highly specific questions. As they visualize the scene, they make their appearances in the courtroom surrounded by an aura of science; they state their pieces with great authority; and then they depart. The effect is dramatic and impressive because difficult controversies are resolved by their testimony.

But, things being the way they are, the drama often plays out altogether differently. Industrial hygienists may blame the attorneys for any mishaps and the attorneys may blame the industrial hygienists. A major part of the problem is a communications gap, if not a communications barrier. This is partly due to increasing specialization within the sciences. Even the field of industrial hygiene has become so compartmentalized that its practitioners may suffer from a sort of professional myopia.

Sometimes an industrial hygienist chafes under the usual requirement that evidence must be given by question and answer. The result thus may depend more on the skill of the questioner than that of the witness. The questioner is dominant. The form of the questions conditions the form and often the content of the answers. Unless the questioner really understands the subject matter, he or she may phrase questions in ways that restrict the scientific witness or misplace the emphasis. At worst, he or she may ask questions that are unanswerable because, to the witness, they have no rational content.

To get an intelligent answer, of course, we have to ask an intelligent question. If the answer is to be highly technical, the question itself must be technical in content and must be clearly phrased to bring out the pertinent answer. With all due respect to the legal skill of the attorney, he or she may find it difficult to feel at home with the highly specialized language of another field. It is a most distressing experience for an industrial hygienist to be subjected, from a supposedly friendly attorney, to a barrage of questions that really do not mean what the attorney thinks they mean. In fact, they may be technologically unintelligible, while at the same time sounding like the best English. This brings us to the obvious solution—good communication.

In any communication between an attorney and an industrial hygienist, in or out of court, precise language must be employed, or at the very least there must be enough of exchange of views so that the language has an agreed meaning. Since the attorney and the industrial hygienist view problems from different vantage points, it is quite possible that they will see different objectives. These can be reconciled if recognized in time, but the danger is that they may go their separate ways, leaving both unaware that they are working at cross purposes. To be effective, the precision of the industrial hygienist and the legal skill of the attorney should focus on the same target.

It should go without saying that the attorney should seek the advice of the industrial hygienist on the wording of questions before they are asked in court, and the attorney should thoroughly understand and heed this advice. Then both will be on target. Too often, however, time limitations on the part of the attorney or the industrial hygienist, or both, prevent such careful collaboration. But both should insist on it as far as they can.

3 TYPES OF LAWSUITS

There are two general types of lawsuits: civil and criminal. A civil action is a lawsuit that is brought by a person or group of persons against another person or group of persons for losses, injuries, or damages they have suffered as the result of another's acts. The causes of these losses can run the gamut from negligence, intentional tort, breach of contract, libel, slander, or invasion of one's right of privacy to the violation of local, state, or federal law. In a civil case the suit is brought in the name of the person (John Smith) who alleges that he has suffered a loss at the hands of another (James Jones). Hence, the case is identified as Smith v. Jones.

A criminal case is one that involves the breach of some local, state, or federal law. In a criminal case the suit is filed in the name of the "people," and not by the individual against whom the crime was committed. This is because our legal system is based on the principle that all crimes are said to be against the public interest. The local, state, or federal legislative bodies that represent the "people" enact laws to protect these "people." Therefore, any breach of such a law is against the "people" and not the individual. If a local law has been violated (a crime has been committed), the lawsuit is brought in the name of the "People of the City" by the local officer in charge of this type of case. Usually this is the city attorney or county prosecutor from the jurisdiction where the crime allegedly occurred. If the offense upon which the case is founded is a violation of state law, the lawsuit is brought in the name of the "People of the State" by the state's attorney of the county in which the crime is alleged to have been committed. If violation of a federal law has been charged, the case is brought in the name of the "People of the United States" by the U.S. attorney for the Federal District Court in which the crime is alleged to have taken place. The defendant in either a civil or criminal case is the person, persons, or company charged with the infraction of violating a particular law.

3.1 Civil Cases

When an attorney determines that an individual has been harmed, for whatever reason, the first step is to draw up certain papers required by law. The first and most important paper is called a *complaint* in most states. It informs the opposing party (defendant) of the facts involved in the particular case, tells why the plaintiff is suing, and lists the amount of damages (i.e., money sought, etc.). After the complaint is filed in court, a summons is issued by the sheriff, which is then served on the defendant. The defendant then has a time period in which to respond. This response is called an *answer*. In his or her answer the defendant either admits or denies each paragraph of the complaint. Depending on how the defendant answers the complaint determines what becomes of the *issues*, which are the facts to be decided by the court and jury. If the defendant admits any charge in the complaint, this does not come before the court or jury since it is *no issue* on that particular point and does not need to be proved by the plaintiff.

Once the issues have been established, the next step is the trial itself. The time it takes for a case to come to trial depends on the individual case and the jurisdiction where the case has been filed. Each of the parties has a right to request a jury trial. Usually it is the plaintiff who will make this request. If neither party requests a jury, then the case will be heard by the judge alone, who is then the sole judge of the facts and the law and makes the decisions concerning all the issues of trial. In a jury trial, the jury determines the facts and the judge has the responsibility of interpreting the law. The witnesses present the facts to the jury whereas the judge gives them the law of the jurisdiction where the case is being heard. It is the responsibility of the jury to decide the reliability of the witnesses and weigh all the evidence presented during the trial. They must then return only one of two verdicts: in favor of the plaintiff or in favor of the defendant. If the verdict is for the plaintiff, the jury must assess the amount of the damages, which is included on the verdict sheet.

In addition to instructing the jury what law should be applied by them to the case, the judge has several other duties. The judge must overturn the jury's verdict if one of the following reasons is present:

1. If the verdict resulted from passion and prejudice (which is seldom found)
2. If there were insufficient facts presented to justify a verdict under the law
3. Because of legal error during the trial

The granting of a new trial is the exception rather than the rule. Another of the judge's responsibilities is to see that the trial is conducted by accepted rules and regulations and that the attorneys and witnesses strictly adhere to the rules of the court. In this day and age, an industrial hygienist could be called to testify in a wide variety of types of cases ranging from criminal or civil cases before a judge to an OSHA hearing before an administrative law judge. For those who have never been to a trial and whose only courtroom experience is exposure to scenes from television's "Peoples' Court" or "LA Law" we might outline the proceeding in the conduct of a trial.

After the issues have been formed in a case, the first step of a trial is the selection of a jury if one has been requested. In most jurisdictions 12 persons form a jury; in some, 6-person juries are used. The jury is picked in most instances by the attorneys for the respective parties and is done after questioning the jurors to determine fairness and absence of prejudice. Actually, each attorney tries to pick jurors who will be sympathetic to his or her side of the case, but it cannot be predetermined that a favorable jury has been chosen.

Once the jury has been selected, the trial actually starts. The attorney who represents the plaintiff/prosecution (the side that has the burden of proof) will make his or her opening statement first. After the opening statement has been made by the plaintiff, or prosecution, then the attorney for the defendant will make his or her opening statement. The purpose of the opening statement is to have the attorneys clarify the issues in the particular case by telling the jury what they believe their respective evidence will show. This enables the jury to hear something about the case before the testimony begins. The judge will instruct the jury that the attorney's opening remarks should not be considered as evidence by the jury. The only evidence that is official is that testimony that comes to the jury from witnesses who have been sworn to tell the truth and testify from the witness stand. Documentary evidence that has been properly submitted to the court may be used as additional evidence in conjunction with the oral testimony.

After the opening statement, there is generally a motion by one of the attorneys to exclude witnesses from the courtroom. This means that all witnesses other than the plaintiff or defendant must leave and cannot hear testimony until each has personally testified. Thus, an expert seldom hears the testimony in any case in which he or she is a witness. Most attorneys prefer that witnesses not listen either before or after testifying because, if they do, the jury may receive the impression that the witnesses are interested in the outcome of the case and may be prejudiced in their testimony.

The next step in a trial is for the side that has the burden of proof, namely the plaintiff in a civil case or the prosecution in a criminal case, to begin presentation of its case through its witnesses. The attorney for the side that has the burden of proof must keep in mind what the particular issues are in the case so that the necessary proof that is required under the law is presented. If he or she fails to prove the elements of his or her case, it is within the prerogative of the judge to enter a directed verdict or a finding in favor of the defendant. A good attorney will always attempt to present witnesses in a good logical and chronological sequence to make it easy for the jury to understand the evidence and to follow the case.

After the evidence has been presented for the side that has the burden of proof, the attorney for the defendant may present a written motion for a verdict directed in his or her client's favor. This is done when the attorney feels that the plaintiff/prosecutor has not met the burden of proof necessary for a favorable verdict. If an attorney does this, it is done outside the presence of the jury. If the judge sustains this motion, the trial is over. In a civil case the term *directing a verdict* is the one most commonly used. In a criminal case the term *finding for the defendant* is the term that is generally used. In essence they both mean the same thing and have the same effect. If the judge overrules the motion for a directed verdict, it means that the attorney for the defendant must present his or her side of the case.

After the plaintiff/prosecutor has presented all of his evidence, and either no motions were presented or if presented they were denied by the judge, the attorney for the defendant presents his or her evidence. It should be pointed out that the same rules of evidence apply equally to both plaintiff/prosecutor and defendant witnesses. If a case involves more than one defendant, the defendant whose name appears first on the complaint will present witnesses, who will be followed by the witnesses for the later named defendant(s).

After all evidence has been presented, the plaintiff/prosecutor and/or defendant have an opportunity to present a written motion for a verdict directed in his or her favor. This is another time that the phrase "directing a verdict," comes into being.

This term can be confusing for most laypeople. In directing a verdict, the judge takes the case away from the jury's consideration and directs the jury to find the decision that he or she tells them. As mentioned earlier, each attorney is given the opportunity to present a written motion requesting that the judge direct the verdict in favor of his or her client. To have the judge take the case away from the jury when it is the duty of the jury to decide issues of the case does not make a whole lot of sense to the average layperson. The judge does not do this very often, but it does occur often enough so that one should be aware of it.

Why would a judge direct a verdict? It occurs if the plaintiff/prosecutor has not proved his or her case under the law or the defendant has not made a satisfactory defense under the law. Remember, each side must prove its case by proving certain elements through the evidence presented through its witnesses. As an example, in a civil personal injury case the plaintiff must prove:

1. Negligence on the part of the defendant; a duty must be owed and a breach must be shown
2. Freedom from negligence on the plaintiff's part
3. That the injuries involved had a direct causal connection with the defendant's negligence

If any of these essential elements is not proven through the evidence presented by witnesses, the plaintiff is not entitled to recover, and the judge will direct a verdict for the defendant and order the jury to sign a verdict finding the defendant not guilty. Conversely, if the defendant fails to provide a defense to the plaintiff's action in which the essential elements were proven, the judge will direct a verdict for the plaintiff. In the latter case, the jury will not decide whether the defendant was liable or not. Their duty as a jury will be to arrive at a decision involving only the issue of damages, which in most cases deals with the amount of money the plaintiff is entitled to receive.

Before a judge directs a verdict in favor of either of the parties, he or she is required, under the law, to construe the evidence most strongly in favor of the party against whom judgment is sought. Generally speaking, the evidence must be overwhelming before the judge will enter a directed verdict. The law is very specific about this and, where possible, wants to give each citizen the benefit of the doubt before the judge directs the verdict. This action is taken only when the case is unfounded under the law.

Considering the number of cases that are heard by judges and juries, there are not very many directed verdicts. If a judge enters a directed verdict, the party against whom the verdict has been directed has the right to appeal to a higher court.

Once all the evidence has been presented, and if there is no directed verdict, the next step in the trial is the final arguments. Attorneys for both sides are allotted the same amount of time, but the plaintiff's/prosecutor's argument usually is in two parts, as he or she has the burden of proof. The plaintiff's/prosecutor's attorney goes first. He or she decides how much time to use initially in the final argument; what is not used can be reserved for use as part of the rebuttal after the defendant's attorney presents his or her final argument. This rebuttal argument of the plaintiff's/prosecutor's attorney is an answer to the final argument of the defendant's attorney.

When the final arguments for both sides have been given, the court will instruct the jury as to the law in that particular case. Sometimes the jury is given written instructions, which they take to the jury room as an aid in their deliberations. After the judge gives the jury his or her instructions, the jury will retire to the jury room where they will choose a foreperson and then deliberate the verdict. After they have reached their decision, they will complete a form that lists all the issues that had to be decided in the case. This is given to the clerk of the court who gives it to the judge to review. When this review is finished, the judge will instruct the clerk to read the verdict to everyone present in the courtroom.

The procedure applies to a case tried before a jury. Some cases are tried before only a judge. The procedures are the same for both with the only difference being that at the end of the trial the judge gives his or her decision without input from anyone else. The evidence is the same in both cases; the main difference is how the attorneys present the evidence since, in the absence of a jury, they do not have to worry about how the jury is going to perceive the witnesses and the arguments.

3.2 Criminal Cases

The trial procedures in a criminal case are a little different from those in a civil case. The action is brought by the "People of the State" in which the action is originated or in the name of the federal government if it is filed in a federal court. One difference is in the document that originates the case. In a criminal case this document is called an *indictment*, or the *information*, either of which is comparable to a complaint in a civil action. This document gives the court the authority to have the defendant apprehended. Once the defendant is taken into custody, he or she is arraigned. At the time of the arraignment the defendant will either plead guilty to the charge, plead not guilty, or may remain mute. If the defendant remains mute, the court will enter a plea of not guilty. When a plea of not guilty is entered, it creates the issue of denial of the charges in the indictment, or the information. No paper or written pleading is filed denying the charges as in a civil case.

Criminal cases may be tried by a court or a court and a jury. Only the defendant is allowed to ask for a jury in a criminal case. The big difference between a criminal case and a civil case is the level of the burden of proof. In a criminal case the burden of proof is greater than that in a civil case because the prosecution must prove beyond a reasonable doubt in order to sustain its verdict. The remainder of the trial is basically conducted in the same way as a civil trial as discussed above.

4 EXPERT WITNESS

An expert witness is one who possesses special knowledge and experience on matters in issue in a lawsuit. More and more industrial hygienists are being asked to perform this role. For those who have never experienced this, it is not surprising that they are hesitant to do so. One basic reason for this hesitating is the fear of not knowing what might confront them when they come to court.

An expert witness is in a special class. The court has placed the expert witness in a little different position than the average witness in an effort to be more scientific.

The selection of an expert witness involves trying to match the education and experience of an individual with the specific facts associated with the case in question. The function of the expert witness is to assist the court and the jurors in arriving at a correct conclusion upon matters that are not familiar to their everyday experiences so that as part of their deliberations they can arrive at an intelligent understanding of the issues that are before them.

An expert may testify only about a matter that is a proper subject for expert testimony. The trial court is entrusted with the responsibility of determining whether the proffered testimony is a proper subject for an expert. As the Supreme Court has noted, the trial court has "broad discretion in the matter of the admission or exclusion of expert evidence, and his action is to be sustained unless manifestly erroneous" (5).

Federal Rule of Evidence 702 provides that expert testimony is admissible if "scientific, technical, or other specialized knowledge will assist the trier of fact to understand the evidence or to determine a fact in issue" (6). According to the federal drafters:

> The rule is broadly phrased. The fields of knowledge which may be drawn upon are not limited merely to the "scientific" and "technical" but extend to all specialized knowledge . . . Thus within the scope of the rule are not only experts in the strictest sense of the word, e.g., physicians, physicists, and architects, but also the large group sometimes called "skilled" witnesses, such as bankers or landowners testifying to land values (7).

The standard adopted by Federal Rule of Evidence 702—whether expert testimony will "assist the trier of fact"—is a more liberal formulation of the subject matter requirement than that found in many common law opinions, which often phrased the requirement as whether the subject was beyond the comprehension of a layperson (8). Under Rule 702, "the test is not whether the jury could reach some conclusion in the absence of the expert evidence, but whether the jury is qualified without such testimony 'to determine intelligently and to the best possible degree the particular issue without enlightenment from those having a specialized understanding of the subject.'" (9). This test is consistent with Wigmore's formulation of the test for expert testimony: "On *this subject* can a jury receive from *this person* appreciable help?" (10).

With respect to expert witnesses the trial judge has the burden of determining whether or not expert testimony should be admitted to prove an issue as well as determining whether or not the expert witness possesses the knowledge and experience to permit him or her to testify on the particular issues involved in that trial. The judge must also rule on whether or not an expert witness can give an opinion upon a given subject matter and whether or not the opinion is relevant and material to the issues in the case.

4.1 Types of Testimony

An expert witness can provide testimony that can be divided into two classifications: (1) where he or she has personal knowledge of the facts and (2) where all or part of his or her knowledge of the case comes from other sources. In the latter situation the evidence will have to be given through the use of hypothetical questions. Courts have relied principally on two different tests to determine the admissibility of innovative scientific evidence. One approach treats the validity of the underlying principle and the validity of the technique as aspects of relevancy. The relevancy approach is to treat scientific evidence in the same way as other evidence, weighing its probative value against countervailing dangers and considerations. In a 1954 text, Professor C. McCormick wrote:

"General scientific acceptance" is a proper condition upon the court's taking judicial notice of scientific facts, but not a criterion for the admissibility of scientific evidence. Any relevant conclusions which are supported by a qualified expert witness should be received unless there are other reasons for exclusion. Particularly, its probative value may be overborne by the familiar dangers of prejudicing or misleading the jury, unfair surprise and undue consumption of time (11).

This approach, which dovetails with the Federal Rules of Evidence, requires a three-step analysis: (1) ascertaining the probative value of the evidence, (2) identifying any countervailing dangers or considerations, and (3) balancing the probative value against the identified dangers.

The second approach, which required the proponent of a novel technique to establish its general acceptance in the scientific community, was based on *Frye v. United States* (12), decided in 1923. In *Frye* the D.C. Circuit Court considered the admissibility of polygraph evidence as a case of first impression. The court wrote:

Just when a scientific principle or discovery crosses the line between the experimental and demonstrable stages is difficult to define. Somewhere in this twilight zone the evidential force of the principle must be recognized, and while the courts will go a long way in admitting expert testimony deduced from a well recognized scientific principle or discovery, the thing from which the deduction is made must be sufficiently established to have gained general acceptance in the particular field in which it belongs (13).

In a unanimous decision on June 28, 1993 (*Daubert v. Merrel Dow Pharmaceuticals, Inc.*, US SupCt, No. 92-102, Blackmun, J.) (14), the U.S. Supreme Court held that the "general acceptance" standard for the admission of scientific evidence adopted in *Frye v. U.S.* no longer governs trials conducted under the Federal Rules of Evidence. The Supreme Court held that Fed. R. Ev. 702, which provides a more liberal rule for admitting relevant scientific evidence, supersedes the common law rule announced in *Frye*.

In the *Daubert* case, two children alleged that they suffered from birth defects caused by their in utero exposure to the drug Bendectin. They presented a number of well-qualified experts who were prepared to testify based on animal studies and reanalysis of epidemiological studies that Bendectin causes the type of birth defects suffered by the plaintiffs. The defendant, Merrell Dow, had equally well-credentialed experts who would have testified, based solely on epidemiological studies, that Bendectin is not a teratogen capable of causing malformations in fetuses.

The U.S. District Court for the Southern District of California granted summary judgment for the defendant, saying that scientific evidence is admissible only if the principle on which it is based is generally accepted in the field to which it belongs. The court found that the plaintiffs' evidence did not meet this standard. Given the vast body of epidemiological data concerning Bendectin, the court held, expert opinion that is not based on epidemiological evidence is not admissible to establish causation.

The U.S. Court of Appeals for the Ninth Circuit affirmed. Citing *Frye*, the court stated that expert opinion based on a scientific technique is inadmissible unless the technique is "generally accepted" as reliable in the relevant scientific community. Reanalysis, the court contended, is generally accepted by the scientific community only when it is subjected to verification and scrutiny by others in the field. Since the reanalysis on which the plaintiffs' experts based their opinions had not been subjected to such peer review, the court concluded that this evidence was not admissible and, therefore, that the plaintiffs could not satisfy their burden of proving causation at trial.

Although the *Frye* standard has dominated the admissibility of scientific evidence for 70 years, its merits have been the subject of much debate. However, the plaintiffs in this case did not attack the standard's scope or application; rather, they argued that the *Frye* test was superseded by the adoption of the Federal Rules of Evidence in 1975. The U.S. Supreme Court agreed. Fed.R.Ev. 402 provides the "baseline" for admissibility. "All relevant evidence is admissible . . . Evidence which is not relevant is not admissible." Evidence is relevant if it has "any tendency to make the existence of any fact that is of consequence to the determination of the action more probable than it would be without the evidence." Rule 401.

In *U.S. v. Abel* (15), the court previously considered the pertinence of background common law in interpreting the Federal Rules of Evidence. The court noted that the rules occupy the field, but that the common law could still serve as an aid to their application. The common law precept at the issue in that case was found to be consistent with Rule 402's general requirement of admissibility. In *Bourjaily v. U.S.* (16), on the other hand, the court was unable to find a particular common law doctrine in the rules, and so held it superseded.

In *Daubert*, the court found, there is a specific rule governing the contested issue. Rule 702 provides:

> If scientific, technical, or other specialized knowledge will assist the trier of fact to understand the evidence or to determine a fact in issue, a witness qualified as an expert by knowledge, skill, experience, training, or education, may testify thereto in the form of an opinion or otherwise.

The court found that there was nothing in the text of this rule that establishes "general acceptance" as an absolute prerequisite to admissibility. Nor was there any evidence presented that the rule was intended to incorporate a general acceptance standard. In fact, the rigid general acceptance test is at odds with the "liberal thrust" of the rules and their "general approach of relaxing the traditional barriers to 'opinion' testimony," the court ruled, quoting *Beech Aircraft Corp. v. Rainey* (17).

The court held that "Given the Rules' permissive backdrop and their inclusion of a specific rule on expert testimony that does not mention 'general acceptance,' the assertion that the Rules somehow assimilated *Frye* is unconvincing. *Frye* made 'general acceptance' the exclusive test for admitting expert scientific testimony. That austere standard, absent from and incompatible with the Federal Rules of Evidence, should not be applied in federal trials," the court concluded (18).

Where this is all headed is open to conjecture. Everyone generally agrees that *Daubert* changed the rules for introducing expert scientific evidence. But agreement pretty much stops there. Justice Harry A. Blackmun, who wrote the opinion for the unanimous court, also wrote in a part joined by only six other justices that the federal rules do place some restrictions on expert testimony. The rules—especially Rule 702—assign to trial judges the task of ensuring that an expert's testimony rests on a reliable foundation and also is relevant. According to this majority of the court, a trial judge, when faced with expert proof, must determine whether the expert's proffered testimony is based on scientific knowledge that will assist the trier of fact to understand or determine a fact in issue. Rule 702, "which clearly contemplates some degree of regulation of the subjects and theories about which an expert may testify (19), is the primary locus of this obligation, Blackmun wrote. Parsing the rule, he said the subject of an expert's testimony must be "scientific . . . knowledge." The adjective "scientific" implies a grounding in the methods and procedures of science, the court found, while the word "knowledge" connotes more than subjective belief or unsupported speculation. "Proposed testimony must be supported by appropriate validation—i.e., 'good grounds,' based on what is known. In short, the requirement that an expert's testimony pertain to 'scientific knowledge' establishes a standard of evidentiary reliability," Blackmun wrote. Additionally, "Rule 702's 'helpfulness' standard requires a valid scientific connection to the pertinent inquiry as a precondition to admissibility" (20).

It is the position of the majority that the trial judge, when faced with a proffer of expert scientific testimony, must determine at the outset whether the expert is proposing to testify in accordance with these principles. While not "presum[ing] to set out a definitive checklist or test," the majority made some "general observations" about how a court should go about this task.

The majority said that the lower courts should look at whether the scientific knowledge being presented has been tested, whether it has been subjected to peer review and publication, what the evidence's known rate of error is, and whether the evidence has a particular degree of acceptance in the relevant community. The majority further cautioned that the inquiry must remain a flexible one. The traditional measures of judicial control—such as vigorous cross-examination, screening unqualified experts or irrelevant evidence, and granting summary judgment and directed verdicts where warranted—are adequate to contain unreliable testimony.

Chief Justice William H. Rehnquist and Justice John Paul Stevens concurred with the majority's holding that the *Frye* rule is no longer good law. In their dissent from the remainder of the court's opinion, they said that the majority should not have gone beyond its determination that the Federal Rules of Evidence has supplanted *Frye*. They recognized that "[g]eneral observations by this Court customarily carry great weight with lower federal courts," and pointed out that "the ones offered here suffer from the fatal flaw common to most such observations—they are not applied to deciding whether or not particular testimony was or was not admissible, and therefore they tend to be not only general, but vague and abstract" (21).

Chief Justice Rehnquist went even further as he wrote, "I do not doubt that Rule 702 confides to the judges some gatekeeping responsibility in deciding questions of the admissibility of proffered expert testimony. But I do not think it imposes on them either the obligation or the authority to become amateur scientists in order to perform that role. I think the Court would be far better advised in this case to decide only the questions presented, and to leave the further development of this important area of the law to future cases" (22).

According to the seven-member majority, the rules contain no general acceptance test. The "austere standard" of *Frye* "absent from and incompatible with the Federal Rules of Evidence, should not be applied in federal trials" (23). The court also said that there is a reliability component in the scientific knowledge component. "Under the rules the trial judge must ensure that any and all scientific testimony or evidence admitted is not only relevant, but also reliable. The requirement that an expert's testimony pertain to 'scientific knowledge' establishes a standard of evidentiary reliability," the opinion said. The judge then must determine whether the reasoning or methodology underlying the testimony is scientifically valid. The court cautioned that the judge's examination must be on the expert's methods, not conclusions.

The following are the key points of the *Daubert* decision:

- The *Frye* Rule, or "general acceptance standard," has been supplanted by the Federal Rules of Evidence, which were enacted in 1975.
- The Federal Rules of Evidence do not contain a requirement that scientific evidence be generally accepted in the field in order to be admissible at trial.
- The inquiry envisioned by Fed. R. Evid. 702 is a flexible one.
- Fed. R. Evid. 702 requires that a trial judge ensure that an expert's testimony is based on scientific knowledge and will assist the trier of fact.
- A standard of evidentiary reliability is established by the requirement that an expert's testimony relate to "scientific knowledge."
- A trial judge, under Fed. R. Evid. 104(a), must determine whether the expert is proposing to testify to scientific knowledge that will assist the trier of fact.
- Factors to be applied in arriving at the above include:
  - whether a theory or technique can be and has been tested;
  - whether it has been subject to peer review or publication;
  - the known or potential error rate; and,
  - whether the theory of technique has gained general acceptance in the field.

Where do we go from here? The "general acceptance" standard is gone. The court's ruling in *Daubert* in favor of a "reliability" standard emphasizes accuracy, validity and trustworthiness of the data and conclusions. The question now, for those involved in the selection of expert witnesses, is "what impact does this ruling have with regard to witness preparation and jury persuasion?" In the past, attorneys chose experts based on their credentials and conclusions. *Daubert* shifts the emphasis to the reliability of the expert's data and conclusions.

This ruling promotes the novel and innovative. It will result in expert witnesses prepared to clearly explain how their testimony relates to the dispute. Will *Daubert* have an effect on the way attorneys prepare experts to testify? Currently, experts undergo preparation for two distinct stages of the trial: the deposition and courtroom testimony. There is a significant difference.

There are four major differences: (1) the purpose for the testimony, (2) the audience for the testimony; (3) the rules governing the testimony considering the purpose and audience; and (4) the tools employed to achieve the purpose. In deposition, the expert sits on a pile of information, which the one taking the deposition wants. The *purpose* for the testimony is to give out *only* the information directly requested. The *audience* is the opposition who wants to secure information that can be used to either disqualify the expert or to challenge the credibility of the conclusions or the person. The *rules* are specific that the deponent only has to answer the questions asked. When a question is asked, however, the answer given is to be done cooperatively, fully, and honestly but that doesn't mean that the deponent has to volunteer any answers over and above the questions asked. The *tools* are supplied by the one taking the deposition and usually are documents.

At trial, the *purpose* of the testimony is to aid the fact-finder (the *audience*) to understand certain key elements. The *rules* are relatively simple—the witness testifies solely for the benefit of the fact-finder (judge or jury) whether he or she is questioned on direct or cross-examination. The expert's sole purpose in testifying is to be the objective supplier of the truth and not an advocate, which means that the expert talks to the judge/jury freely, openly, and honestly. Preparation of the experts should emphasize the need to understand the judge/jurors, their backgrounds, their basic level of understanding, and the best way to present information to help them do their job. The *tools* in this arena can include demonstrative exhibits, documents, drawings, photographs, tables of data, graphs, and so on—whatever it takes to aid their understanding. Clearly, the expert has both a right and an obligation to convince the judge that reliable principles, derived from a reliable, valid, and trustworthy scientific method, form the basis for testimony.

What effect *Daubert* will have on future litigation remains to be seen. Trial courts probably will continue to consider carefully the expert's qualifications during the preliminary stage in preparation for offering the expert to the court as an expert in order to give opinion testimony to the fact-finder. The testimony on "qualifications" very likely will, or at least should, provide the required predicates under *Daubert*, i.e., "relevance" and "reliability."

4.2 Evidence

In general, a witness is permitted to testify to anything that he or she has actually seen or experienced, which is material, that is, pertinent, to the issues involved in the case. The witness is permitted to testify to anything he or she has heard in the presence of the opposing party. He or she can testify to the statements made by one of the parties to the lawsuit that may be against the interest of the party making the statement. The witness is allowed to testify to anything within his or her knowledge that concerns the issues. The witness is *not* permitted to testify to anything he or she heard from a third party or to facts learned from someone other than the parties to the case. This is what is known as *hearsay evidence*, which is inadmissible.

There are many exceptions to the general rules of evidence, but the focus in this discussion will be on those that pertain to the expert witness presenting testimony in the average case. Opinion evidence and hypothetical questions are both exceptions to the hearsay rule. If a witness says, ''It is my opinion that . . . ,'' or ''I think . . . ,'' there will be an immediate objection by the opposing attorney. This objection will be sustained by the court. The reason is that when a witness tries to give an opinion or tell what he or she thinks, he or she is trying to decide one of the issues of the case. This is the role of the jury, and thus such evidence invades the province of the jury, which the court will not permit. The ordinary witness is entitled to give the facts and let the jury decide the case from the facts admitted into evidence. The expert witness is given some latitude as the law permits the testimony of the witness to include opinions upon a given set of facts in evidence, and then permits the jury to decide whether or not it wishes to adopt the opinions given by the expert witness. Expert testimony is not utilized to substitute the opinion of the expert for that of the jury. It is just an aid, but in some cases it can be the deciding factor in the outcome of the particular lawsuit.

4.3 Opinion Questions

As was mentioned, the expert witness is allowed to render opinions in two instances: (1) where he or she has personal knowledge of the facts and (2) where he or she knows none of the facts. An expert witness can render an opinion on facts about which he or she has personal knowledge and on which there is sufficient data on which to base the opinion. This is always done in response to questions presented by the attorney. This could include tests or experiments performed by the witness. It should be pointed out that facts must be within the witness' personal knowledge and not on what someone else did or told him. If he uses other information his opinion will be stricken from the consideration of the jury as being invalid.

Typically, the expert witness will testify to all the facts that he or she has personal knowledge of and upon which he or she can form an opinion. The witness will then be asked an opinion question that must be in his or her specialty or it is not admissible in evidence. The opinion question must be based on a *reasonable* degree of certainty to be admissible in law. It cannot be speculative. If the witness is an industrial hygienist, the question must be based on a *reasonable degree of scientific certainty*. It is not necessary that it be a certainty. The word *reasonable* is all that is required upon which to formulate an opinion. This is far different than the requirements needed to form an opinion in other fields. It is much less stringent and less exacting. It would be unjust for an expert witness to apply a more strict interpretation to the particular problem than is required under law. What is this thing called *reasonable degree of certainty*? Plain and simple, it is a legal fiction. It has been interpreted differently on various occasions. The key word is *reasonable*. A typical definition would be that which would induce a person of ordinary prudence to believe it under the circumstances.

For some to use this standard might prove uncomfortable. However, it might help to view the alternative. Typically the people who make up the jury will not have any technical or scientific knowledge on which to base an opinion, and they are looking to the expert witness to provide them with some assistance. The role of the expert witness is to give his or her opinion based on the facts as presented, using the training and experience that he or she has to arrive at conclusions.

4.4 Hypothetical Questions

The hypothetical question is actually an opinion question. It was stated above that opinion questions were asked of the expert witness where he or she has personal knowledge of the facts. In a hypothetical question situation the expert witness is giving an opinion based on facts from other sources. This is due to the fact that the expert witness was not present when the situation occurred so he or she could not have all the facts needed to express an opinion. Since the expert witness cannot render an opinion unless he or she knows certain facts, the law permits the attorney to ask a *hypothetical question* in which all the necessary facts *that have been proven by witnesses* during the trial will be given. This will be done in a hypothetical form so that the expert witness can give his or her opinion on these facts as they are applicable to his or her special knowledge. From this hypothetical question the expert witness may give as many opinions as are applicable to the issues of the particular case. It is up to the jury to use this information to come to its decision.

The hypothetical question has two principal advantages. First, it informs the jury of the facts upon which the expert's opinion is based. Second, if the question contains assumed facts that are not supported by evidence admitted at trial, it provides the opposing party with an opportunity to object before an opinion is expressed.

The witness must be made to appreciate the underlying premise of a hypothetical question. He or she must understand that the evidence postulated in the question is assumed to be true.

Therefore, a hypothetical question is one in which an attorney gives the expert witness an assumed set of facts that have been given by other witnesses in the case so that the expert witness can give an opinion on that particular set of facts. Generally, most hypothetical questions will begin with "assume," "suppose," or "consider." More often than not, the attorney will generally start the question with "Assume that a hypothetical person . . ." This should alert the expert witness that the question being asked falls into this special category.

How the hypothetical question will be put together must begin long before it is put to the expert in the courtroom. The length of the question is governed by the atmosphere of the trial and the attorney's own sense of effective timing. It would be poor practice to propound a question so long that it needlessly confuses either the expert or the jury. Hopefully, the question will be framed so that it will not become clouded in the jury's minds by frequent objections from opposing counsel.

The expert witness must be able to determine what are the salient facts from among the many that will be presented in the hypothetical question. Typically, it will contain facts that are not needed to render an opinion, but have been included so that the attorney can impress the jury with certain facts that have been brought out in the testimony. Many times the attorney will take this opportunity to present the facts logically and lucidly in a narrative form. The expert witness must keep in mind that regardless of what is being presented, his or her answer must be limited to the facts given in the hypothetical question. It is the expert witness's responsibility to assimilate all the data in the hypothetical question, apply his or her special learning and experience, that is, expertise, to the facts, and then render an opinion based on *reasonable certainty*.

After an expert witness has given his or her opinion, the attorney should ask the reasons why he or she arrived at this opinion. This gives the expert witness the opportunity to discuss all the facts used to arrive at his or her decision and is one of the most important parts of the testimony. Since the purpose of his or her testimony is to instruct the jury about a technical matter that they do not fully understand, the expert witness should clearly itemize and specify the basis for the opinion so that the jury can follow the logic of the analysis. The importance of this part of the testimony cannot be overemphasized. The expert witness is given all the time needed to thoroughly explain his or her position. If the jury is to responsibly perform its duties, then it is up to the expert witness to educate them so that they can understand the technical problem that is before them.

5 WORK PRODUCT

Federal Rule 26(b)(3) sets out the general principles governing the discoverability of materials prepared in anticipation of litigation or for trial. It states that relevant "documents and tangible things" that were prepared by or for a party or his or her "representative," "in anticipation of litigation or for trial," are not discoverable, except on a showing of "undue hardship" and "substantial need" (24).

As the Sixth Circuit Court of Appeals stated in *In re Grand Jury Subpoena dated November 8, 1979*:

> Work product consists of tangible and intangible material which reflects an attorney's efforts at investigating and preparing a case, including one's pattern of investigation, assembling of information, determination of the relevant facts, preparation of legal theories, planning of strategy, and recording of mental impressions (25).

A wide variety of persons have been held to fall within the work product immunity. Materials prepared for an attorney are, of course, covered (26). When a person, such as an industrial hygienist, does work specifically at the direction of an attorney, that person should assume that what he or she is doing will be protected under the work product privilege and should conduct him or herself accordingly. If there are any changes to be made in this relationship, let them be left to the attorney.

The Supreme Court in *United States v. Nobles* (27) held the work product privilege applicable to both the pretrial and trial stages. In cases involving scientific evidence, the work product rule may preclude discovery of reports prepared by expert witnesses. In *Nobles* the Court wrote:

> At its core, the work product doctrine shelters the mental processes of the attorney, providing a privileged area within which he can analyze and prepare his client's case. But the doctrine is an intensely practical one, grounded in the realities of litigation in our adversary system. One of those realities is that attorneys often must rely on the assistance of investigators and other agents in the compilation of materials in preparation for trial. It is therefore necessary that the doctrine protect material prepared by agents for the attorney as well as those prepared by the attorney himself (28).

The Court, however, also held that the calling of a witness at trial is a waiver of the privilege (29).

6 ATTORNEY–CLIENT PRIVILEGE

The attorney–client privilege is designed to protect confidential communications between a client and his or her attorney. The Supreme Court has noted: "its purpose is to encourage full and frank communication between attorneys and their clients and thereby promote broader public interests in the observance of law and administration of justice. The privilege recognizes that sound legal advice or advocacy depends upon the attorney being fully informed by the client" (30).

Application of the attorney–client privilege to expert witnesses sometimes arises in cases involving industrial hygienists. It is important to distinguish between the two ways in which an industrial hygienist might be used as an expert witness. First, he or she may be retained for the purpose of testifying at trial. When this happens, the privilege is waived (31). A "party ought not to be permitted to thwart effective cross-examination of a material witness whom he will call at trial merely by invoking the attorney–client privilege to prohibit pretrial discovery" (32).

Second, an industrial hygienist might be retained as a consultant to provide the attorney with information concerning a particular case. This could be for the plaintiff or the defendant. Many times an attorney will rely on this information to determine whether or not it is feasible to continue with a particular case. A number of courts have held that the attorney–client privilege covers communications made to an attorney by an expert retained for the purpose of providing information necessary for proper representation (33). There are times when the attorney needs to obtain expert advice because of the technical issues of a particular case.

If an industrial hygienist is retained by an attorney, whether his or her input will be covered under the attorney–client privilege is not that industrial hygienist's concern. However, the industrial hygienist should conduct him or herself as though the attorney–client privilege does apply and not discuss the case outside the attorney's office. The confidentiality of the industrial hygienist's role must be honored. That way the industrial hygienist will not inadvertently waive the privilege if, in fact, one does exist.

7 DISCOVERY

Discovery is a pretrial process whereby an attorney attempts to gather information pertaining to the trial for the following purposes:

1. One of the most basic purposes is simply the obtaining of information. By exchanging facts through the discovery process, each side is provided with a great deal more information to help prepare their cases than they would have if required to obtain all information on their own.
2. Another purpose is to create a record of testimony prior to the trial by using depositions and document production. Sometimes attorneys use interrogatories and requests for admissions to aid them in trying to determine what their opponent does or does not know or to locate documents or people who have relevant knowledge.
3. Generally the requests for admissions helps to define and limit the issues at trial. Discovery, more than likely, will eliminate certain issues and focus the parties attention on others.
4. As a result of the exchange of information, all parties are able to evaluate the strengths and weaknesses of their cases for potential settlement discussions. It gives the attorneys an opportunity to evaluate the kind of witness an individual will make at trial in terms of demeanor, credibility, sympathy, and the other intangible factors, which, absent the discovery process, would leave the parties to their peril in the courtroom.
5. Sometimes the pretrial discovery gives the parties additional information to allow them to supplement their pleadings.
6. Sometimes the information gathered during the pretrial discovery process will form the basis of motions for summary judgment and other fact-based motions that may be dispositive of the case in whole or in part.

What can be discovered? Any matter relevant to the subject matter involved in the pending litigation so long as it is not privileged and not trial preparation material. As a matter of right, the discovery pertaining to expert witnesses that the attorney is entitled to is through the use of interrogatories [Federal Rule of Civil Procedure 26(b)(4)]. The information available under the rules pertaining to interrogatories includes the following:

1. The identity of the expert witness who is expected to be used at trial
2. The subject matter of that expert's testimony
3. The substance of the facts and the opinions to which the expert will testify
4. A summary of the grounds for each opinion that the expert will give

8 DEPOSITIONS

A *deposition* is a record of testimony, and the one giving a deposition is a witness. Testimony is a statement made or evidence given by a witness under oath or affirmation in a legal proceeding and more properly refers to oral evidence than to documentary evidence. The reasons for taking a deposition include:

1. To discover the theory of the case
2. To "freeze" the testimony of the witnesses on some crucial points
3. To discover information that could be used at trial or may lead to information that can be used at trial
4. To secure admissions from adverse witnesses
5. To perpetuate the testimony of a friendly witness or an eyewitness who may not be available for trial

Generally, a motion must be filed to take the deposition of an expert although in some cases it may be done by agreement. The need for a deposition is especially manifest if the answers to interrogatories posed to an expert regarding his or her opinion are not complete or are evasive.

If you, as an industrial hygienist, are ever called to give a deposition, you should make sure that your attorney adequately prepares you. The following points may be used as your guide:

1. Your attorney should discuss with you the manner in which you answer the questions. You must remember that what you have to contribute is your credibility.
2. You should not admit that any material with which you are not totally familiar is authoritative.
3. There may be occasions when your attorney will need to object to a question or may instruct you to not answer a question. In order to give him or her time to do so, you need to learn to pause before you answer a question.
4. Be sure to take with you to the deposition only those documents or materials that you have seen before and with which you are totally familiar. If you use a document at a deposition, assume that it will be turned over to the other side.
5. Know that the other attorney will question your credentials.
6. If interrogatories were propounded to you, be familiar with your responses to them because the other attorney will question you about them. He or she will look for what might have been left out of the answers or to supplement them.
7. The opposing attorney will probably ask you who you recognize as authoritative in your field.

8. Your attorney should conduct a full cross-examination of you prior to the deposition. This means that he or she simulates the type of questioning you can anticipate from the opposing attorney.

9. Tell the truth at all times.

10. Only answer the questions that are asked. There is a natural tendency to think out loud and run on with explanations. Do not volunteer any information beyond the scope of inquiry.

9 WITNESS PREPARATION

The attorney should tell the expert witness what his or her expected role will be in the case in which he or she will be involved. The expert witness should be told who he or she is representing and what evidence he or she will be expected to give in court. If you, as an industrial hygienist, are called to be an expert witness in a trial, most attorneys will give you this information, and they will meet with you and discuss the following:

1. Why your testimony is necessary
2. When you will be expected to testify
3. How long the attorney thinks that your testimony will take
4. Where your testimony fits into the overall picture
5. What procedure will be followed in court
6. What testimony you will be expected to give during the trial
7. What you might expect from the opposing attorney during cross-examination

If the attorney who engages you does not give you this information, be sure to question him or her about it. If you are not satisfied with what you think is expected of you and what you should expect while you are testifying, do not hesitate to ask questions of the attorney. On direct examination, most trial attorneys will attempt to anticipate matters that are likely to be raised on cross-examination. The attorney can avoid many stressful moments for the industrial hygienist by placing an abundance of information in the record concerning the sufficiency of the industrial hygienist's qualifications and the adequacy of the examination upon which the opinion, if there is one, will be based. Matters concerning the compensation of the expert, or any other possible motives he or she may have in testifying, should also be brought out in direct examination in order to avoid the impact such facts may have on the trier of fact if they are forced out of the witness by opposing counsel.

9.1 Clarity of Testimony

One of the problems of communication in expert testimony occurs because experts are not used to talking to laypersons about their knowledge and judgments. An expert witness must remember that he or she is more familiar with the terminology than the attorney, and especially the jurors. Even if the terms are clear, an expert witness must be careful that the explanation is not too confusing, which it can be if the witness makes it too complex and it involves situations unfamiliar to the jury. A good rule of thumb to remember is that an expert witness is, in essence, teaching the jury, often from scratch.

9.2 Qualifications

The attorney must qualify an expert witness carefully in order to demonstrate the credibility of his or her testimony (opinion) to the judge and jury. The witness's background and standing in his or her profession should be set out in detail, including university degrees, affiliation with professional societies, writings in the field, tests and experiments conducted, and relevant experience generally.

The following is a sample qualification of an industrial hygienist in an intentional tort case involving a former employee who has charged his employer with intentionally causing his injury:

Q. Mr. _____ what is your profession?

A. I am an industrial hygienist.

Q. Where did you receive your formal education?

A. I took my degree, Bachelor of Science in Chemical Engineering, at the University of _____ in 1975, and I received a Master of Science in Public Health degree from the University of _____ in 1977.

Q. Do you hold a license to practice engineering in this state?

A. Yes, I am licensed as a professional engineer in this and four neighboring states.

Q. Are you certified in industrial hygiene?

A. Yes, I am a Certified Industrial Hygienist (CIH) by the American Board of Industrial Hygiene. I received my certification in 1981.

Q. Upon completion of your formal education, did you become associated with any firm in the practice of your profession?

A. Yes. The _____ Consulting Company in New York City.

Q. What was your position with that firm and how long were you associated with it?

A. I have been with that firm since 1977. I started as Industrial Hygienist I and have been promoted several times. I presently am Manager of the Industrial Hygiene Department.

Q. Of what professional societies are you a member?

A. I am a member of the American Industrial Hygiene Association, the American Academy of Industrial Hygiene, the American Public Health Association, and the American Institute of Chemical Engineers.

Q. Have you written any papers, given any speeches, or taught any classes in your area of practice?

A.   Yes I have.
Q.   Would you please identify them?
A.   In September 1977, I . . .

10 RECORDKEEPING

10.1 General

Many practicing industrial hygienists collect samples and prepare reports on a regular basis. In determining what and how many reports and records should be kept, it is prudent to consider legal requirements and liability matters. Some safety and health regulations require recordkeeping as a form of accountability. These recordkeeping requirements include such documents as fire extinguishing equipment inspection records, local exhaust ventilation system testing records, water quality testing records, and hazardous waste removal records. Some recordkeeping is not obligatory but is considered to be a state-of-the-art requirement in a particular industry. For example, deluge shower and eye wash equipment testing records are considered necessary in the chemical industry.

One of the most important responsibilities of an industrial hygienist is to keep good records. Generally, industrial hygienists probably maintain records on assessment of hazards, exposure measurements, development and maintenance of exposure controls, audit procedures, training and education, to name a few. It is now becoming increasingly important to be able to track where an employee worked so that his or her exposures can be calculated based on places of employment for various time periods. Maintaining a file of job descriptions that can be linked to job title records of individual employees can be of immeasurable value.

An industrial hygienist's records not only serve the primary function of helping document efforts and activities in the recognition, evaluation, and control of employee exposures, but they also could serve as evidence in a workmen's compensation claim, health-related grievance, or arbitration case or might be used by the plant physician to evaluate an environment that might be suspect as related to an employee's exposure and illness. In the long term these records are needed in epidemiological studies. Finally, they may be of great importance in defense of a professional liability lawsuit.

A logical question that needs to be addressed is: "What constitutes adequate records?" Suffice it to say there is not one "adequate" recordkeeping system that will work for every situation. Typically, there should be sufficient detail that if an industrial hygienist has to go back to them in the future he or she can generate the information that the records represent. For example, if the records are of a sampling activity, it is important to record not only the sampling data completely and accurately but also to document the methods used to obtain the results. An industrial hygienist should assume that his or her records will be reviewed as historical data by someone else and that they should be self-explanatory.

In this day and age it makes good sense to develop a system of recordkeeping that can be loaded into a computer database. Regardless of how the data are maintained, the system should allow for a systematic storage and retrieval of the data. Data collected in the field should be kept as a permanent record. It is extremely important to include all notes that are pertinent. It may be of some value to develop a standardized form for collecting the data. When conducting the field work, be sure to write down every detail that could have any value so that when the data are being used some time later all the information needed to do a proper evaluation, make reasonable recommendations, and prepare a comprehensive report will be available. It is just as important to record all data pertinent to a sample as it is to collect the sample. A good rule of thumb is to provide enough information so another industrial hygienist could duplicate the exercise without further assistance from the industrial hygienist making the record.

Be sure to include any calibration information that is pertinent. This should be part of the permanent record. Specific forms can be developed or adopted for the various types of data collection. Forms with labeled spaces for the basic information help to ensure that essential information is documented at the time the work is done. The use of standard forms will promote consistent data collection and documentation.

10.2 Reports

As a general rule an industrial hygiene report should present the facts, show analysis of the data, interpret the findings, and contain conclusions and recommendations. All of the ways in which a report may be used cannot be anticipated at the time it is prepared; an industrial hygienist should make sure that each report includes sufficient information that it can be understood by others who read it later.

The makeup of a report can take various forms, but a common makeup for reports that will be useful in future testimony or litigation would include the following sections:

1. A summary at the beginning that concisely describes the work performed, why it was done, sufficient background information so that a reader can understand what was done, the conclusion(s) drawn, and the recommendations made.

2. A description of the reported exercise or project, which could include illustrations, tables, graphs, diagrams, and photographs. It should tell what was done, why it was done, what was observed at the workplace or otherwise that would be pertinent to the reported exercise, what data were collected, and how they were collected.

3. There should be complete documentation of the sampling and analytical procedures and equipment calibration.

4. There should be discussion of the results and an interpretation of them. Reference should be made to any other pertinent studies.

5. Conclusions should be drawn based on the findings.

6. Recommendations should be identified. All proposed changes or actions should be listed and supported by the conclusions drawn.

7. Detailed data should be included as an addendum or appendix.

8. The report author(s) and principal participants in the project should be identified.

Do not rely on memory. Reports written from memory tend to be incomplete, and there may be a question as to their admissibility should they be needed as evidence in a court proceeding. Generally, unless the original notes from a survey or other project can be produced, the testimony may not be allowed.

Every inspection program must define how inspection results will be handled, filed, classified, and followed up. A plant or area inspection generally results in a determination of what should be done and is very seldom a complete result in and of itself. Usually further action, assistance, reinspection, or correction of some sort is called for. The results of the inspection must be forwarded to the people who have an interest—the supervisor of the area, especially if something should be done, the person in charge of the industrial hygiene program, the medical department, and so forth.

Because supervisors are key persons in safety and health programs, it is important to communicate to them the results, and, if possible, recommendations arising from any industrial hygiene project. For example, in an industrial hygiene survey it is determined that the ventilation hood at a grinding operation has a capture velocity that is less than the minimum recommended for good practice. The mere recording of the potential hazard represents communication. Chances are that the supervisor does not know that this is a problem and may need assistance in process change and design as well as budget support. Whoever inspects the operation should make every effort to communicate at the supervisory level as well as at the management level. Some mechanism should be set up so that problems can be addressed as soon as they are recognized.

There should be a system established for setting priorities for safety and health needs and corrections, and for ways to communicate these needs to management. Sometimes corrections require considerable expenditures of funds that are not immediately available. This situation can sometimes be dealt with by implementing interim programs such as the use of effective personal protective equipment.

One question an industrial hygienist may ask is: "How will the information I have gathered and reported stand up to scrutiny in a courtroom?" Whenever an industrial hygienist is involved in any project that involves sampling, interpreting, and report writing, he or she should proceed on the basis that whatever is being done is going to end up in trial. That way he or she can be prepared for most contingencies.

There are numerous rules that control the admissibility of records during a trial. These are generally found in the Federal Rules of Evidence. The details of this are beyond the scope of this discussion and suffice it to say that your attorney should be very familiar with how the records can be used.

11 CHAIN OF CUSTODY

Regardless of the type of case at issue, the term *chain of custody* basically has the same meaning. When one is dealing with scientific evidence, it is often necessary to show that the sample in question is the sample that was collected at a particular location at a particular time, was brought to the laboratory for analysis, and is the same sample that produced the result that is being introduced at trial. When a chain of custody is required, it is necessary to establish where the chain begins and ends. Between these points are the "links" in the chain of custody.

11.1   Links in the Chain

The links in the chain of custody are those persons who have had physical custody of the object. Failure to account for the sample(s) during possession by a custodian may constitute a break in the chain of custody. The critical point is that while a custodian in the chain of possession need not testify under all circumstances, the sample(s) should be accounted for while in that custodian's control. The most important part of this function is to set up standard operating procedures with sufficient documentation to be able to reproduce the steps involved from the beginning to the end.

11.2   Burden and Standard of Proof

The party offering the evidence has the burden of proving the chain of custody (34). Prior to having the Federal Rules of Evidence adopted, the courts described the standard of proof in a variety of ways. There were several phrases used to describe the standard: "reasonable certainty" (35), "reasonable assurance" (36), and the most common expression in which the offering party had to establish the identity and condition of the exhibit by a "reasonable probability" (37). The reasonable probability standard appears to require no more than the "preponderance of evidence" or "more probable than not" standard (38), and some courts have explicitly expressed the standard in those terms (39).

Contrast this with Federal Rule 901(a), which requires only that the offering party introduce "evidence sufficient to support a finding that the matter in question is what its proponent claims." Thus, the trial court does not decide finally or exclusively whether the item has been identified; rather, the court decides only whether sufficient evidence has been introduced from which a reasonable jury could find the evidence identified (40).

12   STANDARDIZED PROCEDURES

As documentation for possible future use in litigation, and for other purposes as well, every routine function or operation that an industrial hygienist performs should be described in detail in writing as standard operating procedures (SOPs). These procedures should describe the minimum acceptable requirements for the function or operation. For example, if a routine function is to evaluate performance of ventilation systems, SOPs are needed that describe what equipment is to be used, describe how it is to be used (including calibration), describe what information is needed (including forms for standardizing the information), describe in detail the entire procedure to be conducted, and define what reports are required.

Standard operating procedures must be in writing and show date or dates of approval and be signed by someone who has authority and responsibility for industrial hygiene activities. Periodic review is required to determine that the procedures are still adequate and that they reflect the way operations are performed.

13 SUMMARY

Even though the words "professional liability" may cause an industrial hygienist to feel uncomfortable, there is little need to be concerned if he or she conducts his or her affairs as a true professional. Industrial hygienists should use procedures and test methods generally recognized in the profession as reliable and accurate. As a general rule, if industrial hygienists live by the American Academy of Industrial Hygiene Code of Ethics for the Professional Practice of Industrial Hygiene (41), everything will take care of itself with regard to professional liability. The code is divided into four categories:

>   Professional responsibility
>   Responsibility to employees
>   Responsibility to employers and clients
>   Responsibility to the public

Listed under "professional responsibility" is an issue that states that the industrial hygienist must "maintain the highest level of integrity and professional competence." Another issue states that the industrial hygienist must "be objective in the application of recognized scientific methods and the interpretation of findings." Finally, another issue states that the industrial hygienist must "avoid circumstances where compromise of professional judgment or conflict of interest may arise." Implicit in all of this is the responsibility that any industrial hygienist has to workers, namely, that the industrial hygienist must "recognize that the primary responsibility of the Industrial Hygienist is to protect the health of employees."

NOTES AND REFERENCES

1. *United States v. Lee*, 106 U.S. 196, 220 (1882). See also *Carey v. Piphus*, 435 U.S. 247 (1978) ("[O]ver the centuries the common law of torts has developed a set of rules to implement the principle that a person should be compensated fairly for injuries caused by the violation of his legal rights.")
2. The word "actor," in legal terms, defines a person or thing that causes a set of circumstances to occur either by action or inaction. Actor here refers to the industrial hygienist.
3. See also *Privacy Act 1974 Implementation*, 40 F.R. 47405 (Oct. 8, 1975).
4. *General Motors Corporation v. Finklea*, 442 F. Supp. 821 (S.C. W. Va. 1977).
5. *Salem v. United States Lines Co.*, 370 U.S. 31, 35 (1962).

6. See, generally, 3 D Louisell & Co. Mueller, *Federal Evidence*§382 (1979); 3 J. Weinstrin & Berger, *Weinsteins' Evidence*, 702[01] (1982).

7. Advisory Committee's Note, Fed. R. Evid. 702.

8. For example, *Fineberg v. United States*, 393 F.2d 417, 421 (9th Cir. 1968) ("beyond the knowledge of the average layman"); *Jenkins v. United States*, 307 F.2d 637, 643 (D.C. Cir. 1962) ("beyond the ken of the average layman").

9. *State v. Chapple*, 135 Ariz. 281, 660 P.2d 1208, 1219–20 (1963)[quoting Ladd, "Expert Testimony," 5 Vand. L. Rev. 414, 418 (1952)].

10. 7 J. Wigmore, *Evidence*§1923. at 29 (Chadboum rev. 1978). See also Ladd, supra note 9, at 418: "There is no more certain test for determining when experts may be used than the common sense inquiry whether the untrained laymen would be qualified to determine intelligently and to the best possible degree the particular issue without enlightenment from those having a specialized understanding of the subject involved in the dispute."

11. McCormick, C. (1954). *Evidence*, pp. 363–364, West Publishing Co., St. Paul, MN.

12. 293 F. 1013 (D.C. Cir. 1923).

13. 293 F. at 1014.

14. 61 Law Week 4805.

15. 460 US 45 (1984).

16. 483 US 171 (1987).

17. 488 US 169 (1988).

18. 61 Law Week 4808.

19. 61 Law Week 4808.

20. 61 Law Week 4808.

21. 61 Law Week 4810.

22. 61 Law Week 4810.

23. 61 Law Week 4808.

24. Rule 25(b)(3) of the *Federal Rules of Civil Procedure* reads as follows:
    Trial Preparation: Materials. Subject to the provisions of subdivision (b)(4) of this rule, a party may obtain discovery of documents and tangible things otherwise discoverable under subdivision (b)(1) of this rule and prepared in anticipation of litigation or for trial by or for another party or by or for that other party's representative (including his attorney, consultant, surety, indemnitor, insurer, or agent) only upon a showing that the party seeking discovery has substantial need of the materials in the preparation of his case and that he is unable without undue hardship to obtain the substantial equivalent of the materials by other means. In ordering the discovery of such materials when the required showing has been made, the court shall protect against disclosure of the mental impressions, conclusions, opinions, or legal theories of an attorney or other representative of a party concerning the litigation.

25. 622 F.2d at 935.

26. *United States v. Nobles*, 422 U.S. 225, 95 S. Ct. 2160; 45 L. Ed.2d 141 (1975) (work product doctrine "necessarily" applies to materials prepared for an attorney).

27. 422 U.S. 225 (1975). See generally Feldman, "Work Product in Criminal Practice and Procedure," 50 U. Cin. L. Rev. 495 (1981).

28. Id. See also *People v. Collie*, 30 Cal.3d 43, 59, 634 P.2d 534, 543, 177 Cal. Rptr. 458, 467 (1981) (work product doctrine applies to criminal cases and protects the work product of defense investigators).

29. 422 U.S. at 239.

30. *Upjohn Co. v. United States*, 449 U.S. 383, 389 (1981), See also *Fisher v. United States*, 425 U.S. 391, 403 (1976) ("The purpose of the privilege is to encourage clients to make full disclosure to their attorneys").

31. See *United States v. Alvarez*, 519 F.2d 1036, 1046–47 (3d Cir. 1975); *Pouncy v. State*, 353 So.2d 640, 642 (Fla. App. 1977); *State v. Mingo*, 77 N.J. 576, 584, 392 A.2d 590, 595 (1975); see also *United States v. Nobles*, 422 U.S. 225, 239 (1975)("Respondent, by electing to present the investigator as a witness, waived the [work product] privilege with respect to matters covered in his testimony"); *Miller v. District Court*, 737 P.2d 834 (Colo. 1987)(psychiatrist).

32. Friedenthal, "Discovery and Use of an Adverse Party's Expert Information," 14 Stan. L. Rev. 455, 464–65 (1962).

33. See *United States v. Alvarez*, 519 F.2d 1036, 1046–47 (3d Cir. 1975)(psychiatrist); *United States v. Kovel*, 296 F.2d 918, 921–22 (2d Cir. 1961)(accountant); *United States v. Layaton*, 90 F.R.D. 520, 525 (N.D. Cal. 1981); *Baily v. Mesiter Brau, Inc.*, 57 F.R.D. 11, 13 N.D. (Ill. 1972)(financial expert); *Houston v. State*, 602 P.2d 784, 791 (Alaska 1979)(psychiatrist); *People v. Lines*, 13 CAI. 3d 500, 614–15, 531 P.2d 793, 802–03, 119 Cal. Rptr. 225, 234–35 (1975); *Pouncy v. State*, 353 So.2d 640, 642 (Fla. App. 1977); *People v. Knippenberg*, 66 Il.2d 276, 283–84, 362 N.E.2d 681, 684 (1977)(investigator); *State v. Pratt*, 284 Md. 516, 520–22, 396 A.2d 421, 423–24 (1979)(psychiatrist); *People v. Hilliker*, 29 Mich. App. 543, 546–47, 185 N.W.2d 831, 833–34 (1971); *State v. Kociolek*, 23 N.J. 400, 129 A.2d 417 (1957); *State v. Hitopoulus*, 297 S.C. 549, 309 S.E.2d 747 (1983).

34. See *United States v. Santiago*, 534 F.2d 768. 770 (7th Cir. 1976); I. Wigmore, *Evidence* §18, at 841 (Tillers rev. 1983).

35. See *United States v. Jones*, 404 F. Supp. 529, 543 (E.D. Pa. 1975); *Sorce v. State*, 88 Nev. 350, 352–53, 497 P.2d 902, 903 (1972); *State v. Tillman*, 208 Kan. 954, 958–59, 494 P.2d 1178, 1182 (1972).

36. See *State v. Cress*, 344 A.2d 57, 61 (Me. 1975); *State v. Baines*, 394 S.W.2d 312, 316 (Mo. 1965), *cert. denied*, 384 U.S. 992 (1966); *People v. Julian*, 41 N.Y.2d 340, 344, 360 N.E.2d 1310, 1313, 392 N.Y.S.2d 610, 613 (1977).

37. For example, *United States v. Brown*, 482 F.2d 1226, 1228 (8th Cir. 1973)("reasonable probability the article has not been changed in any important respect"); *United States v. Robinson*, 447 F.2d 1215, 1220 (D.C. Cir. 1971), *rev'd on other grounds*, 414 U.S. 218 (1973); *United States v. Capocci*, 433 F.2d 155, 157 (1st Cir. 1970); *Gass v. United States*, 416 F.2d 767, 770 (D.C. Cir. 1969); *West v. United States*, 359 F.2d 50, 55 (8th Cir.), *cert. denied*, 385 U.S. 867 (1966); *Gallego v. United States*, 276 F.2d 914, 917 (9th Cir. 1960); *United States v. S.B. Penick & Co.*, 136 F.2d 413, 415 (2d Cir. 1943); *State v. Johnson*, 162 Conn. 215, 232, 292 A.2d 903, 911–12 (1972); *Doye v. State*, 16 Md. App. 511, 519, 299 A.2d 117, 121 (1973).

38. See *People v. Riser*, 47 Cal.2d 566, 580–81, 305 P.2d 1, 10 ("The requirement of reasonable certainty is not met when some vital link in the chain of possession is not accounted for, because then it is as likely as not that the evidence analyzed was not the evidence originally received"), *appeal dismissed*, 358 U.S. 646 (1959); *State v. Serl*, 269 N.W.2d 785, 788–89 (S.D. 1978).

39. See *State v. Henderson*, 337 So.2d 204, 206 (La. 1976); *State v. Sears*, 298 So.2d 814, 821 (La. 1974); *State v. Williams*, 273 So.2d 280, 281 (La. 1973)("clear preponderance").

40. See *Zenith Radio Corp. v. Matsusshita Elect. Indus. Co.*, 505 F. Supp. 1190, 1219 (E.D. Pa. 1980) ["The Advisory Committee Note to Rule 104(b) makes plain that preliminary questions of *conditional relevancy* are not determined solely by the judge, for to do so would greatly restrict the function of the jury."], *rev'd on other grounds*, 723 F.2d 238 (3d Cir. 1983).

41. The American Academy of Industrial Hygiene, *Code of Ethics for Professional Practice of Industrial Hygiene* (appears in annual membership roster), Suite 101, 4000 W. Saginaw St., Lansing, MI 48917-2737.

Index

RC 967 .P37 1991 v. 3

Patty's industrial hygiene
and toxicology